Approaches to Plant Stress and their Management

R.K. Gaur • Pradeep Sharma
Editors

Approaches to Plant Stress and their Management

Springer

Editors
Dr. R.K. Gaur
Department of Science
Faculty of Arts, Science and Commerce
Mody Institute of Technology and Science
Sikar, India

Dr. Pradeep Sharma
Division of Crop Improvement
Directorate of Wheat Research
Karnal, India

ISBN 978-81-322-1619-3 ISBN 978-81-322-1620-9 (eBook)
DOI 10.1007/978-81-322-1620-9
Springer New Delhi Heidelberg New York Dordrecht London

Library of Congress Control Number: 2013954991

Printed on acid-free paper

Springer is part of Springer Science+Business Media (www.springer.com)

Preface

Plants have evolved to adapt to adverse conditions through cross-wired metabolic pathways to reprogram the progression of development. Stress-mediated responses are triggered to reestablish homeostasis and to repair damaged proteins. In contrast to plant resistance to biotic stresses, which is mostly dependent on monogenic traits, the genetically complex responses to abiotic stresses are multigenic and, thus, more difficult to control and engineer.

Within the framework of genetic background, plant productivity is dependent on this constant adjustment of gene expression in response to environmental cues. The genome-environment interaction is an essential focus for the elucidation of the nature of phenotypic variations leading to successful stress tolerance responses. This interaction is also a key determinant of plant tissue composition related to crop quality factors, as well as plant anatomy, morphology, and development.

Measuring plant stress is useful for many reasons, including the following:

- Finding plants that are resistant to plant stress for breeding.
- Developing optimal plant growth protocols – knowing what types of nutrients to add to the soil and how much should be used, can save money, and enhance crop yields.
- Determining the growing characteristics and limitations of plants under different stress conditions. For example, the effects of different amounts of herbicides and pesticides on plant health and growth are very valuable and can be used to reduce pollution.
- Studying the climate range for specific types of plants. The effects of heat, cold, winter, drought, and light level can be studied for all types of plants.
- Determining optimal and the most economical use of water resources.

The book comprises 22 chapters highlighting physiological, biochemical, and molecular changes due to a particular stress and its management. It describes the interaction between different biotic and abiotic stresses and their economic impact. The management part deals with and compares all aspects of tolerance mechanisms and breeding methods for plant stresses which is useful to understand the pathways or genes important for rendering more tolerance to a certain stress, and to bring forward new ideas for improving the tolerance. This book depicts illustrative tables, colored figures, and complete latest references at the end of each chapter.

Chapter 1 intends to understand the interaction of different pollutants with soil constituents; their impact on soil quality, crop growth, and produce quality; as well as appropriate soil management to counteract the chemical stresses on agricultural crops through analysis of information generated in the area by researchers of different countries.

Chapter 2 summarizes the recent progress in utilizing transgenic plant technology for the improvement of abiotic stress tolerance using research targeted at drought, salinity, and other abiotic stresses, focusing on engineering of stress-specific genes involved in different metabolic pathways in sub-stressed plants.

Chapter 3 briefly describes the transcription factors, leading to the expression of early response transcriptional activators, which then activate downstream stress tolerance effector genes responsible for stress mitigation. This chapter also highlights the role of each organic and inorganic molecule in modern-day stress mitigation strategy.

Chapter 4 attempts to summarize the major findings about the regulatory role of CaM (Calcium-modulated) and its target proteins in abiotic and biotic stress response. These studies employing genetic and molecular biology and biochemical techniques have yielded interesting insights into the function of calmodulin in modulating its various targets to provide stress resistance.

Chapters 5 *and* 14 describe the weeds which provide a tough competition to the crop plants. Once they succeed in doing so, they can easily capture the other resources like water, space, and more importantly photosynthetically active radiation.

Chapter 6 conceptualizes the approaches underlying simulation of age-stage structured populations using the cohort-updating and rate summation principle and the use of geostatistical algorithms integrated in geographic information system (GIS) for risk mapping.

Chapter 7 underlines the effects of abiotic stresses on pre- and postharvest stress susceptibility which is important since they limit the storage and shelf life potential of fruits and vegetables.

Chapter 8 provides a systemic glimpse of integrated cellular and whole plant responses to water stress. It also deals with water stress–associated hormones like ABA that is found to play a central role in orchestrating the molecular and physiological responses leading to protective responses in plants.

Chapter 9 describes the production status, impact of abiotic stresses, and the opportunities for genetic improvement of tolerance to abiotic stresses in major pulses. The chapter highlights marker traits conferring tolerance to such stress(es) which can be used in breeding programs for improving tolerance.

Chapter 10 systematizes current knowledge on the complex network of interactions and regulation of photosynthesis in plants exposed to abiotic stresses. The chapter brings updated information emphasizing on the regulation of photosynthesis and associated aspects that are affected by various abiotic stresses.

Chapter 11 describes the effects of drought on the vegetation of the major plant community types of the desert rangelands in Tunisia with emphasis on cover, species richness, and diversity.

Chapter 12 particularly emphasizes on physiological parameters and the regulation of cold-induced photosynthetic processes that occur after exposure to low temperatures, leading to cold acclimation. This chapter mainly emphasizes on the various molecules and pigments synthesized to acclimatize during low temperature exposure.

Chapter 13 describes the strategies to develop crops which can be resistant to effects of various oxidative stresses. One such way is to develop transgenic plants overexpressing one or more antioxidants, which can confer resistance toward particular stresses. Another way is to develop mutants which are resistant toward certain stresses.

Chapter 15 examines the effects of heavy metals on anatomical traits and molecular machinery that are responsible for their accumulation and tolerance in poplar. Beginning with this deeper molecular information, this chapter provides new ideas for improving poplar trees with traits conferring heavy metal tolerance.

Chapter 16 elucidates the molecular mechanisms that result from treatment of plants with benign microbes under stress conditions, which will then help understand better the full benefits of plant-microbe interaction.

Chapter 17 describes the role of WRKY gene which often responds to several stress factors, after which its proteins may participate in the regulation of several seemingly disparate processes as negative or positive regulators. WRKY genes are shown to be functionally connected forming a transcriptional network composed of positive and negative feedback loops and feed-forward modules. Within a web of partially redundant elements, some WRKY factors hold central positions mediating fast and efficient activation of defense programs.

Chapter 18 provides key insights into the complex, intricate machinery of diverse RNA silencing mechanisms; describes various evolutionary diverse strategies of viral silencing suppressors at various steps; offers a broader view of host recovery following virus infection; and finally suggests the possible applications of RNA silencing to generate virus-resistant plants.

Chapter 19 reviews three strategies for the production of fungus-resistant transgenics: (i) pathogenesis-related proteins, (ii) hypersensitive response, and (iii) RNA interference.

Chapter 20 elaborates on the intracellular, intercellular, and long-distance movement of potyvirus, focusing on the interaction of host cellular factors with movement proteins involved.

Chapters 21 *and* 22 drive the bioinformatics resources like a database of annotated tentative orthologs from crop abiotic stress transcripts, MIPS PlantsDB, GreenPhylDB, Gramene, GCP Comparative Stress Gene Catalog, Plant Stress Gene Database, PASmiR, QlicRice, Rice Stress Gene Catalog, Arabidopsis Stress Responsive Gene Database, STIFDB, and STIFDB2 to facilitate multi-omics research in this field. They highlight the major findings from omics-based studies in response to climate change factors. This offers some perspectives on the need for integrated omics approaches and realistic field-level studies of stresses.

This book presents a range of responses and adaptations that help bring about abiotic and biotic stress tolerance. Some of these involve structural or chemical changes, while others involve restriction of the growing period according to the conditions. To survive under several stresses, symbiotic relationships have been developed by the plants as a response to these stresses.

Contents

Contributors

S. Adam Department of Biochemistry, Sri Venkateswara University, Tirupati, India

P.K. Agrawal National Institute of Abiotic Stress Management, Indian Council of Agricultural Research, Malegaon, Baramati, Pune, India

Andrea Ariani BioLabs, Institute of Life Sciences, Scuola Superiore Sant'Anna, Piazza Martiri della Libertá, Italy

S.K. Bal National Institute of Abiotic Stress Management, Indian Council of Agricultural Research, Malegaon, Baramati, Pune, India

Saumik Basu Molecular Virology Laboratory, School of Life Sciences, Jawaharlal Nehru University, New Delhi, India

Kiran P. Bhagat National Institute of Abiotic Stress Management, Indian Council of Agricultural Research, Malegaon, Baramati, Pune, India

Surajit Bhattacharjee Department of Molecular Biology and Bioinformatics, Tripura University (A Central University), Suryamaninagar, India

Dhriti Bhattacharyya Molecular Virology Laboratory, School of Life Sciences, Jawaharlal Nehru University, New Delhi, India

Supriya Chakraborty Molecular Virology Laboratory, School of Life Sciences, Jawaharlal Nehru University, New Delhi, India

Shivangi Chamoli Department of Biochemistry, College of Basic Sciences and Humanities, G.B.P.U.A&T, Pantnagar, India

Partha Pratim Choudhary Directorate of Weed Science Research, Jabalpur, India

A.K. Choudhary IIPR RRC cum Off-season Nursery, Dharwad, India

D.K. Choudhary Department of Science, Faculty of Arts, Science & Commerce (FASC), Mody Institute of Technology & Science (MITS), Laxmangarh, India

R.L. Choudhary National Institute of Abiotic Stress Management, Malegaon, Baramati, Pune, India

Jaipal S. Choudhary ICAR Research Complex for Eastern Region, Research Centre Plandu, Ranchi, India

Ritika Das Department of Biology, University of Rochester, Rochester, USA

Babasaheb B. Fand School of Atmospheric Stress Management, National Institute of Abiotic Stress Management (Indian Council of Agricultural Research), Malegaon, Baramati, Pune, India

Alessandra Francini BioLabs, Institute of Life Sciences, Scuola Superiore Sant'Anna, Piazza Martiri della Libertá, Italy

Mouldi Gamoun Laboratory of Range Ecology, Arid Regions Institute (IRA), Medenine, Tunisia

A.K. Gaur Department of Molecular Biology and Biotechnology, G.B.P.U.A.T, U.S. Nagar, Pantnagar, India

Rajarshi Kumar Gaur Department of Science, Faculty of Arts, Science & Commerce (FASC), Mody Institute of Technology & Science (MITS), Laxmangarh, Sikar, India

Priyanka Giri Molecular Biology and Genetic Engineering, College of Basic Science and Humanities, G.B. Pant University of Agriculture & Technology, Uttarakhand, India

V. Govindasamy National Institute of Abiotic Stress Management, Malegaon, Baramati, Pune, India

Shekhar Jain Department of Science, Faculty of Arts, Science & Commerce (FASC), Mody Institute of Technology & Science (MITS), Laxmangarh, Sikar, India

Amrita Kasotia Department of Science, Faculty of Arts, Science & Commerce (FASC), Mody Institute of Technology & Science (MITS), Laxmangarh, Sikar, India

Vivek Keshri National Bureau of Agriculturally Important Microorganisms, Indian Council of Agricultural Research, Kushmaur, Maunath Bhanjan, India

Taqi Ahmed Khan Department of Biochemistry, Faculty of Life Sciences, Aligarh Muslim University, Aligarh, India

Anil Kumar Molecular Biology and Genetic Engineering, College of Basic Science and Humanities, G.B. Pant University of Agriculture & Technology, Uttarakhand, India

Satish Kumar National Institute of Abiotic Stress Management, Indian Council of Agricultural Research, Malegaon, Baramati, Pune, India

P. Suresh Kumar National Institute of Abiotic Stress Management, Malegaon, Baramati, Pune, India

Mahesh Kumar School of Drought Stress Management, National Institute of Abiotic Stress Management (Indian Council of Agricultural Research), Malegaon, Baramati, Pune, India

R. Arun Kumar National Institute of Abiotic Stress Management, Indian Council of Agricultural Research, Malegaon, Baramati, Pune, India

Bhumesh Kumar Directorate of Weed Science Research, Jabalpur, India

Sarita Kumari Department of Science, Faculty of Arts, Science & Commerce (FASC), Mody Institute of Technology & Science (MITS), Laxmangarh, Sikar, India

Yong-Fang Li Department of Biochemistry and Molecular Biology, Oklahoma State University, Stillwater, USA

Ramamurthy Mahalingam Department of Biochemistry and Molecular Biology, Oklahoma State University, Stillwater, USA

B. Mandal Bidhan Chandra Krishi Viswavidyalaya, Mohanpur, India

Mohd Mazid Department of Botany, Post Graduate College, Bisalpur, Pilibhit, India

P.S. Minhas National Institute of Abiotic Stress Management, Malegaon, Baramati, Pune, India

Antonio Minnocci BioLabs, Institute of Life Sciences, Scuola Superiore Sant'Anna, Piazza Martiri della Libertá, Italy

Ritesh Mishra Department of Science, Faculty of Arts, Science & Commerce (FASC), Mody Institute of Technology & Science (MITS), Laxmangarh, India

S.D.S. Murthy Department of Biochemistry, Sri Venkateswara University, Tirupati, India

Meenakumari Muthuramalingam Department of Biochemistry and Molecular Biology, Oklahoma State University, Stillwater, USA

A.K. Pal Bihar Agricultural University (BAU), Sabour, India

Amita Pandey Department of Plant Molecular Biology, University of Delhi South Campus, New Delhi, India

Girdhar K. Pandey Department of Plant Molecular Biology, University of Delhi South Campus, New Delhi, India

Ratna Prabha National Bureau of Agriculturally Important Micro-organisms, Indian Council of Agricultural Research, Kushmaur, Maunath Bhanjan, India

B.D. Prasad Bihar Agricultural University (BAU), Sabour, India

Saima Quddusi Amity Institute of Biotechnology, Amity University, Lucknow, India

Pasala RatnaKumar National Institute of Abiotic Stress Management, Indian Council of Agricultural Research, Malegaon, Baramati, Pune, India

A. Subba Rao Director, Indian Institute of Soil Science, Bhopal, India

Meenal Rathore Directorate of Weed Science Research, Jabalpur, India

Stefania Romeo BioLabs, Institute of Life Sciences, Scuola Superiore Sant'Anna, Piazza Martiri della Libertá, Italy

Ajay Krishna Saha Department of Botany, Tripura University (A Central University), Suryamaninagar, Tripura, India

J.K. Saha Principal Scientist, Division of Environmental Soil Science, Indian Institute of Soil Science, Bhopal, India

K.B. Saxena International Crops Research Institute for the Semi-Arid Tropics, Patancheru, India

Luca Sebastiani BioLabs, Institute of Life Sciences, Scuola Superiore Sant'Anna, Piazza Martiri della Libertá, Italy

Sevyaa National Bureau of Agriculturally Important Microorganisms, Indian Council of Agricultural Research, Kushmaur, Maunath Bhanjan, India

Veerendra Kumar Sharma Molecular Virology Laboratory, School of Life Sciences, Jawaharlal Nehru University, New Delhi, India

Arun K. Sharma National Bureau of Agriculturally Important Microorganisms, Indian Council of Agricultural Research, Kushmaur, Maunath Bhanjan, India

Pooja Sharma Department of Science, Faculty of Arts, Science & Commerce (FASC), Mody Institute of Technology & Science (MITS), Laxmangarh, India

Ravigopal Singh Bihar Agricultural University (BAU), Sabour, India

Deepali Singh School of Biotechnology, Gautam Buddha University, Greater Noida, India

Dhananjaya P. Singh National Bureau of Agriculturally Important Microorganisms, Indian Council of Agricultural Research, Kushmaur, Maunath Bhanjan, India

Raghwendra Singh Directorate of Weed Science Research, Jabalpur, India

R. Sultana Bihar Agricultural University (BAU), Sabour, India

Gohar Taj Molecular Biology and Genetic Engineering, College of Basic Science and Humanities, G.B. Pant University of Agriculture & Technology, Uttarakhand, India

Mohd Tasleem Molecular Biology and Genetic Engineering, College of Basic Science and Humanities, G.B. Pant University of Agriculture & Technology, Uttarakhand, India

Sachin Teotia School of Biotechnology, Gautam Buddha University, Greater Noida, India

M.K. Tripathi Agro Produce Processing Division, CIAE, ICAR (Govt of India), Bhopal, India

Anukool Vaishnav Department of Science, Faculty of Arts, Science & Commerce (FASC), Mody Institute of Technology & Science (MITS), Laxmangarh, Sikar, India

Rakesh Kumar Verma Department of Science, Faculty of Arts, Science & Commerce (FASC), Mody Institute of Technology & Science (MITS), Laxmangarh, India

A.K. Verma Department of Biochemistry, College of Basic Sciences and Humanities, G.B.P.U.A&T, Pantnagar, India

Jamila Zammouri Laboratory of Range Ecology, Arid Regions Institute (IRA), Medenine, Tunisia

Integrated Management of Polluted Soils for Enhancing Productivity and Quality of Crops

J.K. Saha, A. Subba Rao, and B. Mandal

Abstract

Soil quality is severely affected due to contamination with salts, toxic metals, non-metals and organic pollutants generated from mainly urban and industrial activities and therefore needs to be managed appropriately for sustaining agricultural productivity. Deterioration in soil quality in the polluted agricultural land can be ascertained through measurement of different physical, chemical and biological indicators. While salts affect crop productivity by degrading rhizosphere environment, heavy metals express toxicity on plant growth and on activities of agriculturally important microflora and fauna and also contaminate food. Although adverse effect of organic pollutants in soil on plant growth and produce quality has not been found significant, these are reported to affect soil microbe activity and therefore are required to be decontaminated. The role of different agricultural operations on countering the adverse effects of soil pollution has been discussed, and different soil and crop management, tillage, nutrient management, water management and soil conservation measures have been suggested for improving productivity of crops, quality of food and environment.

Keywords

Integrated management • Soil quality • Crop • Pollution • Heavy metals • Organic pollutants • Remediation

J.K. Saha (✉)
Division of Environmental Soil Science,
Indian Institute of Soil Science, Nabibagh, Berasia Road,
Bhopal 462038, India
e-mail: jk_saha12000@yahoo.com

A.S. Rao
Indian Institute of Soil Science, Nabibagh, Berasia Road,
Bhopal 462038, India

B. Mandal
Bidhan Chandra Krishi Viswavidyalaya, Mohanpur, India

Introduction

During the course of civilization and in the context of crop production, soil management objectives and technologies have evolved considerably. Before green revolution period, it aimed mainly to ease tilling of land for facilitating seed sowing, conservation of water in the land and in some cases

R.K. Gaur and P. Sharma (eds.), *Approaches to Plant Stress and their Management*,
DOI 10.1007/978-81-322-1620-9_1,

restoration of soil productivity using organics. During and after green revolution period, soil management objectives were reoriented towards augmenting food production per unit area of land for feeding the increasing population, and technologies were developed towards increasing supply of nutrients (mainly through fertilizers) and water to crop plant and intensive tillage operations to facilitate root growth of crops for absorption of water and nutrients. Hence, soil management research focussed on the improvement in the native soil fertility through enhancement of organic matter status, biological activity and physical conditions using integrated nutrient management and integrated soil management technologies (NAAS 2012). However, due to increased human-specific activities after industrial revolution like mining, fossil-fuel-based energy production, industrialization and synthesis of xenobiotics for numerous uses, intentional and unintentional entry of different chemical pollutants into agricultural land is occurring through various routes. As a result of these chemical pressures, native physico-chemical and biological environment in the rhizosphere is altered affecting food production. Increasing urbanization and industrialization is going to enhance chemical stresses on food crops in terms of coverage and intensity through expansion of polluted area in near future. This, coupled with increasing demand for food production, puts an immense challenge before scientists and planners to sustain quality food production in the affected area. As management of a system requires knowledge on its interaction with external agents/forces, this chapter intends to understand the interaction of different pollutants with soil constituents, their impact on soil quality, crop growth and produce quality as well as appropriate soil management to counteract the chemical stresses on agricultural crops through analysis of information generated in the area by researchers of different countries.

Evaluation of Soil Quality in the Context of Degrading Forces

Deterioration in soil quality has been reported as one of the major factors for yield decline under degraded system (Dawe et al. 2000). Soil quality is generally expressed as its capacity to perform different functions which the biosphere depends upon for its continued survival (Karlen et al. 1997); and these functions can be 'support for plant growth', 'production of quality and quantity of food and fibre', 'buffering and cycling of nutrients', 'filtration of water and pollutants', 'breaking down of organic pollutants', etc. A framework for evaluating soil quality has been developed involving integration of physical, chemical and biological functions with indicators that emerge at the system/process level (Andrews and Carroll 2001; Varvel et al. 2006).

Evaluation for quality of polluted soil has two additional purposes: detection of the presence of pollutants and assessment of the extent of damage caused by these. There are several special considerations for the assessment of hazards due to pollutants, including the examination of effects on different components of the environment like humans, animals, plants and microorganisms. While total contents of pollutants may show their build-up, it does not essentially indicate the type of threat that soil poses to the plant and environment, due to highly variable sink capacity. Plants essentially take up heavy metals from the soil via soil solutions (Schindler and Sposito 1991), and phytotoxicity of a metal is related to its free ion activity in the solution to which the plant root is exposed (Parker and Pedler 1997) and also to soluble metal-organic complexes and metal-inorganic ligand complexes (Huang et al. 1997; Jones 1998). Recent studies have also shown that less aggressive reagents that extract predominantly soluble and exchangeable fractions are frequently better at predicting plant availability of excess heavy metals in soils than the traditional more aggressive tests involving the use of reagents such as hydrochloric acid, diethylenetriamine pentaacetic acid and ethylenediamine tetraacetic acid (McBride et al. 2009; Menzies et al. 2007; Meers et al. 2007).

Soil microorganisms are responsible for transformation and mineralization of natural compounds and xenobiotics. Filip (1998) considered biological and biochemical processes linked indicators as important for ecotoxicological evaluation, because these sensitively respond to anthropogenic and/or environmental stresses on soil. All the biochemical reactions are catalysed by enzymes, which are proteins with catalytic properties owing to their power of specific activation (Tabatabai 1982). As heavy metals have been found to impede soil enzyme activities, their assay is considered important in assessing the impact of soil pollution (Tyler 1981; Kuperman and Carreiro 1997; Yang et al. 2006; Saha et al. 2013). In a Russian soddy-podzolic soil, N_2-fixing bacteria, dehydrogenase and respiration activities were found sensitive indicators of lead pollution (Filip 1998). Besides microorganisms, soil invertebrates have also been recognized as indicators of metal pollution. In the short term, species diversity and abundance decrease (Battigelli and Marshall 1993; Eitminaviciute 2006; Gongalsky et al. 2010), and in the long term, the increase in the number of tolerant individuals in the community and the replacement of pollution-sensitive species by less sensitive ones can lead to a different species assemblage inside the community (Salminen et al. 2001). Several invertebrate communities like *Collembolans* (springtails), *Eisenia andrei* (Oligochaeta) and *Eisenia fetida* (earthworm) were reported as biomonitor for testing the chemical toxicity of soil (Peijnenburg et al. 1999; Hirano and Tamae 2011; Santorufo et al. 2012).

Pollutants and Their Ecotoxicity and Remedial Measures

Inorganic Pollutants

Sodium, Salinity and Sodicity

Industries, particularly those associated with chlor-alkali, textiles, glass, rubber production, animal hide processing and leather tanning, metal processing, pharmaceuticals, oil and gas drilling, pigment manufacture, ceramic manufacture and soap and detergent production, are the major consumer of salts (mainly NaCl) produced in the world today. Due to their high mobility in the soil, salt ions present in the industrial effluents when released into the environment percolate through the soil profile and contaminate the groundwater. Most of the effluent treatment plants do not remove salts from the effluent water. As a result of this, salinity of groundwater has been found elevated in and around many industrial clusters of India deteriorating drinking and irrigation water quality (Table 1). As crop production in most of the countries rely considerably on groundwater and surface water, salinity build-up in soil is inevitable around areas of high industrial activity through 'industry → effluent → soil → groundwater → soil' route (Photo 1). Soils of agricultural land surrounding industrial areas of several cities recorded high electrical conductivity (EC) and exchangeable Na indicating considerable accumulation of salts due to irrigation with contaminated surface and groundwater (Saha 2005; Panwar et al. 2010).

Impact of Salts on Soil Quality and Crop Productivity: At higher concentration in soil, Na causes damage to normal plant growth, and such damage is also associated with high salinity and chloride. Thus, injury due to salinity and Na excess occur together (Bergmann 1992). The growth inhibition associated with high concentration of Na and salinity is often not attributed to specific toxic effect, but to the reduced water availability (physiological water stress). Moreover, when Na accounts for over 10–15 % of the cation exchange capacity of clayey and loamy soil or over 20 % in sandy soils, the situation (known as sodicity) results in destruction of soil structure and consequently affects root growth due to compaction, poor permeability and aeration (Maliwal and Somani 2010). As soil pollution due to high salts affects plant growth (consequently, crop-cover and root anchorage), erodibility of land increases and, hence, accelerates soil erosion.

A review by Silva and Fay (2012) indicated detrimental effects of salinity on soil microbial community, as well as on their activities and genetic diversity. Activities of soil enzymes, responsible for mineralization of nutrients, were found severely affected (Pathak and Rao 1998; Tripathi et al. 2007), thereby imparting additional dimension to the plant stress in the salt-affected soil.

Table 1 Impact of industries on electrical conductivity (mS/cm) of groundwater

Location	Nature of industries	Polluted area			Unpolluted area		
		Range	Mean	Reference	Range	Mean	Reference
Gajraula (distr.: Jyotiba Phule Nagar, UP)	Distillery	0.71–0.88	0.79	Jain et al. (2005)	0.46–0.48	0.47	Jain et al. (2005)
Kancheepuram (distr.: Kancheepuram, TN)	Textile dyeing	1.76–4.21	3.34	Balakrishnan et al. (2008)	0.23–3.42	1.28	CGWB[a]
Mettupalayam (distr.: Coimbatore, TN)	Textiles, paper and pulp	0.14–10.38	3.91	Mukherjee and Nelliyat (2007)	0.76–4.15	1.87	Mukherjee and Nelliyat (2007)
Dindigul (distr.: Dindigul, TN)	Tannery industry	0.52–21.2	3.72	Mondal et al. (2005)	0.125–6.52	1.9	CGWB[a]
Chennai (distr.: Chennai, TN)	Tannery, textile	0.48–3.76	2.02	Somasundaram et al. (1993)	1.17–1.93	1.75	CGWB[a]
Tiruppur (distr.: Tiruppur, TN)	Textile, dye	0.41–15.95	4.02	Sellamuthu et al. (2011)	0.66–4.08	1.93	CGWB[a]
		5.9–8.4	6.78	Panwar et al. (2010)	2.4–2.9	2.62	Panwar et al. (2010)
Patancheru (district: Medak, AP)	Miscellaneous	0.70–10.2	2.6	Panwar et al. (2010)	0.18–4.85	1.78	CGWB[a]
Ratlam (distr.: Ratlam, MP)	Chemicals, dye, pharmaceuticals, distillery	1.49–4.50	2.84	Saha and Sharma (2006)	0.28–3.08	1.03	CGWB[a]
Udaipur (distr.: Udaipur, Rajasthan)	Zn smelter	3.14–6.79	5.43	Panwar et al. (2010)	0.36–6.44	1.33	CGWB[a]
Pali (distr.: Pali, Rajasthan)	Textile	5.91–9.54	7.36	Panwar et al. (2010)	0.35–7.95	2.73	CGWB[a]

[a]Range and mean values have been compiled for the district in which the study areas belong (Source: Ground Water Information System. Ministry of Water Resources, Govt. of India. http://gis2.nic.in/cgwb/Gemsdata.aspx), assuming most of the information belonged to unpolluted area

Photo 1 Salt accumulation on the surface of agricultural land irrigated with textile industry effluent affecting germination and growth of wheat crop (Source: Saha and Sharma 2006)

Remediation Technologies: Soil and crop management strategies in such type of polluted soils have been described in details by Maliwal and Somani (2010) and can be summarized as follows:

(a) Leaching of salts with irrigation water containing low salt.
(b) Subsurface drainage for removing the leached-out water (wherever situation permits) so as to protect groundwater from contamination.
(c) Rainwater conservation in the field.
(d) Adoption of soil moisture conservation measures like mulching.
(e) Growing salt tolerant crops and varieties. A partial list of crops having different tolerance level to salinity is given in Table 2.

Heavy Metals

Heavy metals are naturally present in the deeper part of the earth from where they are being mined out and released into the biosphere in the form of waste either during the manufacturing process or after human use of the manufactured products. There are several industries and sources which release heavy metals in the environment, like mines, smelters, thermal power plant, metallurgical industries, electronics, textiles, phosphatic fertilizers, municipal solid wastes (MSW) and sewage and industrial sludge. Latest global assessment indicated a large quanta of metal release into the atmosphere (about 0.6 million tonne/year during mid-1990s) by different anthropogenic activities (Pacyna and Pacyna 2001). Burning of fossil fuel at stationery sources (as in coal burning at power plants) contributed maximum emissions of Cr, Hg, Ni, V and Se; vehicular traffic contributed maximum for Pb; copper production contributed maximum for As, Cd and Cu; and zinc production contributed maximum for Zn. After being emitted to the atmosphere, trace metals are subject to transport within air masses and a large part of this gets deposited on the land mass not only around the source but also in area far away, even crossing boundaries of the country/continent (Pacyna and Pacyna 2001).

Significant part of the metal-loaded effluents, generated particularly from small-scale industries in developing and underdeveloped countries, is released untreated into land and water bodies. An assessment study near some industrial clusters of India has indicated heavy metal build-up in surface layer of soils due to activity of different types of industry (Table 3). In most of cases, metals are present in dilute and small quantities in polluted water bodies and may not cause any harm to plant growth immediately when used for irrigation. However, their immobility and consequent persistence imply that concentrations may become elevated in the long run to such an extent that they begin exhibiting toxic effect on plant, soil microorganisms and food chain. Long-term exposure to heavy metals has been reported to affect human and animal health adversely (ATSDR 2005). Among the heavy metals, Ni, Co, Cr and Cu are relatively more toxic to plants, and As, Cd, Pb and Hg are relatively more toxic to higher animals (McBride 1994).

Compost is the most commonly used organic amendment for improving soil productivity. However, such material prepared from MSW and sewage sludge contains several harmful heavy metals in high concentrations (Sullivan and Miller 2001; Saha et al. 2010). Also, some of the metals are impurities/constituents of extensively used agrochemicals like fertilizers (e.g. Cd through phosphatic fertilizer) and

Table 2 Relative tolerance of selected grain and vegetable crops and fruit plants to soil salinity (Maas 1993)

	Tolerant	Moderately tolerant	Moderately sensitive	Sensitive
Cereals, pulses, oilseeds and others	Barley, rapeseed, cotton, sugar beet, wheat-durum	Pearl millet, safflower, sorghum, soybean, sunflower, wheat	Chickpea, corn, peanut/groundnut, sugarcane	Rice, sesame
Fruits and vegetables		Red beet, cowpea, squash	Broccoli, cabbage, cauliflower, cucumber, brinjal/eggplant, lettuce, pepper, potato, pumpkin, radish, spinach, tomato, turnip, watermelon	Bean, mungbean, carrot, okra, onion pea
Fruits	Date palm, natal plum	Coconut, guava, pineapple	Grape, papaya, pomegranate	Apple, apricot, banana, grapefruit, mango

Table 3 Impact of industries on heavy metal contents in soil (Source: Panwar et al. 2010)

Location	Nature of industries	Heavy metals
Pithampur (Dhar), Madhya Pradesh	Automobile manufacturing, food processing, chemical processing, distilleries, textile industries and others manufacturing industries	Cr, Zn, Co
Debari (Udaipur), Rajasthan	Zinc smelter	Zn, Cd, Pb
Korba, Chhattisgarh	Thermal power plant, metallurgical (Al), textiles, engineering workshops, tyre retreading and others	Cd, Cr
Coimbatore, Tamil Nadu	Electroplating, textile, dye	Ni, Pb, Cd, Cr

pesticides (e.g. Zn, Cu, Sn, Hg, organic pollutants) and contaminate rhizosphere when these are used in intensive agriculture.

Impact of Heavy Metals on Soil and Plant: The presence of heavy metals affects the edaphological environment in several ways like contamination of food as well as a decrease in crop productivity (Rooney et al. 2007) and diminution of soil microbial activity (Chaudri et al. 1993; McGrath 1993; Akerblom et al. 2007). Heavy metals have a strong affinity for chelating substances, including the enzymes implicated in cellular metabolism, and consequently by blocking these enzymes, they cause adverse physiological changes leading to even death of cell. These were found to affect soil microbial activity much earlier than their adverse effect on plant growth were expressed (Saha et al. 2013). The metal tolerance level in spinach (metal concentration in leaf tissue corresponding to 20 % reduction in aboveground biomass) followed the order Cr < Ni = Cu = Cd < Zn, which indicates that Cr is highly toxic to plant followed by Ni, Cu and Cd (Table 4). The order of toxicity to soil microbes as expressed by eco-toxicological dose, ED_{20} values (metal concentration in soil corresponding to 20 % reduction in their activity) was Cd > Pb > Cr > Ni > Cu > Zn. In long-term experiments, adverse effects on soil microbial biomass, as well as symbiotic N_2 fixation by blue-green algae and *rhizobium*, were observed due to heavy metal accumulation from sludge application (Brookes and McGrath 1984; Brookes et al. 1986; Mårtensson and Witter 1990). As soil microorganisms play important role related to plant growth like nutrient mineralization from organic matter, native P solubilization, N_2 fixation from atmosphere and indirect adverse effect of heavy metal accumulation on plant growth

Table 4 Protective heavy metal limits in soil and plant vis-à-vis normal concentration range observed in soils

Metals	PT_{20} (mg/kg soil)	Tolerance level in spinach leaf (μM/g)	Transfer coefficient (TC)	ED_{20} for soil microorganisms (mM/kg soil)
Cd	17.8	1.17	7.39	0.08
Cr	176.4	0.58	0.17	0.60
Cu	409	1.03	0.16	2.82
Ni	153	1.02	0.39	0.87
Pb	Toxicity not shown	–	0.08	0.39
Zn	852	37.0	2.84	5.99

Adapted from Saha et al. (2013)

can be inferred. A comprehensive review indicated inhibitory effect of Cr on uptake of several essential nutrients by plants resulting in their deficiency and crop growth (Shanker et al. 2005).

Contamination of food with heavy metals is another serious issue in management of polluted soils. Transfer of metals to edible part of plant depends on their concentration in soil, interaction with soil constituents and ability of root to exclude or inactivate metals and their translocation from root to edible parts. Transfer coefficient (TC, ratio between concentration in plant tissue and concentration in soil) varies widely with type of metals (Bergmann 1992). In an experiment on MSW compost amended soil, the values of TC for spinach followed the order Cd > Zn > Ni > Cr = Cu > Pb (Table 4). Thus, due to high toxicity to soil microorganisms and higher animals as well as to high mobility from soil to plant, Cd contamination in soil poses maximum threat to the environment. Fine texture, higher pH, organic matter and CEC favoured lower transfer coefficients for Cd, Cr and Hg in flowering Chinese cabbage, resulting from their high sorption capacities (Liu et al. 2007). Due to higher pH and clay content, Vertisol had higher capacity to adsorb Cd and Pb and, as a result, expressed toxicity (PT_{20}, phytotoxicity level corresponding to 20 % growth reduction) to soybean crop at higher levels of contamination as compared to Inceptisol and Alfisol (Fig. 1).

Remediation Technologies: Soils have varying capacity to immobilize metals (through sorption, complexation, precipitation, etc.) due to variable contents of clay types, organic matter, oxides, carbonates, phosphates, sulphides, etc. as well as due to prevailing chemical conditions (pH and Eh), and therefore, tolerable level for metals depends considerably on soil properties. Expression of toxicity to organisms, therefore, depends on degree of contamination and properties of soil. Remediation technologies to counter the toxicity can be grouped into engineering, chemical and biological approaches.

Engineering Approaches: Such technologies can be excavation and landfilling, in situ vitrification, ex situ solidification/stabilization, ex situ soil washing and soil flushing, creating subsurface barrier to protect groundwater from contamination, thermal treatment, electrokinetic method, etc. (Vangronsveld and Cunningham 1998). All these technologies, though quicker and provide a relatively long-term solution, are cost and energy intensive and applicable to a limited volume and area of soil body.

Chemical Approaches: Most of the chemical approaches aim at reducing both total and free ion activity of the metals in soil solution so that their uptake by plant and toxicity to organisms are reduced. Among the chemical methods, application of amendments like phosphates, liming material, Fe/Mn oxyhydroxides, organic materials, zeolites and modified aluminosilicates (beringite) has been advocated (Vangronsveld and Cunningham 1998).

Biological Approaches: Certain hyperaccumulator plants like *Thlaspi caerulescens, Haumaniastrum robertii, Ipomoea alpina, Macadamia neurophylla, Psychotria douarret, Thlaspi rotundifolium, Cistus ladanifer and Salix* sp. are employed to remove

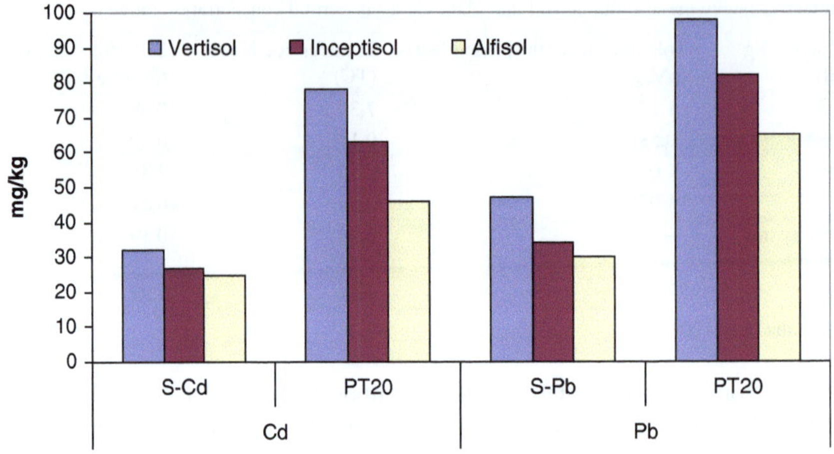

Fig. 1 Effect of soil types on sink capacity of Cd and Pb and their toxicity expressions to soybean crop (Adhikari et al. 2012). [S-Cd and S-Pb represent Langmuir adsorption maxima for Cd and Pb, respectively, and PT20 represents soil metal concentration resulting in 20 % growth reduction of soybean crop]

heavy metals from soils. Such plants can tolerate high metal levels in soil and accumulate 10–500 times higher levels than other plants and crops. Application of chelates like EDTA has been found to enhance metal extraction by the hyperaccumulators (Huang et al. 1997; Nowack et al. 2006). Rock phosphate has been found to accelerate arsenic removal by hyperaccumulator *Pteris vittata* (Fayiga and Ma 2006). Most of the phytoremediating plants capable of accumulating high concentration metals also produce less biomass, which limits their overall phytoextraction efficiency. Using modern techniques of biotechnology, several high-biomass-producing phytoaccumulators have been developed by introducing relevant genes (from hyperaccumulator, bacteria, animals) into non-accumulator plants (Singh et al. 2003). Some of high biomass hyperaccumulators for which regeneration protocols are developed include Indian mustard (*Brassica juncea*), sunflower (*Helianthus annuus*), tomato (*Lycopersicon esculentum*) and yellow poplar (*Liriodendron tulipifera*) (Mello- Farias et al. 2011). Plants have also been used to remove certain metals (most commonly Se) from soil by converting them to volatile forms (termed as phytovolatilization) (Zayed and Terry 1994). Several plants have also been used for phytostabilization purpose (inactivation of soluble forms of metals in the rhizosphere) in order to prevent metal contamination of deeper soil layers and groundwater (Cunningham et al. 1995). Some researchers have demonstrated the potential of microbes for removal of metals from soil using 'Bio metal slurry reactor' technique (Vangronsveld and Cunningham 1998) and indirectly by microbially generated biosurfactant (Wang and Mulligan 2004). However, feasibility of such technologies is yet to be proved at field level.

Organic Pollutants

Organic pollutants polycyclic aromatic hydrocarbons (PAHs), polychlorinated biphenyls (PCBs) and other persistent organic pollutants (POPs) cause diseases like cataracts, kidney and liver damage, jaundice (due to PAH) and cancer as well as have adverse effects on immune system, reproductive system, nervous system and endocrine system (due to PCBs, POPs) of human beings and therefore are widely researched upon nowadays. Some azo-dyes have been found to cause human bladder cancer, hepatocarcinomas and chromosomal aberrations in mammalian cells. Increased use of PCBs in industrial applications and organic/organometallic dyes in textile industries, emission of PAHs during burning of fossil fuels and wide adoption of chemical methods for eliminating pests and weeds in crop production are causing entry of organic pollutants in the environment, including agricultural land. Selected but important organic pollutants, emanating from industrial and fossil-fuel burning activities, are discussed in this chapter.

PAHs

Most of the PAHs are introduced into soil from atmospheric deposition after local and long-range transport, which is supported by their presence in soil of regions remote from any industrial activity (Thomas 1986). Other potential sources of PAHs in soil include disposal from public sewage treatment, irrigation with coke oven effluent and use of MSW compost (Santodonato et al. 1981; White and Lee 1980). In Brazil, Grossi et al. (1988) found increased concentrations of PAHs, PCBs and PCDD/Fs in composts prepared from unsegregated MSW from metropolitan areas. Plants are important sinks for atmospheric PAHs, playing a role in the annual cycling of PAHs, and over 50 % of PAHs emitted to the atmosphere from local sources are deposited to soils either through direct deposition or from plant litter (Simonich and Hites 1994). As PAHs from contaminated soil can enter animal and human body through multiple routes, considerable research is being carried out on minimization of risks associated with this pollutant.

Impact of PAHs on Soil Quality: Most of the toxic organic chemicals like PAHs, PCBs and pesticides are highly hydrophobic and rapidly react with soil-organic matrix and become less available to crops (O'Connor 1996). From a number of experiments, it has been concluded that contamination of crop plants via root absorption with PAHs is negligible (Ellwardt 1977; O'Conor 1996). PAHs do not have any adverse effect on the germination of several crops even at higher concentration (Ghanem et al. 2010).

Changes in soil microbial activity and biodiversity have been reported due to accumulation of persistent and toxic organic pollutants. In the presence of PAHs, biomass-C, respiration, protease activity and heterotrophic counts were significantly enhanced, while urease activity was depressed. N-mineralization was initially, however, reversibly inhibited in the presence of oil and PAHs (Margesin et al. 2000). Adverse effect of PAHs on enzymatic activity, however, depends significantly on organic matter content and pH of soil (Baran et al. 2004). PAHs affected nodulation of alfalfa significantly, and the effect was more in absence of humic matter (Wetzel and Werner 1995). In some of the soil samples, Baran et al. (2004) also observed stimulating effect of PAH on dehydrogenase activity, which was attributed to an adaptation of the soil microflora and the use of the pollutant as a C and energy source.

Degradation of PAHs in Soil: Although soil contaminated with moderate level of PAHs may not pose significant threat to plant, their intake from contaminated soil may occur via ingestion (by children) and inhalation or dermal (skin) exposure to contaminated soil/dust (during tilling) during grazing (by animals) or through soil → water → animal → human route, and hence, these require degradation. Due to strong interaction with non-aqueous phases and soil-organic matter, PAHs become potentially unavailable for microbial degradation since bacteria are known to degrade chemicals only when they are dissolved in water (Weissenfels et al. 1992). As the aqueous solubility of PAHs decreases almost logarithmically with increasing molecular mass, biodegradation of high molecular weight PAHs (having five to seven rings) becomes difficult (Johnsena et al. 2005). While individual isolates exhibited a relatively good capacity to degrade more water-soluble PAHs (acenaphthene, fluorene, phenanthrene, fluoranthene), consortia of microbial culture (*Pseudomonas putida*, *Flavobacterium* sp. and *Pseudomonas aeruginosa*) removed less water-soluble PAHs (anthracene and pyrene) more efficiently compared to isolated cultures (Trzesicka-Mlynarz and Ward 1995). Degradation of PAH in coal tar-contaminated soil was found to be enhanced after pretreatment with solvents like acetone and ethanol (Lee et al. 2001). Increasing temperature has also been found to enhance degradation of lower molecular weight PAH (up to three rings), while high molecular weight PAHs were unaffected by temperature (Coover and Sims 1987).

Polychlorinated Biphenyls (PCBs)

Contamination of soils and waters with PCBs often results from the manufacture, handling, use and disposal of these chemicals. Moreover, their extreme persistence in the environment and ability to bioconcentrate in the food chain make them great environmental and human health risks that need remedial action (Cousins et al. 1998; Hickey 1999). Several experiments with increased sewage-sludge application rates also demonstrated an accumulation of PCBs in the treated soils (Folch et al. 1996; Delschen 1999).

Impact of PCBs on Soil Quality: Several researches have concluded that uptake and translocation of PCBs within the plant is very small (Webber et al. 1990; Gan and Berthouex 1994). Turrio-Baldassarri et al. (2007) found significant level of contamination in forage plant which was assumed to be from airborne PCBs. Soil contamination with PCB (Aroclor 1248; tetrachlorbiphenyl) resulted significant reduction in root growth of several plants (Chekol et al. 2004). This group of organic pollutant, particularly higher-chlorinated congeners, has considerable adverse effect on soil microorganisms (Cámara et al. 2004; Correa et al. 2010).

Degradation of PCBs: This group of chemical is extremely resistant to decomposition due to very low solubility in water and the lack of a structural site of enzymatic attack for degradation. However, several studies have shown that PCBs can be degraded to some extent by enhancing activity of soil microorganisms; however, biodegradation rates decreased with the higher degree of chlorination (Correa et al. 2010). Selected known PCB-degrading bacterial groups (β-*proteobacteria* and acidobacteria) become more abundant in soil contaminated with PCBs, especially higher-chlorinated congeners (Correa et al. 2010). Compounds secreted from roots of photosynthetic plants were shown to support the growth of PCB-degrading bacteria in the rhizosphere of some plant species (Donnelly et al. 1994; Chekol et al. 2004), and therefore, phytoremediation may be an effective strategy for their breakdown. Synergistic interactions between arbuscular mycorrhizal fungus and *rhizobium* enhanced phytoremediation by alfalfa of an agricultural soil contaminated with weathered PCBs (Teng et al. 2010).

Photo 2 Groundwater near Ratlam industrial area has turned red, suspected to be due to contamination with azo-dyes

Dye Pollutants

Several textile, pharmaceutical and printing industries use large volume of azo-dyes, which is a group of synthetic organic colourants (generally derivatives of benzene, toluene, naphthalene, phenol and aniline) containing N as azo group –N = N. A considerable fraction (about one-tenth) of azo-dyes used by textile industries does not bind the fabrics and therefore is released through wastewater (Puvaneswari et al. 2006). Further, some azo-dyes are bonded with heavy metals like Cr and Cu in order to impart shades and resistance to washing. As a result, wastewaters from textile industries, unless treated properly, pose considerable threat to environment from both organic and metal pollutants. Groundwater of several villages near industrial area of Ratlam and Bichhri (India) has been turned red due to contamination from dye industry effluents (DTE 1999; Saha and Sharma 2006, Photo 2). Several types of azo-dyes and their breakdown products are recalcitrant and toxic (lethal, mutagenic and carcinogenic) to aquatic organisms as well as animals (Brown and De Vito 1993; Puvaneswari et al. 2006).

Impact of Dye Pollutants on Soil Quality: As a result of irrigation with effluent from dye industries, considerable accumulation of total organic dyes in cultivated soil had been observed, and such accumulation has also been transported in plant tissue (Zhou 2001). Phytotoxicity of dyes has been reported by some workers (Kalyani et al. 2008; Ayed et al. 2011). Certain azo-dye had been found to affect adversely on growth of atmospheric N-fixing cyanobacterium *Anabaena* sp. (Hu and Wu 2001). Sulphonated azo-dye had significant adverse effect on urease activity, ammonification and nitrification rates in soil and therefore reduces nitrogen use efficiency in crop production (Topaç et al. 2009).

Degradation of Dye Pollutants: Microorganism-mediated biodegradation remained the most promising and cost-effective method of remediation of soil contaminated with dye pollutants (McMullan et al. 2001; Keharia and Madamwar 2003; Khehra et al. 2006; Saratale et al. 2010), and in most cases, degradation products had no or less phytotoxicity. Ligninolytic enzymes exuded by white rot fungi have been found to degrade synthetic dye and other organic pollutants effectively (Pointing 2001). Agricultural residues were used as sorbent material for azo-dyes from effluent which were subsequently used as substrate for growing white rot fungi without having any toxicity effect (Nigam et al. 2000), and the spent composted material was used as soil conditioner. Polycyclic anthraquinone and triphenylmethane groups of azo-dyes are, however, recalcitrant to biodegradation process.

Integrated Management of Polluted Soil

Generally, engineering methods of decontamination and immobilization are adopted for soils having very high levels of toxic metals so as to prevent their spread in unpolluted area and in living organisms, and the threat is required to be brought down to a acceptable level within a short span of time. Also, successful execution of these technologies requires intervention by State due to involvement of high cost which may not be bearable by farmers. On the other hand, agricultural land affected with low to moderate level of pollutants may require modifications/interventions in the existing soil, crop and input managements in order to improve and sustain soil quality, crop productivity and produce quality. For sustainable agriculture on unpolluted land, integrated soil management involves a combined strategy of effective nutrient, crop, water, soil and land management fulfilling the objectives of improving soil fertility, water use efficiency, conservation of soil and water and increasing cropping intensity. In the context of polluted agricultural land, however, soil management should additionally address the issues like protection from pollutant build-up and its remediation and improving soil biological environment. Most often, such management strategies are dependent on:

(a) Type of pollutants and their mode of toxicity expression
(b) Contamination level
(c) Purpose of land use during and after remediation process
(d) Soil type (that determine their interaction with pollutants), depth of profile and topography
(e) Climate of the area
(f) Cropping pattern
(g) Availability of resources
(h) Economics

There exist complex interrelationships among the above determining factors as well as agricultural management technologies (Fig. 2). As a result, arriving at an appropriate strategy for combating threats in polluted soil is often not so simple. The task becomes more complex because of two reasons: firstly, ownership of land (and consequently decision making process) lies with farmers and, secondly, threat from pollutants is occasionally more to consumers rather than to the growers of food due to contamination.

Soils vary widely in respect of physical (depth, texture, structure, compactness, etc.),

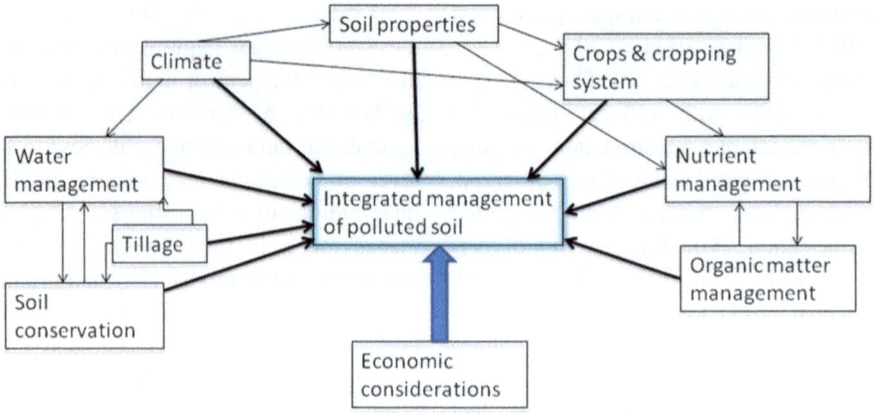

Fig. 2 Different controllable and uncontrollable factors affecting management of polluted soils and their interrelationships

Properties	Process influenced	Effect on pollution process and soil management
Climate	Infiltration and evapotranspiration	Accumulation of Na, Cl and other mobile pollutants on the surface. Leaching of salts and metals down the profile during rainy season
	Available soil moisture	Solubility and bioavailability of pollutants and their uptake by plants
	Soil temperature	Activity of soil microorganisms, degradation of organic pollutants, inactivation of inorganic pollutants (e.g. Cr^{6+} to Cr^{3+})
	Length of growing period of crops	Phytoremediation potential under rainfed situations
Geology of underground rocks	Groundwater recharge	Fractured underground rock formation enhances rate and magnitude of groundwater contamination
Land topography	Runoff, soil erosion	Spread of hazards to nearby land and water bodies
Texture and structure of soil	Infiltration	Contamination of groundwater, removal of salts from rhizosphere through subsurface drainage
	Aeration	Biodegradation of organic pollutants
	Compactness	Soil pulverization and photochemical degradation of organic pollutants during land farming
Depth of soil profile	Vertical and horizontal mobility of water	Spread of polluted area due to industrial effluents
Clay types and contents	Cation exchange capacity (CEC)	Sink capacity of soils for cationic pollutants, removal efficiency during decontamination
Organic matter content	CEC, coordinate complex formation	Sorption and chelation of heavy metals and sorption of organic pollutants influencing their bioavailability
	Microbe population and diversity	Degradation of organic pollutants
Soil reaction	Solubility of minerals and humus, strength of sorption	Expression of heavy metal toxicity, selection of amendment materials for immobilization and phytoextraction
Exchangeable Na	Dispersion of clay, solubilization of humus	Suspended particles carrying contaminated soil particles during runoff and contaminate water bodies
Oxides of Fe, Al	Sorption of anions, soluble Fe and Al ions	Sorption and solubility of anionic pollutants like As, Se, Cr and organic pollutants
	Hardness of soil upon drying (lateritic)	Ease of land farming to degrade organic pollutants

chemical (mineralogical composition, clay type, oxides, organic matter, base saturation, CEC, soil solution composition, nutrient availability, etc.) and biological properties (microbial population, diversity and activity). These soil properties determine effectiveness and adoptability of some of the management options over others. While in situ soil washing and removal of metal ions may be convenient in light-textured soils, these may not be easily achieved in Vertisols due to high clay content. On the contrary, manipulation of redox potential for reducing metal toxicity may be easier in latter soil type due to its slower infiltration capacity. Climatic factors, particularly magnitude and distribution pattern of rainfall and temperature, also influence mobility, transformation, transfer and degradation of pollutants in the rhizosphere. Roles of some of predominant soil and climatic factors on the processes and management for remediation of polluted soils are summarized below:

Agricultural Operations Influencing Remediation of Polluted Land

Fertilization: Fertilizers are considered essential in both types of usage of polluted agricultural land: (1) for higher production of food and other economic crops and (2) for higher removal of metals by phytoremediating crops through higher biomass production. While fertilizer management should enhance mobilization of metals in case of phytoremediation, reverse is desired during growing of food and fodder crops. Application of chloride-containing fertilizers like NH_4Cl and muriate of potash (KCl) should preferably be avoided in Cd-contaminated soil for growing food and fodder crops as these enhance mobility of the metal through formation of soluble $CdCl_n^{2-n}$ complex and promote crop contamination (Sparrow et al. 1994; Lo'pez-Chuken et al. 2012). Fertilizers like ammonium sulphate and monoammonium phosphate lower the soil pH during their continuous application and consequently may increase the availability of metal pollutants (Levi-Minzi and Petruzzelli 1984; Willaert and Verloo 1992) and hence may be avoided in metal-polluted neutral and calcareous soils. In a laboratory experiment, monoammonium phosphate decreased soil pH and the amount of Cd adsorbed by two soils (thereby increasing the availability), while diammonium phosphate precipitated Cd particularly in low organic matter containing soil (Levi-Minzi and Petruzzelli 1984). Contamination of several food crops with arsenic (As) is widely observed in lower Gangetic plains of India and Bangladesh due to irrigation with contaminated groundwater (Meharg and Rahman 2003; Huq et al. 2006). Phosphorus application has been found to reduce its contamination of wheat grain (Pigna et al. 2010). In rice plants (grown on flooded soil), however, added P suppressed uptake of arsenate, not arsenite indicating different uptake mechanism for different forms of As species (Abedin et al. 2002). High available calcium in soils inhibits accumulation of several heavy metals by crop (Kurtyka et al. 2008; Suzuki 2005), and hence, frequent application of Ca-containing fertilizers (like SSP, CAN gypsum) and lime may be advisable in noncalcareous soils in high-rainfall areas.

Soil-Organic Matter Management: Due to its role in maintaining soil fertility and crop productivity, improvement and maintenance of organic matter status in agricultural soil is essentially advocated, particularly in tropical region where rate of C mineralization is quite high. Application of organic matter has been found to reduce availability of heavy metals in contaminated soil (Walker et al. 2004; Skłodowski et al. 2006). However, strength of metal complexes with humic acid increases with increasing pH and decreasing ionic strength (McBride 1994), and therefore, combined application of lime and organic matter may be the most effective strategy in reducing/eliminating the adverse impacts (i.e. phytotoxicity and food contamination) in heavy metal-polluted acidic soils.

Due to their availability, sewage-sludge and MSW compost are widely used as source of organic matter in agricultural land in peri-urban areas. As contents of heavy metals and organic

Table 5 Classification of MSW composts for their marketability and use in different area

Class		Organic matter and nutrient content	Heavy metal content	Quality control compliance	Overall quality and area of application
Marketable classes	A	High	Very low	Complying for all heavy metal parameters	*Best quality* High manurial value and low heavy metal content. Can be used for high value crops and in organic farming
	B	Medium	Very low		*Very good quality* Medium fertilizing potential and low heavy metal content
	C	High	Low		*Good quality* High fertilizing potential and medium heavy metal content
	D	Medium	Low		*Medium quality* Medium fertilizing potential and medium heavy metal content
Restricted use classes	RU-1	Low	–		Low fertilizing potential but safe for environment. Can be used as soil conditioner
	RU-2	High	Low	Not complying for all heavy metal parameters	Can be used for growing non-food crops. Requires periodic monitoring of soil quality if used repeatedly
	RU-3	Medium	Low		Can be used only for developing lawns/gardens, tree plantation (with single application)

Adapted from Saha et al. (2010)

pollutants are generally high in these organic materials, their continuous application is likely to build up these toxicants in soil (Santodonato et al. 1981; Grossi et al. 1988; Saha et al. 2010). However, due to having higher mineral oxides content, toxicity due to repeated sludge application may appear later as compared to MSW compost at similar level of heavy metal input (McBride 1995). In order to prevent heavy metal build-up in soil beyond the level of posing threat to ecosystem, several countries have formulated guidelines regarding use of such contaminated organic matter in agricultural land like 'maximum concentrations of metals allowed in agricultural soils', 'maximum metal concentrations in sewage-sludge and MSW compost' and 'annual metal loading limits for agricultural soils' (McGrath et al. 1994; Düring and Gäth 2002). A scheme has been proposed for utilization of MSW composts with variable plant nutrients and heavy metal contents for different land use systems which can protect soil as well as harvest benefits from its organic matter (Table 5).

Application of stable compost along with maintaining soil pH near neutral and growing of plants has been found to enhance the clean-up of PAH-contaminated soils during in situ remediation (Reilley et al. 1996; Oleszczuk and Baran 2003; Sayara et al. 2010). Enhanced biological activity in the rhizosphere appears to be the reason for such accelerated degradation. High pH and salinity reduces PAHs degradation in saline-alkali soil, and such adverse effect can be reduced by application of organic matter (Luqueño et al. 2008).

Tillage Operations: Tillage methods influence root proliferation (hence rhizosphere depth and volume), aeration, partitioning of rainwater and soil erosion and therefore can be important for the management of polluted soils. Limited studies have indicated role of tillage on amelioration and management of polluted soil. Conservation tillage has been recommended in case of polluted land as it reduces runoff and soil erosion during high intensity rainfall and thereby protects nearby unpolluted area and aquatic life from toxicants (Holland 2004). Reduced tillage or no-tillage promotes higher metal uptake from surface horizon due to restricted root proliferation in the deeper layer (Oliver et al. 1993) and therefore may be advisable during phyto-extraction of surface-contaminated land. On the contrary, enhanced aeration and exposure of below-surface soils to sunlight due to repeated tillage operations can accelerate biodegradation of organic pollutants (Rhykerd et al. 1999).

Soil and Water Conservation and Management: Highly polluted agricultural land is also a source of contamination for nearby land and surface water bodies, and therefore, management strategies warrant appropriate measures to check soil erosion. In high-rainfall area, soil tillage across slope, contour farming, cover cropping, mulching, etc. may be adopted to reduce impact of raindrops and soil loss through runoff water. Growing grasses and cover crops having high phytoremediation potential can be good alternative in such situation as these can reduce runoff soil loss as well (Cook et al. 2009). However, trees (like *Populus* spp. and *Salix* spp.) may be grown along with grasses so as to phytostabilize metals in deeper soil layer and prevent downward metal flux (Quinn et al. 2001). In arid region, contaminated soil particles carried away by wind may have severe health implications on human and animal populations when it enters the body through respiratory route. Capping of polluted land with fertile soils and growing grasses can prevent air pollution with contaminated dust.

Often groundwater of the area surrounding industrial clusters records elevated salinity (Table 1), and use of such water for irrigation is likely to increase osmotic potential in the rhizosphere causing physiological water stress to crop. Moreover, elevated Cl^-(normally associated with salinity of the groundwater) in soil solution may enhance the mobility and uptake of heavy metals (particularly Cd) in contaminated land (Lo'pez-Chuken et al. 2012). Therefore, water management strategy for cultivated land around the industrial area should focus on increasing water use efficiency through efficient irrigation methods and also on reducing dependency on groundwater through conservation of rainwater and soil moisture.

Selection of Crops and Cropping Systems: Crops differ widely in respect of their ability in uptake of metals and in combating heavy metal stress by adopting various mechanisms like exclusion, sequestration and metal homeostasis (Manara 2012). Selection of crops for the polluted land can be decided based on the perceived threat(s) to the farming. A comparatively lower level of pollution poses threat only to the quality of food and fodder due to contamination with toxic metals (Saha et al. 2013), and therefore, appropriate strategy in such situation would be to minimize their transfer to aboveground biomass. Results indicate that bioconcentration is generally the highest in leafy vegetables followed by root vegetables and minimum in grain crops (Page et al. 1987). Therefore, cultivation of leafy and root vegetables (like spinach, lettuce, cabbage, potato, radish, beet and coriander) in soils having elevated levels of metals should be avoided, and growing of other crops can only be taken up after ensuring that national requirement for food and fodder quality in respect of heavy metal contents is met. In an experiment under controlled condition, different cereals, oilseeds, sugarcane and fibre crops exhibited tolerance to high soil Cd levels (Wang 2002). Though grain of cereal crops got contaminated, edible parts of other crops remained free from contamination. Even varieties differ in the metal contamination in grain due to difference in uptake pattern, upward and basipetal translocation in plant parts (Chan and Hale 2004) indicating the importance of genotype selection in the contaminated soil. Crop rotation can also influence heavy metal mobilization and uptake by plants due to residual effect of several organic acids in the root exudates from previous crops (Nigam et al. 2001). Concentration Cd in grain was more in wheat grown after lupins and lowest in wheat grown after cereal, particularly under zero tillage (Oliver et al. 1993).

A higher level of heavy metal pollution may affect productivity of common food crops as well as severely contaminate the food chain, and therefore, the situation warrants decontamination of crop land. The most appropriate strategy in such situation can be avoidance of non-food and non-fodder crops and selection of appropriate phytoremediating plants. Several oilseed crops (particularly *Brassica* sp.), flowering plants, fibre crops and short-duration trees (like eucalyptus, poplar and willow) have exhibited their phytoremediation potential while generating economic produce during the process (Broadley et al. 1999; Angelova et al. 2004; Su and Wong 2004; Indoria and Poonia 2006; Ramana et al. 2009; Ruttens et al. 2011). Use of metal-contaminated lands for production of biofuel crops like sugar (sugarcane, sugar beet), starch (maize, wheat, rice) and oilseeds (rapeseed, sunflower, jatropha) can be a viable option. Incineration of contaminated biomass can generate energy as well as reduce the volume for appropriate disposal or for metal recovery. Meer et al. (2010) estimated that cultivation of maize in a moderately contaminated soil could produce 33,000–46,000 kW h of renewable energy (electrical and thermal) per hectare per year, though its phytoremediation potential is very low. Hemp (*Cannabis sativa*), flax (*Linum usitatissimum*) and peanut (*Arachis hypogaea*) were also found suitable for both biodiesel production and phytoextraction of Cd from contaminated soil (Shi and Cai 2009). Biomass residues after biodiesel/bioethanol production can further be incinerated for energy production with consequent reduction in volume for disposal of hazardous waste (ash).

Management of Soil Contaminated with Organic Pollutants: Degradability of organic pollutants depends primarily on their nature and concentration level, bioavailability (mostly solubility in aqueous phase) and activity of microorganisms (both general and specific). For degradation of toxic and recalcitrant organic pollutants, indigenous microflora may not be effective. Under such situation, bioaugmentation with proper microbials or their consortia capable of degrading target pollutant may be essential. Organic matter application may be counterproductive during degradation strategy for hydrophobic pollutants as they decrease their bioavailability through sorption. On the other hand, compost or manure application may be highly effective in soils with low organic matter contents for degrading relatively soluble simple hydrocarbons, 2–3-ring aromatic hydrocarbons, as it provides substrate for microbial proliferation and hence accelerate degradation rate. Slightly alkaline pH, abundance of electron acceptors (like oxygen, nitrate, Mn and Fe oxides), appropriate moisture level and moderate temperature usually favour microbial growth and consequently contaminant degradation. An integrated remediation system composed of physical (volatilization), photochemical and microbial remediation and phytoremediation has been found superior in degrading 16 priority PAHs in soil over individual techniques (Huang et al. 2004). The techniques applied for this integrated degradation process were land farming (facilitating aeration and light exposure through repeated tillage), introduction of contaminant degrading bacteria, plant growth promoting rhizobacteria (PGPR) and growing contaminant tolerant plant (*Festuca arundinacea*). The investigators used plant species that have the ability to proliferate in the presence of high levels of contaminants and strains of PGPR that increase plant tolerance to contaminants and accelerate plant growth in heavily contaminated soils. Fertilization and clipping of aboveground biomass (which enhanced root turnover) were also found to enhance degradation of PAHs (Olson et al. 2008).

Conclusion

Safe, but profitable, crop production on polluted land can be a challenge due to involvement of large number of controllable and uncontrollable variables. Successful management of such land requires proper understanding of the interrelationships among these variables. On the contrary, most of the researches in understanding pollutant-soil-crop interactions, their impact on crop productivity and food quality and amelioration/management of polluted soils have been carried out taking fewer variables under considerations. This is expected due to obvious constraints in carrying out complex experiments with limited resources and funds. In this chapter, effort has been made to document management options available to reduce the impact of pollutants on crop productivity and produce quality and also to relate different information generated through research to predict recommendable practices in situations where information is scanty. As magnitude of threats from the pollutants also depends on their levels as well as soil types and nature of crops, integrating different remediation technologies can address the above constraints in adoption and has been discussed to a limited extent based on available information. Success and feasibility of these technologies, however, depend largely on their cost of adoption and income generation from concurrent use of polluted land under treatment. Unless contamination reaches to an alarming level, soil pollution in most cases remains unnoticed, and this is more so in developing and underdeveloped countries where facilities for testing of pollutants are grossly inadequate due to high cost. In such a situation, prescribing a package of soil and input management technologies for vulnerable area (e. g. those near industrial and urban areas) can be a step towards sustaining productivity and quality of agricultural produce. However, more researches are required on site-specific integrated management of such vulnerable areas addressing following important issues:

- Conservation of natural soil functions
- Optimization of functionality of urban water-soil system
- Restoration of soil functions after ecosystem damage
- Integration of soil-water system at urban and industrial scale with the irrigation requirement for agriculture
- Harmonization of solid and liquid waste generation with the nutrient requirement of cropping system for better protection environment and food production system

There is also a need to develop decision support system for site-specific management of polluted soils through use of available resources and technologies.

References

Abedin MJ, Feldmann J et al (2002) Uptake kinetics of arsenic species in rice plants. Plant Physiol 128:1120–1128

Adhikari T, Biswas AK et al (2012) Heavy metal pollution in soil-plant system and its remediation. IISS technical bulletin. Indian Institute of Soil Science, Bhopal, pp 1–57

Akerblom S, Baath E et al (2007) Experimentally induced effects of heavy metals on microbial activity and community structure of forest mor layers. Biol Fertil Soils 44:79–91

Andrews SS, Carroll CR (2001) Designing a decision tool for sustainable agroecosystem management: soil quality assessment of a poultry litter management case study. Ecol Appl 11:1573–1585

Angelova V, Ivanova R et al (2004) Bio-accumulation and distribution of heavy metals in fibre crops (flax, cotton and hemp). Ind Crop Prod 19:197–205

ATSDR (2005) Toxicological profile for nickel. Agency for Toxic Substances and Disease Registry, U.S. Department of Health And Human Services, Public Health Service, Atlanta

Ayed L, Mahdhi A et al (2011) Decolorization and degradation of azo dye methyl red by an isolated *Sphingomonas paucimobilis*: biotoxicity and metabolites characterization. Desalination 274:272–277

Balakrishnan M, Antony SA et al (2008) Impact of dyeing industrial effluents on the groundwater quality in Kancheepuram (India). Indian J Sci Technol 1:1–8

Baran S, Bielińska JE et al (2004) Enzymatic activity in an airfield soil polluted with polycyclic aromatic hydrocarbons. Geoderma 118:221–232

Battigelli JP, Marshall VG (1993) Relationship between soil fauna and soil pollutants. In: Proceedings of the forest ecosystem dynamics workshop. FRDA II repost 210. Government of Canada Province of British Columbia, p 31, 10–11 Feb 1993

Bergmann W (1992) Nutritional disorders of plants. Gustav Fischer, Jena/Stuttgart/New York

Broadley MR, Willey NJ et al (1999) A method to assess taxonomic variation in shoot caesium concentration among flowering plants. Environ Pollut 106:341–349

Brookes PC, McGrath SP (1984) The effects of metal toxicity on the size of soil microbial biomass. J Soil Sci 35:341–346

Brookes PC, McGrath SP et al (1986) Metal residues in soils previously treated with sewage-sludge and their effects on growth and nitrogen fixation by blue green algae. Soil Biol Biochem 18:345–353

Brown MA, De Vito SC (1993) Predicting azo dye toxicity. Crit Rev Env Sci Tec 23:249–324
Cámara B, Herrera C et al (2004) From PCBs to highly toxic metabolites by the biphenyl pathway. Environ Microbiol 6:842–850
Chan DY, Hale BA (2004) Differential accumulation of Cd in durum wheat cultivars: uptake and retranslocation as sources of variation. J Exp Bot 55:2571–2579
Chaudri AM, McGrath SP et al (1993) Enumeration of indigenous *Rhizobium leguminosarum* biovar trifolii in soils previously treated with metal –contaminated sewage-sludge. Soil Biol Biochem 25:301–309
Chekol T, Vough LR et al (2004) Phytoremediation of polychlorinated biphenyl-contaminated soils: the rhizosphere effect. Environ Int 30:799–804
Cook LL, Inouye RS et al (2009) Evaluation of four grasses for use in phytoremediation of Cs-contaminated arid land soil. Plant Soil 324:169–184
Coover MP, Sims RC (1987) The effect of temperature on polycyclic aromatic hydrocarbon persistence in an unacclimated agricultural soil. Hazard Waste Hazard Mat 4:69–82
Correa PA, Lin L et al (2010) The effects of individual PCB congeners on the soil bacterial community structure and the abundance of biphenyl dioxygenase genes. Environ Int 36:901–906
Cousins IT, Mclachlan MS et al (1998) Lack of an aging effect on the soil – air partitioning of polychlorinated biphenyls. Environ Sci Technol 32:2734–2740
Cunningham SD, Berti WR et al (1995) Phytoremediation of contaminated soils. Biotech 13:393–397
Dawe D, Dobermann A et al (2000) How widespread are yield declines in long-term rice experiments in Asia? Field Crop Res 66:175–193
Delschen T (1999) Impacts of long-term application of organic fertilizers on soil quality parameters in reclaimed loess soils of the Rhineland lignite mining area. Plant Soil 213:43–54
Donnelly PK, Hegde RS, Fletcher JS (1994) Growth of PCB-degrading bacteria on compounds from photosynthetic plants. Chemosphere 28:981–988
DTE (1999) Bloody waters. Down to earth, Issue 1–15 Apr 1999. Society for Environmental Communications, New Delhi. http://www.downtoearth.org.in/node/19679
Düring RA, Gäth S (2002) Utilization of municipal organic wastes in agriculture: where do we stand, where do we go? J Plant Nutr Soil Sci 165:544–556
Eitminaviciute I (2006) Microarthropod communities in anthropogenic urban soils. 1. Structure of microarthropod complexes in soils of roadside lawns. Entomol Rev 86:128–135
Ellwardt P (1977) Variation in content of polycyclic aromatic hydrocarbons in soil and plants by using municipal waste compost in agriculture. In: Proceedings of the symposium on soil organic matter studies, vol II. International Atomic Energy Agency, Vienna, p 291
Fayiga AO, Ma LQ (2006) Using phosphate rock to immobilize metals in soil and increase arsenic uptake by hyperaccumulator *Pteris vittata*. Sci Total Environ 359:17–25
Filip Z (1998) An internationally based approach to assess soil quality by biological methods. In: Proceedings of the 16th world congress of soil science, Paper no 2041, Montpellier
Folch I, Vaquero MT et al (1996) Extraction and clean-up methods for improvement of the chromatographic determination of polychlorinated biphenyls in sewage sludge-amended soils: elimination of lipids and sulphur. J Chromatogr A 719:121–130
Gan R, Berthouex P (1994) Disappearance and crop uptake of PCBs from sludge amended farming. Water Environ Res 66:54–69
Ghanem A, D'Orazio V et al (2010) Phytotoxicity assay of selected plants to Pyrene contaminated soil. In: Proceedings of the nineteenth world congress of soil science, soil solutions for a changing world, Brisbane. Published on DVD, pp 74–77, 1–6 Aug 2010
Gongalsky KB, Filimonova ZV et al (2010) Relationship between soil invertebrate abundance and soil heavy metal contents in the environs of the Kosogorsky metallurgical plant, Tula Oblast. Russ J Ecol 41:67–70
Grossi G, Lichtig J et al (1998) PCDD/F, PCB, and PAH content of Brazilian compost. Chemosphere 37:2153–2160
Hickey WJ (1999) Transformation and fate of polychlorinated biphenyls in soil and sediment. In: Adriano DC et al (eds) Bioremediation of contaminated soils, vol 37, Agronomy Monograph. ASA, CSSA and SSSA, Madison, pp 213–237
Hirano T, Tamae K (2011) Earthworms and soil pollutants. Sensors 11:11157–11167
Holland JM (2004) The environmental consequences of adopting conservation tillage in Europe: reviewing the evidence. Agr Ecosyst Environ 103:1–25
Hu T, Wu SC (2001) Assessment of the effect of azo dye RP2B on the growth of a nitrogen fixing cyanobacterium – Anabaena sp. Bioresour Technol 77:93–95
Huang JW, Chen J et al (1997) Phytoremediation of lead-contaminated soils: role of synthetic chelates in lead phytoextraction. Environ Sci Technol 31:800–805
Huang X, El-Alawi Y et al (2004) A multi-process phytoremediation system for removal of polycyclic aromatic hydrocarbons from contaminated soils. Environ Pollut 130:465–476
Huq SMI, Joardar JC et al (2006) Arsenic contamination in food-chain: transfer of arsenic into food materials through groundwater irrigation. J Health Popul Nutr 24:305–316
Indoria AK, Poonia SR (2006) Phytoextractability of lead from soil by some oilseed crops as affected by sewage sludge and farmyard manure. Arch Agron Soil Sci 52:667–677
Jain N, Bhatia A et al (2005) Impact of post-methanation distillery effluent irrigation on groundwater quality. Environ Monit Assess 110:243–255
Johnsena AR, Wickb LY et al (2005) Principles of microbial PAH-degradation in soil. Environ Pollut 133:71–84

Jones DL (1998) Organic acids in the rhizosphere – a critical review. Plant Soil 205:25–44

Kalyani DC, Patil PS et al (2008) Biodegradation of reactive textile dye Red BLI by an isolated bacterium *Pseudomonas* sp. SUK1. Bioresour Technol 99:4635–4641

Karlen DL, Mausbach MJ et al (1997) Soil quality: a concept, definition, and framework for evaluation. Soil Sci Soc Am J 61:4–10

Keharia H, Madamwar D (2003) Bioremediation concepts for treatment of dye containing wastewater: a review. Indian J Exp Biol 41:1068–1075

Khehra MS, Saini HS et al (2006) Biodegradation of azo dye C.I. Acid red 88 by an anoxic–aerobic sequential bioreactor. Dyes Pigments 70:1–7

Kuperman RG, Carreiro MM (1997) Soil heavy metal concentrations, microbial biomass and enzyme activities in a contaminated grassland ecosystem. Soil Biol Biochem 29:179–190

Kurtyka R, Małkowski E et al (2008) Effect of calcium and cadmium on growth and accumulation of cadmium, calcium, potassium and sodium in maize seedlings. Pol J Environ Stud 17:51–56

Lee PH, Ong SK et al (2001) Use of solvents to enhance PAH biodegradation of coal tar. Water Res 35:3941–3949

Levi-Minzi R, Petruzzelli G (1984) The influence of phosphate fertilizers on Cd solubility in soil. Water Air Soil Pollut 23:423–429

Liu Y, Kong GT et al (2007) Effects of soil properties on heavy metal accumulation in flowering Chinese cabbage (*Brassica campestris* L. ssp. chinensis var. utilis Tsen et Lee) in Pearl River Delta, China. J Environ Sci Heal B 42:219–227

Lo'pez-Chuken UJ, Lo'pez-Domı'nguez U et al (2012) Implications of chloride-enhanced cadmium uptake in saline agriculture: modeling cadmium uptake by maize and tobacco. Int J Environ Sci Technol 9:69–77

Luqueño FF, Marsch R et al (2008) Remediation of PAHs in a saline–alkaline soil amended with wastewater sludge and the effect on dynamics of C and N. Sci Total Environ 402:18–28

Maas EV (1993) Testing crops for salinity tolerance. In: Proceedings of the workshop on adaptation of plants to soil stresses, Maranville JW, Baligar BV, Duncan RR, Yohe JM (eds) INTSORMIL. Pub no 94–2, Univ of Ne, Lincoln, pp 234–247, 1–4 Aug 1993

Maliwal GL, Somani LL (2010) Nature, properties and management of saline and alkali soils. Agrotech Publishing Academy, Udaipur. ISBN 976-81-8321-177-2

Manara A (2012) Plant responses to heavy metal toxicity. In: Furini A (ed) Plants and heavy metals, SpringerBriefs in molecular science. Springer, Netherlands pp 27–53

Margesin R, Walder G et al (2000) The impact of hydrocarbon remediation (diesel oil and polycyclic aromatic hydrocarbons) on enzyme activities and microbial properties of soil. Acta Biotechnol 20:313–333

Mårtensson AM, Witter E (1990) Influence of various soil amendments on nitrogen fixing soil organisms in a long-term field experiments, with special reference to sewage sludge. Soil Biol Biochem 22:977–982

McBride MB (1994) Environmental chemistry of soils. Oxford University Press, New York

McBride MB (1995) Toxic metal accumulation from agricultural use of sludge: are U.S. EPA regulations protective? J Environ Qual 24:5–18

McBride MB, Pitiranggon M et al (2009) A comparison of tests for extractable copper and zinc in metal-spiked and field-contaminated soil. Soil Sci 174:439–444

McGrath SP (1993) Soil quality in relation to agricultural use. In: Eijsackers HJP, Hamers T (eds) Integrated soil and sediment research: a basis for proper protection. Kluwer, Dordrecht, pp 187–200

McGrath SP, Chang AC et al (1994) Land application of sewage sludge: scientific perspectives of heavy metal loading limits in Europe and the United States. Environ Rev 2:108–118

McMullan G, Meehan C et al (2001) Microbial decolourisation and degradation of textile dyes. Appl Microbiol Biotech 56:81–87

Meers E, Samson R et al (2007) Phytoavailability assessment of heavy metals in soils by single extractions and accumulation in *Phaseolus vulgaris*. Environ Exp Bot 60:385–396

Meers E, Slycken SV et al (2010) The use of bio-energy crops (Zea mays) for 'phytoattenuation' of heavy metals on moderately contaminated soils: a field experiment. Chemosphere 78:35–41

Meharg AA, Rahman MM (2003) Arsenic contamination of Bangladesh paddy field soils: implications for rice contribution to arsenic consumption. Environ Sci Technol 37:229–234

Mello- Farias PC, Chaves ALS et al (2011) Transgenic plants for enhanced phytoremediation – physiological studies, genetic transformation. In: Alvarez M (ed) ISBN: 978-953-307-364-4, InTech, Available from http://www.intechopen.com/books/genetictransformation/transgenic-plants-for-enhanced-phytoremediation-physiological-studies

Menzies NW, Donn MJ et al (2007) Evaluation of extractants for estimation of the phytoavailable trace metals in soils. Environ Pollut 145:121–130

Mondal NC, Saxena VK et al (2005) Assessment of groundwater pollution due to tannery industries in and around Dindigul, Tamilnadu, India. Environ Geol 48:149–157

Mukherjee S, Nelliyat P (2007) Groundwater pollution and emerging environmental challenges of industrial effluent irrigation in Mettupalayam Taluk, Tamil Nadu. In: Comprehensive assessment of water management in agriculture discussion paper 4, International Water Management Institute, Colombo, p 51

NAAS (2012) Sustaining agricultural productivity through integrated soil management. Policy paper no 56, National Academy of Agricultural Sciences, New Delhi, p 24

Nigam P, Armour G et al (2000) Physical removal of textile dyes from effluents and solid-state fermentation of dye-adsorbed agricultural residues. Bioresour Technol 72:219–226

Nigam R, Srivastava S et al (2001) Cadmium mobilisation and plant availability – the impact of organic acids commonly exuded from roots. Plant and Soil 230:107–113

Nowack B, Schulin R et al (2006) Critical assessment of chelant-enhanced metal phytoextraction. Environ Sci Technol 40:5225–5232

O'Connor GA (1996) Organic compounds in sludge-amended soils and their potential uptake by crop plants. Sci Total Environ 185:71–81

Oleszczuk P, Baran S (2003) Degradation of individual polycyclic aromatic hydrocarbons (PAHs) in soil polluted with aircraft fuel. Pol J Environ Stud 12:431–437

Oliver DP, Schultz JE, Tiller KG, Merry RH (1993) The effect of crop rotations and tillage practices on cadmium concentration in wheat grain. Aust J Agr Res 44:1221–1234

Olson PE, Castro A et al (2008) Effects of agronomic practices on phytoremediation of an aged PAH-contaminated soil. J Environ Qual 37:1439–1446

Pacyna JM, Pacyna EG (2001) An assessment of global and regional emissions of trace metals to the atmosphere from anthropogenic sources worldwide. Environ Rev 9:269–298

Page AL, Logan TJ et al (1987) Land application of sludge: food chain implications. Lewis Publications, Chelsea

Panwar NR, Saha JK et al (2010) Soil and water pollution in India: some case studies. IISS technical bulletin. Indian Institute of Soil Science, Bhopal

Parker DR, Pedler JF (1997) Reevaluating the free-ion activity model of trace metal availability to higher plants. Plant Soil 196:223–238

Pathak H, Rao DLN (1998) Carbon and nitrogen mineralisation from added organic matter in saline and alkali soils. Soil Biol Biochem 30:695–702

Peijnenburg WJGM, Baerselman R et al (1999) Relating environmental availability to bioavailability: soil-type-dependent metal accumulation in the Oligochaete *Eisenia Andrei*. Ecotox Environ Safe 44:294–310

Pigna M, Cozzolino V et al (2010) Effects of phosphorus fertilization on arsenic uptake by wheat grown in polluted soils. J Soil Sci Plant Nutr 10:428–442

Pointing SB (2001) Feasibility of bioremediation by white rot fungi. Appl Microbiol Biot 57:20–33

Puvaneswari N, Muthukrishnan J et al (2006) Toxicity assessment and microbial degradation of azo dyes. Indian J Exp Biol 44:618–626

Quinn JJ, Negri CM et al (2001) Predicting the effect of deep-rooted hybrid poplars on the groundwater flow system at a large scale phytoremediation site. Int J Phytoremediat 3:41–60

Ramana S, Biswas AK et al (2009) Phytoremediation of cadmium contaminated soils by marigold and chrysanthemum. Natl Acad Sci Lett 32:333–336

Reilley KA, Banks MK et al (1996) Dissipation of polycyclic aromatic hydrocarbons in the rhizosphere. J Environ Qual 25:212–219

Rhykerd RL, Crews B et al (1999) Impact of bulking agents, forced aeration, and tillage on remediation of oil-contaminated soil. Bioresour Technol 67:279–285

Rooney CP, Zhao F et al (2007) Phytotoxicity of nickel in a range of European soils: influence of soil properties, Ni solubility and speciation. Environ Pollut 145:596–605

Ruttens A, Boulet J et al (2011) Short rotation coppice culture of willows and poplars as energy crops on metal contaminated agricultural soils. Int J Phytoremediat 13:194–207

Saha JK (2005) Changes in salinity and sodicity of soils with continuous application of contaminated water near industrial area. J Indian Soc Soil Sci 53:612–617

Saha JK, Sharma AK (2006) Impact of the use of polluted irrigation water on soil quality and crop productivity near Ratlam and Nagda industrial area, Agricultural bulletin IISS-1. Indian Institute of Soil Science, Bhopal

Saha JK, Panwar N et al (2010) An assessment of municipal solid waste compost quality produced in different cities of India in the perspective of developing quality control indices. Waste Manag 30:192–201

Saha JK, Panwar N et al (2013) Risk assessment of heavy metals in soil of a susceptible agro-ecological system amended with municipal solid waste compost. J Indian Soc Soil Sci 61(1):15–22

Salminen J, Van Gestel CAM et al (2001) Pollution-induced community tolerance and functional redundancy in a decomposer food web in metal stressed soil. Environ Toxicol Chem 20:2287–2295

Santodonato J, Howard P et al (1981) Health and ecological assessment of polynuclear aromatic hydrocarbons. J Environ Pathol Tox 5:1–364

Santorufo L, Cornelis AM et al (2012) Soil invertebrates as bioindicators of urban soil quality. Environ Pollut 161:57–63

Saratale RG, Saratale GD et al (2010) Decolorization and biodegradation of reactive dyes and dye wastewater by a developed bacterial consortium. Biodegrad 21:999–1015

Sayara T, Pognani M et al (2010) Anaerobic degradation of PAHs in soil: impacts of concentration and amendment stability on the PAHs degradation and biogas production. Int Biodeter Biodegr 64:286–292

Schindler PW, Sposito G (1991) Surface complexation at (hydr)oxide surfaces. In: Bolt GHEA (ed) Interactions at the soil colloid-soil solution interface. Kluwer, Dordrecht, pp 115–145

Sellamuthu KM, Mayilswami C et al (2011) Effect of textile and dye industrial pollution on irrigation water quality of Noyyal River Basin of Tamil Nadu. Madras Agric J 98:129–135

Shanker AK, Cervantes C et al (2005) Chromium toxicity in plants. Environ Int 31:739–753

Shi G, Cai Q (2009) Cadmium tolerance and accumulation in eight potential energy crops. Biotechnol Adv 27:555–561

Silva CMMS, Fay EF (2012) Effect of salinity on soil microorganisms. In: Hernandez-Soriano MC (ed) Soil health and land use management, ISBN 978-953-307-614-0, doi: 10.5772/28613. Intech (Available from http://www.intechopen.com/books/soil-health-and-land-usemanagement/effect-of-salinity-on-soil-microorganisms)

Simonich SL, Hites RA (1994) Importance of vegetation in removing polycyclic aromatic hydrocarbons from the atmosphere. Nature 370:49–51

Singh OV, Labana S et al (2003) Phytoremediation: an overview of metallic ion decontamination from soil. Appl Microbiol Biot 61:405–412

Skłodowski P, Maciejewska A et al (2006) The effect of organic matter from brown coal on bioavailability of heavy metals in contaminated soils. Soil Water Pollut Monit Protec Remed 69:299–307

Somasundaram MV, Ravindran G et al (1993) Groundwater pollution of the Madras urban aquifer, India. Ground Water 31:4–11

Sparrow LA, Salardini AA et al (1994) Field studies of cadmium in potatoes (Solanum tuberosum L.). III. Response of cv. Russet Burbank to sources of banded potassium. Aust J Agr Res 45:243–249

Su DC, Wong JWC (2004) Selection of mustard oilseed rape (Brassica juncea L.) for phytoremediation of cadmium contaminated soil. Bull Environ Contam Toxicol 72:991–998

Sullivan DM, Miller RO (2001) Compost quality attributes, measurements, and variability. In: Stoffella PJ, Kahn BA (eds) Compost utilization in horticultural cropping systems. Lewis Publishers, New York, pp 97–120

Suzuki N (2005) Alleviation by calcium of cadmium-induced root growth inhibition in *Arabidopsis* seedlings. Plant Biotech 22:19–25

Tabatabai MA (1982) Soil enzymes. In: Page AL, Miller RH, Keeney DR (eds) Methods of soil analysis, part 2, Chemical and microbiological properties, Agronomy Monograph no 9. ASA-SSSA, Madison, pp 903–947

Teng Y, Luo Y et al (2010) Influence of arbuscular mycorrhiza and *Rhizobium* on phytoremediation by alfalfa of an agricultural soil contaminated with weathered PCBs: a field study. Int J Phytoremediat 12:516–533

Thomas W (1986) Accumulation of airborne trace pollutants by arctic plants and soil. Water Sci Technol 18:47–57

Topaç FO, Dindar E, Uçaroğlu S, Başkaya HS (2009) Effect of a sulfonated azo dye and sulfanilic acid on nitrogen transformation processes in soil. J Hazard Mater 170:1006–1013

Tripathi S, Chakrabarty A et al (2007) Enzyme activities and microbial biomass in coastal soils of India. Soil Biol Biochem 39:2840–2848

Trzesicka-Mlynarz D, Ward OP (1995) Degradation of polycyclic aromatic hydrocarbons (PAHs) by a mixed culture and its component pure cultures, obtained from PAH-contaminated soil. Can J Microbiol 41:470–476

Turrio-Baldassarri L, Abate V et al (2007) A study on PCB, PCDD/PCDF industrial contamination in a mixed urban-agricultural area significantly affecting the food chain and the human exposure. Part I: Soil and feed. Chemosphere 67:1822–1830

Tyler G (1981) Heavy metals in soil biology and biochemistry. In: Paul EA, Ladd JN (eds) Soil biochemistry, vol 5. Marcel Dekker, New York, pp 371–414

Vangronsveld J, Cunningham SD (eds) (1998) Metal-contaminated soils: in situ inactivation and phytorestoration. Springer-Verlag/R G Landes, Georgetown

Varvel GE, Riedell WE et al (2006) Great plains cropping system studies for soil quality assessment. Renew Agric Food Syst 21:3–14

Walker DJ, Clemente R et al (2004) Contrasting effects of manure and compost on soil pH, heavy metal availability and growth of *Chenopodium album* L. in a soil contaminated by pyritic mine waste. Chemosphere 57:215–224

Wang KR (2002) Tolerance of cultivated plants to cadmium and their utilization in polluted farmland soils. Acta Biotechnol 22:189–198

Wang S, Mulligan CN (2004) Rhamnolipid foam enhanced remediation of cadmium and nickel contaminated soil. Water Air Soil Pollut 157:315–330

Webber MD, Pietz RI et al (1990) Plant uptake of PCBs and other organic contaminants from sludge-treated coal refuse. J Environ Qual 23:1019–1026

Weissenfels WD, Klewer H et al (1992) Adsorption of polycyclic aromatic hydrocarbons (PAHs) by soil particles: influence on biodegradability and biotoxicity. Appl Microbiol Biot 36:689–696

Wetzel A, Werner D (1995) Ecotoxicological evaluation of contaminated soil using the legume root nodule symbiosis as effect parameter. Environ Toxic Water Qual 10:127–133

White CM, Lee ML (1980) Identification and geochemical significance of some aromatic components of coal. Geochim Cosmochim Acta 44:1825–1832

Willaert G, Verloo M (1992) Effect of various nitrogen fertilizers on the chemical and biological activity of major and trace elements in a cadmium contaminated soil. Pedologie 43:83–91

Yang Z, Liu S et al (2006) Effects of cadmium, zinc and lead on soil enzyme activities. J Environ Sci 18:1135–1141

Zayed AM, Terry N (1994) Selenium volatilization in roots and shoots: effects of shoot removal and sulphate level. J Plant Physiol 143:8–14

Zhou Q (2001) Chemical pollution and transport of organic dyes in water-soil-crop system of the Chinese coasts. Bull Environ Contam Toxicol 66:784–793

Targeting of Metabolic Pathways for Genetic Engineering to Combat Abiotic Stress Tolerance in Crop Plants

Shivangi Chamoli and A.K. Verma

Abstract

Abiotic stress is a serious threat to sustainable agriculture. Plant adaptation to suboptimal environmental conditions is controlled by cascades of molecular networks involved in stress perception, signal transduction, activation of new biochemical pathways, and repression of others. Protective metabolic adaptations alter physiological homeostatic of the whole plant. Use of modern molecular biology tools for elucidating abiotic stress tolerance relies on expression of specific stress-related gene and gene encoding enzymes present in biosynthetic pathways of functional and structural metabolites. Paramount among the mechanisms are reactive oxygen species scavenging, maintenance of ion uptake and water balance, and accumulation of compatible solutes such as betaines, proline, and alcohol sugars. Instead of single gene manipulation approach, targeting the regulatory machinery involving transcription factors has emerged as new potent tool for developing stress-tolerant transgenic crops. Under this chapter we highlight recent advances to our knowledge that emphasize the development of transgenic crops with improved stress tolerance by targeting different genes of various metabolic pathways.

Keywords

Abiotic stress • Reactive oxygen species • Transcription factors (TFs) • Metabolic engineering in plants

Introduction

Plants in natural environment encounter a wide range of unfavorable conditions such as flooding, drought, salinity, freezing, chilling, high temperature, and strong light or shade which collectively termed as abiotic stress. Abiotic stresses negatively influence growth and productivity of

S. Chamoli • A.K. Verma (✉)
Department of Biochemistry, College of Basic Sciences and Humanities, G.B.P.U.A&T, Pantnagar, Uttarakhand 263145, India
e-mail: akv72@rediffmail.com

R.K. Gaur and P. Sharma (eds.), *Approaches to Plant Stress and their Management*,
DOI 10.1007/978-81-322-1620-9_2, © Springer India 2014

crop and, in extreme cases, cause death of the plant (Tony and Norio 2002). Among these abiotic stresses, drought and salinity are most severe limiting factors affecting the productivity of most field crops to variable degrees, depending on the onset time, duration, intensity of the stress, and degree of susceptibility of a plant species. It is expected that by the year 2050, more than 50 % of agriculture land of different geographical regions would be under drought and salinity (Wang et al. 2001).

Stress perceptions are translated into cascade of altered, morphological, physiological, biochemical, and molecular changes in plants (Shinozaki and yamaguchi-shinozaki 1997a). Plants have evolved to adapt to adverse conditions through cross-wired metabolic pathways to reprogram the progression of development (Hirt 1997). Stress-mediated responses are trigged to reestablish homeostasis and to repair damaged protein. In contrast to plant resistance to biotic stresses, which is mostly dependent on monogenic traits, the genetically complex responses to abiotic stresses are multigenic and thus more difficult to control and engineer. Plant engineering strategies for abiotic stress tolerance rely on the expression of genes that are involved in signaling and regulatory pathways or genes that encode proteins conferring stress tolerance or enzymes present in pathways leading to the synthesis of functional and structural metabolites (Park et al. 2004; Apse and Blumwald 2002; Rontein et al. 2002). Although the conventional approaches in both plant breeding and physiology are of great importance (Vinocur and Altman 2005; Flowers 2004), the genetic engineering of key regulatory genes that govern a subset of stress-related genes appears to be one of the most promising strategies for enabling scientists to minimize the deleterious effects associated with abiotic stress. The complex plant response to abiotic stress, which involves many genes and biochemical-molecular mechanisms, is schematically represented in Fig. 1.

A wide range of metabolites that can prevent these detrimental changes have been identified, including amino acids (e.g., proline), quaternary and other amines (e.g., glycine betaine and polyamines), and a variety of sugars and sugar alcohols (e.g., mannitol and trehalose). Similarly, overexpression of certain enzymes such as superoxide dismutase, ascorbate peroxidase, and glutathione reductase has been implicated in detoxification and scavenging of free radical under oxidative stress (Allen et al. 1997). The use of aquaporins for developing transgenic plants with improved tolerance to abiotic stress resulted in contrasting results. Arabidopsis plants expressing the wild soybean (*Glycine soja*) tonoplast intrinsic protein (TIP), GsTIP2;1, showed more sensitivity to salt and dehydration presumably due to enhanced water loss of the transgenic plants (Wang et al. 2011).

This chapter summarizes the recent progress in utilizing transgenic plant technology for improvement of abiotic stress tolerance using research targeted at drought, salinity, and other abiotic stress with focusing on engineering of stress-specific genes involved in different metabolic pathways in sub-stressed plant.

The Single Gene Target Approach

Most common strategy for improving abiotic tolerance in crop is manipulating single gene encoding enzymes associated with the accumulation of osmolytes, proteins, and enzymes that function scavenging reactive oxygen species (ROS), molecular chaperones, and ion transporters.

Betaines

Betaines are fully N-methyl-substituted quaternary ammonium compounds. The most common betaines in plants include glycine betaine (GB; the most widely studied betaine), as well as proline, betaine, β-alanine betaine, choline-*O*-sulfate, and 3-dimethylsulfoniopropionate (McNeil et al. 1999). Glycine betaine is a widely distributed osmo-protectant in higher plants, accumulating in chloroplast and other plastids of many species at elevated rates induced under various stress conditions. In vitro studies have exhibited unusually strong stabilizing effect of GB on enzymes and complex protein both

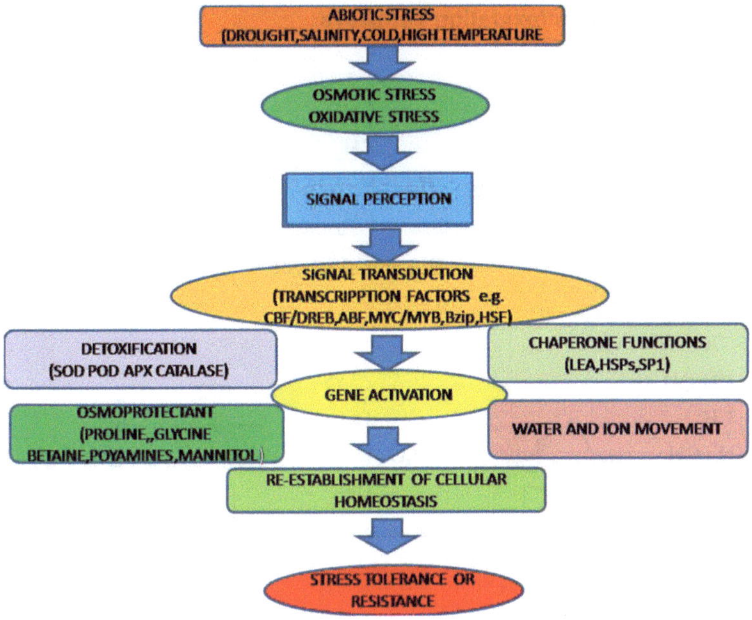

Fig. 1 Complex plant response to abiotic stress. Secondary stress such as osmotic and oxidative stress is a result of different interconnected primary stress (drought, salinity, heat cold, etc.) triggering downstream signaling resulting in activation of stress-responsive mechanisms to reestablish cellular homeostasis and protect plant under stress conditions. Inadequate responses might lead to cell death. *ABF* ABRE binding factor, *AtHK1* Arabidopsis thaliana histidine kinase-1, *APX* ascorbate peroxidase, *bZIP* basic leucine zipper transcription factor, *CBF/DREB* C-repeat-binding factor/dehydration-responsive binding protein, *Hsp* heat-shock protein, *LEA* late embryogenesis abundant, *POD* peroxidase, *SOD* superoxide dismutase, *SP1* stable protein 1

structurally and functionally of photosynthetic and translational machinery, as well as involved in protecting cell membrane mainly at salt and cold stress(Chen and Murata 2002). But some species such as *Arabidopsis*, rice (*Oryza sativa*), tobacco (*Nicotiana tabacum*), potato, and tomato are considered to be non-accumulators of GB (McNeil et al. 1999). Therefore, these plants are potential candidate for introduction of GB biosynthesis pathway, increasing their tolerance of various abiotic stress (McCue and Hanson 1990). Choline and glycine act as precursor for biosynthesis of GB via two distinct pathways: dehydrogenation of choline and N-methylation of glycine.

Choline-Dehydrogenation/Oxidation Pathway

It is two-step oxygenation reaction where choline is oxidized to glycine betaine via the unstable intermediate betaine aldehyde, present in almost all biological system including most animals, plants, and microorganisms (Fig. 2). In higher plants, glycine betaine is synthesized in the chloroplast from choline in two-step reaction: the first step (choline to betaine aldehyde) is mediated by choline monooxygenase (CMO) induced by drought and salinity (McNeil et al. 2000). The second step (betaine aldehyde to glycine betaine) is catalyzed by betaine aldehyde dehydrogenase (BADH), an NAD-dependent dehydrogenase. In *E. coli*, CMO is replaced with enzyme choline dehydrogenase (CDH); rest mechanism of biosynthesis is same to higher plants. Whereas in microorganisms *Arthrobacter globiformis* and a closely related strain *Arthrobacter pascens*, GB is synthesized in single-step reaction catalyzed by only one enzyme choline oxidase (COD,E.C. 1.1.3.17) (Ikuta et al. 1977). Recently, two extremely halophilic microorganisms, *Actinopolyspora halophila* and *Ectothiorhodospira halochloris, are*

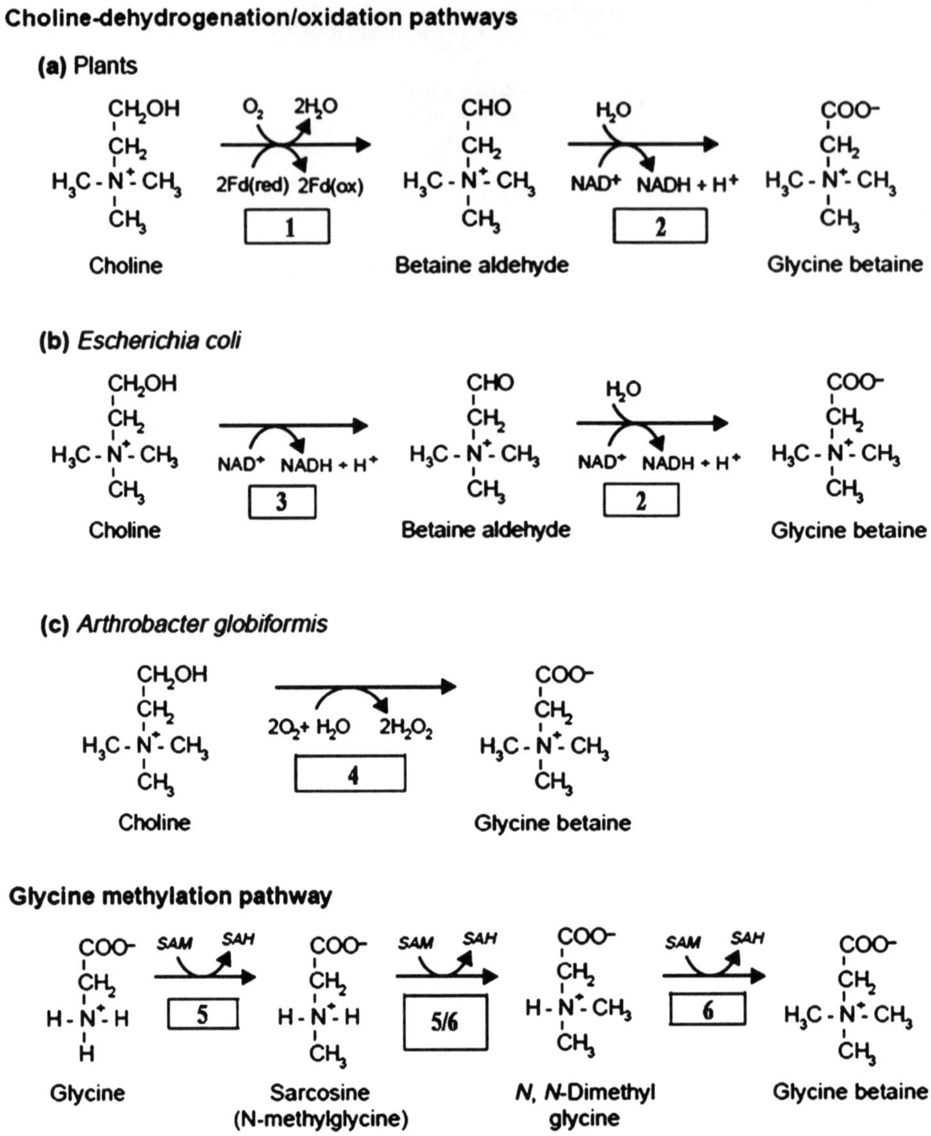

Fig. 2 Biosynthetic pathway for glycine betaine in different organisms. *1*-choline monooxygenase (CMO), *2*-betaine aldehyde dehydrogenase(BADH), *3*-choline dehydrogenase (CDH), *4*-choline oxidase (COD), *5*-glycine sarcosine methyltransferase (GSMT), *6*-sarcosine dimethyltransferase (SDMT); Fd(red) and Fd(ox), ferredoxin in reduced oxidized forms, respectively; *SAH* S-adenosylhomocysteine, *SAM* S-adenosylmethionine

discovered with novel biosynthetic pathway for GB. It is a three-step methylation series where glycine is successively methylated at its amino residue by two methyltransferase, glycine sarcosine methyltransferase (GSMT) and sarcosine dimethylglycine methyltransferase (SDMT), having partially overlapping substrates specificity. *S*-adenosylmethionine acts as the methyl group donor above reactions (Nyyssola et al. 2000). Genes encoding different enzymes (CMO, BADH,CDH,GSMT, and SDMT) of biosynthetic pathway of GB from different sources have been cloned in GB non-accumulating species and have exhibited enhanced tolerance at high-salt concentration and extreme temperature (Tony and Norio 2002).

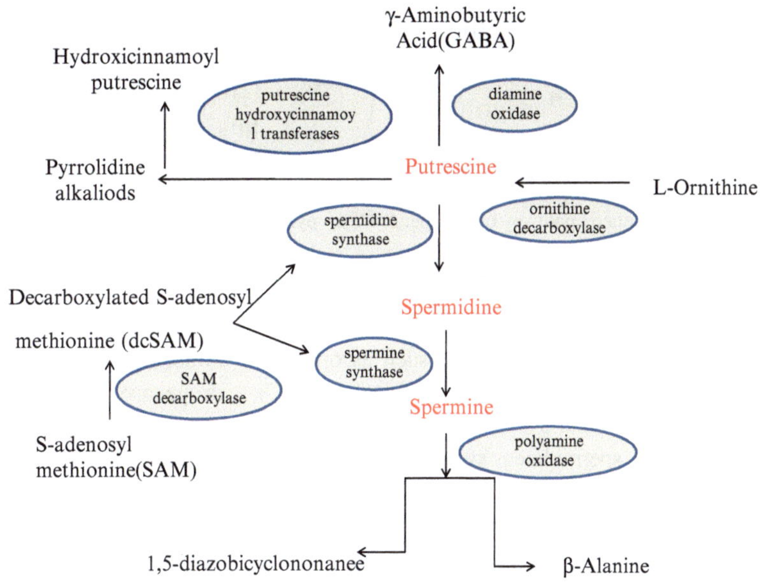

Fig. 3 Polyamine metabolic pathway

However, the observed concentrations of GB in such transgenic plants are generally lower than (<5 μmol g^{-1}) fresh weight (fw) compared with the levels observed in stressed plants of species that normally accumulate GB when under stress (4–40 μmol g^{-1} fw) (Rhodes and Hanson 1993). There are two main constraints limiting accumulation of GB in transgenic plants; endogenous choline source and the transport of choline across the chloroplast envelope (McNeil et al. 1999; Nuccio et al. 2000). Phosphoethanolamine N-methyltransferase (PEAMT; EC 2.1.1.103) is the key enzyme of choline biosynthetic pathway catalyzing all three methylation reactions converting phosphoethanolamine to phosphocholine. Transgenic tobacco plants, co-expressing both spinach CMO and beet BADH, were transformed with PEAMT gene isolated from spinach, significantly synthesize GB in their chloroplast (McNeil et al. 2001). In a research, chloroplastic expression of spinach CMO gene in tobacco plants showed low level of CMO activity and produced little GB (≤ 70 nmol g^{-1} fw) (Nuccio et al. 2000.) Moreover, increasing CMO activity of up to 100-fold among individual transformants did not produce any significant effects on the GB concentration (Nuccio et al. 2000). However, cytosolic expression of CMO resulted in about fivefold more GB accumulation (430 nmol g^{-1} fw) than chloroplastic expression in transformant. A detailed model of the labeling kinetics of choline metabolites with [^{14}C]-choline to transgenic plants demonstrated that the import of choline into chloroplasts limits the synthesis of GB in the chloroplasts.

Polyamines

Polyamines (PAs) are low molecular weight polycationic compounds present in all living organisms. Most commonly found polyamines are the diamine putrescine (Put), the triamine spermidine (Spd), and the tetramine spermine (Spm) (Fig. 3). In most of living organisms, major pathway of PA synthesis is decarboxylation of ornithine via ornithine decarboxylase, and the first PA synthesized is putrescine (Put). Plants and bacteria contain an alternative route to Put production by decarboxylation of arginine by arginine decarboxylase. By sequential addition of aminopropyl moieties to the Put skeleton, higher molecular weight PAs are produced through enzymes spermidine and spermine synthases. Decarboxylated S-adenosylmethionine (dcSAM) acts as donor of aminopropyl groups in

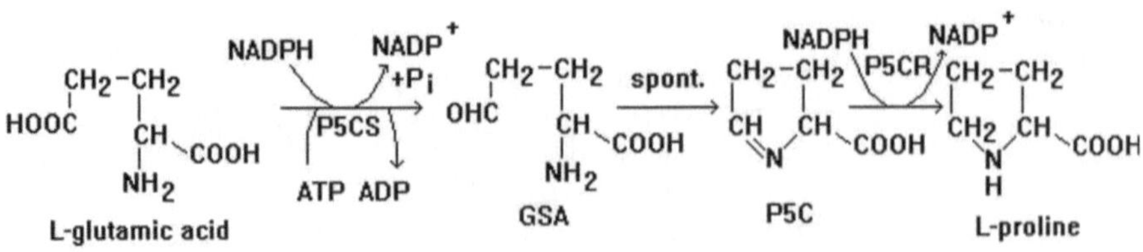

Fig. 4 Biosynthetic pathway of proline. *P5CS* Δ^1-pyrroline-5-carboxylase synthase, *P5CR* Δ^1-pyrroline-5-carboxylate reductase

above reactions. The modification of PA levels by the overexpression of genes such as ornithine or arginine decarboxylases (ODC, ADC), S-adenosylmethionine (SAM) decarboxylase (SAMDC), spermidine synthase (SPDS) in Arabidopsis, tobacco rice (*Oryza sativa*) (Capell et al. 2004), potato (*Solanum tuberosum*) (Kasukabe et al. 2006), and eggplant (*Solanum melongena*) (Prabhavathi and Rajan 2007) was reported to result in enhanced tolerance of these species to different abiotic stresses (Fig. 3).

Proline

Proline is most common osmoticum present in high concentration in glycophytes and halophytes in response to osmotic stress such as drought and high salinity. It acts as carbon and nitrogen sink, functioning as molecular chaperone stabilizing the structure of proteins, and regulator of the redox potential of cells (Csonka 1989; Hare et al. 1997). Promotion of biosynthesis of proline and repression of its degradation pathway jointly result in proline accumulation in plants under stress, establishing "proline homeostasis cycle," depending on the physiological state of tissue (McNeil et al. 1999). Two key enzymes of proline biosynthetic pathways, i.e., Δ^1-pyrroline-5-carboxylate synthetase (P5CS) and Δ^1-pyrroline-5-carboxylate reductase (P5CR), are cloned and overexpressed in several plants to achieve increment in proline level under various abiotic stress (Fig. 4). Overexpression of P5CS in tobacco plant was found to elevate proline content approximately by 2.5-fold after water stress (Konstantinova et al. 2002). Another important enzyme involved in proline catabolism is proline dehydrogenase. For upregulation of proline anabolism, a different approach in engineering Arabidopsis either via knockout of proline dehydrogenase enzyme or transformation with antisense of this enzyme was done and observed significant accumulation of free proline enhancing tolerance to stress (Mani et al. 2002; Nanjo et al. 2003). Δ^1-Pyrroline-5-carboxylase synthase (P5CS) is a rate-limiting enzyme subject to feedback inhibition by proline. A mutated enzyme (P5CSF129A), generated through site-directed mutagenesis via replacing the Phe residue at position 129 in P5CS with an Ala residue, in *V. aconitifolia,* remained no longer inhibited. Transformant expressing P5CSF129A grown in salinity of up to 200 mM NaCl accumulated higher concentration of proline about twice than wild-type expressing non-mutated P5CS gene and observed better growth (Hong et al. 2000). Monitoring the MDA production accentuated the correlation of elevated proline concentration and reduction of free radical levels. These findings indicate that, in addition to serving as an osmolyte, proline might play a role in reducing the oxidative stress that is brought on by osmotic stress.

Mannitol

Mannitol is another sugar alcohol scavenging hydroxyl radicals and stabilization of macromolecular structures (Abebe et al. 2003). Wheat, tobacco, and Arabidopsis are not natural synthesizer of mannitol. However, expression of the *E. coli mtlD* gene for mannitol-1-phosphate dehydrogenase resulted in the biosynthesis of mannitol in transgenic of these species and alleviates abiotic stress. It is reported that more than 6 μmol g^{-1} fw of mannitol accumulated in the leaves of transgenic tobacco plants and a maximum concentration of 3 μmol g^{-1} fw in the leaves of transgenic *Arabidopsis* (Tarczynski et al. 1992; Thomas et al. 1995). In spite of an increase in biomass, plant height and number of tillers (secondary stems in grasses) transgenic wheat plant accumulate considerably low amount of mannitol which elucidated it a relatively minor osmolyte and suggested that it might indirectly enhanced osmotic adjustments and salt tolerance acting as a ROS scavenger (Abebe et al. 2003).

Trehalose

Trehalose, also known as tremalose or mycose, a nonreducing disaccharide, is widely spread in biological systems contributing desiccation tolerance against multiple abiotic stresses (Bartels and Sunkar 2005). Trehalose accumulation in response to adverse conditions in higher plants is rare phenomenon. The possible elucidation for this is presence of sucrose in fairly large quantity and also functioning as preservative against desiccation over trehalose (Oscar et al. 1999). The low levels of trehalose in transgenic plants can be explained by specific trehalase activity, which degrades trehalose; hence, it might be possible to increase trehalose accumulation by downregulating trehalase activity. However, trehalose is found in some resurrection plants involved in effectively stabilizing dehydrated enzymes and lipid membranes. Reversible water-absorption capacity of trehalose protects biomolecules from desiccation- induced damage. Therefore, drought tolerant plants can be produced cost-effectively either by regulating trehalase activity (Goddijn et al. 1997) or expression of trehalose synthesis-related genes (Jang et al. 2003) (Shujun et al. 2010). In positively transformed tobacco with the gene for the trehalose-6-phosphate synthase (TPS1) subunit of yeast trehalose synthase, by the promoter of the *rbcS* gene (Rubisco small subunit) from *Arabidopsis*, 0.8–3.2 mg g^{-1} dw trehalose is quantified, but growth rate of plants reduced up to 30–50 % (Holmström et al. 1996). Potato (*Solanum tuberosum*) is also engineered with yeast gene for TPS1, driven by the 35S promoter of CaMV rendered tolerance along some sever morphological changes including severely retarded growth to yellowish, lancet-shaped leaves and the aberrant development of roots (Yeo et al. 2000). In general, ABA-inducible (rd29A) promoter was used to avoid stunted growth of the transgenic plants (Jang et al. 2003; Suarez et al. 2009). Pleiotropic effects (e.g., necrosis and growth retardation) are result of disturbance in endogenous pathways of primary metabolites. So overproduction of compatible solutes on cost of primary metabolites should be avoided. Overproduction of targeted gene should be strictly tissue specific and stress inducible (Garg et al. 2002).

Fructans

Fructans are highly polymerized polyfructose molecules stored in vacuoles in soluble form mainly in species experiencing water stress conditions. Construct containing Sac B gene from *Bacillus subtilis* encoding levansucrase enzyme fused to yeast vacuolar sorting signal of carboxypeptidase Y placed downstream CaMV35S promoter used to transform *N. tabacum* produced fructans from fructose (Pilon et al. 1995). Same construct also used to transform sugar beets which accumulated fructan both in root and shoot up to 5 % of their dry weight (Pilon et al. 1999). Fructans synthesizing species show better survival rate under unfavorable environment conditions. Tolerance and better growth in plants might be due to pronounced development of rooting system.

Protection Mechanism Against Damaging Effect of ROS

Production of reactive oxygen species is another inevitable aspect of photosynthetic organism exposed to environmental stress. Under stress, uncontrolled ROS increase leads to production of singlet oxygen, superoxide, H_2O_2, and OH^- radicals which are highly reactive species causing large cellular damage. ROS detoxification system comprises of two components existing in all plants – enzymatic (e.g., superoxide dismutase (SOD), ascorbate peroxidase (ASX),

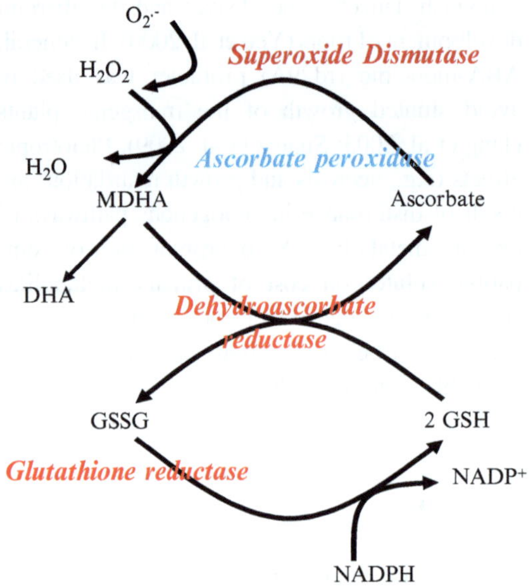

Fig. 5 The Halliwell–Asada pathway for detoxification of reactive oxygen species

glutathione cycle (GST/GPX)) and nonenzymatic (flavonones, anthocyanins, carotenoids, ascorbic acid, etc.). *The Halliwell–Asada pathway* is important in detoxifying reactive oxygen intermediates (Fig. 5). These are produced naturally by the electron-transport chains of mitochondria and especially chloroplasts. Superoxide radicals are eliminated by superoxide dismutase in a reaction that yields hydrogen peroxide. H_2O_2 is consumed through its conversion to oxygen and water by catalase or to water or water alone through the oxidation of ascorbate. Ascorbate is generated by two mechanisms. The enzymatic reduction of monodehydroascorbate takes place in plastids. Alternatively, monodehydroascorbate that is spontaneously dismutated to dehydroascorbate can react with glutathione (GSH) to produce ascorbate and oxidized glutathione (GSSG) in a reaction catalyzed by dehydroascorbate reductase. GSSG is reduced by glutathione reductase, requiring NADPH. Singlet oxygen and hydroxyl ions are eliminated in glutathione pathway.

Genes of enzymes involved in above pathways are targeted to generate transgenic plants of improved abiotic tolerance.

Targeting Ion Transporters at Gene Level to Reestablish Ion Homeostasis in Plants Under Abiotic Stress

Osmotic stress, ion toxicity, and high-salt content in the soil (especially Na^+ and Cl^-), significantly impair plant growth. Ion transporters selectively transport ions and maintain them at physiologically relevant concentrations while Na^+/H^+ antiporters also play a crucial role in maintaining cellular ion homeostasis, thus permitting plant survival and growth under saline conditions. The Na^+/H^+ antiporters catalyze the exchange of Na^+ for H^+ across membranes and have a variety of functions, such as regulating cytoplasmic pH, sodium levels, and cell turgor (Serrano et al. 1999). Plant Na^+/H^+ antiporters have been isolated from Arabidopsis (AtNHX1, SOS1; Shi et al. 2000) and rice plants (Fukuda et al. 1999) and from the halophytic plants *Atriplex gmelini* (Hamada et al. 2001) and *Mesembryanthemum crystallinum* (Chauhan et al. 2000). Overexpression of the vacuolar Na^+/H^+ antiporter AtNHX1 in Arabidopsis plants (Apse et al. 1999) promoted growth and development in potting medium irrigated with up to 200 mM sodium chloride. This salinity tolerance was positively correlated with elevated levels of AtNHX1transcript and with protein and vacuolar Na^+/H^+ antiporter activity. Transgenic *Brassica napus* plants overexpressing AtNHX1 were able to grow flower and produce seeds, in the presence of 200 mM sodium chloride, even though they accumulated sodium at a rate of up to 6 % dry weight. Moreover, their seed yields and seed oil quality were not altered by the high soil salinity (Zhang et al. 2001). Similarly, transgenic tomato plants overexpressing this gene were able to grow flower and produce fruit in the presence of 200 mM sodium chloride (Zhang and Blumwald 2001). Although the tomato leaves accumulated high sodium concentrations, the fruits displayed very low sodium content, demonstrating the potential to maintain fruit yield and quality at high-salt levels. The *A. thaliana* plasma membrane Na^+/H^+ antiporter, encoded by the SOS1 gene, was suggested to be essential for salt tolerance (Shi et al. 2002) and recently (Shi et al. 2003) reported that overexpression of SOS1 improves salt tolerance in transgenic Arabidopsis. Shoots from transgenic melon plants expressing the HAL1 gene showed some level of salt tolerance in vitro (Bordas et al. 1997), and transgenic tomato lines expressing the HAL1gene were found to be more salt tolerant than the wild-type plants, as judged by both callus and plant growth in short-term experiments (Gisbert et al. 2000). Such transgenic lines also demonstrated better fruit yield under salt stress (Rus et al. 2001). In plants, protons are used as coupling ions for ion transport systems, and the proton gradient, generated by proton pumps found in the cell membrane, is the driving force for nutrient uptake (Serrano et al. 1999). Three distinct proton pumps are responsible for the generation of the proton electrochemical gradients (Sze et al. 1999): (i) the plasma membrane H^+ATPase pump (PM H^+ATPase) which extrudes H^+ from the cell and thus generates a proton motive force, (ii) the vacuolar-type H^+ATPase pump (V-ATPase), and (iii) the vacuolar H^+ pumping pyrophosphatase pump (H^+ATPase). The latter two acidify the vacuolar lumen and other endomembrane compartments. Arabidopsis plants were transformed with a vacuolar H^+ATPase pump that is encoded by a single gene, AVP1 (Gaxiola et al. 2001), which can generate an H^+ gradient across the vacuolar membrane, similar in magnitude to that of the multisubunit vacuolar H^+ATPase pump. These transgenic plants expressed higher levels of AVP1 and were more resistant to salt and drought than wild-type plants. It was also found that the resistant phenotypes had an increased vacuolar proton gradient, resulting in increased solute accumulation and water retention.

Manipulation of Stress-Induced Protein

To cope with environmental stress, plants activate a large set of genes leading to the accumulation of specific stress-associated proteins (Vierling 1991; Ingram and Bartels 1996). Heat-shock proteins (Hsps) and late embryogenesis-abundant (LEA)-type proteins are two major types of stress-induced proteins that accumulate upon water, salinity, and extreme temperature stress. They have been shown to play a role in cellular protection during the stress (Ingram and Bartels 1996).

Heat-Shock Protein

Abiotic stress leads to dysfunctioning of enzymes and proteins. For survival under stress, prevention of aggregation of nonnative protein and maintenance of protein in their functional conformation are two important aspects. Hsps act as molecular chaperones, which are responsible for controlling proper folding and conformation of both structural and functional protein (Toäroäk et al. 2001). Among five conserved families of Hsps (Hsp100, Hsp90, Hsp70, Hsp60, and sHsp), the small heat-shock proteins (sHsps) are found to be most prevalent in plants, varying in size from 12 to 40 kDa (Veinger et al. 1998). Recently, it was suggested that sHsps might act as antioxidants in protecting Complex-I electron transport in mitochondria during NaCl stress (Hamilton and Heckathorn 2001). In addition, sHsps are involved in many developmental processes, such as embryo development, seed germination, somatic embryogenesis, pollen development, and fruit maturation (Waters et al. 1996). The results lead us to suggest that Hsps overproduction may also protect plants from oxidative stress. It has demonstrated that the expression of Arabidopsis AtHSP17.6A is regulated by heat shock and osmotic stress and is induced during seed maturation (Sun et al. 2001).

LEA Protein

LEA proteins are high molecular weight proteins that are abundant during late embryogenesis and play crucial roles in cellular dehydration tolerance resulting from desiccation, cold, and osmotic stress (Galau et al. 1987). Hydrophilicity is a common characteristic of LEA-type and other osmotic stress-responsive proteins. Heat stability is another notable feature of LEA proteins, i.e., they do not coagulate upon boiling (Close et al. 1989). Another common characteristic of LEA-type proteins is that, in most cases, their related gene expression is transcriptionally regulated and responsive to ABA (Mundy and Chua 1988). It has been suggested that LEA-type proteins act as water-binding molecules in ion sequestration and in macromolecule and membrane stabilization (Ingram and Bartels 1996). Dehydrins are a subfamily of group 2 LEA proteins (Battaglia et al. 2008) that accumulate in vegetative tissues subjected to drought, salinity, and cold. Strawberry (*Fragaria* × *anassassa*) overexpressing a wheat dehydrin WCOR410 gene showed improved leaf freezing tolerance. Recently, it has been demonstrated that the dehydrin gene Lti30 is involved in cold stress tolerance by interacting electrostatically with vesicles of both zwitterionic (phosphatidylcholine) and negatively charged phospholipids (phosphatidyl glycerol, phosphatidylserine, and phosphatidic acid) (Eriksson et al. 2011).

Role of ABA in Stress-Mediated Responses

Phytohormones regulate every aspect of plant growth, development, and the responses of plants to environmental cues. The hormonal response machinery rapidly alters gene expression by inducing, preventing, or controlling the degradation of regulators as TFs via the ubiquitin-–proteasome pathway (Santner and Estelle 2010). One of the primary plant responses to stress is the accumulation of ABA which results in stomatal closure and reduced water loss via transpiration. The phytohormone (+)-abscisic acid (ABA) plays a key role in plant adaptation to adverse environmental conditions including drought stress. Numerous studies have shown that ABA accumulation is a key factor in controlling downstream responses essential for adaptation to stress. However, molecular and genomic analyses have suggested that both ABA-dependent and ABA-independent regulatory systems are involved in stress-responsive gene expression (Shinozaki and Yamaguchi-Shinozaki 1997b, 2000).

In a previous study, it was established that at least 14 % of Arabidopsis genes were ABA regulated (Huang et al. 2007). In that study, an ABA analogue, 8 %-acetylene ABA (PBI425), was used to reveal and confirm weakly ABA-regulated genes based on its ability to hyperinduce ABA-related gene expression. Transgenic rice plants overexpressing LOS5/ABA3, a key enzyme in the last step of ABA biosynthesis, showed improved yield in the field under drought stress (Xiao et al. 2009). In tomatoes, overexpression of LeNCED1 (a drought-inducible gene encoding a rate-limiting enzyme in ABA biosynthesis) resulted in increased ABA accumulation and improved drought tolerance (Thompson et al. 2007). ERA1 encodes the β-subunit of farnesyl transferase, an enzyme associated with ABA signaling. Transgenic canola carrying era1 antisense (driven by the drought-inducible rd29A promoter) displayed enhanced yield under mild drought conditions in the field (Wang et al. 2005). These results further highlight the need of specific promoters to control gene expression and to avoid negative effects (Thompson et al. 2007) (Fig. 6).

Manipulation of Multiple Genes Simultaneously

Simultaneous Targeting of Different Components of Complexed Stress-Induced Pathways

Plants grown in field face heterogenous conditions at different developmental stages and for varying duration. Tolerance to abiotic stress is a complex combination of genetic and environmental interactions that transduce physiological, molecular, and biochemical responses. Simultaneous expression modification of various components is a promising new approach to generate potent response against combinations of stresses. There are only few examples where the simultaneous co-expression of different components of the same pathway has been tried. Increase in biosynthesis of proline was achieved by co-expression of *E. coli* P5C biosynthetic enzymes gamma-glutamyl kinase 74 (GK74) and gamma-glutamyl phosphate reductase (GPR) and the antisense transcription of proline dehydrogenase (ProDH) in Arabidopsis and tobacco (Stein et al. 2011). The transgenic plants displayed improved tolerance to heat stress associated with the accumulation of cell wall

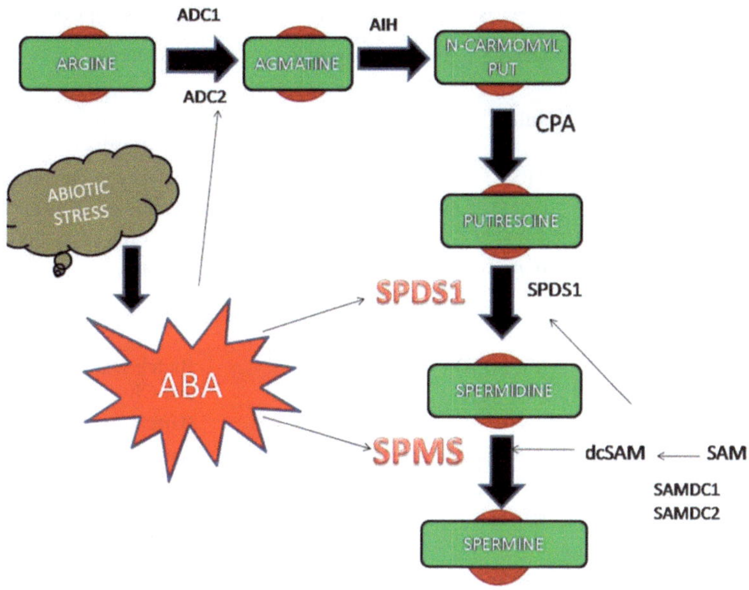

Fig. 6 Scheme of the transcriptional regulation of PA biosynthesis by ABA. Drought stress leads to an increase in ABA levels which enhances the expression of ABA-responsive *ADC2*, *SPDS1*, and *SPMS* genes. The increase in *ADC2* expression leads to Put accumulation, whereas increases in the expression of *SPDS1* and *SPMS* do not lead to accumulation of Spd or Spm

proline-rich proteins (Stein et al. 2011). Simultaneous co-expression of dehydroascorbate reductase (DHAR), glutathione reductase (GR) or glutathione-S-transferase (GST), and glutathione reductase (GR) in tobacco plants also resulted in the increased tolerance of the transgenic plants to a variety of abiotic stresses (Martret et al. 2011). In tobacco seeds, higher antioxidant enzymes activity driven by the simultaneous overexpression of the CuZnSOD and APX genes in plastids allowed the increase of germination rates and longevity of long-term stored seeds under combined stress conditions (Lee et al. 2010), demonstrating the enormous potential of simultaneous gene expression in plant engineering.

Targeting Regulatory Genes

Often crops are exposed to multiple stresses and respond to different environmental factors in complexed metabolic interactions. Therefore, single gene manipulation approach to improve crop tolerance to abiotic stress may not be sufficient. To overcome such constraints, targeting a gene encoding, a stress-inducible transcription factor that regulates a number of other genes, increases the probability of success to generate stress-tolerant plants (Yamaguchi-Shinozaki and Shinozaki 1994). Different families of TF such as ERF/AP2, HSF, bZIP, MYB, MYC, NFY, NAC, WRKY, Cys2His2, MADS box, and zinc finger have been shown to regulate the expression of stress-responsive genes (Hirayama and Shinozaki 2010). Transgenic maize constitutively expressing ZmNF-YB2 showed enhanced tolerance to severe drought stress in field trials (Nelson et al. 2007). Under water-limiting conditions, transgenic plants displayed improved grain yield, as well as reduced wilting, lower leaf temperature, etc. The NAC [NAM (no apical meristem), ATAF1-2, and CUC2 (cup-shaped cotyledon)] TF have been reported to be associated with abiotic stress. Transgenic rice overexpressing SNAC1 (stress-responsive NAC 1) showed increased yield when grown under drought stress field conditions, throughout the control of stomata movement and maintenance of photosynthetic activity (Hu et al. 2006). Likewise, the overexpression of two NAC genes, OsNAC5 and OsNAC6, resulted in stress-tolerant rice via the upregulation of the expression of stress-inducible genes such as OsLEA3 (Takasaki et al. 2010). Recently, expression of OsNAC10 under control of a root-specific promoter (RCc3) yielded more grain in the field under drought conditions (Jeong et al. 2010). The yield advantage of PRCc3::OsNAC10 transgenic rice plants was associated with a larger root diameter (Jeong et al. 2010).

Table 1 Classification, genes, and genetically modified plant species implicated in plant response to abiotic stress

Classification	Genes	Species	References
Compatible solutes			
Mannitol	mtlD	Wheat	Bartels and Sunkar (2005), Wang et al. (2003); Chaves and Oliveira (2004)
Trehalose	TPS1	Tobacco	Bartels and Sunkar (2005)
Glycine betaine	GSMT + DMT	*Arabidopsis*	Waditee et al. (2005)
Polyamines	SPDS	*Arabidopsis*	Kasukabe et al. (2004)
Proline	P5CS	Rice	Bartels and Sunkar (2005)
		Tobacco	Hong et al. (2000)
Antioxidants			
Detoxification	Mn-SOSD	Alfalfa	Bartels and Sunkar (2005); Wang et al. (2003); Chaves and Oliveira (2004)
Ion transport	AVP1	*Arabidopsis*	Bartels and Sunkar (2005); Wang et al. (2003)
ABA biosynthesis	AtNCED3	*Arabidopsis*	Bartels and Sunkar (2005)
Heat-shock protein	AyHsp17.6A	*Arabidopsis*	Bartels and Sunkar (2005)
LEA protein	HVA1	*O. sativa*	Xu et al. (1996)
Transcription control			
	CBF1	*Arabidopsis thaliana*	Jaglo-Ottosen et al. (1998)
	DREB1A	*A. thaliana*	Kasuga et al. (1999)
	CBF3	A. thaliana	Gilmour et al. (2000)
	CBFs	*Brassica napus*	Jaglo et al. (2001)
	AtMYC2 and AtMYB2	*A. thaliana*	Abebe et al. (2003)

Dehydration-responsive element (DRE)/C-repeat (CRT) proteins have been indentified to play important roles in drought, cold, and salinity response (Yamaguchi-Shinozaki and Shinozaki 1994). Overexpression of CBF1/DREB1B genes resulted in improved tolerance to drought, salinity, and temperature stress in model plants (Gilmour et al. 2000) and in crop plants such as rice, wheat, and canola (Dubouzet et al. 2003; Jaglo et al. 2001). At the same time, the transgenic plants showed negative phenological abnormalities such as severe growth retardation under control condition (Ito et al. 2006). This problem was reduced when using more specific promoter, such as the ABA-inducible (rd29a) promoter (Yang et al. 2010). The DRE and Ethylene responsive element binding factors belong to a large family of TFs, APETALA2/EREBP, which mediate signal transduction pathways in response to environmental cues.

Conclusion

Complex traits of abiotic stress phenomena in plants make genetic modification for efficient stress tolerance difficult to achieve. The use of transgenes to improve the tolerance of crops to abiotic stresses remains an attractive option. However, the modification of a single trait (e.g., TFs, antiporters, and others) resulted in several cases in significant improvements in stress tolerance, as discussed earlier. A well-focused approach combining the molecular, physiological, and metabolic aspects of abiotic stress tolerance is required for bridging the knowledge gaps between short-and long-term effects of the genes and their products and between the molecular or cellular expression of the genes and the whole plant phenotype under stress (Table 1).

References

Abebe T, Guenzi AC, Martin B et al (2003) Tolerance of mannitol-accumulating transgenic wheat to water stress and salinity. Plant Physiol 131:1748–1755

Allen RD, Webb RP, Schake SA (1997) Use of transgenic plants to study antioxidant defenses. Free Radic Biol Med 23:473–479

Apse MP, Blumwald E (2002) Engineering salt tolerance in plants. Curr Opin Biotechnol 13:146–150

Apse MP, Aharon GS, Snedden WA et al (1999) Salt tolerance conferred by overexpression of a vacuolar Na^+/H^+ antiport in Arabidopsis. Science 285:1256–1258

Bartels D, Sunkar R (2005) Drought and salt tolerance in plants. Crit Rev Plant Sci 24:23–58

Battaglia M, Olvera-Carrillo Y, Garciarrubio A et al (2008) The enigmatic LEA proteins and other hydrophilins. Plant Physiol 148:6–24

Bordas M, Montesinos C, Debauza M et al (1997) Transfer of the yeast salt tolerance gene HAL1 to *Cucumis melo* L. cultivars and in vitro evaluation of salt tolerance. Transgenic Res 6:41–50

Capell T, Bassie L, Christou P (2004) Modulation of the polyamine biosynthetic pathway in transgenic rice confers tolerance to drought stress. Proc Natl Acad Sci 101:9909–9914

Chauhan S, Forsthoefel N, Ran Y et al (2000) Na^+/myo-inositol symporters and Na^+/H^+-antiport in *Mesembryanthemum crystallinum*. Plant J 24:511–522

Chaves MM, Oliveira MM (2004) Mechanisms underlying plant resistance to water deficits: prospects for water-saving agriculture. J Exp Bot 55:2365–2384

Chen TH, Murata N (2002) Enhancement of tolerance of abiotic stress by metabolic engineering of betaines and other compatible solutes. Curr Opin Plant Biol 5:250–257

Close TJ, Kortt AA, Chandler PM (1989) A cDNA-based comparison of dehydration-induced proteins (dehydrins) in barley and corn. Plant Mol Biol 13:95–108

Csonka L (1989) Physiological and genetic responses of bacteria to osmotic stress. Microbio Rev 53:121–147

Dubouzet JG, Sakuma Y, Ito Y et al (2003) OsDREB genes in rice, *Oryza sativa* L., encode transcription activators that function in drought-, high-salt-and cold-responsive gene expression. Plant J 33:751–763

Eriksson SK, Kutzer M, Procek J et al (2011) Tunable membrane binding of the intrinsically disordered dehydrin Lti30, a cold-induced plant stress protein. Plant Cell 23:2391–2404

Flowers TJ (2004) Improving crop salt tolerance. J Exp Bot 55:307–319

Fukuda A, Nakamura A, Tanaka Y (1999) Molecular cloning and expression of the Na^+/H^+ exchanger gene in *Oryza sativa*. Biochim Biophys Acta 1446:149–155

Galau GA, Bijaisoradat N, Hughes DW (1987) Accumulation kinetics of cotton late embryogenesis-abundant (Lea) mRNAs and storage protein mRNAs: coordinate regulation during embryogenesis and role of abscisic acid. Dev Biol 123:198–212

Garg A, Kim J, Owens T et al (2002) Trehalose accumulation in rice plants confers high tolerance levels to different abiotic stresses. Proc Natl Acad Sci 99:15898–15903

Gaxiola RA, Li J, Undurraga S et al (2001) Drought-and salt-tolerant plants result from overexpression of the AVP1 H^+-pump. Proc Natl Acad Sci 98:11444–11449

Gilmour SJ, Sebolt AM, Salazar MP et al (2000) Overexpression of the Arabidopsis CBF3 transcriptional activator mimics multiple biochemical changes associated with cold acclimation. Plant Physiol 124:1854–1865

Gisbert C, Rus AM, Boları'n MC et al (2000) The yeast HAL1 gene improves salt tolerance of transgenic tomato. Plant Physiol 123:393–402

Goddijn O, Verwoerd TC, Voogd E et al (1997) Inhibition of trehalase activity enhances inhibition of trehalase activity enhances trehalose accumulation in transgenic plants. Plant Physiol 113:181–190

Hamada A, Shono M, Xia T et al (2001) Isolation and characterization of a Na^+/H^+ antiporter gene from the halophyte *Atriplex gmelini*. Plant Mol Biol 46:35–42

Hamilton EW III, Heckathorn SA (2001) Mitochondrial adaptations to NaCl. Complex I is protected by antioxidants and small heat shock proteins, whereas complex II is protected by proline and betaine. Plant Physiol 126:1266

Hare PD, Cress WA, Staden J (1997) The involvement of cytokinins in plant responses to environmental stress. Plant Growth Regul 23:79–103

Hirayama T, Shinozaki K (2010) Research on plant abiotic stress responses in the postgenome era: past, present and future. Plant J 61:1041–1052

Hirt H (1997) Multiple roles of MAP kinases in plant signal transduction. Trends Plant Sci 2:11–15

Holmström K-O, Mäntylä E, Welin B et al (1996) Drought tolerance in tobacco. Nature 379:683–684

Hong Z, Lakkineni K, Zhang Z et al (2000) Removal of feedback inhibition of 1 pyrroline-5-carboxylate synthetase (P5CS) results in increased proline accumulation and protection of plants from osmotic stress. Plant Physiol 122:1129–1136

Hu H, Dai M, Yao J et al (2006) Overexpressing a NAM, ATAF, and CUC (NAC) transcription factor enhances drought resistance and salt tolerance in rice. Proc Natl Acad Sci 103:12987–12992

Huang D, Jaradat MR, Wu W et al (2007) Structural analogs of ABA reveal novel features of ABA perception and signalıng in Arabidopsis. Plant J 50:414–428

Ikuta S, Mamura S, Misaki H et al (1977) Purification and characterization of choline oxidase from *Arthrobacter globiformis*. J Biochem 82:1741–1749

Ingram J, Bartels D (1996) The molecular basis of dehydration tolerance in plants. Annu Rev Plant Biol 47:377–403

Ito Y, Katsura K, Maruyama K et al (2006) Functional analysis of rice DREB1/CBF-type transcription factors involved in cold-responsive gene expression in transgenic rice. Plant Cell Physiol 47:141–153

Jaglo KR, Kleff S, Amundsen KL et al (2001) Components of the Arabidopsis C-repeat/dehydration-responsive element binding factor cold-response pathway are conserved in *Brassica napus* and other plant species. Plant Physiol 127:910–917

Jaglo-Ottosen KR, Gilmour SJ et al (1998) Arabidopsis CBF1 overexpression induces COR genes and enhances freezing tolerance. Science 280:104–106

Jang I-C, Oh S-J, Seo J-S et al (2003) Expression of a bifunctional fusion of the *Escherichia coli* genes for trehalose-6-phosphate synthase and trehalose-6-phosphate phosphatase in transgenic rice plants increases trehalose accumulation and abiotic stress tolerance without stunting growth. Plant Physiol 131:516–524

Jeong JS, Kim YS, Baek KH et al (2010) Root-specific expression of OsNAC10 improves drought tolerance and grain yield in rice under field drought conditions. Plant Physiol 153:185–197

Kasuga M, Liu Q, Miura S et al (1999) Improving plant drought, salt, and freezing tolerance by gene transfer of a single stress inducible transcription factor. Nat Biotechnol 17:287–291

Kasukabe Y, He L, Nada K et al (2004) Overexpression of spermidine synthase enhances tolerance to multiple environmental stresses and up-regulates the expression of various stress-regulated genes in transgenic *Arabidopsis thaliana*. Plant Cell Physiol 45:712–722

Kasukabe Y, He L, Watakabe Y et al (2006) Tachibana, improvement of environmental stress tolerance of sweet potato by introduction of genes for spermidine synthase. Plant Biotechnol 23:75–83

Konstantinova T, Parvanova D, Atanassov A et al (2002) Freezing tolerant tobacco, transformed to accumulate osmoprotectants. Plant Sci 163:157–164

Le Martret B, Poage M, Shiel K et al (2011) Tobacco chloroplast transformants expressing genes encoding dehydroascorbate reductase, glutathione reductase, and glutathione-S-transferase, exhibit altered anti-oxidant metabolism and improved abiotic stress tolerance. Plant Biotechnol J 9:661–673

Lee YP, Baek K-H, Lee H-S et al (2010) Tobacco seeds simultaneously over-expressing Cu/Zn-superoxide dismutase and ascorbate peroxidase display enhanced seed longevity and germination rates under stress conditions. J Exp Bot 61:2499–2506

Mani S, van de Cotte B, van Montagu M et al (2002) Altered levels of proline dehydrogenase cause hypersensitivity to proline and its analogs in Arabidopsis. Plant Physiol 128:73–83

McCue K, Hanson A (1990) Drought and salt tolerance: towards understanding and application. Trends Biotechnol 8:358–362

McNeil SD, Nuccio ML, Hanson AD (1999) Betaines and related osmoprotectants targets for metabolic engineering of stress resistance. Plant Physiol 120:945–949

McNeil S, Rhodes D, Russell B et al (2000) Metabolic model identifies key constraints on an engineered glycine betaine synthetic pathway in tobacco. Plant Physiol 124:153–162

McNeil SD, Nuccio ML, Ziemak MJ et al (2001) Enhanced synthesis of choline and glycine betaine in transgenic tobacco plants that overexpress phosphoethanolamine N-methyltransferase. Proc Natl Acad Sci 98:10001–10005

Mundy J, Chua NH (1988) Abscisic acid and water stress induce the expression of a novel rice gene. EMBO J 7:2279–2286

Nanjo T, Fujita M, Seki M et al (2003) Toxicity of free proline revealed in an Arabidopsis T-DNA tagged mutant deficient in proline dehydrogenase. Plant Cell Physiol 44:541–548

Nelson DE, Repetti PP, Adams TR et al (2007) Plant nuclear factor Y (NF-Y) B subunits confer drought tolerance and lead to improved corn yields on water-limited acres. Proc Natl Acad Sci 104:16450–16455

Nuccio ML, McNeil SD, Ziemak MJ et al (2000) Choline import into chloroplasts limits glycine betaine synthesis in tobacco: analysis of plants engineered with a chloroplastic or a cytosolic pathway. Metab Eng 2:300–311

Nyyssola A, Kerovuo J, Kaukinen P et al (2000) Extreme halophiles synthesize betaine from glycine by methylation. J Biol Chem 275:22196–22201

Oscar J, Goddijn M, van Kees D (1999) Trehalose metabolism in plants. Trends Plant Sci 4(8):315–319

Park EJ, Jeknic Z, Sakamoto A et al (2004) Genetic engineering of glycine betaine synthesis in tomato protects seeds, plants, and flowers from chilling damage. Plant J 40:474–487

Pilon-Smits EAH, Ebstamp MJM, Paul MJ et al (1995) Improved performance of transgenic fructan-accumulating tobacco under drought stress. Plant Physiol 107:125–130

Pilon-Smits EAH, Terry N et al (1999) Enhanced drought resistance in fructan-producing sugar beet. Plant Physiol Biochem 37:313–317

Prabhavathi VR, Rajam MV (2007) Polyamine accumulation in transgenic eggplant enhances tolerance to multiple abiotic stresses and fungal resistance. Plant Biotechnol 24:273–282

Rhodes D, Hanson AD (1993) Quaternary ammonium and tertiary sulfonium compounds in higher plants. Annu Rev Plant Physiol Plant Mol Biol 44:357–384

Rontein D, Basset G, Hanson AD (2002) Metabolic engineering of osmoprotectant accumulation in plants. Metab Eng 4:49–56

Rus AM, MT E˜ o, Gisbert C, Garcia-Sogo B et al (2001) Expressing the yeast HAL1 gene in tomato increases fruit yield and enhances K+/Na + selectivity under salt stress. Plant Cell Environ 24:875–880

Santner A, Estelle M (2010) The ubiquitin–proteasome system regulates plant hormone signaling. Plant J 61:1029–1040

Serrano R, Mulet JM, Rios G et al (1999) A glimpse of the mechanisms of ion homeostasis during salt stress. J Exp Bot 50:1023–1036

Shi H, Ishitani M, Kim C et al (2000) The Arabidopsis thaliana salt tolerance gene SOS1 encodes a putative Na^+/H^+antiporter. Proc Natl Acad Sci USA 97:6896–6901

Shi H, Quintero FJ, Pardo JM et al (2002) Role of SOS1 as a plasma membrane Na^+/H^+ antiporter that controls long distance Na^+ transport in plant. Plant Cell 14:465–477

Shi H, Lee BH, Wu SJ et al (2003) Overexpression of a plasma membrane Na^+/H^+ antiporter gene improves salt tolerance in *Arabidopsis thaliana*. Nature Biotech 21:81–85

Shinozaki K, Yamaguchi-Shinozaki K (1997a) Gene expression and signal transduction in water-stress response. Plant Physiol 115:327–334

Shinozaki K, Yamaguchi-Shinozaki K (1997b) Molecular responses to drought and cold stress. Curr Opin Biotechnol 7:161–167

Shinozaki K, Yamaguchi-Shinozaki K (2000) Molecular responses to dehydration and low temperature: differences and cross-talk between two stress signaling pathways. Curr Opin Plant Biol 3:217–223

Shujun Y, Barbara V, Jiangxin W et al (2010) Narrowing down the targets: towards successful genetic engineering of drought-tolerant crops. Mol Plant 3(3):469–490

Stein H, Honig A, Miller G et al (2011) Elevation of free proline and proline-rich protein levels by simultaneous manipulations of proline biosynthesis and degradation in plants. Plant Sci 181:140–150

Suarez R, Calderon C, Iturriaga G (2009) Enhanced tolerance to multiple abiotic stresses in transgenic alfalfa accumulating trehalose. Crop Sci 49:1791–1799

Sun W, Bernard C, van de Cotte B, Van Montagu M et al (2001) At-HSP17.6A, encoding a small heat-shock protein in Arabidopsis, can enhance osmotolerance upon overexpression. Plant J 27:407–415

Sze H, Lia X, Palmgrenb MG (1999) Energization of plant cell membranes by H-pumping ATPases: regulation and biosynthesis. Plant Cell 11:677–690

Takasaki H, Maruyama K, Kidokoro S et al (2010) The abiotic stress-responsive NAC-type transcription factor OsNAC5 regulates stress-inducible genes and stress tolerance in rice. Mol Genet Genomics 284:173–183

Tarczynski MC, Jensen RG, Bohnert HJ (1992) Expression of a bacterial *mtlD* gene in transgenic tobacco leads to production and accumulation of mannitol. Proc Natl Acad Sci 89:2600–2604

Thomas JC, Sepahi M, Arendall B et al (1995) Enhancement of seed germination in high salinity by engineering mannitol expression in *Arabidopsis thaliana*. Plant Cell Environ 18:801–806

Thompson AJ, Andrews J, Mulholland BJ et al (2007) Overproduction of abscisic acid in tomato increases transpiration efficiency and root hydraulic conductivity and influences leaf expansion. Plant Physiol 143:1905–1917

Toäroäk Z, Goloubinoff P, Horvath I et al (2001) Synechocystis HSP17 is an amphitropic protein that stabilizes heat-stressed membranes and binds denatured proteins for subsequent chaperone-mediated refolding. Proc Natl Acad Sci U S A 98:3098–3103

Tony H, Norio M (2002) Enhancement of tolerance of abiotic stress by metabolic engineering of betaines and other compatible solutes. Curr Opin Plant Biol 5:250–257

Veinger L, Diamant S, Buchner J et al (1998) The small heat-shock protein IbpB from *Escherichia coli* stabilizes stress denatured proteins for subsequent refolding by a multichaperone network. J Biol Chem 273:11032–11037

Vierling E (1991) The roles of heat-shock proteins in plants. Annu Rev Plant Biol 42:579–620

Vinocur B, Altman A (2005) Recent advances in engineering plant tolerance to abiotic stress: achievements and limitations. Curr Opin Biotechnol 16:123–132

Waditee R, Bhuiyan MN, Rai V et al (2005) Genes for direct methylation of glycine provides high levels of glycinebetaine and abiotic-stress tolerance in *Synechococcus* and *Arabidopsis*. Proc Natl Acad Sci 102:1318–1323

Wang W, Vinocur B, Shoseyov O (2001) Biotechnology of plant osmotic stress tolerance: physiological and molecular considerations. Acta Hort 560:285–292

Wang W, Vinocur B, Altman A (2003) Plant responses to drought, salinity and extreme temperatures: towards genetic engineering for stress tolerance. Planta 218:1–14

Wang Y, Ying J, Kuzma M et al (2005) Molecular tailoring of farnesylation for plant drought tolerance and yield protection. Plant J 43:413–424

Wang X, Li Y, Ji W et al (2011) A novel *Glycine soja* tonoplast intrinsic protein gene responds to abiotic stress and depresses salt and dehydration tolerance in transgenic *Arabidopsis thaliana*. J Plant Physiol 168 (11):1241–1248

Waters ER, Lee GJ, Vierling E (1996) Evolution, structure and function of the small heat shock proteins in plants. J Exp Bot 47:325–338

Xiao B-Z, Chen X, Xiang C-B et al (2009) Evaluation of seven function-known candidate genes for their effects on improving drought resistance of transgenic rice under field conditions. Mol Plant 2:73–83

Xu D, Duan X, Wang B et al (1996) Expression of a late embryogenesis abundant protein gene HVA1, from barley confers tolerance to water deficit and salt stress in transgenic rice. Plant Physiol 110:249–257

Yamaguchi-Shinozaki K, Shinozaki K (1994) A novel cis-acting element in an Arabidopsis gene is involved in responsiveness to drought, low-temperature, or high-salt stress. Plant Cell 6:251–264

Yang S, Vanderbeld B, Wan J (2010) Narrowing down the targets: towards successful genetic engineering of drought-tolerant crops. Mol Plant 3:469–490

Yeo ET, Kwon HB, Han SE et al (2000) Genetic engineering of drought-resistant potato plants by introduction of the trehalose-6-phosphate synthase (TPS1) gene from *Saccharomyces cerevisiae*. Mol Cells 10:263–268

Zhang HX, Blumwald E (2001) Transgenic salt-tolerant tomato plants accumulate salt in foliage but not in fruit. Nature Biotech 19:765–768

Zhang HX, Hodson JN, Williams JP (2001) Engineering salt-tolerant *Brassica* plants: characterization of yield and seed oil quality in transgenic plants with increased vacuolar sodium accumulation. Proc Natl Acad Sci 98:12832–12836

Role of Organic and Inorganic Chemicals in Plant-Stress Mitigation

Taqi Ahmed Khan, Mohd Mazid, and Saima Quddusi

Abstract

All kinds of abiotic stress factors are probably the most limiting for crop quality and productivity, comprising economical output and human food supply. Among them, drought, salinity, heavy metal, ultraviolet radiations, and floods are the multidimensional stress factors affecting plants at various levels of their organization. Thus, the effects of these stresses are often manifested at morphophysiological, biochemical, and molecular level, such as inhibition of growth, accumulation of compatible organic solutes, and changes in phytohormones endogenous contents. A number of phospholipid systems are activated by abiotic stress factors, generating a diverse array of messenger molecules, some of which may function upstream of the osmotic stress-activated protein kinases. Today, the role of organic chemicals (e.g., proline, plant growth regulators, and ascorbic acid) and inorganic chemicals (e.g., nitric oxide, hydrogen peroxide, and nutrient, e.g., sulfur) are recently emerged and also widely used approaches for stress mitigation in plant systems. All these molecules modify constitutively expressed transcription factors, leading to the expression of early response transcriptional activators, which then activate downstream stress tolerance effector genes, responsible for stress mitigation. This chapter briefly highlights the role of each organic and inorganic molecule in modern day stress mitigation strategy.

Keywords

Nitric oxide • Hydrogen peroxide • Sulfur • Proline • Plant growth regulators

T.A. Khan (✉)
Department of Applied Biotechnology, Sur College of Applied Sciences, Sur 411, Oman
e-mail: taqikhan34@yahoo.com

M. Mazid
Department of Botany, Post Graduate College, Bisalpur, Pilibhit 262001, India

S. Quddusi
Amity Institute of Biotechnology, Amity University, Lucknow 227106, India

R.K. Gaur and P. Sharma (eds.), *Approaches to Plant Stress and their Management*,
DOI 10.1007/978-81-322-1620-9_3, © Springer India 2014

Introduction

Plants constantly monitor their surroundings and make appropriate metabolic, structural and physiological adjustments to accommodate environmental changes. Within the framework of genetic background, plant productivity is dependent on this constant adjustment of gene expression in response to environmental cues. The genome-environment interaction is an essential focus for the elucidation of the nature of the phenotypic variations leading to successful stress tolerance responses. This interaction is also a key determinant to plant tissue composition related to crop quality factors, as well as plant anatomy, morphology, and development. Plant integrates a diverse range of environmental and metabolic signals via a network of interacting signal transduction pathways that together regulate gene expression during biotic and abiotic stresses. Plants make use of an interacting network of common pathway and components to optimize the stress tolerance responses called cross-tolerance. A common signaling system involving hormones, oxidant, and antioxidants has evolved to provide adequate defense and protection against hazardous amounts of chemicals like heavy metals.

Also, plants are autotrophic organisms powered by photosynthesis. It is therefore no surprise that redox signals from the light reactions of photosynthesis initiate profound changes in gene function. It is now widely accepted that redox signal exerts control on nearly every aspect of plant biology from chemistry to development, growth, and eventual death. It is most probable that redox signaling was the first type of sensory regulation that evolved in nature, since it prevented uncontrolled changes in energy production, utilization, and exchange. The system in the chloroplast that sense redox changes and control redox homeostasis has been intensively studied; these include posttranslational modification of assimilatory reactions and control of gene transcription and translation. Moreover, redox controls and signals also ensure that the expression of plastid-encoded photosynthetic proteins is precisely coordinated with that of nuclear-encoded chloroplastic components. However, about 2 billion years ago, molecular oxygen became intimately involved with the essential energy exchange reactions on which life is based.

Abiotic stresses disrupt the cellular redox homeostasis which leads to the oxidative stress or generation of ROS (reactive oxygen species) (Asada 2006). It is now well established that virtually all kinds of abiotic stresses induce or involve oxidative stress to some degree, and the ability of plants to control oxidant levels is highly correlated with stress tolerance. For this purpose, they are equipped with complex processes such as perception, transduction, and transmission of stress stimuli (Kopyra and Gwozdz 2004). ROS play an active role in intracellular redox signaling, activating various antioxidant resistance mechanisms. Thus, it is a surviving response for plants to control the concentration of ROS. Also, a plethora of evidence indicates that ROS are involved in the DNA damage, cell death, and signal transduction. ROS generated under heavy metal stress especially OH^-, $^1O_2$, H_2O_2, ascorbic acid (AA), plant growth regulators (PGRs), nitric oxide (NO), and proline reaction with sugars, purines, and pyrimidines. However, no direct evidence is observed in study of He and Hader (2002) in respect to lipid peroxidation and DNA damage. In addition, they are also produced as second messengers during growth, tropic, and movement responses as results of hormone action.

Rapid component-specific differences in redox state can be achieved either by modifying the rates of O_2^- and H_2O_2 production or by repression or activation of the antioxidants defense or both. For example, H_2O_2 accumulation in the apoplast is controlled by regulation of the activity of enzymes that generate H_2O_2 potential NADPH oxidases. This allows very precise control of the generation of H_2O_2 either in localized microenvironments or in more global bursts. Much remains to be resolved concerning the components of the H_2O_2-induced signaling cascade and the mechanisms by which information on redox status is used to modify gene expression. H_2O_2-responsive elements are present in the promoters of a number of plants genes. One of the most important developments in the next few years will be the elucidation of transcription factors that are involved in the H_2O_2 signaling cascade. This will be important as a signal transcription factor can orchestrate the expression of many genes to improve tolerance against abiotic stress especially induced by heavy metals.

Stress Mitigation Strategies and Organic Molecules

(a) *Heavy Metal Stress*

The effects of heavy metals on plants resulted in growth inhibition, structural damage, and decline in physiological and biochemical activities as well as the function of plants. Plants have a remarkable ability to take up and accumulate heavy metal from their external, for example, aquatic environment. Heavy metals such as Cd, Pb, Hg, Cu, Zn, and Ni at supraoptimal concentrations affect plants growth, development, and yield. In particular, heavy metal stress results in the production of O_2^-, H_2O_2, and OH^-, which affect various cellular processes mostly the functioning of membrane systems (Weckx and Clijsters 1997). In general, plant expresses an incomplete set of remediating features. Cells are normally protected against free oxy-radicals by the operation of intricate antioxidant systems, comprising both enzymic systems such as SOD, CAT, and APXs and nonenzymatic systems, acting as free radical scavengers such as AA and GSH and phenolic compounds (Foyer et al. 1994). AA is the best well-known compound used for antioxidant and most effective compound which increases the tolerance of the plants to abiotic oxidative stress. The results obtained by using the transgenic plants and mutants confirmed the role of AA in oxidative stress or scavenging free oxy-radicals, but in addition, it affects the physiological activities of the plants. Also, there is evidence that the tolerance of plants is correlated with increasing amount of AA.

The antioxidant defense system in the plant cells includes both enzymatic antioxidants such as SOD, CAT, and APX and nonenzymatic antioxidants like AA, GSH, and tocopherol. Of which, AA is involved as co-factors in the detoxifying enzymatic processes. The oxidative damage to different cellular components by H_2O_2 could be minimized either by CAT and peroxidase activities or by a reaction sequence known as ascorbate-glutathione cycle involving the redox pairs of ascorbate-dehydroascorbate and glutathione-glutathione disulfide. Much information is available on the effect of redox heavy metals on various antioxidant processes in plants. But, however, the effect of excess concentrations of some metals on antioxidative processes is rare, but they are found to be useful to plant at lower concentration and affect drastically at elevated concentrations. Also, studies of Madhava and Sresty (2000) showed that among the two metals treatments (Zn and Ni), the Ni-treated pigeon pea seedlings registered lower value of AA content. It is demonstrated that tolerant plants accumulate more amount of heavy metals in their roots. The knowledge gained in such investigations could facilitate both selection and the breeding of heavy metal tolerant plants. Therefore, an additional research is necessary to provide further insight concerning the specific relationship between metal stress and AA response.

(b) *Salinity Stress*

Salt stress can affect several physiological processes, from seed germination to final plant development, and therefore act as a one of the most limiting factor to plant productivity. Excessive irrigation and poor drainage facilities are the major contributing factors for soil salinity in agricultural lands, and about one-third of the world's irrigated land is being affected by soil salinity (El-Saidi 1997). Decrease in photosynthesis under saline conditions is considered as one of the most important factors responsible for reduced plant growth and productivity and found to be due to internal CO_2 partial pressure and stomatal closure that affect the gaseous exchange. Under natural conditions, high salinity is the major cause of osmotic stress to plant. In addition to imposing water stress and ionic stress, hyperosmolarity also generates secondary stress, particularly, oxidative stress, which caused by excessive ROS. The capacity to scavenge ROS and to reduce their damaging effects on macromolecules appears to represent and important stress tolerance trait. Foliar application of AA at 200 mg/l counteracted the adverse effect of salinity that was accompanied by significant increase in plant growth of flux cultivars (El-Hariri et al. 2010). Also, AA has effects on physiological processes including the regulation of growth, differentiations and metabolism of plants under saline conditions, and increasing physiological availability of water and nutrient. In addition, AA protects metabolic processes, minimizes the damage caused by oxidative processes through synergistic function with other antioxidants, and stabilizes membranes.

Studies of Hassanein et al. (2009) suggested that AA increase the content of IAA, which stimulates cell division and cell enlargement, and this, in turn, improves plant growth. Also, AA is more effective at 400 mg/l and attributed to the increase in nutrient uptake and assimilation. One possibility is that additional AA would inhibit stress-induced increase in the leakage of essential electrolytes following peroxidative damage to plasma membranes. Additional AA did not inhibit increases in leakage of electrolytes from roots of salt-stressed tomato seedlings nor did additional AA significantly reduce the undesirable accumulation of Na^+ in stress of salt-stressed plants. The remarkable protective effect of exogenous AA appeared to be specially related to its antioxidant activity, rather than its possible utility as an organic substrate for respiratory metabolism (Shalata and Neumann 2001). Studies of Shalata and Neumann (2001) also support that an additional exogenous supply of AA to seedlings might decrease the buildup of ROS and thereby increase resistance to salt stress, and this increase in resistance to salt stress is associated with the antioxidant activity of AA and a partial inhibition of salt-induced increase in lipid peroxidation by ROS. Possibly, the protective effect of AA is more related to reduce ROS damage to essential proteins and/or nucleic acids. The fact that new roots and leaves are produced by the seedlings which recovered from salt treatment with AA suggests that additional AA may have affected meristematic cells in the salt-stressed root and shoot tissues.

Moreover, studies of El-Hariri et al. (2010) suggested the effect of salinity on decreasing chlorophyll synthesis which ultimately reduces the biosynthesis of carbohydrates, and also, foliar application of AA counteracted the adverse effect of salinity, this accompanied by significant increase in plant growth, IAA contents, and yield components while decrease in total phenol contents. In respect to foliar spray of AA at 400 mg/l, this caused an increase in fiber diameter compared with control plants or the corresponding salinity levels. Also, foliar spray of AA increases cellulose% and cellulose/lignin % and in the meantime decreases lignin%, attributed to the synthesis of chlorophyll which is involved in increasing photosynthetic metabolites and leads to the accumulation of different fractions of soluble sugars and nitrogen content in plant tissues under saline conditions. This evidence points that AA being an important part of the plant defense machinery maintaining the integrity and normal function of the photosynthetic apparatus and acts directly to neutralize O_2^{-2} or $^1O_2$ in plant cells.

(c) *Ultraviolet Stress*

UV-B is an important environmental signal that regulates diverse responses in plants and promotes UV protection and survival in sunlight and influences metabolism, development, and defense responses. UV-B generates oxidative stress in plant cells due to excessive generation of ROS and also reported that stimulation of activities of SOD, CAT, APX, and GR is observed at initial growth stage, while the activities of CAT and SOD decreased at later stage, but there is no definite trend of change observed for AA. UV radiations induced degradation of AA in a model apple juice and in apple juice where AA degradation is more rapid at higher-dose levels and the reaction accelerated with increasing exposure time. Tikekar et al. (2011) show the effect of UV processing on AA and suggest that process developers and researchers can use this study as a model for designing experiments to identify factors that influence the stability of AA and other bioactive compounds during UV processing. In the combined conditions of high light and low temperature where metabolism is slowed relative to photochemistry, the Mehler reaction may occur at increased rates. Enhanced operation of the Mehler reaction may thus diminish the extent of photoinhibition, the slowly reversible reduction in photosynthetic efficiency and capacity which occurs when light energy is in excess. In addition, AA/DHA and GSH/GSSG ratio may protect the thio-modulated enzymes of the Benson-Calvin cycle from oxidation by H_2O_2 and allow photosynthesis to proceed at relatively high rates even during oxidative stress. In the chloroplast, MDHA is rapidly reduced to ascorbate by reduced ferredoxin. Other membrane-associated ETSs such as those on the plasmalemma may also be instrumental in re-reducing MDHA nonenzymatically.

When MDHA is not reduced, it rapidly disproportionate to AA and DHA, the divalent oxidized product. Although DHA is reduced to AA by DHAR, DHA is always detectable in plant tissues, and AA/DHA ratios are relatively low compared to GSH/GSSG ratios, particularly under field conditions. AA is perhaps the most important antioxidant in plants and plays a pivotal role in the destruction of ROS and the regeneration of α-tocopherol (Foyer et al. 1994). The exogenous addition of ascorbate to the culture of green algae *Chlamydomonas reinhardtii* prevented the ROS increase and, therefore, the ROS-mediated downregulation of large subunits of Rubisco translation induced by excess light stress. In addition, studies of He and Hader (2002) suggest that AA also exhibited a significant protective effect on lipid peroxidation and DNA strand break. The protective effect of AA indicates that survival of the irradiated organisms by UV-B is associated with the extent of DNA damage (He and Hader 2002).

Proline

Proline is a proteinogenic amino acid with an exceptional conformational rigidity and is essential for primary metabolism. It acts as a signaling molecule to modulate mitochondrial functions, influence cell proliferation or cell death, and trigger specific gene expression essential for plant recovery against various types of abiotic stresses including drought, salinity, extreme temperatures, chemical toxicity, and oxidative stress. Although, all the plants did not accumulate proline in sufficient amount that it can avert the adverse effects of abiotic stresses. The use of traditional protocols of plant genetics and breeding program is to develop the cultivars with natural abilities that it can produce high levels of proline under stress conditions. Some progress has been made in introducing genes for the production of these compounds in naturally non-accumulating or low-accumulating plant species; levels of accumulation in transgenic plants have often been low or insufficient to improve plant-stress tolerance. Today's research to determine specific roles of proline in plant-stress tolerance is expected to help improving their application as exogenous treatments to improve growth of plants and productivity under stress conditions. Indeed, there are several traits whose corrective association with resistance has been tested in transgenic plants. Several osmotic stresses cause detrimental changes in cellular components. It is believed that osmoregulation would be the best strategy for abiotic stress tolerance, especially if osmoregulatory genes could be triggered in response to drought, salinity, and high temperature. Therefore, a widely adopted strategy has been to engineer certain osmolytes or by overexpressing such osmolytes in plants as a potential route to breed stress-tolerant crops (Fig. 1).

Moreover, exogenous application of proline can play significant role in enhancing plant-stress tolerance. This role can be in the form of either osmoprotection or cryoprotection. Exogenous application of proline markedly inhibits shoot elongation and root growth of rice seedlings as well as growth of *Brassica napus* callus (Chen and Kao 1995). If under normal conditions, continuous cycling occurs between proline and its ultimate precursor glutamate, and this is blocked by high level of proline; exogenously applied

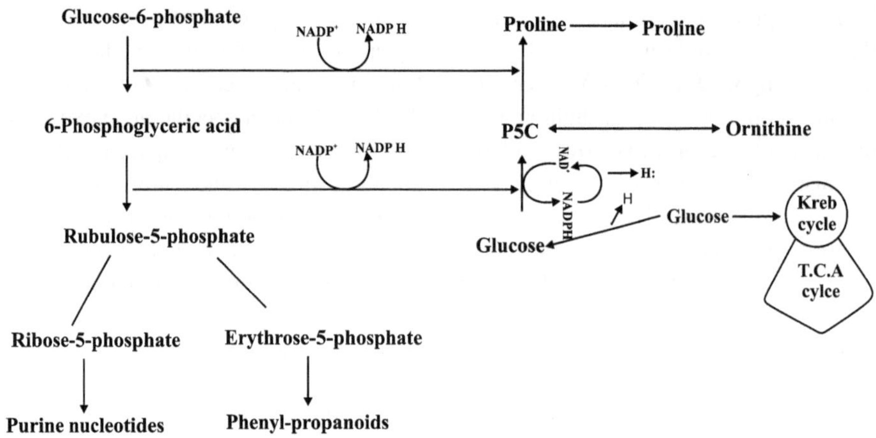

Fig. 1 An intracellular signaling system between proline and P5C by interconversion. Since proline degradation generates reducing equivalents needed during TCA cycle activity, proline metabolism might transfer metabolic information between tissues with different metabolic requirements in the form of redox potential and activate metabolic pathways that do not involve any of the intermediates of metabolism

proline may inhibit C-flux through the OPPP and ADP phosphorylation, both of which are likely to be of importance in sustaining a rapid growth rate.

This metabolic coupling of bacterial and exosymbiont-derived proteins seems to be fundamental to diatom physiology because the compounds affected include the major diatom osmolyte proline and the precursors for long-chain polyamines required for silica precipitation during cell wall formation. Moreover, an efficient system was found for the production of an active xylanase in tobacco plants fused to a proline-rich domain (Zera) of the maize storage protein γ-zein, which is a self-assembling domain able to form protein aggregates in vivo packed in newly formed endoplasmic reticulum-derived organelles known as protein bodies and aggregate of xylanase fused to Zera in plants used as an become efficient and low-cost bioreactors for industrial purposes (Llop-Tous et al. 2011).

Although exogenous application of proline to plants expressed to abiotic stresses generally provides a stress-preventing or recovery effect, high concentrations of proline may be harmful to plants also. Besides beneficial effects, a few reports suggest that foliar application of proline to rice plants growing under saline conditions did not change concentration of either Na^+ and or Cl^- in the leaves (Krishnamurthy and Bhagwat 1993). In addition, in case of Arabidopsis, proline was suggested to cause damage to ultrastructure of chloroplast and mitochondria. These findings bolster the argument that the synthesis of proline from glutamate and did merely its presence is of importance in counteracting the effects of osmo-abiotic stress. The observation that the proline content in wheat seeds increased more than sevenfold 48 h after inhibition but thereafter declined and had stabilized after 120 h ends more credence to the idea that proline biosynthesis may activate OPPP activity during germination.

Proline treatment inhibited Arabidopsis seed germination, restricted growth of petunia plants, and arrested root growth of *Thellungiella halophila*. In mung bean (*Vigna radiata* L.), for example, it was determined that while addition of 20–33 mM of proline to cell cultures mitigated the adverse effects of NaCl stress, concentrations of 50 mM or higher were inhibitory to the growth of both salt-stressed and nonstressed cultures (Kumar and Sharma 1989). In rice, while 30 mM proline was the most effective concentration in improving germination and seedling growth under salt stress, higher concentration (40 or 50 mM) resulted in reduced seedling growth and lowered K^+/Na^+ ratio. Therefore, it is essential to determine optimal concentration of proline that produces beneficial effects in different plant species. In addition, growth of tobacco suspension cells under salt stress was promoted by exogenous application of 10 mM proline, which was proposed to be due to proline action as a protectant of enzymes and membranes. In barley embryo cultures, under saline conditions, exogenous application of proline resulted in a decrease in Na^+ and Cl^- accumulations and an increase in growth. Such ameliorative effects of proline were indicated to be due to plasma-membrane stabilization.

Plant Growth Regulators

Phytohormones (PGRs) are in a prominent position, playing important regulatory roles in plant physiology affecting the responses to a wide range of abiotic and biotic stresses. These hormones generate a signal transduction network that leads to a cascade of events responsible for the physiological adaptation of the plant to stress. It should be noted that the degree of drought tolerance varies with developmental stages in most plant species. Drought during the vegetative phase of sunflower plants affects both biological and economic yields. In maize, water deficit in the late developmental stage tends to reduce kernel size rather than number. Under stressful growing conditions to characterize the agronomic and ecological traits related to environmental tolerance of switch grass, it was found that drought treatments reduced tiller length and number, leaf area, and biomass production by up to 80 %.

Abscisic Acid

The molecular basis of ABA biosynthesis and catabolism was established by genetic and biochemical approaches. ABA action is one of the most studied topics in abiotic stress response research. An increase in ABA content in response to water-deficit stress may arise from an increase in ABA biosynthesis and/or a decrease in ABA breakdown. In *Arabidopsis thaliana* seedlings, drought enhanced both ABA biosynthesis and catabolism, resulting in an increase in ABA and catabolites. Likewise, drought-treated plants of *Laurus azorica* showed an increase in leaf ABA concentrations with respect to that of the control. Exogenous application of ABA enhances the tolerance of plants or plant cells to drought. In relation to endogenous ABA, different reports showed that drought-tolerant cultivars have more ABA than susceptible ones.

Salicylic Acid

Application of exogenous SA improves the plant performance under water. Low concentrations of exogenous SA provided tolerance against the damaging effects of drought in tomato and bean plants, whereas higher concentrations did not show the same results. Enhanced tolerance to drought and dry matter accumulation was also observed in plants of wheat raised from grains soaked in acetylsalicylic acid aqueous solution. Wheat seedlings subjected to drought and treated with SA exhibited higher moisture content and dry matter accumulation, carboxylase activity of Rubisco, SOD, and total chlorophyll content compared to untreated control. The SA treatment also provided a considerable protection to the enzyme nitrate reductase thereby maintaining the level of diverse proteins in leaves. In addition, the treatment of water-stressed plants with SA low concentrations significantly enhances the photosynthetic parameters, membrane stability index, leaf water potential, and activities of the enzymes nitrate reductase and carbonic anhydrase, thus improving tolerance to drought. SA is also involved in the promotion of drought-induced leaf senescence in *Salvia officinalis* plants grown under drought in Mediterranean field conditions. In addition, SA applied exogenously was effective in providing resistance to the plants against the excessive water stress in cell suspensions. Exogenous application of SA and glycine-betaine enhanced the yield of sunflower hybrids under different degrees of water stress. Under stress, diameter of the head, number of achene, and seed oil content were reduced. However, applications of SA and GB improved these parameters. SA increases the activity of the oxidative enzymatic system as is the case of CAT and SOD. In plants of Brassica, exogenous application of SA increased CAT and SOD activity.

Jasmonic Acid

The participation of JA in response to abiotic stress, such as drought and salinity, has been reported in several species. The treatment of barley leaves with sorbitol or mannitol increased JAs endogenous contents, followed by synthesis of jasmonate-induced proteins. Sorbitol treatment enhanced octadecanoids and JAs content, and this threshold was necessary and sufficient to initiate JA-responsive gene expression. Under water stress, endogenous JA content increased in maize root cells, and this compound was able to elicit betaine accumulation in pear leaves. Tomato cultivars differing in salt tolerance differed in basal JA content. Steady-state amounts of JA and related compounds were higher in salt tolerant compared to the salt sensitive.

Nutrient (Sulfur)

Sulfur (S) is an essential macronutrient for all living organisms. Plants are able to assimilate inorganic sulfur and incorporate it into organic compounds. In the last decades, sulfate availability in soils has become the major limiting factor for plant production in many countries due to significant reduction of anthropogenic S emission forced by introducing stringent environmental regulation. Sulfur deficiency stress causes decrease in plant growth and productivity by altering in biochemical processes, especially photosynthesis and nitrogen fixation. Under S starvation, the sequestration of heavy metals in intracellular compartments induces oxidative stress, which seems to affect regulatory mechanisms of basic bioenergetic processes of photosynthesis and N fixation.

Regulation of photosynthetic ETS is a critical aspect of tailoring the metabolism of the cell to nutrient availability. This redox regulation based on disulfide-dithiol conversion catalyzed by thioredoxins is an important component of chloroplast function. In addition, chloroplasts are equipped with a peculiar NADPH-dependent thioredoxin reductase, termed NTRC, with a joint thioredoxin domain at the carboxyl terminus. NTRC is important to maintain the chloroplast redox homeostasis under light limitation. Phylogenetic analysis suggests that chloroplast NTRC originated from an ancestral cyanobacterial enzyme. While the biochemical properties of plant NTRC are well documented, little is known about the cyanobacterial enzyme (Pascual et al. 2011).

High concentration of heavy metals in the soil is toxic to most plant (Macnair 1993). Based on their solubility under physiological conditions, heavy metals may be available to living cells and have significance for the plant and animal communities with in various ecosystems. At metabolic level, the heavy metal stress combination had a profound effect on central metabolic pathways such as the tricarboxylic acid (TCA) cycle, glycolysis, pentose phosphate cycle, and large parts of amino acid metabolism. Under these conditions, central metabolites, such as sugars and their phosphates, increased, while S-containing compounds are decreased (Wulff-Zottele et al. 2010). Among the heavy metals, As, Hg, Ag, Sb, Cd, Pb, and Al have no known functions as nutrients and seem to be more or less toxic to plants and microorganisms. Of the known metals, Cd, Ni, Zn, and Cu are toxic to plants at elevated levels, whereas Pb has generally observed to cause phytotoxicity.

Cd in plants interrupts numerous metabolic processes and reduces the water and nutrient uptake that cause chlorosis, growth retardation, and ultimately plant death. Consequently, both the plant growth and activity of diurnal photosynthetic system remain the least altered under Cd-provoked toxicity stress (Wan et al. 2011). There are two types of causal relationships existing between the high concentration of heavy metals on soil and the expression of toxicity symptoms. On the one type, heavy metals compete with essential mineral nutrients for uptake thereby disturbing the mineral nutrition of plants, and on the other type, after uptake by the plant, it accumulates in plant tissues and cell compartments and hampers the general metabolism of the plant. Moreover, heavy metal accumulation in plants has multiple direct and indirect effects on the plant growth and alters many physiological functions by forming complexes with O, N, and S ligands. They interfere with mineral uptake, water relations, and seed germination. Furthermore, they cause metabolic disturbance by altering essential biochemical reactions. The demand of plant species for S as an essential major nutrient is very species specific, with enhanced need of plant species belonging to the family Brassicaceae. When standard nutrient solution is diluted, the SO_4^{2-} concentration decreases to levels of S deficiency with impact on gene expression and enzyme activities. Because of concomitant changes in metal sensitivity, it is difficult to compare plant reactions at extreme low SO_4^{2-} supply with those of plants grown at 18-fold higher SO_4^{2-} concentration. In contrast to SO_4^{2-}, the exposure to heavy metals, especially Cd, surpasses environmentally relevant concentration in many physiological experiments. Investigation with high Cd levels can only show how a heavily disturbed metabolism of a nearly dead plant reacting to an extreme, environmentally never occurring plants. Cd and Pb exposure causes high mutation rates in *Arabidopsis thaliana* with floral anomalies, poor seed production, and malformed embryos. Moreover, other nutrients have also an impact on the uptake of heavy metals especially that of Cd and Pb.

Inorganic Molecules

Hydrogen Peroxide (H_2O_2)

H_2O_2 is the two-electron reduction product of O_2. The potential of H_2O_2 as a stress-tolerant signal to abiotic stress in plants has positioned for much attention. As it is well cleared in this chapter, recent studies on H_2O_2 functionality to induce tolerance in plants to abiotic stresses have been unraveled and very impressive. In plants, H_2O_2 has been implicated to posses antioxidant properties, regulate the antioxidant defense system, and might act as a signal in activating a number of defense gene expression pathways through various sensing mechanisms under abiotic stress. H_2O_2 plays an important role in resistance to temperature, drought, heavy metal, UV-B, and salt stress. H_2O_2 generation via electron transport is increased in response to environmental stresses such as excess excitation energy, drought, and cold and also induced in plants following exposure to a wide variety of abiotic stimuli. Also H_2O_2 has been implicated as a key factor mediating programmed cell death. Plants exposed to abiotic stresses can produce a systemic signal, a component of which may be H_2O_2 which sets up an acclamatory response in unstressed regions of plants (Bhattacharjee 2005). These include extremes of temperatures, UV irradiation, O_3 exposure, heavy metal accumulation, and drought. Therefore, H_2O_2 is considered as a fundamental fact of life in an aerobic environment (Fig. 2).

Exogenous H_2O_2 possibly functioned directly or may have induced intracellular H_2O_2 generation to act as a signaling molecule under stress. Pretreatment of wheat seeds with H_2O_2 has been shown to improve the subsequent salt tolerance of the seedlings. Avsian et al. (2004) have shown that salt stress induced oxidative stress in the form of H_2O_2, the production of which occurred in intracellular spaces. The control of Na^+ involvement across the PM and tonoplast to maintain a low Na^+ concentration in the cytoplasm is a key factor of cellular adaptation to salt stress. It has been initially hypothesized that H_2O_2 might be downstream signal molecules to regulate the activity of PM H^+-ATPase. Further results indicated that H_2O_2 content increased greatly under salt stress. Since H_2O_2 might be the candidate downstream signal molecule, H_2O_2 induced an increased K to Na ratio. Therefore, it is clear from summing up this new assay that NO

Fig. 2 H_2O_2 signaling and representation of potential factors responsible for synthesis and its effects

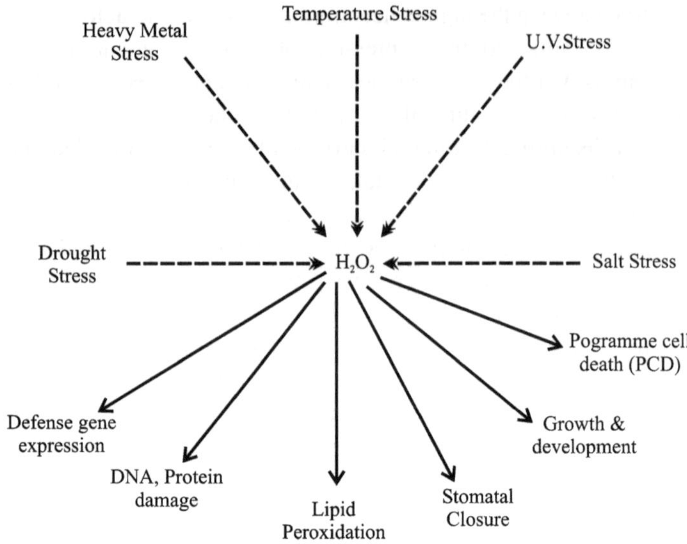

may be regulated by the H_2O_2 generation, since H_2O_2 is involved in downstream signal molecule of NO, PM NADPH oxidase.

As a consequence of a general stress response, cytotoxic H_2O_2 can get accumulated in the cells and can act as a secondary messenger. High H_2O_2 and O_2^- had been reported earlier in the case of various other plants under Cr, Zn, Pb, etc. (Panda et al. 2003a, b). The increase in SOD activity indicated higher H_2O_2 level seen by the increase in total H_2O_2 content in leaves, which tallies with those observed in the case of *Brassica juncea* and *Vigna radiata* under Zn and Al treatment (Panda et al. 2003b). Moreover, H_2O_2 increase usually occurred after Cu, Cd, and Hg treatment of *Arabidopsis thaliana* and *Solanum tobacum* plants, respectively. However, in barley plants only Mn increased the H_2O_2 content after 5 days but not Cu. This difference may indicate that H_2O_2 accumulation developed differently during a larger stress action. After a long time of Cd action, decrease in SOD activity observed. However, this effect was connected with attenuation of the enzymatic antioxidative system, and increased per oxidation of lipids may have not resulted in H_2O_2 level decrease. Cu can act through changes in H_2O_2-dependent peroxidase activity followed by cell wall stiffening due to the formation of cross-linking among the cell wall polymers. Also, Cd enhances H_2O_2 accumulation. Taking all these observation together, a hypothetical framework may be suggested that Cd induces a transient loss in antioxidative capacity perhaps accompanied by a stimulation of oxidant-producing enzymes, which results in intrinsic H_2O_2 accumulation. H_2O_2, then, would act as a signaling molecule triggering secondary defenses. NADPH oxidase is involved in plant growth and plant response to Cu. Increased accumulation of H_2O_2 usually connected with changes in the cellular redox status alerts the plant cell against environmental stresses and may enhance the plants antioxidant response through calcium signaling in the expression of glutathione transferase gene.

Sergi et al. (2003) reported that accumulation of H_2O_2 in the walls of mesophyll cells of *Cistus clusii and Cistus albidus* plants increased at the onset of drought and also recently have shown that H_2O_2 accumulates in senescing leaves of drought-susceptible plants and conclude that drought stress causes the H_2O_2 accumulation in test species. The accumulation of H_2O_2 observed at the onset of drought in the mesophyll cell walls of *C. albidus* and *C. clusii* may be associated with its function in cellular signaling at the first stage of drought or with drought-induced changes in cell wall structure or both. Accumulation of H_2O_2 in mesophyll cell walls occurred as the first symptom of drought in both species, indicating that H_2O_2 may play a role in inter- or intracellular signaling or both. Drought, oxidative stress, and H_2O_2 in particular can enhance the expression of several genes, and H_2O_2 can play a role in inter- and intracellular signaling. Thus, the possibility of an H_2O_2-dependent regulation of stress response genes cannot be excluded. Studies of Apostolova et al. (2008) reported that levels of endogenous H_2O_2 strongly increased in the spring wheat cultivar in response to cold hardening and to a lesser extent in the winter wheat. However, H_2O_2 pretreatment reduced production of H_2O_2 under further chilling stress, postponing oxidative damage. Results of Wang et al. (2010) showed that pretreatment with H_2O_2 at appropriate concentration may improve the tolerance of warm-season Zoysia grasses to chilling stress and that Manila grass had better tolerance to chilling, as evaluated by lower MDA and EL, and better turfgrass quality, regardless of the pretreatment applied. Pretreatment with H_2O_2 has been shown to induce chilling tolerance in normally chill-sensitive maize seedlings. Similarly regenerated potato nodal explants treated with H_2O_2 became significantly more thermotolerant compared with untreated control.

Nitric Oxide

Environmental stress like heavy metal constitutes the most significant factor leading to substantial and unpredictable decrease in crop yield in agriculture. NO was produced as primary signals in stress signal cascade. Alternatively, it can initiate programmed cell death, particularly when NO is also produced depending on the intensity of the oxidative signal or oxidative load exerted by heavy metals toxicity on the tissues. H_2O_2 along with NO functions as stress-signaling molecules in plants (Mazid et al. 2011a, b), mediating a range of defensive mechanisms in plants under stressful conditions. Similarly, NO is an important signaling molecule in plants, which can induce the decrease of ROS accumulation when plants differ from abiotic stress. Also, NO is a widespread intracellular and intercellular messenger with a broad spectrum of regulatory functions in many physiological processes. In plants NO was reported to be involved in ET emission, response to drought, growth and cell proliferation, metabolism and senescence, PCD, and stomatal closure. Today, it has been considered that NO also plays a vital role in diverse physiological processes in plants like induction of seed germination, reduction of seed dormancy, regulation of plant metabolism and senescence induction in cell death, regulation of stomatal movement, photosynthetic regulation, and floral regulation. In many recent researches, high levels of NO have the capacity to damage membranes and DNA fragmentation and to reduce photosynthesis in oat and alfalfa, as well as regulating the multiple plant responses towards a variety of abiotic stresses.

Moreover, previous researches also confirm that NO involves various physiological processes and prolongs the storage life of fruit. However, little attention has been paid to the effects of NO on the storage metabolism in fruit during storage. Decrease of firmness and accumulation of sugar and acid to sugar ratio in peach fruit during storage is significantly inhibited by treatment with NO. Treatment with NO could promote fructose and glucose metabolism during the first few days of storage and increase the content of sucrose and the activities of sorbitol dehydrogenase, sorbitol oxidase, and sucrose phosphate synthase in fruit during storage. Interestingly, NO signaling is based on interactions with plant hormones like auxin and jasmonic acid. NO inhibits the plants from oxidative damage by the regulation of general mechanisms for cellular redox homeostasis and also promoting the transformation of O_2 to H_2O_2 and O^- and also by enhancing the $H_2O_2^-$scavenging enzyme activities (Zheng et al. 2009).

Heavy metal toxicity can elicit a variety of adoptive responses in plants. Cells are normally protected against free oxy-radicals by the operation of intricate antioxidant systems, comprising both enzymatic systems such as SOD, CAT, and APXs and nonenzymatic systems, acting as free radical scavengers such as AA and GSH and phenolic compounds.

The oxidative damage to different cellular components by H_2O_2 could be minimized either by CAT and APX activities or by a reaction sequence known as ascorbate-glutathione cycle involving the redox pairs of ascorbate-dehydroascorbate and glutathione-glutathione disulfide. Enzymatic antioxidants such as selenium-dependent GPX, GST, GR, and SOD, as well as the concentration of H_2O_2 and MDA, an indicator of LPX determined to identify which antioxidant enzymes participate in the efficient scavenging of ROS, were generated upon exposure to high doses of Cd^{2+} exposure (Zhang et al. 2011).

LPX and H_2O_2 are measured as oxidative stress markers while antioxidant defenses are measured as CAT, GST, and AA in order to understand their dissimilarity with respect to pollution levels. They also reported that variations of oxidative stress indices in response to accumulation of heavy metals within *Padina tetrastromatica* commonly found in tropics could be used as molecular biomarkers in assessment and monitoring environmental quality of ecologically sensitive marine habitats. In the case of several plants such as *Phytolacca americana L.*, responsiveness to Cd stress is unknown, but, however, in the case of a 6-week-old seedlings of *Phytolacca americana* exposed to half strength Hoagland solution with 200 μmol/L $CdCl_2$ or 400 μmol/L $CdCl_2$ for 4 days, the content of H_2O_2 and MDA and electrolyte leakage increased, while the photosynthetic rate decreased, indicated that the oxidative damage induced by Cd stress in *Phytolacca americana* is one of the metal toxicity mechanism. Moreover, the activities of SOD and POD increased rapidly with elevated Cd concentration and exposure time, and CAT activity was stable in response to 200 μmol/L $CdCl_2$ stress and increased only at 3 days later upon 400 μmol/L $CdCl_2$ treatment (Zhang et al. 2011). Thus, the enzymatic antioxidation capacity played an important role in Cd tolerance of a hyperaccumulator plant. Therefore, much information is available on the effect of redox heavy metals on various antioxidant processes in plants. But, however, the effect of excess concentrations of some metals (e.g., Zn and Ni) on antioxidative processes is rare, but found to be useful to plant at lower concentration and affect drastically at elevated concentrations. In addition, the symptoms of Zn and Ni toxicity appeared as reduction in seedling growth. The growth of the main root is affected much, and as a result, it exhibits the function of fibrous roots.

Conclusion

AA can act efficiently in plants as immunomodulators when applied at the appropriate concentration and the current stage of plant development. Ascorbate is implicated in plant responses to abiotic environmental stresses and regulates stress response as a result of a complex sequence of biochemical reactions such as activation or suppression of key enzymatic reactions, induction of stress responsive protein synthesis, and the production of various chemical defense compounds. Due to its essential function as a cofactor for the biosynthesis of various plant hormones, AA appears to influence not only the endogenous level but also signaling of these plant hormones, and thus affect responses against the abovementioned environmental stresses. Also, redox status of AA may play a role in the signaling of this interconnected phytohormone network. However, there are obviously still large gaps to fill in order to elucidate the precise role of AA in enhancing the tolerance of plant to environmental stresses during development as well as normal conditions.

Presently, NO and H_2O_2 have emerged as central players in the field of abiotic stress-induced signaling in plant system. Since, it is also clear that both signaling molecules can mediate the transcription of number of stress responsive genes. However, some most interesting questions remain to be elucidated till date is as follows: (a) to identify probable mechanisms through which H_2O_2 and NO have direct/indirect effects on transcription factors, (b) the recognition of full range of biological functions for H_2O_2 and NO remain to be catalogued, and (c) determining the ways in which they interact. So far, it is still under research whether H_2O_2 is situated at a common center for the signaling pathways providing responses to various signals triggered by heavy metal stress like production of NO.

Acknowledgment We thank eminent authorities, whose works have been consulted and whose ideas and insights have richly contributed to this work, and my research partners who have shared productively my interest in the study. Financial support from the Indian Council of Medical Research, New Delhi to Taqi Ahmed Khan in the form of S.R.F (45/14/2011-BIO/BMS) is gratefully acknowledged.

References

Apostolova P, Yordanova R, Popova L (2008) Response of antioxidative defence system to low temperature stress in two wheat cultivars. Gen Appl Plant Physiol 34:281–294

Asada K (2006) Production and scavenging of reactive oxygen species in chloroplasts and their functions. Plant Physiol 141:391–396

Avsian KO, Gueat DY, Lev YS, Gollop R, Ben HG (2004) The salt stress signal transduction pathway that activates the gpxl promoter is mediated by intracellular H_2O_2, different from the pathway induced by extracellular H_2O_2. Plant Physiol 135(3):1685–1696

Bhattacharjee S (2005) Reactive oxygen species and oxidative burst: roles in stress senescence and signal transduction in plants. Current Sci 89:7–15

Chen SL, Kao CH (1995) Cd induced changes in proline level and peroxidase activity in roots of rice seedlings. Plant Growth Regul 17:67–71

El-Hariri DM, Sadak MS, El-Bassiouny HMS (2010) Response of flax cultivars to ascorbic acid and A-tocopherol under salinity stress conditions. Int J Acad Res 2(6):101–109

El-Saidi TM (1997) Salinity and its effect on growth, yield and some physiological process of crop plant. In: Jaiswal PK, Singh RP (eds) Strategies for improving salt tolerance. Oxford/IBH, New Delhi, pp 111–127

Foyer CH, Descourvieres P, Kunert KJ (1994) Protection against oxygen radicals: an important defence mechanism studied in transgenic plants. Plant Cell Environ 17:507–523

Hassanein RA, Bassony FM, Barakat DM, Khalil RR (2009) Physiological effects of nicotinamide and ascorbic acid on *Zea mays* plant grown under salinity stress. 1-changes in growth, some relevant metabolic activities and oxidative defense systems. Res J Agric Biol Sci 5(1):72–81

He YY, Hader DP (2002) UV-B-induced formation of reactive oxygen species and oxidative damage of the cyanobacterium *Anabaena* sp: protective effects of ascorbic acid and *N*-acetyl-L-cysteine. J Photochem Photobiol B 66:115–124

Kopyra M, Gwozdz EA (2004) The role of nitric oxide in plant growth regulation and responses to abiotic stresses. Acta Physiol Plant 26:459–472

Krishnamurthy R, Bhagwat KA (1993) Effect of foliar application of proline on the salt stressed rice seedlings. Acta Agron Hung 42:267–272

Kumar V, Sharma DR (1989) Effect of exogenous proline on growth and ion content in NaCl stressed and nonstressed cells of mungbean, Vigna radiata var. radiata. Indian J Exp Biol 27:813–815

Llop-Tous I, Ortiz M, Torrent M, Ludevid MD (2011) The expression of a xylanase targeted to ER-protein bodies provides a simple strategy to produce active insoluble enzyme polymers in tobacco plants. PLoS One 6(4):e19474

Macnair MR (1993) The genetics of metal tolerance in vascular plants. New Phytol 124:541–559

Madhava Rao KV, Sresty TVS (2000) Antioxidative parameters in the seedlings of pigeonpea (Cajanus cajan (L.) Millspaugh) in response to Zn and Ni stresses. Plant Sci 157:113–128

Mazid M, Khan TA, Mohammad F (2011a) Potential of NO and H2O2 as signaling molecules in tolerance to abiotic stress in plants. J Indus Res Technol 1(1):56–68

Mazid M, Khan TA, Mohammad F (2011b) Role of nitric oxide in regulation of H2O2 mediating tolerance of plants to abiotic stress: a synergistic signaling approach. J Stress Physiol Biochem 7(2):34–74

Panda SK, Chaudhury I, Khan MH (2003a) Heavy metal induced lipid peroxidation and affects antioxidant in wheat leaves. Biol Plant 46:289–294

Panda SK, Singa LB, Khan MH (2003b) Does aluminium phytotoxicity induce oxidative stress in green gram (*Vigna radiata*)? Bulgarian J Plant Physiol 29:77–86

Pascual MB, Mata-Cabana A, Florencio FJ, Lindahl M, Cejudo J (2011) A comparative analysis of the NADPH thioredoxin reductase C-2-Cys peroxiredoxin system from plants and cyanobacteria. Plant Physiol 155(4):1806–1816

Sergi MB, Tana JM, Leonor A (2003) Enhanced photo and antioxidative protection and hydrogen peroxide accumulation in drought stresses cistus clusii and cistus albidus plants. Tree Physiol 23:1–12

Shalata A, Neumann PM (2001) Exogenous ascorbic acid (vitamin C) increases resistance to salt stress and reduces lipid peroxidation. J Exp Bot 52 (364):2207–2211

Tikekar RV, Anantheswaran RC, LaBorde LF (2011) Ascorbic acid degradation in a model apple juice system and in apple juice during ultraviolet processing and storage. J Food Sci 76(2):62–71

Wan G, Najeeb U, Jilani G, Naeem MS, Zhou W (2011) Calcium invigorates the cadmium-stressed Brassica

napus L. plants by strengthening their photosynthetic system. Environ Sci Pollut Res Int 18(9):1478–1486
Wang L, Yang L, Yang F, Li X, Song Y, Wang X, Hu X (2010) Involvements of H_2O_2 and metallothionein in NO-mediated tomato tolerance to copper toxicity. J Plant Physiol 167(15):1298–1306
Weckx JEJ, Clijsters H (1997) Zinc phytotoxicity induces oxidative stress in primary leaves of *Phaseolus vulgaris*. Plant Physiol Biochem 35:405–410
Wulff-Zottele C, Gatzke N, Kopka J, Orellana A, Hoefgen R, Fisahn J, Hesse H (2010) Photosynthesis and metabolism interact during acclimation of *Arabidopsis thaliana* to high irradiance and sulphur depletion. Plant Cell Environ 33(11):1974–1988
Zhang YX, Zhang HM, Huang ZB, Li LF, Liu JG, Li X (2011) Antioxidative enzymes play key roles in cadmium tolerance of *Phytolacca americana*. Huan Jing Ke Xue 32(3):896–900
Zheng C, Jiang D, Liu F, Dai T, Liu W, Jing Q, Cao W (2009) Exogenous nitric oxide improves seed germination in wheat against mitochondrial oxidative damage induced by high salinity. Environ Exp Bot 67:222–227

Role of Calcium/Calmodulin in Plant Stress Response and Signaling

Ritika Das, Amita Pandey, and Girdhar K. Pandey

Abstract

Calcium is a common intracellular second messenger in all the eukaryotes, regulating plethora of cellular processes. Many effects of calcium are mediated by calmodulin superfamily of calcium-binding regulatory protein. Calmodulin is a ubiquitously found and highly conserved calcium sensor throughout eukaryotes; it plays a critical role in calcium-mediated signaling involved in a myriad of cellular processes and responses. Calmodulin works by binding to short peptide sequences in target proteins, bringing about structural changes, which alter the activity of target proteins in response to changes in intracellular calcium level. Plants have evolved a complex network of calmodulin and calmodulin-binding target proteins that serve to play an important role in mediating stress-signaling pathways. Many of the calmodulin-binding proteins include transcription factors, ion channels, and metabolic enzymes that assist plant to effectively cope with environmental stress and pathogens. Extensive research over the past decade has been focused on understanding the function of calmodulin in biotic and abiotic stress response. Several studies employing genetic, molecular biology and biochemical techniques have yielded interesting insights into the function of calmodulin in modulating its various targets to provide stress resistance. In this chapter, we attempt to summarize the major findings about the regulatory role of CaM and its target proteins in abiotic and biotic stress response.

R. Das
Department of Biology, University of Rochester,
Rochester, NY 14627, USA

A. Pandey • G.K. Pandey (✉)
Department of Plant Molecular Biology, University of Delhi South Campus, Benito Juarez Road, Dhaula Kuan,
New Delhi 110021, India
e-mail: gkpandey@south.du.ac.in

R.K. Gaur and P. Sharma (eds.), *Approaches to Plant Stress and their Management*,
DOI 10.1007/978-81-322-1620-9_4,

Keywords
CaM- Calcium-modulated protein • CML- CaM-like protein • EF-hand • AtSRs/CAMTAs- (CaM-binding transcription activator) • CBP- CaM-binding protein • cDNA expression library • HRP- Horseradish peroxidase • CaMBD- CaM-binding domain

Introduction

Calmodulin was first discovered by Ebashi and Kodama (1965) who demonstrated calcium requirement of calcium-binding protein troponin C in skeletal muscle. Later studies (Cheung 1971) also showed requirement of calcium by an activator for cyclic nucleotide phosphodiesterase in regulating cAMP levels. The "activator"/calcium-binding protein in both cases was found to be structurally similar and named as calmodulin (CaM) or calcium-modulated protein, present in all eukaryotes. The first CaM sequence reported in bovine showed it to be an acidic, single polypeptide containing protein, 148 amino acid long, having a molecular weight close to 16.7 kDa (Watterson et al. 1980), and an isoelectric point (pI) of 3.9–4.3. Survey of genomes of different organisms revealed a greater diversity in sequence of CaM displayed by plants unlike animals that have only few genes encoding calmodulin (Yang and Poovaiah 2003).

Calmodulin and CMLs as a Major Calcium Sensor

In response to a variety of stimuli, such as hormones, gravity, light, and abiotic and biotic stresses, plants generate a "Ca^{2+} spike," a transient increase in intracellular calcium level. Given that a plethora of stimuli can generate a "Ca^{2+} spike," specificity in signaling is achieved by downstream transducers of Ca^{2+} signal, the "EF-hand"-containing proteins called "Ca^{2+} sensors" or Ca^{2+}-binding proteins (Bouche et al. 2005). Presence of a large repertoire of Ca^{2+}-transducing proteins in plants points to the existence of a wide variety of cellular responses regulated by calcium.

Calcium-binding domain or "EF-hand" is basically a helix-loop-helix structure, which can bind a single "Ca^{2+} ion." An "EF-hand" contains approximately 40 residues and is found in single or multiple pairs giving the protein various structural and functional variations. EF-hand-containing proteins can be divided into two functional categories: regulatory and structural. The regulatory EF-hand-containing proteins upon binding to Ca^{2+} undergo a conformational change, which is transmitted to their target protein, whereas the structural EF-hand-containing proteins do not undergo conformational change upon binding to Ca^{2+} and are involved in buffering intracellular Ca^{2+} levels. Proteins like calbindin and parvalbumin belong to the latter category. The regulatory EF-hand-containing proteins based on the number and organization of EF-hands have been divided into three major classes, including calmodulin (CaM) and calmodulin-like proteins (CMLs), calcium-dependent protein kinase (CDPK), and calcineurin B-like protein (CBL). CaM, CMLs, and CBL differ from CDPK by the absence of an effector kinase domain; CDPK contains a kinase domain in addition to "EF-hands." Calmodulin (CaM) is one of the major calcium transducer conserved across species. Unlike yeast and animals, plants have a large repertoire of CaM isoforms and CaM-like proteins (CMLs); a possible explanation could be the inherent complexity of plants and their sessile nature, which requires prompt response to diverse environmental stimuli, mediated by different calcium-sensing proteins.

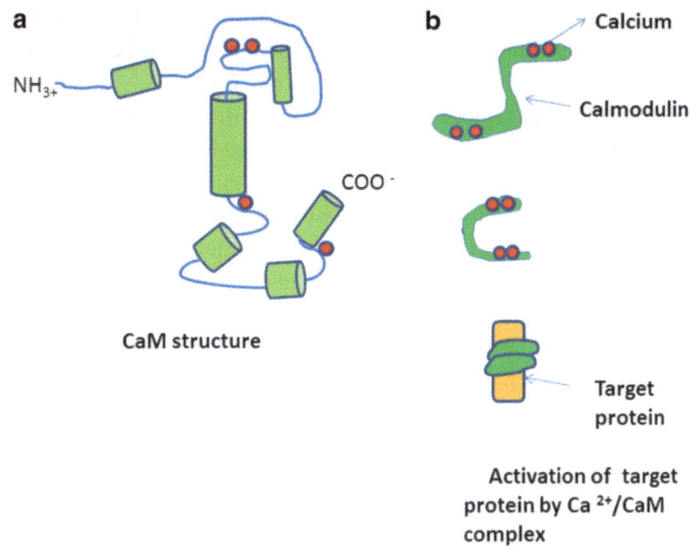

Fig. 1 *Calmodulin structure and function.* (**a**) General structure of calmodulin showing "EF-hand" motif comprising of α-helices (*shown in green*) connected by β-loops (*shown in blue*) that binds calcium ions. (**b**) Mechanism of calmodulin activation by calcium ions (*shown in red*) involves alterations in CaM conformation upon binding of calcium that then "wraps" around target protein to bring about its regulation

Plasticity in the Structure of CaM and CMLs

Calmodulin is an exceptionally versatile molecule at the structural level; calcium-saturated calmodulin has four helix-loop-helix (EF-hands), residing in the two globular domains, the N- and C-terminal globular domains, connected by a long α-helix also called the "central helix." Evidenced by NMR studies, the middle portion of the "central helix" is non-helical and flexible in solution, allowing it to extend in solution and facilitating accommodation of targets of different sizes. Each "EF-hand" is comprised of β-sheets with loops exposed to bind calcium ions (Fig. 1) and binds one calcium ion and exhibits cooperative binding, which results in a change in conformation of apocalmodulin (apoCaM) or calcium-free CaM, from a more compact structure to a dumbbell-shaped structure of calcium CaM (CCaM) as determined from biophysical studies involving NMR and X-ray diffraction studies. Binding of Ca^{2+} to CaM results in change in conformation from a closed to an open state thereby acting as a molecular switch; the binding energy released during this process is utilized to allow activation of downstream target proteins.

Comparison of X-ray structure of Ca^{2+}-saturated CaM and NMR structure of Ca^{2+}-free CaM has revealed presence of a large hydrophobic cavity/surface in the former upon binding to Ca^{2+}. In plants a variety of stimuli are transduced in the form of specific "calcium signatures," defined with respect to its spatial-, temporal-, and concentration-based parameters. A specific "calcium signature" induces conformational changes in calmodulin upon binding to Ca^{2+}, exposing hydrophobic methionine and phenylalanine residues in the hydrophobic cavity that bind to the aliphatic or aromatic amino acids in the target peptide thereby regulating target protein activation (Perochon et al. 2011). Besides hydrophobic interactions, calmodulin also utilizes electrostatic forces for interacting with target proteins. Electrostatic forces are contributed by a rim of negatively charged residues surrounding the hydrophobic surface also called the polar rim. These residues interact with the positively charged residues in the target protein further contributing to the binding energy.

The specificity in CaM target regulation arises from the Ca^{2+}-CaM complex specific target interactions with variable numbers of bound Ca^{2+} ions. Ca^{2+} binding to each EF-hand causes conformational transitions in the CaM molecule leading to multiple conformational states in complex with variable numbers of Ca^{2+} ions. CaM may regulate its targets with one, two, or three Ca^{2+} ions as well as in the apo- or fully loaded state. Recent experimental evidences suggest that the concentration-dependent profiles for several Ca^{2+}-/CaM-dependent protein targets exhibit quite a diverse range of behavior, for example, the ACII isoform is inhibited by increasing Ca^{2+} concentration.

Besides CaM, there are CaM-like proteins (CMLs) that have at least 16 % identity with CaM at the amino acid level with variable number of EF-hands (2–6) and no other identified functional domain (McCormack and Braam 2003). The EF-hand helices similar to CaM including 12 residues in a looped structure binds Ca^{2+} on their first, third, fifth, seventh, and twelfth amino acid position in the 12 amino acid long calcium-binding loop. Most of the CMLs have a strict conservation in Ca^{2+}-binding residues except several CMLs that have an E to D substitution in at least one of their EF-hands that might reduce affinity for calcium but may increase affinity for another divalent cation such as magnesium ions, though this remains to be experimentally confirmed.

Apart from the typical calcium-/CaM-binding proteins, calmodulin-binding proteins with IQ motif have been identified (Xie et al. 1994), where "IQ" represents the first two amino acids of the motif: isoleucine and glutamine. IQ motif, represented by the sequence IQXXXRGXXXR where X is any amino acid sequence and is about 20–25 amino acids long and assumes an α-helical conformation, stabilized in the presence of CaM. IQ motif introduces further diversity in CaM action. It has been found in calcium-independent CaM-binding protein such as AtBAG6 (Kang et al. 2006) and calcium-dependent CaM-binding protein such as IQD1 (Levy et al. 2005). Presence of large number of CaM and CaM-like proteins indicates that it serves important functions in a variety of developmental and stress-signaling pathways.

Mechanisms of Regulatory Action of CaM on Downstream Target Proteins

The regulatory action of CaM on its target occurs by more than one means such as relieving autoinhibition, dimerization, scaffolding, and remodeling of active site (Fig. 2). Activation by releasing autoinhibition involves binding of CaCaM to an autoinhibitory domain, adjacent to the CaM-binding domain, thus relieving the inhibition. GAD (glutamate decarboxylase) enzyme involved in GABA synthesis is activated by binding of CaM to its C-terminal autoinhibitory domain (Yap et al. 2003). Similar mechanism exists in activation of Ca^{2+}/calmodulin-dependent kinase (CCaMK). Activation by dimerization has been studied in K^+ channels where CaM binds to two channels and binding of Ca^{2+} leads to activation promoting inward K^+ ion movement (Li et al. 2005). Scaffolding or remodeling of an enzyme's active site involves a conformational change upon CaCaM binding that can promote its activation. Lastly, CaM may competitively bind to an overlapping ligand-binding site on a target protein thereby modulating its activity such as in the case of KCBP (kinesin CaM-binding protein), where it binds to the microtubule-binding facing domain of kinesin (Vinogradova et al. 2008).

Calmodulin Gene Family in Plants

CaM Gene Family in Arabidopsis and Rice

Unlike yeast genome, which encodes for a single CaM protein (Davis et al. 1986), animals and plant genomes encode highly conserved CaM gene family (McCormack et al. 2005). One of the striking differences between animal and plant calmodulin gene family is that the number of calmodulin proteins far exceeds in plants than in animals. Recent data provide evidence for their involvement in Ca^{2+}-/CaM-mediated signaling pathways. Moreover, in human and rat, CaM gene family is composed of three nonallelic

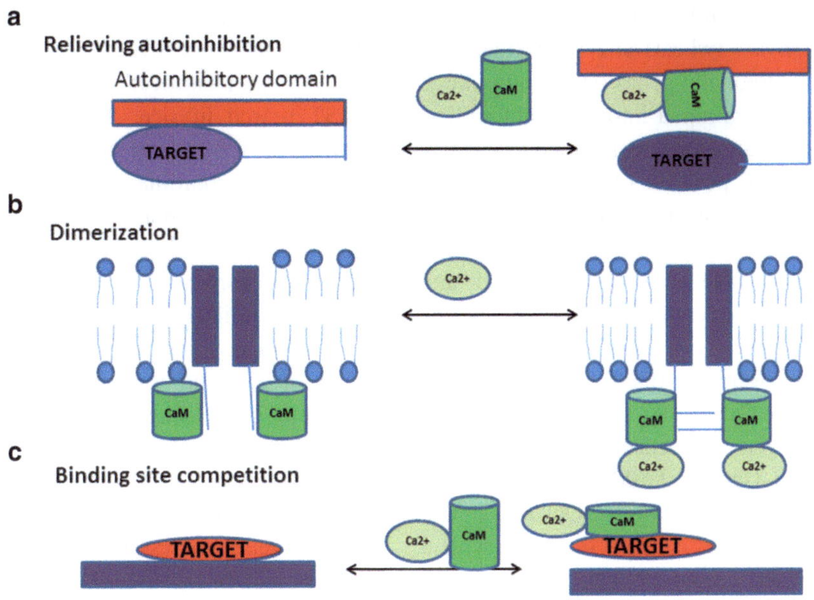

Fig. 2 *Mechanisms of CaM-mediated activation of target proteins.* CaM can bind to its putative targets and induce various conformational changes that activate the target protein by relieving autoinhibition, causing dimerization, acting as a binding-site competitor, and scaffolding and performing binding site remodeling

genes encoding for identical calmodulin protein (Ranty et al. 2006). Development of genomic tools and sequencing has led to characterization of CaM and CMLs in several plant species such as *Arabidopsis*, rice, wheat, maize, and potato. Out of a total of 250 "EF-hand"-containing genes identified in *Arabidopsis*, 7 genes (AtCaM1 to AtCaM7) encoding 4 isoforms of CaM and 50 genes encoding CML have been identified (McCormack et al. 2005). Seven CaM genes in *Arabidopsis* have nearly identical (97–99 %) amino acid sequence to animal CaM (typical CaM) (McCormack et al. 2005); CaM1 to CaM4 differ from CaM7 by four amino acids while CaM 2, 3, 5, and 6 differ from CaM7 by just one amino acid, such highly identical sequence conservation in multiple genes encoding the same protein points to a selection pressure to avoid sequence divergence that might have important functional consequence. By functional analysis of BLAST searches of TIGR rice database, a maximum of 243 genes encoding "EF-hand"-containing proteins were identified (Boonburapong and Buaboocha 2007). Bioinformatics analyses of rice (*Oryza sativa*) genome using BLAST search and Interpro scan revealed the existence of 5 CaM and 32 CML genes (Boonburapong and Buaboocha 2007). Phylogenetic analysis based on amino acid sequence revealed that OsCam1-1, OsCaM1-2, and OsCaM1-3 are exactly identical to each other in their amino acid sequence and shared 98.7 % identity with OsCaM2 and OsCaM3. OsCaM1 displayed a very high degree of similarity to typical CaM protein in barley and wheat suggesting some degree of conservation of CaM in monocot cereal plants. Hydrophobic residues, required for interaction with downstream target proteins, were found to be highly conserved in all the isoforms. Also, lysine 116, which regulates CaM activity by posttranslational modification and is important for CaM functionality, exhibited conservation across *Arabidopsis* CaMs and rice CaMs.

In addition to rice and *Arabidopsis*, CaM genes have been identified and characterized to some extent in wheat *(Triticum aestivum)* and oak (*Quercus petraea* Liebl). A wide repertoire of specific CaM and CML proteins provides an evolved system to deal with different types of environmental stimuli in higher plants. Yang et al. (1996) first isolated and characterized the calmodulin gene family in hexaploid crop plant *Triticum aestivum* (wheat) and found the existence of ten calmodulin genes representing three distinct isoforms grouped into four subfamilies. Southern blot analysis was performed to identify the chromosomal position of CaM genes, and each subfamily was concluded to form a homoallelic loci.

CaM genes have also been isolated and characterized in oak plant (*Quercus petraea* Liebl) where three CaM genes important during flooding stress were identified. QpCaM-1 that is one of the three CaM genes identified in oak had a very divergent sequence as compared to QpCaM-2 and QpCaM-3 (90.6 % and 89.9 % identity, respectively), while QpCaM-2 and QpCaM-3 had a very similar sequence with a difference of just one amino acid.

Presence of a multigene CaM family in plants may be attributed to the complexity of plants itself. Accumulating evidence suggests that each CaM gene may have distinct and significant function (Heo et al. 1999) supported by reports showing highly conserved CaM isoforms that modulate target proteins differently (Ishida et al. 2008). Besides, induced expression of some but not all CaM isoforms in response to certain stimuli has been reported (Duval et al. 2002), and competition among CaM isoforms for target protein might exist.

CML (CaM-Like Proteins) Genes in Arabidopsis and Rice

Like calmodulin, the number of calmodulin-like proteins (CMLs) far exceeds in plants than in animals. Besides, animal CMLs are very similar to typical CaM; plant CMLs are vastly diverse proteins (McCormack et al. 2005; Boonburapong and Buaboocha 2007). Studies reporting genome-wide identification of CaM genes in *Arabidopsis* and rice have also identified multiple CML genes in these plant species (McCormack et al. 2005; Boonburapong and Buaboocha 2007). Fifty CML genes have been identified in *Arabidopsis* genome most of which have four EF-hand motifs. CML1 has only one EF-hand, while CML12 has six EF-hands, the maximum number of EF-hands present in any of the *Arabidopsis* CMLs.

Similar to identification of CaM genes by BLAST search in rice, 32 CML genes were identified in rice genome that show high conservation in Ca^{2+}-coordinating residues with some degree of variation in amino acid positions such as those of OsCML7, OsCML8, and OsCML13 that have an aspartate (D) instead of glutamate (E) as a residue in the twelfth calcium-coordination position believed to be reducing affinity of EF-hand for calcium by over 100-fold while increasing its affinity for Mg^{2+} (Boonburapong and Buaboocha 2007; Cates et al. 2002). However, the physiological significance of Mg^{2+} binding remains unknown. Ten rice CML genes possessed an extra EF-hand that does not pair with any of the other EF-hand motifs. Pairing of EF-hand motifs helps increase its affinity for calcium, thus it is possible that this extra motif may bind calcium with lower affinity or may even be nonfunctional. OsCML7 and OsCML9 also appeared to be distinct in the sense that they had significant identity with OsCaM1-1 but have only 2 and 1 EF-hands, respectively.

Differential Expression of CaM Genes

A transcriptome analysis of *Arabidopsis* CaM during different developmental stages and under stress conditions using microarray was performed. The CaM transcripts were expressed in almost all developmental stages (callus, cell suspension, seedling, inflorescence, rosette, flowering, fruit, silique, and senescence stages). No significant change in CaM transcript was detected under tested stress conditions (salt, ozone, methyl jasmonate) as evident in the microarray data from the Geneinvestigator (https://www.genevestigator.ethz.ch). This could be explained by the fact that the basal expression of CaM transcripts is itself very high and comparable to that of housekeeping gene *TUB4* encoding tubulin. However, a previous study has shown upregulation of *CaM2* (also known as *TCH1*) by touch stimulus (Lee et al. 2005). Another recent report has shown the upregulation of seven CaM genes in *A. thaliana* by RT-PCR upon heat stress treatment suggesting possible involvement of CaM genes in heat stress signaling (Al-Quraan et al. 2010). Transcriptome analysis by Wang et al. (2008) also showed enrichment of some CaM/CML genes during pollen tube growth and germination in *Arabidopsis*. Twelve CaM genes such as *CaM7*, *CaM3*, and *CaM16* had higher expression during pollen growth hints towards their possible role in a complex signaling network activated during pollen growth and germination.

Differential Expression of CML Genes

CML11 and *CML33* had highest expression in cell suspension, while others such as CML*2*, CML*6*, CML*21*, and CML*35* have highest expression in floral organs specifically stamens. Four CML genes (*CML3*, CML*16*, CML*39*, and CML*49*) also showed enrichment in pollen germination and growth stage suggesting their possible role in signaling pathways regulating pollen growth and development.

Most of the CML genes showed at least five-fold change in expression in response to a variety of stresses such as ozone, salt, and cycloheximide. *CML37*, *CML39*, and *CML40* were among the ones most strongly regulated by stimuli. Divergent *CML* sequence and distinct expression pattern in different plant organs as well as differential regulation under stress seems to suggest duplication and subsequent mutation in the *CML* genes to impart novel functions to these calcium-binding proteins.

Recently, expression of CaM and CML genes was analyzed in rice during different developmental stages and under a variety of abiotic stress stimuli (Boonburapong and Buaboocha 2007). Significant differences could be found in the expression pattern of CaM isoforms and CML genes providing interesting insights into possible functions and tissues where these genes might be functionally active. OsCaM genes (*OsCaM1-1*, *OsCaM1-2*, *OsCaM1-3*, *OsCaM2*, and *OsCaM3*) were found to be expressed in almost all the developmental stages (root, leaf, inflorescence, and seed) examined. *OsCaM1-1* was highly expressed in all tissues and showed strong induction under salt stress. *OsCaM1-1* was studied in further detail by the authors who made several *OsCaM1-1::GUS* reporter lines that showed induction of GUS activity under different stress conditions such as salt and drought. Consistent with the stress stimulus-induced expression, analysis of *OsCaml-1* promoter element revealed the existence of several stress response elements such as DRE (drought response element) motif (ACCGAC) in the upstream region that suggests possible regulation of *OsCaml-1* expression by DREB class of transcription factors that recognize DRE *cis*-element under salt, cold, and drought stresses.

OsCML7 and *OsCML13* were expressed at high levels in all the tissues, whereas *OsCML1*, *OsCML4*, and *OsCML8* showed similar expression pattern with higher levels expressed in the leaf and root and lower expression in young inflorescence. Real-time PCR was performed to examine the expression of CML genes showing differential expression under salt and drought stress. *OsCML4*, *OsCML5*, *OsCML8*, and *OsCML11* showed strong induction under salt stress, whereas *OsCML7* and *OsCML9* showed weak induction. *OsCML4* and *OsCML8* exhibited highest increase in transcript levels upon PEG (drought stress-mimicking agent) treatment that was maintained till 48 h after treatment, while *OsCML15* was found to be showing early upregulation that return to normal levels after 5 h. Similar to *OsCaml-1*, *OsCML4*, *OsCML5*, *OsCML7*, and *OsCML11* exhibited the presence of DRE *cis*-element in the promoter region that supports real-time PCR data showing the respective upregulation of these genes under osmotic and salt stresses (Boonburapong and Buaboocha 2007).

Thus, identification of multiple CaM and CML genes in these plant species together with the development and stress-induced expression highlights the importance of these calcium sensors in Ca^{2+}/CaM signaling in plant growth and function. Also, conservation of identical CaM genes with no amino acid difference hints towards a positive selection pressure that has maintained these sequences within the genome. Unlike CaM, CML sequences have undergone some divergence in their respective sequences that suggests distinct functions may be performed by these paralogous genes. Biochemical functional characterization of these different isoforms will further help elucidate the precise roles performed by them.

Subcellular Localization of CaM

Local calcium signature generates signal at specific sites/compartment within the cell, which alters the subcellular localization of CaM and CML proteins. This intracellular trafficking of calmodulin has been shown to play an important role in its functional regulation. CaM is the primary Ca^{2+} decoder in the nucleus. Subcellular localization has revealed the presence of CaM in the nucleus. Immunocytochemical observations conducted using CaM-specific antibodies in several plant species including pea and barley revealed the presence of CaM in the nucleus (Dauwalder and Roux 1986). Another study showed PhCaM53 from petunia to be either membrane or nuclear localized depending on its prenylation status (Rodriguez-Concepcion et al. 1999). CAAX motif characterized by the presence of C cysteine residue, A an aliphatic amino acid, and X serine, methionine, alanine, glutamine, and leucine serves as a prenylation site for GGTase (geranyl geranyl transferase) enzyme. In addition to the CAAX motif, a C-terminal region rich in arginine and lysine residues was also found to be important for both membrane and nuclear localization. Prenylation acts as a signal for membrane attachment while inhibiting this signal moves PhCaM53 into the nucleus. This nucleocytoplasmic shuttling was regulated by light exposure and carbon source. Functional relevance of this localization was proven using mutant lines expressing constitutively nuclear PhCaM53 that displayed growth defects, explained by the fact that CaM53 might interact with a variety of target proteins in the nucleus to alter expression of genes involved in development (Rodriguez-Concepcion et al. 1997).

Using CaM-conjugated quantum dot (QD) system, sensitized to a single molecule on the surface of plant cells, it was shown that QD-CaM binds selectively on the outer surface of the plasma membrane as confirmed by TEM. A recent study by Wang et al. (2009a) used this technique to show the modulation of calcium fluxes across plasma membrane by an apoplasm (extracellular matrix)-localized CaM. In addition to QD technique, electron microscopy and measurement of calcium fluxes in pollen protoplast using microelectrodes were used to demonstrate the localization of conjugated CaM on the membrane and its downstream effects on intracellular calcium levels, respectively. This study provided direct evidence for extracellular CaM in acting as a signaling molecule.

In addition to nucleus and membrane, CaM proteins have also been detected in peroxisome and chloroplast. AtCat3 (catalase), regulated by CaM, was shown to be localized inside the peroxisome using antihuman CaM antibody from fractions isolated using sucrose density centrifugation (Yang and Poovaiah 2002). The authors did not address how this CaM may be localized in the peroxisome. But as observed in the case of CaM53, it is possible that posttranslational modifications could act as a signal to influence its localization. CaM has been also found to be localized in the chloroplast in a study that identified it to be the NAD kinase "activating" factor required for photosynthetic light reaction converting NAD to NADP (terminal electron acceptor of photosystem I) (Jarrett et al. 1980). Use of calmodulin inhibitors like phenothiazine inhibited this reaction, suggesting requirement of CaM that was also supported by the presence of CaM in isolated chloroplast fractions.

Subcellular Localization of CML

Like CaM, a CML protein AtCML19 (AtCENTRIN2) has been found to be showing dynamic subcellular localization depending on stress condition. AtCML19 translocates from the cytoplasm to the nucleus upon UV exposure and provides protection from DNA damage (Liang et al. 2006). Immunoblotting of AtCENTRIN2 revealed increased accumulation of the protein in response to UV exposure. EGFP fusion constructs of AtCML19 showed enhanced nuclear localization after UV-C exposure or cisplatin treatment that induces bulky DNA lesions in the genome. This study provided evidence for the critical role played by nuclear-localized CML19 in DNA damage response. Another CML protein, CML42, has been shown to be localized to both cytoplasm and nucleus where it functions to regulate responses to herbivore attack (Vadassery et al. 2012).

Thus, studies showing different subcellular localization of CaM/CML provide important information about functions performed by them in different organelles under diverse developmental/stress cues.

Increasing Repertoire of CaM-Binding Proteins

Calmodulin by itself does not have any activity so the signal has to be transduced through another set of proteins containing a CaM-binding domain. In the past, extensive use of proteomics and genomics has led to the identification of calmodulin-binding target proteins. Yeast two-hybrid screens and protein microarray revealed that the target proteins lack a consensus sequence in the CaM-binding motif, which provide flexibility in CaM regulation. Also, sequencing and annotation of *Arabidopsis* genome has increased the repertoire of CaM-binding proteins. To date ~50 CaM-binding proteins have been identified, and the number is expected to increase further.

CaM-binding proteins can be broadly categorized into three groups. One of the groups comprises plant-specific CaM-binding proteins, such as auxin-responsive SAURs and pollen-specific MPCBP. Interestingly, the CaM-binding proteins in plants are more diversified, greater in number, and some are plant specific. This suggests existence of wide and diverse plant-specific CaM signaling networks.

Approaches to Identify CaM-Binding Proteins

A variety of molecular biology, biochemical, and computational methods have been utilized to identify potential CaM-binding proteins to gain insight into pathways regulated by calmodulin (Reddy et al. 2011). One of the most commonly employed methodologies to study downstream targets of CaM has been screening of cDNA expression library with CaM labeled with horseradish peroxidase (HRP) enzyme, radioactive S^{35}, or biotin. The technique involves screening plaques induced for protein expression with IPTG and probing them with labeled CaM. Positive clones are further characterized by in vitro assays such as CaM overlay assay, CaM electrophoretic mobility shift assay, and CaM sepharose assay. cDNA expression library prepared from a variety of development stages and hormone-treated plant material has been used to identify potential CaM-binding proteins (CBPs). Initial screening studies identified a wide range of proteins as CBPs, few of which include GABA (gamma-aminobutyric acid) synthesizing enzyme GAD (glutamate decarboxylase), stress-activated transcription factor, CAMTA3 (calmodulin-binding transcription activators), and KCBP (kinesin-like calmodulin-binding protein), a cytoskeletal protein.

Recently, protein microarray studies have been employed to identify CaM/CML targets where 1,133 *Arabidopsis* ORF clones were used to express fusion proteins that were then probed on an array with alexa-fluor conjugated CaM1, CaM6, CaM7, CaM8, CML10, and CML12 (Popescu et al. 2007). Of all the clones tested, 173 proteins were identified to be interacting with at least one of these CaMs/CMLs. The proteins identified to be binding to CaM/CML in this study could be grouped into a variety of classes such as transcription factors, intracellular and receptor protein kinases, cell cycle specific proteins, and cytochrome P450s. In vitro immunoprecipitation studies of 83 candidate target proteins revealed only about 50 % of these interactions are positive, suggesting that protein array studies like yeast two-hybrid may yield interaction artifacts and, therefore, need to be treated with caution when studying these target proteins individually as putative CaM targets.

Besides the biochemical approaches described, some computational tools have also been designed for in silico prediction of putative CaM targets. However, it is important to note that computational methods have a high false positive rate of prediction and still needs to be improved for a more accurate prediction. One such approach involves domain/motif method that is based on the idea that "proteins with similar domains may exhibit similar interactions." Comparative genomics can also be employed as a method to identify putative CaM-interacting plant proteins based on the hypothesis that evolutionarily important protein interactions are conserved across species (Pazos and Valencia 2001). Thus, PPI (protein-protein interaction) dataset from more extensively studied organisms such as yeast can be utilized to create similar protein network maps. Co-expression and co-regulation of genes encoding potential interacting proteins has also been utilized to create interaction network. Protein interaction requires that the two proteins need to be expressed in the same cell and at the same time. DNA microarray studies revealing genes showing similar expression pattern and similar regulation by same transcription factors have been considered to be useful source for determining protein-protein interaction sets (Lu 2005).

Calmodulin and Its Role in Abiotic Stress Response

Plants possess extraordinary plasticity in many of their growth and developmental processes in response to a variety of stresses, achieved by adoption of intricate mechanisms to perceive and respond to various stimuli in order to adapt to their environment. CaM and downstream effectors of CaM play a major role in regulating responses to various stresses. This is achieved partly by presence of CaM and proteins with CaMBD in various cellular compartments. This section basically discusses the recent advances in the field of CaM- and CML-regulated abiotic stress responses.

Regulation by CaM and CMLs of Nuclear CaM-Binding Proteins

Transcriptional activation of genes promoting stress tolerance is one of the major mechanisms by which signaling pathways act to evoke response to various abiotic stress conditions. Besides activating gene transcription via calcium sensors, a direct role of CaM and CML in controlling the activity of transcription factors has been found.

Regulating Transcription by DNA-Binding Transcription Factors

AtSRs/CAMTAs (*Arabidopsis thaliana* signal responsive/CaM-binding transcription activator) family of transcription factor is a conserved family of transcription factors in multicellular eukaryotes, identified in a cDNA library screen by Bouche et al. (2005) as a CaM-binding protein. Structural characterization of AtSRs/CAMTAs revealed presence of an N-terminus DNA-binding CG-1 domain, transcriptional activation domain (TAD), a transcription factor immunoglobulin (TIG)-like nonspecific DNA-binding domain, ankyrin repeats (ANK), and a noncanonical CaM-binding IQ motifs connected to a canonical calmodulin-binding domain (CaMBD). Arabidopsis genome encodes for six CAMTA genes (AtCAMTA1–AtCAMTA6).

The molecular interaction between CaM and AtSR/CAMTA was primarily hydrophobic-polar in nature like most other CaM interactions. AtSR2/CAMTA1 in *Arabidopsis* negatively regulates auxin response, determined by transcriptome profiling of *camta1* mutant lines (Galon et al. 2010). Of the 63 genes observed to be upregulated in *camta1* mutant, 27 % were auxin-induced genes such as F-box proteins, auxin-responsive protein, and cytochrome P450. Further analysis using BAR expression browser and AtGene Express data revealed that transcriptome changes in *camta1* mutant are most similar to that of changes in mRNA profile by gibberellins, cytokinin, and auxin. Consistent with this observation, hypocotyls of *camta1* T-DNA insertion mutant and repressor lines exhibit a hypersensitive auxin response (Galon et al. 2010). Significant proportion of genes involved in abiotic stress response such as cold, heat, and drought were downregulated in *camta1* loss-of-function plants suggesting its importance in abiotic stress signaling. Also, *AtCaMTA1*::GUS reporter assays revealed induction under stress conditions such as heat (in leaf trichome and root cortex) and salinity (Galon et al. 2010).

AtSR1/CaMTA3 has been also observed in regulating cold stress acclimation and freezing tolerance together with CaMTA1 by serving as activator of CBF (C-repeat/DRE-binding factor) genes that encodes AP2/ERF family of transcription factors involved in expression of a variety of cell metabolites such as raffinose and sucrose, which act as cytoprotective agents (Doherty et al. 2009) in response to cold stress. Another study revealed a conserved DNA motif 2 (CM 2) with a consensus sequence of CCGCGT as a typical AtSR/CaMTA recognition motif in promoter of CBF2. Expression of endogenous CBF2 was reduced in *atsr1/camta3* null mutants; transcript levels were restored to normal level either by complementation of AtSR1/CaMTA3 under control of 35S promoter or by introducing another null mutation in AtSR2/CaMTA1. These results implied that CBF2 was positively regulated by AtSR1/CaMTA3 and negatively regulated by AtSR2/CaMTA1. Interestingly, no difference in freezing tolerance was found in single mutants, *camta1camta3* double mutant plants exhibited reduced cold acclimation upon prior exposure to mild cold stress before being treated with freezing temperature suggesting the specific role of these genes in response to cold stress (Doherty et al. 2009). Electrophoretic mobility shift assay (EMSA) revealed CaMTA1, CaMTA2, CaMTA3, and CaMTA5 binds to CM2 sequence confirming the prediction that CaMTA transcription factors might be acting via this upstream sequence to regulate gene transcription. Indeed, promoter analysis using various GUS constructs with mutations in the promoter regions identified CM2 sequence as a critical sequence for CaMTA3-mediated activation of CBF gene expression (Doherty et al. 2009). This study established CaMTA1 and CaMTA3 as a critical link between calcium signaling and freezing tolerance. However, exact nature of CaM mediating this effect remains to be identified.

Studies in a member of GTL (GT-2-LIKE1) family of transcription factor in Poplar (*Populus species*), namely, PaGTL1 with respect to its role in water use efficiency and drought tolerance (Weng et al. 2012), revealed it to be regulated by Ca^{2+}/CaM under water deficit condition posttranscriptionally. This regulation is achieved by direct binding of its C-terminal DNA-binding helix to CaM in a calcium-dependent manner exerting a negative effect on transcriptional repressive activity of PaGTL1 on *PtaSdd1* (*stomatal density and distribution1*) expression that upon activation is believed to promote improved water use efficiency and drought tolerance via the stomatal lineage pathway (Weng et al. 2012). Enhanced survival under drought conditions of *gtl1 Arabidopsis* mutants could be reversed by expression of PaGTL1 under AtGTL1 promoter indicating that PaGTL1 structure and function may be well conserved across species. Poplar is commonly cultivated as a hybrid species of *Populus tremula* x *P. alba* for biofuel production that is greatly reduced by water deficit conditions. This study highlights the importance of calmodulin-regulated PaGTL1 in creation of more drought-resistant variety of this species without compromising its biomass (Weng et al. 2012).

AtGTL is another CaM-binding transcription factor that was first identified using S^{35}-labeled CaM as a probe for cDNA expression library prepared from cold stress-treated seedlings (Yoo et al. 2010). AtGTL structure resembles GT subfamily of transcription factors that consist of a tri-helix DNA-binding domain that bind GT element. Sequence analysis revealed the existence of a putative CaM-binding domain which was further verified by site-directed mutagenesis of methionine^{506} and leucine^{507} residue that abolished interaction of AtGTL with bovine CaM in vitro and also compromised its ability to bind DNA as concluded from EMSA studies (Yoo et al. 2010). GFP fusion construct of AtGTL transformed into *Nicotiana* epidermal cells confirmed its nuclear localization, and transgenic lines overexpressing AtGTL had increased transcript levels of cold- and salt-responsive genes such as *RD29A* and *ERD10*.

However, this increased transcription of stress-responsive genes could not be correlated with increased resistance to stress probably because of the fact that additional factors might be required downstream of AtGTL1 to bring about enhanced stress tolerance (Yoo et al. 2010).

Perruc et al. (2004) identified a plant-specific 25 kDa CaM-interacting protein, AtCaMBP25, in a screen using osmotically stressed *Arabidopsis* seedlings. CaM overlay assay found the interaction between AtCaMBP25 and AtCaM1 in a calcium-dependent manner and not with other isoforms such as AtCaM8 that suggests involvement of a specific CaM isoform in binding and regulation of distinct subset of target proteins. Northern blotting revealed AtCaMBP25 transcript to be induced by cold, osmotic, and salt stresses. GFP fusion protein confirmed its nuclear localization, which upon overexpression resulted in increased sensitivity to mannitol-induced osmotic stress. On the other hand, transgenic lines expressing knockdown construct showed enhanced resistance to osmotic stress suggesting involvement of AtCaMBP25 as a negative regulator of osmotic stress response (Perruc et al. 2004). Multiple transgenic lines expressing knockout construct of AtCaMBP25 displayed increased germination rates and root length upon mannitol and NaCl treatment in comparison to wild type. However, the authors have not investigated molecular mechanism of negative regulation, and further experiments are required for detailed understanding of the process. Analysis of AtCaMBP25 mRNA in ethylene and ABA synthesis mutants suggested it to be regulated independently of these two pathways (Perruc et al. 2004).

Regulating Transcription by Phosphorylation and Dephosphorylation

Plants being sessile are constantly exposed to fluctuating temperature; exposure to temperature above optimal leads to heat stress and results in retarded growth and development. Accumulation of heat stress proteins (HSPs) under the control of heat stress transcription factors (HSFs) is known to play crucial role in heat stress response (HSR) and provides thermotolerance to plants. AtCaM3 has been characterized in detail with respect to its role in HSF1 (heat shock transcription factor 1)- and AtPP7 (serine/threonine phosphatase)-mediated thermotolerance in *Arabidopsis* (Zhang et al. 2009a). In a forward genetics approach, different T-DNA mutants available for *CaM* genes were analyzed, and *AtCBK3 (AtCRK1, CDPK-related kinase1)* mutants were observed to be impaired in thermotolerance. This phenotype was correlated to decreased expression levels of heat shock proteins in the mutant lines. On the other hand, plants overexpressing AtCBK3 were observed to be more resistant to heat stress showing its sufficiency in activating HSF1 to mediate thermotolerance. Data gathered from in vitro phosphorylation assays, AtCBK3 knockout, and overexpression of AtCBK3 suggested that Ca^{2+}/CaM regulates the activity of protein kinase AtCBK3, which phosphorylates transcription factor AtHSFA1a to increase its interaction with HSE and stimulate the expression of HSPs.

Similar to AtCBK3, AtPP7 a novel serine/threonine phosphatase imparted thermotolerance upon overexpression while loss of function of this gene resulted in compromised thermotolerance. Increased transcript level of heat shock genes AtHSP70 and AtHSP101 as assayed by quantitative PCR was observed in transgenic lines overexpressing AtPP7, while the reverse was true for knockout lines (Liu et al. 2007). CaM-binding motifs were identified in AtPP7, and it was activated in Ca^{2+}-/CaM-dependent manner to regulate AtHSF1 (AtHSFA1a) transcriptional activity (Li et al. 2004). Studies in other systems also suggest the existence of a PP2C-type phosphatases and their interaction with CaM; in moss, *Physcomitrella patens*, PsCaMPP, has a CaM-binding domain (Takezawa 2003), and in maize, Li et al. (2004) reported that HSF1 binds to DNA upon activation by calmodulin under heat stress. EMSA assay revealed a calcium dependence of HSF1 binding to heat shock element under heat stress. Blocking CaM activity by W7 inhibitor inhibited

DNA-binding activity of HSF1 suggesting a possible conserved role of calcium/calmodulin in activation of HSF1 across different plant species. However, exact function of this interaction remains to be explored.

Soybean genome encodes for number of CaM isoforms (SCaM1–SCaM5), out of which only SCaM4 exhibited rapid and differential induction in response to salt stress as shown by northern blot and in vivo GUS assay. This supports the hypothesis that different CaM isoforms can activate different Ca^{2+}-/CaM-mediated signaling pathways. Study by Park et al. (2009) found the existence of GT1 *cis*-element (GAAAAA) upstream of SCaM4 gene sequence that mediates its regulation by GT family of transcription factors under stress. In order to find which putative GT member was acting to regulate its expression, genome-wide sequence analysis of *Arabidopsis* genome was conducted that found four GT1-related transcription factors, namely, AtGT1, AtGT3a, AtGT3b, and AtGT4. Only AtGT3b expression was induced significantly under NaCl or pathogen stress, which hints towards its possible role in regulation by SCaM4. EMSA studies using AtGT3b-GST fusion construct showed it to be directly binding to the GT element oligonucleotide probe. This was further confirmed in vivo in yeast cells where AtGT3b fused to yeast GAL activation domain specifically induced the expression of HIS and lacZ reporter gene bearing GT element in their promoter region.

AtMYB2 transcription factor has also been studied in context of its activation by calmodulin and its role in salt and dehydration tolerance. A study by Yoo et al. (2005) used GmCaM4 to probe a cDNA expression library prepared from salt-treated Arabidopsis and identified MYB2 as its interacting partner. MYB2 is a R2R3 DNA-binding domain containing transcription factor that has previously been shown to regulate the expression of various salt and dehydration inducible genes such as *RD22* and *P5CS1*. The CaM-binding region in AtMYB2 was found to be in its DNA-binding domain. Detailed analyses with CaM overlay assay and site-directed mutagenesis confirmed the direct in vitro interaction of AtMYB2 with CaM. Further, overexpression of GmCaM4 in Arabidopsis led to upregulation of MYB2-dependent genes such as those encoding proline-synthesizing enzyme (pyrroline carboxylate synthetase 1), accompanied by enhanced salt tolerance. Interestingly, GmCaM1 and GmCaM4 differentially regulated DNA-binding activity where GmCaM4 was observed to increase the DNA-binding activity of MYB2, while GmCaM1 was found to be inhibiting it. However, how this binding activity is modulated in different ways at the structural level remains to be answered.

CBP60g is a CaM-binding transcription factor and plays a role in ABA response and drought stress in addition to its previously identified role in biotic stress (Wan et al. 2012; Wang et al. 2009b). CBP60g overexpression transgenic lines were observed to be more resistant to drought conditions as compared to wild type and also have a hypersensitive response to ABA that was further linked to a downstream activation of SA biosynthetic enzyme *ICS1* and *EDS5*. Specifically which genes cause these abiotic stress-resistant phenotypes needs further investigation since no change at the transcript level of ABA synthesis genes could be detected.

Regulation of Membrane CaM-Binding Proteins by CaM and CML

Regulating Ion Channels

Creation of Ca^{2+} current/fluxes in the cell forms an initial step in calcium signaling occurring in response to various abiotic and biotic stimuli. This alteration in calcium levels within the cell is brought into effect by ion channels that have been characterized to some extent in *Arabidopsis* Boursiac and Harper 2007. Some studies have suggested the involvement of cyclic nucleotide-gated channel (CNGC) in calcium uptake and currents within the cell under different stimuli (White et al. 2000; Hetherington and Brownlee 2004). Of the 20 CNGC genes present in *Arabidopsis*, CNGC2 has been shown to be conducting calcium/potassium current in heterologous system. In fact, the first CNGCs were originally identified as CaM-binding protein in a cDNA library screen in tobacco (NtCBP4) and barley (HvCBT1) (Demidchik and Maathuis 2007) implicating them in calcium/CaM signaling. However, precise roles performed by CNGC family of ion channels in heteromeric arrangement and calcium-signaling cascade require further investigation. Other ion channels such as sodium/proton exchanger (NHX) family of transporters that are located on vacuolar membranes in plant cells serve to act as proton pumps helping maintain Na^+ levels in the cell providing protection from salinity stress (Jiang et al. 2010). As described below some of these NHX classes of transporters have been regulated by CaM at some level.

AtNHX1 is the most abundant Na^+/H^+ antiporter, a cation/anion exchanger present on plant vacuole, and plays an important role in Na^+ sequestration in the vacuole and vacuolar pH regulation, regulated by AtCaM15 (Yamaguchi et al. 2005). AtCaM15 was identified as an AtNHX1-interacting protein in a cDNA library screen using C-terminal NHX1 as bait. C-terminal NHX1 harbors a hydrophilic region present entirely in the vacuole, which upon deletion doubled the Na^+/K^+ selectivity of the channel suggesting an important regulatory role of this domain (Yamaguchi et al. 2005). Testing this interaction with another CaM isoform, CaM81 showed it to be occurring specifically with AtCaM15. As in yeast, GFP fusion construct showed AtNHX1 to be localizing in the vacuole. Co-immuno-precipitation using FLAG-tagged AtNHX1 and HA-tagged AtCaM15 further confirmed the physical interaction between CaM and NHX (Yamaguchi et al. 2005). Further assays measuring the transport activity of AtNHX1 showed the interaction to be repressing the Na^+/H^+ activity of the channel and also showed the interaction to be governed by pH as well as Ca^{2+} levels (Yamaguchi et al. 2005). Collectively all the experiments performed suggest a possible role for CaM15-mediated control over NHX1 activity, relieved under stress conditions such as high salinity (high pH) that would abolish this interaction increasing Na^+/H^+ exchange activity facilitating increased storage of sodium ions in the vacuole providing protection from salt stress.

Other multidrug transporters in plant such as AtMRP1 also possess a CaMBD hint towards the possible role of CaM in protecting plant from chemical toxicity (Giesler et al. 2004). This knowledge of cell membrane channels, which uptake heavy metal ions together with other ions in the soil, provides an important avenue by which plants having differential capacity to uptake ions. This can be exploited for phytoremediation or to generate crop varieties that are resistant to accumulation of such heavy metals and are hence suitable for human consumption.

ACA4 is an *Arabidopsis* Ca^{2+}-ATPase located in the vacuolar membrane possessing an N-terminus autoinhibitory and CaM-binding domain. This particular Ca^{2+}-ATPase is necessary for conferring protection against high NaCl, KCl, and mannitol stress as confirmed by expression experiments in yeast calcium transport mutant K616 (Geisler et al. 2000). Expression of a truncated version of ACA4 lacking the N-terminal domain was able to rescue yeast mutant growth under low calcium conditions suggesting that CaM probably activates ACA4 by relieving autoinhibition by the N-terminal region of ACA4 (Geisler et al. 2000). In addition, ACA4 transcript was observed to accumulate under high salt conditions, and increased expression of ACA4 in yeast conferred it with enhanced tolerance to osmotic stress suggesting the existence of Ca^{2+}-/CaM-mediated activation of ACA4 under high salt conditions (Geisler et al. 2000).

Regulation of Enzymes During Stress Tolerance

CaM modulates the activity of ion transport channels and metabolic enzymes as a rapid response mechanism to deal with abiotic stress conditions. Gamma-aminobutyric acid (GABA) is a nonprotein four-carbon amino acid that is involved in stress response involving changes in temperature and light conditions (Wallace et al. 1984; Bouche et al. 2005). Later studies linked the increased accumulation of GABA- to CaM-mediated binding and activation of glutamate decarboxylase (GAD) enzyme. X-ray scattering studies showed Ca^{2+}/CaM pair binding to hexamer GAD enzyme in a 1:3 stoichiometry exhibiting a relief of autoinhibition mode. C-terminal lobe of CaM binds to a CaM-binding domain defined by 25 amino acid residues close to or within the autoinhibition domain of GAD that activates the enzyme (Gut et al. 2009).

Redox status of cell is regulated by a host of enzymes such as NADP (nicotinamide adenine dinucleotide phosphate) kinase and detoxifying enzymes such as catalase and superoxide dismutase (SOD). Reactive oxygen species (ROS) generation under abiotic and biotic stress serves as a crucial cue for downstream protective signaling cascades. Ca^{2+}/CaM has been found to be contributing to oxidative stress signaling by activation of a variety of such enzymes. Study conducted by Yang and Poovaiah (2002) reported a twofold increases in catalase (AtCAT3) activity upon stimulation by Ca^{2+}/CaM complex. Similarly, SOD another ROS-scavenging enzyme has been reported to have CaM-binding sites. Increase in NADPH levels upon a variety of stimuli including mechanical, osmotic, and incompatible bacteria has been shown to be the outcome of CaM-mediated activation of NAD kinase. NADPH is possibly used as a reductant by NADH oxidase to generate active oxygen species that further activate downstream signaling pathways (Harding et al. 1997). Another calmodulin-binding protein isolated by screening tobacco cDNA expression library, NtCBP4, has a putative CNGC (cyclic nucleotide-gated cation channel) domain and is important for heavy metal tolerance (Sunkar et al. 2000). Structurally, NtCBP4 is similar to invertebrate and vertebrate nonselective K^+ channel. This is the first report demonstrating the existence of a plasma membrane CaM-binding protein channel involved in tolerance to metal toxicity. Interestingly, expression of full length NtCBP4 leads to increased accumulation of Pb^{2+}in seedlings, while expression of just the calmodulin-binding domain removing part of the CNGC channel domain produces enhanced resistance to Pb^{2+}. Consistent with this, T-DNA insertion mutant lines of *Arabidopsis* AtCNGC1 a homolog of NtCBP4 had increased resistance to lead (Sunkar et al. 2000).

Regulation of Membrane Proteins and Receptor Kinases

Apyrase (nucleoside triphosphatase or NTPase) is a membrane-localized hydrolytic enzyme; pea psNTP9 which possesses a CaMBD and is activated by Ca^{+2}/CaM overexpression of pea apyrase in a heterologous system like *Arabidopsis* was observed to provide increased resistance to cycloheximide, cytokinin, and herbicides (Hsieh 2000).

Recent studies have revealed that plant genome encodes for putative receptor-like kinase (RLK) proteins, which like their animal counterparts may play a key role in growth and development of plants. Role of a *C*aM-regulated *r*eceptor-*l*ike *k*inase (CRLK1) was studied by Yang et al. (2004, 2010) who characterized this plasma membrane protein to play an important role by acting as a bridge between calcium and cold signaling. *crlk1* mutant plants were found to be more susceptible to cold stress as compared to wild type, and this phenotype could be further correlated with delayed expression of cold response genes such as *RD29A* and *COR15a*. A follow-up study by the same group found CRLK1 to be interacting both in vitro and *in planta* with MAP kinase pathway member MEKK1 in regulating cold stress response (Yang et al. 2004). Mass spectrometric analysis of CRLK1-immunoprecipitated protein complex led to identification of several candidate proteins as its molecular interactors. MEKK1, a member of the MAPK pathway previously implicated in low temperature response, was also identified in the pull-down assays and selected for further studies. MEKK1 was activated in response to cold stress by CRLK1, and this activation was abolished in *crlk1* knockout mutants as demonstrated by in-gel phosphorylation assay (Yang et al. 2004). BiFC (bimolecular fluorescence complementation assay) using transfection of fluorescently labeled CRLK1 and MEKK1 in the protoplast confirmed in vivo interaction of the two proteins and also showed this association to be occurring both on cell membrane and endosomes. MEKK1 itself is a MAPK cascade protein believed to be cytoplasmic; however, a recent study has shown it to be localized in the nucleus in connection with its role in leaf senescence (Miao et al. 2007). Thus, MEKK1 together with CRLK1 may localize at both membrane and endosomal vesicles to regulate downstream pathways involved in stress acclimation. GmCaMK1, a soybean *crlk1* ortholog, carries a stretch of 24 amino acids in its C-terminal domain that forms the CaMBD overlapping with a kinase domain (DeFalco et al. 2010). Therefore, the existence of such kinases that depend on CaM for their activity highlights the role of calcium/CaM as an upstream effector of stress-response signaling mediated by kinases that further act to relay the signal via sequential phosphorylation and activation of proteins involved in stress tolerance.

Another calcium-/calmodulin-binding receptor-like kinase isolated from EST-based sequencing from salt-tolerant plant *Glycine soja*, namely, GsCBRLK1, has been studied by Yang et al. (2010). Cold, salinity, drought, and ABA induced GsCBRLK1 transcript levels, suggesting its importance as a linker between stress and calcium-induced gene expression. Overexpression of GsCBRLK1 increased resistance to salinity and ABA associated with increased expression of stress and ABA gene markers including *RD29A*, *RD22*, *COR15A*, and *KIN1*. GFP fusion construct showed GsCBRLK1 to be a membrane-localized protein that could be attributed to the existence of an N-myristoylation motif MGXXXS/T(K) that serves as an anchorage signal to the plasma membrane. CaM-binding assay using synthetic peptide (GsCBRLK from Tyr147 to Asn169) revealed CaM binding of GsCBRLK1 and calcium dependent, supporting the in vivo interaction of GsCBRLK1 to calmodulin as predicted by yeast two-hybrid assay. Further, in vitro enzyme assays showed GsCBRLK1 to be exhibiting autophosphorylation as well as myelin basic protein substrate phosphorylation in a calcium-dependent manner.

Other CaM/CML During Abiotic Stress Tolerance

A variety of other CaM/CML members have been identified to be acting in stress response pathways and their respective interacting proteins required for mediating this response as described below for some of them have been identified.

AtCML9 has been reported as a negative regulator of ABA signaling pathway and response to abiotic stress stimuli as concluded from *cml9* mutant lines that have increased resistance to salt and drought conditions and exhibit an ABA sensitive phenotype in germination assays (Magnan et al. 2008). However, expression of stress response genes such as *RD29A* and *RD20* was reduced in the *cml9* mutants suggesting that the stress resistance conferred by loss of *cml9* was via a different mechanism. Recently, using a protein microarray approach (Popescu et al. 2007) led to identification of ABF4 (ABRE-binding factor 4) transcription factor to be interacting with AtCML9, and thus, a possible role of AtCML9 in ABA signaling by virtue of this interaction could be postulated. As noted by the authors Magnan et al. (2008), validating the significance of this interaction would further help establish the role of AtCML9 as an important node between calcium and ABA signaling. In addition to stress tolerance, AtCML9 has also been shown to be interacting with PPR2 (pseudo-response regulator) in a cDNA expression library screen by Perochon et al. (2010). This interaction was shown to be specific in nature as concluded from the absence of PPR2 interaction with other CaM isoforms. PPR2 exhibits features of transcription factors such as GARP DNA-binding domain (a signature motif of type B response regulators involved in histidine to asparagine phosphorylation signaling system), a proline-rich domain and golden C-terminal domain. Fluorescent-tagged fusion constructs showed both CML9 and PPR2 to be localizing in the nucleus suggesting a plausible function of CML9 in regulating PPR2 and other transcription factors (Perochon et al. 2010). Although direct evidence is lacking, as speculated by the authors, PPR2 interaction with AtCML9 could be important under stress and development conditions since PPR2 expression appears to be induced during seed germination and osmotic stress conditions.

Another CML member, AtCML19 (AtCENTRIN2) has been found to be crucial for resistance to UV exposure by virtue of its ability to translocate to nucleus upon UV-C exposure and help in DNA damage repair via its interaction with DNA repair protein AtRAD4 (Liang et al. 2006). *cml19* mutant plants displayed reduced resistance to UV-C exposure due to reduced efficiency of DNA repair. Overexpression of AtCML19, on the other hand, increased DNA repair efficiency as measured by in vitro DNA synthesis rate in plants treated with UV or cisplatin. Immunoblotting using anti-CENTRIN2 antibody showed strong induction of the protein upon UV-C exposure, which did not increase using other DNA-damaging agents such as bleomycin suggesting AtCENTRIN2 to be specifically responding to UV stress (Liang et al. 2006). CENTRIN2 has been found to be interacting with AtRAD4 in mammals, and investigation by the authors using GST pull-down assay showed this to be true for AtCENTRIN2 and AtRAD4 as well, suggesting this to be a conserved mechanism for DNA repair (Liang et al. 2006). Deletion of the EF-hand domain of AtCENTRIN2 abolished its interaction with AtRAD4 and its improved efficiency of DNA synthesis-based repair providing further evidence for the importance of calcium regulation in mediating this interaction (Liang et al. 2006).

A novel calmodulin-like gene, namely, *OsMSR2* (*Oryza sativa L.* multi-stress-response gene 2) has been recently studied and characterized with respect to its role and function in positively regulating ABA signaling and abiotic stress response (Xu et al. 2011). Expression of OsMSR2 was strongly upregulated by a wide range of stresses including cold, drought, and heat in different tissues and during different developmental stages as evidenced by microarray and real-time PCR studies. Transgenic lines overexpressing this gene were found to be more resistant to high salt and drought conditions as well as exhibiting hypersensitive response to ABA (Xu et al. 2011). Recombinant protein exhibited electrophoretic shift upon binding to calcium. Enhanced salt and drought stress tolerance could be correlated with increased expression of stress response genes such as *RD29A* and *P5CS1* in transgenic lines using real-time PCR to measure the transcript levels. However, the exact protein targeted by OsMSR2 to mediate ABA stress response has not been identified and needs further investigation (Xu et al. 2011).

A study by Felzon et al. (2005) focused on isolation and characterization of CaM genes in woody plant oak (*Quercus petraea* Liebl). This was a first report attempting to identify CaM genes important under drought stress in oak, which is economically and silviculturally important plant species. RT-PCR in 6-week-old oak seedlings subjected to flooding stress using primer mix designed from known CaM genes in other plant species was used as a strategy to identify three CaM genes, namely, QpCaM-1, QpCaM-2, and QpCaM-3 (Felzon et al. 2005). Further analysis using northern blot analysis revealed a downregulation of QpCaM-1 and QpCaM-3 upon submergence stress, while that of QpCaM-2 showed downregulation between 6 and 24 h after flooding stress and a robust upregulation at 48 h (Felzon et al. 2005). This increase in QpCaM-2 mRNA levels might reflect importance of this CaM member in tolerance to hypoxic environment generated during flooding.

Role of CaM proteins in modulating downstream calcium-sensing protein kinases such as CCaMK has been studied in some details in the past (Patil et al. 1995). CCaMK is a class of Ca^{2+}-/CaM-binding proteins that have been studied mostly with respect to their function in development. However, some studies have also shown their importance in regulating abiotic stress response. A recent study in maize identified ABA-induced ZmCCaMK activation as a part of antioxidant defense response (Ma et al. 2012). ZmCCaMK YFP fusion constructs were found to be localized to nucleus, cytoplasm, and plasma membrane and reported to be acting as a common node between NO, hydrogen peroxide, and calcium signaling during oxidative stress response. Gene expression analysis by quantitative PCR as well as measurement of enzymatic activity in maize mesophyll cells by transient expression of

ZmCCaMK revealed an upregulation in transcription level as well as activity of antioxidant enzymes such as SOD4 and cAPX (ascorbate peroxidase). Consistent with this observation, knockdown of *ZmCCaMK* by RNAi impaired ABA-induced increase in SOD and cAPX enzyme activity providing further evidence of this calcium-/CaM-regulated kinase in maize as an antioxidant defense mechanism by upregulating the enzymes which are involved in free radical scavenging (Ma et al. 2012).

Hydrogen sulfide is believed to be third endogenous gaseous molecule after nitric oxide (NO) and carbon monoxide (CO) that performs different physiological roles in plant and animal cells. Hydrogen sulfide (H_2S) previously characterized as a potential plant toxin has been recently found to be imparting heat stress tolerance in tobacco suspension culture cells in a Ca^{2+}-/CaM-dependent manner (Li et al. 2012). The precise mechanism of how hydrogen sulfide provides heat tolerance remains to be elucidated; however, a direct function of this molecule in providing protection from oxidative damage has been found via activation of various enzymes such as catalase and superoxide dismutase that are involved in free radical scavenging and detoxification (Zhang et al. 2009b). Initial studies by Li et al. (2012) show the heat stress tolerance imparted by hydrogen sulfide donor sodium hydrogen sulfide (NaHS) increases in a concentration-dependent manner up to 50 micromolar H_2S. Addition of calcium source in the form of $CaCl_2$ improves this heat tolerance, while using calcium chelators such as EGTA or CaM antagonists such as chlorpromazine inhibits this thermotolerance imparted by H_2S (Li et al. 2012). However, which specific CaM genes are involved in imparting hydrogen sulfide-mediated heat stress resistance require further experiments.

Thus, collectively all the data described above (summarized in Fig. 3) reflects diverse means via which CaM and CMLs promote or inhibit abiotic stress tolerance by influencing various cellular processes such as those involved in transcription, transport, and enzymatic reactions in different subcellular compartments. Existing reports and further characterization of CaM isoforms involved in these pathways will provide valuable knowledge for creation of transgenic plants or screening of specific *cam* mutants with enhanced tolerance to abiotic stress conditions.

CaM/CML During Biotic Stress Response

Besides coping with environmental stress, calcium signaling mediated by calmodulin has been shown to be important in providing protection from pathogens and various other forms of biotic stresses (shown in Fig. 4). Initial studies showing changes in cytosolic calcium levels in cowpea epidermal cells in response to incompatible interaction with fungus pathogen *Uromyces phaseoli* and other similar reports in suspension cultures challenged with fungal elicitors have indicated calcium signaling as an important early event in plants cellular response to pathogen (Xu and Heath 1998). However, the exact nature of downstream pathways forming a part of this response to biotic stresses has not been explored in very detail until recently. Various transcription factors, ion channels, and target proteins have been identified as a part of CaM-modulated machinery to mount response against pathogen attack. As described below, some of these CaM target proteins have been observed to be modulating hypersensitive response (HR) and cell death pathway that serves as an important means of restricting pathogen infection and colonization (Iakimova et al. 2005). In brief, hypersensitive response is a genetically controlled pathway that involves a cell suicide mechanism brought into effect by recognition of pathogen by a plant receptor and activation of caspase like proteases that is accompanied by fragmentation of DNA and dismantling of protoplast that ultimately results in cell death inhibiting further pathogen growth (Iakimova et al. 2005). Hypersensitive response also strengthens plant immunity via secretion of toxic compounds called phytoalexins and other antimicrobial proteins called pathogen-related (PR) proteins by dying cells, providing

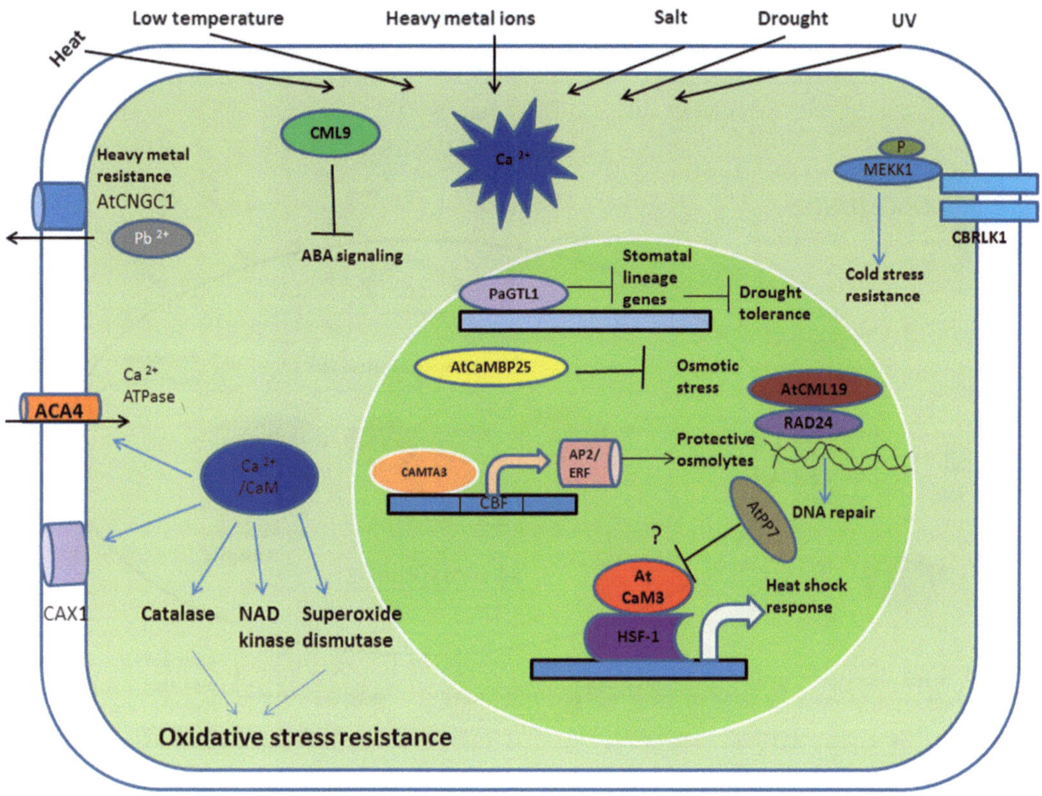

Fig. 3 *Model depicting different CaM members and their respective cellular targets in abiotic stress response.* Different abiotic stress stimuli such as temperature stress, high salt, and drought evoke distinct calcium signature in the cell, which is decoded by various CaM proteins to help plants cope with stressful conditions. Some CaM members modulate transcription factors such as HSF1 and CaMATA3 that further bring about changes in expression of downstream genes involved in stress resistance such as heat shock proteins or others involved in cold acclimation such as *RD22*. Other CaM target proteins include ion channels such as ACA4 and CAX1 that regulate abiotic stress response positively and negatively, respectively, suggesting dual role for calcium/CaM in abiotic stress signaling. AtCNGC1 which is a CaM-binding cyclic nucleotide-gated ion channel has been found to be providing heavy metal resistance by exporting lead ions out of the cell. Different cellular enzymes involved in ROS detoxification such as catalase and superoxide dismutase have also been found to be activated by CaM that might serve as a protection against oxidative damage by free oxygen radicals generated in the cell under different stress conditions and serve as stress-signaling molecules. Other proteins such as CML9 and CBRLK1 (CaM-binding receptor-like kinase) have also been associated with specific stress response pathways suggesting the importance of CaM-related pathways in a broad range of abiotic stress stimuli

protection against pathogen infection (Dangl et al. 1996).

Regulating Transcription Factors During Biotic Stresses

WRKY transcription factors characterized by WRKYGQK peptide sequence and a zinc finger motif form an important class of transcription factor network involved in plant defense response (Pandey and Somssich 2009). Protein-protein-interaction-based cDNA expression library screening identified AtWRKY7 as a CaM interactor (Park et al. 2005). CaMBD of AtWRKY7 was found to be conserved among the group IID of the WRKY protein family, which includes ten members, out of all the members tested WRKY7, WRKY11, and WRKY17 bind to CaM in calcium-dependent

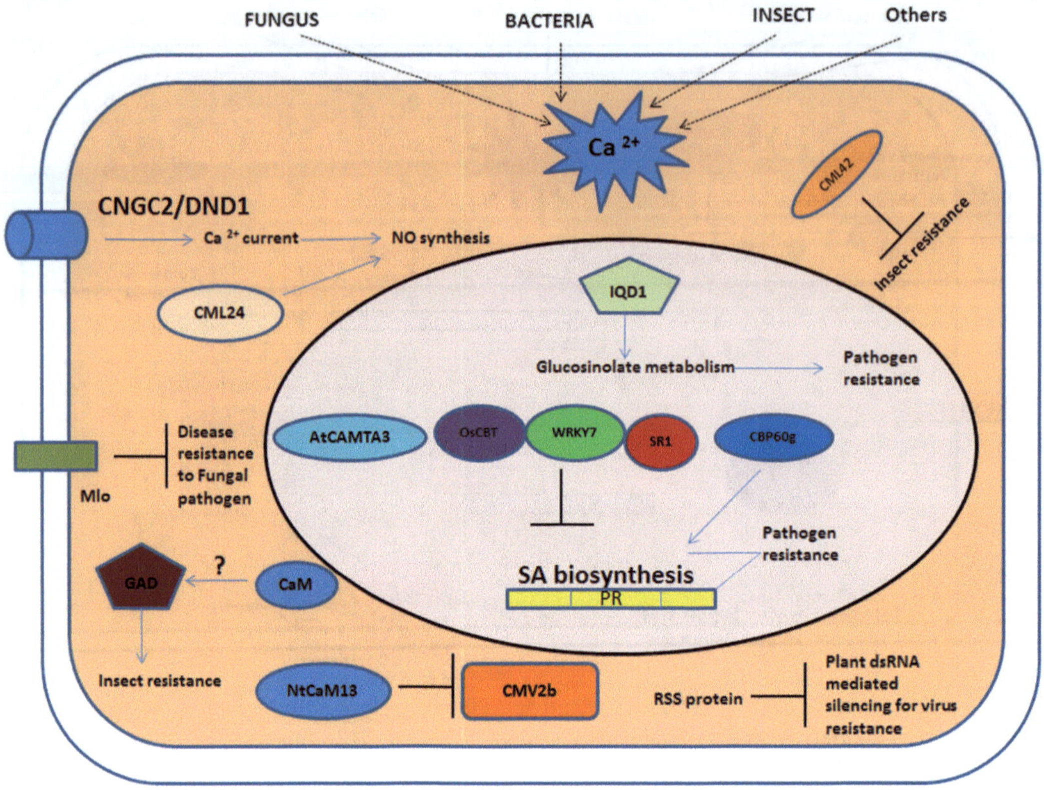

Fig. 4 *Role of CaM and CaM-regulated cellular targets in biotic stress response induced by pathogen as a stimulus.* Different types of pathogen including but not restricted to bacteria, viruses, fungi, and insects evoke a calcium spike/current in the plant cell upon physical contact. This calcium signature is further relayed by a variety of CaM members and their specific interacting partners to ultimately regulate downstream pathways involved in pathogen resistance. A variety of transcription factors such as OsCBT, WRKY7, and AtCaMTA3 act to negatively regulate pathogen response by having an inhibitory effect on salicylic acid (SA) synthesis, which serves as a key molecule to activate SA-mediated innate immunity pathway. Other CaM-modulated transcription factors such as IQD1 and CBP60g have an opposite role where they act to upregulate glucosinolate metabolic and SA biosynthesis pathway, respectively, that provide resistance against pathogens. Various other cellular proteins, enzymes, and ion channels as depicted in the model have also been implicated in CaM-regulated pathways for biotic stress resistance

manner. Further analysis using gel overlay assay, site-directed mutagenesis, and split ubiquitin assay confirmed this interaction suggesting the possibility of this interaction occurring in vivo to help plants cope with pathogen attack (Park et al. 2005). WRKY7 has also been studied by Kim et al. (2006) who have found it to be acting as a transcriptional repressor of basal immunity upon pathogen challenge. WRKY7-GFP construct showed nuclear localization when bombarded into onion epidermal cells. Further, transgenic lines overexpressing WRKY7 were more susceptible to infection by virulent strains of *Ps. syringae,* whereas loss-of-function mutants showed increased resistance. EMSA studies were used to demonstrate the binding of WRKY7 to W-box DNA sequence TTGACC. The negative regulation of plant resistance to pathogen by WRKY7 could be further linked to SA signaling where lower accumulation of SA-signaling target gene *PR1* was observed upon infection in transgenic plants overexpressing WRKY7 (Kim et al. 2006). In addition to WRKY7, WRKY 11, and WRKY17 have also been found to be negatively regulating resistance to both virulent as well as avirulent strains *of Ps. syringae* in *Arabidopsis*. Pathogen resistance phenotype of *wrky11* was enhanced in *wrky11wrky17* double mutants suggesting possible separate roles of these transcription factors in immunity (Kim et al. 2006). Transcriptome and expression analysis of the single and double mutant plants under pathogen challenge revealed shared target genes suggesting a partially redundant role of these transcription factors in plant immune response (Journot-Catalino et al. 2006). Numerous other reports have established link between calcium/CaM signaling and defense response in plants. AtSR1/AtCaMTA3 has been shown to be negatively regulating salicylic acid (SA)-controlled adaptive immunity in Arabidopsis plants (Du et al. 2009). Mutations in CaMBD of *Atcamta3* failed to rescue the disease resistance phenotype of *atsrs1* mutant plants providing the evidence for direct role of CaM binding in regulation of AtCaMTA3 in disease resistance to *Pseudomonas syringae*. Further, the authors could show this inhibitory role of AtCaMTA3 in SA-induced immunity to be mediated via negative regulation of SA biosynthesis genes such as *EDS1*.

NAC family of transcription factors that play a wide variety of role in plant development, abiotic, and biotic stress responses have also been found to be CaM regulated in some reports (Olsen et al. 2005; Kim et al. 2007). NAC family of transcription factors are defined by a conserved N-terminal NAC domain that is derived from *NO-APICAL MERISTEM* (*NAM*) gene in *Petunia* and *ATAF1*, *ATAF2*, and *CUP-SHAPED COTYLEDON* (*CUC*) gene in *Arabidopsis*. NAC binding to CaM named as CBNAC was identified by Kim et al. (2007) in a cDNA expression library screen using HRP-labeled AtCaM2. This CaM interaction was further confirmed using yeast split ubiquitin assay and CaM overlay assay. The CaM-binding specificity was demonstrated using W487R mutant of CBNAC that as expected did not bind CaM concluded from split ubiquitin assay in yeast (Kim et al. 2007). The subcellular localization was studied using GFP and RFP fusion constructs both of which showed CBNAC to be constitutively localized in the nucleus. Further experiments utilizing random site selection method identified a GCTT core DNA sequence to be bound by CBNAC. Transient expression assay using GUS reporter plasmids carrying CBNAC-binding DNA sequence in the upstream region showed CBNAC to be acting as a transcriptional repressor upon binding CaM (Kim et al. 2007). Recently, following up on this study, the same group of authors found CBNAC to be actually acting as a repressor of PR1 together with SN1 (suppressor of nonexpressor of PRgenes1, inducible 1) (Kim et al. 2007). *cbnac* mutants showed increased resistance to *Ps. syringae,* while plants overexpressing *cbnac* had the opposite phenotype in terms of pathogen resistance that could be correlated with *PR1* expression level (Kim et al. 2007). SN1 that is also known to be negatively regulating PR1 expression but lacks a DNA-binding domain was examined by yeast two-hybrid for its interaction with CBNAC, and this was indeed found to be true. Further, *cbnac*/*sni1* double mutants displayed a greater resistance to pathogen than either *cbnac* or *sni1* mutant alone suggesting a synergistic interaction between the two proteins in acting as repressor of plant response to infection (Kim et al. 2007).

OsCBT (*Oryza sativa* CaM-binding transcription factor) is a CaM-modulated transcription factor, first identified by Choi et al. (2002) in a rice cDNA library screen using HRP-labeled Gm CaM. Structural analyses revealed this protein to be very similar to AtCAMTAs in *Arabidopsis* with DNA-binding CG-1 domain, ankyrin repeats, five CaM-binding domains, and a putative transcriptional activation domain. Gel overlay assays identified CaMBDI and CaMBDII as the two CaM-binding domains, which bind to calmodulin in a Ca^{+2}-independent and Ca^{+2}-dependent manner, respectively (Choi et al. 2002). The authors also mapped the critical residues within OsCBT required for its interaction with CaM. Site-directed mutagenesis revealed Ile^{763} in CaMBDI and Trp^{829} and Val^{836} in CaMBDII to be important for this interaction. GFP fusion constructs of OsCBT transfected into *Arabidopsis* protoplasts showed it to be nuclear localized (Choi et al. 2002). Further, random site selection method (RBSS) looking for enrichment of OsCBT binding to a population of oligonucleotides revealed TWCG (C/T)GT*KKKK*TKCG (W is A or C and K is T or C) as a consensus sequence that was bound by OsCBT, which was further validated using electrophoretic mobility shift assays (Choi et al. 2002). Effect of CaM on OsCBT transcriptional activity was monitored using GUS assay in *Arabidopsis* protoplasts that revealed OsCaM to downregulate OsCBT-mediated transcription. Recently, the exact role of OsCBT in stress response was determined, and it was found to be negatively regulating defense response against fungal pathogen *Magnaporthe grisea* and bacterial pathogen *Xanthomonas oryzae* (Koo et al. 2009). *oscbt-1* mutant plants were isolated and found to be showing increased resistance to these pathogens that was correlated with increased hypersensitive response (HR) and upregulation of PR (pathogen-related) genes that provide protection from pathogens (Koo et al. 2009).

SR1 (signal responsive 1) a CaM-binding transcription factor has been studied for its role in resistance to powdery mildew disease (Nie et al. 2012). Sr1-4D gain-of-function mutation was found to be suppressing the increased resistance to powdery mildew pathogen *Golovinomyces cichoracearum* observed in *edr1* mutants. *edr1*-mediated increased resistance to powdery mildew has been attributed to activated SA signaling. Sr1-4D mutation was observed to suppress the enhanced expression of *PR1* gene upon pathogen infection as determined by real-time PCR. Genetic mapping revealed the Sr1-4D mutation to be in the calmodulin-binding region within *SR1* gene hint towards the possibility of alteration in calmodulin regulation of SR1 in these mutants that might lead to a more enhanced transcriptional repression of its target genes (Nie et al. 2012). Further experiments such as chromatin immunoprecipitation and EMSA confirmed SR1 to bind in the promoter region of pathogen resistance gene *ndr1*, thereby exerting a negative effect on its expression as concluded from decreased level of *ndr1* expression in *SR1* gain-of-function mutant plants (Nie et al. 2012).

CBP60g, a CaM-binding transcription factor, has been found to positively regulating SA-mediated defense response to *Ps. syringae* (Wang et al. 2009b). CBP60g harbors a CaM-binding domain in its N-terminal and was initially identified as an early response gene induced upon *Ps. syringae* infection. Overexpression of CBP60g-enhanced resistance to *Ps. syringae*, increased SA accumulation, and pathogen response genes such as *PR1* and *PR2* that accumulated at earlier time points in CBP60g overexpressing transgenic lines upon pathogen inoculation as compared to wild type. Increased SA accumulation in CBP60g transgenic lines could be attributed to an increase at the transcript level in SA biosynthetic enzymes *ICS1* and *EDS1* (Wang et al. 2009b). Consistent with the critical role played by CBP60g in biotic stress response, T-DNA insertion mutant lines of *cbp60g* display compromised defense response. GFP fusion constructs of CBP60g made to identify subcellular localization of the protein showed constitutive nuclear localization consistent with its role as a transcription factor (Wan et al. 2009). As discussed before, the same study also found a role for CBP60g in drought stress response suggesting the involvement of CBP60g as a versatile CaM-regulated protein acting in biotic as well as abiotic stress response.

Regulating Ion Channels and Membrane Proteins

An interesting study has suggested possible role of Ca^{2+}/CaM in innate immunity response by regulation of cyclic nucleotide-gated channel 2 (CNGC2/DND1) to activate downstream NO (nitric oxide) signaling resulting in HR response (Ali et al. 2007). Various genetic and biochemical assays established a link between pathogen-elicited activation of CNGC2 and generation of calcium current and synthesis of NO. Loss of HR response to avirulent pathogen in *dnd1* mutants could be rescued by application of NO-generating agent sodium nitroprusside clearly demonstrating the importance of NO as a downstream signal for defense response. Further, blocking CaM function using W7(N-(6-aminohexyl)-5-chloro-1-naphthalenesulfonamide) as an antagonist prevented NO generation implicating CaM as a crucial link between

CNGC2 activation and NO signaling providing protection against pathogen (Ali et al. 2007). Taken together, this is one of the few reports confirming the previously speculated role of Ca^{2+}/CaM function in innate immune response. Another recent study reported *cml24-4* to be serving as a possible link between upstream calcium/CaM signaling triggered by pathogen and downstream NO generation (Ma et al. 2008). NO fluorescence monitored using NO-specific dye 4,5-diaminofluorescein diacetate showed NO generation to be significantly reduced in *cml24-4* mutant compared to wild-type epidermal guard cells. In addition to this, *cml24-4* mutant plants were found to be defective in hypersensitive response as compared to wild type when challenged with bacterial pathogen *Ps. syringae* (Ma et al. 2008). Future experiments studying how exactly CaM/CML24 participates in NO generation would yield interesting insights into cross talk between calcium and NO signaling in innate immunity response.

Novel Roles of CaM and CaM-Binding Proteins in Pathogen Resistance

Besides microbial pathogens, insect herbivory is also a commonly encountered form of biotic stress faced by plants that elicits the production of phytohormones and antimicrobial peptides to provide an effective defense mechanism against such attacks. Methyl jasmonate is one of the hormones induced under such conditions that activates defense-signaling cascade to provide protection from herbivores. Recognition of herbivore-secreted elicitors by the plant has been shown to be triggering an increase in calcium influx into the cell, which forms an early event in such defense responses suggesting the involvement of calcium-regulated sensor proteins in activating downstream defense-signaling cascade. CML42 in *Arabidopsis* has been found to be negatively regulating defense response against insect herbivory by *Spodoptera littoralis* (Vadassery et al. 2012). CML42 upregulation was first identified in microarray studies, looking into transcriptional changes brought about upon application of *S. littoralis* oral secretion, confirmed by real-time PCR. Further analysis revealed a rise in calcium levels using transgenic *Arabidopsis* leaves expressing the Ca^{2+} sensor aequorin upon application of oral secretion (Vadassery et al. 2012). The role of CML42 in defense signaling however was concluded to be inhibitory in nature as suggested by enhanced resistance to *S. littoralis* in *cml42* loss-of-function plants. In addition, enhanced resistance in *cml42* mutants could be linked to enhance calcium response to jasmonic acid (known hormone involved in defense against herbivory) in the mutants correlated with increased expression of JA signaling pathway response genes (Vadassery et al. 2012). *cml42* was itself shown to be negatively regulated at the transcript level by COI1 (jasmonate receptor) (Vadassery et al. 2012). CML42-GFP constructs showed both cytosolic as well as nuclear localization hint towards possibly diverse cellular targets of this calcium sensor in mediating calcium-signaling cascade and response (Vadassery et al. 2012). As speculated by the authors, CML42 and other similar proteins might form part of a complex network to balance defense response upon attack by insect or may themselves be induced by specific elicitors in insect secretion to downregulate defense signaling so as to provide a more conducive environment facilitating insect infection.

A study by Levy et al. (2005) found CaM-binding IQ domain containing nuclear-localized protein, namely, IQD1, to be important for protection against herbivory. Overexpression of IQD1 was sufficient to provide enhanced resistance to herbivory-associated infestation, and this resistance imparted by IQD1 was linked to glucosinate metabolism that is a class of secondary metabolites known to be required for plant defense (Levy et al. 2005).

OsMlo (rice mildew locus O), a modulator of plant defense response, was identified by Kim and colleagues (2002) in a cDNA library screen using cDNA prepared from rice suspension culture treated with fungal elicitor using HRP-labeled CaM. Mlo family of protein comprises of a plant-restricted family of transmembrane proteins first identified in barley that was shown to negatively regulating disease resistance to powdery mildew (Kim et al. 2002). Mlo protein in rice was found to be binding CaM via its 20 amino acid C-terminal tail sequence as determined by deletion and site-directed mutagenesis studies (2002). Regulation of Mlo proteins by CaM might be a general feature as speculated by the authors that could have important implications in pathogen resistance and cell death. However, details regarding the role of calmodulin and calcium in modulating Mlo to impact downstream pathways to control biotic stress signaling remains to be elucidated.

GAD as previously described is known to aid in synthesis of GABA which has also been reported to be protective against phytopathogen *Heliothis virescens* (tobacco budworm) (Macgregor et al. 2003). Exposure of the insect neuromuscular junction to GABA results in paralysis and forms an important line of defense to insect herbivory. Overexpression of GAD in tobacco was sufficient to impart enhanced pathogen resistance suggesting that increase in level of GABA is an important mechanism to deal with invertebrate pathogen (Macgregor et al. 2003). As GAD has also been documented to be a CaM-binding protein, it is possible that calcium-/CaM-mediated activation of GAD is critical for protection from herbivore pathogen. However, further studies exploring this link between calmodulin and GAD function in biotic stress are required. For instance, whether GAD-mediated resistance is calmodulin dependent and if so, which CaM proteins are involved in this function would help address this question in greater detail.

A very unique role of CML protein in tobacco has been uncovered very recently in a study that showed viral RNA-silencing suppressor protein to be targeted by CML protein, which binds to their dsRNA-binding domain and prevents the viral mechanism to block plant dsRNA-mediated defense response (Nakahara et al. 2012). Various in vitro and in vivo studies by the authors showed NtCaM13 interaction with RNA-silencing suppressor (RSS) protein, namely, cauliflower mosaic virus protein CMV2b. CaM binding to the RSS protein was eventually concluded to be promoting autophagy-mediated proteolytic degradation of the viral protein, thereby allowing plant RNAi response to mount against the viral pathogen (Nakahara et al. 2012). Thus, CaM and CMLs may also protect plants from viral pathogens by playing a role in dsRNA-mediated plant response to pathogen.

Thus, diverse mechanisms of CaM control over biotic stress response have been revealed using genetic and molecular biology techniques that show important functions performed by CaM and CML proteins in modulating various transcription factors, ion channels, enzymes, and cytosolic proteins. These interactions may exert either positive or negative control over biotic stress signaling suggesting dual role of CaM in regulation of such response pathways depending on the stimulus and proteins involved.

Cross Talk Between Calmodulin and Other Signaling Pathways

Calcium-/CaM-binding proteins have been identified that have often been found to be performing important functions in pathways such as hormone, phospholipid, and MAPK-signaling cascades suggesting the existence of critical cross talk nodes where calcium modulates other signaling pathways important for plant development and function (summarized in Fig. 5).

Brassinosteroid Signaling

Brassinosteroid is a recently discovered class of hormones that plays an important role in plant growth and development. Studies in *Arabidopsis* identified mutants in BR biosynthesis or receptor genes display dwarfism whereas gain-of-function mutants overexpressing BR biosynthetic genes such as DWF4 and DET2 results in increased growth (Vert et al. 2005). Initial hints for the role of calcium/calmodulin signaling in BR signaling came from studies of dwarf *det3* mutants that display altered calcium oscillations as compared

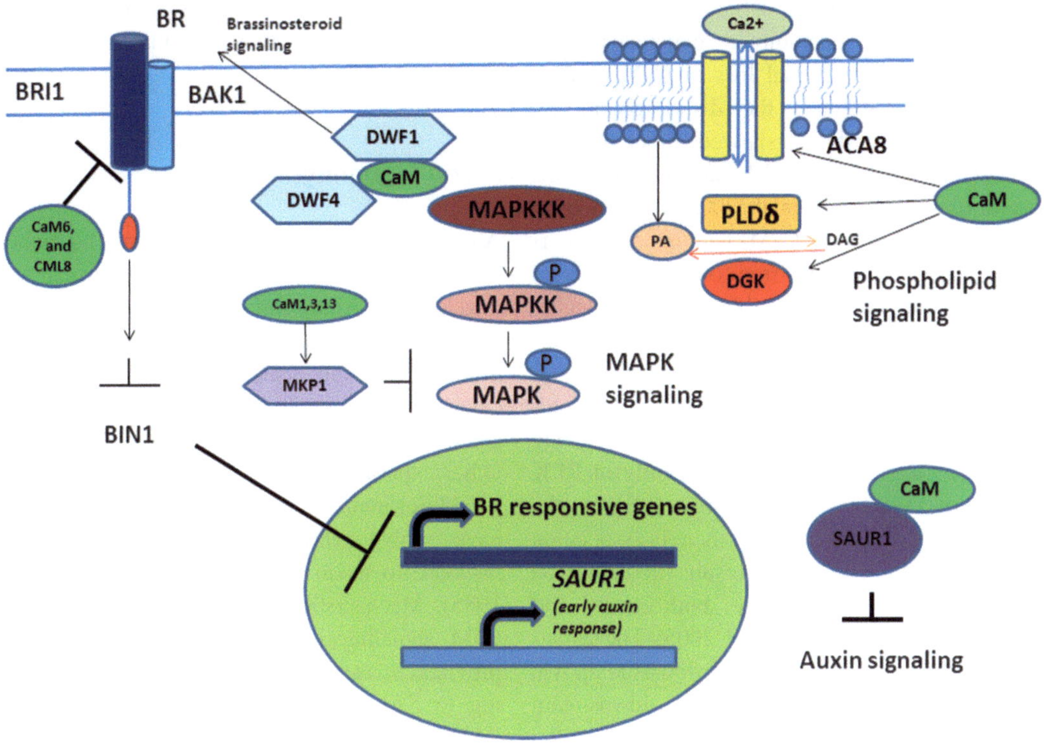

Fig. 5 *Model depicting cross talk of calcium/CaM signaling with other signaling pathways.* Calcium-/CaM-binding proteins have been identified that provide evidence for its role in modulating other signaling pathways in plant growth, development, and stress tolerance. Brassinosteroid signaling mediated by receptor BRI1 is relayed by various downstream effectors that act to inhibit the activity of BIN1 resulting in activation of BR responsive genes that function in cell growth and elongation. CaM has been found to be playing dual role in BR signaling where some CaM isoforms as shown in this model are negative regulators of the pathway acting to inhibit the autophosphorylation of BRI1. Other CaMs have been identified to be required for activity of BR biosynthesis enzymes such as DWF1 and DWF4, thus acting as potential positive regulators of the pathway. MAPK signaling has also been found to be linked to CaM via the negative regulation of MAPK activity by activation of MAPK phosphatase, MKP1, that dephosphorylates MAPK to inhibit the pathway during wound stress. Membrane phospholipids are hydrolyzed during stress and senescence-related processes that intersects with calcium/CaM requirement as revealed by activation of PLDδ, DGK, and ACA8, all of which act in phospholipid signaling. Auxin, which is a central hormone in plant growth and development, has also been shown to be regulated at some level by CaM via SAUR1 (identified as CaM-binding protein) that acts to negatively regulate the pathway. *BR* brassinosteroid, *BRI1* BR insensitive, *BAK1* BR-associated kinase1, *BIN1* brassinosteroid insensitive 1, *DWF* dwarf1, *MAPK* mitogen-activated protein kinase, *PLD* phospholipase D, *DGK* diacylglycerol kinase, *MKP1* MAPK phosphatase1, *PA* phosphatidic acid, *SAUR1*-small auxin up RNA, *ACA8* autoinhibited Ca^{2+} ATPase

to wild-type plants after treatment with various stimuli such as cold, ABA, and oxidative stress (Allen et al. 2000). Also, the dwarf phenotype of these mutants could be at least partially rescued upon treatment with BR. Previous studies had also implicated CaM to be important for plant growth as shown by altered growth and apical dominance in transgenic plants with upregulated expression of potato CaM isoform PCM1(Poovaiah et al. 1996). However, a direct function of calcium/CaM in the BR signaling pathway was identified in a study that reported direct interaction between BR biosynthesis enzyme DWF1 and CaM (Du and Poovaiah 2005). Screening of cDNA library in Arabidopsis with S^{35}-labeled calmodulin uncovered a novel interaction between CaM and BR synthesizing enzyme. Genetic analyses carried out using deletion constructs and site-directed mutagenesis revealed CaM binding to be important for DWF1 function as concluded from experiments where partial loss of CaM binding only partially rescued the dwarf phenotype. Co-immunoprecipitation showed direct physical interaction between DWF1 and various CaM isoforms (Du and Poovaiah 2005). As noted by the authors, similar CaM-binding domain is found in DWF1 proteins from other plant species as well as suggesting a possible conserved mechanism of calcium/CaM regulation over BR signaling. BR signaling is initiated by the brassinosteroid insensitive 1/BRI1 receptor and its co-receptor BRI1-associated kinase (BAK1) that both belong to the leucine-rich repeat (LRR) family of RLK family of kinases. Upon binding of BR to the extracellular domain of BRI1, there is association of BRI1 with BAK1 that results in activation of its kinase domain by both trans and autophosphorylation (Vert et al. 2005). The signaling cascade is further relayed downstream via intracellular kinases to activate downstream effectors such as transcription factors that function in cell growth and elongation (Vert et al. 2005). Another recent study by Oh et al. (2009) showed negative regulation of BR receptor kinase (brassinosteroid insensitive 1) by Ca^{2+}/CaM, providing an interesting new link between the two pathways (Oh et al. 2009). Pull-down assay involving His tagged CaM and W980 peptide (Ala^{974} to Phe^{996} of BR1 kinase) showed direct binding of CaM7 to kinase subdomain VIa, and this binding was seen to be increasing in a Ca^{2+}-dependent manner (Oh et al. 2009). Co-expression of BRI1 and different CaM isoforms in *E. coli* revealed a decrease in BRI1 autophosphorylation by phosphoprotein staining (Oh et al. 2009). This decrease in BRI1 autophosphorylation was more pronounced at the tyrosine residues than serine/threonine residues and was more robust in case of certain CaMs such as CaM6, CaM7, and CML8 (Oh et al. 2009). Unlike auxin, BR is not transported over long distances between plant tissues; hence the role of calcium/CaM as a second messenger in modulating this signaling pathway may serve as a mechanism of relaying this signaling cascade beyond its site of production. This interplay between calcium and BR signaling could provide avenues for crop size manipulation via alteration of different regulators of these pathways.

Auxin Signaling

Auxin is one of the main plant hormones performing central functions in plant growth and development. Many early studies have indicated a possible link between auxin and calcium signaling. Studies utilizing calcium fluorophores and calcium-sensitive microelectrodes showed changes in calcium oscillations upon auxin treatment. Other experiments using calcium inhibitors such as chlorpromazine and chelators such as EGTA have shown reduced epicotyls elongation in response to auxin treatment (Raghothama et al. 1985). Mechanistic insight into how auxin and CaM signaling might be connected was first provided in a study that reported Ca^{2+}/CaM binding to SAUR (small auxin up RNA) protein. SAUR represents a class of early auxin response genes that is upregulated upon auxin treatment (Kant et al. 2009). *SAUR39* was characterized as a negative regulator of auxin signaling by Kant et al. (2009) who showed that plants overexpressing this gene exhibited lower auxin transport and downregulated expression of auxin transporters and auxin biosynthesis genes. Biochemical analyses of the transgenic plants overexpressing *SAUR39* showed lower chlorophyll content, increased anthocyanin accumulation, and premature leaf senescence that correlated with decreased auxin signaling (Kant et al. 2009). The study implicated SAUR genes to be acting as a negative regulator of auxin signaling. ZmSAUR1 from maize was found to be encoding a calcium-/CaM-binding protein (Yang and Poovaiah 2000). Maize root cDNA expression library was screened with S^{35}-labeled CaM that identified SAUR1 as a putative binding partner (Yang and Poovaiah 2000). Further biochemical experiments using CaM affinity chromatography also found SAUR1 to be forming a stable complex with CaM (Yang and Poovaiah 2000). Deletion analysis identified amino acid sequence from amino acid 20–45 as the putative CaM-binding region. Synthetic peptide bearing the putative CaM-binding sequence revealed stable binding of CaM to this peptide in a calcium-dependent manner. Expression analysis upon use of CaM inhibitors revealed no change at the transcript level of *ZmSAUR1* suggesting that regulation via CaM occurs probably at the posttranscriptional stage (Yang and Poovaiah 2000). Sequence analysis of SAUR proteins in rice and Arabidopsis revealed existence of a stretch of basic amphiphilic helix in the N-terminal region similar to the sequence found in ZmSAUR1 (Yang and Poovaiah 2000). However, how CaM binding regulates SAUR function remains to be deciphered.

MAPK Signaling

MAPK or mitogen-activated protein kinase cascade has been studied and found to be responding to a variety of environmental and developmental cues in plants. The MAPK-signaling cascade involves three protein kinases (MAPKKKs, MAPKKs, and MAPKs) acting in a sequential manner to activate various downstream responses in the form of activated/phosphorylated proteins such as transcription factors, ion channels, and enzymes. The initial link between MAPK signaling and calmodulin was uncovered in tobacco where NtMKP1 (an MAPK-inactivating phosphatase) was identified as a CaM-binding protein. Yamakawa et al. (2004) identified NtMKP1 as a CaM-binding protein in a cDNA library screen using phage expression library prepared from wounded leaves. Bacterially expressed NtMKP1 was found to be binding to three different CaM isoforms, namely, NtCaM1, NtCaM3, and NtCaM13. Further, it was shown that *NtMKP1* transcript accumulates in response to viral (TMV) infection hint towards its possible role in modulation of MAPK signaling under pathogen stress (Yamakawa et al. 2004). NtMKP1 overexpressing plants displayed reduced activation of wound activated MAPKs such as WIPK and SIPK providing evidence for negative regulation of MAPK signaling by NtMKP1. Amino acids of NtMKP1 from position 436 to 453 were found to be forming the CaM-binding domain providing direct evidence for NtMKP1 as a CaMBP (Yamakawa et al. 2004). Other studies in Arabidopsis and rice also revealed AtMKP1 and OsMKP1, respectively, to be binding CaM suggesting a conserved function of CaM in regulating MAPK signaling pathway via MAPK phosphatase (Yamakawa et al. 2004). A recent study by Bartels et al. (2009) revealed regulation of MPK6 by MAPK phosphatase1 and a protein tyrosine phosphatase 1(PTP1). A detailed analysis of *ptp1* and *mpk1* mutant plants showed constitutive activation of defense responses and reduced growth confirming the previous reports of negative regulation of MAPK signaling by these phosphatases, which may in turn be modulated by calmodulin in a complex cross talk between calcium and other signaling pathways.

Phospholipid Signaling

Phospholipid is a major constituent of the plasma membrane that can also serve as an important signaling molecule involved in a variety of plant functions such as environmental stress and senescence. Ca^{2+}-ATPase transporters present in the cell membrane extrude or trap calcium ions to maintain appropriate cellular calcium concentration. ACA8, a Ca^{2+}-ATPase, was found to be activated by both phospholipid and calmodulin because of overlap of phospholipid-binding site with that of calmodulin (Meneghelli et al. 2008). Biochemical assay revealed increased V_{max} and lower K_m of CaM binding to ACA8 in presence of acid phospholipids. This provided the first evidence for a complex interplay between calcium/CaM and phospholipid in plant function. Another means of phospholipid and calcium/CaM cross talk is via the step of phospholipid hydrolysis that serves a critical role in generating secondary messengers, regulating ion fluxes, vesicle trafficking, and membrane remodeling (Meneghelli et al. 2008). Different classes of phospholipases have been identified that perform initial steps of phospholipid hydrolysis and these include phospholipase D (PLD), PLC, PLA, and PLA_1. PLD has been shown to be activated by calmodulin to release phosphatidic acid (PA) that serves as a signal during stress and in response to hormone (Bargmann and Munnik 2006). Different PLD isoforms act during varied stress conditions to influence biophysical status of lipid membrane. Six classes (PLD α, β, γ, δ, ε, and ζ) of PLDs have been characterized based on their pH requirement and range of calcium concentration for in vitro activity (Bargmann and Munnik 2006). ABA, cold, salt, and drought conditions were found to be activating *AtPLD δ* at the transcript level, and loss of *AtPLD δ* has been associated with induction of PLD activity under dehydration conditions (Bargmann and Munnik 2006). Overexpression of *AtPLD δ* increases tolerance to freezing, whereas loss of *AtPLD δ* decreases survival under cold conditions that is correlated with increased and decreased PA, respectively. However, the identification of exact CaM isoforms involved in PLD activation is lacking and needs further detailed investigation. Diacylglycerol (DAG) that is generated upon hydrolysis of PA also serves as a signaling molecule is produced by a calmodulin-binding phosphatase (Snedden and Blumwald 2000). The reverse reaction converting DAG to PA is also achieved by a CaM-binding kinase diacylglycerol kinase (DGK), first studied in tomato (Snedden and Blumwald 2000). LeCBDGK (*Lycopersicon esculentum* calmodulin-binding diacylglycerol kinase) was identified in a cDNA library screen as a calmodulin-binding protein, and the CaM-binding domain was found to be located in the last 29 amino acid residue of the C-terminal region. Co-immunoprecipitation of LeCBDGK detected calmodulin in the soluble extracts suggesting in vivo interaction (Snedden and Blumwald 2000). Immunoblotting revealed existence of LeCBDGK in the plasma membrane fraction. Interestingly, addition of CaM antagonists disrupted this interaction of LeCBDGK with membrane suggesting functional relevance of CaM binding in attaching of LeCBDGK to the membrane that may be critical for its function in generating PA and downstream signaling (Snedden and Blumwald 2000).

Thus, studies of CaM-binding proteins acting in other hormonal/developmental pathways suggest far greater role performed by CaM in plant function by modulation of diverse signaling cascades.

Conclusion

A large number of studies have shed light on CaM-modulated target proteins expanding our understanding of calcium/CaM function in both environmental and biotic stress response. Both rapid responses such as regulation of ion channels as well as downstream transcriptional response mediated by CaM helps plants cope with a variety of stress conditions. Plants unlike other species have evolved multiple CaM isoforms; however, the exact differences in functions of these isoforms need more investigation. Development of both biochemical and computational methods by different research groups has facilitated identification of novel proteins regulated by CaM in response to different stimuli. However, mechanistic details of how CaM is regulating the activity of such target proteins within the cell in cross talk with other pathways are just beginning to emerge. Taken together, work over the last decade has provided us with the knowledge of many novel roles played by calcium signaling and calmodulin in vast majority of abiotic and biotic stress pathways. Future studies attempting to discover many more such targets of calmodulin and their precise role in stress-signaling pathway would be a major research focus that would also provide a platform for utilizing this basic knowledge in creation of various stress-resistant crop varieties.

References

Ali R, Ma W, Lemtiri-Chlieh F et al (2007) Death don't have no mercy and neither does calcium: Arabidopsis cyclic nucleotide gated channel2 and innate immunity. Plant Cell 19:1081–1095

Allen GJ, Chu SP, Schumacher K et al (2000) Alteration of stimulus-specific guard cell calcium oscillations and stomatal closing in Arabidopsis det3 mutant. Science 289:2338–2342

Al-Quraan NA, Locy RD, Singh NK (2010) Expression of calmodulin genes in wild type and calmodulin mutants of *Arabidopsis thaliana* under heat stress. Plant Physiol Biochem 48(8):697–702

Bargmann BO, Munnik T (2006) The role of phospholipase D in plant stress responses. Curr Opin Plant Biol 9:515–522

Bartels S, Anderson JC, Gonzalez Besteiro MA et al (2009) MAP kinase phosphatase1 and protein tyrosine phosphatase1 are repressors of salicylic acid synthesis and SNC1-mediated responses in Arabidopsis. Plant Cell 21:2884–2897

Boonburapong B, Buaboocha T (2007) Genome wide identification and analyses of the rice calmodulin and related potential calcium sensor proteins. BMC Plant Biol 7:4

Bouche N, Yellin A, Snedden WA et al (2005) Plant specific calmodulin-binding proteins. Annu Rev Plant Biol 56:435–466

Boursiac Y, Harper JF (2007) The origin and function of calmodulin regulated Ca2+ pumps in plants. J Bioenerg Biomembr 39:409–414

Cates MS, Teodoro ML, Phillips GN Jr (2002) Molecular mechanisms of calcium and magnesium binding to parvalbumin. Biophys J 82:1133–1146

Cheung WY (1971) Cyclic 3',5'-nucleotide phosphodiesterase. Evidence for and properties of a protein activator. J Biol Chem 246(9):2859–2869

Choi JY, Lee SH, Park CY et al (2002) Identification of calmodulin isoform-specific binding peptides from a phage-displayed random 22-mer peptide library. J Biol Chem 277:21630–21638

Dangl JL, Dietrich RA, Richberg MH (1996) Death don't have no mercy: cell death programs in plant–microbe interactions. Plant Cell 8:1793–1807

Dauwalder M, Roux SJ (1986) Distribution of calmodulin in corn seedlings: immunocytochemical localization in coleoptiles and root apices. Adv Space Res 6 (12):67–70

Davis TN, Urdea MS, Masiarz FR et al (1986) Isolation of the yeast calmodulin gene: calmodulin is an essential protein. Cell 47(3):423–431

DeFalco TA, Bender KW, Snedden WA (2010) Breaking the code: Ca2+ sensors in plant signalling. Biochem J 425:27–40

Demidchik V, Maathuis FJ (2007) Physiological roles of nonselective cation channels in plants: from salt stress to signalling and development. New Phytol 175 (3):387–404

Doherty CJ, Van Buskirk HA, Myers SJ et al (2009) Roles for Arabidopsis CAMTA transcription factors in cold-regulated gene expression and freezing tolerance. Plant Cell 21(3):972–984

Du L, Poovaiah BW (2005) Ca2+/calmodulin is critical for brassinosteroid biosynthesis and plant growth. Nature 437:741–745

Du L, Ali GS, Simons KA et al (2009) Ca2+/calmodulin regulates salicylic-acid-mediated plant immunity. Nature 457:1154–1159

Duval FD, Renard M, Jaquinod M et al (2002) Differential expression and functional analysis of three calmodulin isoforms in germinating pea (Pisum sativum L.) seeds. Plant J 32(4):481–493

Ebashi S, Kodama A (1965) A new protein factor promoting aggregation of tropomyosin. J Biochem 58(1):107–108

Felzon H, Capelli N, Dat J et al (2005) Molecular cloning and characterization of calmodulin genes in young oak seedlings (*Quercus petraea* L.) during early flooding stress. Biochim Biophys Acta 1727:213–219

Galon Y, Nave R, Boyce JM et al (2010) Calmodulin-binding transcription activator 1 mediates auxin signaling and responds to stresses in Arabidopsis. Planta 232(1):165–178

Geisler M, Frangne N, Gomès E et al (2000) The *ACA4* gene of Arabidopsis encodes a vacuolar membrane calcium pump that improves salt tolerance in yeast. Plant Physiol 124(4):1814–1827

Giesler M, Marjolaine G, Sabine B et al (2004) *Arabidopsis* immunophilin-like TWD1 functionally interacts with vacuolar ABC transporters. Mol Biol Cell 15(7):3393–3405

Gut H, Dominici P, Pilati S et al (2009) A common structural basis for pH- and calmodulin-mediated regulation in plant glutamate decarboxylase. J Mol Biol 392:334–351

Harding SA, Oh SH, Roberts DM (1997) Transgenic tobacco expressing a foreign calmodulin gene shows an enhanced production of active oxygen species. EMBO J 16:1137–1144

Heo WD, Lee SH, Kim MC et al (1999) Involvement of specific calmodulin isoforms in salicylic acid-independent activation of plant disease resistance responses. Proc Natl Acad Sci USA 96:766–771

Hetherington AM, Brownlee C (2004) The generation of Ca(2+) signals in plants. Annu Rev Plant Biol 55:401–427

Hsieh H-L (2000) Regulation of a recombinant pea nuclear apyrase by calmodulin and casein kinase II. Biochim Biophys Acta 1494:248–255–542

Iakimova ET, Michalczuk L, Woltering EJ (2005) Hypersensitive cell death in plants – its mechanisms and role in plant defence against pathogens. J Food Ornament Plant Res 13:135–158

Ishida H, Huang H, Yamniuk AP et al (2008) The solution structures of two soybean calmodulin isoforms provide a structural basis for their selective target activation properties. J Biol Chem 283(21):14619–14628

Jarrett HW, Charbonneau H, Anderson JM et al (1980) Plant calmodulin and the regulation of NAD kinase. Ann N Y Acad Sci 356:119–129

Jiang X, Leidi EO, Pardo JM (2010) How do vacuolar NHX exchangers function in plant salt tolerance? Plant Signal Behav 5(7):792–795

Journot-Catalino N, Somssich IE, Roby D et al (2006) The transcription factors WRKY11 and WRKY17 act as negative regulators of basal resistance in Arabidopsis thaliana. Plant Cell 18:3289–3302

Kang CH, Jung WY, Kang YH et al (2006) AtBAG6, a novel calmodulin-binding protein, induces programmed cell death in yeast and plants. Cell Death Differ 13(1):84–95

Kant S, Bi YM, Zhu T et al (2009) SAUR39, a small auxin-up RNA gene, acts as a negative regulator of auxin synthesis and transport in rice. Plant Physiol 151:691–701

Kim MC, Lee SH, Kim JK et al (2002) Mlo, a modulator of plant defense and cell death, is a novel calmodulin-binding protein. J Biol Chem 277:19304–19314

Kim KC, Fan B, Chen Z (2006) Pathogen-induced Arabidopsis WRKY7 is a transcriptional repressor and enhances plant susceptibility to *Pseudomonas syringae*. Plant Physiol 142(3):1180–1192

Kim HS, Park BO, Yoo JH et al (2007) Identification of a calmodulin-binding NAC protein as a transcriptional repressor in Arabidopsis. J Biol Chem 282:36292–36302

Koo S, Choi M, Chun H et al (2009) The calmodulin-binding transcription factor OsCBT suppresses defense responses to pathogens in rice. Mol Cells 27:563–570

Lee D, Polisensky DH, Braam J (2005) Genome-wide identification of touch- and darkness-regulated Arabidopsis genes: a focus on calmodulin-like and XTH genes. New Phytol 165(2):429–444

Levy M, Wang Q, Kaspi R et al (2005) Arabidopsis IQD1, a novel calmodulin-binding nuclear protein, stimulates glucosinolate accumulation and plant defense. Plant J 43(1):79–96

Li B, Liu HT, Sun DY et al (2004) Ca(2+) and calmodulin modulate DNA-binding activity of maize heat shock transcription factor in vitro. Plant Cell Physiol 45 (5):627–634

Li X, Borsics T, Harrington HM, Christopher DA (2005) Arabidopsis AtCNGC10 rescues potassium channel mutants of E. coli, yeast and Arabidopsis and is regulated by calcium/calmodulin and cyclic GMP in *E. coli*. Funct Plant Biol 32:643–653

Li Z-G, Gong M, Xie H et al (2012) Hydrogen sulfide donor sodium hydrosulfide-induced heat tolerance in tobacco(*Nicotiana tabacum* L) suspension cultured cells and involvement of Ca2+ and calmodulin. Plant Sci 185–196. doi:10.1016

Liang L, Flury S, Kalck V et al (2006) CENTRIN2 interacts with the Arabidopsis homolog of the human XPC protein (AtRAD4) and contributes to efficient synthesis-dependent repair of bulky DNA lesions. Plant Mol Biol 61(1–2):345–356

Liu HT, Li GL, Chang H et al (2007) Calmodulin-binding protein phosphatase PP7 is involved in thermotolerance in Arabidopsis. Plant Cell Environ 30(2):156–164

Lu HP (2005) Single-molecule study of protein-protein and protein-DNA interaction dynamics. Methods Mol Biol 305:385–414

Ma W, Smigel A, Tsai YC et al (2008) Innate immunity signaling: cytosolic Ca2+ elevation is linked to downstream nitric oxide generation through the action of calmodulin or a calmodulin-like protein. Plant Physiol 148(2):818–828

Ma F, Lu R, Liu H et al (2012) Nitric oxide-activated calcium/calmodulin-dependent protein kinase regulates the abscisic acid-induced antioxidant defence in maize. J Exp Bot. doi:10.1093/jxb/ers161

MacGregor KB, Shelp BJ, Peiris S et al (2003) Overexpression of glutamate decarboxylase in transgenic tobacco plants deters feeding by phytophagous insect larvae. J Chem Ecol 29:2177–2182

McCormack E, Braam J (2003) Calmodulins and related potential calcium sensors of Arabidopsis. New Phytol 159:585–598

McCormack E, Tsai Y-C, Braam J (2005) Handling calcium signaling: Arabidopsis CaMs and CMLs. Trends Plant Sci 10:383–388

Magnan F, Ranty B, Charpenteau M et al. (2008) Mutations in AtCML9, a calmodulin-like protein from Arabidopsis thaliana, alter plant responses to abiotic stress and abscisic acid. Plant J 56(4):575–578.

Meneghelli S, Fusca T, Luoni L et al (2008) Dual mechanism of activation of plant plasma membrane Ca2 + −ATPase by acidic phospholipids: evidence for a phospholipid binding site which overlaps the calmodulin-binding site. Mol Membr Biol 25:539–546

Miao Y, Laun TM, Smykowski A (2007) Arabidopsis MEKK1 can take a short cut: it can directly interact with senescence-related WRKY53 transcription factor on the protein level and can bind to its promoter. Plant Mol Biol 65(1–2):63–76

Nakahara KS, Masutaa C, Yamada S et al (2012) Tobacco calmodulin-like protein provides secondary defense by binding to and directing degradation of virus RNA silencing suppressors. PNAS 109:10113–10118

Nie H, Zhao C, Wu G et al (2012) SR1, a calmodulin-binding transcription factor, modulates plant defense and ethylene-induced senescence by directly regulating NDR1 and EIN31. Plant Physiol 158:1847–1859

Oh MH, Wang X, Kota U et al (2009) Tyrosine phosphorylation of the BRI1 receptor kinase emerges as a component of brassinosteroid signaling in Arabidopsis. Proc Natl Acad Sci U S A 106:658–663

Olsen AN, Ernst HA, Leggio LL et al (2005) NAC transcription factors: structurally distinct, functionally diverse. Trends Plant Sci 10:79–87

Pandey SP, Somssich IE (2009) The role of WRKY transcription factors in plant immunity. Plant Physiol 150:1648–1655

Park CY, Lee JH, Yoo JH et al (2005) WRKY group IId transcription factors interact with calmodulin. FEBS Lett 579:1545–1550

Park HC, Kim ML, Kang YH et al (2009) Functional analysis of the stress-inducible soybean calmodulin isoform-4 (GmCaM-4) promoter in transgenic tobacco plants. Mol Cells 27(4):475–480

Patil S, Takezawa D, Poovaiah BW (1995) Chimeric plant calcium/calmodulin-dependent protein kinase gene with a neural visinin-like calcium-binding domain. Proc Natl Acad Sci U S A 92:4897–4901

Pazos F, Valencia A (2001) Similarity of phylogenetic trees as indicator of protein-protein interaction. Protein Eng 14(9):609–614

Perochon A, Aldon D, Galaud JP et al (2011) Calmodulin and calmodulin-like proteins in plant calcium signaling. Biochimie 93(12):2048–2053

Perruc E, Charpenteau M, Ramirez BC et al (2004) A novel calmodulin-binding protein functions as a negative regulator of osmotic stress tolerance in *Arabidopsis thaliana* seedlings. Plant J 38(3):410–420

Poovaiah BW, Takezawa D, An G et al (1996) Regulated expression of a calmodulin isoform alters growth and development in potato. J Plant Physiol 149:553–558

Popescu SC, Popescu GV, Bachan S et al (2007) Differential binding of calmodulin-related proteins to their targets revealed through high-density Arabidopsis protein microarrays. Proc Natl Acad Sci U S A 104 (11):4730–4735

Raghothama KG, Mizrahi Y, Poovaiah BW (1985) Effect of calmodulin antagonists on auxin induced elongation. Plant Physiol 79:28–33

Ranty B, Aldon D, Galaud J-P (2006) Plant calmodulins and calmodulin-related proteins multifaceted relays to decode calcium signals. Plant Signal Behav 1(3):96–104

Reddy AS, Ben-Hur A, Day IS (2011) Experimental and computational approaches for the study of calmodulin interactions. Phytochemistry 72(10):1007–1019

Rodriguez-Concepcion M, Yalovsky S, Zik M et al (1999) The prenylation status of a novel plant calmodulin directs plasma membrane or nuclear localization of the protein. EMBO J 18:1996–2007

Snedden WA, Blumwald E (2000) Alternative splicing of a novel diacylglycerol kinase in tomato leads to a calmodulin-binding isoform. Plant J 24:317–326

Sunkar R, Kaplan B, Bouche N et al (2000) Expression of a truncated tobacco NtCBP4 channel in transgenic plants and disruption of the homologous Arabidopsis CNGC1 gene confer Pb2+ tolerance. Plant J 24:533

Takezawa D (2003) Characterization of a novel plant PP2C-like protein Ser/Thr phosphatase as a calmodulin-binding protein. J Biol Chem 278(39):38076–38083

Vadassery J, Reichelt M, Hause B et al (2012) CML42-mediated calcium signaling coordinates responses to Spodoptera herbivory and abiotic stresses in Arabidopsis. Plant Physiol 159(3):1159–1175

Vert G, Nemhauser JL, Geldner N et al (2005) Molecular mechanisms of steroid hormone signaling in plants. Annu Rev Cell Dev Biol 21:177–201

Vinogradova MV, Malanina GG, Reddy VS et al (2004) Structural dynamics of the microtubule binding and regulatory elements in the kinesin-like calmodulin binding protein. J Struct Biol 163(1):76–83

Wallace W, Secor J, Schrader LE (1984) Rapid accumulation of gamma-aminobutyric acid and alanine in soybean leaves in response to an abrupt transfer to lower temperature, darkness, or mechanical manipulation. Plant Physiol 75:170–175

Wan D, Li R, Zou B et al (2012) Calmodulin-binding protein CBP60g is a positive regulator of both disease resistance and drought tolerance in Arabidopsis. Plant Cell Rep 31(7):1269–1281

Wang Y, Zhang WZ, Song LF et al (2008) Transcriptome analyses show changes in gene expression to accompany pollen germination and tube growth in Arabidopsis. Plant Physiol 148(3):1201–1211

Wang Q, Chen B, Liu P, Zheng M et al (2009a) Calmodulin binds to extracellular sites on the plasma membrane of plant cells and elicits a rise in intracellular calcium concentration. J Biol Chem 284(18):12000–12007

Wang L, Tsuda K, Truman W et al (2009b) Arabidopsis CaM binding protein CBP60g contributes to MAMP-induced SA accumulation and is involved in disease resistance against *Pseudomonas syringae*. PLoS Pathog 5(2):e1000301

Watterson DM, Sharief F, Vanaman TC (1980) The complete amino acid sequence of the Ca2 + − dependent modulator protein (calmodulin) of bovine brain. J Biol Chem 255(3):962–975

Weng H, Yoo CY, Gosney MJ et al (2012) Poplar GTL1 is a Ca2+/calmodulin-binding transcription factor that functions in plant water use efficiency and drought tolerance. PLoS One 7(3):e32925

White PJ, Piñeros M, Tester M et al (2000) Cation permeability and selectivity of a root plasma membrane calcium channel. J Membr Biol 174:71–83

Xie X, Harrison DH, Schlichting I et al (1994) Structure of the regulatory domain of scallop myosin at 2.8 A resolution. Nature 368(6469):306–312

Xu H, Heath MC (1998) Role of calcium in signal transduction during the hypersensitive response caused by basidiospore-derived infection of the cowpea rust fungus. Plant Cell 10(4):585–598

Xu GY, Pedro SCF, Rocha et al (2011) A novel rice calmodulin-like gene, OsMSR2, enhances drought and salt tolerance and increases ABA sensitivity in Arabidopsis. Planta 234(1):47–59

Yamaguchi T, Aharon GS, Sottosanto JB et al (2005) Vacuolar Na+/H + antiporter cation selectivity is regulated by calmodulin from within the vacuole in a Ca2 + − and pH-dependent manner. Proc Natl Acad Sci U S A 102:16107–16112

Yamakawa H, Katou S, Seo S et al (2004) Plant MAPK phosphatase interacts with calmodulins. J Biol Chem 279:928–936

Yang T, Poovaiah BW (2000) Molecular and biochemical evidence for the involvement of calcium/calmodulin in auxin action. J Biol Chem 275:3137–3143

Yang T, Poovaiah BW (2002) Hydrogen peroxide homeostasis: activation of plant catalase by calcium/calmodulin. Proc Natl Acad Sci U S A 99(6):4097–4102

Yang T, Poovaiah BW (2003) Calcium/calmodulin-mediated signal network in plants. Trends Plant Sci 8 (10):505–512

Yang T, Segal G, Abbo S et al (1996) Characterization of the calmodulin gene family in wheat: structure, chromosomal location, and evolutionary aspects Mol. Gen. Genet 252(6):684 94

Yang T, Chaudhuri S, Yang L et al (2004) Calcium/calmodulin up-regulates a cytoplasmic receptor-like kinase in plants. J Biol Chem 279:42552–42559

Yang T, Chaudhuri S, Yang L et al (2010) A calcium/calmodulin-regulated member of the receptor-like kinase family confers cold tolerance in plants. J Biol Chem 285:7119–7126

Yap KL, Yuan T, Mal TK et al (2003) Structural basis for simultaneous binding of two carboxy-terminal peptides of plant glutamate decarboxylase to calmodulin. J Mol Biol 328(1):193–204

Yoo JH, Park CY, Kim JC et al (2005) Direct interaction of a divergent CaM isoform and the transcription factor, MYB2, enhances salt tolerance in Arabidopsis. J Biol Chem 280:3697–3706

Yoo CY, Hasegawa PM, Mickelbart MV et al (2010) Regulation of stomatal density by the GTL1 transcription factor for improving water use efficiency. Plant Signal Behav 6(7):1069–1071

Zhang W, Zhou RG, Gao YJ et al (2009a) Molecular and genetic evidence for the key role of AtCaM3 in heat-shock signal transduction in Arabidopsis. Plant Physiol 149(4):1773–1784

Zhang H, Ye YK, Wang SH et al (2009b) Hydrogen sulfide counteracts chlorophyll loss in sweet potato seedling leaves and alleviates oxidative damage against osmotic stress. Plant Growth Regul 58(2009):243–250

Herbicide Resistance in *Phalaris minor* and Genetic Medication in Crop

M.K. Tripathi and A.K. Gaur

Abstract

Introduction of dwarf wheat varieties through revolutionized wheat production had made India the second largest wheat grower in the world, but these varieties could not cope up with the weeds. Weed problem has been continuously increasing along with development of high-yielding cultivars, which require more fertilizers and irrigation for their growth and development. The major weed growing in wheat fields is *Phalaris minor*, and it is being controlled since 1982, by using a single herbicide, isoproturon. Continuous application of single herbicide exerted a selection pressure, and it had been noticed that this weed has developed resistance against isoproturon. The problem of resistance is increasing, leading to development of resistance day by day. Currently it has become a very serious problem in some parts of Haryana and Punjab and has also been observed in the Terai region of Uttaranchal.

Keywords

Herbicide • Isoproturon • P. minor • PSII • Cytochrome • D1 protein

Phalaris minor as a Weed of Wheat

Little seed canary grass (*Phalaris minor* Ritz.) is a monocot weed in the Poaceae family. A few years ago, it was a minor weed but with the adoption of rice-wheat cropping system, which provided favorable environment for the seed production of this weed, it has become a major weed of the wheat crop. Due to its similarity to wheat, it is detected only at the time of reverence of ear head in wheat crops.

M.K. Tripathi (✉)
Agro Produce Processing Division, CIAE, ICAR (Govt of India), NabiBagh, Baresia Road, Bhopal, MP 462038, India
e-mail: mktripathi@ciae.res.in

A.K. Gaur
Department of Molecular Biology and Biotechnology, Govind Ballabh Pant University of Agriculture & Technology, U.S. Nagar, Pantnagar, UK, India

Origin and Distribution

At present, 22 species of *Phalaris* are native to the Mediterranean, including *Phalaris minor* Retz. and four other in Southern and Western USA. *P. minor* was reported to be a major weed

R.K. Gaur and P. Sharma (eds.), *Approaches to Plant Stress and their Management*,
DOI 10.1007/978-81-322-1620-9_5, © Springer India 2014

in Latin America and probably reached India through the import of Mexican wheat, and it was recognized to be problematic in 1970s (Bhan and Chaudhary 1976). Surveys of wheat crops in the states of Punjab (Bir and Sidhu 1979) and Haryana (Malik et al. 1985; Singh et al. 1995a) established *P. minor* as the most dominant weed of wheat in northwest India. *P. minor* is widely distributed in all the continents of world.

Taxonomy and Morphology

P. minor is self-pollinated (2n = 28, rarely 29) with C-3 photosynthetic pathway, similar to wheat. Morphologically, *P. minor* is similar to wheat until the flowering stage. There are, however, significant differences in the leaf characteristics and growth habits between wheat and *P. minor*. Both tillering and branching occur in *P. minor*, whereas wheat has no branching habit. *P. minor* has a purplish pigmentation at the base of the stem and internodes; leaves are narrow and dropping and a purplish orange sap oozes out when the leaf is broken. The length of the *P. minor* ear head is normally half (6 vs. 12 cm) that of wheat. Maturity of *P. minor* is asynchronous and starts earlier than wheat. Inflorescence maturity of *P. minor* starts at the top and progresses towards the base with continuous shattering of seeds before harvesting of the crop. The weight of *P. minor* seeds is 20-fold lower than that of wheat.

Biology

Germination Periodicity

A large proportion of *P. minor* seeds germinate between mid-November and mid-December (Ray et al. 1982). Germination progresses rapidly between 10 °C and 20 °C (Mehra and Gill 1988). Dark brown coated seeds germinate better than light colored seed (Mehra and Gill 1988). Germination of *P. minor* sown at various soil depths was found to spread over a 9-week period and could last up to 10–12 weeks (Walia and Gill 1985a). The emergence pattern of *P. minor* was affected by irrigation and hoeing (Tewari and Bisen 1982). Okerekea et al. (1981) observed the germination of *P. minor* to be unaffected by soil moisture. Bhan and Chaudhary (1976) found that plants emerging in late December had more tillers than plants emerging in November. Singh and Ghosh (1982) found the highest number of seeds on the soil surface, and this decreased with an increase in depth to 15 cm.

Freshly harvested mature seeds remained dormant for a period of 6 months (Singh 1998) and germination increased to 88–96 % after 12 months. The test weight (1,000 seed weight) varied from 1.5 to 2.1 g depending upon soil type and growth conditions. No dormancy was observed 13 months after harvesting of *P. minor*. Similarly, Jimenez-Hidalgo et al. (1993) found considerable variations in germination of *Phalaris spp.* from different locations and years, but no differences were found between seeds aged 6 and 18 months.

P. minor seeds tolerated anaerobic conditions by entering into secondary dormancy and by avoiding anaerobic decomposition. The chemical status of seeds under anaerobic conditions suggested that *P. minor* resisted oxygen stress probably due to formation of chemical metabolites and changes in membrane permeability (Parashar and Singh 1985). This could be the reason for the high infestation of *P. minor* under rice-wheat rotation areas (Singh et al. 1995b). The seed coat of *P. minor* is hard and may not be affected by anaerobic conditions under rice cultivation prior to wheat sowing.

Growth and Development

P. minor has more tillers/plant than wheat under noncompetitive conditions, and its branching habit contributes to greater seed production (300–450 seeds per panicles). Flowering initiation (heading) was observed in *P. minor* at 66–98 days after sowing (DAS). On average, *P. minor* matured 20 days earlier than wheat under the December and normal November sowings, respectively. Uptake of nitrogen (N), phosphorus (P), and potash (K) by *P. minor* was 6, 7, and

54 % higher than wheat, respectively, at the 20th December sowing (Anonymous 1990).

The rate of biomass accumulation of *P. minor* is slow during the initial growth stages, dry weight increases being greater after 90 days of sowing compared to the first 60 days (Malik and Singh 1993), and plant height varies depending upon the growing conditions, generally growing 50–70 cm at 90 DAS (Malik and Singh 1995). At maturity, the *P. minor* may be taller than the dwarf wheat varieties.

Crop Weed Interaction

P. minor thrives under high-fertility and high-moisture conditions (Singh and Malik 1992; Singh et al. 1995a) and competes vigorously with wheat reducing its yield by up to 80 % depending upon weed intensity, cultivation practices, soil, and environmental factors. On average, a population of 200–400 plants/m^2 of *P. minor* is normally observed under field conditions in Haryana and Punjab. Differential effects of weed populations on wheat yields were also observed by Cudney and Hill (1979) in the USA, Godinho and Costa (1981) in Portugal, Montazeri (1993) in Iran, and Afentouli and Eleftherohorinos (1996) in Greece.

Depletion of the soil moisture was greater from the soil profile in the presence of *P. minor* at higher levels of N (Khera et al. 1995). Under controlled environmental conditions, water requirement of wheat and *P. minor* was comparable when 20–25-day-old plants were placed in nutrient solution for 2 weeks (Singh et al. 1998a).

Surveys conducted in Haryana state revealed that under high-fertility conditions, *P. minor* occurred in 92 % fields compared to 67 % and 32 % under medium- and low-fertility soils, respectively (Singh and Malik 1992), indicating that *P. minor* is a vigorous competitor of wheat under improved nutrient and moisture conditions. Weed competition with wheat was severe up to 80 kgN/ha and decreased at 120 and 160 kg/ha due to vigorous crop growth, reduced weed population, and dry matter production (Singh and Dhaliwal 1984). Walia and Gill (1985a, b) also reported that higher levels of nitrogen (120 and 160 kg/ha) suppressed the population of *P. minor*; however, effects on wheat grain yield were nonsignificant between 120 and 160 kgN/ha.

Weed Control

Mechanical Weed Control

Bidirectional (cross-rows) sowing of wheat reduced the dry weight of *C. album* but had no effect on grass weeds (Malik et al. 1985). Cross-row sowing combined with isoproturon treatment had a complementary effect and was found more desirable. Sharma et al. (1985) found that closer row spacing (15 cm) and cross-rows (22.5 cm) resulted in a significant reduction in *P. minor* density compared to unweeded check plots. Narrow or cross-row spacing coupled with 0.5 kg/ha of isoproturon at 2 weeks after sowing (WAS) significantly reduced *P. minor* dry weight and increased tiller humblers and yield of wheat (Prakash et al. 1996). Increased seed rate (150 kg/ha) and bidirectional sowing significantly reduced the N, P, and K uptake by weeds compared with wider rows and normal seed rate (100 kg/ha), resulting in higher nutrient uptake by wheat (Johri et al. 1992).

Ray et al. (1982) observed that pre-sowing irrigation helped to germinate the first major flush of *P. minor* which was effectively killed by soil scarification or by application of paraquat (0.5 kg/ha). Singh et al. (1985) observed that delayed wheat sowing decreased the density and growth of *P. minor*. Kolar and Mehra (1992) observed that November-sown wheat produced a higher grain yield (383 kg/ha) compared to October (2,050 kg/ha) or December (3,050 kg/ha) sown.

The competitive nature of wheat was found to improve with increase in seed rate from 100 to 150 kg/ha even in light soils under optimum fertilizer and irrigation conditions. High seed rates of 150 kg/ha with herbicide use can provide weed-free conditions and good yields. Increasing the seed rate from 100 to 175 kg/ha in the wheat variety WH-473 decreased the dry weight of weeds from 135 to 96 g/m^2 under unweeded plots and increased average wheat yield from 4.57 to 5.44 t/ha under higher seed rates (Panwar et al. 1995).

Chemical Control

Several herbicides have been evaluated and used for the control of *P. minor* and associated weeds in wheat under different agroclimatic conditions. It is always desirable to have several candidate herbicides to allow choice of chemical weed control under different cropping systems. Herbicides are effective and efficient tools of weed management, and their selective use not only provides economic yields but also helps to avoid/delay weed resistance.

Chemical Weed Control by Isoproturon

Isoproturon applied at the two leaf stage showed a maximum reduction in dry weight (DW) of *P. minor* and *A. fatua,* whereas application at the six leaf stage had no effect on *A. fatua* (Bhan et al. 1985a); Balyan et al. (1987) evaluated the effect of time of application of isoproturon on *P. minor*-, *A. ludoviciana*-, *C. album*-, and *L. aphaca*-infesting wheat (cv. WH-283). All the weeds were most susceptible to isoproturon (1.0 kg/ha) when applied at 20–30 DAS; tolerance was observed with delayed application. In wheat, isoproturon treatment can reduce photosynthetic activity, but recovery occurred within 2 weeks (Singh et al. 1997a). Application of isoproturon at 25 DAS not only resulted in saving of 25 % in herbicide rates but also increased the efficacy of weed control (Singh and Malik 1994). The corresponding yields of wheat were 5,345 and 4,084 kg/ha following application of isoproturon (0.75 kg/ha) at 25 DAS compared to 5,010 and 4,020 kg/ha with 1.0 kg/ha isoproturon applied at 35 DAS (Singh and Malik 1994).

Sharma et al. (1985) also found that application of isoproturon (0.50 kg/h) at 2 WAS (weeks after sowing) had a similar effect on *P. minor* to that of 1.0 kg/ha, applied 5 WAS under late showing conditions. Lower application rates of isoproturon were required when applied at the 2–3 leaf stages (Singh and Malik 1992). Interaction of sowing time and application of isoproturon significantly influenced the populations of *P. minor* and *A. ludoviciana*. Susceptibility of *P. minor* to isoproturon was similar at the 2–4 leaf stages, but at the six leaf stage, *P. minor* showed a marked tolerance to isoproturon (Yaduraju 1991).

Pandey et al. (1996) conducted field experiments during the winter seasons of 1991–1993 at New Delhi with wheat. The applications of pendimethalin (preemergence) or isoproturon (powder) after first irrigation were highly effective against both grassy (*P. minor*) and broad-leaved weeds.

Dixit and Bhan (1998) reported that isoproturon applied at 0.50, 0.75, and 1.0 kg/ha just before irrigation provided better weed control and higher wheat yields as compared to its application at 30 days after sowing. The greatest weed control efficiency was obtained with 1.0 kg/ha isoproturon applied just before irrigation.

Herbicides

The term herbicides as used in agriculture applies to a heterogeneous group of chemicals with the property of eradicating all vegetation or of selectively killing weeds without seriously injuring the cultivated crops. The actions of herbicides on the plants include auxin herbicides, inhibitors of growth regulator action, inhibitors of cell division, inhibitors of photosynthesis, inhibitor of metabolic synthesis, inhibitors of normal chloroplast development, and inhibitors of more than a single process (Andreaws 1963). Classification of herbicides based on risk of resistance development is shown in Table 1.

In vitro studies suggested that herbicides that inhibit or modify photosynthesis can be classified as electron transport inhibitors, uncouplers, energy transfer inhibitors, inhibitory uncouplers, or electron acceptors inhibitor (Moreland 1980).

Table 1 Classification of herbicides based on risk of resistance development

Risk of resistance	Mode of action	Example
High	Acetolactate synthase (ALS) inhibitors	Sulfonylureas, imidazolinones
	Lipid synthesis inhibitors	Thiocarbamates, benzofuran
	Cell membrane disrupters	Dinitrophenols
Medium	Contact photosynthesis inhibitors	Paraquat, Basagran Dinitroanilines
	Root growth inhibitors	
	Pigment inhibitors	Pyridazinones, triazoles triazines, phenylureas, and amides
	Systemic photosynthesis inhibitors	
Low	Amino acid derivatives	Glyphosate, Touchdown
	Growth regulators	2,4-D, MCPA, glufosinate
	Shoot growth inhibitors	Alachlor

Molecular Mechanism of Herbicide Action with Photosynthetic Electron Transport

More than 50 % of all herbicides used interact with the process of photosynthesis. Most of them are inhibitors of photosystem II-dependent electron flow (Van Rensen 1971).

Photosynthetic electron transport occurs in chloroplast thylakoid membrane. Thylakoid membranes are either stacked (referred to as apprised or grana lamellae) or unstacked (referred to as stroma lamellae). The thylakoid membranes contain four membrane-spanning protein complexes: the PSII complex, the cyt b_6f complex, the PSI complex, and ATP synthase complex. The PSII is primarily but no exclusively localized in the apprised lamellae, while the PSI and ATP synthase complex are present only in the stroma lamellae. The ATP synthase complex converts the potential energy of proton gradient, developed during electron transport, into high-energy phosphate bond energy in the form of ATP. The cytochrome b_6/f complex is distributed approximately equally between the apprised and unapprised membrane (Anderson and Anderson 1988). The separation of PSII and PSI complex is thought to optimize the relative amounts of light energy transferred to each reaction center. Photosynthetic electron transport involves all components except for the ATP synthase complex.

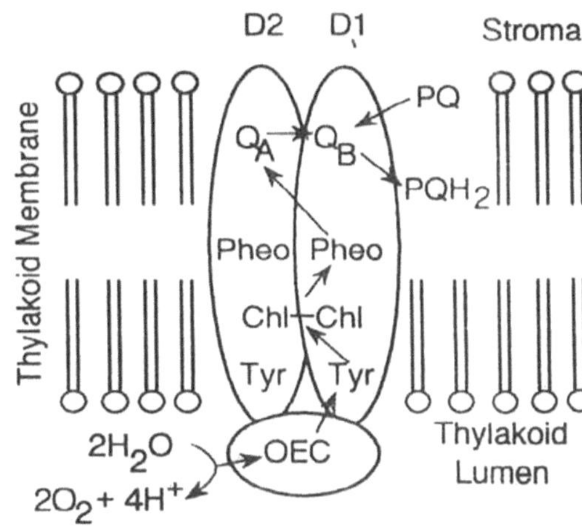

Fig. 1 Photosystem II reaction center complex. *OEC* oxygen-evolving complex, *Tyr.* tyrosine amino acid residue 160, *Chl-Chl* chlorophyll a dimer (P680 pheo, pheophytin; Q_A and Q_B, bound plastoquinone molecules, *PQH_2* plastohydroquinone. Electron transport carriers bound D1 and D2 proteins

The electron carriers are cofactors that are bound to chloroplast proteins. Thus, the proteins comprising the various complexes do not carry the electron directly. The only exception to this is the tyrosine amino acid residue of the D1 protein that is an electron carrier in PSII (Fig. 1) (Vermaas 1988).

PSII Electron Transport

The PSII complex includes the oxygen-evolving (water-splitting) complex, a reaction center complex, and light-harvesting chlorophyll antenna proteins. The model for the PSII reaction center complex is based on homologies between PSII and the bacterial photosynthetic reaction center (Michel and Deisenhofer 1988; Sinning et al. 1989a, b). The light-harvesting chlorophyll molecules associated with PSII and transfer excitation energy to the PSII reaction center, a chlorophyll dimer known as P_{680}. When excitation energy is transferred to the chlorophyll, a dimer charge separation takes place and the excited electron is transferred to pheophytin. An electron derived from the splitting of water neutralizes the residual positive charge of the chlorophyll dimer. From pheophytin, the electron is transferred to Q_A and QB. Q_A and Q_B are plastoquinone molecules bound in a special niche of the D2 and D1 proteins, respectively. Q_B accepts two electrons from Q_A, then accepts two protons from the stroma side of the membrane, and then leaves its binding niche as plastohydroquinone. Another plastoquinone molecule then binds to the D1 protein, replacing the molecule that has left, and when bound, is called "Q_B."

Electron Transport Between PSII and PSI

Plastoquinone donates its electrons to the cytochrome b_6/f complex. The Q-cycle is thought to utilize plastohydroquinone and the cytochrome b_6/f complex to transport two protons across the membrane per electron utilized in linear electron transport from PSII to PSI (Haehnel 1984). Plastocyanin accepts electrons from cytochrome b_6/f and shuttles the electrons along the lumen side of the thylakoid membrane to the PSI reaction center.

PSI Electron Transport

The PSI complex can be defined as the components of photosynthetic electron transport that catalyze the photoreduction of ferredoxin with plastocyanin as the electron donor (Reilly and Nelson 1988). PSI is composed of reaction center complex and light-harvesting chlorophyll antenna proteins which transfer absorbed light energy to the PSI reaction center, known as P_{700}. It is estimated that eight protein subunits are associated with the PSI complex (Polos et al. 1987). Two 70 kDa polypeptides, designated A_1 and A_2, are associated with the reaction center (Haehnel 1984).

P_{700} is generally considered to be chlorophyll, a dimer (Lagoutte and Mathis 1989) which undergoes a light-induced charge separation resulting in the transfer of an excited electron to A_0. A_0 is generally considered to be a chlorophyll monomer.

The membrane-bound acceptors, Fx and Fa/Fb, are protein-bound iron-sulfur center. FX has one 2 Fe−2S center (Andreasson and Vaungard 1988) and Fa/Fb contains two 4 Fe−4S centers (Haehnel 1984). Fa/Fb is probably the intermediate from which paraquat and diquat accept electrons (Parrett et al. 1989).

The membrane-bound acceptors, Fx and Fa/Fb, are protein-bound iron-sulfur centers. Fx has one 2 Fe−2S center (Andreasson and Vaungard 1988) and Fa/Fb contains two 4 Fe−4S centers (Haehnel 1984). Fa/Fb is probably the intermediate from which parquet and diquat accept electrons (Parrett et al. 1989).

Ferredoxin transfer electrons from Fa/Fb to ferredoxin: $NADP^+$ oxidoreductase (FNR). FNR is restricted to the stromal surface of non-apprised lamellae and catalyzes the reduction of $NADP^+$ to NADPH.

Isoproturon Resistance in *P. minor*

Chemical Nature of Isoproturon

IUPAC name of isoproturon is 3-(4-isopropylphenyl)-1, 1-dimethyl urea. It is solid crystalline powder with molecular formula of $C_{12}H_{18}N_2O$ and has molecular weight 206.29; chemical structure of isoproturon is shown in Fig. 2 (Fedtke 1982). Melting point of isoproturon is 155 C and its solubility in water is 70 mg/100 ml at 20 C. Its LC-50 value for goldfish is calculated 100 mg/l and for guppy it is 90 mg/l, but in case of bee, it is found nontoxic. Its optimum pH is 6–8. It can be used as pre- and

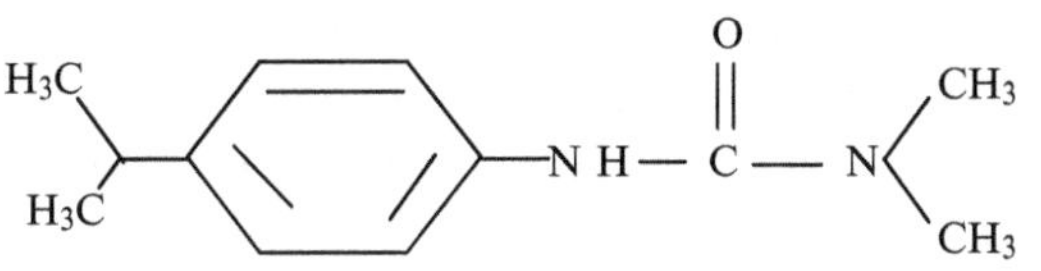

Fig. 2 Chemical structure of isoproturon

postemergence control of black grass, wild oats, annual meadow grass, ryegrass, and many broadleaf weeds in spring and winter wheat, spring and winter barley, and winter rye (www.hclrss.demon.co.uk/isoproturon.html).

Action of Isoproturon

Isoproturon is expected to have a mode of action similar to that of other substituted ureas, where the identified target site is usually in the photosynthetic apparatus, especially at D1 protein of PSII (Pfistar et al. 1981). Phenyl urea herbicides having binding site (Q_B of the D1 protein) between PSII and PSI in chloroplast electron transport system block electron transport (Velthuys 1981 and Fuerst and Norman 1991) and result in destruction of PSII reaction center (Pallett and Dodge 1980; Barry et al. 1990). Isoproturon application decreases the protein and chlorophyll contents and RUBP carboxylase activity which result in disorganization of grana and intergrana in wheat (De Felipe et al. 1989).

PSII Electron Transport Inhibitors

PSII electron transport inhibitors, such as urea derivative herbicide and triazine, bind to the Q_B-binding niche on the D1 protein. The D1 protein has also been referred to as the Q_B-binding protein and the 32 kDa herbicide-binding protein. PSII inhibitors bind to the D1 reaction center protein and inhibit electron transport by acting as nonreducible analogs of plastoquinone (Gardner 1989).

Q_B is bound to the D1 protein by two hydrogen bonds between the protein and the two carbonyl groups of plastoquinone. A hydrogen bond is formed between one carbonyl and the hydroxyl groups of serine 264; an amino acid adjacent to serine 264 can also hydrogen bond to the same carbonyl on Q_B (Trebst 1987). A hydrogen bond is formed between the second carbonyl of Q_B and histidine 215 (Sinning et al. 1989). PSII herbicides such as atrazine bind to the D1 protein due to hydrogen bonds, van der Waals forces, and hydrophobic interactions (Michel et al. 1986). Hydrogen bonds between triazine herbicides and the amino acids serine 264 and phenyl amine 265 are essential for binding (Trebst 1986). Phenylamine 255 contributes to hydrophobic interactions in herbicide binding (Shigematsu et al. 1989).

PSII electron transport inhibitors belong to a variety of chemical families. Three of these families, the nitrophenols, nitriles, and pyridazinones, inhibit photosynthesis by preventing Q_B binding in vitro, but another mode of action may also be involved in their herbicidal activity in vitro. This diverse group of herbicides binds to overlapping, but not identical, binding sites on the D1 protein (Pfister and Arntzen 1979). Herbicides in the nitrophenol and nitrile family probably bind to the D1 protein due to interaction with histidine 215 rather than serine 264 (Trebst 1987).

Treatment of plants with PSII herbicides blocks the flow of electrons through PSII and thus also indirectly blocks the transfer of excitation energy from chlorophyll molecules to the PSII reaction center. Excited chlorophyll molecules (singlet chlorophyll) spontaneously form triplet chlorophyll through a non-radiative energy transformation of chlorophyll known as intersystem crossing. The triplet chlorophyll reacts with molecular oxygen to form singlet oxygen. Lipid peroxidation is then initiated by triplet chlorophyll and singlet oxygen.

Isoproturon Resistance

It is the ability of an organism to cope with adverse environmental conditions along with continuation of life processes. Thus an ability of a weed to continue life cycle at recommended doses of the weedicide, which must kill the weed. Persistent herbicide application to weed plant populations created strong selection pressure resulting in herbicide resistance. Herbicide-resistant little seed canary grass is present globally. In India, since 1991, it infests 10,001–1,00,000 sites, covering 1,000,001–2,000,000 acres, possessing resistance to ureas and amides (www.weedscience.org). Herbicide resistance trait is heritable, and therefore it became difficult to control its spread. The production of seed being too high (500–1,000 per plant) and size being too small (1,000 seeds weigh about 2 g), the seeds can get distributed to newer areas easily along with seed of main crop wheat, machinery, wind, irrigation, etc. (Yaduraju 1999). Increasing infestation of little seed canary grass and development of resistance to isoproturon substantially decreased wheat production in rice-wheat system in India (Malik and Singh 1993). During the winters of 1991–1992 and 1992–1993, some biotypes of little seed canary grass were acquired resistance to isoproturon in India, but cross-resistance to pendimethalin and diclofop-methyl was not much pronounced. Resistant biotype required a higher dose of diclofop-methyl than the susceptible biotype. Resistant biotypes required 2–8 times more isoproturon doses compared to susceptible biotype for the same level of control. Control of little seed canary grass with isoproturon dropped from 78 % to 21 % from 1990 to 1993 (Malik and Singh 1995).

Malik (1995) reported that efficacy falls as resistance develops and crop failure is now being reported. Malik et al. (1998) studied herbicide-resistant weed problem in cereal crops (mainly wheat and rice) in developing countries. The main resistance problems were reported to be in *P. minor* against isoproturon in India, *P. minor* and *P. paradoxa* against diclofop in Mexico, etc. Singh et al. (1997) reported that *P. minor* resistance to isoproturon in wheat is increasing rapidly in the rice-wheat cropping zones of northwest India. Walia et al. (1997) conducted investigation during 1993–1994 and 1994–1995 in different *Phalaris* biotypes. Most of the biotypes were not controlled by isoproturon even at double the recommended dose (1.88 kg/ha). These studies clearly showed that *Phalaris* populations had developed resistance to isoproturon. Sharma and Pandey (1997a) reported isoproturon resistance in different biotypes of *P. minor*. The Delhi biotype was completely killed by 1.5 kg/ha isoproturon application at preemergence and at all the doses applied at postemergence, whereas the Haryana biotype was only killed at 1.5 kg/ha of isoproturon at postemergence.

Malik et al. (1998) reported widespread resistance in Haryana and Punjab which might have affected adjoining areas in other cropping systems. Some resistant biotypes required more than 6 times greater doses of isoproturon than sensitive biotypes for the same level of control. Dhaliwal et al. (1998) conducted field studies in Ludhiana, India, during 1996–1997 and 1997–1998 on 24 isoproturon-resistant biotypes of *P. minor*. Isoproturon at even double the recommended concentration (1.88 kg/ha) failed to control *P. minor* biotypes.

Mechanisms of Isoproturon Resistance in *P. minor*

Herbicide resistance could be acquired due to three basic mechanisms: (i) prevention of herbicide from reacting to the site of action, (ii) metabolic detoxification of the herbicide, and (iii) resistance at the site of action (Kearney and Kaufman 1969).

Herbicide Translocation

Isoproturon is usually applied as foliar spray. Before it could reach its target site, it encounters many barriers, such as the waxy cuticle layer of leaf, the cell wall, plasma membrane, and chloroplast membranes (Preston 1994). It is also likely that some membrane-bound protein carriers may be involved in its translocation. Although isoproturon is a nonpolar molecule and it may freely diffuse through the lipidous barriers, the involvement of a protein carrier (soluble or membrane bound) cannot be ruled out. Selective activity of isoproturon to wheat was attributed to differential uptake also (Archireddy and Kirkwood 1986; Giessbuhler et al. 1975). Sensitive biotypes of *P. minor* showed greater membrane permeability after herbicide application than tolerant (Dhawan and Malik 1999). Yaduraju (1998) reported that uptake and distribution of ^{14}C radioactivity in the foliage treatment with ^{14}C isoproturon was almost identical.

Metabolic Detoxification of Herbicides

Metabolites, besides affecting plant metabolism, themselves get metabolized within the plant. Plant species differ in their ability to react with certain herbicides, and in some cases, the observed difference in response of susceptible and resistant plant species is due to the differences in their metabolic activities. The biochemical mechanisms involved are quite often similar to those already established in animal tissues (Andreaws 1963).

Transformation of urea herbicides in plants generally involves two basic reactions: phase I oxidation, reduction, or hydrolysis and phase II acetylation of anilines, glucosidation, and binding to nitrogen constituents (Giessbuhler et al. 1975). Wheat degrades isoproturon by two primary reactions, ring alkyl hydroxylation and N-demethylation (Cabanne et al. 1987). de Prado et al. (1991) studied the effect of several substituted ureas including isoproturon on leaf fluorescence and Hill reaction activity in herbicide-resistant biotypes of *Alopecurus myosuroides* (a weed found in wheat fields in Spain) and concluded that resistance against herbicides was not due to herbicides' insensitivity of the chloroplast target site; rather it was due to metabolic detoxification. Devine (1997) reviewed several weed biotypes in which resistance conferred through enhanced herbicide detoxification, primary through elevated expression or activity of $cytP_{450}$.

Yaduraju (1998) reported a substantial difference between species in the metabolism of ^{14}C-isoproturon, which indicated that *P. minor* contained 2–3 times greater concentration of ^{14}C in the form of undegraded isoproturon than *P. paradoxa*. The latter had a greater proportion of ^{14}C in the form of conjugates and metabolites. The faster recovery in photosynthesis after treatment with isoproturon of *P. paradoxa* plants was also indicative of a better detoxification mechanism in that species. Kulshreshta et al. (1999) examined that isoproturon was completely degraded in 8 days by resistant biotypes and wheat but persisted up to 18 days in the susceptible biotypes. The toxic monodesmethyl and didesmethyl analogs were found in the susceptible biotypes for up to 18 days. In the resistant biotype and wheat, two additional nontoxic metabolites, p-isopropyl aniline and a hydroxyl derivative, were detected from 8 days onwards, confirming that metabolism of isoproturon in resistant *P. minor* was similar to that in wheat.

These degradations are believed to be carried out by cytochrome P_{450} monooxygenase since the $cytP_{450}$ inhibitor significantly increased the activity of isoproturon and inhibited its degradation in wheat (Cabanne et al. 1985). Two inhibitors of $cytP_{450}$ monooxygenases, ABT and piperonyl butoxide (PBO), enhanced the phytotoxicity of chlorotoluron in wheat (Gaillardon et al. 1985). It had been demonstrated that the effect of ABT was due to an inhibition of chlorotoluron oxidative metabolism (Cabanne et al. 1989). Oxidative detoxification of a number of xenobiotics in plant was also achieved by the cyt P_{450}-dependent monooxygenase system (Riviere and Cabanne 1987; Donaldson and Luster 1991; Hatzios 1991). Several cyt P_{450} inhibitors had been reported to act as herbicide synergists including PBO (Varsano et al. 1992; Feng et al. 1995; Leah et al. 1997; Singh et al. 1998b, c). Detoxification decreased through inhibition of cyt P_{450} monooxygenases as evidenced from mixed-function oxidase (MFO) inhibitors, which have shown binding to the heme group of the cyt P_{450}, thus forming a stable inactive cyt P_{450} complex (Kemp et al. 1990). The synergistic effects of PBO had been found to vary with different herbicides and species. However, little informations on the nature of metabolites of isoproturon formed in *P. minor* including role of cyt P_{450} inhibitors had been reported.

Resistance at the Site of Action

An alternation in the herbicide target site or by having many more copies of the target site, though very uncommon mechanisms may cause resistance at the site of action of herbicide, and thus, the vital process mediated by the target site could not be inhibited by the herbicide (Kearney and Kaufman 1969). Isoproturon, a PSII inhibitor, binds to the D1 protein of reaction center and inhibits electron transport through action as a nonreducing analog of plastoquinone Q_B (Gardner 1989). S-triazine and other urea herbicides also act as nonreducible analogs of semiquinone anion of plastoquinone.

The D1 polypeptide is also known as the 32-kDa protein or the QB protein. Responding to light, the D1 protein undergoes a continuous cycle of degradation and resynthesis ranging from 30 min to a few fours. The QB protein was first proposed by Renger (1976) to regulate electron flow from QA to QB and found to contain the DCMU-binding site. The DCMU-binding site was shown to be proteinaceous in nature which was concluded from trypsin treatment experiments. Binding studies demonstrated that various herbicides bind partly to overlapping site at the same protein.

The decrease in sensitivity for herbicides due to trypsin treatment could be correlated with a change in several polypeptides of which the 32-kDa polypeptide was the most important (Croze et al. 1979; Steinback et al. 1981). Mullet and Arntzen (1981) found that selective depletion of the 32-kDa polypeptide from photosystem II particles caused loss of triazine inhibition. The sensitivity to phenolic herbicides was not altered, whereas sensitivity to ureas showed a slight decrease. Therefore, they proposed that the Q_B protein is made up of at least two polypeptides.

Pfister et al. (1981) studied the radioactive photoaffinity labeling to determine the protein to which herbicides bind. ^{3}H-labeled azido-1-dinoseb specially labels 30–40-kDa polypeptides. He and his co-worker demonstrated that ^{14}C-azidoatrazine covalently binds to the 32–34-kDa protein.

The D1 protein is encoded by a chloroplast gene psbA (Tripathi et al. 2005) which is conserved from cyanobacteria to higher plants (Morden and Golden 1989). The location of the psbA gene encoding D1 polypeptide of PSII is also highly conserved in chloroplast genome, and in most cases, it has been shown to be present as a single uninterrupted copy in the large single copy (LSC) region near inverted repeat (IR) (Erickson et al. 1984). Exceptions to the psbA gene organization and copy number are reported in green alga *C. reinhardtii* where the psbA gene is present in the invert repeat as two copies containing four large introns ranging in size from 1.1 to 1.8 kbp are observed. In cyanobacteria, the psbA gene is present as a small multigene family of usually three copies, e.g., in Synechocystis PCC6803, but sometimes these may be four (e.g., Anabaena 7120) (Trivedi and Sane 1994).

The deduced amino acid sequence of psbA gene product was relatively found conserved and reflected hydrophobic nature of protein, while the protein consisted of 353 amino acids in plants and 360 in cyanobacteria (Erickson et al. 1985). A remarkable degree of amino acid sequence homology had been noticed across the evolution of the chloroplast (Trivedi and Sane 1994).

In all cases of target site resistance, point mutation in the chloroplast psbA gene was found, resulting in the substitution of glycine for serine 264 (Sinning et al. 1989). This modification greatly reduced affinity. As a result of this modification, there was a 1,000-fold reduction in affinity at the Q_B-binding site and 100-fold increase in triazine resistance at the whole plant level (Metz and Thiel 1989).

Several herbicide-resistant cell lines were isolated from photosynthetic cell suspensions of soybean with different levels of herbicide resistance, photosystem II activity, and chlorophyll a/b ratio. Sequencing of the psbA gene (coding for the D1 polypeptide of photosystem II) from STR7 mutant of soybean revealed a single change serine 268 to proline, in the D1 protein, which was not previously described in any photosynthetic organism (Alfonso et al. 1996). In addition to affecting atrazine resistance, this single amino acid change resulted in a decrease in the electron transfer rate between the secondary acceptors Q_A and Q_B and a stabilization of the S_2Q_B and S_3Q_3 states.

Two herbicide-resistant strains of the cyanobacteria *Spherchococcus sp.* PCC7002 were compared to the wild type with respect to DNA change which resulted in herbicide resistance. The mutations were mapped to a region of the cyanobacterial genome which encodes 1 of 3 copies of psbA gene. A point mutation at codon 211 in the psbA coding locus (TTC to TCC) resulted in an amino acid change from phenylalanine to serine in the D1 protein. This mutation conferred resistance to atrazine and diuron at 7 times and at 2 times the minimal inhibitory concentration (MIC) for the wild type, respectively. A second point mutation at codon 219 in the psbA coding locus (GTA to ATA) resulted in an amino acid change from valine to isoleucine in the D1 protein. This mutation conferred resistance to diuron and atrazine at 10 times and 2 times the MIC for the wild type, respectively (Gingrich et al. 1988).

Properties of Photosystem II Reaction Center

Purified PSIIRC preparations from spinach and pea contain five polypeptide subunits, as revealed by SDS-PAGE. The two polypeptides of about 30 kDa are named D1 and D2, respectively, because of their diffuse migration pattern on SDS-PAGE (Satoh 1983). Cytb559 is constituted of two subunits: α-subunits of about 10 kDa and β-subunits of about 4 kDa (Cramer et al. 1985). Another 4.8-kDa component was identified as a psb I gene product (Ikeuchi and Inoue 1988; Webber et al. 1989). Larger components of about 60 kDa were observed in SDS-PAGE of the PSIIRC (Nanba and Satoh 1987). Western blotting analysis identified this band to be homodimer of D1 or D2 proteins (Satoh et al. 1987) or heterodimers of D1 and D2 proteins (Marder et al. 1987).

Spinach D1 protein has been shown to be rich in alanine, glycine, leucine, and isolucione but lacks lysine. Alanine also has been reported one of the major components of the D2 protein, a psbI gene product enriched in phenylalanine (16.7 mol% in the case of spinach), but lacks alanine (Satoh 1983).

The genes for the five subunits of PSIIRC are present in chloroplast DNA (Okyama et al. 1986; Shinozaki et al. 1986). The N- and C-terminal amino acid sequences of mature proteins in isolated complex have been identified and sequenced (Michel et al. 1988; Takahashi et al. 1988, 1990). The existence of a less but significant amino acid sequence homology, as deduced from the nucleotide sequence of the chloroplast genome, has been documented to exist between the D1 and D2 proteins of the PSIIRC and the L and M subunits of the purple bacterial RC (Deisenhofer et al. 1985; Allen et al. 1987). Although the sequence homology is only about 10–20 % which is significant in the region of five putative transmembrane helices predicted on the basis of hydropathy plots (Trebst 1986; Michel and Deisenhofer 1988) and site-directed immunological analysis (Sayre et al. 1986).

D1 and D2 proteins are homologous to each other, thus originated from a single gene by a mechanism such as gene duplication in the early stage of evolution of photosynthetic systems. Interestingly, within an area of high homology, a tyrosine residue (Tyr 161) serves as the secondary donor (to P_{680}) of PSII which is responsible for ERR signal II that is situated on the D1 protein (Debus et al. 1989). Symmetrically, another tyrosine is also responsible for the dark stable EPR signal II, along with residues of the D2 protein (Debus et al. 1988a; Vermass et al. 1988). Rochaix et al. (1984) reported that significant homology also exists between D1 (psbA gene product) and D2 protein (psbD gene product) of PSII. Low-fluorescent LF1 mutant of the green alga *Scenedesmus obliquus* (Metz et al. 1980) is known to have an active PSIIRC and a normal reducing-side function but is deficient in O_2- evolution capability and bound decreased amount of Mn compared with wild-type cells (Metz et al. 1980; Metz and Bishop 1980). However, the mutant is affected in only one azidoatrazine-binding PSII core protein of about 34 kDa. Metz et al. (1980) unambiguously identified the affected protein as D1 using D1 antibodies (from *Amaranthus hybridus)* and thus showed that D1, known to bind Q_B (the secondary PSII quinine acceptor), had both a donor and acceptor side role in PSII. This evidence strongly supported the prediction that D1 was part of the PSIIRC. The binding niche of the D1 protein for the physiological electron carrier Q_B and the PSII-binding herbicides was located between the D and E helices and the parallel DE helix, which was located in the stroma. This helix involved the amino acids 211 Phe-275 Leu of the D1 protein (Svensson et al. 1991). The D2 protein, like the D1 protein, has five hydrophobic membrane-spanning helices of which the D and E helices appear to be the most important ones for its function on the electron acceptor side of PSII. A fifth polypeptide of low molecular mass has recently been identified as a component of this core reaction center, which is encoded by the chloroplast gene psb I (Webber et al. 1989). Structural homologies between the D1 and D2 polypeptides and the L and M subunits of the photosynthetic bacterial reaction center have also been reported. It is now generally accepted that D1 and D2 comprise the heart of the reaction center: a D1/D2 heterodimer appears to favor formation of a bound chlorophyll dimer, P_{680}, to bind the quinine and pheophytin electron acceptors and to contain tyrosine residues which have been identified as the Z (Debus et al. 1988b) and D (Vermaas et al. 1988) electron donor sites to P_{680}. D1 and D2 may also function in manganese binding. Two transmembrane chlorophyll-binding proteins of approximately 47 and 43 kDa are functionally associated with the PSII reaction center core, and these are present in equal stoichiometric amounts with D1 and D2.

Several other polypeptides have been co-purified with PSII under various conditions. These include three nuclear-encoded, intrinsic polypeptides of approximately 24, 10, and 22 kDa (Ljungberg et al. 1984). These may serve to anchor the luminal polypeptides to PSII and possibly modulate their function. The gene for this 10-kDa nuclear genome-encoded polypeptide has been identified as psbR. A chloroplast-encoded phosphoprotein of approximately 10-kDa gene has been identified as psb H which may alter the conformation of the D1/D2 dimer by serving a role analogous to that of the H subunit of the bacterial reaction center (Deisenhofer et al. 1985). Several low molecular weight polypeptides of cytosolic and chloroplast origin are also found to be associated with PSII (Delepelaire 1984).

PSII Genes

Genes encoding PSII polypeptides are located in one or two distinct genetic compartments, the nucleus or the chloroplast. Genes located on nuclear chromosomes are transcribed in the nucleus, their mRNAs are translated in the cytoplasm on 80S ribosome, and the protein products are imported across the chloroplast envelope and targeted for a specific suborgenellar location. Genes located on the multi-copy, circular chloroplast genomes are transcribed in the chloroplast, and their mRNAs are translated in the chloroplast on 70S ribosome (Fig. 3). A system of nomenclature proposed for the PSII chloroplast genes designated psb (photosystem b), followed by a letter of the alphabet assigned in the general order of gene identification.

PSII gene expression can be regulated at all stages that intervene between the presence of the gene on a DNA molecule in a cell and the presence of the mature, functional gene product in the cell.

psbA Gene and D1 Protein

Nomenclature

The chloroplast gene, psbA, codes for the D1 core reaction center polypeptide of PSII. The nomenclature for the D1 and D2 polypeptides is derived from the fact that these polypeptides, when labeled in vitro under conditions that block cytoplasmic translocation in *C. reinhardtii* which were then separated from other thylakoid membrane proteins by gel electrophoresis, were initially visualized on autoradiogram broad, diffuse bands 1 and 2, called D1 and D2 obtained (Chula and Gillham 1977).

Apparent molecular weights of D1 and D2 were calculated in the range of 30–34 kDa. D1 migrated more slowly than D2 in a standard SDS-PAGE system, but the relative mobility of these two polypeptides reversed in polyacrylamide gels containing urea (Delepelaire 1984). D1 and D2 also stained poorly with Coomassie brilliant blue. D1 has alternately been referred to as the rapidly synthesized 32-kDa polypeptide of

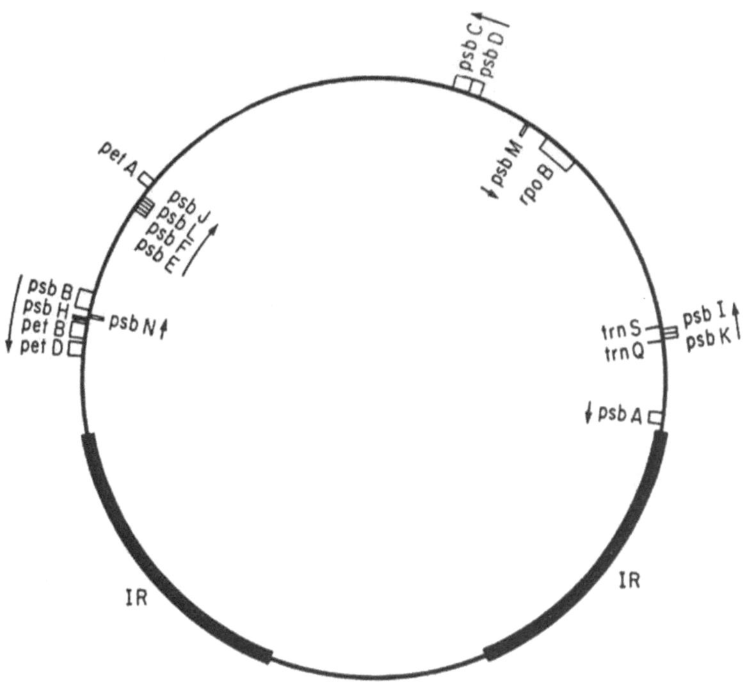

Fig. 3 Circular chloroplast genome map of tobacco (Shinozeki et al. 1986)

PSII, the herbicide-binding protein, or the Q_B protein (Hirschberg et al. 1984). D1 is found to have binding niche for both quinine and herbicides.

Nucleotide Sequence of psbA

Sequence homologous to D1 was initially localized by southern hybridization using labeled RNA from light-grown seedlings to restriction fragments from the chloroplast genome of maize and spinach (Bedbrook et al. 1978). The nucleotide sequencing of the D1 gene from spinach and *Nicotiana debneyi* was the first to be performed (Zurawski et al. 1982) and thus honored with name psbA.

Point mutations in the psbA genes of plants, algae, Euglena, and cyanobacteria resulted in a resistance to several classes of herbicides which competed with plastoquinone for binding to the D1 polypeptide and hence blocked electron transfer between Q_A and Q_B. DNA sequence analysis of psbA isolated from herbicide-resistant mutants had shown that an alteration in psbA codons produced an herbicide-resistant phenotype. The amino acid sequence encoded by psbA gene predicts a chloroplast translation product of 353 amino acid residues (352 in *Chlamydomonas* and 345 in *Euglena*) was identified with a predicated Mr. of 38,950 D. This was larger than the apparent molecular weight of 32,000. Posttranslational processing of D1 had been shown to take place at the C-terminus (Marder et al. 1984), and the amino acid sequence of the ten C-terminal residues of spinach D1 showed that mature D1 polypeptide's (Takahashi et al. 1988) last nine residues form the precursor D1 polypeptide.

D1 Is Highly Conserved

Deduced D1 polypeptide sequence is highly conserved among all species for which psbA is becoming available. D1 polypeptides are found 93 % homologous with the *Chlamydomonas* D1, 90 % homologous with *Cyanophora* D1, and 84–87 % homologous with the *Euglena* and cyanobacterial D1 polypeptides. It is noticed that the tyrosine residue 161 in the core PSII reaction center with the quinine (Q_B) binding site is responsible for the Z^+ EPR signal and binding sites for chlorophyll special pair, pheophytin, and possibly manganese. Hydropathy analysis predicted five membrane-spanning helices (Trebst 1987). The N-terminus appeared to be in the stroma, while the C-terminus is probably in the lumen where it most likely interacts with the OEE polypeptides (Sayre et al. 1986).

D1 Is a Photogene

A 32–34-kDa protein (D1) is observed as one of the most prevalent products of light-driven protein synthesis in chloroplasts isolated from pea (Eaglesham and Ellis 1974) and spinach (Bottomley et al. 1974). It is also noticed that light appeared to affect the mRNA levels for this protein, since translatable message for the 32-kDa polypeptide could be isolated from light-grown but not from dark-grown maize seedlings (Bedbrook et al. 1978). These and subsequent studies with spinach, mustard (Link and Langridge 1984), spirodela (Reisfeld et al. 1982) pea, and mung bean lead to the classification of psbA as one of several plastid "photogenes." Evidence for transcriptional regulation of psbA expression was obtained from the experiment from "run-on" assays in which transcripts initiated in vivo could be elongated in lysed plastid extracts in vitro. Transcriptional activity of psbA declined steadily with continued illumination, but at 72 h, it still exceeded transcription of psbA in dark-grown seedling by threefold. psbA gene expressed with own promoter and did not appear to be co-transcribed with any other PSII genes. In spinach, Northern blot analysis and 3′ S1 protection studies of spinach RNA suggested that psbA and trnH, located downstream, may be co-transcribed (Zurawski et al. 1984).

A single major transcript is seen in pea. In *C. reinhardtii* (Erickson et al. 1984) and Euglena, psbA contains four introns, group II introns in Euglena, and group I introns for Chlamydomonas which are spliced from the precursor mRNA. None of the other psbA genes characterized to date contain introns.

Diurnal rhythms have been shown to affect the steady-state levels of several plastid mRNA, including psbA. Piechula and Gruissem (1987) clearly showed that psbA mRNA declines steadily during the day between 8 a.m. and 4 p.m. to reach the minimum level (50 % of maximum), then goes to a maximum at 8 p.m., and maintained this level until the following morning when levels again started dropping to a low by 4 p.m. Similar diurnal differences were seen during the developmental cycle from day 7 to day 35, although overall levels of psbA mRNA were maximal at 15 days, 2–3 times that seen at day 7 or day 35.

Translation of psbA Message Is Light Dependent

Although psbA mRNA used to be present in dark-grown plastids as well as in developing plastid and chloroplasts, synthesis of D1 polypeptide is tightly regulated by light. The D1 synthesis that is undetectable in dark-grown barley seedlings can be seen 15 min after transfer of seedlings to light. The psbA mRNA levels increased after transfer of seedlings to the light, and this increase was not observed until at least 16 h of illumination and could account for the rapid induction of D1 synthesis (Klein and Mullet 1990).

Herbicide-Resistant Mutants

Herbicides such as s-triazines, substituted ureas, and phenolic derivatives block photosynthesis on the reducing side of PSII (Trebst 1980), compete with quinines for binding to the D1 protein, and thereby block electron transport (Steinback et al. 1981). The role of D1 protein in determining resistance to several PSII-specific herbicides was revealed by a comparative analysis of psbA from herbicide-resistant and herbicide-sensitive algae and plants. Resistance to the herbicide DCMU (3–3, 4-dichlorophenyl-dimethyl urea) and atrazine was correlated in all these cases with single amino acid substitution at Ser 264 which was changed to Ala in *C. reinhardtii* (Erickson et al. 1984) and Euglena and glycine in vascular plants (Bettini et al. 1987).

Analysis of other mutants of *C. reinhardtii* with distinctive patterns of resistance to atrazine and other herbicides has revealed five additional and distinct changes on the D1 protein localized in the Q_B-binding niche in and between helix IV and helix V. These changes involve Val219-Tle, Phe 255-Tyr, Gly256-Asp, His275-Phe, and Ala251-Val (Erickson et al. 1985). psbA mutation sites affecting D1 residues 219, 215, 255, and 264 have also been characterized from cyanobacteria (Hirschberg et al. 1984). Two additional amino acid residues of D1 altered in herbicide-resistant mutants of cyanobacteria were identified as Phe211-Ser (Gingrich et al. 1988) and Asn^{266}-Thr or Asn^{266}-Asp (Ajlani et al. 1989).

Large numbers of herbicide-resistant mutants of various plants, cyanobacteria, and green algae have been analyzed. The analyses of the sites of these mutants also helped to define the Q_B-binding pocket in PSII and the bacterial reaction center. All of these mutations localized in (i) the C-terminal end of the putative helix D of D1, (ii) the N-terminal end of the putative helix E of D1, or (iii) the exposed region connecting the D and E helices. The majority of the herbicide-resistant mutations identified to date are found to be located between helices D and E. They include Ala251D_1-Val, Phe 255D1-Tyr, Gly 256D1-Asp, and Ser264D1-Ala (Gly/Asn/Thr; Asn266D1-Thr/Asp). Among these, the Ser264D1 residue was analogous to Ser223 residue in the bacterial reaction center. Thus, it seems that all of the mutations resulting in herbicide resistance might be clustered around the putative non-heme iron and Q_B-binding site.

Development of Herbicide Resistance

Development of herbicide resistance in weeds is an evolutionary process. There are two prerequisites for the evolution of herbicide resistance in plant populations: the occurrence of heritable variation in genetic composition for herbicide resistance and natural selection for increased resistance to herbicides. In response to repeated treatment with particular herbicide or class of herbicides, weed populations change in genetic composition such that the frequency of resistant alleles and resistant individuals increases. The selection pressure for herbicide resistance contributes by three factors: efficiency of the herbicide, frequency of herbicide use, and duration of herbicide effect. In general, gene mutations conferring resistance to a herbicide class are not induced by application of the herbicide but rather occur spontaneously. Spontaneous mutations at gene loci occur with characteristic frequency such that new mutations are continuously generated in natural populations of weeds. Mutations at some loci, particularly that encoding herbicide site of action, may confer resistance. Typical spontaneous mutation rates in biological organisms are often cited as 1×10^5 or 1×10^6 gametes per locus per generation.

There are two modes of inheritance of herbicide resistance: nuclear inheritance and cytoplasmic inheritances. In nuclear inheritance, the resistance alleles are transmitted through pollen and ovules. In majority of cases, resistance is controlled by a single, major gene. Cytoplasmic inheritance of resistance occurs with triazine herbicides in several weed species. The gene conferring triazine resistance is located in the chloroplast genome (Hirschberg and McIntosh 1983). Transmission of the chloroplast-resistant gene mostly occurs by pollen, the paternal parent. The mutation that confers maternally inherited triazine resistance involves a single-base substitution in the psbA chloroplast gene which codes for photosystem II membrane protein to which triazine herbicide binds.

Strategies for Avoidance of Herbicide Resistance

Herbicide Rotation

As a remedy control of isoproturon-resistant *P. minor* with fenoxaprop-ethyl in limited pot studies (Singh et al. 1995; Kirkwood et al. 1997) had been confirmed which could be extended in field conditions. Some variations in the activity of fenoxaprop reported by Malik and Yadav (1997) were observed due to its application at various stages at few locations. Montazeri (1993) and Mirkamali (1987) also found fenoxaprop to provide good control of *P. minor* and wild oat along with tralkoxydim and diclofop-methyl. Clodinafop applied at 48 g/ha was less effective, and a higher application rate was required for satisfactory control of *P. minor* and other grass weeds, whereas a mixture of clodinafop (18 g/ha) with mineral oil provided satisfactory control of *P. minor* when applied at the two leaf to second tiller stage.

Contrary to expectation, chlorotoluron (with the same mode of action as isoproturon) was found to provide excellent control of the resistant and susceptible biotypes of *P. minor* prior to evolution of isoproturon resistance and was found to be equally effective to that of isoproturon.

Improved control of *A. ludoviciana*, *P. minor*, and some broadleaf weeds was achieved by tralkoxydim + isoproturon (Singh et al. 1992a). Chlorotoluron can be substituted for isoproturon where resistance to the latter has been observed. Other promising herbicides which provided effective control of the resistant *P. minor* include terbutryne, metachlor, propachlor, trifluralin, atrazine, and pendimethalin (Singh et al. 1995a; Kirkwood et al. 1997).

Crop Rotation

Rotation of crops integrates new agronomic practices and more competitive crops to help suppress the weed flora. Resistance to isoproturon in *P. minor* was observed in 67 % of fields under rice-wheat rotations compared to 8 %, 9 %, and 16 % when rice-berseem, sunflower wheat, sugarcane, vegetable wheat, cotton, and pigeon pea, respectively, were rotated (Malik and Singh 1995). Sugarcane is one of the important crops to break the dominance of *P. minor*, not only because of its smothering effect during the later stage of growth but also due to rotation of herbicides, atrazine, and simazine to which resistant *P. minor* is sensitive (Singh et al. 1995; Yaduraju and Ahuja 1995; Kirkwood et al. 1997).

Cultivation of oilseed crops with large biomass suppresses the growth and seed production of *P. minor*. Pulse crops are not good competitors with *P. minor*, but the use of an alternative herbicide helps in checking *P. minor* growth and infestation.

Economic Impact of Losses

Weeds not only reduce the yield and quality of crops but also utilize essential nutrients and moisture. In India, annual losses inflicted by pests including weeds are estimated to be US$ 175 million, out of which weeds alone contribute a loss of 40 % (Singh and Malik 1992). Grain yield of wheat, averaged from 11 locations in India, ranged between 3,119 and 4,367 kg/ha under unweeded and herbicide-treated plots, respectively. Nonavailability of manual labors for weeding at critical time for weed removal and poor efficiency with grass weeds are other constraints in achieving optimum yield targets. The increased yield obtained in herbicide-treated fields was a function of 59 %, 50 %, and 46 % higher uptake of N, P, and K/ha by wheat compared with unweeded conditions. On average, weeds utilized 40, 7, and 35 kg/ha of N, P, and K, respectively, costing $ 8/ha; application of herbicides increased nitrogen efficiency of wheat from 50 % to 90 % (Singh and Malik 1992).

Singh and Singh (1996) found that the sowing method in wheat influences the financial returns. Among weed control methods, isoproturon + 2, 4-D treatment provided the highest cost-benefit ratio compared with two hand weedings and weedy check plots. Isoproturon + 2, 4-D gave the highest gross margin followed by isoproturon and pendimethalin. Herbicides are thus the most effective tools in managing the weeds and harvesting a good crop.

Future Aspects of Photosystem II Herbicides

PSII herbicides are not being sought for at the moment, and very few commercial products have been introduced to the market during the last decade. One reason may be that the dosage necessary for PSII herbicide to control weeds in general is much higher than the dosage for herbicides which interfere with carotenoid or amino acid biosynthesis. However, there may be a renascence of PSII herbicides for three reasons: (i) the growing interest for naturally occurring substances as biocides (ii), the design of new herbicides by molecular modeling, and (iii) crops with mutants of D1 proteins conferring herbicide resistance.

Another aspect of a renascence of PSII herbicides is the possibility of molecular modeling. Sequencing of new mutants allows us to identify additional amino acids in the herbicide-binding pocket. New mutants can be generated by site-directed mutagenesis in algae and cyanobacteria. In addition, the photoaffinity labeling technique in conjugation with sequencing can identify additional amino acids. The ultimate goal would be to obtain crystals of PSII, which would allow an X-ray structure to be determined.

In Canada, atrazine-resistant oilseed rape is used as a crop and is obtained by a sexual cross with the atrazine-resistant *Brassica campestris*. Construction of herbicide-resistant crop plants would lead to a higher demand of PSII herbicides. However, the sexual cross, as in oil rapeseed, is possible with a few crop plants only. The most universal approach would be to engineer a psbA gene, which codes for a resistant D1 protein. A modified psbA gene has already been successfully introduced into *Chlamydomonas reinhardtii*, and thus, new herbicide-resistant crops seem to be feasible in the future.

References

Afentouli CG, Eleftherohorinos IG (1996) Little seed canary grass (*Phalaris minor*) and short spiked canary grass (*Phalaris brachystachys*) interference in wheat and barley. Weed Sci 44:560–565

Ajlani G, Meyer I, Vernotte C, Astier C (1989) Mutation in phenol-type herbicide resistance maps within the psbA gene in *Synechocystis* 6714. FEBS Lett 246:207–210

Alfonso M, Pulyo JJ, Gaddour K, Etienne AL, Kirilovsky D, Picorel R (1996) Induced new mutation of D1 serine-268 in soybean photosynthetic cell cultures produced atrazine resistance, increased stability of $S2Q_B$ and $S3Q_B$-states, and increased sensitivity to light stress. Plant Physiol 112:1499–1508

Allen JP, Feher G, Yeates TR, Komiya H, Rees DC (1987) Structure of the reaction center from *Rhodobacter sphaeroides* R-26. The cofactors. Proc Natl Acad Sci U S A 84:5730–5734

Anderson JM, Anderson B (1988) The dynamic photosynthetic membrane and regulation of solar energy conversion. Trends Biochem Sci 13:351–355

Andreasson LE, Vaungard T (1988) Electron transport in photosystems I and II. Annu Rev Plant Physiol 39:379–411

Andreaws WA (1963) Herbicides. In: Hochster RM, Questel JH (eds) Metabolic inhibitors. Academic Press, New York and London pp 243–261

Anonymous (1990) Final technical report, all India coordinated research programme on weed control. Haryana Agricultural University, Hisar, 180 pp

Archireddy NR, Kirkwood RC (1986) The foliar uptake of isoproturon by wheat and *Alopecurus myosuroides* huds. Pestic Sci 17:53–56

Balyan RS, Malik RK, Panwar RS, Bhan VM (1987) Evaluation of herbicides for weed control in wheat. Haryana Agric Univ J Res 17:170–173

Barry P, Young AJ, Britton G (1990) Photodestruction of pigments in higher plants by herbicides action. I. The effect of DCMU (diuron) on isolated chloroplasts. J Exp Bot 91:123–127

Bedbrook JR, Link G, Coen D, Bogard L, Rich A (1978) Maize plastid gene expressed during photoregulated development. Proc Natl Acad Sci 75:3060–3064

Bettini PS, McNally S, Sevignac M, Darmency H, Gasquez J, Dron M (1987) Atrazine resistance in *Chenopodium album*. Plant Physiol 84:1442–1446

Bhan VM, Chaudhary DB (1976) Germination, growth and reproductive behaviour of *Phalaris minor* Retz. as affected by date of planting. Indian J Weed Sci 8:126–130

Bhan VM, Gupta VK, Malik RK (1985) Effect of isoproturon at different stages of wheat and associated weeds. In: Indian Society of weed science annual conference, Anand, Gujarat, Abstract, 42 pp

Bir SS, Sidhu M (1979) Observation in the weed flora of cultivable lands in Punjab-wheat fields in Patiala district. New Bot 6:79–89

Bottomley W, Spencer D, Whitfeld PR (1974) Protein synthesis in isolated spinach chloroplasts: comparison of light-driven and ATP-driven synthesis. Arch Biochem Biophys 164:106

Cabanne F, Gaillardon P, Scalla R (1985) Phytotoxicity and metabolism of chlorotoluron in two wheat varieties. Pestic Biochem Physiol 23:212–215

Cabanne F, Hubby D, Gaillardon P, Scalla R, Durst F (1987) Effect of cytochrome P_{450} inhibitor 1-aminobenzotriazole on metabolisms of chlorotoluron and isoproturon in wheat. Pestic Biochem Physiol 28:371–375

Cabanne F, Huby D, Gaillardon P, Scalla R, Durst F (1989) Effect of the cytochrome P_{450} inactivator ABT on the metabolisms of chlorotoluron and isoproturon in wheat. Pestic Biochem Physiol 23:212–215

Chula NH, Gillham NW (1977) The sites of synthesis of the principal thylakoid membrane polypeptides in *Chlamydomonas reinhardtii*. J Cell Biol 74:441

Cramer WA, Widger WR, Herrmann RG, Trebst A (1985) Topography and function of thylakoid membrane proteins. Trends Biochem Sci 10:125–129

Croze E, Kelly M, Horton P (1979) Loss of sensitivity to diuron after trypsin digestion of chloroplast photosystem II particles. FEBS Lett 103:22–26

Cudney DW, Hill JE (1979) The response of wheat grown with three population levels of canary grass to various herbicide treatments. Proc West Weed Sci Soc (USA) 32:55–56

De Felipe MR, Golvanao MP, Lucas MM, Lang P, Pozuele JM (1989) Differential effects of isoproturon on the photosynthetic apparatus and yield of two varieties of wheat and *L. rigidum*. Weed Res 28 (2):85–89

de Prado R, Menendez J, Tena M, Caseley JC, Taberner A (1991) Response of substituted ureas, triazines and chloro acetanilides in a biotype of *Alopecurus myosuroides* resistant to chlorotoluron. Proc Brighton Crop Prot Conf Weeds 3:1065–1070

Debus R, Barry BA, Sithole I, Babcock GT, Mclnotsh L (1988a) Directed mutagenesis indicates that the donor to $P680^+$ in photosystem II in tyrosine-161 of the D1 polypeptide. Biochemistry 27:9071–9074

Debus RJ, Barry BA, Babcock GT, McIntosh L (1988b) Site-directed mutagenesis identifies a tyrosine radical involved in the photosynthetic oxygen-evolving system. Proc Natl Acad Sci U S A 85:427–430

Debus RJ, Barry BA, Sithole I, Babcock GT, McIntosh L (1989) Directed mutagenesis indicates that the donor to P^+680 in photosystem II is Tyr-161 of the D1 polypeptide. Biochemistry 26:9071–9074

Deisenhofer J, Epp O, Miki K, Huber R, Michel H (1985) Structure of the protein subunits in the photosynthetic reaction centre of *Rhodopseudomonas viridis* at 3A resolution. Nature (London) 318:618–624

Delepelaire P (1984) Partial characterization of the biosynthesis and integration of the photosystem II reaction centers in the thylakoid membrane of *Chlamydomonas reinhardtii*. EMBO J 3:701

Devine MD (1997) Mechanisms of resistance to acetyle coenzyme a carboxylase inhibitory – a review. Pestic Sci 51(3):259–264

Dhaliwal BK, Walia US, Brar LS (1998) Response of *Phalaris minor* Retz biotypes to various herbicides. Indian J Weed Sci 30(3–4):116–120

Dhawan RS, Malik RK (1999) Effect of isoproturon on leaf membrane permeability in some biotypes of *Phalaris minor* Retz. Indian J Plant Physiol 4 (2):125–126

Dixit A, Bhan VM (1998) Influence of isoproturon application for controlling *P. minor* plant biotypes. Weed News 2(1–2):33–35

Donaldson RP, Luster DG (1991) Multiple forms of plant cytochrome P_{450}. Plant Physiol 96:669–674

Eaglesham ARJ, Ellis RJ (1974) Protein synthesis in chloroplasts. II. Light driven synthesis of membrane proteins by isolated pea chloroplasts. Biochim Biophys Acta 335:396–402

Erickson JM, Rahire M, Rachaix JD (1984) *Chlamydomonas reinhardtii* gene for the Mr 32,000 protein of photosystem II contains four large introns and is located entirely within the chloroplast inverted repeat. EMBO J 3:2753–2756

Erickson JM, Rahire M, Rochaix JD, Mets L (1985) Herbicide resistance and cross-resistance: changes at three distinct sites in the herbicide-binding protein. Science 228:204–207

Fedtke C (1982) Biochemistry and physiology of herbicide action. Springer, Berlin, p 145

Feng PC, Rao SR, Schafer DE (1995) Inhibition of thiazopyro metabolism in plant seedlings by inhibitors of monoxygenases. Pestic Sci 45:203–206

Fuerst EP, Norman MA (1991) Interaction of herbicides with photosynthetic electron transport. Weed Sci 39:458–464

Gaillardon P, Cabanne F, Scalla R, Durst F (1985) Effect of mixed function oxidase inhibitors on the toxicity of chlorotoluron and isoproturon to wheat. Weed Res 25:397–402

Gardner G (1989) A stereochemical model for the active site of photosystem II herbicides. Photochem Photobiol 49:331–336

Giessbuhler H, Martin H, Voss G (1975) The substituted ureas. In: Kearney PC, Kaufmann DD (eds) Herbicides, chemistry, degradation and mode of action. Dekker, New York, pp 203–291

Gingrich JC, Buzby JS, Stirwalt VL, Brayant DA (1988) Genetic analysis of two new mutations resulting in herbicide resistance in the *Cyanobacterium synechococcus* sp. PCC7002. Photosynth Res 16:1–2

Godinho I, Costa JCA (1981) Competition of Phalaris minor in wheat. In: 1st congress Portuguese de Fitiatria e de Fitofarmacologia e III. Oeiras: Portugal, 3:237–243

Haehnel W (1984) Photosynthetic electron transport in higher plants. Annu Rev Plant Physiol 35:659–693

Hatzios KK (1991) Biotransformation of herbicides in higher plants. In: Grover R, Cessna AJ (eds) Environmental chemistry of herbicides, vol 12. CRC Press, Boston, pp 141–185

Hirschberg J, Bleecker A, Kyle DJ, McIntosh L, Arntzen CJ (1984) The molecular basis of triazine-herbicide resistance in higher-plant chloroplasts. Z. Naturforsch. 39c 412–420

Hirschberg J, McIntosh L (1983) Molecular basis of herbicide resistance in *Amaranthus hybridus*. Science 222:1346–1349

Ikeuchi M, Inoue Y (1988) A new photosystem II reaction centre component (4.8 kDa) encoded by chloroplast genome. FEBS Lett 241:99–104

Jimenez-Hidalgo MJ, Saavedra M, Garcia-Torres L (1993) Germination of *Phalaris* species as affected by temperature and light. In: Proceeding congress of the Spanish weed science society, Lugo, 1–3 Dec 1993, pp 98–102

Johri AK, Singh G, Sharma D (1992) Nutrient uptake by wheat and associated weeds as influenced by management practice. Trop Agric 69:391–393

Kearney PC, Kaufman DD 1969 The substituted ureas. In: Degradation of herbicides. Marcel Dekker, New York, p 79

Kemp MS, Moss SR, Thomas TH (1990) Herbicide resistance in *Alopecurus myosuroides*. In: Green MB, LeBaron HM, Moberg WK (eds) Managing research to agrochemicals from fundamental research to practical strategies. American Chemical Society, Washington, DC, pp 376–393

Khera KL, Sandhu BS, Aujla TS, Singh CB, Kumar K (1995) Performance of wheat (*Triticum aestivum*) in relation to small canarygrass (*Phalaris minor*) under different levels of irrigation, nitrogen and weed population. Indian J Agric Sci 65:717–722

Kirkwood RC, Singh S, Marshall G (1997) Mechanisms of isoproturon resistance in *Phalaris minor*, implications and control measures. In: Proceeding 16th Asian Pacific weed science society conference integrated weed management towards sustainable agriculture, Kuala Lumpur, 8–12 Sept, pp 204–207

Klein RR, Mullet JE (1990) Control of gene expression during higher plant chloroplast biogenesis. J Biol Chem 262:4341–4348

Kolar JS, Mehra SP (1992) Changing scenario of weed flora in agroecosystem of Punjab. In: Dhaliwal GS, Hansra BS, Jerath N (eds) Changing scenario of our environment. Punjab Agricultural University Ludhiana, India, pp 252–262

Kulshrestha G, Singh SB, Yaduraju NT (1999) Proceeding of an international conference, Brighton, 15–18 Nov 1999, 1, pp 167–172

Lagoutte B, Mathis P (1989) The photosystem I reaction center: structure and photochemistry. Photochem Photobiol 49:833–844

Leah JM, Caseley JC, Riches CR, Valverde B (1997) Effect of monoxygenase inhibitors on uptake, metabolism and phytotoxicity of propanil in resistant biotype of jungle rice, *Echinochloa colona*. Pestic Sci 49:141–145

Link G, Langridge U (1984) Structure of chloroplast gene for the precursor of the Mr 32,000 photosystem II protein from mustard. Nucleic Acids Res 10:593–596

Ljungberg U, Akerlund HE, Andersson B (1984) The release of a 10 kDa polypeptide from everted photosystem II thylakoid membranes by alkaline Tris. FEBS Lett 175:255–257

Malik RK, Singh S (1993) Evolving strategies for herbicide use in wheat: resistance and integrated weed management. In: Proceeding Indian society of weed science international symposium on integrated weed management for sustainable agriculture, 18–20 Nov 1993, Hisar, 1, pp 225–238

Malik RK, Singh S (1995) Little seed canary grass (*Phalaris minor*) resistance to isoproturon in India. Weed Technol 9:419–425

Malik RK, Yadav A (1997) Potency of alternate herbicides against isoproturon-resistant littleseed canarygrass. In: Proceeding 16th Asian-Pacific weed science society conference integrated weed management towards sustainable agriculture, Kuala Lumpur, 8–12 Sept, pp 208–210

Malik RK, Panwar RS, Katiyal SK, Bhan VM (1985) Weed seed distribution in wheat – a case study. Haryana J Agron 1:17–20

Malik RK, Yadav A, Singh S, Malik YP, Tyagi BS (1998) Development of resistance to herbicides in *Phalaris minor* and mapping of variations in weed flora. Wheat: research needs beyond 2000 AD. In: Proceedings of an international conference, Karnal, 12–14 Aug 1997, pp 291–298

Marder JB, Golobinoff P, Edelman M (1984) Molecular architecture of the rapidly processing of a chloroplast membrane polypeptide. J Biol Chem 259:3900

Marder JB, Chapman DJ, Telfer A, Nixon PJ, Barber J (1987) Identification of *psbA* and *psbD* gene products, D1 and D2, as reaction centre proteins of photosystem 2. Plant Mol Biol 9:325–333

Mehra SP, Gill HS (1988) Effect of temperature on germination of *Phalaris minor* Retz. and its competition in wheat. J Res Punjab Agric Univ 25:529–534

Metz J, Bishop NI (1980) Identification of a chloroplast membrane polypeptide associated with the oxidizing side of photosystem II by the use of select low fluorescent mutants of *Scenedesmus*. Biochem Biophys Res Commun 94:560–566

Metz L, Thiel A (1989) Biochemistry and genetic control of the photosystem II herbicide target site. In: Boger P, Sandmann G (eds) Target sites of herbicide action. CRC Press, Boca Raton, pp 1–24

Metz JG, Wong J, Bishop NI (1980) Changes in electrophoretic mobility of a chloroplast membrane polypeptide associated with the loss of the oxidizing side of photosystem II in low fluorescent mutants of *Scenedesmus*. FEBS Lett 114:61–66

Michel H, Deisenhofer J (1988) Relevance of the photosynthetic reaction centre from purple bacteria to the structure of photosystem II. Biochemistry 27:1–7

Michel H, Epp O, Deisenhofer J (1986) Pigment-protein interactions in the photosynthetic reaction center from *Rhodopseudomonas viridis*. EMBO J 5:2445–2451

Michel H, Hunt DF, Shabanowitz J, Bennett J (1988) Tandem mass spectrometry revels that three photosystem II proteins of spinach chloroplasts contain *N-acetyl-O*-phosphothreonine at their NH_2 termini. J Biol Chem 263:1123–1130

Mirkamali H (1987) Control of *Phalaris brachystachys* and *Phalaris minor* in wheat grown in northern Iran. Proc Br Crop Prot Conf Weeds 2:407–412

Montazeri M (1993) Evolution of a few wheat selective grass herbicides at two times of application on canarygrass (*Phalaris minor*) and wild oat (*Avena sterilis*). In: Proceeding Indian society of weed science international Symposium on integrated weed management for sustainable agriculture, Hisar, 18–20 Nov, 2, pp 166–167

Moreland DE (1980) Mechanisms of action of herbicides. Ann Rev Plant Physiol 31:597–638.

Morden CW, Golden SS (1989) psbA genes indicate common ancestry of prochlorophytes and chloroplasts. Nature 337:382–385

Mullet JE, Arntzen CJ (1981) Identification of a 32–34 kilodalton polypeptide as a herbicide receptor protein in photosystem II. Biochim Biophys Acta 635:236–248

Nanba O, Satoh K (1987) Isolation of a photosystem II reaction center consisting of D-1 and D-2 polypeptides and cytochrome b-559. Proc Natl Acad Sci U S A 84:109–112

Okereka OU, Blair AM, Caseley JC (1981) Effects of depth of planting, temperature and soil moisture on seed emergence and survival in soil of *Bromus sterilis* and *Phalaris minor*. In: Proceeding conference on grass-weeds in cereals in the UK, Reading, pp 41–46

Okyama K, Fukuzawa H, Kohchi T, Shirai H, Sana T, Sano S, Umesono K, Shiki Y, Takeuchi M, Chang Z, Aota S, Inokuchi H, Ozeki H (1986) Chloroplast gene organization deduced from complete sequence of liverwort *Marchantia polymorpha* chloroplast DNA. Nature (London) 322:572–574

Pallett KE, Dodge AD (1980) Studies into the action of some photosynthetic inhibitor herbicides. J Expt Bot 31:1051–1055

Pandey J, Sharma R, Singh RP, Chander S (1996) Effect of time and yield of wheat (*Triticum aestivum*). Indian J Agron 41(4):573–576

Panwar RS, Malik RK, Balyan RS, Singh DP (1995) Effect of isoproturon, sowing method and seed rate on weed and yield of wheat *(Triticum aestivum)*. Indian J Agric Sci 65:109–111

Parashar V, Singh OS (1985) Mechanism of anoxia induced secondary dormancy in canarygrass (*Phalaris minor* Retz.) and wild oat (*Avena fatua* L.). Seed Res 13:91–97

Parrett KG, Mehari T, Warren PG, Golbeck JH (1989) Purification and properties of the intact P-700 and Fx-containing photosystem I core protein. Biochim Biophys Acta 973:324–332

Pfister K, Arntzen CJ (1979) The mode of action of photosystem II specific inhibitors in herbicide-resistant weed biotypes. Z Naturforsch 34c:996–1009

Pfistar K, Radosevich SR, Gardeney G, Arntzen CJ (1981) Photoaffinity labeling of a herbicides receptor protein in chloroplast membranes. Proc Natl Acad Sci U S A 78:981–986

Pfister K, Steinback KE, Arntzen CJ (1981) Photoaffinity labelling of an herbicide receptor protein in chloroplast membranes. Proc Natl Acad Sci 78:981

Piechula B, Gruissem J 1987 Diurnal mRNA fluctuations of nuclear and plasmid genes in developing tomato fruits. EMBO J 6:3593–3599

Polos E, Laskay G, Szigeti Z, Pataki S, Lehoczki E (1987) Photosynthetic properties and cross-resistance to some urea herbicides of trizine-resistant *Conyza Canadensis Cronq. (L.)*. Z Naturforsch 42c:783–793

Prakash A, Shara KC, Mishra RD (1996) Control of *Phalaris minor* and wild oats in wheat by cultural and chemical methods. Indian J Agron 31:411–413

Preston C (1994) Resistance of PS I disrupting herbicides. In: Powels SB, Holms JAM (eds) Herbicide resistance in plants. Lewis Publishers, Boca Raton, pp 61–82

Ray B, Seth AK, Tyagi SD, Jindal DL (1982) Effect of gramoxone (paraquat) for pre-plant reduction of canary grass (*Phalaris minor* Retz.) in wheat in stale seed technique. In: Indian society of weed science annual conference, Hisar, Haryana, Abstract (cf. Weed Abstract 1984: 2460)

Reilly P, Nelson N (1988) Photosystem I complex. Photosynth Res 19:73–84

Reisfeld A, Mattoo AK, Edelman M (1982) Processing of chloroplast-translated membrane protein *in vitro*. Analysis of the rapidly synthesized 32,000 Da shield protein and its precursor in. Spirodela oligorrhiza. Eur J Biochem 124: 125

Renger G (1976) Studies on the structural and functional organization of system II of photosynthesis; the use of trypsin as a structurally selective inhibitor at the outer surface of the thylakoid membrane. Biochem Biophys Acata 440:287–300

Riviere JL, Cabanne F (1987) Animal and plant cytochrome P_{450} systems. Biochimie 69:669–674

Rochaix JD, Dron M, Rahire M, Malnoe P (1984) Sequence homology between the 32 Kdalton and the D2 chloroplast membrane polypeptide of *Chlamydomonas reinhardtii.* Plant Mol Biol 3:363–370

Satoh K (1983) Photosystem II reaction center complex purified from higher plants. In: Inoue Y, Crofts AR, Govindjee NM, Reger G, Satoh K (eds) The oxygen evolving system of photosynthesis. Academic Press, Tokyo, pp 27–38

Satoh K, Fujii Y, Aoshima T, Tado T (1987) Immunological identification of the polypeptide bands in the SDS-polyacrylamide gel electrophoresis of photosystem II preparations. FEBS Lett 216:7–10

Sayre RT, Anderson B, Bogorad L (1986) The topology of a membrane protein: the orientation of the 32 kd Qb-binding chloroplast thylakoid membrane protein. Cell 37:601–608

Sharma R, Pandey J (1997) Effect of various formulations of isoproturon on weeds and yield of wheat. Ann Agric Res 18(2):246–247

Sharma KK, Verma SP, Singh CM (1985) Cultural and chemical manipulations for weed management in wheat with reference to grassy weeds. Trop Pest Manage 31:133–138

Shigematsu Y, Sato F, Yamada Y (1989) A binding model for phenylurea herbicides based on analysis of a Thr264 mutation in the D-1 protein of tobacco. Pestic Biochem Physiol 35:33–41

Shinozaki K, Ohme M, Tanaka M, Wakasugi T, Hayashida N, Matsubayashi T, Zaita N, Chunwongse J, Obokata J, Yamaguchi-Schinozaki K, Ohto C, Tarazawa K, Meng BY, Sugita M, Deno H, Kamogashira T, Yamada K, Kusuda J, Takaiwa F, Katoh A, Tohdoh N, Shimada H, Sugiura M (1986) The complete nucleotide sequence of the tobacco chloroplast genome: its gene organization and expression. EMBO J 5:2043–2049

Singh S (1998) Studies on the mechanism of isoproturon resistance in *Phalaris minor* (littleseed canarygrass). Ph.D. thesis University of Strathclyde, Glasgow

Singh D, Dhaliwal HS (1984) Control of *Phalaris minor* Retz. and broad-leaved and weeds in wheat with selective herbicides. Pesticides 18:45–47

Singh RD, Ghosh AK (1982) Soil profile distribution and effect of temperature and soil depth on germination of *Phalaris minor* Retz. In: Indian society of weed science annual conference, Hisar, Abstract, pp 41–42

Singh S, Malik RK (1992) Weed management and fertilizer utilization. Fertil News 37:59–63

Singh S, Malik RK (1994) Effect of application methods of chlorotoluron on the control of *Phalaris minor* in wheat (*Triticum aestivum*). Indian J Agron 39:23–26

Singh G, Singh OP (1996) Response of late sown wheat (*Triticum aestivum*) to seeding methods and weed control measures in flood prones areas. Indian J Agron 41:237–242

Singh G, Singh D, Singh NK (1985) Effect of dates of sowing of wheat on *Phalaris minor* and wheat yield. In: Indian society of weed science. Annual conference, Anand, Gujarat, Abstract, pp 42

Singh S, Kirkwood RC, Marshall G (1995a) Evaluation of isoproturon resistant littleseed canarygrass (*Phalaris minor*) to range of graminicides. In: Annual meeting weed science, Society of America, Seattle, Abstract 162, 30 Jan–2 Feb, p 54

Singh S, Malik RK, Balyan RS, Singh S (1995b) Distribution of weed flora of wheat in Haryana. Indian J Weed Sci 27:114–121

Singh S, Kirkwood RC, Marshall G (1997a) Brighton crop protection conference: weeds. In: Proceedings of an international conference, Brighton, 17–20 Nov 1997, 1, pp 357–362

Singh S, Kirkwood RC, Marshall G (1997b) Effects of isoproturon on photosynthesis in susceptible and resistant biotypes of *Phalaris minor* and wheat. Weed Res 37:315–324

Singh S, Kirkwood RC, Marshall G (1998a) Effect of monooxygenase inhibitor piperonyl butoxide on the herbicidal activity and metabolism of isoproturon in herbicide resistant and susceptible biotypes of *Phalaris minor* and wheat *Pestic*. Science 53 (2):123–132

Singh S, Kirkwood RC, Marshall G (1998b) Effect of the monooxygenase inhibitor piperonyl butoxide on the herbicidal activity and metabolism of isoproturon in herbicide resistant and susceptible biotypes *P. minor* and wheat. Pest Biochem Physiol 59:143–153

Singh V, Singh S, Malik RK (1998c) Performance of oxyfluorfen in mixture with isoproturon, diclofop or tralkaoxyolin and control of *Phalaris minor* in wheat. Indian J Weed Sci 30(1–2):62–64

Sinning I, Koepke J, Schiller B, Michel H (1989a) First glance on the three-dimensional structure of the photosynthetic reaction center from a herbicide-resistant *Rhodopseudomonas viridis* mutant. Z Naturforsch 45:455–458

Sinning I, Michel H, Mathis P, Rutherford AW (1989b) Characterization of four herbicide-resistant mutants of *Rhodopseudomonas viridis* by genetic analysis, electron paramagnetic resonance, and optical spectroscopy. Biochemistry 28:5544–5553

Steinback KE, Gardner G, Arntzen CJ (1981) Photoaffinity labeling of a herbicides receptor protein in chloroplast membranes. Proc Natl Acad Sci USA 78:981–985

Svensson B, Vass I, Styring S (1991) Sequence analysis of the D1 and D2 reaction centre proteins of photosystem II. Z Naturforsch 46:765–776

Takahashi M, Shiraishi T, Asada K (1988) COOH-terminal residues of D1 and the 44 kDa CPa-2 at spinach photosystem II core complex. FEBS Lett 240:6–8

Takahashi Y, Nakane H, Kojima H, Satoh K (1990) Chromatographic purification and determination of the carboxy terminal sequences of photosystem II reaction center protein D1 and D2. Plant Cell Physiol 31:273–280

Tewari JP, Bisen CR (1982) Studies on emergence and population dynamics of *Phalaris minor* and other weeds. In: Indian society of weed science annual conference, Hisar, Abstract, pp 40–41

Trebst A (1980) Inhibitors in electron flow: tools for the functional and structural localization of carriers and energy conservation sites. Methods Enzymol 69:675

Trebst A (1986) The topology of plastoquinone and herbicide binding peptides of photosystem II in the thylakoid membrane. Z Naturforsch 41c:240–245

Trebst A (1987) The three-dimensional structure of the herbicide binding niche on the reaction center polypeptides of photosystem II. J Naturforsch 42c:742–750

Tripathi MK, Yadav MK, Gaur AK, Mishra DP (2005) PCR based isolation of psbA (herbicide binding protein encoding) gene using chloroplast and genomic DNA for *Phalaris minor* biotype (S). Physiol Mol Biol Plants 11(1):161–163

Trivedi PK, Sane PV (1994) Characterization of the psbA gene from chloroplasts of *Populus deltoides*. J Plant Biochem Biotechnol 3:97–102

Van Rensen JJS (1971) Action of some herbicides in photosynthesis, as studied by their effects on oxygen evolution and cyclic photophosphorylation. Meded Landbouwhogesch Wageningen 71(9):1–80

Varsano R, Rabinowitch HD, Rubin B (1992) Mode of action of piperonyl butoxide as herbicide synergists of atrazine and terbutryne. Pestic Biochem Physiol 44:174–177

Velthuys BR (1981) Electron dependent competition between plastoquinone and inhibitors for binding to photosystem II. FEBS Lett 126:277–281

Vermaas WFJ, Rutherford AW, Hansson O (1988) Site-directed mutagenesis in photosystem II of the cyanobacterium *Synechocystis* sp. PCC 6803: donor D is a tyrosyl residue in the D2 protein. Proc Natl Acad Sci U S A 85:8477–8481

Walia US, Gill HS (1985a) Interactions between herbicides and nitrogen in the control of *Phalaris minor* in wheat. Trop Pest Manage 31:226–231

Walia US, Gill HS (1985b) Influence of nitrogen and substituted urea herbicides on the uptake of N, P and K by *Phalaris minor* Retz. and wheat. Indian J Weed Sci 17:12–17

Walia US, Brar SS, Kher DS (1997) Uptake of nutrients by wheat and *Phalaris minor* as influenced by crop residue management systems and weed control methods in rice wheat. Environ Ecol 15(3):613–620

Webber AN, Packman L, Chapman DJ, Barber J, Gray JC (1989) A fifth chloroplast encoded polypeptide is present in the photosystem II reaction centre complex. FEBS Lett 242:259–262

www.weedscience.org (2001) Group C2/7 resistant little seed canary grass (*Phalaris minor*)

Yaduraju NT (1991) Effect of plant-growth stage and environmental factors on the activity of isoproturon against small canary grass (*Phalaris minor*). Indian J Agron 36:202–206

Yaduraju NT (1998) Physiological basis of selectivity of isoproturon in *Phalaris* species. Indian J Exp Biol 36 (6):599–604

Yaduraju NT (1999) Control of herbicides resistant *Phalaris minor* need for a sound weed management system. Pestology Spec Issue 23(2):264–266.

Yaduraju NT, Ahuja KN (1995) Response of herbicide resistant *Phalaris minor* to pre-and post-emergence herbicides, herbicide mixtures and adjuvants. Proc Brighton Crop Prot Conf Weeds 1:225–230

Zurawski G, Bohnert HJ, Whitfeld PR, Bottomley W (1982) Nucleotide sequence of the gene for the M, 32,000 thylakoid membrane protein from *Spinacia oleracea* and *Nicotiana debneyl* predicts a totally conserved primary translation product of M, 38,500. Proc Natl Acad Sci 79:7699

Zurawski G, Clegg M T, Brown AHD (1984) The nature of nucleotide sequence divergence between barley and maize chloroplast DNA. Genetics 106:735–749

Phenology Modelling and GIS Applications in Pest Management: A Tool for Studying and Understanding Insect-Pest Dynamics in the Context of Global Climate Change

Babasaheb B. Fand, Jaipal S. Choudhary, Mahesh Kumar, and Santanu K. Bal

Abstract

Intensification of agricultural yield losses due to pest aggravation in the context of global climate change has been the key focus of ecological research. In this regard, interest in forecasting models is now days growing radically among entomologists to predict the environmental suitability for new and invading agricultural insect pests. This chapter describes the approaches for development of temperature-based phenology models that helps in understanding insect behaviour and physiology under diverse environmental conditions. A few suitable illustrations are provided on how phenology models can be used for simulating variability in insect development times through stochastic and deterministic simulation functions with inclusion of temperature as a main predictor of insect development. Further, discussions were also included on linking of phenology models with geographic information systems (GIS) for mapping pest population growth potentials according to real-time or interpolated temperature data, as a tool for pest risk assessments in different agro-ecological regions and to support the development of management strategies. The concepts and approaches underlying simulation of age-

B.B. Fand (✉) • S.K. Bal
School of Atmospheric Stress Management, National Institute of Abiotic Stress Management (Indian Council of Agricultural Research), Malegaon, Baramati, Pune 413 115, Maharashtra, India
e-mail: babasahebfand@gmail.com

J.S. Choudhary
ICAR Research Complex for Eastern Region, Research Centre Plandu, Ranchi 834 010, Jharkhand, India
e-mail: choudhary.jaipal@gmail.com

M. Kumar
School of Drought Stress Management, National Institute of Abiotic Stress Management (Indian Council of Agricultural Research), Malegaon, Baramati, Pune 413 115, Maharashtra, India
e-mail: mahesagrawal@gmail.com

R.K. Gaur and P. Sharma (eds.), *Approaches to Plant Stress and their Management*,
DOI 10.1007/978-81-322-1620-9_6, © Springer India 2014

stage-structured populations using cohort-updating and rate summation principle and the use of geostatistical algorithms integrated in GIS for risk mapping are described briefly.

Keywords
Climate adaptation • Geographic information system • Global climate change • Pest risk assessment • Phenology models

Introduction

Setting the Scene

Climate change is now days scientifically established and globally acknowledged fact (Abrol et al. 1996; IPCC 2007). Considering the declining production efficiency of agro-ecosystems due to depleting natural resource base and serious consequences of climate change, food security for twenty-first century is the major challenge for human kind in years to come (Fig. 1). Abiotically stressful environment in changing climate is predicted to impact negatively the diversity and abundance of insect pests and ultimately the extent of damage caused in economically important agricultural crops. Being a tropical country, India is more challenged with impacts of looming climate change. The Indian climate has undergone significant changes with an average temperature increase of 0.56 °C over the last 100 years (IPCC 2007; Rao et al. 2009; IMD 2010). According to various climate change scenarios, the temperature in India is predicted to increase by 1–5 °C by the end of 2100 (Lal 2003; IPCC 2007). Already, the productivity of Indian

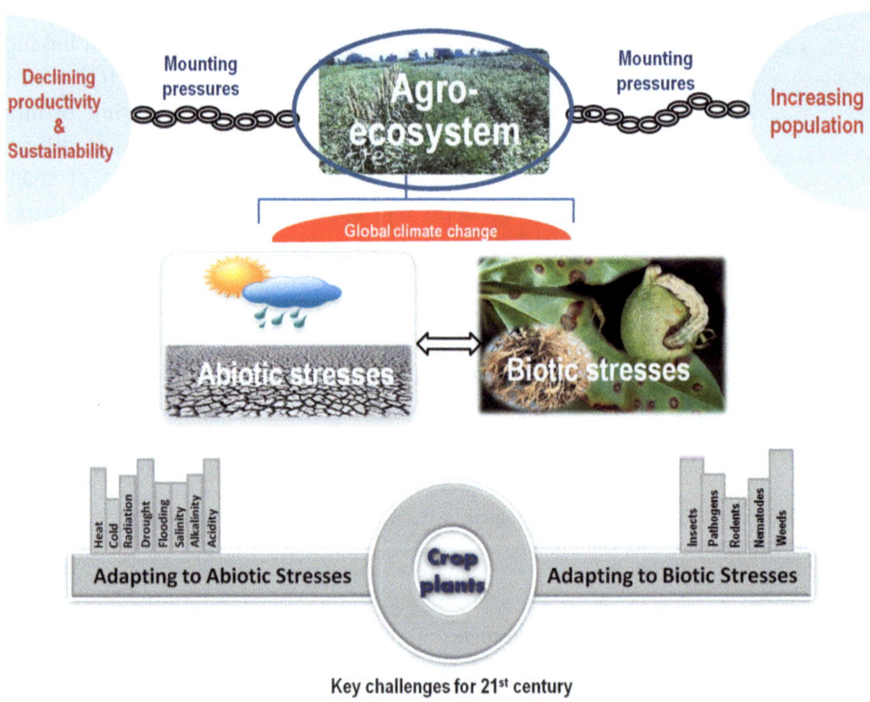

Fig. 1 Schematic presentation of key challenges of agriculture in the context of global climate change

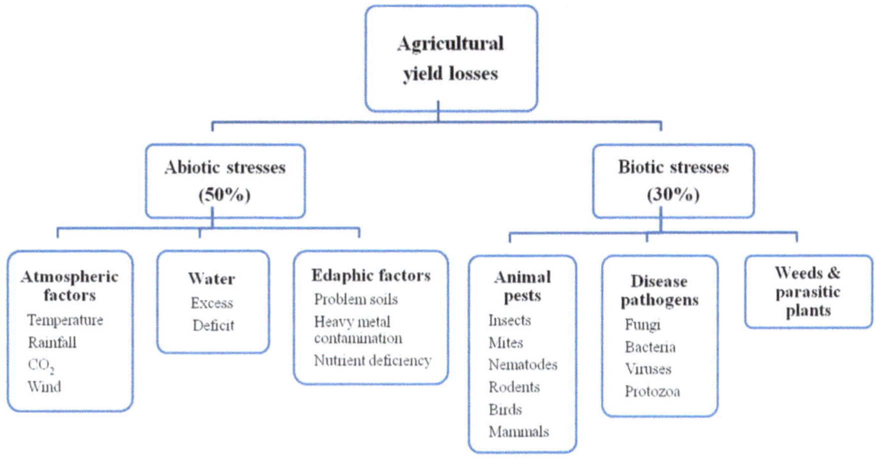

Fig. 2 Agricultural yield losses due to abiotic & biotic stresses

agriculture is limited by its high dependency on monsoon rainfall which is most often erratic and inadequate in its distribution (Chand and Raju 2009). The country is experiencing declining trend of agricultural productivity due to fluctuating temperatures (Samra and Singh 2004; Aggarwal 2008), frequently occurring droughts and floods (Samra 2003), problematic soils and increased outbreaks and resurgences of insect pests (Joshi and Viraktamath 2004; IARI News 2008; Nagrare et al. 2009, 2011) and diseases. These problems are likely to be aggravated further by changing climate which may put forth major challenge to attain a goal of food security. It will have serious environmental and socio-economic impacts on rural farmers whose livelihoods depend directly on the agriculture and other climate-sensitive sectors.

Dealing with the impacts of climate change on biological systems is really tedious task owing to its complexity, uncertainty, unpredictability and differential impacts over time and place. Understanding abiotic stress responses in crop plants, insect pests and their natural enemies is an important and challenging topic ahead in agricultural research. There is urgent need that the impacts of climate change on crop production mediated through changes in populations of serious insect pests should be given careful attention for planning and devising adaptation and mitigation strategies for future pest management programmes.

Agricultural Yield Losses Due to Abiotic and Biotic Stresses

Agro-ecosystem environment is largely governed by the interactions between abiotic and biotic components to which plants are continuously exposed, and these factors influence the crop growth and productivity to a great extent. Among these, abiotic stress factors which include temperature, rainfall, drought and soil fertility share major proportion of yield losses in agricultural crops and are most harmful when occur in combination of several other stress factors (Mittler 2006). The abiotic stress factors also modulate the effects of biotic stresses such as pests, diseases and weeds and in combination can substantially reduce the crop yield potential to the extent of 80 % (Fig. 2) (Oerke 2006; Theilert 2006). Poor crop performance due to stresses exerted by abiotic and biotic environmental factors leads to a lower attainable yield, i.e. yield actually realised under practical growth conditions. When the potential yields i.e. yields obtained under ideal growth conditions are compared with attainable yields of crops, the impact of the environmental stresses on plant growth and productivity becomes apparent. These losses are predicted to be intensified further due to future climate changes which may pose serious threat to the food security and economic development of the human societies (IPCC 2007).

Emerging Pest Problems in India

India during the last decade has witnessed severe outbreaks of different insect pests in many crops of economic importance (Table 1) (Fand et al. 2012) resulting in heavy yield losses and increased cost of crop protection. Certain changes in climatic variables have led to increased frequency and intensity of outbreaks of insect pests. Outbreak of sugarcane woolly aphid *Ceratovacuna lanigera* Zehntner in sugarcane belt of Karnataka and Maharashtra states during 2002–2003 resulted in 30 % yield losses (Joshi and Viraktamath 2004; Srikanth 2004, 2007; Tripathi et al. 2008; Rafee 2010). Recent abnormal weather patterns coupled with insecticide misuse lead to the widespread outbreaks of rice plant hoppers, namely, *Nilaparvata lugens* (Stal) and *Sogatella furcifera* (Horvath), resulting in crop failure over more than 33,000 ha paddy area in North India (Fig. 3a) (IARI News 2008; IRRI News 2009). Mealybug, *Phenacoccus solenopsis* Tinsley, which was earlier not a major pest of economic importance has emerged as a serious insect pest in India since 2006 affecting cotton-growing belt across the country resulting in heavy yield loss to the cotton growers (Jhala et al. 2008; Nagrare et al. 2009, 2011). Besides cotton, the pest has been recorded on more than 194 plant species that included agricultural and horticultural crops and weeds (Fig. 3b) (Vennila et al. 2011). Papaya mealybug *Paracoccus marginatus* Williams and Granara de Willink has assumed the status of a major pest and caused havoc in papaya and other horticultural crops of economic importance during 2007–2009 in the parts of Tamil Nadu, Karnataka and Maharashtra states (Tanwar et al. 2010a). Prolonged drought conditions followed by heavy rains resulted in severe outbreak of swarming caterpillars, *Spodoptera mauritia* (Boised), on rice in Orissa over an area of 1.25 m ha leading to 90 % crop damage (Anonymous 2010; Tanwar et al. 2010b). Choudhary et al. (2013) reported an outbreak of stink bug *Tessaratoma javanica* (Tessaratomidae) on litchi crop in Chotanagpur region of Jharkhand state, resulting in approximately 80 % yield loss to litchi fruits (Fig. 3c). The possible reason for outbreak was the water-stressed litchi plants at flowering stage due to scanty rainfall received during the previous year. Zanskar and Chanthang valleys of Ladakh region of Jammu and Kashmir state experienced serious problem of migratory locust, *Locusta migratoria migratorioides* (Reiche & Fairmaire), during August 2006 (Kumar et al. 2009). Extreme weather fluctuations and subsequent loss of natural enemies caused an outbreak of pine lappet moth, *Kunugia latipennis* Walker, on agroforestry plants in the mid-altitude hills of Meghalaya during May–June 2011 (Firake et al. 2012). Severe incidence of leaf-eating caterpillar *Spodoptera litura* (Fabricius) in soybean (Fig. 3d) and pod borer *Helicoverpa armigera* (Hubner) in pigeon pea and chick pea due to abnormal weather during kharif season is a regular feature in semi-arid cropping regions of the country. These situations of increased and frequent pest damage to the crops have made another big hole in the pockets of already distressed farmers by increasing the cost of plant protection and reducing the margin of profit.

Predicted Impacts of Climate Change on Arthropod Diversity and Abundance in Agro-ecosystems

Pest menace at various stages of crop growth is one of the limiting factors to the productivity of agricultural crops. Despite the timely crop protection measures, arthropod pests have a potential to reduce crop production substantially, with an estimated worldwide average of 16–18 % yield losses (Oerke et al. 1994; Oerke 2006). Considerable variation in the pest damage in different agro-climatic regions is attributed mainly to the differential impacts of several abiotic factors such as temperature, humidity, rainfall and light intensity (Reed and Pawar 1982). Being poikilotherms, insects have limited ability of homeostasis with external temperature changes, and hence temperature is considered as the most dominant environmental factor influencing their behaviour, distribution, development, survival and reproduction (Andrewartha and Birch 1954;

Table 1 Recorded instances of recent insect-pest outbreaks in relation to changing climate scenario in India (Fand et al. 2012)

Insect pest	Order/family	Host plant(s)	Region/location	Probable reason(s)	Impact of pest outbreak	Reference
Sugarcane woolly aphid *Ceratovacuna lanigera* Zehntner	Hemiptera: Aphididae	Sugarcane	Sugarcane belt of Karnataka and Maharashtra states during 2002–2003	Recent abnormal weather patterns	30 % yield losses	Joshi and Viraktamath (2004)
				Insecticide misuse	Reduced cane recovery	Srikanth (2007)
Rice plant hoppers *Nilaparvata lugens* (Stal) and *Sogatella furcifera* (Horvath)	Hemiptera: Delphacidae	Rice	North India	-do-	Crop failure over more than 33,000 ha paddy area	IARI News (2008)
						IRRI News (2009)
Mealybug, *Phenacoccus solenopsis* Tinsley	Hemiptera: Pseudococcidae	Cotton, vegetables and ornamentals	Cotton-growing belt of the country	Recent abnormal weather patterns	Heavy yield (30–40 %) loss to the cotton	Dhawan et al. (2007)
				Insecticide misuse	Increased cost of crop protection due to overuse of pesticides	
				Changed cropping environment (introduction of Bt cotton)		
Papaya mealybug *Paracoccus marginatus*	Hemiptera: Pseudococcidae	Papaya	Tamil Nadu, Karnataka, Maharashtra	Recent abnormal weather patterns	Significant yield loss to the papaya growers	Tanwar et al. (2010b)
				Insecticide misuse		
Litchi stink bug *Tessaratoma javanica* (Thunberg)	Hemiptera: Tessaratomidae	Litchi	Jharkhand	Drought conditions and resultant water stress in plants	Severe infestation on inflorescence and developing fruits resulted in up to 80 % yield loss to the litchi	Choudhary et al. (2013)

a. Suddenly broken out plant hopper infestation on rice (inset) in North India during 2008 leading to hopper burn symptoms (Photograph by M. Sujithra, Scientist (Entomology), IARI, New Delhi.)

b. Severe infestation of *Phenacoccus solenopsis* on pomegranate fruit (Photograph by Babasaheb B. Fand, Scientist (Agril. Entomology), NIASM, Baramati)

c. Outbreak of litchi stink bug in Jharkhand. Photo in inset shows litchi flowers and tender fruits attacked by stink bug (Source: Choudhary et al. 2013)

d. *Spodoptera litura* larvae feeding gregariously on soybean foliage (Photograph by Mahesh Kumar, Scientist (Plant Physiology), NIASM, Baramati)

Fig. 3 Emerging pest problems in economically important agri-horticultural crops in India

Yamamura and Kiritani 1998; Bale et al. 2002). The global climate change has been predicted to raise the mean surface temperature of earth by 1.5–5.8 °C by the end of 2100 (Govindasamy et al. 2003; Hijmans et al. 2005a; IPCC 2007). This is expected largely to aggravate the already serious pest problems and yield losses further, which may pose serious threat to food security and economic development of the human societies (IPCC 2007; Chahal et al. 2008).

Impact of climate change on interactions between crop plants and insect pests has been extensively reviewed by many workers from different parts of the world (Rao et al. 2006; Petzoldt and Seaman 2010; Fand et al. 2012). The major predictions on changing pest scenario due to climate change included (1) the expansion of the geographic ranges due to shift in cultivation areas of host crops at higher latitudes and altitudes (Porter et al. 1991; Cannon 1998; Parmesan and Yohe 2003), (2) increased overwintering survival due to milder winters (Harrington et al. 2001; Bale et al. 2002; Bale and Hayward 2010), (3) increased number of generations due to accelerated growth rates and shorter generation times (Yamamura and Kiritani 1998; Bale et al. 2002; Ward and Masters 2007), (4) increased risk of introducing invasive alien species (Sutherst et al. 1991; Sutherst 2000; Mooney and Hobbs 2000; Ward and Masters 2007), (5) breakdown of host resistance to insect pests due to increased environmental stress and reduced vigour (Rhoades 1985; Kaiser 1996; Hilder and Boulter 1999; Kranti et al. 2005; Sharma et al. 2005) and (6) disruption of synchrony between life cycles of insect pests and their natural enemies leading to failure of biological control (Bale et al. 2002; Gutierrez et al. 2008). The tropical and subtropical regions of the world are more challenged with the impacts of looming climate change and resultant pest outbreaks because of year-round favourable climatic conditions for pest multiplication and food availability.

Ecological Models for Forecasting Pest Activity

Early predictions on the future pest distribution and abundance through forecasting models could facilitate better preparedness to combat outbreaks of serious insect pests by developing effective pest management strategies well in advance. In that way, models are considered as important analytical tools that allow better understanding and prediction of insect population dynamics and growth potential under an array of environmental conditions (Baker 1991; Jarvis and Baker 2001a, b). Improved knowledge on basic physiological aspects of insect growth and development in relation to environmental factors forms a sound basis for successful pest management in varied agro-ecological conditions (Sporleder et al. 2004, 2008; Kroschel et al. 2012). In addition, coupling of phenology models with geographic information system (GIS) allows the prediction of pest population dynamics in response to climate change in different agro-ecological regions on a spatial mode (Sporleder et al. 2008; Kroschel et al. 2012; Fand et al. 2013) and also helps in delineation of agro-ecological hotspots and future areas of pest risk (Liebhold et al. 1993; Yadav et al. 2010).

This chapter describes the existing modelling approaches employed in applied insect ecology with major focus on describing process-based climatic response models, i.e. phenology models with inclusion of temperature as a main predictor of insect development. We tried to explain with suitable illustrations how phenology models can be used for simulating variability in insect development times through stochastic and deterministic simulation functions. Moreover, coupling of model outputs with a geographic information system for pest hotspot zonation and analysis of climate change impacts is also discussed.

Temperature-Based Phenology Modelling: A Tool to Understand and Predict Pest Activity

Approaches to Insect Modelling

Models are the simplified mathematical expressions that mimic a real-world phenomenon and which can be used for making predictions on numerical changes in insect populations in time and space (Odum and Barrett 2008). For modelling the risk of establishment and geographic distribution of insect pests under array of environmental conditions, there exist following two most frequently used modelling approaches: climatic pattern matching functions and process-based models.

Climatic Pattern Matching Functions

These models describe the species' potential geographic distribution and risk based on their response to climatic factors. For analysis of pest risk with such models, minimal inputs are pest occurrence data from different locations and climate data relevant to those occurrence points. The risk for potential establishment and invasion of new species in an area can be estimated based on the relationship between long-term meteorological data and the presence/absence of the species for each location (Sutherst 2000; Sutherst and Maywald 1990). Execution of models by this approach involves the use of computer-aided tools, e.g. CLIMEX (Peacock and Worner 2006; Wilmot Senaratne et al. 2006), BIOCLIM (Kohlmann et al. 1988; Steinbauer et al. 2002), MAXENT (Phillips et al. 2004, 2006; Elith et al. 2006, 2011) and recent developments like GARP and HABITAT (Venette et al. 2010). These

BOX 1: Potential Geographic Distribution of Cotton Mealybug *Phenacoccus solenopsis* Based on MAXENT Ecological Niche Model (for Details, Please Refer to Fand 2012)

The potential geographic distribution of newly emerged invasive mealybug of cotton Phenacoccus solenopsis Tinsley (Hemiptera: Pseudococcidae) in India has been predicted using MAXENT ecological niche model. The probability for pest distribution and abundance at both current and future climatic conditions was estimated using presence-only data on P. solenopsis occurrence and corresponding climate data relevant to the study area.

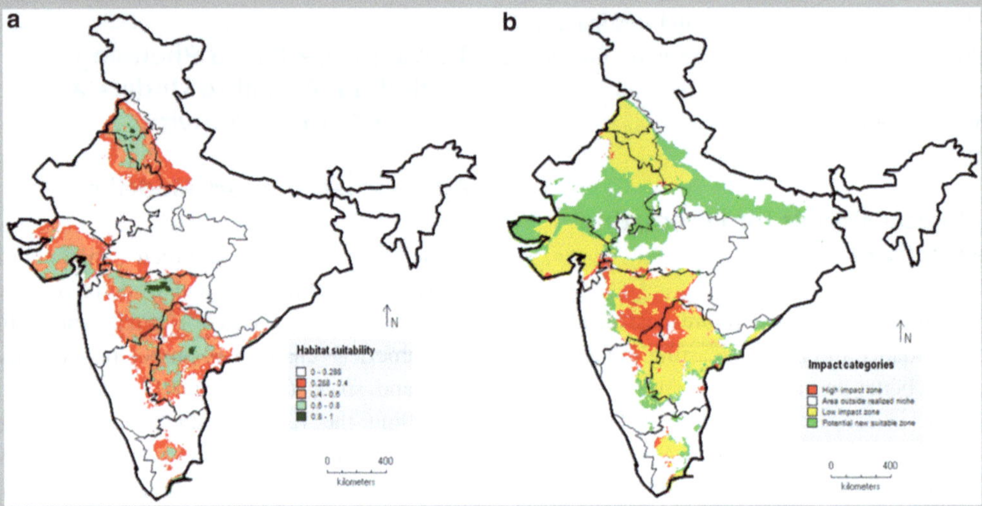

Fig. 4 (**a**) Habitat suitability for potential geographic distribution of *P. solenopsis* at current climatic conditions {values ranging from low (0) to high (1)} (**b**) impact of climate change on the future geographic distribution of *P. solenopsis*

The model predicted extensive areas for P. solenopsis occurrence throughout the north, central and south cotton-growing area of India (Fig. 4a). *It was observed that P. solenopsis will extend its present geographic range, especially at higher altitudes and latitude, as area presently unsuitable may become more conducive for its distribution and abundance due to future climate warming* (Fig. 4b).

models have been extensively used for pest risk analysis (Sutherst and Maywald 1990; Zalucki and Furlong 2005) and predicting possible impacts of future climatic changes on geographic distribution and spread (Sutherst and Maywald 1990; Wang et al. 2010a, b; Fand 2012; Fand et al. 2013). This modelling approach is useful in situations where detailed information about insect species is lacking; however, it does not take into account biological characteristics of the species in modelling framework. Hence, the resultant maps only show information about potential distribution of species but do not give any idea about its population dynamics and damage potential (Venette et al. 2010). For illustration on prediction of environmental suitability using climate pattern matching functions, see Box 1.

Phenology Models or Process-Based Climatic Response Models

These models are more sophisticated and progressive models than pattern matching functions described in earlier section. They use non-linear functions of higher biological significance and include stochastic functions for simulating variability in development times within a population based on detailed laboratory assessments of life table parameters (Logan et al. 1976; Sharpe et al. 1981; Wagner et al. 1984). Phenology models are important analytical tools that help to evaluate, understand and predict insect population dynamics in agro-ecosystems under wide range of environmental conditions and management practices and can very well be used in phytosanitary risk assessments (Baker 1991; Jarvis and Baker 2001a, b). For execution of these models, several computer-aided tools have been developed and tested, e.g. DYMEX (Kriticos et al. 2003), ECAMON (Trnka et al. 2007) and ILCYM (Sporleder et al. 2009, 2012).

Why Phenology Models?

When we are speaking about the insect pests, it is a matter of numbers and not merely the presence of solitary individual that annoys us. The insect numbers are governed by particular environment inhabitant for the kind of insect species. The environment plays major role in controlling physiology and behaviour of insects which in turn governs the dynamics of insect numbers in time and space and also affects the timings of biological events, i.e. 'phenology' of insects. Hence, basic understanding of how environment influences diversity, abundance and phenology (timings of hatching, diapause, mating, migration, etc.) of insect species forms a sound basis for any ecologically based pest management (Pedigo 2006). Insects pass through different developmental stages before becoming fully grown adults and require certain amount of heat energy to develop from one stage to the other, e.g. from egg stage to larval stage or so on. Because of seasonal and intra-annual climatic variations, judgement of management decisions based on calendar dates does not hold always realistic (Ascerno 1991). Contrarily, measurement of amount of heat units accumulated over time represents the physiological age of the species and thus explains timings of biological events in insects' life cycle more accurately than calendar days. Thus, insect phenology, a direct relationship between insect development and prevailing weather conditions, can be a precise method to time the treatment. In that way, development of phenology models helps in predicting time of events in insects' life and thereby making better management decisions (Sporleder et al. 2004, 2009, 2012; Kroschel et al. 2012).

Temperature Inclusion in Insect Phenology Models

As stated earlier, insects being poikilothermic organisms, temperature is considered as single most dominating abiotic factor affecting their development, survival, reproduction, behaviour and geographic abundance (Andrewartha and Birch 1954; Yamamura and Kiritani 1998; Bale et al. 2002). Because of acceleration of biochemical reactions or metabolic activities at increasing temperatures within a favourable range, temperature can be used as a main predictor of rates of insect development and thus timings of phenological events in insects' life cycle (Sharpe and DeMichele 1977; Pedigo 2006). This temperature dependency of insect growth and development can be applied in process-based modelling framework to describe basic physiological aspects of insect species, i.e. development, survival and reproduction (Sporleder et al. 2004, 2009, 2012).

The critical importance of temperature to development of arthropod populations has long been recognised, and good deal of research efforts has been made in determining temperature dependence of life-history parameters of insect species (Stinner et al. 1974; Logan et al. 1976; Sharpe et al. 1981; Wagner et al. 1984; Lactin et al. 1995). Linear degree-day method based on the accumulation of temperature above lower-temperature threshold has been used extensively in estimating physiological age and thus to predict the rate of insect development (Allen 1976; Pedigo 2006: Nietschke et al. 2007). The following expression of linear

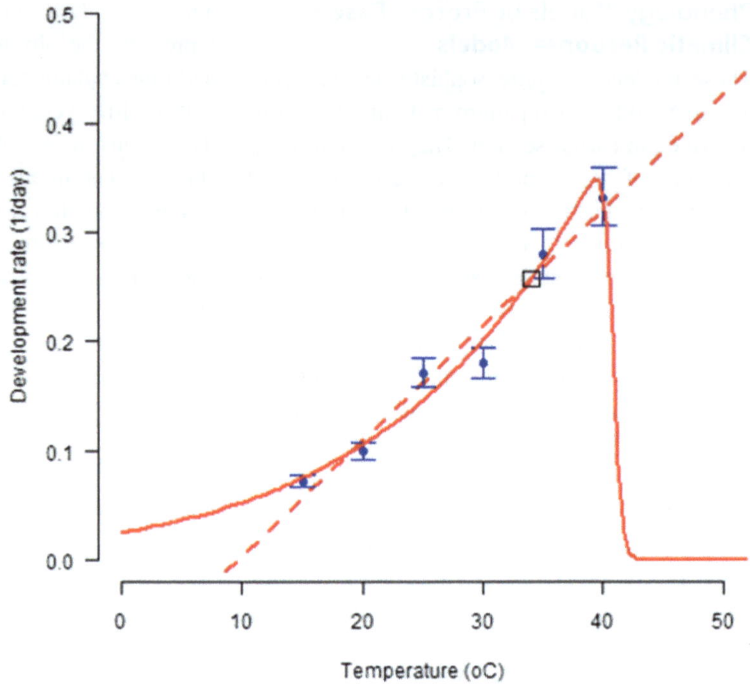

Fig. 5 Theoretical linear (*dashed line*) and non-linear (*solid line*) relationship between temperature and development rate of third instar nymphs of mealybug *P. solenopsis* (Fand et al. 2013)

degree-day model is used for estimating number of degree-days (DD) accumulated by insect on any particular day:

$$\mathrm{DD} = (\mathrm{T} - \mathrm{K}) \quad (1)$$

where T = daily mean temperature (°C) and K = threshold temperature (°C). As the particular life stage *(j)* of insect extends over a specific time period, say, days (*i*), the cumulative number of degree-days required to complete *j*th instar after n days is given by following formula:

$$\mathrm{D}n = \sum_{i=1}^{n} dj \quad (2)$$

where i = 1, 2, 3...., n, the cumulative degree-days required to complete *j*th instar.

However, due to non-linear relationship existing between development rates and temperature, especially under fluctuating temperature condition, the linear models produce errors at temperature extremes (Fig. 5). In other words, linear models work well for intermediate temperatures but do underestimate development rates at low temperatures and lead to overestimation of development at higher temperatures. Hence, linear models are considered as poor predictors of insect development times (Stinner et al. 1974; Worner 1992). On the other hand, non-linear relationship between rate of insect development and temperature extremes can be used to estimate the optimum temperature requirement for insect development (Briere et al. 1999). The phenology models are based on non-linear functions for predicting development rate as a function of temperature. The non-linear functions that estimate development rates account for biological processes taking place in poikilotherms such as retardation or acceleration of biochemical reactions involved in metabolic activities according to temperature conditions (Logan et al. 1976; Sharpe and DeMichele 1977; Wagner et al. 1984; Pedigo 2006; Worner 1992). The most frequently used non-linear function for estimating insect development rates because of its biological relevance is the Sharpe and DeMichele model (Sharpe and DeMichele 1977; Schoolfield et al. 1981), the equation of which is as shown below:

$$r(T) = \frac{RH025 \times \frac{T}{298.16} \times \exp\left[\frac{\Delta Ha}{R}\left(\frac{1}{298.16} - \frac{1}{T}\right)\right]}{1 + \exp\left[\frac{\Delta Hl}{R}\left(\frac{1}{Tl} - \frac{1}{T}\right)\right] + \exp\left[\frac{\Delta Hh}{R}\left(\frac{1}{Th} - \frac{1}{T}\right)\right]} \quad (3)$$

BOX 2: Temperature-Dependent Reproduction of Cotton Mealybug *P. solenopsis* (for Further Details, Please Refer to Fand et al. 2013)

The phenology model for estimating reproductive potential of P. solenopsis revealed that the reproductive period of P. solenopsis shortened with increasing temperatures as female lifespan abridged. A second-order polynomial function described the temperature dependence of P. solenopsis fecundity showing a curvilinear response with a maximum fecundity at 30 °C and dropping off at temperatures below and above it (Fig. 6a). *The exponential function described relationship between oviposition (cumulative oviposition rate) and female age. The 50 % oviposition was completed by the time female reached a normalized age of 0.607* (Fig. 6b).

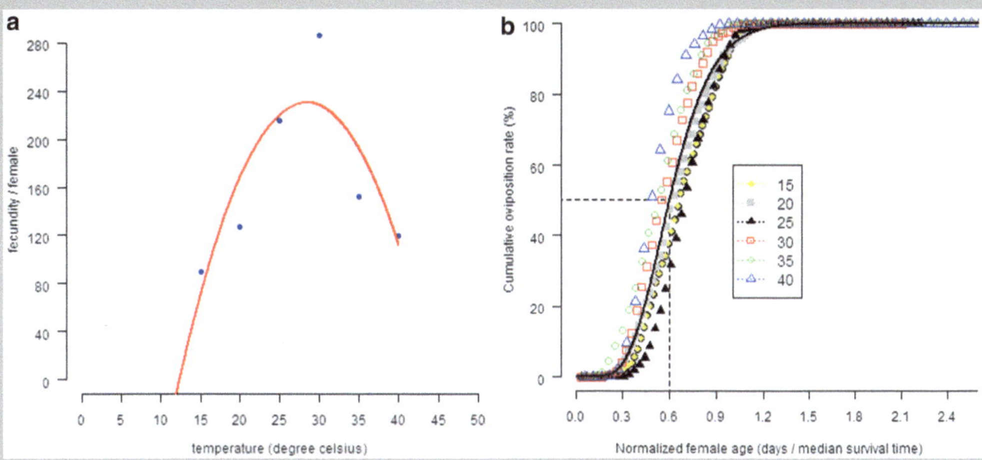

Fig. 6 (**a**) Temperature-dependent total egg production (polynomial regression model) (**b**) Cumulative proportion of egg production in relation to female age expressed as normalised time (exponential function)

where r (T) is development rate at temperature T (°K); R is universal gas constant (1.987 *cal degree*$^{-1}$ *mol*$^{-1}$); RHO_{25} is development rate at 25 °C temperature (298.16 °K), assuming no enzyme inactivation; ΔH_a is enthalpy of activation of reaction catalysed by enzyme (*cal mol*$^{-1}$); ΔH_l is change in enthalpy at low temperature (*cal mol*$^{-1}$); ΔH_h is change in enthalpy at high temperature (*cal mol*$^{-1}$); T_l is low temperature at which enzyme is half active; and T_h is high temperature at which enzyme is half active.

In order to model the effects of temperature on insect reproduction, the non-linear modelling functions allow describing different biological processes like temperature-dependent sex ratio and age-dependent and temperature-dependent oviposition frequencies. Generally, polynomial models of second and third order gave good estimate of temperature-dependent immature mortality and female fecundity as shown below:

$$m\,(T) = aT^2 + bT + c \tag{4}$$

where *m (T)* is the mortality or fecundity at temperature *T* (°C).

Age-related oviposition frequency of females as a function of temperature can be described by exponential or cumulative distribution models (Sporleder et al. 2004). The equation of exponential model is given below:

$$y = \left(1 - \exp\left(-\left(ax + bx^2 + cx^3\right)\right)\right) \tag{5}$$

where *y* is the cumulative oviposition rate and *x* is the normalised female age. For illustrations on modelling temperature-dependent reproductive behaviour of insects, see Box 2.

Cohort-Updating and Rate Summation Approach for Age-Stage Structured Life Table Simulations

As the insects have different development stages before becoming a matured adult ready for reproduction, the insect populations are mostly heterogeneous (except univoltine species), in their age-stage structure at any particular time and space. This is mainly because of continuous reproduction and overlapping generations that are produced by the species. In order to describe the temperature-dependent development and mortality in each immature life stage and senescence and reproduction in adults of the insect species under study, Sporleder et al. (2004) suggested a cohort-updating and rate summation approach that simulates stochastically the variability in development within a population. The cohort-updating algorithm is based on scheme proposed by Curry et al. (1978) and further described by Wagner et al. (1985) and Logan (1988). In cohort-based approach, each cohort, i.e. group of individuals of same age, is tracked independently for development, mortality and transfer of one life stage to the next. The concept of cohort-updating approach allows accounting for individual's responses to the variety of conditions experienced by them, and hence phenology models based on this approach are more realistic to actual field conditions (Yonow et al. 2004).

The developmental response of insects to temperature is usually determined by exposing insect cohorts to constant temperatures. Usually, the insect development shows linear relationship with increasing temperatures within the favourable range. However, development at temperature extremes (lower and upper limits) may show non-linear response due to retardation or acceleration of developmental rates, especially when insects are exposed to fluctuating temperatures. This deviation in development rates from linearity at constant and fluctuating temperature conditions has been described as rate summation effect (Behrens et al. 1983; Ratte 1985; Worner 1992). Hence, in order to reproduce similar insect behaviour under test conditions as that of under natural environment, the development is approximated by rate summation over discrete time increments, say, daily or hourly for precise simulation (Worner 1992). The proportion of development taking place at each time increment is summed till the insect has completed its development and passed on to the next stage as shown in the following equation:

$$D = \Sigma r[T(t)]\, dt \quad (6)$$

where D = development as a function of temperature T at time t and r = temperature-dependent development rate.

GIS Applications in Pest Management

Geographic Information System (GIS): An Enabling Technology for Entomologists

The historical studies on insect population biology were focused mainly to address temporal changes in insect dynamics with very little attention on exploring spatial patterns of their spread and abundance (Iwao 1972; Taylor et al. 1978; Taylor 1984). The major obstruction to study spatial processes in applied insect ecology has been the lack of sophisticated analytical tools. However, advent of the technology of geographic information system (GIS) and applied geostatistics has opened up new avenues for analysing spatial patterns in insect populations (Liebhold et al. 1993).

GIS is a computer-aided programme that allows management of georeferenced spatial data on insects and has been proved as an emerging technology in area-wide pest management because many dimensions of insect ecology have a spatial component. Recent advances in computational sciences made it easier to process huge amount of available ecological and climate data using algorithms (geostatistics) integrated in GIS. This helps better understanding of the pattern of species spread (Stockwell and Noble 1992; Peterson 2003; Ganeshaiah et al. 2003; Sporleder et al. 2008). Earlier, the simple statistical functions (dispersion indices) such as Iwao's patchiness index (Iwao 1972) and Taylor's power law (1978, 1984) have been used extensively to describe spatial patterns in insects. However, these indices only address spatial variability based on relationship between sample variance and mean and do not account for spatial locations of samples (Jumars et al. 1977; Sawyer 1989; Hurlbert 1990). On the other hand, modern geostatistical algorithms that are integrated in present-day GIS softwares use both sample value and geographic location simultaneously to quantify and model spatial patterns in insect species.

GIS is such an enabling technology for entomologists as well as ecologists, which helps in relating insect-pest outbreaks to biographic and physiographic features of the landscape and hence can best be utilised in area-wide pest management programmes for effective forecasting and forewarning of the probable pest outbreaks (Van Sickle 1989; Liebhold et al. 1993). How climatic changes will affect development, incidence and population dynamics of insect pests can be studied through GIS by predicting and mapping trends of potential changes in their geographic distribution (Liebhold et al. 1993) and delineation of agro-ecological hotspots and future areas of pest risk (Sporleder et al. 2008; Yadav et al. 2010; Kroschel et al. 2012). Certain open-source programmes such as DIVA-GIS and Quantum-GIS can successfully be employed in wide range of situations for ecological modelling and predictions (Ganeshaiah et al. 2003; Hijmans et al. 2005b; Scheldeman and van Zonneveld 2010). The recent developments in GIS and geostatistics have made it easier to analyse complex spatial patterns of insect species and will landmark a major breakthrough in applied insect ecology.

Coupling of Phenology Models with GIS Platform for Risk Mapping

Most of the currently available computer-aided modelling tools such as CLIMEX (Peacock and Worner 2006; Wilmot Senaratne et al. 2006), DYMEX (Kriticos et al. 2003), ECAMON (Trnka et al. 2007), HABITAT (Venette et al. 2010) and ILCYM (Sporleder et al. 2012) are coupled with basic GIS platforms which allow for spatial simulations of estimated pest risk indices. The spatial simulations in GIS on different scales, i.e. regional or worldwide, help in visualising pest risk potentials for a geographic area of interest and thereby making a better judgement on pests' abundance and damage potential. The minimal input requirement for spatial simulations, i.e. data on minimum and maximum temperatures from WorldClim database (freely downloadable at www.worldclim.org), is already incorporated within most of the modules (Sporleder et al. 2012) or can be imported easily while doing analysis (Hijmans et al. 2005b). It further allows analysis of the impact of climate change on future pest distribution, abundance and incidence by generating future pest risk maps using temperature forecasts from atmospheric general circulation model described by Govindasamy et al. (2003). For illustrations on how GIS platform helps in pest risk assessment, see Box 3.

GIS-Based Risk Maps as a Decision-Making Tool in Pest Management

The risk maps allow a detailed analysis of climate change impact on the future distribution and damage potential of many invasive insect pests and diseases of economically important agricultural crops under projected climate changes. Such predictions on probable future pest activity will facilitate better preparedness to combat outbreaks of serious insect pests well in advance. The pest risk maps prepared in GIS are effective and meaningful communication and decision-making tool for adaptation planning to climate change. They help in making strategic pest management decisions such as design of pest surveys, undertaking area-wide pest management programmes, region-specific risk assessments and enforcement of quarantine measures to restrict entry of invasive insect pests and diseases of crops based on the risk of pest species under investigation. The information generated also helps policymakers and national pest management authorities in adapting, planning and improving national pest management and quarantine programmes.

BOX 3: Risk Mapping of Cotton Mealybug *P. solenopsis* Based on Phenology Model Coupled with GIS Platform (for Further Details, Please Refer to Fand et al. 2013)

In present study, a process-based climatic response phenology model established for P. solenopsis was employed in a geographic information system for mapping its population growth potentials. The risk index for pest activity was computed using interpolated temperature data (1950–2000) from WorldClim database (www.worldclim.org) and used to visualise the potential population abundance. The downscaled temperature data from SRES A1B scenario for the year 2050 from WorldClim database were used to highlight the changes in P. solenopsis activity due to future climate change. Western semiarid regions, Deccan plateau and Eastern Ghats of India, were predicted optimal for P. solenopsis activity under current climatic conditions (Fig. 7a), *whereas drastic change in P. solenopsis population growth and abundance is predicted throughout the country, barring Northern and North-Eastern Himalayas where low temperatures limit the pest survival, under future climate conditions* (Fig. 7b).

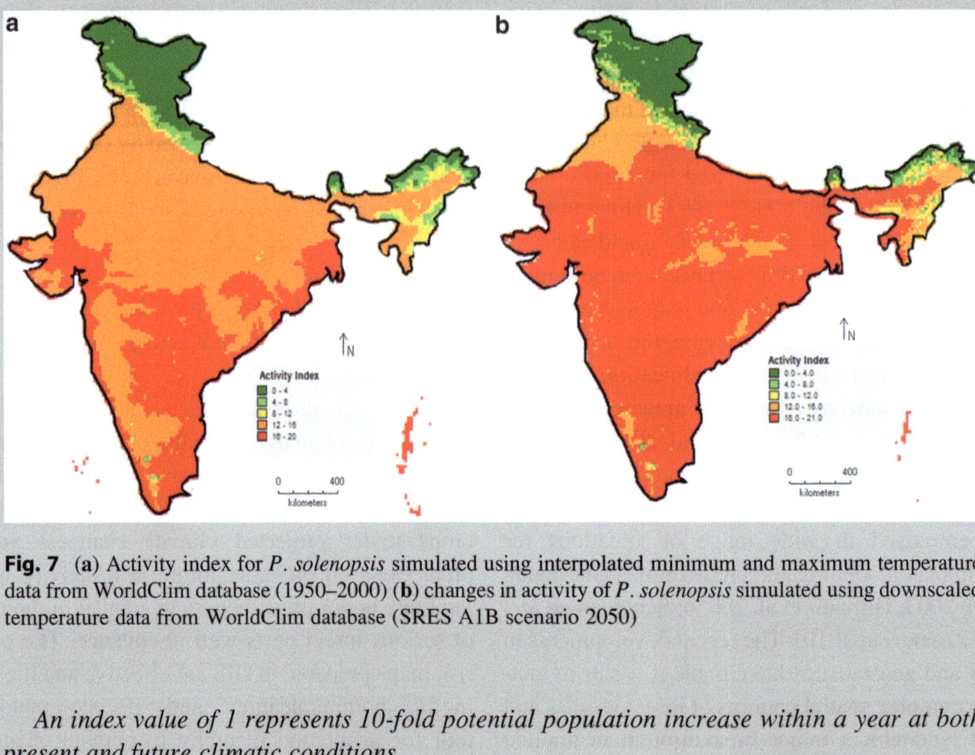

Fig. 7 (**a**) Activity index for *P. solenopsis* simulated using interpolated minimum and maximum temperature data from WorldClim database (1950–2000) (**b**) changes in activity of *P. solenopsis* simulated using downscaled temperature data from WorldClim database (SRES A1B scenario 2050)

An index value of 1 represents 10-fold potential population increase within a year at both present and future climatic conditions.

Conclusion

The mathematical models based on detailed life-history phenology of insect species and GIS-aided tools for delineation of pest hotspots could be an important and useful strategy for assessing the impacts of climate change on insect abundance, geographic distribution and severity of incidence under an array of environmental conditions. The modelling framework described in this chapter has immense ecological applications to address various insect-related problems. However, model is a tool only and hence cannot be a panacea to all the pest problems; rather it can be a useful means for getting fist-hand knowledge about the intricacies of the issue and to develop an approach to solve it. Further, uncertainties in climate change predictions and accuracy of interpolated climate data used in spatial risk assessments are of major concerns.

Acknowledgements The authors gratefully acknowledge the Director, National Institute of Abiotic Stress Management, Baramati, for extending his cooperation and support in bringing out compilation of this document. We are grateful to M. Sujithra, Scientist, Division of Entomology, Indian Agricultural Research Institute, New Delhi, for providing photographs on brown plant hopper incidence in paddy fields. All the online sources of information in public domain referred in this chapter have been duly acknowledged.

References

Abrol Y, Gadgil S, Pant GB (eds) (1996) Climate variability and agriculture. Narosa Publishing House, New Delhi

Aggarwal PK (2008) Climate change and Indian agriculture: impacts, adaptation and mitigation. Indian J Agric Sci 78:911–919

Allen JC (1976) A modified sine wave method for calculating degree days. Environ Entomol 5:388–396

Andrewartha HG, Birch LC (1954) The innate capacity for increase in numbers. In: Andrewartha HG (ed) The distribution and abundance of animals, 3rd impression. University of Chicago Press, Chicago, pp 31–54

Anonymous (2010) Training manual on e-Pest surveillance (awareness and surveillance programme for management of major pests of paddy) under Rashtriya Krishi Vikas Yojna-2nd green revolution. Directorate of Agriculture & Food Production, Bhubaneswar, p 54

Ascerno ME (1991) Insect phenology and integrated pest management. J Arboric 17:13–15

Baker CRB (1991) The validation and use of a life-cycle simulation model for risk assessment of insect pests. Bull OEPP 21:615–622

Bale JS, Hayward SAL (2010) Insect overwintering in a changing climate. J Exp Biol 213:980–994

Bale J, Masters GJ, Hodkins ID et al (2002) Herbivory in global climate change research: direct effects of rising temperature on insect herbivores. Global Change Biol 8:1–16

Behrens W, Hoffmann KH, Kempa S, Gabler S, Merkel-Wallner G (1983) Effects of diurnal thermoperiods and quickly oscillating temperatures on the development and reproduction of crickets, *Gryllus bimaculatus*. Oecologia 59:279–287

Briere JF, Pracros P, Le Roux AY, Pierre JS (1999) A novel rate model of temperature-dependent development for arthropods. Environ Entomol 28:22–29

Cannon RJC (1998) The implications of predicted climate change for insect-pests in the UK, with emphasis on non-indigenous species. Global Change Biol 4:785–796

Chahal SK, Bains GS, Dhaliwal LK (2008) Climate change: mitigation and adaptation. In: Abstracts of international conference on climate change, biodiversity and food security in the South Asian region. Punjab state council for science and technology. Chandigarh and United Nations Educational, Scientific and Cultural Organization, New Delhi, 3–4 Nov 2008

Chand R, Raju SS (2009) Instability in Indian agriculture during different phases of technology and policy. Indian J Agric Econ 64:187–207

Choudhary JS, Prabhakar CS, Moanaro DB, Kumar S (2013) Litchi stink bug (*Tessaratoma javanica*) outbreak in Jharkhand, India, on litchi. Phytoparasitica 41:73–77

Curry GL, Feldman RM, Smith KC (1978) A stochastic model for a temperature dependent population. Theor Popul Biol 13:197–213

Dhawan AK, Singh K, Saini S, Mohindru B, Kaur A, Singh G, Singh S (2007) Incidence and damage potential of mealybug, *Phenaccous soloenposis* Tinsley on cotton in Punjab. Indian J Ecol 34:110–116

Elith J, Graham CH, Anderson RP, Dudik M, Ferrier S, Guisan A, Hijmans RJ, Huettmann F, Leathwick JR, Lehmann A, Li J, Lohmann LG, Loiselle BA, Manion BA, Manion G, Moritz C, Nakamura M, Nakazawa Y, Overton JMC, Peterson AT, Phillips SJ, Richardson KS, Scachetti-Pereira R, Schapire RE, Soberson J, Williams S, Wisz MS, Zimmermann NE (2006) Novel methods improve prediction of species' distributions from occurrence data. Ecography 29:129–151

Elith J, Phillips SJ, Hastie T, Dudik M, Chee YE, Yates CJ (2011) A statistical explanation of MaxEnt for ecologists. Diver Distrib 17:43–57

Fand BB (2012) Modeling the impact of climate change on potential geographic distribution of polyphagous mealybug *Phenacoccus solenopsis* in India. In: Abstracts of the "IVth national symposium on plant protection in horticultural crops-emerging challenges and sustainable pest management. Association for the advancement of pest management in horticultural ecosystems", Bangalore, 25–28 Apr 2012

Fand BB, Kamble AL, Kumar M (2012) Will climate change pose a serious threat to crop pest management? Int J Sci Res Pub 2:1–15. http://www.ijsrp.org/research-paper-1112.php?rp=P11336

Fand BB, Tonnang HEZ, Kumar M, Kamble AL, Bal SK (2013) Temperature based phenology modeling and GIS-based risk mapping: a tool for forecasting potential changes in the abundance of mealybug *Phenacoccus solenopsis* Tinsley (Hemiptera: Pseudococcidae). In: Proceedings of the IVth international insect science congress. University of Agricultural Sciences, Bangalore, 14–18 Feb 2013

Firake DM, Behere GT, Firake PD, Azad Thakur NS, Dubal ZB (2012) An outbreak of pine lappet moth, *Kunugia latipennis*, in mid-altitude hills of Meghalaya state, India. Phytoparasitica 40:231–234

Ganeshaiah KN, Barve N, Nath N, Chandrashekara K, Swamy M, Uma Shaanker R (2003) Predicting the potential geographical distribution of the sugarcane woolly aphid using GARP and DIVA-GIS. Curr Sci 85:1526–1528

Govindasamy B, Duffy PB, Coquard J (2003) High-resolution simulations of global climate, part 2: effects of increased greenhouse gases. Climate Dyn 21:391–404

Gutierrez AP, Pointi L, d'Oultremont T, Ellis CK (2008) Climate change effects on poikilotherm tritrophic interactions. Climate Change 87:67–92

Harrington R, Fleming R, Woiwood IP (2001) Climate change impacts on insect management and conservation in temperate regions: can they be predicted? Agric For Entomol 3:233–240

Hijmans RJ, Cameron SE, Parra JL, Jones PG, Jarvis A (2005a) Very high resolution interpolated climate surfaces for global land areas. Int J Climatol 25:1965–1978

Hijmans RJ, Guarino L, Jarvis A, O'Brien R, Mathur P, Cruz M, Barrantes I, Rojas Eb (2005b) DIVA-GIS (Version 5.2) manual. p 79

Hilder VA, Boulter D (1999) Genetic engineering of crop plants for insect resistance – a critical review. Crop Prot 18:177–191

Hurlbert SH (1990) Spatial distribution of the montane unicorn. Oikos 58:257–271

IMD (2010) Annual climate summary 2010, India Meteorological Department (Government of India, Ministry of earth sciences). Pune, p 27

IPCC (2007) Climate Change 2007 – Impacts, adaptation and vulnerability contribution of Working Group II to the Fourth Assessment Report of the IPCC. In: Parry ML, Canziani OF, Palutikof JP, van der Linden PJ, Hanson CE (eds) Cambridge University Press, Cambridge, p 976

IARI News (2008) Brown plant hopper outbreak in rice. 24:1–2

IRRI News (2009) Pest outbreaks in India. Rice today 6. www.ricenews.irri.org

Iwao S (1972) Application of the m-m method to the analysis of spatial patterns by changing the quadrat size. Res Popul Ecol 14:97–128

Jarvis CH, Baker RHA (2001a) Risk assessment for non-indigenous pests: 2. Accounting for inter-year climate variability. Diver Distrib 7:237–248

Jarvis CH, Baker RHA (2001b) Risk assessment for non-indigenous pests: 1. Mapping the outputs of phenology models to assess the likelihood of establishment. Diver Distrib 7:223–235

Jhala RC, Bharpoda TM, Patel MG (2008) *Phenacoccus solenopsis* (Hemiptera: Pseudococcidae), the mealybug species recorded first time on cotton and its alternate host plants in Gujarat, India. Uttar Pradesh J Zool 28:403–406

Joshi S, Viraktamath CA (2004) The sugarcane woolly aphid, *Ceratovacuna lanigera* Zehntner (Hemiptera: Aphididae): its biology, pest status and control. Curr Sci 87:307–316

Jumars PA, Thistle D, Jones ML (1977) Detecting two-dimensional spatial structure in biological data. Oecologia 28:109–123

Kaiser J (1996) Pests overwhelm Bt cotton crop. Nature 273:423

Kohlmann B, Nix H, Shaw DD (1988) Environmental predictions and distributional limits of chromosomal taxa in the Australian grasshopper *Caledia captiva* (F.). Oecologia 75:483–493

Kranthi KR, Naidu S, Dhawad CS et al (2005) Temporal and intraplant variation in Cry1Ac expression of Bt cotton and its influence on the development of cotton bollworm, *Helicoverpa armigera* (Hubner), (Noctuidae, Lepidoptera). Curr Sci 89:291–298

Kriticos DJ, Brown JR, Maywald GF, Radford ID, Nicholas DM, Sutherst RW, Adkins SW (2003) SPAnDX: a process-based population dynamics model to explore management and climate change impacts on an invasive alien plant, *Acacia nilotica*. Ecol Model 163:187–208

Kroschel J, Sporleder M, Tonnang HEZ, Juarez H, Carhuapoma P, Gonzales JC, Simon R (2012) Predicting climate-change-caused changes in global temperature on potato tuber moth *Phthorimaea operculella* (Zeller) distribution and abundance using phenology modeling and GIS mapping. Agric For Meteorol 170:228–241, http://dx.doi.org/10.1016/j.agrformet.2012.06.017

Kumar R, Khan ZH, Ramamurthy VV (2009) Topical outbreak of migratory locust, *Locusta migratoria migratorioides* (Reiche & Fairmaire) in Ladakh valley of Jammu and Kashmir. Biol Forum 1:107–109

Lactin DJ, Holliday NJ, Johnson DL, Craigen R (1995) Improved rate model of temperature-dependent development by arthropods. Env Entomol 24:68–75

Lal M (2003) Global climate change: India's monsoon and its variability. J Environ Stud Policy 6:1–34

Liebhold AM, Rossi RE, Kemp WP (1993) Geostatistics and geographic information systems in applied insect ecology. Ann Rev Entomol 38:303–327

Logan JA (1988) Toward an expert system for development of pest simulation models. Environ Entomol 17:359–376

Logan JA, Wollkind DJ, Hoyt SC, Tanigoshi LK (1976) An analytical model for description of temperature dependent phenomenon in arthropods. Environ Entomol 5:1133–1140

Mittler R (2006) Abiotic stress, the field environment and stress combination. Trends Plant Sci 11:15–19

Mooney HA, Hobbs RJ (eds) (2000) Invasive species in a changing world. Island Press, Washington, DC

Nagrare VS, Kranthi S, Biradar VK, Zade NN, Sangode V, Kakde G, Shukla RM, Shivare D, Khadi BM, Kranthi KR (2009) Widespread infestation of the exotic mealybug species, *Phenacoccus solenopsis* Tinsley (Hemiptera: Pseudococcidae) on cotton in India. Bull Entomol Res 99:537–541

Nagrare VS, Kranthi S, Kumar R, Dhara Jothi B, Amutha M, Deshmukh AJ, Bisane KD, Kranthi KR (2011) Compendium of cotton mealybugs. Central Institute for Cotton Research, Nagpur, 42 p

Nietschke BS, Magarey RD, Borchert DM, Calvin DD, Jones E (2007) A developmental database to support insect phenology models. Crop Prot 26:1444–1448

Odum EP, Barrett GW (2008) Fundamentals of ecology, 2nd edn. Cengage Learning India Pvt. Ltd., New Delhi, pp 1–10

Oerke EC (2006) Crop losses to pests. J Agric Sci 144:31–43

Oerke EC, Debne HW, Schonbeck F, Weber A (1994) Crop production and crop protection. Elsevier, Amsterdam

Parmesan C, Yohe G (2003) A globally coherent fingerprint of climate change impacts across natural systems. Nature 421:37–42

Peacock L, Worner SP (2006) Using analogous climates and global insect distribution data to identify potential sources of new invasive insect pests in New Zealand. N Z J Zool 33:141–145

Pedigo LP (2006) Entomology and pest management, 4th edn. Dorling Kindersley (India) Pvt. Ltd., New Delhi, pp 175–210

Peterson AT (2003) Indo-US forum workshop on eco-informatics. ATREE, Bangalore, 9–11 June

Petzoldt C, Seaman A (2010) Climate change effects on insects and pathogens, climate change and agriculture: promoting practical and profitable responses, New York State IPM Program, 630 W. North St. New York State Agricultural Extension Station, Geneva, NY 14456, pp 6–16. http://www.climateandfarming.org/pdfs/FactSheets/III.2Insects.Pathogens.pdf

Phillips SJ, Dudikm M, Schapire RE (2004) A maximum entropy approach to species distribution modeling. In: Proceedings of the 21st international conference on machine learning. Banff

Phillips SJ, Anderson RP, Schapire RE (2006) Maximum entropy modeling of species geographic distributions. Ecol Model 190:231–259

Porter JH, Parry ML, Carter TR (1991) The potential effects of climatic change on agricultural insect-pests. Agric Forest Meteorol 57:221–240

Rafee CM (2010) Status of sugarcane woolly aphid, *Certovacuna lanigera* Zehntner in Thunga bhadra command area. Karnataka J Agric Sci 23:127

Rao MS, Khan MA, Srinivas K, Vanaja M, Rao GGS, Ramakrishna YS (2006) Effects of elevated carbon dioxide and temperature on insect-plant interactions – a review. Agric Rev 27:200–207

Rao GGSN, Rao AVMS, Rao VUM (2009) Trends in rainfall and temperature in rainfed India in previous century. In: Aggarwal PK (ed) Global climate change and Indian agriculture case studies from ICAR network project. ICAR Publication, New Delhi, pp 71–73

Ratte HT (1985) Temperature and insect development. In: Hoffman KH (ed) Environmental physiology and biochemistry of insects. Springer, New York, pp 33–66

Reed W, Pawar CS (1982) Heliothis: a global problem. In: ICRISAT proceedings of the international workshop on Heliothis Management, international crops research institute for the semi-arid tropics, Hyderabad, 15–20 Nov 1981

Rhoades DF (1985) Offensive-defensive interactions between herbivores and plants: their relevance in herbivore population dynamics and ecological theory. Am Nat 125:205–238

Samra JS (2003) Impact of climate and weather on Indian agriculture. Indian Soc Soil Sci 51:418–430

Samra JS, Singh G (2004) Heat wave of March 2004: impact on agriculture. Indian Council of Agricultural Research, New Delhi

Sawyer AJ (1989) Inconstancy of Taylor's b: simulated sampling with different quadrat sizes and spatial distributions. Res Popul Ecol 31:11–24

Scheldeman X, van Zonneveld M (2010) Training manual on spatial analysis of plant diversity and distribution. Bioversity International, Rome, p 179

Schoolfield RM, Sharpe PJH, Magnuson CE (1981) Non-linear regression of biological temperature-dependent rate models based on absolute reaction-rate theory. J Theor Biol 88:719–731

Sharma HC, Dhillon MK, Kibuka J, Mukuru SZ (2005) Plant defense responses to sorghum spotted stem borer, *Chilo partellus* under irrigated and drought conditions. Int Sorghum Millets Newsl 46:49–52

Sharpe PJH, DeMichele DW (1977) Reaction kinetics of poikilotherm development. J Theor Biol 64:649–670

Sharpe PJH, Schoolfield RM, Butler GD (1981) Distribution model of *Heliothis zea* (Lepidoptera: Noctuidae) development times. Can Entomol 113:845–856

Sporleder M, Kroschel J, Gutierrez Quispe MR, Lagnaoui A (2004) A temperature-based simulation model for the potato Tuberworm, *Phthorimaea operculella* Zeller (Lepidoptera; Gelechiidae). Environ Entomol 33:477–486

Sporleder M, Simon R, Juarez H, Kroschel J (2008) Regional and seasonal forecasting of the potato tuber moth using a temperature-driven phenology model linked with geographic information systems. In: Kroschel J, Lacey L (eds) Integrated pest management for the potato Tuber Moth, *Phthorimaea operculella* Zeller – a potato pest of global importance. Tropical agriculture, advances in crop research. Margraf Publishers, Weikersheim, pp 15–30

Sporleder M, Simon R, Gonzales J, Carhuapoma P, Juarez H, De Mendiburu F, Kroschel J (2009) ILCYM – Insect life cycle modeling: a software package for developing temperature-based insect phenology models with applications for regional and global pest risk assessments and mapping. In: proceeding of international society for tropical root crops (ISTRC). Lima

Sporleder M, Tonnang HEZ, Juarez H, Carhuapoma P, Gonzales JC, Simon R, Kroschel J (2012) ILCYM – insect life cycle modeling (version 3.0): a software package for developing temperature-based insect phenology models with applications to regional and global analysis of insect population and mapping (user manual). International Potato Center, Lima, p 124

Srikanth J (2004) The epidemic of sugarcane woolly aphid Ceratovacuna lanigera Zehntner (Homoptera: Aphididae) in western India: an appraisal. In: Balasundaram N (ed) Proceedings of national seminar on use of appropriate varieties and management practices for improving recovery of sugarcane, Sugarcane Breeding Institute, Coimbatore

Srikanth J (2007) World and Indian scenario of sugarcane woolly aphid. In: Mukunthan N, Srikanth J,

Singaravelu B, Rajula Shanthy T, Thiagarajan R, Puthira Prathap D (eds) Woolly aphid management in sugarcane, vol 154. Extension Publication, Sugarcane Breeding Institute, Coimbatore, pp 1–12

Steinbauer MJ, Yonow T, Reid IA, Cant R (2002) Ecological biogeography of species of *Gelonus, Acantholybas* and *Amorbus* in Australia. Aust Ecol 27:1–25

Stinner RE, Gutierrez AP, Butler GD (1974) An algorithm for temperature-dependent growth rate simulation. Can Entomol 106:519–524

Stockwell DRB, Noble IR (1992) Induction of sets of rules from animal distribution data: a robust and informative method of data analysis. Math Comp Simul 33:385–390

Sutherst R (2000) Climate change and invasive species: a conceptual framework. In: Mooney H, Hoobs R (eds) Invasive species in a changing world. Island Press, Washington, DC, pp 211–240

Sutherst RW, Maywald GF (1990) Impact of climate change on pests and diseases in Australasia. Search (Sydney) 21:230–232

Sutherst RW, Maywald GF, Bottomly W (1991) From CLIMEX to PESKY, a generic expert system for risk assessment. EPPO Bull 21:595–608

Tanwar RK, Anand P, Panda SK, Swain NC, Garg DK, Singh SP, Sathya Kumar S, Bambawale OM (2010a) Rice swarming caterpillar (Spodoptera mauritia) and its management strategies, vol 24, Technical bulletin. National Centre for Integrated Pest Management, New Delhi

Tanwar RK, Jeyakumar P, Vennila S (2010b) Papaya mealybug and its management strategies, vol 22, Technical bulletin. National Centre for Integrated Pest Management, New Delhi, p 26

Taylor LR (1984) Assessing and interpreting the spatial distributions of insect populations. Ann Rev Entomol 29:321–357

Taylor LR, Woiwood IP, Perry JN (1978) The density-dependence of spatial behaviour and the variety of randomness. J Animal Ecol 47:383–406

Theilert W (2006) A unique product: the story of the imidacloprid stress shield. Pflanzenschutz-Nachrichten Sci Forum Bayer 59:73–86

Tripathi GM, Singh SK, Kumar M (2008) Role of biotic and abiotic factors on the population dynamics of sugarcane woolly aphid, *Ceratovacuna lanigera* Zehntner and its natural enemies in sugarcane. Curr Sci 94:718–720

Trnka M, Muska F, Semeradova D, Dubrovsky M, Kocmankova E, Zalud Z (2007) European corn borer life stage model: regional estimates of pest development and spatial distribution under present and future climate. Ecol Model 207:61–84

Van Sickle GA (1989) GIS-a tool in forest pest management. In: Proceedings of GIS-89: a wider perspective. Forestry, Canada, pp 213–219

Venette RC, Kriticos DJ, Magarey RD, Koch FH, Baker RHA, Worner SP, Gomez NN, Raboteaux DW, McKenney EJ, Dobesberger D, Yemshanov PJ, De Barro WD, Hutchison G, Fowler T, Pedlar J, Kalaris M (2010) Pest risk maps for invasive alien species: a roadmap for improvement. Bio Sci 60:349–362

Vennila S, Prasad YG, Prabhakar M, Rishi K, Nagrare V, Amutha MD, Agarwal M, Sreedevi G, Venkateswarlu B, Kranthi KR, Bambawale OM (2011) Spatio-temporal distribution of host plants of cotton mealybug, *Phenacoccus solenopsis* Tinsley in India, vol 26, Technical bulletin. National Centre for Integrated Pest Management, New Delhi, p 50

Wagner TL, Wu HI, Sharpe PJH, Coulson RN (1984) Modeling distributions of insect development time: a literature review and application of the Weibull function. Ann Entomol Soc Am 77:474–487

Wagner TL, Wu HI, Feldman RM, Sharpe PJH, Coulson RN (1985) Multiple cohort approach for simulating development of insect populations under variable temperatures. Ann Entomol Soc Am 78:691–704

Wang XY, Huang XL, Jiang LY, Qiao GX (2010a) Predicting potential distribution of chestnut phylloxerid (Hemiptera: Phylloxeridae) based on GARP and Maxent ecological niche models. J Appl Entomol 134:45–54

Wang Y, Watson GW, Zhang R (2010b) The potential distribution of an invasive mealybug *Phenacoccus solenopsis* and its threat to cotton in Asia. Agric Fores Entomol 12:403–416

Ward NL, Masters GJ (2007) Linking climate change and species invasion: an illustration using insect herbivores. Global Change Biol 13:1605–1615

Wilmot Senaratne KAD, Palmer WA, Sutherst RW (2006) Use of CLIMEX modelling to identify prospective areas for exploration to find new biological control agents for prickly acacia. Aust J Entomol 45:298–302

Worner SP (1992) Performance of phenological models under variable temperature regimes: consequences of the Kaufmann or rate summation effect. Environ Entomol 21:689–699

Yadav DS, Subhash C, Selvraj K (2010) Agro-ecological zoning of brown planthopper, {*Nilaparvata lugens* (Stal.)} incidence on rice {*Oryza sativa* (L.)}. J Sci Ind Res 69:818–822

Yamamura K, Kiritani K (1998) A simple method to estimate the potential increase in the number of generations under global warming in temperate zones. Appl Entomol Zool 33:289–298

Yonow T, Zalucki MP, Sutherst RW, Dominiak BC, Maywald GF, Maelzer DA, Kriticos DJ (2004) Modeling the population dynamics of the Queensland fruit fly *Bactrocera (Dacus) tryoni* – a cohort based approach incorporating the effects of weather. Ecol Model 173:9–30

Zalucki MP, Furlong MJ (2005) Forecasting *Helicoverpa populations* in Australia: a comparison of regression based models and a bioclimatic based modelling approach. Insect Sci 12:45–56

Influence of Moisture Stress on Growth, Development, Physiological Process and Quality of Fruits and Vegetables and Its Management Strategies

P. Suresh Kumar, P.S. Minhas, V. Govindasamy, and R.L. Choudhary

Abstract

According to world estimates, only 9 % of the area is conducive for crop production. Abiotic stresses like temperature, water, radiation and hazardous chemicals/pollutants are responsible for major reduction in agricultural production. Fruits and vegetables can be potentially exposed to numerous abiotic stresses during production and distribution. Water, an integral part of living systems, is ecologically important because it is a major force in shaping climatic patterns and is biochemically important because it is a necessary component in physiological processes. Plant acclimation to drought stress depends on the process like osmotic adjustment and osmoregulation. Among hydrophilic proteins, several LEA proteins including dehydrins accumulate in various plant parts during the process of osmotic adjustment. ABA signalling pathway plays a vital role in plant stress responses. Moisture stress has also been shown to induce phenolic accumulation through up-regulation of phenylalanine ammonia lyase (PAL) and proline accumulation. Reactive oxygen species (ROS) levels increase when plants are exposed to abiotic stress conditions. Carbon metabolism and the levels of specific sugars are severely affected by abiotic stress. Deficit irrigation (DI) has effects on fruit maturation and ripening. Shallow-rooted vegetables are known to be sensitive to water deficiency. Some of the basic methods to collect and conserve water are harvesting of rainwater, development of catchment area and storing run-off water for recycling and construction of waterways. Crop rotation, conservation agriculture, cover crops, strip cropping, mulching, contour bunds, contour drains, terraces, contour furrow, stabilisation structures, soil conservation and water retention through use of vegetation are some of the agronomical measures used to conserve moisture. Systems biology approach which integrates the series of -omics, i.e. phenomics, genomics,

P.S. Kumar (✉) • P.S. Minhas • V. Govindasamy
R.L. Choudhary
National Institute of Abiotic Stress Management,
Malegaon, Baramati, Pune, Maharashtra, India
e-mail: psureshars@gmail.com

R.K. Gaur and P. Sharma (eds.), *Approaches to Plant Stress and their Management*,
DOI 10.1007/978-81-322-1620-9_7, © Springer India 2014

transcriptomics, proteomics and metabolomics, is getting roots on moisture stress alleviation and evolving new breeding strategies. Growth regulator applications can also potentially enhance stress resistance. Plant growth-promoting rhizobacteria (PGPR) and fungi (mycorrhizae) can facilitate plant growth directly by facilitating the uptake of nutrients from the environment, by influencing phytohormone production and by enzymatic lowering of plant ethylene levels.

Keywords
Fruits • Vegetables • Moisture stress • Quality • Omics • Microbes • Systems biology

Introduction

Since more than 1,000 years ago, research on evolution has furnished abundant experimental evidence on how the genetic variability of organisms (Perezdela 2004) and the selectivity of environmental influences could have led to the present-day diversity of plant, animal and human forms and also to what extent this process continues in the presence and absence of human activity in altering the environment (Gupta 2004; Bossdorf et al. 2005). Fundamentally, plants require energy (light), water, carbon and mineral nutrients for growth. Abiotic stress can be termed as the negative impact of environmental factors on the organisms in a specific situation. It is a natural phenomenon that occurs in multiples and interdependent, and its impact varies across the sectors of agriculture. Unlike a biotic stress that would include such living disturbances as fungi or harmful insects, abiotic stress factors or stressors are naturally occurring, often intangible, factors such as intense sunlight or wind that may cause harm to the plants. Plants are especially dependent on environmental factors, so it is particularly constraining. The most common of the stressors is the easiest for people to identify, but there are many other less recognisable abiotic stress factors which affect environments constantly. The abiotic stresses like temperature (heat, cold chilling/frost), water (drought, flooding/hypoxia), radiation (UV, ionising radiation), chemicals (mineral/nutrient deficiency/excess, pollutants heavy metals/pesticides, gaseous toxins) and mechanical (wind, soil movement, submergence) are responsible for major reduction in agricultural production. The lesser-known stressors generally occur on a smaller scale and so are less noticeable; they include poor edaphic conditions like physical and physico-chemical properties, high radiation, compaction, contamination and rapid dehydration during seed germination.

Climatic variability is the biggest challenging factor that affects agriculture in India and elsewhere (Goyary 2009). Drought, salinity, extreme temperatures and oxidative stress are interconnected and affect the water relations of a plant on the cellular as well as whole plant level causing specific as well as unspecific reactions (Beck et al. 2007a). This leads to a series of morphological, physiological, biochemical and molecular changes that adversely affect plant growth and productivity (Wang et al. 2001). Dealing with the abiotic stresses is in reality a highly onerous task owing to their complexity, uncertainty and differential impacts over the time and place. According to world estimates, an average of 50 % yield losses in agricultural crops are caused by abiotic factors (Oerke et al. 1994; Sanchez-Rodríguez et al. 2011), mostly shared by high temperature (20 %), low temperature (7 %), salinity (10 %), drought (9 %) and other forms of stresses (4 %). FAO report stated that only 3.5 % of the global land area is not affected by some environmental constraint (FAO 2007). Plant responses to abiotic stresses are dynamic and complex (Skirycz and Inze 2010; Cramer 2010); they are both elastic (reversible) and plastic (irreversible). The plant responses to stress are dependent on the tissue or organ affected by the stress. In addition, the level and duration of stress (acute vs. chronic) can have a significant effect on the complexity of the response (Tattersall et al. 2007).

The decade has witnessed the development of horticulture in India, which has resulted in appreciable growth and laudable achievements, owing to technological advancement and policy environment (Singh 2011). Horticulture contributes 29.65 % to agricultural GDP with the growth rate of 5–6 % during the decade. The expansion of horticulture in nontraditional areas resulted in increase in area from 12.77 million hectares (MHa) in 1991–1992 to 20.66 million hectares in 2008–2009 with an increased production of about 12 million tons (MT). India has emerged as the second largest producer of fruits and vegetables and has first position in several horticultural crops. Fruits and vegetables can be potentially exposed to numerous abiotic stresses during production, handling, storage and distribution (Hodges 2003). Hence understanding of effects of abiotic stresses on pre- and postharvest stress susceptibility will become more important since they limit the storage and shelf life potential of fruits and vegetables (Toivonen and Beveridge 2005).

Major Abiotic Stresses Which Influence the Horticultural Crop Growth and Production

Abiotic stresses occurring during production can either be the primary cause (direct) for disorders that exhibit themselves during postharvest handling and storage practices or they can influence the susceptibility of a fruit or vegetable to postharvest conditions that cause abiotic stresses resulting in disorders (indirect) (Ferguson et al. 1999). It is important to characterise the relationship between preharvest abiotic stresses occurring during production and postharvest abiotic stresses that the fruit or vegetable is exposed after harvest and during storage and distribution, since the solution to these different problems will be best resolved by focusing on preharvest or postharvest abiotic stress amelioration, respectively.

Drought

Water stress refers to the situation where cells and tissues are less than truly turgid. It occurs whenever the loss of water in transpiration exceeds the rate of absorption. When plants are unable to absorb enough water to replace that lost by transpiration, water stress develops in the plant system. The results may be wilting, reduction in photosynthesis, disturbances in physiological processes, cessation of growth or even death of the plant or plant parts. In fact water stress practically affects every aspect of plant growth, modifying the anatomy, morphology, physiology and biochemistry which are related with decrease in turgor, water potential and osmotic potential (Chundawat 1990). The occurrence of drought conditions during production of fruit and vegetable crops is becoming more frequent with climate change patterns (Whitmore 2000). Much work has been devoted to understanding of drought effects on production and productivity of horticultural crops (Whitmore 2000). The existing literature provides some insight which may lead to better understanding and perhaps also encourage future research. Water stress during the production phase of some fruits and vegetables may affect their physiology and morphology in such a manner as to influence susceptibility to weight loss in storage. There have been both positive effects reported for field water deficits (stress) in tree fruits and root vegetables. Size of fruit is important, since larger fruit has lower surface area to volume ratios, which confers lower relative water loss (Shibairo et al. 1997). Another negative effect associated with water deficits is the case of root vegetables, such as carrot, where preharvest water stress (watering to 25–75 % of soil water field capacity) can weaken the cells, resulting in higher membrane leakage (i.e. cell damage) and consequently greater weight loss in storage (Shibairo et al. 1998b). Timing of a water stress event can also be very important in determining response to postharvest abiotic stress response. One example is that 'Kensington' mango fruit (*Mangifera indica* L.) will be significantly more susceptible to postharvest chilling injury with exposure to water stress during the cell expansion phase of growth as opposed to being exposed to the stress during cell division or at a time near to harvest maturity (Léchaudel and Joas 2007). Therefore it is critical to avoid water stress until the fruit has reached maximum size in order to minimise incidence of chilling-induced injury in storage. Water stress, particularly at the tuber-forming stage, can also lead to a higher susceptibility of potatoes to postharvest development of black spot disorder (Hamouz et al. 2011). Black spot disorder is correlated primarily to susceptibility of cells in the potatoes to undergo decompartmentation in response to mechanical stress (i.e. bruising) (Stevens and Davelaar 1997).

Temperature Extremes

Susceptibility to high (heat injury inducing) or low (chill injury inducing) temperatures is known to be reduced by prior exposure of the sensitive fruit or vegetable to low ambient temperatures (Saltveit and Morris 1990). However, if the preharvest temperature leads to chilling-induced injury in the field, then susceptibility to postharvest chilling injury can be increased. Therefore, the level of the preharvest temperature extreme will be a determinant as to if the exposure will have positive or negative effects on postharvest stress sensitivity. Extreme high temperatures can occur in the field and apple fruit exposed to direct sunlight can reach in excess of 40 °C (Ferguson et al. 1999). High temperatures during the late season (leading up to harvest) can enhance susceptibility of apples to superficial scald which develops in storage (Bramlage and Weis 1997).

Salinity

Tomatoes grown under high salinity will produce smaller fruit with higher soluble solids (Mizrahi 1982). Smaller fruit will have higher surface area to volume ratios, hence greater susceptibility to postharvest water loss (i.e. desiccation stress) (Shibairo et al. 1997; Datta and de Jong 2002). While there is no direct information in the literature to confirm that smaller tomato fruit from saline growing conditions would be subject to greater desiccation stress postharvest, firmness declines for tomatoes grown under 3 and 6 dS m^{-1} salinity levels were increased by 50–130 %, respectively, at 2 weeks holding at 20 °C compared with control fruit (Mizrahi et al. 1988).

Light

It would be considered logical to assume that effects of exposure to high light are difficult to dissociate from effects exposure to high temperatures. However, research has shown that low light (bagging of apples) in the preharvest interval reduced susceptibility of apples to developing superficial scald in cold storage (Barden and Bramlage 1994) and, in contrast, high ambient temperatures resulted in increased susceptibility (Bramlage and Weis 1997). Generally, only sun-exposed surfaces of susceptible cultivars of apples develop scald in storage (Ferguson et al. 1999). In the case of ambient low light, when lettuce is grown under low light which is suboptimal for photosynthetic activity, shelf life of fresh-cut lettuce (i.e. lettuce subjected to mechanical stress) is much shorter than lettuce produced under optimal light conditions (Witkowska and Woltering 2010). Tomato size is smaller when the crop is grown under ambient low light levels, such as in the early spring season in northern latitudes (Gruda 2005), and since surface area to volume ratio is greater in smaller fruits, susceptibility to postharvest desiccation stress would increase (Shibairo et al. 1997). Low light also results in lower levels of ascorbate in many greenhouse-grown fruits and vegetables (Gruda 2005), which would render them less fit to deal with postharvest stresses since ascorbate contents are general directly proportional to relative levels or stress tolerance (Noctor and Foyer 1998).

Plant Nutrition

There is limited literature regarding the effects of crop nutrition on crop growth and quality. There is one review dealing with the effect of preharvest nutrition on postharvest physiology and disorders of fruits and vegetables (Sams and Conway 2003); however, most of the reviewed literature touches on nutrition effects on postharvest biotic stress effects (i.e. disease resistance). Calcium nutrition during production has been well documented in regard to postharvest disorders of many fruits and some vegetables (Sams and Conway 2003). Calcium is also been suggested as a putative signalling molecule involved in the development of cross-tolerance to abiotic stresses (Bowler and Fluhr 2000). Therefore the role of preharvest calcium nutrition in postharvest stress resistance may be complex and dependent on whether the fruit or vegetable is also exposed to environmental abiotic stresses. Potassium nutrition has been shown to have a few important effects on postharvest abiotic stress susceptibility of vegetables. In carrots, deficiency in potassium is associated with greater weight loss (desiccation stress) in storage (Shibairo et al. 1998a). At levels below 1 mM potassium in the soil medium, weight loss was directly associated with increased membrane leakage (i.e. damaged cells) in the carrot tissues. Above 1 mM potassium, there were no significant differences in weight loss under standardised storage conditions (Shibairo et al. 1998a). Improved potassium nutrition has also been shown to reduce susceptibility of potatoes to internal bruising in response to mechanical stresses imposed during postharvest handling (Maier et al. 1986; McGarry et al. 1996). Relatively high preharvest nitrogen is often associated with poor postharvest quality of many fruits and vegetables (Sams and Conway 2003). In regard to affecting susceptibility to postharvest stress, applying higher than recommended levels of preharvest nitrogen for a specific crop has been linked to storage discoloration susceptibility in both cabbage and potato. In the case of cabbage, it appears that excessive nitrogen fertilisation leads to high accumulations of zinc and aluminium and nitrate-induced manganese deficiency (Berard 1990). High nitrogen applications in the field resulted in increased incidence and severity of black midrib in cold storage, particularly for the susceptible cultivar, 'safe keeper' (Berard 1990). In the case of potatoes, black spot, susceptibility (a consequence of bruising; Stevens and Davelaar 1997) is influenced by nitrogen fertilisation, particularly the balance of nitrogen applied in relation to levels of potassium applied (Horneburg and Wirsing 1995). In contrast nitrogen deficiency or lower than recommended nitrogen application rates will most often result in increased vitamin C content in many fruits and vegetables (Lee and Kader 2000). Vitamin C content has been tightly linked with storage life potential (Hodges et al. 2001) which is likely a consequence of the importance of this antioxidant nutrient in forestalling oxidative injury that leads to quality losses in storage (Noctor and Foyer 1998). Among the above-mentioned different abiotic stressors, stress due to moisture especially drought stress is most predominant in the context of Indian condition. Following few pages are dedicated to explain the intricacies of moisture stress, and we tried to infer some conclusion with available literature on moisture stress.

Insight on Moisture Stress: Effect on Physiology, Growth, and Development of Horticultural Crops

Worldwide, 6,510 million hectares (mha) of land is under rainfed agriculture of which approximately 60 % is in the developing countries. India ranks first among the dry land agricultural countries in terms of both extent and value of produce (DAC 2013). Out of a total 142.1 million hectares of cultivated area in India, dry land accounts for 91.0 million hectares and in the foreseeable future also nearly 60 % of our population will still continue to depend on dry land/ rainfed agriculture. Water, an integral part of living systems, is ecologically important because it is a major force in shaping climatic patterns and is biochemically important because it is a necessary component in physiological processes (Brown 1995). Water plays a key role in transpiration and photosynthesis and regulates the stomata and thus is crucial to growth and leaf expansion of plants. In addition, it is the primary solvent in physiological processes by which gases, minerals and other materials enter plant cells and by which these materials are translocated to various parts of the plant.

Too much rain, which can drown the crop, delay harvest and accelerate soil erosion, can be just as serious as too little rain. It is apparent that annual rainfall averages alone are not a dependable gauge of the rainfall in an area, although it gives a good general indication of the amount of moisture available for crop production. The amount of rainfall that actually ends up stored in the soil for crop use depends on other factors such as water run-off and evaporation from the soil surface. The rate at which crops use water is highest under hot, dry conditions and the lowest when it is very humid. In contrast, under flood condition, all pores are filled with water; so the oxygen supply is almost completely deprived (water logging) and plant roots cannot obtain oxygen for respiration to maintain their activities for nutrient and water uptake. Plants weakened by lack of oxygen are much more susceptible to diseases caused by soilborne pathogens. Flood tolerance is dependent upon crop species, prior plant stress (e.g. freezing weather, drought), crop load, air temperatures (warm temperatures are more detrimental), soil type, flooding depth and duration. When plants are unable to absorb enough water to replace that lost by transpiration, water stress develops in the plant system (Whitmore 2000). The results may be wilting, cessation of growth or even death of the plant or plant parts. In India, irrigation is available for only 40 % of the cultivated area and the remaining 60 % depends on scanty rains.

Physiological Aspects of Drought Stress

Plant response to osmotic stress caused by drought is at the morphological, anatomical, cellular and molecular levels (Greenway and Munns 1980; Bohnert et al. 1995; Yeo 1998; Hasegawa et al. 2000; Pessarkli 2011). These changes include developmental changes such as a life cycle, inhibition of shoot growth and enhancement of root growth, adjustment in ion transport through uptake, extrusion and sequestration of ions and metabolic changes such as carbon metabolism and the synthesis of compatible solutes. Some of these signals are triggered by the primary osmotic stress signals, whereas others may result from secondary stresses caused by primary signals. These secondary signals may be phytohormones mainly ABA, reactive oxygen species (ROS) and intercellular second messengers such as phospholipids. In general plant responses are of three kinds: *maintenance of homeostasis, detoxification of harmful elements and recovery of growth.* ABA signalling pathway plays a vital role in plant stress responses as evidenced by the fact that many of drought-inducible genes studied to date are also induced by ABA (He et al. 2005). Similarly, *drought tolerance is dependent upon crop species, prior plant stress and crop load.* In addition, water deficit alters the cell wall nonenzymatically, for example, by the interaction of pectate and calcium (Boyer 2009). Furthermore, water conductance to the expanding cells is affected by aquaporin activity and xylem embolism (Parent et al. 2009; Nardini et al. 2011). The initial growth inhibition by water deficit occurs prior to any inhibition of photosynthesis or respiration (Vandeleur et al. 2008; Hummel et al. 2010).

Plant acclimation to drought stress which tries to eliminate excessive water loss lies in the osmotic adjustment, i.e. a decrease of cell water potential in order to diminish the difference in water potential between the plant cell and the ambient soil. This decline in osmotic potential as a response to water deficit can be achieved by solute accumulation within the plant cells or by a decreased cell volume leading to an increased concentration of osmotic solutes as water leaves from vacuole. These phenomena are described as osmoregulation and osmotic adjustment. *Osmoregulation* has been defined as the regulation of osmotic potential within a cell by the addition or removal of solutes from solution until the intracellular osmotic potential is approximately equal to the potential of the medium surrounding the cell (Turner and Jones 1980). *Osmotic adjustment* refers to the lowering of the water potential due to the net accumulation of solutes in response to water deficits. Osmotic adjustment is an important mechanism in drought tolerance because it enables (1) a continuation of cell expansion (Wyn Jones and Gorham 1983), (2) stomatal and photosynthetic adjustments (Ludlow 1980), (3) better plant growth and (4) yield production (Morgan 1983). The degree of the osmoregulatory process is affected by the rate of stress, the stress preconditioning, the organ type and the genetic variation between and within species. *Osmotic adjustment* is associated with accumulation of low-molecular soluble metabolites collectively called compatible solutes like low-molecular saccharides, monosaccharides, glucose and fructose; disaccharides, sucrose; oligosaccharides like raffinose, stachyose and verbascose (Garcia et al. 1997; Goddijn and Van Dun 1999; Garcia-Orenes et al. 2005); organic acids and sugar alcohols, mannitol, pinitol and sorbitol (Smith et al. 1987); nitrogen-containing compounds such as amino acids and amides such as glutamine and asparagine; quaternary ammonium compounds called betaines, alanine betaine, glycine betaine and imino acid proline; polyamines spermine, spermidine and putrescine (Rabe 1990; Rhodes and Hanson 1993; Pareek et al. 1997; Mansour 2000; Olvera-Carrillo et al. 2011) and with relatively high-molecular hydrophilic proteins inside the cells. Among hydrophilic proteins, several LEA proteins including dehydrins accumulate to the relatively high extents in various plant parts during the process of osmotic adjustment. Although proline role in plant osmotolerance remains controversial, proline is thought to contribute to osmotic adjustment, detoxification of ROS and protection of membrane integrity. Proline accumulation is believed to play adaptive roles in plant stress tolerance and has been proposed to act as a compatible osmolyte and to be the way to store carbon and nitrogen (Olvera-Carrillo et al. 2011). Proline has also been proposed to function as a molecular chaperone stabilising the structure of proteins, and proline accumulation can provide a way to buffer cytosolic pH and to balance cell redox status.

Effect of Moisture Stress on Metabolic Changes

During drought stress, protoplast volume shrinkage by water loss leads to loss of turgor, osmotic stress and a potential change of membrane potentials. Upon severe loss of water from the cells, membrane disintegration and abolition of metabolic processes occur (Mahajan and Tuteja 2006). Moisture stress can also inhibit the production and accumulation of lycopene in tomato (Hall 1964). Moisture stress has also been shown to induce phenolic accumulation through up-regulation of phenylalanine ammonia lyase (PAL) (Ke and Saltveit 1989). This up-regulation of PAL was associated with wound-induced ethylene production. Some of the metabolic shifts are mediated by stress response messengers (e.g. phenolics, suberin and isocoumarin accumulation), and others are a direct consequence of cellular disruption that occurs during wounding or bruising (e.g. methanethiol, allyl isothiocyanate and dimethyl sulphide accumulations). Enzymes such as polygalacturonase and pectinesterase may increase in activity leading to loss of cell wall structure and concomitant increases in soluble sugars (Inari et al. 2002). This may explain at least a component of the loss of firmness that has been observed with carrots as they lose water (Shibairo et al. 2002; Jacobo-Velázquez DA, Cisneros-Zevallos 2009). This hypothesis is borne out by results of work with cucumbers where water stress resulted in up-regulation of polygalacturonase activity, suggesting that water loss itself was not the only factor in causing softening of stressed fruit (Kubo et al. 2000). Another aspect of water stress is induction of ethylene production (Kubo et al. 2000), which may explain why water stress leads to accelerated ripening in bananas (Burdon et al. 1994) and accelerated senescence in bell peppers (Lurie et al. 1986).

Mechanisms for Moisture Stress Response at the Biochemical and Molecular Levels

Exposure of plants to unfavourable environmental conditions such as alteration of temperature, high light intensity, water availability, air pollutants or salt stress can increase the production of reactive oxygen species (ROS). This phenomenon is called oxidative stress and is known as one of the major causes of plant damage as a result of environmental stresses (Sunkar et al. 2003). ROS include hydrogen peroxide, hydroxyl radicals and superoxide anions. ROS are usually generated by normal cellular activities such as photorespiration and β-oxidation of fatty acids, but their levels increase when plants are exposed to biotic or abiotic stress conditions (Xiong and Zhu 2002). Superoxide radical is regularly synthesised in the chloroplast and mitochondria, though some quantity is also reported to be produced in microbodies. Hydroxyl radical can damage, thus fatally affecting plant metabolism and ultimately growth and yield (Sairam and Tyagi 2004). The capacity to scavenge ROS and to reduce their damaging effects on macromolecules such as protein, DNA, lipids, chlorophyll and other important macromolecules appears to represent an important stress tolerance (Xiong and Zhu 2002). Increase in activities of antioxidant enzymes such as SOD, APX, CAT and GR under abiotic stresses and also in tolerant species/varieties has also been reported by various workers (Sairam and Tyagi 2004). Several reports have shown that over-expression of superoxide dismutases leads to increased tolerance to abiotic stresses such as low temperature and water stress (Bohnert and Sheveleva 1998).

Many reactive oxygen species (ROS), particularly hydrogen peroxide, behave as signalling agents to trigger biochemical changes at the gene expression level (Jaspers and Kangasjärvi 2010). In general, abiotic stressors will induce perturbations in the fruit or vegetable cellular homeostasis which will then result in the increased generation of ROS in the apoplast, mitochondria, peroxisomes, cytoplasm, chloroplasts and endoplasmic reticulum (Jaspers and Kangasjärvi 2010). The ability of the cell to initially cope will depend largely on the endogenous free radical scavenging capacity (Mittler 2002). When free radical generation exceeds the endogenous scavenging capacity, the ROS interact with sensors, for which the full nature is not currently understood, that will initiate mitogen-activated protein kinase (MAPK) cascade reactions and also directly up-regulate transcription factors and calcium/calmodulin kinases (Mittler 2002; Jaspers and Kangasjärvi 2010). The MAPK cascade reaction will activate various transcription factors that enable de novo production of ROS, ROS scavenging systems, and accumulation of heat shock proteins and modulate NADPH supply in the cell (Mittler 2002; Sabehat et al. 1996). Some of the MAPK cascade paths have been also shown to be linked specifically to ethylene production (Jaspers and Kangasjärvi 2010), which is probably why ethylene production seems to be intrinsic to most stress responses. However, not all stressors produce identical response pathways, and so there is still a lot of work to be done in mapping of stress response networks (Jaspers and Kangasjärvi 2010). Water stress will lead to accelerated softening (Kubo et al. 2000), and that response has been associated with the induction of ethylene production in response to water stress.

Effect of Moisture Stress on Synthesis of Secondary Metabolites

Plants produce a huge variety of secondary metabolites with roles in various biological processes, such as pollination, seed dispersal and resistance to biotic and abiotic stresses (Wink 1999). Until recently it was thought that genes for plant metabolic pathways were not clustered, and this is certainly true in many cases. However, five plant secondary metabolic gene clusters have now been discovered, all of them implicated in synthesis of defence compounds, with enzymes for the first committed steps apparently recruited directly or indirectly from primary metabolic pathways involved in hormone synthesis (Chu et al. 2011). The genes and corresponding gene products for the first committed steps in these pathways can be regarded as signature genes/enzymes, as they are required for the synthesis of the skeleton structures of the different classes of secondary metabolite (Osbourn 2010). The signature genes all share homology with genes from plant primary metabolism and so are likely to have been recruited directly or indirectly from primary metabolism by gene duplication and acquisition of new functions (Chu et al. 2011).

In the past decade, several substances that were once considered to be secondary metabolites, such as jasmonic acid, salicylic acid and brassinosteroids, have been shown to be important internal signals (D'Auria and Gershenzon 2005). A wide range of metabolites that can prevent detrimental changes in cellular components by abiotic stresses have been identified such as amino acids (e.g. proline), quaternary and other amines (e.g. glycine betaine and polyamines) and a variety of sugars and sugar alcohols (e.g. mannitol and trehalose). Proline might confer a protective effect by inducing stress-protective proteins. Glycine betaine is a widely studied osmoprotectant, the accumulation of which has been studied with respect to modifications of several metabolic steps. Tomato plants transformed with a bacterial choline dehydrogenase gene (codA gene) targeted to the chloroplast were highly tolerant to chilling and oxidative stress, showing an increase in photosynthetic rate, plant survival, flower retention and fruit set (Park et al. 2004).

Effect of Moisture Stress on Fruit Growth and Quality

There have been both positive and negative effects reported for field water deficits (stress) in tree fruits and root vegetables (Wang and Frei 2011). Size of fruit is important, since larger fruit has lower surface area to volume ratios, which confers lower relative water loss (Shibairo et al. 1997). Timing of a water stress event can also be very important in determining response to postharvest abiotic stress response. Mango var. Kensington fruit was significantly more susceptible to postharvest chilling injury with exposure to water stress during the cell expansion phase of growth as opposed to being exposed to the stress during cell division or at a time near to harvest maturity (Léchaudel and Joas 2007). Carbon metabolism and the levels of specific sugars are severely affected by abiotic stress. Under stress, a decrease in sucrose, starch and soluble sugar content was reported. The shift of metabolism towards sucrose might occur because starch synthesis and degradation are more affected than sucrose synthesis (Silva and Arrabaca 2004). Trehalose is thought to protect biomolecules from environmental stress, as suggested by its reversible water absorption capacity to protect biological molecules from desiccation-induced damage (Penna 2003). Mannitol is another sugar alcohol that accumulates upon salt and water stress and can thus alleviate abiotic stress (Williamson and Coston 1990). Stress can also result in the losses of nutrient constituents in the fruit or vegetable, with vitamin C loss being the most sensitive indicator of stress exposure (Noctor and Foyer 1998; Pignocchi and Foyer 2003; Ioannidi et al. 2009). Any water deficit during growth of banana would retard its growth, and the effects may sometimes be evident only several months after the drought. The soil moisture deficit stress during vegetative stage of banana causes plant to extend its life cycle, poor bunch formation, lesser number of fingers and small-sized fingers (Singh et al. 2010). The cultivars belonging to BB genome appear to be more tolerant to drought. Observations on gas exchange characteristics, root and shoot dry matter showed low shoot/root ratio, i.e. higher root biomass for Bee hee kela (BB) and Kacha kela (ABB) under water stress, indicating higher drought tolerance.

Water stress in certain period of crop growth imparts beneficial effects to plant growth and development. In the case of peaches, it has been shown that lower levels of irrigation result in higher density of fruit surface trichomes and consequent lower weight losses in storage (Crisosto et al. 1994). In addition, two studies have shown that deficit irrigation of apples and pears could reduce water loss of these fruits in subsequent storage (Kilili et al. 1996; Lopez et al. 2011), and this was attributed to reduction in skin permeance of the deficit-irrigated fruit (Kilili et al. 1996). Miller et al. (1998), on kiwi cv. Hayward observed a significant loss in fruit weight, especially in plants exposed to stress in early summer (fruit set period). In contrast, an increase in the total soluble solids occurred. However, in a similar experiment conducted by Reid et al. (1996), kiwi fruit harvested from vines exposed to a less severe drought stress was unaffected in size, and fruit firmness was retained for 30 days longer in comparison to control. Bordonaba and Terry (2010), testing strawberry (cvs. Elsanta, Sonata, Symphony, Florence and Christine) in response to water deficits obtained promising results. Berry size was equivalent (Florence and Christine) or smaller (Sonata and Symphony) than control plants. Considering that the main components of red colour in strawberries are anthocyanins is plausible to presume that these fruits have lower contents of this secondary metabolite. However, in a previous study, Terry et al. (2007) observed the same reduction in red colour, but anthocyanin measurements pointed to a higher content of this metabolite.

Deficit irrigation (DI) has effects on fruit maturation and ripening depending on timing of application in apple (Mpelasoka et al. 2001b; 2001c). All DI treatments increased fruit total soluble solids (TSS) and firmness regardless of maturity but had little or no effect on titratable acidity. Fruit thinning has been proposed as a feasible strategy to compensate the loss in fruit size caused by water stress (Mpelasoka et al. 2001a). Regarding quality parameters, deficit-irrigated plants exhibited higher contents of TSS than fully irrigated plants. Gelly et al. (2003) reported an increase in soluble solid content and coloration of peach fruit when RDI was applied during production. RDI caused fruit peel stress lowering the content of vitamin C and carotenoids while increasing the phenolic content, mainly anthocyanins and procyanidins in peach (Buendía et al. 2008). The effects of RDI and crop load on Japanese plum (*Prunus salicina*) cv. Black Gold were investigated by Intrigliolo and Castel (2010). RDI strategy increased the efficiency of water usage, with 30 % of water savings, having minimal effect on crop yield and fruit growth. The combination of medium crop load and RDI shifted fruit mass distribution towards the low-value categories. García-Tejero et al. (2010) determined the post-harvest fruit quality of oranges (*Citrus sinensis* L. Osbeck, cv. Salustiano) exposed to RDI. Fruit quality parameters as TSS and TA increased in all stressed treatments resulting better organoleptic parameters (Treeby et al. 2007). In a similar study Velez et al. (2007), using the 'maximum daily trunk shrinkage' method, which is used as an indicator of water stress, was able to obtain water savings reaching 18 % without significant decreases in average fruit yield, weight and number. Working with grape (cv. Rizamat), table type, Du et al. (2008) reported in alternate drip irrigation condition, the photosynthetic rate was similar to control whilst the transpiration rate kept in the same level. Stressed grapes presented higher concentrations of both ascorbic acid and total soluble solids and lower titrated acidity, culminating in healthier and sweeter grapes. In a similar study, Santos et al. (2007) compared the effects of partial root zone drying irrigation system (50 % ETc irrigating one side at a time) with the conventional deficit irrigation system (50 % ETc applied on both sides), full irrigation system (100 % ETc applied on both sides) and nonirrigated vines. Partial drying regime showed a decreased vegetative growth, expressed by the smaller values of leaf layer number, percentage of water shoots, shoot weight, pruning weight and total leaf area.

Effect of Water Stress on Vegetable Quality

Shallow-rooted vegetables are known to be sensitive to water deficiency and require frequent supply of moisture for better yield and quality. Water stress, particularly at the tuber-forming stage, can also lead to a higher susceptibility of potatoes to postharvest development of black spot disorder (Hamouz et al. 2011). Black spot disorder is correlated primarily to susceptibility of cells in the potatoes to undergo decompartmentation in response to mechanical stress (i.e. bruising) (Stevens and Davelaar 1997). Another negative effect associated with water deficits is the case of root vegetables, such as carrot, where preharvest water stress (watering to 25–75 % of soil water field capacity) can weaken the cells, resulting in higher membrane leakage (i.e. cell damage) and consequently greater weight loss in storage (Shibairo et al. 1998b). In onion soil water stress caused by withholding irrigation at both 3 and 7 leaf stages reduced yields by 26 % as compared to control. In beans water stress lower than −40 hPa decreased leaf water potential and stomatal conductance, and it also reduced all growth and yield parameters. This effect was significant especially at −150 and −200 hPa. Plants at −40 hPa grew normally while plants at −100 hPa showed drought acclimation (El-Tohamy 1999). In tomato, water stress accompanied by high temperature above 28 C during fruiting induced about 30–45 % flower drop in different cultivars (Sanchez-Rodriguez et al. 2011). The differential growth of root and shoot of tomato was related to osmotic adjustments (Shashidhar et al. 1991). Tomato Arka Meghali has better osmotic adjustment under rainfed conditions, and crop water stress index (CWSI) was low in Arka Meghali (0.76) at flowering and fruit development stages compared to other cultivars (0.80–0.96) (Bhatt et al. 2006).

During rainy season due to excess rains, flooding occurs which most of the vegetables unable to tolerate. Vegetables are mostly sensitive to flooding, and damage is mainly due to reduction of oxygen in root zone which inhibits aerobic process (Heo et al. 2007). Flooded tomato accumulates ethylene which led to cell death (Singh 2011). Low oxygen levels stimulate an increased production of a 1-aminocyclopropane-1-carboxylic acid (ACC) in roots. The rapid epinastic growth of leaves is a characteristic response of tomatoes to waterlogged conditions (Kawase 1981). The severity of flooding symptoms increases with rising temperature; rapid wilting and death of tomato plants are usually observed following a short period of flooding at higher temperatures (Kuo et al. 1982). Although cassava and sweet potato are considered to be tolerant to drought conditions, significant reduction in tuber yield as well as in starch content occurs, and varieties vary in their response to water deficit stress/high-temperature stress conditions (Ravi et al. 1996; Reyes and Cisneros-Zevallos 2003). High drought tolerance and productivity are associated with leaf retention during drought in cassava (El-Sharkawy et al. 1992; Osiru et al. 2009). Bejarano et al. (2000) evaluated the content of glycoalkaloids (GAs, α-solanine and α-chaconine) in drought-tolerant potato (*Solanum tuberosum* L.) under drought stress. GAs concentration increased an average of 43 % and 50 % in the improved and control cultivars, respectively, but never above the recommended food safety limit (200 mg kg^{-1} fresh tubers). GAs are natural toxins synthesised by plants of the *Solanaceae* family and are believed to be associated with resistance to certain insects. Thipyapong et al. (2004) proposed that in tomato plants a decrease in PPO activity lowers the H_2O_2 concentration, reducing lipid peroxidation and improving resistance against water stress. Coelho et al. (2005) investigated the yield and bioactive amine content of American lettuce (*Lactuca sativa* cv. Lucy Brown) grown under greenhouse conditions and drip irrigation. Spermidine was the prevalent amine, followed by putrescine, cadaverine and agmatine. Supplying irrigation to achieve maximum cabbage (*Brassica oleracea* L. Capitata group) yield will also optimise sensory quality by minimising the compounds responsible for pungency. However, glucosinolate concentration will be reduced (Radovich et al. 2005). Glucosinolates are amino acid-derived secondary metabolites that may exhibit antibiotic, anticarcinogenic and organoleptic activity after hydrolysis. Sinigrin and progoitrin are the most important compounds with regard to flavour, since they are the primary determinants of pungency, bitterness and sulphurous aroma in cabbage (Fahey et al. 2001; Talalay and Fahey 2001; Wang and Frei 2011). Low soil water content (0.40 MPa of soil water tension) during broccoli growth leads to leaf size reduction, without affecting weight or yield, and contributes to the maintenance of green colour, possibly due to induced cytokinin synthesis (Wurr et al. 2002; Zaicovski et al. 2008).

Approaches to Ameliorate Moisture Stress Sensitivity

Despite limitations with the supply of freshwater in several regions, considerable amounts of water are lost through one or any combination of the mechanisms such as (1) evaporation from soil surface during conveyance and irrigation, (2) leakage during storage and transport to the fields where crops are grown, (3) run-off and (4) uncontrolled drainage. Under irrigated agriculture, about 30 % of water to be used as irrigation is lost in storage and conveyance. There are also other losses such as run-off and drainage when this remaining 70 % water reaches the farmers' fields. Organisms need to adapt themselves to changes in fluctuating environmental conditions. The plants, since they are not able to escape from adverse environmental conditions, have to rely entirely on their developmental plasticity to survive. Some of the most common ways of inducing moisture resistance and conserving moisture are discussed below.

Agronomical Measure

Recently, there has been a greater application of the inherent dangers involved in the current unsustainable cultural practices, and more emphasis is being laid to alternative approaches, which are ecologically sustainable, economically sound and socially acceptable (Munda and Das 2007). Some of the basic methods to collect and conserve water are harvesting of rainwater, harvesting of water from snow melting (Srinivasa Rao and Bhatt 1992), development of catchment area and storing run-off water for recycling, check dams and construction of waterways (Bhatt and Bujarbaruah 2007). Crop rotation (Yadav 2003), cover crops, strip cropping (Pandey 1992), mulching (Singh et al. 2002, 2003; Suresh Kumar et al. 2012), contour hedgerow intercropping system (Chandrasekharan and Pandian 2009), contouring, contour bunds (Sharma et al. 1991), sloping area land technology (SALT), contour drains (Misra and Ahamed 1993), graded bunding (Sharma et al. 1991; Prihar and Sandhu 1987), terraces, broad bed furrow system for raising crop and conservation of water, contour furrow (Gyawali 1993), stabilisation structures, curved land ploughing for intensive land preparation (Budathoki et al. 1993), soil conservation and water retention through use of vegetation, live check of bamboo pieces, growing leguminous trees (Mishra and Sharma 2003) and loose boulders are some of the other agronomical measures used to conserve moisture (Kaur et al. 2002). Trench planting (0.5–1.0 m) is recommended for ber, *aonla* and custard apple to conserve moisture. The trenches collect rainwater along with silt and organic matter and thus promote tree growth (Lu et al. 1997). Planting in trenches is common for pineapple under dry conditions. Windbreaks help to deflect and to filter through the wind current thereby reducing the velocity of wind resulting lower displacement of wind around tree and cause reduction in transpiration and evaporation. The orientation of windbreak should more or less be at right angles to the prevailing winds. It is believed that the windbreak is effective for a distance equivalent to 3–4 times of tree height (Chundawat 1990). *Acacia tortilis, Cassia siamea* and *Prosopis juliflora* are some of the useful tree species suitable as windbreak under less water conditions.

Water needs for irrigation can be met, in part, by practising uniformity of water application – precise irrigation with microirrigation – that delivers water from piped main lines and laterals directly to the root zone frequently and in small amounts and at rates matched to crop needs. This irrigation strategy has shown to be the best method for saline waters. However, such precise irrigation systems are expensive, but benefits include reduction of hidden costs of water wastage and land degradation, and the environmental costs of drainage and land reclamation. The net benefits of microirrigation improve markedly when such advantages are taken into account. A tax on groundwater withdrawals in a region where demand exceeds the natural rate of recharge will have a similar impact on the relative cost of microirrigation. Thus, there is a need of widespread adoption of policies that motivate farmers to reduce off-farm impacts and encourage entrepreneurs to develop low-cost microirrigation systems that are financially compatible with a wide range of crops and production environments. It has been found that up to 81 % water saving was observed in lemon compared to flood irrigation with the over 35 % increase in yield. Similarly, banana, grapes and pomegranate recorded 45 % saving in water using drip irrigation. Waste water can also be used for horticultural plantations in between 2 and 3 irrigations with normal water. As agriculture is the major industry which consumes maximum water, judicious use of waste water or saline water is alternative strategy to bring down the water consumption and to alleviate moisture stress during drought period.

Water availability, water use and nutrient supply to the plants are closely interacting factors influencing plant growth and yield. Under limited water, excessive application of fertiliser should be avoided as excess foliage leads to higher transpiration. Limited soil moisture influences nutrient availability of plants. However, adequately fertilised plants may show higher drought tolerance. Water use efficiency of guava plants is known to be increased by adequate fertilisation. In the initial stages of tree growth, nutrients influence leaf area, leaf growth rates and senescence and thus transpiration (Stefanelli et al. 2010). Nitrogen and phosphorus deficiencies can reduce cell expansion which results in smaller leaves in tomato. Generally, adequately fertilised soils promote both rapid leaf area expansion, thus increasing transpiration, and reduced evaporation. Nitrogen is the key nutrient and should be applied according to the expected crop load. Application of N fertilisers enhances new leaf growth and delays plant senescence.

Antitranspiration coatings have been shown to be effective for maintaining quality through control of water loss (Chundawat 1990). Antitranspirants are chemicals which when sprayed on plants form a film which increases the diffusion resistance of water from stomata and thus reduces transpiration losses of water. Several chemicals have been successfully used like acropyl in grapes, polycot in banana and kaolinite (3–8 %) in different fruit plants. The energy input can be reduced by increasing plant reflectivity by using effective chemicals like zinc oxide, kaolinite and chalk alone or in combination with other antitranspirants. These chemicals are used to reduce temperature on plant parts. Film-forming compounds like Wilt-Pruf, Mobileaf, clear spray, Vapor Gard and Foli-Coat can be used to reduce transpiration and water loss. Water loss can also be restricted by applying chemicals which facilitate stomatal closure. Some of these are phenylmercuric acetate (PMA), decenylsuccinic acid (DSA), atrazine and sodium azide.

ABA synthesis is one of the fastest responses of plants to water stress, triggering ABA-inducible gene expression and causing stomatal closure, thereby reducing water loss via transpiration and eventually restricting cellular growth. The potent antitranspirant effect of ABA suggests that it would be ideal for such use, and seedlings whose roots were dipped in ABA prior to transplanting showed greater survival than those dipped in water. IAA and some CKs can antagonise ABA-induced stomatal closure (Singh et al. 2001; 2010). 100 μM abscisic acid (ABA) or 10 μM N^6-benzyladenine (BA) was used to avoid water stress in grafted mango plants. There are also cases where foliar applications of these PGRs have increased photosynthesis in sweet oranges. Foliar sprays of 50 mM IAA, GA_3 or benzylaminopurine (BAP) partially counteracted the effect of water deficit on photosynthesis and transpiration. Growth regulator applications can also potentially enhance stress resistance, particularly for fruits and vegetables which are prone to show accelerated ripening or senescence in response to drought stress. Accelerated ripening or senescence is most often mediated by ethylene production in response to the stress. As consequence anti-ethylene products such as aminovinylglycine (AVG) and 1-methylcyclopropene (1-MCP) could be used to mitigate drought stress. One of the inherent difficulties is to know up to what extent the substance of interest enters the leaves and its fate in the leaf. While foliar cytokinin (CK) application can prevent ABA-induced photosynthetic limitation, the effects can be transient and of little consequence in the long term. Paclobutrazol (10 mg/lit) is used to avoid moisture stress in mango.

Conservation Agriculture

The basic goal of conservation agriculture is agricultural sustainability, conserving, improving and making more efficient use of natural reserve through the integrated management of available soil, water and biological resources combined with judicious use of external inputs (Gupta and Rao 1994). The basic principles of conservation agriculture are soil cover, particularly through retention of crop residues on the soil surface as mulch or cover crops; a minimum level of soil disturbance, e.g. reduced or zero tillage practices; and sensible, profitable crop rotations (Panwar and Vyas 2002). Soil moisture can be conserved through mulching either with black polythene or locally available mulches (Singh et al. 1987, 1994), growing cover crops or interculturing in the orchards to check soil erosion and run-off rainwater. If area around tree basin remains covered with mulches, like grass mulch (10–12 kg/basin) and black polythene mulch (100 micron) throughout the growth period, it helps in conserving the soil moisture (Singh et al. 1997; Singh and Bag 2002), saving water for more critical stages during summer and reduced the weed population by 60 % with grass mulch and 100 % with black polythene mulch. The polythene mulch maintained 29 % more soil moisture compared to un-mulched trees on soil available water content basis. Experiments on onion and chilli under rainfed conditions proved the utility of black polythene and dry grass mulch at the rate 5 t/ha for increasing the productivity under moisture stress situations (Sharma 2007). Surface-applied mulches provide several benefits to crop production by controlling evaporation from the soil surface (Pawar et al. 2004), heat energy and nutrient status in soil, buffering drastic changes in soil temperature (Naeini and Cook 2000). Soil temperature, especially at the surface layer, is reported to be important for the translocation of photoassimilates. Straw mulch ameliorates environmental stresses (Macilwain 2004; Suresh Kumar et al. 2012) and improves the product quality and safety. Jat et al. (2006) also have been reported higher (almost double) water productivity of carrot under bed planting over conventional surface broadcasting method of planting. The higher yield and lower irrigation water requirement of cabbage were recorded when seeded with a precision planter or transplanted on 37 cm wide raised beds at the top as compared to traditional planting on sloppy ridges cultivation (Jat et al. 2006). Microirrigation in muskmelon and capsicum resulted in 15–70 % more yield with water saving of about 36 % (Sharma 2007).

Germplasm/Crop Selection

There have been reports on the selection of germplasm and cultivars from breeding programmes that will have greater drought stress resistance and hence better production capability. The fruit crop selected for arid region must be such that its maximum growth period synchronises with the period of maximum water availability and low vapour pressure deficit in the atmosphere. Ideally the period from flowering to fruiting must be completed well before the onset of summers. Fruits like ber, custard apple, phalsa, and *Cordia myxa* are suitable for such condition. Otherwise the crop selected should be such that their reproductive cycle can be monitored to synchronise with maximum moisture availability periods, e.g. guava, pomegranate and acid lime which bear fruits in distinct bahars, and the bahar which coincides with rainy season can be encouraged. Since water is a limiting factor in arids, the crop selected should have drought tolerance mechanism like deep root system to draw water from deeper soil (ber, date palm), leaf shedding (ber, gonda), water-binding mechanism (fig), presence of thorns (karonda), stomata at lower side (Annona), wax coating (ber), leaf orientation (aonla), hairiness and sunken and covered stomata (fig, phalsa, ber and gonda). Similarly intense radiation is another important feature of arid region. A crop like kinnow mandarin can be selected which has tolerance against drought and bear fruits inside well-formed canopy (Chundawat 1990).

Varieties selected or evolved for drought conditions should be short in duration (Bhansali et al. 2003). Some of the fruit and vegetable varieties are developed in various crops especially for drought stress tolerance, namely, pomegranate hybrid Ruby, Annona hybrid Arka Sahan and fig selections like Deanna and Excel. Dogridge (*Vitis champine*) was found promising both for improvement in vigour, yield and quality of seedless grapes as well as tolerance to drought and salinity. Procedures have been standardised to multiply the rootstock plants and to raise vineyards in situ on Dogridge (Singh 2011). Gola, Himida and Seb in ber; Ganesh, Bassein seedless and Jalore seedless in pomegranate; and Halawy and Barhee in date have been found to be promising in arid region. Apple cultivars which have resistance to developing browning in response to cutting have greater levels of apoplastic antioxidant enzymes and lower levels of ROS in the apoplast after cutting than cultivars with lower levels of antioxidant enzymes (Toivonen and Sweeney 1997; Toivonen 2003; Toivonen and Beveridge 2005). In vitro selection is an approach where plant cells of a fruit or vegetable of interest are tissue cultured and exposed to a stressor (Vinson et al. 2001), taking surviving cells to regenerate new plants having a superior level of stress resistance (Rai et al. 2011). This approach has been highly successful to regenerate germplasm of many crop plants which can be regenerated via tissue culture techniques.

Based on the requirements of dry situation, the crops like ber, *Lehsua*, pomegranate, custard apple, guava, *aonla*, acid lime, phalsa, bael, wood apple, jamun, karonda, date palm and fig are best suitable for cultivation. Besides, there are many indigenous drought hardy plants which have characteristics like deep root system (tamarind), scanty foliage (ker (*Capparis decidua*)) and mucilaginous sap in plant parts (ker, gonda, pilu, bael). In addition, there are many high-value fruit crops like grape, sweet orange, papaya, mango and kinnow which require dry weather during flowering and fruiting for developing high-quality fruits.

Use of Molecular Probes for Marker-Assisted Breeding

Metabolic profiling of plants under stress is an important tool to study stress-induced changes in metabolites. This will reveal the exact nature of the metabolites present in the plants and also help in analysis of transgenic plants that might alter due to the insertion of foreign genes. Traditional breeding approaches are limited by the complexity of the stress-tolerant traits, low genetic variance of yield component under stress conditions and the lack of an efficient selection technique, but by including molecular methods to introduce the genes for stress-associated mechanism will give an advantage by surpassing limitation of the former (Rajam 1997, 1998). As tolerance to stress is multifactorial syndrome rather than result of a single reaction or gene (Mittler 2006; Shinozaki and Yamaguchi-Shinozaki 2000), tackling of the primary stress reactions by gene transfer can also alleviate the secondary stress and generate plant with higher stress tolerance (Beck et al. 2007b). There are a large number of genes and proteins associated with stress tolerance in plants, and so the best approach to identifying stress-tolerant lines is apply the stress of interest and perform quantitative trait loci (QTL) analyses (Foolad 1999; Grover et al. 2003). Molecular engineering for stress resistance in fruits and vegetables is limited due to two major factors: (1) the complexity of the stress response network (Toivonen and Beveridge 2005) means that modulating the stress resistance with single gene insertions is unlikely, and (2) methods to successfully transform many important fruit and vegetable crops have yet to be developed (Rai et al. 2011).

Innovative and Improved Postharvest Management Strategies

Stress response and resistance are a very complex matrix of processes and pathways, which are not fully understood at this point in time (Toivonen and Beveridge 2005; Jaspers and Kangasjärvi 2010). Hence, it is very difficult to design effective treatments to achieve the resistance required for any or all abiotic stresses that a fruit or vegetable may encounter. Simple modifications to postharvest handling systems can sometimes result in significant reduction in stress exposure and consequently result in storage and/or shelf life extension.

One of the most successful strategies is the application of plastic film packaging or wraps to prevent desiccation, resulting in significant improvements and shelf life and quality of many fruits and vegetables (Ben-Yehoshua and Rodov 2003). In many cases, modified atmosphere packaging is considered to largely control humidity around product and thus prevent moisture loss of fresh-cut and whole fruits and vegetables (Toivonen 2009). Also, antitranspiration coatings have been shown to be effective for maintaining quality through control of water loss (Baldwin 2003). In regard to maintaining water content on the retail shelf, the application of misting systems can 'recharge' the vegetable and in so doing maintain quality over longer durations at less than ideal storage temperatures (Barth et al. 1990; Shibairo et al. 1998a).

Plant Growth Regulators

Plant hormones regulate every aspect of plant growth, development and the responses of plants to biotic and abiotic stresses (Peleg and Blumwald 2011). Classical phytohormones are abscisic acid (ABA), ethylene, cytokinin (CK), auxin, gibberellin, jasmonate as well as brassinosteroids, salicylic acid, nitric oxide and strigolactone, and it is likely that additional growth regulators are yet to be discovered (Santner and Estelle 2009). ABA synthesis is one of the fastest responses of plants to water stress, triggering ABA-inducible gene expression and causing stomatal closure, thereby reducing water loss via transpiration and eventually restricting cellular growth (Wilkinson and Davies 2010; Yamaguchi-Shinozaki and Shinozaki 2006). Many ABA-mediated physiological processes induced by water deficit, including closure of the stomata and acceleration of leaf senescence, are counteracted by CKs which increase stomatal aperture and/or delay ABA-induced stomatal closure (Stoll et al. 2000). Growth regulator applications can also potentially enhance stress resistance, particularly for fruits and vegetables which are prone to show accelerated ripening or senescence in response to drought stress (Baldwin 2003). Accelerated ripening or senescence is most often mediated by ethylene production in response to the stress. As consequence anti-ethylene products such as aminovinylglycine (AVG) and 1-methylcyclopropene (1-MCP) could be used to mitigate drought stress (Baldwin 2003; Blankenship and Dole 2003). Other growth hormones, such as methyl jasmonate (which promotes leaf senescence), and polyamines can enhance chilling resistance in avocado, grapefruit, bell peppers and zucchini squash (Gill and Tuteja 2010; Wang 1994). Abscisic acid has been demonstrated to reduce chilling-induced injury in some crops (Wang 1993). Other growth regulators have been suggested for use in preventing senescence in leafy vegetables (e.g. 2, 4-D); however, their practical application is limited.

Systems Biology Approach

'Systems biology' approach which integrates the series of -omics, i.e. phenomics, genomics, transcriptomics, proteomics and metabolomics, is getting roots since the early new millennium (Cramer et al. 1994). Future crop improvement programmes using systems biology approach with the integration of experimental results from lab and field conditions will perhaps be the best strategy to address abiotic stresses. Due to advances in systems biology, structural biology and breeding technologies based on genomic information, the ability to design biological functions is coming within reach (Cramer 2010). However, to truly comprehend the dynamics of the entire system, the complexity, which is generated by the interactions of many factors, must also be understood (Nilo et al. 2010). To date, there are limited examples of such research based on understanding life processes as systems due to the lack of life scientists, who cover multiple disciplines. Nevertheless, systems biology, as defined above, will become a central field in the next generation of life sciences in which all the sectors of agriculture will have the major benefit. The recombinant

DNA technologies hold big promise towards the development of crop varieties tolerant to a variety of abiotic stress factors such as drought, heat, cold and salinity.

Beneficial Microbes for Management of Moisture Stress in Horticultural Crops

Microbial activity in soils is stimulated by the release of carbon-rich material in the form of root border cells and/or the selective exudation of specific sugars, carboxylic acids or amino acids that encourage the development of cultivar-specific, plant-beneficial, microbial communities (Bünemann et al. 2004; Barea et al. 2005). Some of these microorganisms benefiting from the availability of fresh energy are also capable of enhancing plant growth and development, which are generally called as plant growth-promoting (PGP) rhizosphere microorganisms (Glick 1995; Vessey 2003). Plant growth-promoting rhizobacteria (PGPR) and fungi (mycorrhizae) can facilitate plant growth directly by facilitating the uptake of nutrients from the environment, by influencing phytohormone production (e.g. auxin, cytokinin or gibberellin) and/or by enzymatic lowering of plant ethylene levels (Glick et al. 1999). In addition to facilitating the growth of plant, these microorganisms can protect plants from the deleterious effects of flooding and drought (Harley and Smith 1983).

The beneficial plant-microbe interactions in the rhizosphere are the primary determinants of plant health and soil fertility (Jeffries et al. 2003). Rhizobacteria include mycorrhization helper bacteria (MHB) and plant growth-promoting rhizobacteria (PGPR), which assist AMF to colonise the plant roots. Synergistic positive interactions have been reported between AMF and plant growth-promoting bacteria (PGPB) such as nitrogen fixers, fluorescent *pseudomonads* and sporulating *bacilli* (Hameeda et al. 2007). The most common bacteria in the mycorrhizosphere are *Pseudomonas* (Vosátka and Gryndler 1999). Schubert and Cravero (1985) isolated natural AMF infection in grapevine roots. For example, through one mechanism, treating of tomato plants with ACC deaminase containing PGPR or genetically engineered plants to express this microbial enzyme significantly decreased the damage caused in these plants as a result of flooding (Grichko and Glick 2001). Mayak et al. (2004) reported that treating peppers and tomato plants with ACC deaminase containing *Archromobacter piechaudii* ARV 8 significantly reduced the growth inhibition caused by drought stress. Similar to rhizobacterial associations, mycorrhizae are also the most common symbiotic species on earth in almost all ecosystems. Several studies have demonstrated that mycorrhizal associations play vital role in plant nutrition. They greatly increase the efficiency of nutrient and water uptake and buffer plant species against several environmental stresses including soil moisture deficit (Rao et al. 1998; Subramanian and Charest 1995; Subramanian et al. 2006; Busso 1997). Some extremely mycorrhizal-dependent plants, including grapes, citrus, melons, oaks and pines, may quite literally starve to death in soils that lack this helpful fungi (Hemalatha et al. 2010). These benefits of mycorrhizal symbioses in vegetables, fruits and tree species, both agronomically by increased growth and yield as well as ecologically by improved fitness, indicate that mycorrhizal plants are often more competitive and better able to tolerate environmental stresses (Mathur and Vyas 1995; Krishna et al. 2006). Science of metagenomics can be utilised for exploring novel microbial genes from various stress environments for abiotic stress tolerance especially for drought, salinity and temperature.

Conclusion

Abiotic stresses are significant determinants of quality and nutritional value of fruits and vegetables during growth, harvest, handling, storage and distribution to consumer. Crop management can have a significant influence on susceptibility to stress. The most important criterion of adaptation in a natural ecosystem is survival in space and time. An important aspect of the breeding and selection approach is that there must be stressors applied in reproducible ways to allow the breeder to identify expression of stress resistance since that characteristic is adaptive, rather than constitutive, in nature. High yield potential has been the most important criterion for man from the domestication during the cultivation. But, this point is clear that mechanism of the adaptation and relationships of plant production is still a dilemma! It needs clarification, but to try them with only one targeted factor is almost impossible. If these points can be adequately answered (in terms of plant domestication, cultivation and agronomic respects), many affected and hidden/invisible obstacles easily can be removed or overcome. Both an early crop cultivar and an earlier sowing may be reducing the irrigation requirements and water amount. The increased water use efficiency caused by increasing CO_2 will compensate only partially for the negative effects of increasing water limitation. Farming practices like 'conservation tillage and zero tillage' and 'optimum land use' practices for the agricultural application lead to decomposition of organic materials and more carbon stored in the soil. Agricultural practices are among the biggest water-consuming activities, considering that alternatives to reduce water use in agricultural practices are of special interest in the present moment. Irrigation practices often aim at total replacement of evapotranspiration in order to obtain maximum yield. The use of regulated water stress (RDI) is a feasible strategy to enhance accumulation of health-promoting compounds in food.

References

Baldwin EA (2003) Coating and other supplemental treatments to maintain vegetable quality. In: Bartz JA, Brecht JK (eds) Postharvest physiology and pathology of vegetables, 2nd edn. Marcel Dekker, New York, pp 413–456

Barden CL, Bramlage WJ (1994) Separating the effects of low temperature, ripening, and light on loss of scald susceptibility in apples before harvest. J Am Soc Hort Sci 119(1):54–58

Barea JM, Pozo MJ, Azcon R et al (2005) Microbial cooperation in the rhizosphere. J Exp Bot 56:1761–1778

Barth MM, Perry AK, Schmidt SJ et al (1990) Misting effects on ascorbic acid retention in broccoli during cabinet display. J Food Sci 55(4):1187–1188

Beck EH, Fetitig S, Knake C et al (2007a) Specific and unspecific responses of plants to cold and drought stress. J Bio Sci 32(3):501–510

Beck EH, Heim R, Hansen J (2007b) Plant resistance to cold stress: mechanisms and environmental signals triggering frost hardening and dehardening. J Bio Sci 29:449–459

Bejarano L, Mignolet E, Devaux A et al (2000) Glycoalkaloids in potato tubers: the effect of variety and drought stress on the α-solanine and α-chaconine contents of potatoes. J Sci Food Agric 80 (14):2096–2100

Ben-Yehoshua S, Rodov V (2003) Transpiration and water stress. In: Bartz JA, Brecht JK (eds) Postharvest physiology and pathology of vegetables, 2nd edn. Marcel Dekker, New York, pp 111–159

Berard LS (1990) Effect of nitrogen fertilization on stored cabbage. I. Development of physiological disorders on tolerant and susceptible cultivars. J Hort Sci 65 (3):289–296

Bhansali RJ, Manjit S, Jain SM et al (2003) Micro-propagation of arid zone fruit trees of India. In: Bhansali RJ, Manjit S (eds) Micropropagation of woody trees and fruits. Springer, Dordrecht, pp 381–432

Bhatt BP, Bujarbaruah KM (2007) Integrated intensive farming system-a sustainable approach of land use in Eastern Himalayas. Shaping agrarian prosperity through integrated intensive farming. ICAR RC NEH Region, Umiam, pp 1–20

Bhatt RM, Srinivasa Rao NK, Gowda VR (2006) Response of bulb onion to water stress: photosynthesis, stomatal conductance and osmotic adjustment. Indian J Hort 63:276–280

Blankenship SM, Dole JM (2003) 1-Methylcyclopropene: a review. Postharvest Biol Technol 28(1):1–25

Bohnert HJ, Nelson DE, Jensen RG (1995) Adaptation to environmental stresses. Plant cell 7:1099–1111

Bohnert HJ, Sheveleva E (1998) Plant stress adaptations, making metabolism move. Curr Opin Plant Biol 1:267–274

Bordonaba JG, Terry LA (2010) Manipulating the taste-related composition of strawberry fruits (Fragaria x ananassa) from different cultivars using deficit irrigation. Food Chem 122(4):1020–1026

Bossdorf OH, Auge L, Lafuma WL et al (2005) Phenotypic and genetic differentiation between native and introduced plant populations. Oecologia 144:1–11

Bowler C, Fluhr R (2000) The role of calcium and activated oxygens as signals for controlling cross-tolerance. Trends Plant Sci 5(6):241–246

Boyer JS (2009) Evans review: cell wall biosynthesis and the molecular mechanism of plant enlargement. Funct Plant Biol 36(5):383–394

Bramlage WJ, Weis SA (1997) Effects of temperature, light and rainfall on superficial scald susceptibility in apples. Hort Sci 32(5):808–811

Brown RW (1995) The water relations of range plants: adaptations of water deficits. In: Bedunah DJ, Sosebee RE (eds) Wildland plants: physiological ecology and developmental morphology. Society for Range Management, Denver, pp 291–413

Budathoki et al (1993) Vegetable crops: indigenous knowledge and technology in the western hills of Nepal. In: Tamang D, Gill GJ, Thapa GB (ed) Indigenous management of natural resources in Nepal: proceedings of the workshop on indigenous management of Agriculture, held on 8–9 June,1992 at Dhulikhel. Winrock International Institute for Agricultural Development, Kathmandu

Buendía B, Allende A, Nicolás E et al (2008) Effect of regulated deficit irrigation and crop load on the antioxidant compounds of peaches. J Agric Food Chem 56 (10):3601–3608

Bünemann EK, Bossio DA, Smithson PC et al (2004) Microbial community composition and substrate use in a highly weathered soil as affected by crop rotation and P fertilization. Soil Biol Biochem 36:889–901

Burdon JN, Dori S, Lomaniec E et al (1994) The post-harvest ripening of water stressed banana fruits. J Hort Sci 69(5):799–804

Busso CA (1997) Towards an increased and sustainable production in semiarid rangelands of central Argentina: two decades for research. J Arid Environ 36:197–210

Chandrasekharan B, Pandian BJ (2009) Rainwater harvesting and water saving technologies. Indian J Agron 54(1):90–97

Chu HY, Wegel E, Osbourn A (2011) From hormones to secondary metabolism: the emergence of metabolic gene clusters in plants. Plant J 66(1):66–79

Chundawat BS (1990) Arid fruit culture. Oxford & IBH, New Delhi

Coelho AFS, Gomes EP, Sousa AD, Gloria MB (2005) Effect of irrigation level on yield and bioactive amine content of American lettuce. J Sci Food Agri 85 (6):1026–1032

Cramer GR (2010) Abiotic stress and plant responses from the whole vine to the genes. Aust J Grape Wine Res 16:86–93

Cramer GR, Alberico GJ, Schmidt C (1994) Leaf expansion limits dry matter accumulation of salt-stressed maize. Aust J Plant Physiol 21:663–674

Crisosto CH, Johnson RS, Luza JG et al (1994) Irrigation regimes affect fruit soluble solids concentration and rate of water loss of 'O'Henry' peaches. Hort Sci 29 (10):1169–1171

D'Auria JC, Gershenzon J (2005) The secondary metabolism of Arabidopsis thaliana: growing like a weed. Curr Opin Plant Biol 8(3):308–316

DAC (2013) Department of agriculture and cooperation, agricultural statistics at a glance 2012. http://agricoop.nic.in/agristatistics.htm. Accessed 3 Mar 2013

Datta KK, de Jong C (2002) Adverse effect of waterlogging and soil salinity on crop and land productivity in northwest region of Haryana India. Agric Water Manag 57:223–238

dos Santos TP, Lopes CM, Rodrigues ML, de Souza CR et al (2007) Effects of deficit irrigation strategies on cluster microclimate for improving fruit composition of *Moscatel* field-grown grapevines. Sci Hort 112 (3):321–330

Du T, Kang S, Zhang J et al (2008) Water use efficiency and fruit quality of table grape under alternate partial root-zone drip irrigation. Agric Water Manag 95 (6):659–668

El-Tohamy W, Schnitzler WH, El-Behairy U, El-Betagy MS (1999) Effect of VA mycorrhiza on improving drought and chilling tolerance of bean plants. J App Bot 73:178–183

El-Sharkawy MA, De Tafur SM, Cadavid LF (1992) Potential photosynthesis of cassava as affected by growth conditions. Crop Sci 32:1336–1342

Fahey JW, Zalcmann AT, Talalay P (2001) The chemical diversity and distribution of glucosinolates and isothiocyanates among plants. Phytochem 56(1):5–51

FAO (2007) Food and agriculture organization. http://www.fao.org/docrep/010/a1075e/a1075 e00.htm. Accessed 26 Feb 2013

Ferguson IB, Volz R, Woolf A (1999) Preharvest factors affecting physiological disorders of fruit. Postharvest Biol Technol 15(3):255–262

Foolad MR (1999) Genetics of salt quantitative and cold analysis tolerance in tomato and QTL mapping. Plant Biotechnol 16(1):55–64

Garcia A, Rizzo CA, Ud-Din J, Bartos SL, Senadhira D, Flowers TJ et al (1997) Sodium and potassium transport to the xylem are inherited independently in ice and the mechanism of sodium: potassium selectivity differs between rice and wheat. Plant Cell Environ 20:1167–1174

Garcia-Orenes F, Guerrero C, Mataix-Solera J et al (2005) Factors controlling the aggregate stability and bulk density in two different degraded soils amended with biosolids. Soil Tillage Res 82:65–76

García-Tejero I, Jimenez-Bocanegra JA, Martinez G et al (2010) Positive impact of regulated deficit irrigation on yield and fruit quality in a commercial citrus orchard Citrus sinensis (L.)Osbeck, cv. salustiano. Agric Water Manage 97(5):614–622

Gelly M, Recasens I, Mata M, Arbones A, Rufat J, Girona J, Marsal J (2003) Effects of water deficit during stage II of peach fruit development and postharvest on fruit quality and ethylene production. J Hort Sci Biotechnol 78:324–330

Gill SS, Tuteja N (2010) Polyamines and abiotic stress tolerance in plants. Plant Signal Behav 5(1):26–33

Glick BR (1995) The enhancement of plant growth promotion by free living bacteria. Can J Microbiol 41:109–117

Glick BR, Patten CL, Holguin G et al (1999) Biochemical and genetic mechanisms used by plant growth promoting bacteria. Imperial College Press, London, p 267

Goddijn OJ, Van Dun K (1999) Trehalose metabolism in plants. Trends Plant Sci 4:315–319

Goyary D (2009) Transgenic crops, and their scope for abiotic stress environment of high altitude: biochemical and physiological perspectives. DRDO Sci Spectr, March 2009 :195–201

Greenway H, Munns R (1980) Mechanisms of salt tolerance in nonhalophytes. Ann Rev Plant Physiol 31:149–190

Grichko VP, Glick BR (2001) Amelioration of flooding stress by ACC deaminase-containing plant growth-promoting bacteria. Plant Physiol Biochem 39:11–17

Grover A, Aggarwal PK, Kapoor A et al (2003) Addressing abiotic stresses in agriculture through transgenic technology. Curr Sci 84(3):355–367

Gruda N (2005) Impact of environmental factors on product quality of greenhouse vegetables for fresh consumption. Crit Rev Plant Sci 24(3):227–247

Gupta A (2004) Origin of agriculture and domestication of plants and animals linked to early Holocene climate amelioration. Curr Sci 87:54–59

Gupta RK, Rao DLN (1994) Potential of wasteland for sequestering carbon by reforestation. Curr Sci 66:378–380

Gyawali BK (1993) Integrated pest management through indigenous techniques in Nepal. In: Tamang D, Gill GJ, Thapa GB (ed) Indigenous management of natural resources in Nepal: proceedings of the workshop on indigenous management of Agriculture, held on 8–9 June,1992 at Dhulikhel. Winrock International Institute for Agricultural Development, Kathmandu

Hall CB (1964) The effect of short periods of high temperature on the ripening of detached tomato fruits. Proc Am Soc Hortl Sci 84:501–506

Hameeda B, Srijana M, Rupela OP, Reddy G (2007) Effect of bacteria isolated from composts and macrofauna on sorghum growth and mycorrhizal colonization. World J Microbiol Biotechnol 23 (6):883–887

Hamouz K, Bečka D, Morava J (2011) Effect of environmental conditions on the susceptibility to mechanical damage of potatoes. In: Agris on-line papers in economics and informatics. Available from http://www.agris.cz/vyhledavac/detail.php?id=116921&iSub=518&sHighLight=agria&PHPSESSID=997764de5509230f23dc06268b7b8344. Accessed 30 Mar 2011

Harley JL, Smith SE (1983) Mycorrhizal symbiosis. Academy Press, London

Hasegawa PM, Bressan RA, Zhu JK, Bohnert HJ (2000) Plant cellular and molecular responses to high salinity. Annu Rev Plant Physiol Plant Mol Biol 51:463–499

He C, Yan J, Shen G, Fu L, Holaday AS, Auld D, Blumwald E, Zhang H (2005) Expression of an Arabidopsis vacuolar sodium/proton antiporter gene in cotton improves photosynthetic performance under salt conditions and increase fibre yield in the field. Plant Cell Physiol 46:1848–1854

Hemalatha P, Velmurugan M, Harisudan C, Davamani V (2010) Importance of mycorrhizae for horticultural crops. In: Mycorrhizal biotechnology. CRC press, New York, pp 39–150

Heo HJ, Kim YJ, Chung D, Kim DO (2007) Antioxidant capacities of individual and combined phenolics in a model system. Food Chem 104:87–92

Hodges DM (2003) Postharvest oxidative stress in horticultural crops. Food Products Press, New York

Hodges DM, Wismer WV, Forney CF (2001) Antioxidant responses in leaves of two cultivars of spinach (*Spinacia oleracea* L.) differing in their senescence rates. J Am Soc Hort Sci 126(5):611–617

Horneburg B, Wirsing F (1995) The tendency of potato tubers to black spot. 2. Influence of fertilizer application and irrigation on the expression of black spot. Kartoffelbau 46(8):318–320. http://www.rcub.bg.ac.yu/monograph/2003/rad04.htm

Hummel I, Pantin F, Sulpice R et al (2010) Arabidopsis plants acclimate to water deficit at low cost through changes of carbon usage: an integrated perspective using growth, metabolite, enzyme, and gene expression analysis. Plant Physiol 154(1):357–372

Inari T, Yamauhi R, Kato K, Takeuchi T (2002) Changes in pectic polysaccharides during the ripening of cherry tomato fruits. Food Sci Tech Res 8(1):55–58

Intrigliolo DS, Castel JR (2010) Response of plum trees to deficit irrigation under two crop levels: tree growth, yield and fruit quality. Irrigation Sci 28(6):525–534

Ioannidi E, Kalamaki MS, Engineer C et al (2009) Expression profiling of ascorbic acid-related genes during tomato fruit development and ripening and in response to stress conditions. J Exp Bot 60 (2):663–678

Jacobo-Velázquez DA, Cisneros-Zevallos L (2009) Correlations of antioxidant activity against phenolic content revisited: a new approach in data analysis for food and medicinal plants. J Food Sci 74:R107–R113

Jaspers P, Kangasjärvi J (2010) Reactive oxygen species in abiotic stress signalling. Physiol Plant 138 (4):405–413

Jat ML, Gupta RK, Erenstein O, Ortiz R (2006) Diversifying the intensive cereal cropping systems of the Indo-Ganges through horticulture. Chronica Hortic 46:16–20

Jeffries PS, Gianinazzi S, Perotto KK, Turnau A, Barea JM (2003) The contribution of arbuscular mycorrhizal fungi in sustainable maintenance of plant health and soil fertility. Biol Fertility Soils 37:1–16

Kaur B, Gupta SR, Singh G (2002) Bioamelioration of a sodic soil by silvopastoral systems in northwestern India. Agroforest Syst 54:13–20

Kawase M (1981) Anatomical and morphological adaptation of plants to water logging. Hort Sci 16:30–34

Ke D, Saltveit ME (1989) Carbon dioxide-induced brown stain development as related to phenolic metabolism in iceberg lettuce. J Amer Soc Hort Sci 114 (5):789–794

Kilili AW, Behboudian MH, Mills TM (1996) Postharvest performance of 'Braeburn' apples in relation to withholding of irrigation at different stages of the growing season. J Hort Sci 71(5):693–701

Krishna H, Singh SK, Patel VB (2006) Screening of arbuscular-mycorrhizal fungi for enhanced growth and survival of micropropagated grape (*Vitis vinifera*) plantlets. Indian J Agric Sci 76(5):297–301

Kubo Y, Xue YB, Nakatsuka A, Mathooko FM, Inaba A, Nakamura R (2000) Expression of a water stress-

induced polygalacturonase gene in harvested cucumber fruit. J Jpn Soc Hort Sci 69(3):273–279
Kuo DG, Tsay JS, Chen BW, Lin PY (1982) Screening for flooding tolerance in the genus *Lycopersicon*. Hort Sci 17:76–78
Lee SK, Kader AA (2000) Pre-harvest and postharvest factors influencing vitamin C content of horticultural crops. Postharvest Biol Technol 20:207–220
Léchaudel M, Joas J (2007) An overview of preharvest factors influencing mango fruit growth, quality and postharvest behaviour. Braz J Plant Physiol 19 (4):287–298
Lopez G, Larrigaudière C, Girona J, Behboudian MH, Marsal J (2011) Fruit thinning in 'conference' pear grown under deficit irrigation: implications for fruit quality at harvest and after cold storage. Sci Hort 129 (1):64–70
Lu J, Wilson MJ, Yu J (1997) Effect of trench planting and soil chiseling on soil properties and citrus production in hilly ultisols of china. Soil Till Res 43:309–318
Ludlaw MM (1980) Adaptive significance of stomatal responses to water stress. In: Kreeb KH, Richter H, Hinckley TM (eds) Adaptations of plants to water and high temperature stress. Wiley Interscience, New York, pp 123–138
Lurie S, Shapiro B, Ben-Yehoshua S (1986) Effects of water stress and degree of ripeness on rate of senescence of harvested bell pepper fruit. J Am Soc Hort Sci 111(6):880–885
Macilwain C (2004) Is organic farming better for the environment? Nature 428:797–798
Mahajan S, Tuteja N (2006) Cold, salinity and drought stresses: an overview. Arch Biochem Biophys 444:139–158
Maier NA, Dahlenburg AP, Frensharn AB (1986) Potassium nutrition of irrigated potatoes in South Australia. 3. Effect on specific gravity, size and internal bruising of tubers. Aust J Exp Agric 26(6):737–744
Mansour MMF (2000) Nitrogen containing compounds and adaptation of plants to salinity stress. Biologia Plantarum 43:491–500
Mathur N, Vyas A (1995) Changes in enzymes of nitrogen metabolism by VA mycorrhizae in *Ziziphus nummularia*. J Plant Physiol 147:331–333
Mayak S, Tirosh T, Glick BR (2004) Plant growth promoting bacteria that confer resistance to water stress in tomato and pepper. Plant Sci 166:525–530
McGarry A, Hole CC, Drew RLK, Parsons N (1996) Internal damage in potato tubers: a critical review. Postharvest Biol Technol 8(4):239–258
Miller SA, Smith GS, Boldingh HL, Johansson A (1998) Effects of water stress on fruit quality attributes of kiwifruit. Ann Bot 81(1):73–81
Mishra A, Sharma SD (2003) Leguminous trees for the restoration of degraded sodic wastelands in eastern U. P. in India. Land Degr Dev 14:245–261
Misra RD, Ahmed M (1993) Manual on irrigation agronomy. Oxford/IBH Publishing Co. Pvt Ltd, New Delhi
Mittler R (2002) Oxidative stress, Antioxidants and stress tolerance. TRENDS in Plant Sci 7(9):405–410
Mittler R (2006) Abiotic stress, the field environment and stress combination. Trends Plant Sci 11:15–19
Mizrahi Y (1982) Effect of salinity on tomato fruit ripening. Plant Physiol 69(4):966–970
Mizrahi Y, Taleisnik E, Kagan-Zur V, Zohar Y, Offenbach R, Matan E, Golan R (1988) A saline irrigation regime for improving tomato fruit quality without reducing yield. J Am Soc Hort Sci 113(2):202–205
Morgan JM (1983) Osmoregulation as a selection criterion for drought tolerance in wheat. Austral J Agr Res 34:607–614
Mpelasoka BS, Behboudian MH, Ganesh S (2001a) Fruit quality attributes and their interrelationships of 'Braeburn' apple in response to deficit irrigation and to crop load. Gartenbauwissenschaft 66(5):247–253
Mpelasoka BS, Behboudian MH, Mills TM (2001b) Effects of deficit irrigation on fruit maturity and quality of 'Braeburn' apple. Sci Hort 90(3–4):279–290
Mpelasoka BS, Behboudian MH, Mills TM (2001c) Water relations, photosynthesis, growth, yield and fruit size of 'Braeburn' apple: responses to deficit irrigation and to crop load. J Hort Sci Biotech 76(2):150–156
Munda GC, Das A (2007) Increasing cropping intensity in IIFS. Shaping agrarian prosperity through integrated intensive farming. ICAR RC NEH Region, Umiam
Naeini SARM, Cook HF (2000) Influence of municipal compost on temperature, water, nutrient status and the yield of maize in a temperate soil. Soil Use Manag 16:215–221
Nardini A, Lo GMA, Salleo S (2011) Refilling embolized xylem conduits: is it a matter of phloem unloading? Plant Sci 180(4):604–611
Nilo R, Saffie C, Lilley K, Baeza-Yates R, Cambiazo V et al (2010) Proteomic analysis of peach fruit mesocarp softening and chilling injury using difference gel electrophoresis (DIGE). BMC Genomics 11:322–329
Noctor G, Foyer CH (1998) Ascorbate and glutathione: keeping active oxygen under control. Annu Rev Plant Physiol Plant Mol Biol 49:249–279
Oerke EC, Debne HW, Schonbeck F, Weber A (1994) Crop production and crop protection. Elsevier, Amsterdam
Olvera-Carrillo Y, Reyes JL, Covarrubias AA (2011) Late embryogenesis abundant proteins: versatile players in the plant adaptation to water limiting environments. Plant Signal Behav 64:1–4
Osbourn A (2010) Secondary metabolic gene clusters: evolutionary toolkits for chemical innovation. Trends Genet 26(10):449–457
Osiru MO, Olanya OM, Adipala E, Kapinga R, Lemaga B (2009) Yield stability analysis of Ipomoea batatus L. cultivars in diverse environments. Aust J Crop Sci 3 (4):213–220
Pandey DN (1992) Environment and biodiversity. Yojana 36(13):8–12
Panwar JA, Vyas A (2002) AM fungi: a biological approach towards conservation of endangered plants in Thar Desert India. Curr Sci 82(5):576–578
Parent B, Hachez C, Redondo E, Simonneau T, Chaumont F, Tardieu F (2009) Drought and abscisic acid effects on aquaporin content translate into changes in

hydraulic conductivity and leaf growth rate: a trans-scale approach. Plant Physiol 149(4):2000–2012
Park EJ, Jeknic Z, Sakamoto A, Denoma J, Yuwansiri R, Murata N, Chen TH (2004) Genetic engineering of glycine-betaine synthesis in tomato protects seeds, plants, and flowers from chilling damage. Plant J 40:474–487
Pareek A, Singla SL, Grover A (1997) Salt responsive proteins/ genes in crop plants. In: Singh RP, Gulati A, Jaiswal PK (eds) Stretagies for improving salt tolerance in higher plants. Oxford/IBH Publishing Co, New Delhi, pp 365–391
Pawar SN, Divekar SP, Ghule SB, Kadale AS (2004) Effect of mulching on moisture conservation and yield of summer groundnut. J Soil Crops 14:410–413
Peleg Z, Blumwald E (2011) Hormone balance and abiotic stress tolerance in crop plants. Curr Opin Plant Biol 14(3):290–295
Penna S (2003) Building stress tolerance through overproducing trehalose in transgenic plants. Trends Plant Sci 8:355–357
Perezdela VM (2004) Plant genetic adaptedness to climatic and edaphic environment. Euphytica 92:27–38
Pessarkli M (2011) Saltgrass, a high salt and drought tolerant species for sustainable agriculture in desert regions. Inter J Water Resources Arid Environ 1(1):55–64
Pignocchi C, Foyer CH (2003) Apoplastic ascorbate metabolism and its role in the regulation of cell signalling. Curr Opin Plant Biol 6(4):379–389
Prihar SS, Sandhu BS (1987) Irrigation to field crops – principles and practices. Indian Council of Agricultural Research, New Delhi
Rabe E (1990) Stress physiology: the functional signifi cance of the accumulation of nitrogen-containing compounds. J Hortl Sci 65:231–243
Radovich TJK, Kleinhenz MD, Streeter JG (2005) Irrigation timing relative to head development influences yield components, sugar levels, and glucosinolate concentrations in cabbage. J Am Soc Hort Sci 130 (6):943–949
Rai MK, Kalia RK, Singh R, Gangola MP, Dhawana AK (2011) Developing stress tolerant plants through in vitro selection – an overview of the recent progress. Environ Exp Bot 71:89–98
Rajam MV (1997) Polyamines. In: Prasad MNV (ed) Plant ecophysiology. Wiley, New York, pp 343–374
Rajam MV, Dagar S, Wale B, Yadav JS, Kumar A, Shoeb F (1998) Genetic engineering of polyamine and carbohydrate metabolism for osmotic stress tolerance in higher plants. J Biosci 23:473–482
Rao MB, Tanksale AM, Gathe MS, Deshpande VV (1998) Molecular and biotechnological aspects of microbial proteases. Micrbiol Mol Biol Rev 62 (3):597–635
Ravi V, Aked J, Balagopalan C (1996) Review on tropical root and tuber crops. I. Storage methods and quality changes. Crit Rev Food Sci Nutr 36:661–709
Reid JB, Brash DW, Sorensen IB, Bycroft B (1996) Improvement in kiwifruit storage life caused by withholding early-season irrigation. NZ J Crop Hort Sci 24:21–28
Reyes LF, Cisneros-Zevallos L (2003) Wounding stress increases the phenolic content and antioxidant capacity of purple-flesh potatoes (*Solanum tuberosum* L.). J Agric Food Chem 51:5296–5300
Rhodes D, Hanson AD (1993) Quaternary ammonium and tertiary sulfonium compounds in higher plants. Annual Rev Pl Phsiol Pl Molec Biol 44:357–384
Sabehat A, Weiss D, Lurie S (1996) The correlation between heat-shock protein accumulation and persistence and chilling tolerance in tomato fruit. Plant Physiol 110(2):531–537
Sairam RK, Tyagi A (2004) Physiology and molecular biology of salinity stress tolerance in plants. Curr Sci 86:407–421
Saltveit ME, Morris LL (1990) Overview of chilling injury of horticultural crops. In: Wang CY (ed) Chilling injury of horticultural crops. CRC Press, Florida, pp 3–15
Sams CE, Conway WS (2003) Preharvest nutritional factors affecting postharvest physiology. In: Bartz JA, Brecht JK (eds) Postharvest physiology and pathology of vegetables, 2nd edn. Marcel Dekker, New York, pp 161–176
Sanchez-Rodriguez E, Moreno DA, Ferreres F, Rubio-Wilhelmi MDM, Ruiz JM (2011) Differential responses of five cherry tomato varieties to water stress: changes on phenolic metabolites and related enzymes. Phytochemistry 72(8):723–729
Santner A, Estelle M (2009) Recent advances and emerging trends in plant hormone signalling. Nature 459:1071–1078
Schubert A, Cravero MC (1985) Occurrence and infectivity of vesicular-arbuscular mycorrhizal fungi in North-Western Italy. Vitis 24:129–138
Sharma SS (2007) Resource conserving techniques in fruit crops. Lecture delivered in summer school on "resource conserving techniques for improving nutrient use efficiency and crop productivity". IARI, New Delhi, 4–24 Sept 2007, pp 187–206
Sharma DP, Singh KN, Rao KVGK, Kumbhare PS (1991) Irrigation of wheat with saline drainage water on a sandy loam soil. Agril Water Manage 19:223–233
Shashidhar VR, Mervat ES, Hebber BK, Prasad TG, Udayakumar M (1991) Root osmotic adjustment promotes higher water status in shoots. Evidence for a link between drought tolerance and avoidance mechanism in two crop species. In: National seminar of young scientists on environmental physiology, Garhwal, Srinagar, p 22
Shibairo SI, Upadhyaya MK, Toivonen PMA (1997) Postharvest moisture loss characteristics of carrot (*Daucus carota* L.) cultivars during short-term storage. Sci Hort 71(1–2):1–12
Shibairo SI, Upadhyaya MK, Toivonen PMA (1998a) Potassium nutrition and postharvest moisture loss in

carrots (*Daucus carota* L.). J Hort Sci Biotechnol 73(6):862–866

Shibairo SI, Upadhyaya MK, Toivonen PMA (1998b) Influence of preharvest water stress on postharvest moisture loss of carrots (*Daucus carota* L.). J Hort Sci Biotechnol 73(3):347–352

Shibairo SI, Upadhyaya MK, Toivonen PMA (2002) Changes in water potential, osmotic potential, and tissue electrolyte leakage during mass loss in carrots stored under different conditions. Sci Hort 95 (1–2):13–21

Shinozaki K, Yamaguchi-Shinozaki K (2000) Molecular responses to dehydration and cold: differences and cross-talk between two stress signal pathways. Curr Opin Plant Biol 3:217–223

Silva JM, Arrabaca MC (2004) Contributions of soluble carbohydrates to the osmotic adjustment in the C4 grass *Setaria sphacelata*: a comparison between rapidly and slowly imposed water stress. J Plant Physiol 161:551–555

Singh HP (2011) *On going research in abiotic stress due to climate change in horticulture*. http://www.niam.res.in/pdfs/DDG_Hort_lecture.pdf. Accessed 18 July 2012

Singh HP, Batish DR, Kohli RK (2003) Allelopathic interactions and allelochemicals: new possibilities for sustainable weed management. Crit Rev Plant Sci 22:239–311

Singh KA, Bag TK (2002) Some relevant farming practices for Arunachal Pradesh. Arunachal Pradesh environmental planning and sustainable development-opportunities and challenges. Bishen Singh Mahendra Pal Singh, Dehradun

Singh NK, Bracker CA, Hasegawa PM et al (1987) Characterization of osmotin. A thaumatin-like protein associated with osmotic adaptation in plant cells. Plant Physiol 85:529–536

Singh G, Singh NT, Abrol IP (1994) Agroforestry techniques for the rehabilitation of salt-affected lands in India. Land Degrad Dev 5:223–242

Singh G, Dagar JC, Singh NT (1997) Growing fruit trees in highly alkali soils – a case study. Land Degrad Dev 8:257–268

Singh DV, Srivastava GC, Abdin MZ (2001) Amelioration of negative effect of water stress in *Cassia angustifolia* by benzyladenine and/or ascorbic acid. Biol Plant 44:141–143

Singh KA, Singh A, Satapathy KK, Ramchandra Yadav BPS, Verma A (2002) Micro-watershed approach to hill slope management in North East India. In: Singh KA (ed) Resource management perspective of Arunachal agriculture. ICAR RC NEH Region, Umiam, pp 323–365

Singh HP, Singh JP, Lal SS (2010) Challenges of climate change-Indian horticulture. Westville publishing house, New Delhi

Skirycz A, Inze D (2010) More from less: plant growth under limited water. Curr Opin Biotechnol 21 (2):197–203

Smith JAC, Schulte PJ, Nobel PS (1987) Water flow and water storage in Agave deserti: osmotic implications of crassulacean acid metabolism. Plant cell Environ 10:639–648

Srinivasa Rao NK, Bhatt RM (1992) Responses of tomato to moisture stresses: plant water balance and yield. Plant Physiol Biochem 19:36–41

Stefanelli D, Goodwin I, Jones R (2010) Minimal nitrogen and water use in horticulture: effects on quality and content of selected nutrients. Food Res Int 43 (7):1833–1843

Stevens LH, Davelaar E (1997) Biochemical potential of potato tubers to synthesize blackspot pigments in relation to their actual blackspot susceptibility. J Agric Food Chem 45(11):4221–4226

Stoll M, Loveys B, Dry P (2000) Hormonal changes induced by partial rootzone drying of irrigated grapevine. J Exp Bot 51:1627–1634

Subramanian KS, Charest C (1995) Influence of arbuscular mycorrhizae on the metabolism of maize under drought stress. Mycorrhiza 5:273–278

Subramanian KS, Santhanakrishnan P, Balasubramanian P (2006) Responses of field grown tomato plants to arbuscular mycorrhizal fungal colonization under varying intensities of drought stress. Sci Hort 107:245–253

Sunkar R, Bartels D, Kirch HH (2003) Over expression of a stress-inducible dehydrogenase gene from Arabidopsis thaliana in transgenic plants improves stress tolerance. Plant J 35(4):452–464

Suresh Kumar P, Choudhary VK, Bhagawati R (2012) Influence of mulching and irrigation level on water-use efficiency, plant growth and quality of strawberry (Fragaria x ananassa). Indian J Agric Sci 82(2):127–133

Talalay P, Fahey JW (2001) Phytochemicals from cruciferous plants protect against cancer by modulating carcinogen metabolism. J Nutr 131(11):3027S–3033S

Tattersall EA, Grimplet J, Deluc L et al (2007) Transcript abundance profiles reveal larger and more complex responses of grapevine to chilling compared to osmotic and salinity stress. Funct Integr Genomics 7 (4):317–333

Terry LA, Chope GA, Bordonaba JG (2007) Effect of water deficit irrigation and inoculation with *Botrytis cinerea* on strawberry (Fragaria x ananassa) fruit quality. J Agric Food Chem 55(26):10812–10819

Thipyapong P, Melkonian J, Wolfe DW, Steffens JC (2004) Suppression of polyphenol oxidases increases stress tolerance in tomato. Plant Sci 167(4):693–703

Toivonen PMA (2003) Effects of storage conditions and postharvest procedures on oxidative stress in fruits and vegetables. In: Hodges DM (ed) Postharvest oxidative stress in horticultural crops. Food Products Press, New York, pp 69–90

Toivonen PMA (2009) Benefits of combined treatment approaches to maintaining fruit and vegetable quality. Fresh Produce 3(1):58–64

Toivonen PMA, Beveridge THJ (2005) Maturity, ripening and quality relationships. In: Lamikanra O, Imam SH,

Ukuku DO (eds) Produce degradation: pathways and prevention. CRC Press, Boca Raton, pp 55–77
Toivonen PMA, Sweeney M (1997) Differences in chlorophyll loss at 13 °C for two broccoli (*Brassica oleracea* L.) cultivars associated with antioxidant enzyme activities. J Agric Food Chem 46(1):20–24
Treeby MT, Henriod RE, Bevington KB, Milne DJ, Storey R (2007) Irrigation management and rootstock effects on navel orange *Citrus sinensis* (L.) Osbeck fruit quality. Agric Water Manage 91(1–3):24–32
Turner NC, Jones MM (1980) Turgor maintenance by osmotic adjustment: A review and evaluation. In: Turner NC, Kramer PJ (eds) Adaptations of plants to water and high temperature stress. John-Wiley & Sons, New York pp, pp 87–103
Vandeleur RK, Mayo G, Shelden MC et al (2008) The role of plasma membrane intrinsic protein (PIP) aquaporins in water transport through roots: diurnal and drought stress responses reveal different strategies between isohydric and anisohydric cultivars of grapevine. Plant Physiol 149(1):445–460
Velez JE, Intrigliolo DS, Castel JR (2007) Scheduling deficit irrigation of citrus trees with maximum daily trunk shrinkage. Agric Water Manage 90 (3):197–204
Vessey JK (2003) Plant growth promoting rhizobacteria as biofertilizers. Plant Soil 255:571–586
Vinson JA, Su X, Zubik L, Bose P (2001) Phenol antioxidant quantity and quality in foods: fruits. J Agric Food Chem 49:5315–5321
Vosátka M, Gryndler M (1999) Treatment with culture fractions from *Pseudomonas putida* modifies the development of *Glomus fistulosum* mycorrhizae and response of potato and maize plants to inoculation. Appl Soil Ecol 11(2–3):245–251
Wang CY (1993) Approaches to reducing chilling injury of fruits and vegetables. Hort Rev 15:63–95
Wang CY (1994) Reduction of chilling by Methyl Jasmonate. Acta Hort 368:901–907
Wang Y, Frei M (2011) Stressed food – the impact of abiotic environmental stresses on crop quality. Agric Ecosyst Environ 141(3–4):271–286
Wang WX, Vinocur B, Shoseyov O, Altman A (2001) Biotechnology of plant osmotic stress tolerance: physiological and molecular considerations. Acta Hort 560:285–292
Whitmore JS (2000) Drought management on farmland. Kluwer Academic, Dordrecht
Williamson JG, Coston DC (1990) Planting method and irrigation rate influence vegetative and reproductive growth of peach planted at high density. J Am Soc Hort Sci 115(2):207–212
Wilkinson S, Davies WJ (2010) Drought, ozone, ABA and ethylene: new insights from cell to plant to community. Plant Cell Environ 33:510–525
Wink M (1999) Introduction: biochemistry, role and biotechnology of secondary metabolites. In: Wink M (ed) Biochemistry, role and biotechnology of secondary metabolites. Sheffield Academic/CRC Press, Sheffield, pp 1–16
Witkowska I, Woltering EJ (2010) Pre-harvest light intensity affects shelf-life of fresh-cut lettuce. Acta Hort 877:223–227
Wurr DCE, Hambidge AJ, Fellows JR et al (2002) The influence of water stress during crop growth on the postharvest quality of broccoli. Postharvest Biol Technol 25(2):193–198
Wyn Jones RG, Gorham J (1983) Aspects of salt and drought tolerance in plants. In: Kosuge T, Meredith CP, Hollaender A (eds) Genetic engineering of plants. Plenum, New York, pp 355–370
Xiong L, Zhu JK (2002) Molecular and genetic aspects of plant responses to osmotic stress. Plant Cell Environ 25:131–139
Yadav JSP (2003) Managing soil health for sustained high productivity. J Indian Soc Soil Sci 51 (4):448–465
Yamaguchi-Shinozaki K, Shinozaki K (2006) Transcriptional regulatory networks in cellular responses and tolerance to dehydration and cold stresses. In: Delmer DP (ed) Annual review of plant biology. Annual Reviews Inc, New York, pp 781–803
Yeo AR (1998) Molecular biology of salt tolerance in the context of whole-plant physiology. J Exp Bot 49:915–929
Zaicovski CB, Zimmerman T, Nora L et al (2008) Water stress increases cytokinin biosynthesis and delays postharvest yellowing of broccoli florets. Postharvest Biol Technol 49(3):436–439

Plant Water-Stress Response Mechanisms

Surajit Bhattacharjee and Ajay Krishna Saha

Abstract

Water, the central molecule of life, plays a profound role in a number of plant life processes ranging from photosynthesis to macromolecular interaction through hydrophobic bond. Due to imbalances in natural status of the different physiological, environmental conditions and during natural calamities, plants are exposed to either deficit of water (i.e. drought) or excess of water (i.e. flooding). Both of these conditions lead to water stress on plants which in turn results in disruption of agriculture and food supply in different parts of the world. In this chapter, a brief idea on the causes, indicators, responses and adaptation processes to the water stress in plants and the associated molecular mechanisms has been presented. In this chapter, the stresses related to water are expressed as "drought". The cellular and molecular responses of plants to water stress have been studied intensively throughout the world. Understanding the mechanisms by which plants perceive water stress and transmit the subsequent signals to cellular machinery and modulate expression of genes and their products to activate adaptive responses is of fundamental importance to plant biology. Knowledge about water-stress signal transduction is therefore vital for continued development of rational breeding and transgenic strategies to improve stress tolerance in crops. Factors controlling water-stress conditions alter the normal equilibrium and lead to a series of morphological, physiological, biochemical and molecular changes in plants which adversely affect their growth and productivity. However, plants also have developed innate adaptations to water-stress conditions

S. Bhattacharjee
Department of Molecular Biology and Bioinformatics,
Tripura University (A Central University),
Suryamaninagar, TR 799022, India
e-mail: sbhattacharjee@gmail.com

A.K. Saha (✉)
Department of Botany, Tripura University (A Central
University), Suryamaninagar, TR 799022, India
e-mail: aksaha.58@gmail.com

R.K. Gaur and P. Sharma (eds.), *Approaches to Plant Stress and their Management*,
DOI 10.1007/978-81-322-1620-9_8, © Springer India 2014

with an array of biochemical and physiological interventions that involve the function of many stress-associated genes. Water-stress-associated hormones like ABA are found to play a central role in orchestrating the molecular and physiological responses leading protective responses in plants. Overall, this chapter provides a systemic glimpse of integrated cellular and whole plant responses to water stress.

Keywords
Water stress • Drought • Signal transduction • Water-stress resistance • Stress tolerance

Abbreviations

ABA	Abscisic acid
ABREs	Abscisic acid responsive elements
APX	Ascorbate peroxidase
AtHD6	Histone deacetylase 6
ATP	Adenosine triphosphate
BapA	Boiling staple protein
bZIP	Basic leucine zipper domain
CBL	Calcineurin B-like protein
CDSP	Chloroplast drought-stress protein
CIPKs	CBL-interacting protein kinases
CO_2	Carbon dioxide
COR	Cold regulated
CPKs/CDPKs	Calcium-dependent protein kinases
DAG	Diacylglycerol
DREs	Dehydration-responsive elements
ERD	Early response to dehydration
ET	Electron transport
ETR	Electron transport rate
Fs	Steady state of chlorophyll fluorescence
GL	Glycolipids
GRase	Glutathione reductase
HSP	Heat-shock protein
IP3	Inositol 1,4,5-trisphosphate
KIN	Cold inducible
LEA	Late embryogenesis abundant
LEAPs	Late embryogenesis abundant proteins
MAPK	Mitogen-activated protein kinase
MYC	Myelocytomatosis oncogene
NAADP	Nicotinic acid adenine dinucleotide phosphate
NAD	Nicotinamide adenine dinucleotide
NADP	Nicotinamide adenine dinucleotide phosphate
NFYA5	Nuclear factor Y, subunit A5
PL	Phospholipids
PLC	Phospholipase C
PS	Photosystem
RAB	Responsive to ABA
RC	Reaction centre
RD	Responsive to dehydration
RNA	Ribonucleic acid
ROS	Reactive oxygen species
RuBisCO	Ribulose-1,5-bisphosphate carboxylase oxygenase
RWC	Relative water content
SOD	Super oxide dismutase
WD	Water deficit

Introduction

Plants in nature are constantly exposed to various abiotic stresses resulting from unfavourable environmental conditions which adversely affect their growth and development (Atkinson and Urwin 2012). Water stress is one of the main abiotic stresses to which crops are exposed in India. Plant water stress, often caused by drought, can have major impact on plant growth and development (Jaleel et al. 2009). When drought occurs, then it can be the cause of lower yields and possible crop failure. The effects of plant water stress vary between the plant species. Early recognition of water-stress symptoms can be critical to maintain the growth of a crop. The most common symptom of plant water stress is wilt. As the plant undergoes water stress, the water pressure inside the leaves decreases and the plant wilts. Drying to a condition of wilt will reduce the growth of any plant (Kaur and Gupta 2005).

From an irrigator's perspective, managing water to minimise stress means knowing plant water availability, recognising symptoms of water stress and planning ahead. This chapter outlines how water stress impacts plant growth and development and how to anticipate plant water stress to minimise negative consequences. Drought (water stress) is one of the main abiotic stress factors that affect all organisms' lives. Drought occurs when soil moisture level and relative humidity in air are low, while temperature is also high. Almost every plant process is affected directly or indirectly by water supply (Akıncı 1997; Lobell et al. 2013). Plants, as one of the basic food sources, either in nature or in cultivations, in their growing period, require water or at least moisture for germination. It is obvious that most land plants are exposed to short- or long-term water stresses at some times in their life cycle and tend to develop some adaptive mechanisms for adapting to changing environmental conditions. The extent and duration of the water deprivation determines the magnitude of stress response (Pugnaire et al. 1999). Some plants may adapt more easily than others giving them an advantage over competitors. Water stress may range from moderate, and of short duration, to extremely severe and prolonged summer drought (Pereira and Chaves 1993, 1995; Bottner et al. 1995). At the whole plant level, the effect of water stress is usually perceived as a decrease in photosynthesis and growth and is associated with alteration in carbon and nitrogen metabolism (Cornic and Massacci 1996; Mwanamwenge et al. 1999). It is observed that within a few seconds following the onset of water stress, short-term responses which are primarily linked to stomatal regulation appeared. Short-term responses lead to reduction in water loss by transpiration and maximising CO_2 intake. Optimum efficiency of these initial responses is found to be responsible for maintenance of constant ratio of transpiration to photosynthesis (Kozlowski et al. 1991). Midterm responses also known as acclimation comprise of the fine-tuning of the osmotic potential by accumulation of solute, modifications in cell wall elasticity and morphological variations. Long-term adaptation to drought is characterised by variation in gene responses, anatomical modifications of specific organs and acquisition of modified physiological mechanisms with an aim to reduce the overall growth to balance resource utilisation (Chapin 1980, 1991) (Fig. 1). Under field conditions, these responses can be synergistically or antagonistically modified by the superimposition of other stresses.

The most severe form of water deficit is desiccation – when most of the protoplasmic water is lost and only a very small amount of tightly bound water remains in the cell. It is reported that water stress encompasses both destructive and constructive elements and acts as a determining factor as well as a driving force for improving resistance and adaptive evolution (Larcher 1987). Plant resistance to water stress which leads to adaptation results from either tolerance or a mechanism that supports avoidance. Whole plant can contribute to the avoidance of water deficit through an array of mechanisms during the plant's life cycle, and evasion to water stress can also occur at the cellular level. The important determinants of these adaptive responses include the species and genotype, the extent and severity of water loss, the age and phase of development, the organ and cell type and the subcellular compartment. An example of avoidance at the cellular level is the process of osmotic adjustment where the osmotic potential of the cell is lowered in order for the water potential gradient to favour water uptake and maintenance of turgor (Bray 1997) (Fig. 1).

In response to water stress, a plethora of modification occurred in the intracellular milieu of the plant cells. The changes include the

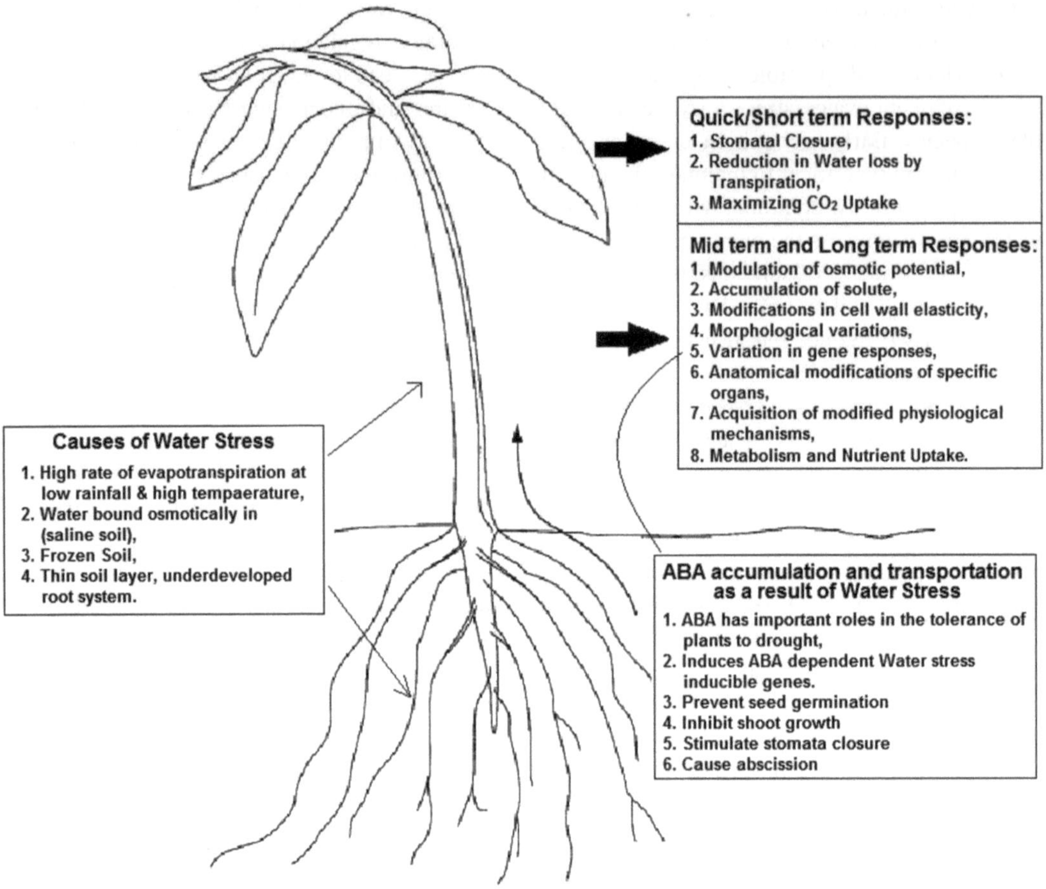

Fig. 1 Causes of water stress and variable responses shown by plants against drought

modification of different intracellular metabolic pathways, changes in the nutrient and ion uptake, synthesis of new proteins, modulation of free radical generation and all these changes found to be preceding the induction of signal transduction pathways. Water stress induces a multiple signal transduction pathway which follows the generation of second messengers (e.g. inositol phosphates, lipid mediators and reactive oxygen species). These second messengers in turn modulate intracellular Ca^{2+} level and activate kinases to initiate protein phosphorylation cascades. These events lead to activation of target proteins which are directly involved in cellular protection or acting as transcription factors controlling genes explicitly involved in regulation of water stress. The activation of these genes is found to be involved in the generation of the plant hormones like abscisic acid (ABA), ethylene and salicylic acid (SA) which in turn initiate a second round of signalling which may be responsible for the adaptive and tolerance responses associated with water stress (Xiong et al. 2002). Specificity in water-stress responses in plants is further determined by a complex regulatory network of molecular mechanisms which include the interaction between transcription factors, kinase cascades, production of reactive oxygen species as well as involvement of heat-shock factors and small RNAs (Atkinson and Urwin 2012).

One of the important adaptive features acquired by the plants to water stress includes sun-type or shade-type chloroplast adaptation which is also induced by many other stress factors including drought (Lichtenthaler et al. 1981). Regions with adequate but non-uniform precipitation also experience water-limiting environments. The general effects of drought on plant growth are fairly well known. However, the primary effect of water deficit at the biochemical and molecular levels is not considerably understood yet, and such understanding is crucial. Knowledge of the biochemical and molecular responses to drought is essential for a holistic perception of plant resistance mechanisms to water-limited conditions in higher plants. The response to abiotic stress results in a dramatic change of the whole plant transcriptome. It is reported that the transcriptomic response to drought can vary with the time of day. These responses seem to interact with hormonal and other stress pathways that naturally vary during the course of the day (Wilkins et al. 2010; Cramer et al. 2011). Sometimes a comparison between cellular response and whole plant response may reveal the level of organisation where the adaptation operates (Kar 2011). In this chapter, we provide an overview of the current understanding of plant responses to drought. In addition, we will describe the cellular signalling mechanisms leading to protect the plant from the deleterious effects of drought.

Origin of Plant Water Stress

Water stress in plants results either from restricted water supply to their roots or due to increased rate of transpiration. Plants growing under arid and semiarid environments frequently experienced the water stress associated with drought. It is reported that high temperatures act as an indirect driver of plant water stress (Lobell et al. 2013). Roots are the primary site of water intake in plants. The extent of force required for a plant to absorb water from the soil is known as the matric potential. In conditions of low soil moisture, more energy is required by the plants to remove water from the soil; thus, the matric potential is greater. Symptoms of water stress have been experienced by plants when the soil is dry and the matric potential is strong (Glyn Bengough et al. 2011). This condition is recognised as the matric effect. It has also been shown that heat is an indirect driver of reduced crop yield through increased plant evapotranspiration (Lobell et al. 2013) (Fig. 1).

Measurement of Water Stress in Plants

The extent of water stress experienced by plants in their habitat can be assessed by measuring the soil moisture and analyses of the distribution of precipitation. Measurement of water potential (ψ) in plants is found to be the most fundamental indicator of water stress. No water stress (small negative water potential values) was found in soils with high water-holding capacity. On the other hand, moderate to high water stress was recorded at the end of the season in those sites with low water-holding capacity. A linear relationship between predawn leaf water potential and stem water potential is also reported (McCutchan and Shackel 1992). Another commonly used indicator of plant water status is relative water content or RWC which at one time had been less accurately termed as relative turgidity. Tissue water content (percent of fresh weight) and fresh weight have also been used as indicators of water status. Unfortunately, water content or fresh weight of tissue at full turgor is normally not given as a reference. Water content can be very misleading because of its superficial resemblance to RWC (Hsiao 1973). In some studies, visual wilting is considered as the sole indicator of water status. Although wilting is dependent on turgor pressure, it is also a function of the mechanical properties of cell wall and tissue (Hsiao 1973; Joly 1985) (Fig. 2).

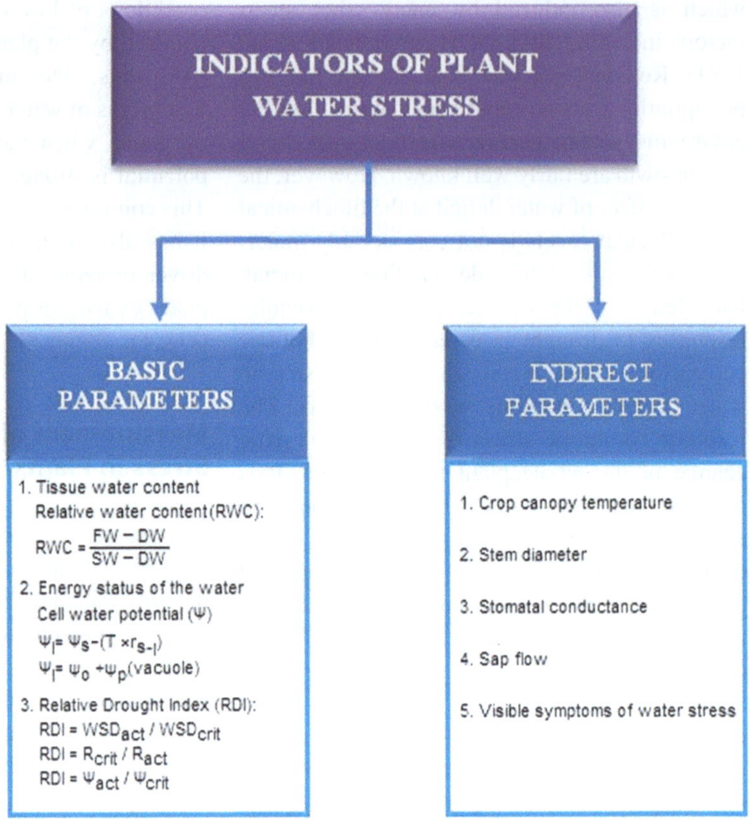

Fig. 2 Indicator of plant water stress

Plant Responses to Water Stress

Photosynthetic Responses to Drought

Reduced rate of photosynthesis is a usual effect of water stress in plants. Water stress reduces photosynthesis by decreasing both leaf area and photosynthetic rate per unit leaf area (McCree 1986). Photosynthesis is severely inhibited and may cease altogether as water deficits increase in crop plants. Water deficiency in plants generates metabolic changes along with functional and structural rearrangements of photosynthesising apparatus. The decrease in leaf growth or increasing senescence of leaves under drought conditions may also inhibit photosynthesis in existing leaves (Boyer 1976). Decreasing water content is accompanied by loss of turgor and wilting, cessation of cell enlargement, closure of stomata, alteration of photosynthesis and interference with many other basic metabolic processes (Kramer and Boyer 1995).

The inhibition of photosynthesis during water stress could be explained as the cause of the stomata closure and the internal CO_2 concentration decrease. Stomatal limitation is more severe when a plant is stressed than when it is not. Therefore, it is rather surprising that photosynthesis often decreases in parallel with or more than stomatal conductance. Limitation of carbon uptake during water stress might be associated with stomatal control of water (Chaves 1991; Cornic and Massacci 1996). Stomata close in response either to a decline in leaf turgor and/or water potential or to a low-humidity atmosphere (Maroco et al. 1997).

The photosynthetic rate in higher plants decreases more rapidly than respiration rate with increased water stress. A nearly effect of water reduction in leaves is usually partial or complete stomatal closures which markedly decrease the movement of carbon dioxide into the assimilating leaves and reduce the photosynthetic rate up to ten times depending on the amount of water removal and the sensitivity of the plant (Ghannoum 2009; Akıncı and Lösel 2012; Chaves et al. 2003).

In C_4 plants, stomatal closure is found to be a major determinant in the inhibition of photosynthesis under water stress, while non-stomatal factors like metabolic impairments are also reported to play the major role in this inhibition. In both C_3 and C_4 plants, the rate of photosynthesis decreases under the drought conditions. It is evidenced that the rate of photosynthesis is more affected in C_4 plants (like corn) than C_3 plants (such as wheat) in conditions of water deficits. This explains the fact that hot arid areas with prevalence of C_4 plants are more susceptible to frequent drought. A number of cofactors like (a) low CO_2 uptake due to stomatal closure and resistance, (b) qualitative and quantitative changes in photosynthesising pigments and (c) poor assimilation rates in photosynthetic leaves are found to be affected under water stress which in turn decreases the rate of photosynthesis in plants. Water stress is also found to inhibit chlorophyll synthesis and subsequently decrease chlorophyll content of leaves. In severe stress, photosynthesis may be more controlled by the chloroplast's capacity to fix CO_2 than by the increased diffusive resistance (Faver et al. 1996; Herppich and Peckmann 1997).

Unlike chlorophyll, other plant pigments like xanthophyll are found to be less sensitive to water stress. During water stress, the synthesis of xanthophyll pigment is shown to be upregulated which supports the finding that xanthophyll pigments have a protective role in plants under stress and also are found to play an inhibitory role on reactive oxygen species (ROS) production (Lisar et al. 2012). The photosynthetic enzymes have been shown to be significantly affected by water stress. In case of C_4, it is difficult to draw a conclusion regarding the specific pattern in the modulation of enzyme activity in response to drought stress, whereas in C_3 cycle enzymes are found to be consistently inhibited in response to water stress. Activity of RuBisCO, the key enzyme for carbon metabolism in leaves, is reported to be strikingly decreased in conditions of water stress. Inhibition of the RuBisCO activity during water stress is found to be associated with acidification of the chloroplast stroma. Furthermore, water-stress-associated suppression in RuBisCO activity is also related to the alterations of the chloroplast structure, conformational change of the RuBisCO, lack of the substrate and reduction in the activity of the coupling factor – ATPase – and sometimes due to damage, the plastids may lose RuBisCO. Activity of other photosynthetic enzymes like NAD-dependent malate dehydrogenase, phosphoenolpyruvate carboxylase, fructose-1,6-bisphosphatase and other related enzymes also is found to be inhibited to different extents (Ramachandra et al. 2004; Lisar et al. 2012) (Fig. 3).

Water stress also disrupts the cyclic and non-cyclic types of electron transport during the light reaction of photosynthesis. The disruption is clear in the oxygen-releasing complex and electron transfer from protochlorophyllide to P_{700}. Lower electron transport rate negatively affects photophosphorylation process and decreases ATP synthesis as well as $NADP^+$ reduction. ATPase inhibition under water deficiency is also responsible for the reduction in ATP levels in chloroplasts. All these factors cumulatively affect the intensity of photo-assimilation and the stability of the photosynthetic apparatus under the conditions of water stress. Both of the PSs in chloroplasts are affected by water deficiency; however, PS1 of some plants is more severely damaged compared to PS2, though there is an opposite concept as well (Ramachandra et al. 2004; Lisar et al. 2012).

Transpiration and Stomata

Stomatal closure is commonly the principal mechanism responsible for restricting transpiration rates in plants during exposure to water stress. Transpiration is directly proportional to the gradient of water vapour concentration from the internal evaporation surface to the bulk air outside the leaf and inversely proportional to the

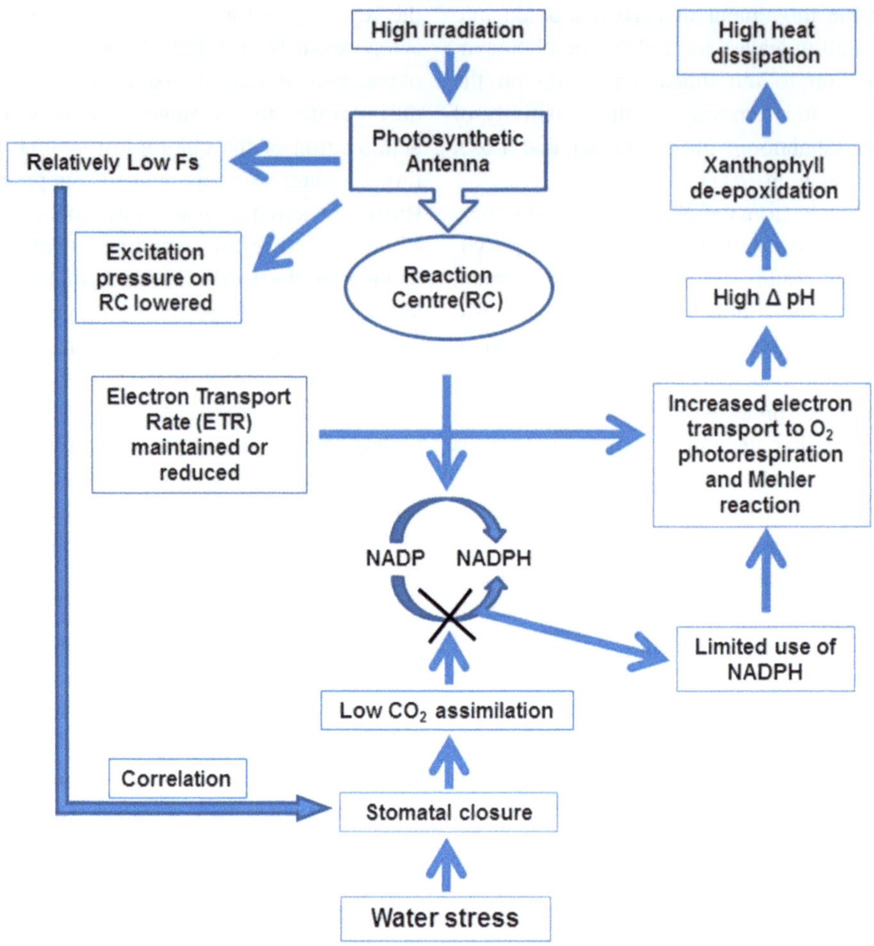

Fig. 3 Schematic presentation of photosynthetic control mechanisms under water stress. *ETR* electron transport rate, *RC* reaction centre, *Fs* steady state of chlorophyll fluorescence

total resistance to water vapour transport of the air boundary layer and of the leaf. In addition, increased stomatal resistance may not cause proportional decreases in transpiration rate because diminished dissipation of heat by vaporisation and the consequent rise in leaf temperature increase the water vapour concentration inside the leaf. In most situations, the rise in leaf temperature accompanying substantial reduction in transpiration has been calculated or measured to be only a few degrees (Hsiao 1973; Chaves et al 2003). Therefore, it would be reasonable to assume that elevation in leaf temperature does not play a general role in water-stress effects. Some other non-stomatal factors in the leaf like "mesophyll" or "wall" resistance cause significant reductions in transpiration as water stress develops. The "wall" resistance is small in turgid leaves and tends to rise with moderate water deficits to a significant level which is nevertheless still minor compared with the expected stomatal resistance (Crafts 1968). Adaxial and abaxial stomata have been observed to differ in response to water stress in some cases but apparently not in others (Wang et al. 1998). The above results indicate that stomata are somewhat insensitive to mild water stress. However, this conclusion probably cannot be generalised, since there are direct or indirect indications that stomata of other species may be sensitive to small water deficits. The stomatal response is found to be dependent on threshold water status. It is observed that the optimum water content for stomatal opening can be actually something less than the tissue water content at full turgor. Full turgor can cause some stomatal closure, presumably because of excessive back pressure from the epidermal cells surrounding the guard cells. Once the threshold water status for stomatal closure is reached, leaf resistance increases sharply, rising 20- or 30-fold. Such large increases in leaf resistance may be taken as indicative of almost complete stomatal closure. Aside from leaf water status, there is some evidence that water vapour content of the air may be very important in determining stomatal opening. In case of maize leaves it is reported that at the same water deficit the diffusive resistance is upto several times as great in dry air (nearly zero humidity) as in moist air. Light may also modify stomatal response to water deficit. At higher light levels, more water deficit seemed to be required to induce closure. It has also been reported that stomatal response to water stress was attenuated by oxygen-free air (Hsiao 1973; Yokota et al. 2006).

Stomatal opening and closing result from turgor differences between guard cells and the surrounding subsidiary or epidermal cells. Stomatal interactions with environmental factors such as light and CO_2 are complex and appear to be mediated by a net gain or loss of guard cell potassium and turgor with the consequent stomatal movement. As the opening of the stomata is turgor dependent, water deficits by reducing leaf turgor would directly reduce opening. It has also been reported that mild water deficit is associated with marked loss of solutes from guard cells which is concurrent with stomatal closure. Thus, a part of the water-stress effect on stomatal closure and associated decrease in the rate of transpiration may not be direct but is linked to the regulation of osmotic solutes in guard cells. Another important determinant in modulating the stomatal opening during water stress is found to be abscisic acid (ABA). It is reported that ABA rises markedly in leaves subjected to water stress and that exogenous ABA is a potent and fast-acting inhibitor of stomatal opening; it is also being hypothesised that stress affects stomata via its effect on ABA levels or on plant hormonal balance, specifically the balance between ABA and cytokinins. It was reported that the rapidity and ready reversibility of the action of ABA on stomata would make it a good modulator of stomatal behaviour. Although stomatal opening is reduced during stress by a concerted effect of depressed cytokinin level and rise in ABA but kinetin, a member of cytokinin family, can promote stomatal opening within a few hours of application. Unfortunately, the stomatal response to kinetin is dependent on the duration of exposure and age of the plants. Stomata of many species and apparently of younger leaves do not respond to kinetin. It is also observed that kinetin is unable to reverse the ABA-mediated inhibition of stomata (Xiong and Zhu 2003; Hsiao 1973; Yokota et al. 2006).

Respiration

Water stress exerts a variable response on plant respiration which ranges from inhibition to stimulation under different water-stress conditions. In different plant organs like leaves, shoots, roots, flowers or whole plants, a decreased rate of respiration in response to water stress has been reported. In contradiction, some other reports have shown that in water-stressed plants the rate of respiration is almost unaffected or even increased. Leaf respiration shows a biphasic response to relative water content (RWC), decreasing in the initial stages of water stress (RWC $>$ 60 %) and increasing as RWC decreases below 50 % (Flexas et al. 2005). Under this hypothesis, the initial decrease in respiration would be related to the immediate inhibition of leaf growth and, consequently, the growth of respiration component. The increase of respiration at lower RWC would relate to an increasing metabolism as the plant triggers acclimation mechanisms to resist water stress. These mechanisms would increase the maintenance component of respiration and, as such, the overall respiration rate. In case of root, the changes in rate of respiration in response to water stress are found to be age dependent. Respiration in the established root and rain root is shown to respond differentially in response to water stress. In established root, the rate of respiration never reached zero in response to water stress and rapidly recovers upon direct rewatering, whereas it has been shown that in rain root the rate of respiration rapidly reached zero and did not recover upon rewatering (Graham and Nobel 1999). It is hypothesised that the differential rate of respiration in response to water stress occurs at a certain threshold of water-stress intensity. It has been reported that dark respiration is generally suppressed, more or less proportionately but not very markedly, by moderate to severe water stress. Similar kind of response was observed in light respiration. It is observed that the effects might be due to plasmolysis rather than water stress. The biphasic response of respiration in whole plants against water stress has also been observed. The initial tendency is for the rate of respiration to decrease probably as a consequence of decreased energy demand for growth. A second trend that appears at severe water stress is the increase of respiration rates, possibly as a consequence of enhanced metabolism (osmoregulation, water-stress-induced senescence processes). It has been reported that the fast-growing plant species show a more pronounced biphasic response than slow-growing species (Flexas et al. 2005) (Fig. 5).

Osmotic Adjustment Mechanisms Under Water Stress

Water is essential in the maintenance of the turgor which is essential for cell enlargement and growth and for maintaining the form of herbaceous plants. Turgor is also important in the opening of stomata and the movements of leaves, flower petals and various specialised plant structures (Kramer and Boyer 1995). The turgor measurements on the lamina have often appeared to show declining rates of leaf growth with decreasing turgor (Kramer and Boyer 1995; Meyer and Boyer 1972; Michelena and Boyer 1982; Westgate and Boyer 1985). The turgor decrease may or may not occur during soil drying, and this is believed to be due to osmotic adjustment, the process in which solutes accumulate in growing cells as their water potential falls of osmotic potential arising from the net accumulation of solutes in response to maintain turgor in tissues (Turner and Jones 1980; Morgan 1984). Osmotic adjustment may allow growth to continue at low water potential. Osmotic adjustment usually depends mainly on photosynthesis to supply compatible solute. Osmotic adjustment has been defined as "the lowering water deficits or salinity" (Turner and Jones 1980). With continued water limitation, osmotic adjustment delays, but cannot completely prevent, dehydration (Kramer and Boyer 1995). Osmotic adjustment has been found in many species and has been implicated in the maintenance of stomatal conductance, photosynthesis, leaf water volume and growth (Turner and Jones 1980; Morgan 1984). In wheat and other cereals, osmotic adjustment leads to rapid responses for decreasing the effect of water stress (Richter and Wagner 1982). It is reported that water stress increases the osmotic pressure of the cell sap, increasing the percentage of sugar in sugar cane and often in sugar beet, although the yield per acre may be reduced (Russel 1976). Solutes known to accumulate with water stress and contribute to osmotic adjustment in non-halophytes include inorganic cations, organic acids, carbohydrates and free amino acids. In some plants, potassium is the primary inorganic cation accumulating during water stress, and it is often the most abundant solute in a leaf (Jones et al 1980; Ford and Wilson 1981). Osmotic adjustment is usually not permanent, and plants often respond rapidly to increased availability of water. Morgan and Condon (1986) showed that such increase in solute concentration gives tissues a temporary advantage, enabling turgor to be maintained at low water potentials by decreasing their osmotic potentials (Morgan and Condon 1986).

Cell Growth and Cell Division

Because plant growth is the result of cell division and enlargement, water stress directly reduces growth by decreasing CO_2 assimilation and reducing cell division and elongation. The effect of water stress is more evident on cell wall expansion because cell enlargement involves the extensibility of the cell wall under turgor pressure. Therefore, any loss in turgor pressure as a consequence of the imbalance in the plant water content could result in reduced growth and even in the total absence of growth under dry environmental conditions.

Cell growth rate, Gr, can be expressed as a function of turgor pressure, P, and the extensibility coefficient, Φ, by the equation $Gr = \Phi (P - Y)$ where Y is the yield threshold pressure. The equation shows that growth rate decreases as P decreases, but it could also be maintained if either Φ increases or Y decreases. Therefore, reduced growth rate may not rely only on reduced turgor caused by desiccation. There is some evidence of reduced growth without loss of turgor in plants subjected to desiccation stress, but this reduction may be part of the osmotic adjustment process. Some mechanism may control cell wall extensibility through the perception of soil dryness, giving rise to smaller plants and hence lower water requirements and higher survival (Hsiao 1973).

Plant Metabolic Response to Water Stress

Plant adaptations to dry environments can be expressed at four levels: phenological or developmental, morphological, physiological and metabolic. The metabolic response is least known where the metabolic or biochemical adaptations are involved (Hanson and Hitz 1982). Physiological and biochemical changes including carbohydrates, proteins and lipids are observed in many plant species under various water-stress levels which may help in better understanding survival mechanisms in drought.

Carbohydrate Changes Under Water Stress

The available reports stated that the content of soluble sugars and other carbohydrates in the leaves of various water-stressed plants is altered and may act as a metabolic signal in the response to drought (Akıncı and Lösel 2009, 2010; Chaves et al. 2003; Koch 1996; Jang and Sheen 1997). Munns et al. (1979) and Quick et al. (1992) showed that sugars are major contributors to osmotic adjustment in expanding wheat leaves (Munns et al. 1979; Quick et al. 1992). The increase of sugar in various plant tissue responses to water stress supports the idea of contribution of solutes while the plants are exposed to different stress levels. The studies have shown that soluble sugars accumulate in leaves during water stress and have suggested that these sugars might contribute to osmoregulation, at least under moderate stress (Morgan 1984; Quick et al. 1992; Jones et al. 1980; Munns and Weir 1981; Ackerson 1981; Kameli and Losel 1993, 1996; Al-Suhaibani 1996).

Increase in total carbohydrate is recorded in cotton by Timpa et al. (1986) and Evans et al. (1992). Total soluble sugar is found to be increased in wheat, alfalfa, lupins, bean and cucumber (Kameli and Losel 1996, 1993; Irigoyen et al. 1992; Quick et al. 1992; Al-Suhaibani 1996; Akıncı and Lösel 2009). But depletion of sucrose and starch content is also recorded in soya bean, grapevine, lupins, bean and cucumber by Westgate et al. (1989), Rodriguez et al. (1993), Quick et al. (1992), Steward (1971) and Akıncı and Lösel (2009).

Plant Proteins: Responses to Water Stress

Many specified proteins synthesised under water scarcity have been isolated and characterised by researches (Singh et al. 1987; Close 1997; Pelah et al. 1997; Claes et al. 1990). The water-stress-specific proteins (stress induced) have been described as dehydrins (polypeptide) and LEA (late embryogenesis abundant), RAB (responsive to ABA) and storage proteins (in vegetative tissues) (Artlip and Funkhouser 1995). Under water-stress conditions, plants synthesise alcohols, sugars, proline, glycine, betaine and putrescine and accumulate that of those molecular weights which are low (Chopra and Sinha 1998; Galston and Sawhney 1990). Dehydrins have been the most observed group among the accumulated proteins in response to loss of water and increased in barley, maize, pea and *Arabidopsis*. Under water stress, LEA proteins play an important role as protection of plants. Osmotin is also an accumulated protein under water stress in several plant species such as tobacco, triplex, tomato and maize (Ramagopal 1993).

Heat-shock proteins (HSPs) and late embryogenesis abundant (LEA)-type proteins are two major types of stress-induced proteins during different stresses including water stress. Protection of macromolecules such as enzymes, lipids and mRNAs from dehydration is the well-known function of these proteins. LEA proteins accumulate mainly in the embryo. The exact functions and physiological roles of these proteins are unknown. HSPs act as molecular chaperones and are responsible for protein synthesis, targeting, maturation and degradation in many cellular processes. They also have important roles in stabilisation of proteins and membranes and in assisting protein refolding under stress conditions. Expression of LEA-type genes under osmotic stress is regulated by both ABA-dependent and independent signalling pathways. Genes encoding LEA-type proteins are diverse – RD (responsive to dehydration), ERD (early response to dehydration), KIN (cold inducible), COR (cold regulated) and RAB (responsive to ABA) genes (Lisar et al. 2012; Wang et al. 2004; Singh et al. 2005).

Changes of amino acids and protein have been mentioned in many reports which have stated that water stress caused different responses depending on the level of stress and plant type. Water stress has a profound effect upon plant metabolism and results in a reduction in protein synthesis. Several protein contents were reduced by stress in maize mesocotyls (Bewley and Larsen 1982; Bewley et al. 1983). Dasgupta and Bewley (1984) pointed out water stress reduced protein synthesis in all regions of barley leaf. Vartanian et al. (1987) mentioned the presence of drought-specific proteins in taproot in *Brassica*.

Various water-stress-induced proteins like dehydrins, LEAs, RABs, osmotins, boiling staple proteins, Beta alanine amino peptidase A (BapA, 87 kDa proteins) and chloroplast proteins (CDSP32 and CDSP 34) are recorded by many scientists. Protein content decrease has been recorded in *Avena* coleoptiles (Xu et al. 1996; Artlip and Funkhouser 1995; Ramagopal 1993; Sinha et al. 1996; Bray 1995; Naot et al. 1995; Pareek et al. 1997; Pelah et al. 1997; Mantyla et al. 1995; Pruvot et al. 1996).

Inhibition and/or decrease in protein synthesis has been recorded in *Avena* coleoptiles (Dhindsa and Cleland 1975), in sugar beet (Shah and Loomis 1965) and in *Pisum sativum* L. nodules (Gogorcena et al. 1995). Water stress inhibits cell division and expansion, consequently leaf expansion, and also halts protein synthesis. The direct significance of the inhibition of protein synthesis by stress to growth and leaf expansion is difficult to assess. Free proline accumulation in response to drought in many plant species tissues is well documented (Andrade et al. 1995; Aspinall and Paleg 1981; Chandrasekhar et al. 2000; Tholkappian et al. 2001; Nair et al. 2006). The functions of many of these proteins have not been established (Hughes et al. 1989). However, water stress may inhibit the synthesis of different proteins equally while inducing the synthesis of a specific stress protein (Dasgupta and Bewley 1984).

Treshow (1970) concluded that water stress inhibits amino acid utilisation and protein synthesis (Treshow 1970). Due to unutilisation of amino acids, they are accumulated, giving a 10–100-fold accumulation of free asparagine, valine and glutamic acid, but alanine levels decreased. Barnett and Naylor (1966) found no significant differences in the amino acid and protein metabolism of two varieties of Bermuda grass during water stress. They have also reported that during water deficit, amino acids were continually synthesised but protein synthesis was inhibited followed by decrease in protein content.

Plant Lipid–Water-Stress Interactions

Along with proteins, lipids are the most abundant component of membranes, and they play a role in the resistance of plant cells to environmental stresses (Kuiper 1980; Suss and Yordanov 1986). Strong water deficit leads to a disturbance of the association between membrane lipids and proteins as well as to a decrease in the enzyme activity and transport capacity of the bilayer (Caldwell and Whitman 1987). In plant cell, polar acyl lipids are the main lipids associated with membranous structures (Harwood 1979; Bishop 1983). Glycolipids (GL) are found in chloroplast membranes (more than 60 %), and phospholipids (PL) are thought to be the most important mitochondrial and plasma membrane lipids (Harwood 1980). Many workers have investigated the effect of different levels of water stress on lipid content and composition in different parts of plants (Kameli 1990; Al-Suhaibani 1996; Pham Thi et al. 1982, 1985, 1987; Navari-Izzo et al. 1989, 1990, 1993; Douglas and Paleg 1981; Liljenberg and Kates 1982). Fatty acid, phospholipid, total lipid, etc., are recorded to be increased in soya bean, cotton, wheat, alfalfa and maize by various workers (Navari-Izzo et al. 1990; Pham Thi et al. 1982; Kameli 1990; Al-Suhaibani 1996; Douglas and Paleg 1981; Quartacci et al. 1994; Poulson et al. 2002). It is observed that for Arabidopsis, polyunsaturated trienoic fatty acids may be an important determinant of responses of photosynthesis and stomatal conductance to environmental stresses such as vapour pressure deficit. When *Vigna unguiculata* plants are submitted to drought, the enzymatic degradation of galacto- and phospholipids increased. The stimulation of lipolytic activities is greater in the drought-sensitive than in drought-tolerant cvs (Sahsah et al. 1998).

Phospholipid and glycolipid decline is recorded in cotton (Wilson et al. 1987; Ferrari-Iliou et al. 1984; El-Hafid et al. 1989), wheat and barley (Chetal et al. 1981), sunflower and maize (Quartacci and Navari-Izzo 1992). Total lipid content decrease is recorded in cucumber and squash by Akıncı (1997). Linoleic, linolenic acid, galactolipid, hexadecenoic acid and diacylglycerol are found to be decreased in cotton (Pham Thi et al. 1982, 1985) and in maize (Navari-Izzo et al. 1989). Investigations on various crop species record a general decrease in phospholipid, glycolipid and linoleic acid contents and an increase in the triacylglycerol of leaf tissues exposed to long periods of water deficits. Enzyme activity and transport capacity are affected by the composition and phase properties of the membrane lipids (Kuiper 1985; Gronewald et al. 1982; Whitman and Travis 1985). Wilson et al. (1987) observed that water deficit caused a significant decline in the relative degree of acyl unsaturation (i.e. FA unsaturation) in phospholipids and glycolipids in two different drought-tolerant cotton plants (Wilson et al. 1987). Pham Thi et al. (1987) pointed out that changes in oleic and linoleic acid during water stress resulted in desaturation and water stress markedly inhibited the incorporation of the precursors into the leaf lipids (Pham Thi et al. 1987).

The study of Navari-Izzo et al. (1989) revealed the responses of maize seedling to field water deficits and found that the diacylglycerol, free fatty acid and polar lipid contents decrease significantly with stress (Navari-Izzo et al. 1989). The dry land conditions induced a decrease of more than 50 % in phospholipid levels, and triacylglycerols increased by about 30 % over the control. Pham Thi et al. (1982) have shown that the most striking effects are a decrease of total fatty acids especially trans-hexadecenoic acid. Water deficits inhibit fatty acid desaturation resulting in a sharp decrease of linoleic and linolenic acid biosynthesis. Wilson et al. (1987) and Navari-Izzo et al. (1993) found that in plasma membranes isolated from sunflower seedlings grown under water stress, there is a reduction of about 24 % and 31 % in total lipids and phospholipids, respectively, and also significant decreases in glycolipids and diacylglycerols.

Drought and Nutrient Uptake

The capacity of plant roots to absorb water and nutrients generally decreases in water-stressed plants, presumably because of a decline in the nutrient element demand (Alam 1999). It is well documented that essential plant nutrients are known to regulate plant metabolism even the plants exposed to drought by acting as cofactor or enzyme activators (Nicholas 1975).

Many reports stated that water stress mostly causes reduction in uptake of nutrients (Levitt 1980), for instance, phosphorus, K^+, Mg^{2+} and Ca^{2+} in some crops (Foy 1983; Abdalla and El-Khoshiban 2007; Bie et al. 2004); Ca^{2+}, Fe^{3+}, Mg^{2+}, nitrogen and phosphorus and potassium in *Spartina alterniflora* (Brown et al. 2006); Fe^{3+}, Zn^{2+} and Cu^{2+} in sweet corn (Oktem 2008); and Fe^{3+}, K^+ and Cu^{2+} in *Dalbergia sissoo* leaves (Nambiar 1977). Gerakis et al. (1975) and Kidambi et al. (1990) stated that nutrient elements increased in forage plant species and alfalfa. An increase in some specific elements such as K^+ and Ca^{2+} was reported in maize (Tanguilig et al. 1987) and K^+ in drought-tolerant wheat varieties (Sinha 1978). In leaves of *Dalbergia sissoo*, nitrogen, phosphorus, Ca^{2+}, Mg^{2+}, Zn^{2+} and Mn^{2+} increased with increasing water stress (Singh and Singh 2004).

It is generally accepted that the uptake of phosphorus by crop plants is reduced in dry soil conditions (Pinkerton and Simpson 1986; Simpson and Lipsett 1973). According to Singh and Singh (2004), availability of soil nutrients decreases with increasing soil drying, with K^+, Ca^{2+}, Mg^{2+}, Zn^{2+}, Fe^{3+} and Mn^{2+} decreasing by 24 %, 6 %, 12 %, 15 %, 25 % and 18 %, respectively.

Drought Perception, Signal Transduction and Response

Plant response to water stress depends on their ability to sense the extent or severity of drought they are exposed. It has been reported that water stress can be sensed by a membrane-bound two-component histidine kinase which is activated by high osmolarity. The increase of a cell osmolarity upon water loss during drought therefore triggers the signal transduction in response to drought. The active signal receptor activates phospholipase C (PLC) which hydrolyses phosphatidylinositol 4,5-bisphosphate to yield the second messengers inositol 1,4,5-trisphosphat (IP3) and diacylglycerol (DAG) (Mahajan and Tuteja 2005). IP3 releases calcium from internal stores, and the Ca^{2+} sensor (calcineurin B-like protein, CBL) activates downstream protein kinases and phosphatases. Drought-inducible genes display characteristic promoter cis-acting elements, the dehydration-responsive elements (DREs) which at least partially resemble those of the cold-induced genes (Bray 1997). Abscisic acid triggers a major signalling pathway in drought-stress response. Activation of the abscisic acid responsive elements (ABREs) by several transcription factors such as the DRE-binding factors and bZIP proteins leads to the expression of drought-stress tolerance effectors such as dehydrins or enzymes catalysing low molecular weight osmolytes. The signal transduction pathway of ABA involves cADP ribose, NAADP and Ca^{2+} as second messenger (Quatrano et al. 1997). Calcium appears as a prime candidate in drought-stress signal transduction resulting in a metabolic or structural mitigation of the effect of the stressor. Therefore, proteins, which sense changes in the cytoplasmic calcium concentrations, are important components of the signal transduction chain. Calcium-dependent protein kinases (CDPKs or, in Arabidopsis, CPKs) act as sensor responders by combining Ca^{2+}-binding and kinase activity in the same polypeptide. CPK4 and CPK11 have also been identified as positive transducers of Ca^{2+}-dependent ABA signalling. Strong ABA insensitivity in stomata closure and increased drought sensitivity were reported in the cpk4 and cpk11 single and double mutants, with opposite phenotypes observed in CPK4 and CPK11 overexpression lines. Calcineurin B-like proteins (CBLs) are sensor relay proteins that, upon Ca^{2+} binding, interact with and modulate the activity of CBL-interacting protein kinases (CIPKs). CBL1 an isoform of CBL was identified as a relay for ABA-mediated responses and can act as a positive regulator of drought signalling, CBL1-overexpressing plants exhibit enhanced drought tolerance and constitutive expression of stress genes. Although not only CBL single mutant is ABA hypersensitive in guard cells but also the cbl1cbl9 double mutant was reported to be more drought tolerant in wilting assays and the stomatal closure response in the double mutant was hypersensitive to ABA. It has been shown that in the vasculature and in guard cells, luciferase reporter expression under the control of ABA-responsive AtHD6 (histone deacetylase 6) promoter was detected in response to drought, suggesting a role for tissue autonomous ABA synthesis in addition to long-distance root-to-shoot movement of ABA in response to water stress. It has been observed that the transcription factors like NFYA5 (nuclear factor Y, subunit A5) in Arabidopsis and the maize NF-YB2 function as positive regulators of drought-stress responses, suggesting a possible role of the CCAAT box element and its binding partner NF-Y in ABA/drought-stress signalling. Besides transcriptional induction by ABA, NFYA5 gene expression is further enhanced by posttranscriptional control of NFYA5 mRNA stability. NFYA5 transcripts contain a target site for the microRNA, miR169, which is downregulated by drought. Furthermore, overexpression of miR169 and a T-DNA insertion mutation in NFYA5 both caused drought sensitivity (Raghavendra et al. 2010; Xiong et al. 2002).

Other intracellular hazards observed in plants in response to drought stress are the generation of reactive oxygen species (ROS), which is being considered as the cause of cellular damage. However, recently, a signalling role of such ROS in triggering the ROS scavenging system that may confer protection or tolerance against stress is emerging. Such scavenging system consists of antioxidant enzymes like SOD, catalase and peroxidases and antioxidant compounds like ascorbate and reduced glutathione; a balance between ROS generation and scavenging ultimately determines the oxidative load. As revealed in case of defences against pathogen, signalling via ROS is initiated by NADPH oxidase-catalysed superoxide generation in the apoplastic space (cell wall) followed by conversion to hydrogen peroxide by the activity of cell wall-localised SOD. Wall peroxidase may also play role in ROS generation for signalling. Hydrogen peroxide may use Ca^{2+} and MAPK pathway as downstream signalling cascade. Plant hormones associated with stress responses like ABA and ethylene play their role possibly via a crosstalk with ROS toward stress tolerance, thus projecting a dual role of ROS under drought stress (Kaur and Gupta 2005; Xiong et al. 2002; Raghavendra et al. 2010).

DNA Elements Controlling Gene Expression During Water Deficit

The most comprehensive information about the mechanism of regulation of gene expression in response to water deficit has been obtained from the investigation of DNA elements and sequence-specific DNA-binding proteins. Presently, two classes of DNA elements have been identified: the ABA-responsive element (ABRE) and the dehydration-responsive element (DRE). The ABRE has been shown to be sufficient for ABA-regulated gene expression during water deficit, but in some genes it must be associated with a coupling element. The dehydration-responsive element from the rd29A gene from Arabidopsis, TACCGACAT, has been shown to be involved in the regulation of this gene by an ABA-independent pathway induced by water deficit. It has been shown that these are insufficient for controlling the genes that are induced by water deficit, and new additional DNA elements and several of these elements are beginning to be defined. In the Arabidopsis gene rd22, which requires protein synthesis for expression, there is a DNA element, CACATG, that is similar to the element bound by the transcription factor MYC (Kaur and Gupta 2005; Xiong et al. 2002).

Mechanisms of Acclimation to Water Deficit and Stress Tolerance

Plants have developed multiple mechanisms in order to protect PSA against different kinds of stresses. At the cellular level, plants attempt to alleviate the damaging effects of stress by altering their metabolism to cope with the stress. Many plant systems can survive dehydration but to a different extent. According to Hoekstra et al. (2001) on the basis of the critical water level, two types of tolerance are distinguished:

1. Drought tolerance can be considered as the tolerance of moderate dehydration, down to moisture content below which there is no bulk cytoplasmic water present – about 0.3 g $H_2O\ g^{-1}$ DW.

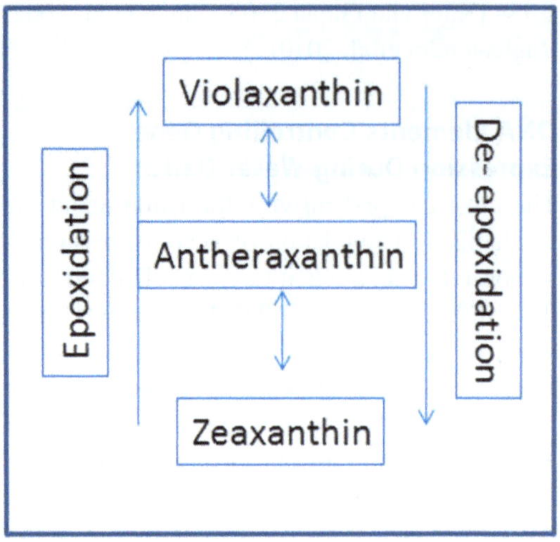

Fig. 4 Water stress induced the synthesis of accessory photosynthetic pigments like zeaxanthin and antheraxanthin

2. Desiccation tolerance refers to the tolerance of further dehydration, when the hydration shell of the molecules is gradually lost. Desiccation tolerance includes also the ability of cells to rehydrate successfully.

According to Bohnert and Shen (1999), a nearly universal reaction under stress conditions, including WD, is the accumulation of "compatible solutes", many of which are osmolytes (i.e. metabolites whose high cellular concentration increases the osmotic potential significantly) considered to lead to osmotic adjustment. These observations indicate that "compatible solutes" may have other functions as well, namely, in the protection of enzyme and membrane structure and in scavenging of radical oxygen species. One of the principal mechanisms employed by plants to prevent or to alleviate damage to the PSA is non-photochemical chlorophyll fluorescence quenching (qN) (Ruban and Horton 1995). In this mechanism, excess light energy is dissipated as heat in the light-harvesting antenna of PS2. This dissipation is primarily controlled by the trans-thylakoid pH gradient (pH) (Gounaris et al. 1984).

When CO_2 fixation and therefore ATP consumption are decreased at low RWC, the functioning electron flow gives rise to an acidification of the thylakoid lumen that induces Zx and Ax synthesis. It has been proposed that the photoprotective process results in the diversion of energy away from the reaction centres (Ruban and Horton 1995; Medrano et al. 2002). According to Tambussi et al. (2002), the non-photochemical fluorescence quenching (qN), as well as the content of zeaxanthin and antheraxanthin after moderate WS, increased significantly. However, at severe WS, a further rise in these xanthophylls was not associated with any increase in qN. In addition, the β-carotene content rose significantly during severe WD, suggesting an increase in antioxidant defence. Besides the above-mentioned mechanisms of energy dissipation, there are also other ways. For example, the energy dissipation in closed stomata can occur via ATP and NADPH, which are used for other metabolic processes, and they are obviously important mechanisms of tolerance and protection against water stress and photooxidative damage (Lichtenthaler 1996) (Fig. 4).

During dehydration, anhydrobiotes pass through hydration ranges that also necessitate protection against drought. The desiccation tolerance programme can be switched on by dehydration and the plant hormone ABA (Ingram and Bartels 1996). Upon water loss, the cellular volume decreases and cell content becomes increasingly viscous and the chance for molecular interactions rises. The danger of protein denaturing and membrane fusion increases. But a range of compatible solutes which do not interfere with cellular structure and function hinder this process. It is considered that at lower water contents, molecular oxidants (glutathione, ascorbate, tocopherol) play a preponderant role in elevating oxidative stress. Hoekstra et al. (1997) showed that desiccation may increase the transfer of these amphiphiles from the polar cytoplasm into the lipid phase of membranes. They thought that this partitioning into membrane might be extremely effective in automatically inserting amphiphilic antioxidant into membranes upon dehydration.

Reduction of metabolism coincides with survival of desiccation (Leprince et al. 1999). In vegetative tissues, genes encoding enzymatic

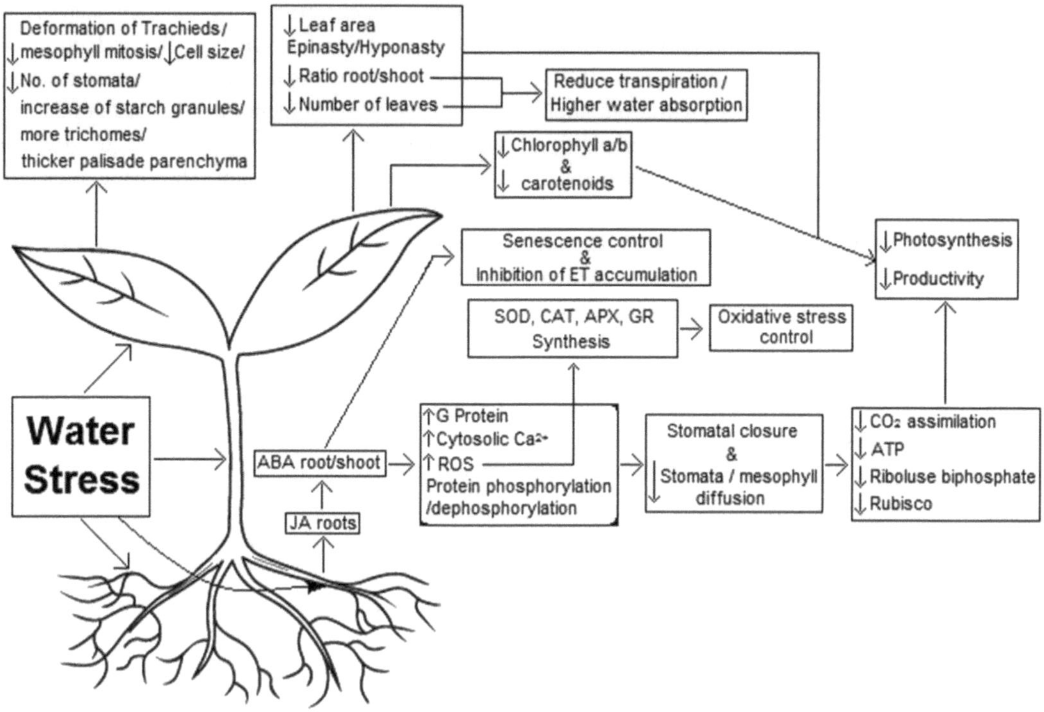

Fig. 5 Schematic presentation showing integrated approach of acclimation to water stress

antioxidants such as APX, SOD and GRase are upregulated during drying or rehydration (Fig. 5). When the bulk water is removed (below 0.3 g H_2O g^{-1} DW), the mechanism keeping the macromolecules preferentially hydrated through amphiphiles fails to work, because there is no water left for preferential hydrations (Crowe et al. 1990). It has been established that during desiccation, soluble sugars interact with the polar head groups and replace the water molecules. Phospholipid molecules largely retain the original spacing between one another. When water dissipates from the water shell of macromolecules at moisture contents below 0.3 g H_2O g^{-1} DW, the hydrophobic effect responsible for structure and function is lost. After bulk water is lost, the hydrogen bonding and glass formation are the mechanisms by which membranes and proteins are structurally and functionally preserved.

Sugars are special in that they allow the removal of the closely associated water from protein without this leading to conformational changes and loss of enzymatic function. According to the water replacement hypothesis, sugars act as a water substitute by satisfying the hydrogen-bonding requirement of polar groups of the dried protein surface (Carpenter and Growe 1988; Wolkers et al. 1998). At around 0.3 g H_2O g^{-1} DW, the cytoplasm vitrifies and exists in a so-called glassy state, an amorphous metastable state, retaining the disorder and physical properties of the liquid state (Franks et al. 1991). This state decreases the probability of chemical reactions and is indispensable for surviving the dry state. A very important role in this process is played by late embryogenesis abundant proteins (LEAPs), especially their Group 1 – dehydrins, in stabilisation and protecting during desiccation. It was observed that their accumulation coincides with the acquisition of desiccation tolerance (Bartels et al. 1988). Group 1 proteins have very high potential for hydration – several times greater than that for "normal" cellular proteins (McCubbin et al. 1985). Because of these special features, LEAPs potentially bind to intracellular macromolecules coating them with a cohesive water layer and preventing their coagulation during desiccation (Close 1996). Upon removal of their own hydration shell, these proteins would still be capable of playing a role in stabilising macromolecular structures. They could provide a layer of their own hydroxylated residues to interact with surface groups of other proteins, acting as "replacement water" (Cuming 1999; Buitink et al. 2002). Wolkers et al. (1999) suggested that LEAPs embedded in the glassy matrix might confer stability on slowly dried carrot somatic embryos.

Another class of proteins associated with desiccation tolerance are low molecular weight HSPs. Coordinated expression of LEAPs and sHSP transcripts is observed during embryo development in response to ABA, indicating the existence of common regulatory elements of LEAPs, sHSPs and desiccation tolerance (Wehmeyer et al. 1996). But so far, there is no direct experimental evidence for a specific role of sHSPs in desiccation tolerance. Satoh et al. (2002) followed recovery of the photosynthetic system during rewatering in a terrestrial, highly drought-tolerant cyanobacterium *Nostoc commune*. With absorption of water, the weight of the Nostoc colony increased. Fluorescence intensities of phycobiliproteins and PS1 complexes recovered almost completely within 1 min, suggesting that their functional forms were restored very quickly. PS1 activity and cyclic ET flow around PS1 recovered within 2 min, while the PS2 activity recovered after a time lag of 5 min. Photosynthetic CO_2 fixation was restored almost in parallel with the first recovery phase of PS2 reaction centre activity (Fig. 5).

There is need to search for valuable approaches in order to identify those metabolic steps that are most sensitive to drought and to elucidate which metabolites and gene products are of primary importance for increasing drought tolerance of plants. Many proteins are involved in damage limitation or the removal of toxic compounds which are induced during water deficit. For example, ubiquitin, chaperones and proteases may all be involved in the recovery of proteins or their building blocks. Genes encoding enzymes that detoxify reactive oxygen species are also induced. It is difficult to ascertain whether the induction of these genes is to repair damage caused directly by reduced water content or if they accumulate to ameliorate damage caused by a secondary stress or to restrict pathogen invasion. The characterisation of genes induced by water deficit has greatly improved our understanding of plant responses to the environment.

Conclusion

The multitude of different stressors, their spatial and temporal character, their variation in intensity and dose and their potential interaction yield an abundance of scientific questions. One of the most interesting aspects of water-stress physiology is how mild or moderate stress is transduced into alterations in metabolism. The foregoing considerations make it seem unlikely that mild stress could, by any of the mechanisms mentioned, damage biochemical components or organelles of the cell; yet mild stress does have pronounced effects. It is more probable that changes in metabolism elicited by mild stress represent plant regulatory responses rather than damage. This in turn implies that many of the changes in plant processes brought about by stress arise indirectly. Among all the changes, the most important aspect of water stress proven to be results reduced cell growth. Inhibition of cell growth during water stress is found to corroborate with inhibition of protein synthesis, cell wall synthesis, membrane proliferation, etc. For maintaining balance of metabolites, the plant has probably evolved controls which slow down synthesis of cell building blocks when low turgor prevents expansion. This may be a likely explanation for the susceptibility of cell wall synthesis and polyribosomes (hence protein synthesis) in growing tissue to very mild water stress. It has also been reported that water stress is associated with impaired lipid synthesis in such tissue. These explain how cell wall synthesis is impaired during water stress, and it may be coupled with suppression of plant growth. Various other changes may also be expected in growing cells under mild to moderate water stress, on the basis of suppressed enlargement of cell volume, for the alteration in volume can be marked. Therefore, a slowing or cessation of growth should result in a quick accumulation of many metabolites, which in turn could affect various processes. Another interesting fact is that water stress affects metabolic processes in a variable fashion. A clear differentiation is made between the changes in metabolism in growing and nongrowing tissue and in the case of young tissue when they are exposed to water stress. There seems to be little doubt that in nongrowing or slowly growing tissue, some metabolic parameters, such as protochlorophyll formation and nitrate reductase levels, are also susceptible to water stress. Regarding possibilities other than turgor changes, the lowering of water activity is the least likely to be a mechanism underlying water-stress effects. Changes in molecular and ionic concentrations and spatial relations may mediate water-stress effects within the limitations.

All the results from research focusing on explaining the mechanism of resistance to water stress of plants have important applications for other fields, such as ecology, forestry, biology and agriculture. At the same time the expanding human population is facing acute food shortage. Therefore, the study of the mechanisms of stress resistance of plants has assumed practical importance over and above academic interest. The greater our understanding of plant response to stressors and stress tolerance, the greater will be our ability to manage natural and human-made ecosystems.

References

Abdalla MM, El-Khoshiban NH (2007) The influence of water stress on growth, relative water content, photosynthetic pigments, some metabolic and hormonal contents of two *Triticum aestivum* cultivars. J Appl Sci Res 3(12):2062–2074

Ackerson RC (1981) Osmoregulation in cotton in response to water stress. II. Leaf carbohydrate status in relation to osmotic adjustment. Plant Physiol 67:489–493

Akıncı S (1997) Physiological responses to water stress by *Cucumis sativus L.* and related species. Ph.D. thesis, University of Sheffield

Akıncı S, Lösel DM (2009) The soluble sugars determination in *Cucurbitaceae* species under water stress and recovery periods. Adv Environ Biol 3 (2):175–183

Akıncı S, Lösel DM (2010) The effects of water stress and recovery periods on soluble sugars and starch content in cucumber cultivars. Fresen Environ Bull 19(2):164–171

Akıncı S, Lösel DM (2012) Plant water-stress response mechanisms. In: Prof. Ismail Md, Rahman M (ed) Water stress. InTech, ISBN: 978-953-307-963-9

Alam SM, (1999) Nutrient uptake by plants under stress conditions. In: Rijeka Croatia, Pessarakli M (ed) Handbook of plant and crop stress, 2 rev and expth edn. Marcel Dekker, New York, pp 285–313

Al-Suhaibani NAR (1996) Physiological studies on the growth and survival of *Medicago sativa* L. (alfalfa) seedlings under low temperature. Ph.D. thesis, Animal and Plant Sciences Department, University of Sheffield

Andrade JL, Larque-Saavedra A, Trejo CL (1995) Proline accumulation in leaves of four cultivars of *Phaseolus vulgaris* L. with different drought resistance. Phyton 57:149–157

Artlip TS, Funkhouser EA (1995) Protein synthetic responses to environmental stresses. In: Pessarakli M (ed) Handbook of plant and crop physiology. Marcel Dekker, New York, pp 627–644

Aspinall D, Paleg LG (1981) Proline accumulation: physiological effects. In: Paleg LG, Aspinall D (eds) The Physiology and biochemistry of drought resistance in plants. Academic, New York, pp 206–225

Atkinson NJ, Urwin PE (2012) The interaction of plant biotic and abiotic stresses: from genes to the field. J Exp Bot 63(10):3523–3543

Barnett NM, Naylor AW (1966) Amino acid protein metabolism in Bermuda grass during water stress. Plant Physiol 41:1222–1230

Bartels D, Singh M, Salamini F (1988) Onset of desiccation tolerance during development of the barley embryo. Planta 175:485–492

Bewley JD, Larsen KM (1982) Differences in the responses to water stress of growing and non-growing regions of maize mesocotyls: protein synthesis on total, free and membrane-bound polyribosome fractions. J Exp Bot 33(134):406–415

Bewley JD, Larsen KM, Papp JEE (1983) Water-stress-induced changes in the pattern of protein synthesis in maize seedling mesocotyls: a comparison with the effects of heat shock. J Exp Bot 34(146):1126–1133

Bie Z, Ito T, Shinohara Y (2004) Effects of sodium sulphate and sodium bicarbonate on the growth, gas exchange and mineral composition of lettuce. Sci Hortic 99:215–224

Bishop DG (1983) Functional role of plant membrane lipids. In: Proceeding of the 6th annual symposium in botany. University of California, Riverside, pp 81–103, 13–15 Jan

Bohnert HJ, Shen B (1999) Transformation and compatible solutes. Sci Hort 78:237–260

Bottner P, Couteaux MM, Vallejo VR (1995) Soil organic matter in mediterranean-type ecosystems and global climatic changes: a case study-the soils of the mediterranean basin. In: Moreno JM, Oechel WC (eds) Global change and mediterranean-type ecosystems. Ecological studies, vol 117. Springer-Verlag, New York, pp 306–325

Boyer JS (1976) Water deficits and photosynthesis. In: Kozlowski TT (ed) Water deficits and plant growth, vol 4. Academic, New York, pp 153–190

Bray EA (1995) Regulation of gene expression during abiotic stresses and the role of the plant hormone abscisic acid. In: Pessarakli M (ed) Handbook of plant and crop physiology. Marcel Dekker, New York, pp 733–752

Bray EA (1997) Plant responses to water deficit. Trends Plant Sci 2(2):48–54

Brown CE, Pezeshki SR, DeLaune RD (2006) The effects of salinity and soil drying on nutrient uptake and growth of Spartina alterniflora in a simulated tidal system. Environ Exp Bot 58(1–3):140–148

Buitink J, Hoekstra F, Leprince O (2002) Biochemistry and biophysics of tolerance systems. In: Black M, Pritchard HW (eds) Desiccation and survival in plants: drying without dying. CAB International, Wallingford, pp 293–318

Caldwell CR, Whitman CE (1987) Temperature-induced protein conformational changes in barley root plasma membrane-enriched microsomes. I. Effect of temperature on membrane protein and lipid mobility. Plant Physiol 84:918–923

Carpenter JF, Growe JH (1988) The mechanism of cryoprotection of proteins by solutes. Cryobiology 25:244–255

Chandrasekhar V, Sairam RK, Srivastava GC (2000) Physiological and biochemical responses of hexaploid and tetraploid wheat to drought stress. J Agron Crop Sci 185(4):219–227

Chapin FS (1980) The mineral nutrition of wild plants. Annu Rev Ecol Syst 11:233

Chapin FS (1991) Effects of multiple environmental stresses on nutrient availability and use. In: Mooney HA, Winner WE, Peli EJ (eds) Response of plants to multiple stresses. Academic, San Diego, p 67

Chaves MM (1991) Effects of water deficits on carbon assimilation. J Exp Bot 42:1–16

Chaves MM, Maroco JP, Pereira JS (2003) Understanding plant response to drought: from genes to the whole plant. Funct Plant Biol 30:239–264

Chetal S, Wagle DS, Nainawatee HS (1981) Glycolipid changes in wheat and barley chloroplast under water stress. Plant Sci Lett 20:225–230

Chopra RK, Sinha SK (1998) Prospects of success of biotechnological approaches for improving tolerance to drought stress in crop plants. Curr Sci 74 (1):25–34

Claes B, Dekeyser R, Villarroel R et al (1990) Characterization of a rice gene showing organ-specific expression in response to salt stress and drought. Plant Cell 2:19–27

Close TJ (1996) Dehydrins: emergence of a biochemical role of a family of plant dehydration proteins. Physiol Plant 97:795–803

Close TJ (1997) Dehydrins: a commonalty in the response of plants to dehydration and low temperature. Physiol Plant 100:291–296

Cornic C, Massacci A (1996) Leaf photosynthesis under drought stress. In: Baker NR (ed) Photosynthesis and environment. Kluwer Academic, Dordrecht/Boston, pp 347–366

Crafts AS (1968) Water deficits and physiological processes. In: Kozlowski TT (ed) Water deficits and plant growth, vol 2., pp 85–133

Cramer GR, Urano K, Delrot S et al (2011) Effects of abiotic stress on plants: a systems biology perspective. BMC Plant Biol 11:163. doi:10.1186/1471-2229-11-163

Crowe JH, Carpenter JF, Crowe LM et al (1990) Are freezing and dehydration similar stress vectors? A comparison of models interaction of stabilizing solutes with biomolecules. Cryobiology 27:219–231

Cuming AC (1999) Lea proteins. In: Shewry PR, Casey R (eds) Seed proteins. Kluwer Academic, Dordrecht/Boston, pp 753–780

Dasgupta J, Bewley D (1984) Variations in protein synthesis in different regions of greening leaves of barley seedlings and effects of imposed water stress. J Exp Bot 35(159):1450–1459

Dhindsa RS, Cleland RE (1975) Water stress and protein synthesis. II. Interaction between water stress, hydrostatic pressure, and abscisic acid on the pattern of protein synthesis in *Avena* coleoptiles. Plant Physiol 55:782–785

Douglas TJ, Paleg LG (1981) Lipid composition of *Zea mays* seedlings and water stress-induced changes. J Exp Bot 32(128):499–508

El-Hafid L, Pham-Thi AT, Zuily-Fodil Y et al (1989) Enzymatic breakdown of polar lipids in cotton leaves under water stress. I. Degradation of monogalactosyl-diacylglycerol. Plant Physiol Biochem 27:495–502

Evans RD, Black RA, Loescher WH, Fellows RJ (1992) Osmotic relations of the drought-tolerant shrub Artemisia tridentata in response to water stress. Plant, Cell and Environment 15:49–59

Faver KL, Gerik TJ, Thaxton PM et al (1996) Late season water stress in cotton: II: leaf gas exchange and assimilation capacity. Crop Sci 36:922–928

Ferrari-Iliou R, Pham Thi AT, Vieira da Silva J (1984) Effect of water stress on the lipid and fatty acid composition of cotton (*Gossypium hirsutum* L.) chloroplasts. Physiol Plant 62:219–224

Flexas J, Galmes J, Ribas-Carbo M et al (2005) The effects of water stress on plant respiration. In: Lambers H, Ribas-Carbo M (eds) Plant respiration, vol 18. Springer, Dordrecht, pp 85–94. doi:10.1007/1-4020-3589-6_6

Ford CW, Wilson JR (1981) Changes in levels of solutes during osmotic adjustment to water stress in leaves of four tropical pasture species. Aust J Plant Physiol 8:77–91

Foy CD (1983) Plant adaptation to mineral stress in problem soils. Iowa J Res 57:339–354

Franks F, Hatley RHM, Mathias SF (1991) Materials science and the production of self-stable biological. BioPharm 4:38–42

Galston AW, Sawhney RK (1990) Polyamines in plant physiology. Plant Physiol 94:406–410
Gerakis PA, Guerrero FP, Williams WA (1975) Growth, water relations and nutrition of three grassland annuals as affected by drought. J Appl Ecol 12:125–135
Ghannoum O (2009) C_4 photosynthesis and water stress. Ann Bot 103:635–644. doi:10.1093/aob/mcn093
Glyn Bengough A, McKenzie BM, Hallett PD et al (2011) Root elongation, water stress, and mechanical impedance: a review of limiting stresses and beneficial root tip traits. J Exp Bot 62(1):59–68
Gogorcena Y, Iturbe-Ormaetxe I, Escuredo PR et al (1995) Antioxidant defences against activated oxygen in pea nodules subjected to water stress. Plant Physiol 108:753–759
Gounaris K, Brain APR, Quinn PJ et al (1984) Structural reorganization of chloroplast thylakoid membranes in response to heat stress. Biochim Biophys Acta 766:198–208
Graham EA, Nobel PS (1999) Root water uptake, leaf water storage and gas exchange of a desert succulent: implications for root system redundancy. Ann Bot 84:213–223
Gronewald JW, Abou-Khalil W, Weber EJ et al (1982) Lipid composition of a plasma membrane enriched fraction of maize roots. Phytochemistry 21(4):859–862
Hanson AD, Hitz WD (1982) Metabolic responses of mesophytes to plant water deficits. Annu Rev Plant Physiol 33:163–203
Harwood JL (1979) The synthesis of acyl lipids in plant tissues. Prog Lipid Res 18:55–86
Harwood JL (1980) Plant acyl lipids. In: Stumpf PK (ed) The biochemistry of plants, vol 4. Academic, New York, pp 1–55
Herppich WB, Peckmann K (1997) Responses of gas exchange, photosynthesis, nocturnal acid accumulation and water relations of *Aptenia cordifolia* to short-term drought and rewatering. J Plant Physiol 150:467–474
Hoekstra FA, Wolkers WF, Buitink J et al (1997) Membrane stabilization in the dry state. Comp Biochem Physiol 117A:335–341
Hoekstra F, Golovina E, Buitink J (2001) Mechanisms of plant desiccation tolerance. Trends Plant Sci 8(9):431–438
Hsiao TC (1973) Plant responses to water stress. Annu Rev Plant Physiol 24:519–570
Hughes SG, Bryant JA, Smirnoff N (1989) Molecular biology: application to studies of stress tolerance. In: Jones HG, Flowers TJ, Jones MB (eds) Plants under stress. Biochemistry, physiology and ecology and their application to plant improvement. Cambridge University Press, Cambridge, pp 131–155
Ingram J, Bartels D (1996) The molecular basis of dehydration tolerance in plants. Annu Rev Plant Physiol Plant Mol Biol 47:377–403
Irigoyen JJ, Emerich DW, Sanchez-Diaz M (1992) Water stress induced changes in concentrations of proline and total soluble sugars in nodulated alfalfa (*Medicago sativa*) plants. Physiol Plant 84:55–60
Jaleel CA, Manivannan P, Wahid A et al (2009) Drought stress in plants: a review on morphological characteristics and pigments composition. Int J Agric Biol 11:100–105
Jang JC, Sheen J (1997) Sugar sensing in plants. Trends Plant Sci 2:208–214
Joly RJ (1985) Techniques for determining seedling water status and their effectiveness in assessing stress. In: Duryea ML (ed) Proceedings: evaluating seedling quality: principles, procedures, and predictive abilities of major tests. Oregon State University, Forest Research Laboratory, Corvallis. ISBN 0-87437-000-0
Jones MM, Osmond CB, Turner NC (1980) Accumulation of solutes in leaves of sorghum and sunflower in response to water deficits. Aust J Plant Physiol 7:193–205
Kameli A (1990) Metabolic responses of durum wheat to water stress and their role in drought resistance. Ph.D. thesis, Animal and Plant Sciences Department, University of Sheffield
Kameli A, Losel DM (1993) Carbohydrates and water status in wheat plants under water stress. New Phytol 125:609–614
Kameli A, Losel DM (1996) Growth and sugar accumulation in durum wheat plants under water stress. New Phytol 132:57–62
Kar RK (2011) Plant responses to water stress. Role of reactive oxygen species. Plant Signal Behav 6(11):1741–1745
Kaur N, Gupta AK (2005) Signal transduction pathways under abiotic stresses in plants. Curr Sci 88 (11):1771–1780
Kidambi SP, Matches AG, Bolger TP (1990) Mineral concentrations in alfalfa and sainfoin as influenced by soil moisture level. Agron J 82(2):229–236
Koch KE (1996) Carbohydrate-modulated gene expression in plants. Annu Rev Plant Physiol Plant Mol Biol 47:509–540
Kozlowski TT, Kramer PJ, Pallardy SG (1991) The physiological ecology of woody plants. Academic, San Diego
Kramer PJ, Boyer JS (1995) Water relations of plants and soils. Academic, San Diego
Kuiper PJC (1980) Lipid metabolism as a factor in environmental adaptation. In: Maliak P et al (eds) Biogenesis and function of plant lipids. Elsevier, Amsterdam, pp 169–176
Kuiper PJC (1985) Environmental changes and lipid metabolism of higher plants. Physiol Plant 64:118–122
Larcher W (1987) Stress bei Pflanzen. Naturwissenschaften 74:158–167

Leprince O, Buitink J, Hoekstra FA (1999) Axes and cotyledons of recalcitrant seeds of *Castanea sativa* Mill. J Exp Bot 50:1515–1524

Levitt J (1980) Responses of plants to environmental stresses, 2nd edn. Academic Press, New York

Lichtenthaler HK (1996) Vegetation stress: an introduction to the stress concept in plants. J Plant Physiol 148:4–14

Lichtenthaler HK, Buschmann C, Doll M et al (1981) Photosynthetic activity, chloroplast ultrastructure, and leaf characteristics of high-light and low-light plants and of sun and shade leaves. Photosynth Res 2:115–141

Liljenberg C, Kates M (1982) Effect of water stress on lipid composition of oat seedling root cell membranes. In: Wintermans JFGM, Kuiper PJC (eds) Biochemistry and metabolism of plant lipids. Elsevier Biomedical, Amsterdam, pp 441–444

Lisar SYS, Motafakkerazad R, Hossain MM et al (2012) Water stress in plants: causes, effects and responses. In: Prof. Rijeka, Croatia, Ismail Md. Mofizur Rahman (ed) Water stress. InTech, ISBN: 978-953-307-963-9

Lobell DB, Hammer GL, McLean G et al (2013) The critical role of extreme heat for maize production in the United States. Nat Climate Change 3:497–501. doi:10.1038/nclimate1832

Mahajan S, Tuteja N (2005) Cold, salinity and drought stresses: an overview. Arch Biochem Biophys 444:139–158

Mantyla E, Lang V, Palva ET (1995) Role of abscisic acid in drought-induced freezing tolerance, cold acclimation, and accumulation of LTI78 and RAB18 proteins in *Arabidopsis thaliana*. Plant Physiol 107:141–148

Maroco JP, Pereira JS, Chaves MM (1997) Stomatal responses of leaf-to-air vapour pressure deficit in *Sahelian* species. Aust J Plant Physiol 24:381–387

McCree KJ (1986) Whole-plant carbon balance during osmotic adjustment to drought and salinity stress. Aust J Plant Physiol 13:33–43

McCubbin WD, Kay CM, Lane BG (1985) Hydrodynamic and optical properties of the wheat germ Em protein. Can J Biochem Cell Biol 63:803–811

McCutchan H, Shackel KA (1992) Stem-water potential as a sensitive indicator of water stress in Prune trees (*Prunus domestica* L. cv. French). J Am Soc Hort Sci 117(4):607–611

Medrano H, Escalona JM, Boto J et al (2002) Regulation of photosynthesis of C3 plants in response to progressive drought: stomatal conductance as a reference parameter. Ann Bot 89:895–905

Meyer RF, Boyer JS (1972) Sensitivity of cell division and cell elongation to low water potentials in soybean hypocotyls. Planta 108:77–87

Michelena VA, Boyer JS (1982) Complete turgor maintenance at low water potentials in the elongating region of maize leaves. Plant Physiol 69:1145–1149

Morgan JM (1984) Osmoregulation and water stress in higher plants. Annu Rev Plant Physiol 35:299–319

Morgan JM, Condon AG (1986) Water-use, grain yield and osmoregulation in wheat. Aust J Plant Physiol 13:523–532

Munns R, Weir R (1981) Contribution of sugars to osmotic adjustment in elongating and expanded zones of wheat leaves during moderate water deficits at two light levels. Aust J Plant Physiol 8:93–105

Munns R, Brady CJ, Barlow EWR (1979) Solute accumulation in the apex and leaves of wheat during water stress. Aust J Plant Physiol 6:379–389

Mwanamwenge J, Loss SP, Siddique KHM et al (1999) Effect of water stress during floral initiation, flowering and podding on the growth and yield of faba bean (*Vicia faba* L.). Eur J Agron 11:1–11

Nair AS, Abraham TK, Jaya DS (2006) Morphological and physiological changes in cowpea (*Vigna unguiculata* L.) subjected to water deficit. Indian J Plant Physiol 11(3):325–328

Nambiar EKS (1977) The effects of drying of the topsoil and of micronutrients in the subsoil on micronutrient uptake by an intermittently defoliated ryegrass. Plant Soil 46(1):185–193

Naot D, Ben-Hayyim G, Eshdat Y et al (1995) Drought, heat and salt stress induce the expression of a citrus homologue of an atypical late-embryogenesis *lea5* gene. Plant Mol Biol 27:619–622

Navari-Izzo F, Quartacci MF, Izzo R (1989) Lipid changes in maize seedlings in response to field water deficits. J Exp Bot 40(215):675–680

Navari-Izzo F, Vangioni N, Quartacci MF (1990) Lipids of soybean and sunflower seedlings grown under drought conditions. Phytochemistry 29(7):2119–2123

Navari-Izzo F, Quartacci MF, Melfi D et al (1993) Lipid composition of plasma membranes isolated from sunflower seedlings grown under water-stress. Physiol Plant 87:508–514

Nicholas DJD (1975) The functions of trace elements in plants. In: Nicholas DJD (ed) Trace elements in soil-plant-animal systems. Academic, New York

Oktem A (2008) Effect of water shortage on yield, and protein and mineral compositions of drip-irrigated sweet corn in sustainable agricultural systems. Agric Water Manage 95(9):1003–1010

Pareek A, Singla SL, Kush AK et al (1997) Distribution patterns of HSP 90 protein in rice. Plant Sci 125 (10):221–230

Pelah D, Wang W, Altman A et al (1997) Differential accumulation of water stress-related proteins, sucrose synthase and soluble sugars in *Populus* species that differ in their water stress response. Physiol Plant 99:153–159

Pereira JS, Chaves MM (1993) Plant water deficits in Mediterranean ecosystems. In: Smith JA, Griffiths H (eds) Water deficits plant responses from cell to community. BIOS Scientific, Oxford, pp 237–251

Pereira JS, Chaves MM (1995) Plant responses to drought under climate change in Mediterranean-type ecosystems. In: Moreno JM, Oechel WC (eds) Global change and Mediterranean-type ecosystems, vol 117, Ecological studies. Springer-Verlag, New York, pp 140–160

Pham Thi AT, Flood C, Vieira da Silva J (1982) Effects of water stress on lipid and fatty acid composition of cotton leaves. In: Wintermans JFGM, Kuiper PJC (eds) Biochemistry and metabolism of plant lipids. Elsevier Biomedical, Amsterdam, pp 451–454
Pham Thi AT, Borrel-Flood C, Vieira da Silva J et al (1985) Effects of water stress on lipid metabolism in cotton leaves. Phytochemistry 24(4):723–727
Pham Thi AT, Borrel-Flood C, Vieira da Silva J et al (1987) Effects of drought on [1-14C]- oleic acid and [1-14C]- linoleic acid desaturation in cotton leaves. Physiol Plant 69:147–150
Pinkerton A, Simpson JR (1986) Interactions of surface drying subsurface nutrients affecting plant growth on acidic soil profiles from an old pasture. Aust J Exp Agric 26(6):681–689
Poulson ME, Edwards GE, Browse J (2002) Photosynthesis is limited at high leaf to air vapor pressure deficit in a mutant of *Arabidopsis thaliana* that lacks trienoic fatty acids. Photosynth Res 72:55–63
Pruvot G, Massimino J, Peltier G et al (1996) Effects of low temperature, high salinity and exogenous ABA on the synthesis of two chloroplastic drought-induced proteins in *Solanum tuberosum*. Physiol Plant 97:123–131
Pugnaire FI, Serrano L, Pardos J (1999) Constraints by water stress on plant growth. In: Pessarakli M (ed) Hand book of plant and crop stress, 2nd edn. Marcel Dekker, Inc., New York, pp 271–283
Quartacci MF, Navari-Izzo F (1992) Water stress and free radical mediated changes in sunflower seedlings. J Plant Physiol 139:621–625
Quartacci MF, Sgherri CIM, Pinzino C et al (1994) Superoxide radical production in wheat plants differently sensitive to drought. Proc R Soc Edinb 102B:287–290
Quatrano RS, Bartels D, Ho THD, Pages M (1997) New insights into ABA-mediated processes. The Plant Cell 9(4):470–475
Quick WP, Chaves MM, Wendler R et al (1992) The effect of water stress on photosynthetic carbon metabolism in four species grown under field conditions. Plant Cell Environ 15:25–35
Raghavendra AS, Gonugunta VK, Christmann A et al (2010) ABA perception and signalling. Trends Plant Sci 15:395–401
Ramachandra RA, Chaitanya KV, Vivekanandan M (2004) Drought-induced responses of photosynthesis and antioxidant metabolism in higher plants. J Plant Physiol 161(11):1189–1202
Ramagopal S (1993) Advances in understanding the molecular biology of drought and salinity tolerance in plants- the first decade. In: Lodha ML, Mehta SL, Ramagopal S et al (eds) Advances in biochemical engineering/biotechnology. Indian Society of Agriculture Biochemists, Kanpur, pp 39–48
Richter H, Wagner SB (1982) Water stress resistance of photosynthesis: some aspects of osmotic relations. Effects of stress on photosynthesis. In: Marcelle R, Clijsters H, Poucke MV (eds) Proceedings of a conference held at the "Limburgs Universtair Centrum" Diepenbeek. Martinus Nijhoff/Dr W. Junk publishers, The Hague, pp 45–53
Rodrigues ML, Chaves MM, Wendler R et al (1993) Osmotic adjustment in water stressed grapevine leaves in relation to carbon assimilation. Aust J Plant Physiol 20:309–321
Ruban AV, Horton P (1995) Regulation on non-photochemical quenching of chlorophyll fluorescence in plants. Aust J Plant Physiol 22:221–230
Russel EW (1976) Water and crop growth. (Soil conditions and plant growth). Longman, London, pp 448–478
Sahsah Y, Campos P, Gareil M et al (1998) Enzymatic degradation of polar lipids in *Vigna unguiculata* leaves and influence of drought stress. Physiol Plant 104:577–586
Satoh K, Hirai M, Nishio J et al (2002) Recovery of photosynthetic systems during rewatering is quite rapid in a terrestrial cyanobacterium, *Nostoc commune*. Plant Cell Physiol 43:170–176
Shah CB, Loomis RS (1965) Ribonucleic acid and protein metabolism in sugar beet during drought. Physiol Plant 18:240–254
Simpson JR, Lipsett J (1973) Effects of surface moisture supply on the subsoil nutritional requirements of lucerne (*Medicago sativa* L.). Aust J Agric Res 24(2):199–209
Singh B, Singh G (2004) Influence of soil water regime on nutrient mobility and uptake by *Dalbergia sissoo* seedlings. Trop Ecol 45(2):337–340
Singh NK, Bracker CA, Hasegawa PM et al (1987) Characterization of osmotin. Plant Physiol 85:529–536
Singh S, Cornilescu CC, Tyler RC et al (2005) Solution structure of a late embryogenesis abundant protein (LEA14) from *Arabidopsis thaliana*, a cellular stress-related protein. Protein Sci 14(10):2601–2609
Sinha SK (1978) Influence of potassium on tolerance to stress. In: Sekhon GS (ed) Potassium in soils and crops. Potash research Institute, New Delhi
Sinha KM, Sachdev A, Johari RP et al (1996) Lathyrus dehydrin-a drought inducible cDNA clone: isolation and characterization. J Plant Biochem Biotechnol 5:97–101
Stewart CR (1971) Effect of wilting on carbohydrates during incubation of excised bean leaves in the dark. Plant Physiol 48(6):792–794
Suss KH, Yordanov I (1986) Biosynthetic cause of in vivo acquired thermotolerance of photosynthetic light reactions and metabolic responses of chloroplasts to heat stress. Plant Physiol 81:192–199
Tambussi EA, Casadesus J, Munne-Bosh SM, Araus JL (2002) Photoprotection in water-stressed plants of durum wheat (*Triticum turgidum* var. durum): changes in chlorophyll fluorescence spectral signature and photosynthetic pigments. Funct Plant Biol 29:35–44
Tanguilig VC, Yambao EB, O' Toole JC, De Datta SK (1987) Water stress effects on leaf elongation, leaf

water potential transpiration and nutrient uptake of rice, maize and soybean. Plant Soil 103:155–168
Tholkappian P, Prakash M, Sundaram MD (2001) Effect of AM-fungi on proline, nitrogen and pod number of soybean under moisture stress. Indian J Plant Physiol 6(1):98–99
Timpa JD, Burke JB, Quisenberry JE et al (1986) Effects of water stress on the organic acid and carbohydrate compositions of cotton plants. Plant Physiol 82:724–728
Treshow M (1970) Disorders associated with adverse water relations. In: Environment & plant response. McGraw-Hill, New York, pp 153–174
Turner NC, Jones MM (1980) Turgor maintenance by osmotic adjustment: a review and evaluation. In: Turner NC, Kramer PJ (eds) Adaptations of plants to water and high temperature stress. Wiley, New York, pp 87–103
Vartanian N, Damerval C, De Vienne D (1987) Drought-induced changes in protein patterns of *Brassica napus* var. oleifera roots. Plant Physiol 84:989–992
Wang XQ, Wu WH, Assmann SM (1998) Differential responses of Abaxial and Adaxial guard cells of broad bean to abscisic acid and calcium. Plant Physiol 118(4):1421–1429
Wang W, Vinocur B, Shoseyov O et al (2004) Role of plant heat-shock proteins and molecular chaperones in the abiotic stress response. Trends Plant Sci 9 (5):244–252
Wehmeyer N, Hernandez LD, Finkelstein RR et al (1996) Synthesis of small heat-shock proteins is part of the developmental program of late seed maturation. Plant Physiol 112:747–757
Westgate ME, Boyer JS (1985) Osmotic adjustment and the inhibition of leaf, root, stem and silk growth at low water potentials in maize. Planta 164:540–549
Westgate ME, Schussler JR, Reicosky DC, Brenner ML (1989) Effect of water deficits on seed development in Soybean: II. Conservation of seed growth rate Plant Physiol 91:980–985
Whitman CE, Travis RL (1985) Phospholipid composition of a plasma membrane-enriched fraction from developing soybean roots. Plant Physiol 79:494–498
Wilkins O, Brautigam K, Campbell MM (2010) Time of day shapes *Arabidopsis* drought transcriptomes. Plant J 63(5):715–727
Wilson RF, Burke JJ, Quisenberry JE (1987) Plant morphological and biochemical responses to field water deficits. II. Responses of leaf glycerolipid composition in cotton. Plant Physiol 84:251–254
Wolkers WF, van Kilsdonk MG, Hoekstra FA (1998) Dehydration-induced conformational changes of poly-L-lysine as influenced by drying rate and carbohydrates. Biochim Biophys Acta 1425:127–136
Wolkers WF, Tetteroo FA, Albedra M et al (1999) Changed properties of the cytoplasmic matrix associated with desiccation tolerance of dried carrot somatic embryos. An in situ Fourier transform infrared spectroscopic study. Plant Physiol 120:153–163
Xiong L, Zhu JK (2003) Regulation of abscisic acid biosynthesis. Plant Physiol 133:29–36
Xiong L, Schumaker KS, Zhu JK (2002) Cell signaling during cold, drought, and salt stress. Plant Cell 14: S165–S183. doi:10.1105/tpc.000596
Xu DP, Duan XL, Wang BY et al (1996) Expression of a late embryogenesis abundant protein gene: HVA1. From barley confers tolerance to water deficit and salt stress in transgenic rice. Plant Physiol 110:249–257
Yokota A, Takahara K, Akashi K (2006) Water stress. In: Rao KVM, Raghavendra AS, Reddy KJ (eds) Physiology and molecular biology of stress tolerance in plants. Springer, Dordrecht, pp 15–40

Abiotic Stresses in Major Pulses: Current Status and Strategies

R. Sultana, A.K. Choudhary, A.K. Pal, K.B. Saxena, B.D. Prasad, and Ravigopal Singh

Abstract

Environmental stresses such as erratic and insufficient rainfall, extreme temperatures, salinity, alkalinity and aluminium toxicity limit the yield and productivity of many cultivated crops including pulses. Pulses are leguminous plants whose grains are used exclusively for food and are generally grown in harsh environments. Therefore, pulses encounter a number of abiotic stresses during various stages of their life cycle. Each type of stress hampers the growth of the plant by disturbing the normal physiology and morphology. The exact mechanisms governing the cause and effect of abiotic stresses in pulses are very complex and difficult to understand. Due to changing environmental conditions, very often referred to as 'climate change', pulses have become more prone to oxidative damage by overproduction of toxic reactive oxygen species (ROS) such as superoxide radicals, hydrogen peroxide and hydroxyl radicals. These radicals disturb the cellular homeostasis of the cell resulting in significant yield losses. In North India, high temperatures (>30 °C) coupled with drought stress during flowering stage produce distinct effect on the grain yield of chickpea and lentil, whereas in pigeon pea, low temperatures (<10 °C) cause severe flower drop resulting in yield losses. However, recent empirical evidence suggests that genotypic variations have been observed for almost all the abiotic stresses in pulses and several genotypes tolerant to heat, drought and waterlogging have been identified. Marker traits conferring tolerance to such stress(es) have also been

R. Sultana (✉) • A.K. Pal • B.D. Prasad • R. Singh
Bihar Agricultural University (BAU), Sabour, BR 813 210, India
e-mail: rafat.hayat@gmail.com

A.K. Choudhary
IIPR RRC cum Off-season Nursery, Dharwad, KA 580 005, India

K.B. Saxena
International Crops Research Institute for the Semi-Arid Tropics, Patancheru, AP, India

R.K. Gaur and P. Sharma (eds.), *Approaches to Plant Stress and their Management*,
DOI 10.1007/978-81-322-1620-9_9, © Springer India 2014

identified which can be used in breeding programmes for improving tolerance. This chapter describes the production status, the impact of abiotic stresses and the opportunities for genetic improvement of tolerance to abiotic stresses in major pulses.

Keywords
Pulses • Pigeon pea • Chickpea • Abiotic stress • Antioxidants • ROS

Introduction

Pulse crops are leguminous plants whose grains are used exclusively for food. It comprises cool and warm season pulses which include chickpea, pigeon pea, mung bean, urd bean, lentil, field pea, rajma and grass pea besides several other pulses of local importance such as cowpea, moth bean, horse gram, faba bean and rice bean. In Asia, Africa and many developing countries, pulses constitute a major source of dietary protein for millions of poor people. Considering the limited breeding effort and development of management packages for pulses, compared with cereals, there is considerable scope for further improvement of pulse production. However, in recent years, fluctuations in pulse yield and frequent crop failure due to global climate change caused by biotic and abiotic stresses are threatening the future expansion of the pulse industry in India. The susceptibility/sensitivity of most pulses to various abiotic and biotic stresses (Kumar et al. 2011) and high genotype × environment (G × E) interactions on the expression of yield (Kumar and Ali 2006) have been identified as the major constraints, leading to only a marginal gain in genetic improvement for yield and stability. For instance, India alone has released more than 550 improved varieties of pulses, but production increased only marginally (from 12.70 million tons in 1960–1961 to 17.79 million tons covering an area of 27.94 million hectare with productivity of 636.5 kg per hectare in 2011) (FAOSTAT 2013). Therefore, these issues require an immediate attention, and overall, a paradigm shift is needed in the breeding strategies to alleviate the effects of such stress on yield and stability pulses.

Pulses: The Major Component of Subsistence and Low-Input Agriculture

In general, pulses in most developing countries form an integral component of the subsistence and low-input agriculture. In subsistence and low-input agriculture, yield in farmers' fields is many folds less than what is realised in well-managed experimental fields. Resource-poor farmers manage subsistence agriculture through strategies based on subdividing the risk of crop failure through interspecific (inter- or mixed cropping) and intraspecific (heterogeneous cultivars) diversity. Such production environments are characterised by harsh and erratically unpredictable climatic conditions with the high risk of crop failure. Such environments are prone to large errors, less differentiation between genotypes and less repeatability across years. Various components of such environments are highly unpredictable and often intractable (Allard 1999). Components of such environments largely include abiotic stresses such as extremes of moisture (waterlogging/drought) and temperature (high/low), salinity and mineral stresses (such as aluminium toxicity in acid soil).

Mechanism of Abiotic Stress in Pulses

Literatures reveal that one abiotic stress is often confounded with several other stresses, making it difficult to identify the exact causes that the crop faced under stress condition. Hence, development of stress-tolerant genotypes is difficult due to the complex nature of mechanism imparting tolerance. It is also a well-known fact that a combination of different morphological, physiological and biochemical traits is controlled by several major and minor genes and the interaction among them leads to stress tolerance in plants (Bohnert et al. 1995). Therefore, for improving breeding efficiency, there is a need to identify specific physiological, biochemical and molecular characteristics that may improve yields under such stresses. In pulses, proper characterisation of abiotic stresses into its different components at physiological, genetic and molecular levels is not well understood, and only limited information is available in pulses. However, some efforts have been initiated and some generic traits including earliness, leaf morphology (wax/pubescence, posture/rolling), seed hardiness, pollen viability and germination, receptivity of stigma, pigments (chl a:b, carotenoids), antioxidant, cool canopy, harvest index and stay green trait have been conceptually considered to be associated with different mechanisms operating for heat tolerance.

Physiology of Abiotic Stresses

The exact mechanisms governing the cause and effect of abiotic stress in pulses are very complex and difficult to understand. However, each type of stress hampers the plant growth by disturbing the normal physiology and morphology. Changing environmental conditions or very often said as 'climate change' can lead to oxidative damage by overproduction of toxic reactive oxygen species (ROS) such as superoxide radicals, hydrogen peroxide and hydroxyl radicals which are significant limiting factors for crop yield (Bowler et al. 1992; Mehlhorn et al. 1990; Foyer et al. 1994; McKersie and Leshem 1994; Price et al. 1989; Moran et al. 1994; Mittler 2002). Production of ROS is an inevitable result of the leakage of electron onto molecular oxygen in chloroplast; mitochondrion and plasma membrane linked electron transport in plant cell (Rubinstein and Luster 1993; Asada 1994; Fridovich 1995). Chloroplast is very sensitive to abiotic stresses which lead to oversaturation of the photosynthetic light reactions, eventually causing photo-inhibitory damage to the photosynthetic apparatus (Powles 1984; Aro et al. 1993). Oxygen is produced by photosystem II during photosynthesis, which increases the internal oxygen concentrations and potentially augments the chances for reactive oxygen species (ROS) formation, especially under stressful conditions (Aro et al. 1993).

The unutilised energy in the electron transport systems can excite oxygen from the triplet to a singlet form. Activation of oxygen may occur by two different mechanisms: first by absorption of sufficient energy to reverse the spin on one of the unpaired electrons and second by monovalent reduction of oxygen to form superoxide ($O_2^{\cdot -}$), hydrogen peroxide (H_2O_2), hydroxyl radical ($OH^{\cdot}$) and finally water. Superoxide acts as either an oxidant or a reductant; it can oxidise sulphur, ascorbic acid or reduce nicotinamide adenosine dinucleotide (NADPH); alternatively, it can reduce cytochrome C and metal ions. The oxidising nature of photosystem II (PSII) facilitates four single-electron transfers from water to the PSII reaction centre releasing triplet or ground state oxygen. Leaking of electrons from this site to molecular oxygen, or release of partially reduced oxygen products, results in the production of activated oxygen species. Reactions associated with photosynthesis and photorespirations are major sources of ROS within plant cells (Foyer and Noctor 2003). The reactions of activated oxygen with organic substrates are complex even in vitro with homogenous solutions, but in biological systems, there are even more complications due to the surface properties of membranes, electrical charges, binding properties of macromolecules and compartmentalisation of enzymes, substrates and catalysts. Thus, various sites even within a single cell differ in the nature and extent of reactions with oxygen.

Under prolonged oxidative stress, active oxygen causes DNA damage, protein denaturation and lipid peroxidation (Scandalios 1993; Asada and Takahashi 1987). ROS play an important role in endonuclease activation and consequent DNA damage (Hagar et al. 1996). Both the sugar and base moieties are susceptible to oxidation by the hydroxyl radical, causing base degradation, single-strand breakage and cross-linking to protein (Imlay and Linn 1986; Oleinick et al. 1986). DNA is effective in binding metals that are involved in Fenton reactions, and secondly, less damage can be tolerated by DNA than other macromolecules (Beyer et al. 1991). Oxidative attack on proteins results in site-specific amino acid modifications, fragmentation of the peptide chain, aggregation of cross-linked reaction products, altered electrical charge and increased susceptibility to proteolysis.

Tolerant plants develop defence mechanism at morphological, biochemical, anatomical and physiological levels. Morphologically, plants under drought condition have deeper root system and waxy leaf with fewer and closed stomata. Membrane integrity is yet another aspect in which a tolerant plant modifies itself at the onset of drought and heat stresses. Severe cold also induces freezing tolerance. Water–ice transitions can cause cellular damage via dehydration and physical disruption of membranes, resulting in loss of chemical gradients and eventually cellular death (Sinclair et al. 2003). Responses to cold should especially prevent membrane disruption due to ice formation. In other stresses such as heavy metal or high salinity stresses, plants follow two type of mechanism to overcome the situation. In the first type, plants check the entry of salt or heavy metal by creating ion gradient. In the second type, plants scavenge the circumstances which results from the entry of those ion or heavy metals. The overall results of different stresses are the generation of ROS as discussed above, and these are scavenged by special mechanism by plants as described below.

The damaging effects of ROS have caused plant cells to develop complex redox homeostatic mechanisms that scavenge active oxygen and keep it below harmful levels. Antioxidants (ROS scavengers) include enzymes such as SOD, APX, CAT, GR, MDAR and DHAR as well as non-enzyme molecules such as ascorbate, glutathione, proline, betaine, carotenoids, anthocyanins and flavanoids (Zagorchev et al. 2013; Talukdar 2013; Garg and Kaur 2013; Rasool et al. 2013; Wu et al. 2007; Shao et al. 2007; Sulpice et al. 2003). Additional compounds, such as osmolytes, proteins (e.g. peroxiredoxin) and amphiphilic molecules (e.g. tocopherol), can also function as ROS scavengers (Bowler et al. 1992; Noctor and Foyer 1998). $O_2^{\cdot-}$ are mainly scavenged enzymatically by SOD in a dismutation reaction, which produces hydrogen peroxide (H_2O_2) and oxygen (O_2) (Gill et al. 2010; Pal et al. 2013; Bansal and Srivastava 2012; Nautiyal and Sinha 2012; Rasool et al. 2013). H_2O_2 is then disposed of enzymatically by APX (in chloroplast) or CAT (in mitochondrion) (Asada 2006; Willekens et al. 1997; Bansal and Srivasta 2012; Rasool et al. 2013). Oxidised ascorbate is then reduced by reactions that are catalysed by MDAR, DHAR and GR in a series of reactions known as the Asada–Halliwell pathway (Bowler et al. 1992; Zagorchev et al. 2013; Talukdar et al. 2012; Garg and Kaur 2013; Rasool et al. 2013). The water–water cycle has been proposed to explain the role of these antioxidant mechanisms at the onset of oxidative stress in chloroplasts (Asada 1999). Apart from these, there are many other genes and transcription factors which are induced in specific stresses, e.g. heat shock transcription factor (HSF) and hypoxia-inducible transcription factor 1 (HIF-1) had a greater heat endurance in plants. In cold, DREB or CBF are larger family of transcription factors involved (Fowler and Thomashow 2002).

Nature of Abiotic Stresses in Pulses

Pulses encounter a number of abiotic stresses during various stages of their life cycle. Among various abiotic stresses, heat in lentil, chickpea and urd bean; waterlogging and low temperature in pigeon pea; and drought in chickpea are the most important. The crop losses due to these stresses largely depend on seasonal variations. It has been observed that the productivity of late sown lentil and chickpea has gone down due to abrupt rise in temperature in the month of February/March in northern India. The increase in night temperature has resulted in low amount of dewfall during winter season, leading to unbalanced growth in the lentil crop. The temperature extremities including heat at reproductive stage adversely affect the

Table 1 Importance of major abiotic stresses affecting pulse crops

	Warm season pulses			Cool season pulses		
Stress	Pigeon pea	Urd bean	Mung bean	Chickpea	Lentil	Field peas
Drought	**	*	**	***	***	***
Waterlogging	***	***	***	*	*	***
Heat	*	***	**	***	***	***
Cold	***	**	**	**	*	*
Salinity/alkalinity	**	*	*	**	**	**
Al toxicity	**	**	**	**	**	**

***very important, **important, *not very important

developing grain and result in low yield in lentil and chickpea. If suitable varieties possessing heat tolerance are developed, we can popularise cultivation of lentil and chickpea in rice fallow (out of estimated 9 m ha of rice fallow, ~0.50 m ha can be brought under lentil in northern India).

Warm Season Pulses

Rainy season pulses such as pigeon pea, mung bean and urd bean often experience temporary water submergence that may vary from hours to a few days. This group of pulses also encounter moisture deficit owing to uneven rainfall pattern. Besides these, this group is relatively sensitive to low temperature stress (Table 1).

Pigeon pea (*Cajanus cajan*) or tur is an important annual food legume crop in the semi-arid regions of the world and second most important pulse crop after chickpeas in India. Over a billion people worldwide rely on pigeon pea as a main source of protein. Pigeon pea is grown as a cash crop by small farmers in different parts of the world including India, Africa and the Caribbean. In India, the total area under pigeon pea is around 4.42 million hectares and has an average crop productivity of 647 kg/ha which is relatively lower compared to some other legumes (FAOSTAT 2013). Pigeon pea, which is perennial by nature but is cultivated as an annual, encounters almost all stresses such as its excessive soil moisture/waterlogging (during seedling stage) and drought and low temperature stresses (during reproductive stage). Hence, improvement of pigeon pea for tolerance to these abiotic stresses is very important for obtaining increases in the harvest index and ultimately the yield.

Waterlogging is one of the most important abiotic stresses, limiting pigeon pea production in northern India where farmers grow short-duration pigeon pea in double-cropping systems. Long-duration or medium-duration pigeon pea also suffers from waterlogging in different parts of the country. Heavy rainfall results in temporary waterlogging in fields particularly those of high water-holding capacity soils, such as vertisols and Indo-Gangetic alluvial soils (Reddy and Virmani 1981). The risk of crop failure or yield reduction due to waterlogging is quite acute in extra-early and early duration varieties of pigeon pea because they have less time to recover from the stress as compared to medium-/long-duration varieties (Matsunaga et al. 1994). There are only two options available to enhance availability of any crop, i.e. enhancement in productivity or horizontal expansion. In pigeon pea, still there is scope for enhancing area in North West plains of India, where after short-duration pigeon pea, wheat can be grown successfully. Thus, development of short-duration varieties possessing waterlogging tolerance is of utmost importance.

Pigeon pea also suffers with different intensities of intermittent and/or terminal drought stress. Pigeon pea can be exposed to intermittent drought stress during dry periods of the rainy season and to terminal drought stress in the post-rainy season. Shorter-duration pigeon pea genotypes which mature within 90 days have been developed

(Vales et al. 2012). However, most of the short-duration genotypes are sensitive to intermittent drought as it is difficult to develop phenotypic screens for intermittent drought tolerance since the timing and intensity of this type of drought are fairly unpredictable, whereas screening for terminal drought has been successful in many crop plants (Turner 1986; Subbarao et al. 1995).

The sensitivity of pigeon pea to temperature is another major constraint leading to its low productivity. In East and North part of the India where traditionally long-duration (200–300 days) pigeon pea genotypes are cultivated (sowing is done in June–July and harvested by March–April), are quite sensitive to low-temperature stress. Low temperate during early flowering phase leads to flower drop in pigeon pea (when temperature drops below 10 °C). Due to this, farmers harvest produce from the second flush of flower; this leads to delay in maturity and significant yield losses. Efforts to know the physiological reason behind flower drop and identifying genotypes tolerance to low temperature would be a great asset for pigeon pea grower of the region.

India had a major shortfall in mung bean production over the last few years, and this created a major export opportunity for the countries like Myanmar and Australia. International prices have been gone up to USD1200/t, making mung bean one of the most profitable summer season crops for export. Mung bean had second highest returns to land and management after chickpea and was twice as profitable as both first- and second-season maize. The availability of short-duration, heat- and waterlogging-resistant mung bean cultivars allows them to slot into the gap between rice–rice crops; they mature on residual moisture in these paddies, countering the degradation risks of continuous cereal cultivation in the Indo-Gangetic Plain. Great opportunity exists for expanding mung bean cultivation into different cropping niches in order to increase food security (as well as income) to increase system productivity through the sustainable intensification of cropping patterns.

Black gram (*Vigna mungo* L. Hepper), popularly known as urd bean, urid or mash is an important self-pollinating diploid grain legume, and it is an important pulse crop of the Indian subcontinent. The main producer of black gram is India, which produces about 1.5 million t of seeds annually. Black gram accounts for more than 40 % of total legume seeds traded in the world (CRN India 2011). The urd bean is another important crop which can be fitted within cereal–cereal cropping system in spring/summer of northern India. However, this crop is highly thermo-sensitive, which makes it unsuitable to fit in the gap of the above cropping system. To meet the growing demand of urd bean in the country, it has become necessary to introduce this crop in other season also. Being a short-duration crop, it can fit well in spring/summer season in northern India and will be an alternate pulse crop available with farmers to fetch additional income in the lean season. The development of heat-tolerant urd bean genotypes offers ample scope for their adoption by farmers. Looking at the variability available for such traits in urd bean and other crops of the similar group like mung bean, horse gram and rice bean and in wild Vigna, the genotypes possessing tolerance to high temperature can be developed. Popularisation of such genotypes will help in ensuring more availability of pulses to the agrarian population of the country and in turn also help in saving huge foreign exchange on import. Although urd bean tolerates waterlogging to a greater extent, yield level is adversely affected in case excess water is not drained out after 2–3 days. Generally, urd bean tolerates drought stress (Arora and Mauria 1989).

Winter Season Pulses

Winter pulses such as chickpea, lentil, field peas and faba beans, which relatively tolerate low temperature, more often experience high temperature stress during their reproductive period. In low-input agriculture, this condition is often intertwined with moisture stress (Table 1).

Chickpea (*Cicer arietinum* L.) is an important pulse crop worldwide and a major source of protein for millions of families in developing countries. Seed yield of chickpea is generally low, unstable and less than its potential as it is most often grown on marginal land with minimum inputs. India grows chickpea on about 9.21 million ha area producing 8.22 million tonnes with an annual productivity of 892.5 kg/ha (FAOSTAT 2013). Despite its high yield potential, low and unstable yields are generally due to prevalence of various biotic and abiotic stresses. On a global basis, annual yield losses in chickpea were estimated to be 6.4 million tonnes due to abiotic stresses and 4.8 million tonnes due to biotic stresses (Ryan 1997). The most common abiotic stresses affecting chickpea production, in order of importance, are drought, heat and cold (Croser et al. 2003). In chickpea, terminal drought is one of the major constraint that limits its productivity and in future due to impact of global warming, this situation will get even more worsen.Therefore, breeding chickpea genotypes tolerance to drought is an important task towards breeder.

Flowering and podding in chickpea are known to be very sensitive to changes in external environment, and exposure to heat stress at this stage is known to lead to the reduction in seed yield (Summerfield et al. 1984). Drastic reductions in chickpea seed yields were observed when plants at flowering and pod development stages were exposed to high (35 °C) temperatures (Summerfield et al. 1984; Wang et al. 2006).

Samineni et al. (2011) have demonstrated that chickpea is sensitive to salinity at both vegetative and reproductive phases, with pod formation being particularly sensitive. The sensitivity occurs even for a reputably tolerant cultivar 'JG 11' even at 20 and 40 mM NaCl levels that may be considered relatively mild for many crops including bread wheat with which chickpea may be grown in rotation.

Lentil (*Lens culinaris* Medik.) is one of the important cool season pulse crop of India which is being grown on an area of about 1.60 m ha with annual production of 0.93 m tones (FAOSTAT 2013). Lentil is of paramount importance for the carbohydrate to protein ratio for the poor population in India, especially eastern Gangetic region. Therefore, in India, there is an ever-increasing annual demand for human consumption, but lentil production in the country is characterised by low yield potential of 697 kg/ha. Terminal heat due to early arrival of summer along with drought after anthesis and grain filling and particularly during seed maturity has emerged as a major threat to lentil production in eastern Gangetic region. Combating with the problem which leads to significant yield loss in lentil can provide sufficient protein in the diet of common people. Drought stress has also been found to affect germination of lentil seed significantly (Salehi 2012). The water deficit was found to reduce plant height by about 20 %, leaf area by 48–81 % and total dry matter by about 60 % compared with well-watered plants. The water deficit reduces flower number by 35–46 % and increased seed abortion (empty pods) by 17–46 % (Shrestha et al. 2005). The 70 % reduction in seed yield induced by the water deficit was primarily due to a reduction in pod and seed numbers (by 59–70 %) rather than individual seed growth rate and seed size.

Sources of Tolerance

Efforts have already been made on identification of important donors for tolerance to different abiotic stresses in pulses. The genotypes ILL6002 and ILL7663 have been identified in lentil for rapid initial growth habit and earliness, respectively. In urd bean, a genotype RBU38 has shown some level of tolerance to heat, which can be a useful donor for breeding programme. In pigeon pea, a genotype tolerant to waterlogging (i.e. ICPL 84023) has been identified (Sarode et al. 2007). Recently, results of screening a large set of materials (n = 272) with different genetic origins for waterlogging tolerance at seed level revealed that significant variability for waterlogging tolerance exists in cultivated pigeon pea genotypes (Sultana et al. 2012). Chauhan et al. (1997) tested ten genotypes and Krishnamurthy et al. (2012) tested 160 accessions (146 mini core pigeon pea germplasm accessions, four control entries and ten previously tested genotypes). Sultana et al. (2012) reconfirmed the reactions of ICP 7035 previously reported by Chauhan et al. (1997) as sensitive and those of ICPH 2671, ICPH 2740, ICPH 3762 and ICPR 2671 as tolerant (Krishnamurthy et al. 2012). Similarly chickpea genotype ICCV 92944 characterised as early flowering type with quick canopy growth enabled it to thrive well under late sown condition before temperature reached to 35 °C. Some other genotypes, namely, ICC 1052, ICC 15614, ICC 12916, ICC 1205, ICC 637 and ICC 8522, have also been identified possessing heat avoidance tendency by virtue of high LAI.

Screening Techniques and Marker Traits

Genotypic variations have been observed for almost all the abiotic stresses wherever a large number of genotypes have been screened. Large-scale screening of germplasm for different abiotic stresses and using them in breeding programme is very much essential for bringing resistance trait in pulses. Marker traits conferring tolerance to such stress(es) have also been identified. In the following section, we will discuss the specific screening technique(s) and the marker traits apt to differentiate genotypes for such a stress.

Drought

Drought refers to a situation wherein available soil moisture is not sufficient to meet the demands of potential evapotranspiration. It occurs where soil moisture and rainfall are inadequate during the growing season to support healthy crop growth to maturity and cause extreme crop stress and wilting of plants (Choudhary and Vijayakumar 2012). Therefore, large-scale screening of germplasm for drought is very much essential to address the anticipated drought condition. Several screening techniques including soil-based and field screening have been suggested to identify drought-tolerant genotypes in pulses which are based on identifying genotypes having long, dense and efficient root systems. Soil-based screening techniques have been criticised as several other constraints are encountered. To overcome these constraints, a new phenotyping technique to screen drought tolerance in lentil based on hydroponic system has been recently suggested by Singh et al. (2013).

The new screening technique as developed by Singh et al. (2013) is based on seedling survivability, drought tolerance score, root and shoot length and fresh and dry weight of roots and shoots of 80 lentil genotypes that were exposed to drought under hydroponic conditions. The effectiveness of this technique was compared with two soil culture techniques. The hydroponic technique involved removing 15-day-old seedlings of each genotype from the nutrient solution and exposing them to air for 5 h daily continuously for 6 days. Three genotypes, namely, 'ILL-10700', 'ILL-10823' and 'FLIP-96-51', which were received from International Centre of Agriculture in the Dry Areas (ICARDA), showed the maximum seedling survivability and minimum reduction in the growth parameters with a drought score of 0.0–0.2, indicating higher tolerance to drought stress than other genotypes. The researchers have claimed that this new phenotyping technique is effective, rapid and easy for screening a large number of genotypes. The same technique with minor modifications has also been suggested to screen mung bean genotypes for drought tolerance (IARI News 2012).

However, despite big claims made by the researchers, it appears that hydroponic technique is based on only partial component of even drought avoidance. They have not corroborated the results of the hydroponic technique with the field-level screening. Furthermore, it is also not clear whether the parameters (taken as the index of drought tolerance) confer only survival advantage to the surviving genotypes. Under field conditions, several variables interact to produce final outcome. Therefore, controlled field screening aided by rainout shelter appears to be more reliable than simply soil or solution culture-based screening techniques.

A number of morphological markers in pulses are known to reduce water loss from soil/aerial portion of the plants. These include plant type (spreading or semi-spreading), morphology (leaf area index; LAI) and orientation of leaves (leaf angle), cuticular waxiness (which results in 2–50 % reduction in transpiration) and leaf reflectance (up to 50 % reduction in light absorption).

In pigeon pea, some physiological parameters such as water retention capacity of leaves (relative water content; RWC), osmotic adjustment (OA), dehydration tolerance and stomatal regulation appear to be equally important in most of the legumes for combating moisture-deficit condition. Agronomic traits such as pods/plant, seeds/pod, seed size and seed yield/plant under actual water deficit condition should be given much importance while breeding for drought its resistance a (Choudhary et al. 2011a).

Waterlogging

Screening technique(s) to discriminate tolerant and sensitive genotypes for waterlogging stress in pulses is only a few and is also not well documented. However, some screening work has been performed in pigeon pea. Identification of key mechanisms of tolerance by designing simple screening protocols that would be useful in identification of tolerant sources is of paramount importance in pigeon pea. Since pigeon pea receives maximum rain during the months of July and August. Sultana et al. (2012) found that the seed (just after sowing) and early seedling stage (15–35 days) in pigeon pea are very sensitive to waterlogging; hence, screening methodology was optimised by taking into account the crop growth stages that were most severely affected by waterlogging. The seeds of the genotypes were subjected to water submergence treatments in 200 ml beakers (100 mm diameter) containing 100 ml of water at 23 ± 1 °C. The submergence treatments were established as a function of the submersion time (S120, S144, S168 and S192 for groups of seeds submerged for 120, 144, 168 and 192 h, respectively). A baseline (S0 = no submergence treatment) germination test was performed by placing 20 seeds of each genotype between two paper towels in plastic Petri dishes and maintaining humidity as necessary. The durations of S120, S144 and S168 were comparable with field observations of soil waterlogging timing at the study site, especially during rainy years. The S192 duration was specifically selected for the present experiment in order to check seed viability under extended submergence (8 days). After completing each stress period, seeds were dried on a filter paper for 4–5 h to drain excess water and then placed on a paper towel in a Petri dish and kept for germination at a constant temperature (25 ± 2 °C) in a dark room. The seeds were considered to have germinated when radicle length reached a minimum of 2 mm. The germinated seeds were counted and percent survival was calculated 5–6 days after completing stress treatment (Sultana et al. 2012).

A set of genotypes in 2–3 replications are grown in pots. Thirty-five-day-old seedlings were placed in a well-levelled tank. Thereafter, 6–8-day water submergence condition was artificially created. After a fixed period of exposure to such stress, water is drained out completely. Data on plant density, chlorophyll content, oxygen content of water, pH of soil and other parameters are taken before and after water treatment. Agronomic data such as days to flowering, plant height, pods/plant, seeds/pod, yield/plant and maturity period are recorded on surviving genotypes in both treated and control treatments. This screening technique may be satisfactory. However, the timing of the experiment must be such that it mimics that of natural condition (high temperature, high humidity, long day, and the like).

Extremes of Temperature

In general, warm season pulses are relatively tolerant to heat stress and sensitive to low temperature. Cool season pulses, on the other hand, tolerate low temperature, but are sensitive to heat stress (>35 °C). Among warm season pulses, pigeon pea is the crop which encounters low temperature stress during winter months in North India. Low temperature adversely affects growth, survival and reproductive capacity of plants if the minimum temperature falls below 10 °C.

Conclusive evidence for the presence of genetic variability vis-à-vis cold tolerance was provided by Sandhu et al. (2007). They screened for cold tolerance in a set of 480 pigeon pea lines at Ludhiana. During the first fortnight of January, minimum temperature more often touches 0 °C, which was good enough to assess cold reaction. As many as 32 genotypes were rated cold tolerant as the plants retained their normal morphology with intact floral buds. They suggested utilising these genotypes to enhance cold tolerance of sensitive varieties and study the genetics of cold tolerance. However, they did not report any data on other reproductive traits.

Low temperature primarily affects development and growth of flower buds and opening of flowers (Choudhary 2007). In some sensitive genotypes such as 'IPA 209' and 'IPA 06-1', filaments of stamens fail to enlarge at low temperature and thus affect opening of flowers. Pollen dehiscence does not occur too, although pollens are fully fertile. As a consequence, unfertilised flowers wither and fall down, resulting in no pod formation in these genotypes under low temperature (IIPR Annual Report 2008–2009). It appears that formation of floral buds, number of blossomed flowers and pod setting at low temperature can be used as selection criterion to identify tolerant genotypes in pigeon pea. These traits need to be investigated also in wild relatives of pigeon pea so as to screen tolerant wild accessions as have been done in wild relatives of mung bean and urd bean, IC 251372 in *V. glabrescens* and IC 251442 in *V. umbellata*, for example (IIPR Annual Report 2011–2012).

Cool season pulses experience terminal heat stress, especially during pod formation and grain-filling period in North India. Screening techniques to identify heat-tolerant genotypes have been developed at Indian Institute of Pulses Research, Kanpur (IIPR Annual Report 2011–2012). Fifty chickpea genotypes were evaluated for heat tolerance under late sown condition. The thermo-tolerance of the selected genotypes involved assessment of their pollen viability at 43 °C, nondestructive assessment of photosynthetic ability through fluorescence imaging system and membrane injury test. Genotypic variation was observed for LAI. Among the test genotypes, ICCV 92944 showed least deviation in the LAI during late sown condition as compared to the LAIs of the same genotype sown under normal condition. The photosynthetic ability as quantified by kinetics of chlorophyll fluorescence differed significantly with increasing temperature. The damaging effect of high temperature on membrane was evident by rise in the minimal fluorescence. The LAI and dry matter accumulation at high temperature beyond 35 °C was found to be largely affected by inhibition of photosynthesis. The membrane injury was likely to be associated with prolonged exposure of photosynthesising leaf at high temperature; however, significant variation in the membrane injury index was observed among genotypes. Pollens of heat-tolerant genotype ICCV 92944 were deeply stained at > 40 °C, while the sensitive genotypes ICC 14077 and PBG 5 showed non-viable pollen without taking stain. The increase in the non-viable pollen decreased fertility in these genotypes substantially at > 40 °C.

Genotypic variation has also been observed among lentil genotypes for heat tolerance under field condition which results in synchronisation of heat stress with reproductive phase in North India (IIPR Annual Report 2011–2012). The biomass decreased significantly in all genotypes under late (January) sown condition compared to under normal (November) plantings. Therefore, lentil genotypes having relatively high biomass under heat stress would be desirable as there appears to be strong association of biomass and seed yield in lentil.

Salinity/Alkalinity Stress

Soil salinity is an ever-increasing production constraint for both cool season and warm season pulses in many parts of the world. Salinity of only 3 dSm^{-1} in field soils was the threshold for reduced shoot growth and yield in chickpea (Rao et al. 2002), although this exceeds the even lower salinity threshold (<1.3 dSm^{-1}) in some pulses like cowpea, soybean and pigeon pea (Keating and Fisher 1985).

In chickpea, germination is less sensitive to salinity than early vegetative stage, and reproductive phase is considered to be even more sensitive than vegetative phase. In chickpea, 40 mM NaCl may be considered optimum level to discriminate tolerant and sensitive genotypes. Upon removal of salinity (NaCl), which may hardly be a case under field condition, chickpea shows excellent recovery with substantial new shoot growth. This could happen presumably in most pulses as they are endowed with the unique adaptive advantage of having indeterminate growth habit. Tissue ion regulation is a key trait for salt tolerance in plants, but whether Na or Cl 'exclusion' contributes to tolerance in chickpea remains uncertain. However, Samineni et al. (2011) reported that sensitivity during the reproductive phase was not caused by changes in pollen viability but was potentially due to toxic accumulation of Na and Cl in flowers, and possibly the sensitivity of pollen tube growth if NaCl entered the stigma. However, in another salinity experiment, no relation was found between final yield and Na (% dry mass) in shoots at the vegetative stage (Vadez et al. 2007). It therefore appears that in chickpea, a combination of mechanisms – ion exclusion and tissue tolerance of excess ions – is likely to contribute to salt tolerance. Under saline condition, symptoms of leaf necrosis, presumably related to the destruction of chlorophyll in leaf cells resulting from ion toxicity when Na + and/or Cl- exceeds threshold level in tissues, have been observed; 'visual scores' of necrosis could be used as an index of salinity tolerance in chickpea (Flowers et al. 2010). Salinity also causes physiological drought so that chickpea is unable to remove as much water from saline soil as from non-saline soil. Although chickpea shows osmotic adjustment (OA), its role in salt-sensitive compared with salt-resistant genotypes is not conclusive and requires further study. Salinity has been shown to decrease the number of pods/plant, seeds/pod and size of seeds; nevertheless, size is relatively less affected. Vadez et al. (2007) reported even no significant difference in seed size of salt-sensitive and salt-resistant genotypes, indicating the possibility to develop salt-resistant cultivars in the market-preferred seed size category. No consistency in performance has been observed when genotypes selected in seedling stage are brought to maturity under saline condition. It indicates that selections for salt resistance are required across the entire life cycle. Further, as genotypes differ in expression of resistance at different stages, it also provides an opportunity to combine sources of resistance for different stages from contrasting parents in chickpea (Flowers et al. 2010).

In warm season pulse like pigeon pea, differential tolerance to salinity vis-à-vis pigeon pea maturity groups has been observed by Dua and Sharma (1996). Late maturing genotypes showed better tolerance than early maturing ones. Like chickpea, no correlation was found between the tolerance at germination and later stages. However, percentage survival showed some association with seed yield under salinity. Low and high accumulation of *Na* and *K*, respectively, in the roots and other plant parts (main stem, branches and leaves) perhaps helped salinity tolerance in pigeon pea. Salinity also delays days to 50 % flowering by 1–2 weeks and prolongs the peak period of flower production and reduces the number and weight of the pods and seeds (Promila and Kumar 1982). Subbarao et al. (1991) studied comparative salinity tolerance among pigeon pea genotypes and their wild relatives. Among the cultivated genotypes, 'ICPL 227' and 'Hy3C' were observed as the most tolerant and the most sensitive genotypes, respectively. Among the wild relatives of pigeon pea, several species including *C. scarabaeoides*, *C. albicans* and *C. platycarpus* showed a wide range of variation in their salinity tolerance. They further clarified that certain physiological attributes that confer salinity tolerance in these wild species include *Na* and *Cl* retention in the roots and limited translocation to the shoots, high *K* selectivity and maintenance of transpiration rate under saline conditions. It has been observed that increasing salt concentrations adversely affects growth of pigeon pea. Among the three test accessions, 'ICPL 151' was superior to the other two accessions in fresh and dry biomass, yield and yield components when assessed at the adult stage. The tolerant accession 'ICPL 151' accumulated significantly lower Na^+ and Cl^- in shoots. By contrast, the accession had higher shoot and root K^+, K/Na ratio, K vs. Na selectivity, soluble sugars, free amino acids and proline compared to the other two accessions.

Srivastava et al. (2006) found that a *NaCl* treatment of 1.01 g/kg alfisol was suitable to salinity screening in pigeon pea. Using that treatment, they found large variations in the salinity susceptibility index (SSI) and the percent relative reduction (RR %) in both cultivated and wild accessions. The amount of *Na* accumulation in shoot showed that more tolerant genotypes accumulated less *Na* in the shoot except the wild species, which followed a different pattern compared to cultivars. Overall, they found that *C. acutifolius*, *C. cajanifolius* and *C. lineata* were mostly sensitive, whereas *C. platycarpus*, *C. scarabaeoides* and *C. sericeus* provided good sources of tolerance. It was interesting to notice that *C. scarabaeoides* also provided a large range of sensitive materials.

Aluminium Toxicity

Aluminium (Al) toxicity is known to limit crop production in 30 % of arable lands (Campbell et al. 1988). It constrains crop productivity in acid soils which is widely prevalent in large areas of the world particularly in the tropics and subtropics. The two most common ways to mitigate Al toxicity are by liming and by using tolerant cultivars. It is possible to detoxify aluminium in surface soil in the field by liming with pH 5.5 or above. However, liming may not always be cost effective and does not remedy subsoil acidity. Such situations call for use of tolerant cultivars as the satisfactory solution to the problem of Al toxicity.

Considerable variation for tolerance to Al toxicity in plant species has been reported. Literature says that most of the pulses, such as chickpea, pigeon pea, pea, mung bean, lentil and alfalfa, are sensitive to Al toxicity. The work on screening for tolerance to Al toxicity in these pulses are only a few and limited to seedling screening.

The four techniques, which have been commonly used for screening of Al toxicity in pulses, are sand and hydroponic assays, haematoxylin staining and root regrowth assay. These four techniques have unconditional advantages over field screening because reliable ranking of tolerance in the field screening is difficult due to the large temporal and spatial variation in acidic soils (Choudhary et al. 2011b).

Choudhary et al. (2011b) studied the effects of five levels of aluminium concentrations (0, 10, 20, 30 and 50 μg ml^{-1} Al) on 32 pigeon pea genotypes by four different methods: hydroponic and sand assays (growth response methods), root regrowth and haematoxylin root staining. Significant variability was noted for tolerance to aluminium toxicity among the pigeon pea genotypes (Fig. 1). The results of all the four screening methods were consistent, suggesting that any one of the four methods could be used for screening purpose. However, due to operative simplicity, reliable and better precision and short test period, the haematoxylin staining at 30 μg ml^{-1} Al concentrations was suggested as the best method to discriminate pigeon pea genotypes for Al tolerance (Fig. 2). The same Al concentration (30 μg ml^{-1}) has also been suggested for screening of other pulses including chickpea (Singh et al. 2009) and pea (Singh and Choudhary 2010). However, the optimum level of concentration may vary in other pulses like lentil and mung bean.

Choudhary and Singh (2011) further assessed tolerant and sensitive genotypes of pigeon pea for phosphorus, potassium, calcium and magnesium contents in their root and shoot. Tolerant genotypes with Al toxicity (IPA 7–10, T 7, 67B and GT 101E) accumulated significantly high amounts of these nutrients (>1.5 times) compared to the sensitive ones (Bahar, Pusa 2002–2002 and Pusa 9) (Figs. 1 and 2). The primary effect of Al toxicity was the restriction

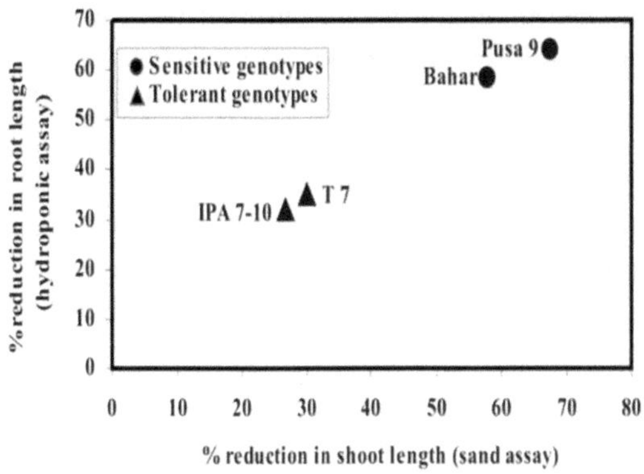

Fig. 1 Relationship between root length reduction in the hydroponics assay (0 compared to 50 μg ml^{-1} Al) and shoot reduction in the sand assay (compared to 50 μg ml^{-1} Al) for four pigeon pea genotypes

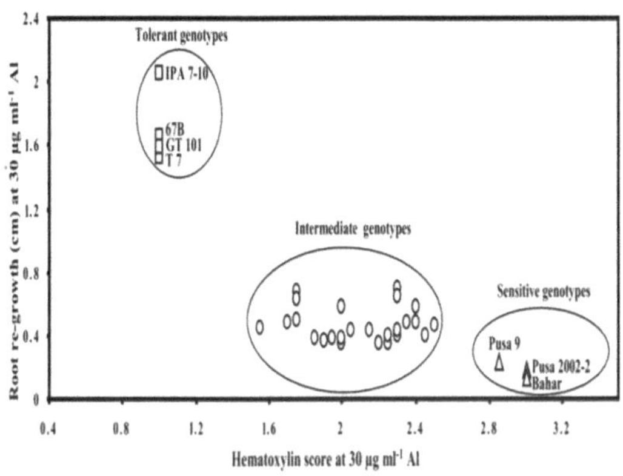

Fig. 2 Relationship between root regrowth and haematoxylin scores of 32 genotypes of pigeon pea at 30 μg ml^{-1} Al concentration

of root growth (rapid inhibition of root elongation by destroying the root apex); injured roots did not branch normally and were shorter at higher levels of aluminium (123 μM Al = 30 μg ml^{-1} Al) than the roots grown in nutrient solution without aluminium (control). Such damage to roots perhaps caused differential nutrient uptake by tolerant and sensitive genotypes. Al toxicity has been reported to affect uptake of water and nutrients in a similar manner in other crops as well (Fageria 1985). In their experiment, Choudhary et al. (2011) indicated that Al exclusion from root could be the possible mechanism for Al tolerance. However, internal detoxification in several other crops has also been reported as the possible mechanism of Al tolerance.

All the experiments conducted for assessing Al tolerance in pulses such as chickpea, pigeon pea, mung bean and lentil were based on sand or solution culture techniques. The results are still to be confirmed under natural acid soils. Furthermore, all experiments were based on seedling response to Al toxicity. Therefore, it seems imperative to expose plants at reproductive stage also to ascertain whether any association indeed occurs between vegetative and reproductive stages for response to Al toxicity. It would further be desirable to include wild accessions of pulses along with cultivated genotypes to find out even higher degree of tolerance to Al toxicity.

Genetics of Traits and Breeding Strategies

Pulses encounter different kinds of abiotic stresses; thus, breeding strategy also varies accordingly. For example, for mitigating drought and terminal heat stress, selection of pulses that take less time to mature (e.g. mung bean, moth bean, cowpea) could be done. In other pulses such as pigeon pea and chickpea, breeding for earliness would be desirable as their reproductive stage will tend to escape such stresses. By extrapolation, pulses are likely to accumulate a specific combination of genes (alleles) if exposed separately to above-mentioned stresses. Therefore, a genotype showing tolerance to drought may react differently if exposed to waterlogging and vice versa. JG 11 – a leading chickpea variety of central and south zones in India – is considered relatively tolerant to heat stress and is not a good performer in north India where atmospheric temperature is comparatively low during winter months (December–January). It has been found that the best breeding strategy appears to be selection of superior genotypes of pulses on the basis of yield under actual field condition. According to Flowers et al. (2010), final assessment of stress tolerance (salinity) of genotypes must be based on grain yield. If stress-tolerant genotypes have been identified based on component traits such as number of pods/plant, it needs further confirmation by evaluating these genotypes for seed yield under actual saline condition. It is often argued that there may be temporal and spatial variation under field condition; therefore, screening should be done under controlled condition. However, under field conditions, a number of variables interact to produce final outcome; thus, field testing of genotypes cannot be ignored. Therefore, we suggest to assess genotypes for such abiotic stresses under actual field condition, and the results may be reconfirmed under controlled condition and vice versa. However, selection must finally be practised for high-yielding genotypes.

Sometimes it has been observed that surviving genotypes up to the stage of maturity are poor yielding. However, such genotypes can be used as donors for specific abiotic stress tolerance. There are instances where traits in question are mono- or oligogenic having high heritability (high regression of offspring on the parents). Such Al tolerance has been shown to be dominant monogenic in pea (Singh and Choudhary 2010) and chickpea (Singh and Raje 2011) and oligogenic in pigeon pea (Singh et al. 2011). Similarly, pollen dehiscence and pod setting under low temperature (IIPR Annual Report 2008–2009) and waterlogging tolerance (Sarode et al. 2007) in pigeon pea; salinity tolerance in certain accessions of *Cajanus albicans*, a wild relative of pigeon pea (Subbarao et al. 1990); and winter hardiness in lentil (Frt gene) are also reported as a dominant monogenic trait. Under such a situation, simple backcross breeding can be used to improve Al or other stress tolerance in pigeon pea or any other concerned grain legume. There are instances when even monogenic/oligogenic traits show high G × E interaction and low heritability; this may be the case for many marker traits conferring abiotic stress tolerance in pulses. This calls for improving tolerance through marker-assisted backcrossing (MABC), which generally involves transfer of a limited number of trait loci including transgenes from one genetic background (donor genotype) to the other genetic background (elite variety) using molecular markers.

However, resistance/tolerance to most abiotic stresses is quantitative in nature. For example, earliness, plant height, plant structure, growth habit and yield in lentil (Kumar et al. 2011) and root traits, drought tolerance score, canopy temperature differential and seed size in chickpea (Varshney et al. 2013) are controlled by several QTLs. Yield and component traits (seed number, seed weight), which should be given due importance while making final assessment of tolerance to such stresses, are also governed by many QTLs. In such cases, retaining desirable gene combinations or pyramiding of several QTLs through MABC approach may be a challenging task. The best approach is to resort to marker-assisted recurrent selection (MARS). In some cases, superior alleles for a given trait (e.g. salinity tolerance in *C. albicans*) are identified and transferred from the wild species to a leading variety/cultivar. Under such situations, advanced backcross QTL (AB-QTL) approach as proposed by Tanksley and Nelson (1996) for simultaneous discovery and transfer of superior alleles from wild species to develop improved lines may be followed as this approach facilitates efficient tracking for desired and non-desired alleles in breeding lines (Choudhary et al. 2013).

Recently, 75 ESTs have been identified from the cDNA libraries of drought stressed plants and 20 ESTs proved to be unique to the pigeon pea (Priyanka et al. 2010). Ectopic expression of *CcHyPRP* (*C. cajan* hybrid proline-rich protein), *CcCYP* (*C. cajan* cyclophilin) and *CcCDR* (*C. cajan* cold and drought regulatory) genes in Arabidopsis showed marked tolerance, higher plant biomass and increased photosynthetic rates under PEG/NaCl/cold/heat stress conditions (Priyanka et al. 2010). Therefore, these candidate genes can be further used for engineering crop plants for enhance tolerance to major abiotic stresses (Priyanka et al. 2010). Bioinformatics searches followed by wet lab experiment will facilitate in identification of more stress-responsive genes in pigeon pea. Newly identified stress-responsive genes in this crop can be exploited for improving other legumes crop.

These candidate genes are useful tool for basic and applied research, especially for drought tolerance in pigeon pea improvement. Recently, presence of ten well-characterised abiotic stress-responsive genes (*AKIN*, *AMADH*, *DHN*, *DREB*, *Myb*, *CAD*, *EREBP*, *LEA*, *SAMS*, *STPK*) in various model crops and other legumes was confirmed (Roorkiwal and Sharma 2012). These candidate genes can be further used in generating superior chickpea varieties with an aim to increase yields under stress conditions using biotechnological approaches. This shows that the application of biotechnological approaches has a potential to contribute efficiently to solve or reduce these problems in the pulses, thereby contributing to sustainable agriculture.

Conclusions

Physio-genetic approach involving efficient screening techniques and evaluation of breeding material/lines under targeted environment for the traits linked to tolerance is likely to lead to the identification of specific component traits (imparting tolerance to above-mentioned stresses) and high-yielding varieties with improved stress tolerance. Recent studies suggest that cultivated and wild species of pulses posses desired traits for a number of abiotic stresses discussed above. Future effort is required to identify desirable genes from these germplasm to transfer it into adapted cultivars by conventional and/or biotechnological approaches.. Understanding the complex network of genes/QTL underlying the targeted traits is also essential for precise utilisation of genetic stocks in breeding programmes. Therefore, proper genetic dissection of specific traits can be worked out with the help of molecular markers, which can lead to the identification of genes/QTLs that are able to control a number of traits imparting tolerance to a target stress. For this purpose, mapping populations can be developed from the donors. These mapping populations will be useful genomic resources for mapping/tagging of genes/QTL for target traits which can be utilised in marker-assisted breeding.

References

Allard RW (1999) Principles of plant breeding, 2nd edn. Wiley, New York

Aro E-M, Virgin I, Anersson B (1993) Photoinhibition of photosystem II: inactivation, protein damage and turnover. Biochim Biophys Acta 1143:113–134

Arora RK, Mauria SS (1989) *Vigna mungo* (L.) Hepper. Record from Proseabase. In: van der Maesen LJG, Somaatmadja S (eds) PROSEA (plant resources of South-East Asia) foundation. PROSEA Project, Bogor

Asada K (1994) Production and action of active oxygen species in photosynthetic tissues. In: Foyer CH, Mullineaux PM (eds) Causes of photooxidative stress and amelioration of defense systems in plants. CRC Press, Boca Raton, pp 77–104

Asada K (1999) The water-water cycle in chloroplasts: scavenging of active oxygen and dissipation of excess photons. Annual Review in Plant Physiology and Plant Molecular Biology 50:601–639

Asada K (2006) Production and scavenging of reactive oxygen species in chloroplast and their function. Plant Physiol 141:391–396

Asada K, Takahashi M (1987) Production and scavenging of active oxygen in chloroplasts. In: Kyle DJ, Osmond CB, Arntzen CJ (eds) Photoinhibition. Elsevier, Amsterdam, pp 227–288

Bansal R, Srivasta JP (2012) Antioxidant defense system in pigeonpea roots under waterlogging stress. Acta Physiol Plant 34(2):515–522

Beyer W, Imlay J, Fridovich I (1991) Superoxide dismutases. Prog Nucleic Acid Res Mol Biol 40:221–253

Bohnert HJ, Nelson DE, Jensen RG (1995) Adaptations to environmental stresses. Plant Cell 7:1099–1111

Bowler C, Montagu MV, Inz'e D (1992) Superoxide dismutase and stress tolerance. Annu Rev Plant Physiol Plant Mol Biol 43:83–116

Campbell TA, Foy CD, Mc Murty E, Elgin JE Jr (1988) Selection alfalfa for tolerance to toxic levels of Al. Can J Plant Sci 68:743–753

Chauhan YS, Silim SN, KumarRao JVDK, Johansen C (1997) A pot technique to screen pigeonpea cultivars for resistance to waterlogging. J Agron Crop Sci 178:179–183

Choudhary AK (2007) Selection criteria for low temperature tolerance in long-duration pigeonpea. In: Abstract published in the national symposium on "legumes for ecological sustainability: emerging challenges and opportunity, Indian Institute of Pulses Research, Kanpur, p 266, 3–5 Nov 2007

Choudhary AK, Vijayakumar AG (2012) Glossary of plant breeding: a perspective. LAP LAMBERT

Academic/AV Akademikerverlag GmbH & Co. KG Heinrich-Böcking, Saarbrücken

Choudhary AK, Sultana R, Pratap A, Nadarajan N, Jha UC (2011a) Breeding for abiotic stresses in pigeonpea. J Food Legume 24:165–174

Choudhary AK, Singh D, Iquebal MA (2011b) Selection of pigeonpea genotypes for tolerance to aluminium toxicity. Plant Breed 130(4):492–495

Choudhary AK, Raje RS, Subhojit D, Sultana R, Timmanna O (2013) Conventional and molecular approaches towards genetic improvement in pigeonpea for insects resistance. Am J Plant Sci 4:372–385

Croser JS, Ahmad F, Clarke HJ, Siddique KHM (2003) Utilization of wild *Cicer* in chickpea improvement-progress, constraint and prospects. Aust J Agric Res 54:429–444

CRN India (2011) http://www.crnindia.com/commodity/urad.html. Accessed on 21 Apr 2013

Dua RP, Sharma PC (1996) Physiological basis of salinity tolerance in pigeonpea (*Cajanus cajan*) and method of testing materials under highly variable soil conditions. Indian J Agric Sci 66:405–412

Fageria NK (1985) Influence of aluminum in nutrient solutions on chemical composition in two rice cultivars at different growth stages. Plant Soil 85:423–429

FAOSTAT (2013) http://faostat.fao.org/site/567/DesktopDefault.aspx?PageID=567#ancor. Accessed 21 Apr 2013

Flowers TJ, Gaur PM, Gowda CLL, Krishnamurthy L, Srinivasan S, Siddique KHM, Turner N, Vadez V, Varshney RK, Colmer TD (2010) Salt sensitivity in chickpea. Plant Cell Environ 33:490–509

Fowler S, Thomashow MF (2002) Arabidopsis transcriptome profiling indicates multiple regulatory pathways are activated during cold acclimation in addition to the CBF cold-response pathway. Plant Cell 14:1675–1690

Foyer CH, Noctor G (2003) Redox sensing and signaling associated with reactive oxygen in chloroplast, peroxisomes and mitochondria. Physiol Plant 119:355–364

Foyer CH, Descourvierse P, Kunert KJ (1994) Protection against oxygen radicals: an important defence mechanism studied in transgenic plants. Plant Cell Environ 17:507–523

Fridovich I (1995) Superoxide radical and superoxide dismutases. Annu Rev Biochem 64:97–112

Garg N, Kaur H (2013) Impact of cadmium-zinc interactions on metal uptake translocation and yield in pigeonpea genotypes colonized by arbuscular mycorrhizal fungi. J Plant Nutr 36:67–90

Gill T, Kumar S, Ahuja PS, Sreenivasulu Y (2010) Over-expression of *Potentilla superoxide dismutase* improves salt stress tolerance during germination and growth in *Arabidopsis thaliana*. J Plant Genet Transgen 1:1–10

Hagar H, Ueda N, Shah SV (1996a) Role of reactive oxygen metabolites in DNA damage and cell death in chemical hypoxic injury to LLC-PK1 cells. American Journal of Physiology 271:209–215

IARI News (2012) 28:1–2

IIPR Annual Report (2008–2009) Indian Institute of Pulses Research, Kanpur

IIPR Annual Report (2011–2012) Indian Institute of Pulses Research, Kanpur

Imlay JA, Linn S (1986) DNA damage and oxygen radical toxicity. Science 240:1302–1309

Keating BA, Fisher MJ (1985) Comparative tolerance of tropical grain legumes to salinity. Aust J Agric Res 36:373–383

Krishnamurthy L, Upadhyaya HD, Saxena KB, Vadez V (2012) Variation for temporary waterlogging response within the mini core pigeonpea germplasm. J Agric Sci 150:357–364

Kumar S, Ali M (2006) GE interaction and its breeding implications in pulses. Botanica 56:31–36

Kumar J, Choudhary AK, Solanki R, Pratap A (2011) Towards marker- assisted selection in pulses – a review. Plant Breed 130:297–313

Matsunaga R, Ito O, Tobita S, Rao TP, Johansen C (1994) Response of short-duration pigeonpea to nitrogen application after short-term waterlogging on a vertisol. Field Crops Res 38:167–174

McKersie BD, Leshem YY (1994) Stress and stress coping in cultivated plants. Kluwer Academic, Dordrecht, pp 79–100

Mehlhorn H, Tabner B, Wellburn A (1990) Electron spin resonance evidence for the formation of free radicals in plants exposed to ozone. Physiol Plant 79:377–383

Mittler R (2002) Oxidative stress, antioxidants and stress tolerance. Trends Plant Sci 7:405–410

Moran JF, Becana M, Iturbe-Ormaetxe I, Frechilla S, Klucas RV, Aparicio-Tejo P (1994) Drought induces oxidative stress in pea plants. Planta 194:346–352

Nautiyal N, Sinha P (2012) Lead induced antioxidant defence system in pigeon pea and its impact on yield and quantity of seeds. Acta Physiol Plant 34 (3):977–983

Noctor G, Foyer CH (1998) Ascorbate and glutathione: keeping active oxygen under control. Annu Rev Plant Physiol Plant Mol Biol 49:249–279

Oleinick NL, Chiu S, Ramakrishman N, Xue L (1986) The formation, identification, and significance of DNA-protein cross-links in mammalian cells. Br J Cancer 55:135–140

Pal AK, Acharya K, Vats SK, Kumar S, Ahuja PS (2013) Over expression of *PaSOD* in transgenic potato enhances photosynthetic performance under drought. Biol Plant 57(2):359–364

Powles SB (1984) Photoinhibition of photosynthesis induced by visible light. Annu Rev Plant Physiol 35:15–24

Price AH, Antherton N, Hedry GAF (1989) Plants under drought-stress generates activated oxygen. Free Radic Res Commun 8:61–66

Priyanka B, Sekhar K, Reddy VD, Rao KV (2010) Expression of pigeonpea hybrid-proline-rich protein

encoding gene (CcHyPRP) in yeast and *Arabidopsis* affords multiple abiotic stress tolerance. Plant Biotechnol J 8:76–87
Promila K, Kumar S (1982) Effect of salinity on flowering and yield characters in pigeonpea. Indian J Plant Physiol 25:252–257
Rao DLN, Giller KE, Yeo AR, Flowers TJ (2002) The effects of salinity and sodicity upon nodulation and nitrogen fixation in chickpea. Ann Bot 89:563–570
Rasool S, Ahmad A, Siddiqi TO, Ahmad P (2013) Changes in growth, lipid peroxidation and some key antioxidant enzymes in chickpea genotypes under salt stress. Acta Physiol Plant 35(4):1039–1050
Reddy SJ, Virmani SM (1981) Pigeonpea and its climatic environment. In: Nene YL (ed) Proceeding of the international workshop on pigeonpea held in Patancheru, India, vol 1. ICRISAT, Patancheru, pp 259–270, 15–19 Dec 1980
Roorkiwal M, Sharma PC (2012) Sequence similarity based identification of abiotic stress responsive genes in chickpea. Bioinformation 8(2):92–97
Rubinstein B, Luster DG (1993) Plasma membrane redox activity: components and role in plant processes. Annu Rev Plant Physiol Plant Mol Biol 44:131–155
Ryan J (1997) A global perspective on pigeon pea and chickpea sustainable production systems: present status and future potential. In: Asthana A, Ali M (eds) Recent advances in pulses research. Indian Society for Pulses Research and Development, Kanpur, pp 1–31
Salehi M (2012) The study of drought tolerance of lentil (lens culinaris Medik) in seedling growth stage. Int J Agron Plant Prod 3(1):38–41
Samineni S, Siddique KHM, Gaur PM, Colmer TD (2011) Salt sensitivity of the vegetative and reproductive stages in chickpea: podding is a particular sensitive stage. Environ Exp Bot 71:260–268
Sandhu JS, Gupta SK, Singh S, Dua RP (2007) Genetic variability for cold tolerance in pigeonpea. E-J SAT Agric Res 5:1–3
Sarode SB, Singh MN, Singh UP (2007) Genetics of waterlogging tolerance in pigeonpea (*Cajanus cajan* L. Millsp.). Indian J Genet Plant Breed 67:264–265
Scandalios JG (1993) Oxygen stress and superoxide dismutase. Plant Physiol 101:7–12
Shao HB, Chu LY, Wum G, Zhang JH, Lu ZH, Hu YC (2007) Changes of some anti-oxidative physiological indices under soil water deficits among 10 wheat *(Triticum aestivum* L.) genotypes at tillering stage. Colloids Surf B Biointerfaces 59:113–119
Shrestha R, Siddique KHM, Turner NC, Turner DW, Berger J (2005) Growth and seed yield of lentil (*Lens culinaris* Medikus) genotypes of West Asian and South Asian origin and crossbreds between two under the rainfed conditions in Nepal. Aust J Agric Res 56:971–981
Sinclair BJ, Vernon P, Klok CJ, Chown SL (2003) Insects at low temperatures: an ecological perspective. Trends Ecol Evol 18:257–262
Singh D, Choudhary AK (2010) Inheritance pattern of aluminum tolerance in pea (*Pisum sativum* L). Plant Breed 129:688–692
Singh D, Raje RS (2011) Genetics of aluminium tolerance in chickpea (*Cicer arietinum*). Plant Breed 130:563–568
Singh D, Rai AK, Panyang O (2009) Hematoxylin staining as a potential screening technique for aluminum tolerance in pea (*Pisum sativum* L.). Curr Sci 96:1029–1030
Singh D, Raje RS, Choudhary AK (2011) Genetic control of aluminium tolerance in pigeonpea (*Cajanus cajan* L.). Crop Pasture Sci 62:761–764
Singh D, Dixit HK, Rajendra S (2013) A new phenotyping technique for screening for drought tolerance in lentil (*Lens culinaris* Medik.). Plant Breed 132(2):185–190
Srivastava N, Vadez V, Upadhyaya HD, Saxena KB (2006) Screening for intra and interspecific variability for salinity tolerance in pigeonpea (*Cajanus cajan*) and its related wild species. J SAT Agric Res 2:1–12
Subbarao GV, Johansen C, Rao JVDKK, Jana MK (1990) Salinity tolerance in F1 hybrids of pigeonpea and a tolerant wild relative. Crop Sci 30:785–788
Subbarao GV, Johansen C, Jana MK, Rao JVDKK (1991) Comparative salinity responses among pigeonpea genotypes and their wild relatives. Crop Sci 31:415–418
Subbarao GV, Johansen C, Slinkard AE, Rao RCN, Saxena NP, Chauhan YS (1995) Strategies for improving drought resistance in grain legumes. Crit Rev Plant Sci 14(6):469–523
Sulpice R, Tsukaya H, Nonaka H, Mustardy L, Chen TH, Murata N (2003) Enhanced formation of flowers in salt-stressed *Arabidopsis* after genetic engineering of the synthesis of glycine betaine. Plant J 36:165–176
Sultana R, Vales MI, Saxena KB, Rathore A, Rao S, Rao SK, Myer M, Kumar RV (2012) Water-logging tolerances in pigeonpea (*Cajanus cajan* L. Millsp.). Genotypic variability and identification of tolerant genotypes. J Agric Sci :1–13. doi: 10.1017/S0021859612000755, Published online
Summerfield RJ, Hadley P, Roberts EH, Minchin FR, Rawsthrone S (1984) Sensitivity of chickpea (*Cicer arietinum* L.) to hot temperatures during the reproductive period. Exp Agric 20:77–93. doi:10.1017/S0014479700017610
Talukdar D (2013) Studies on antioxidant enzymes in *Cana indica* plant under copper stress. J Environ Biol 34:93–98
Tanksley SD, Nelson JC (1996) Advanced backcross QTL analysis: a method for the simultaneous discovery and transfer of valuable QTLs from unadapted germplasm into elite breeding lines. Theor Appl Genet 92:191–203
Turner NC (1986) Adaptation to water deficits: a changing perspective. Aust J Plant Physiol 13(1):175–190
Vadez V, Krishnamurthy L, Serraj R, Gaur PM, Upadhyaya HD, Hoisington DA, Varshney RK, Turner NC, Siddique KHM (2007) Large variation in

chickpea is explained by differences in sensitivity at the reproductive stage. Field Crops Res 104:123–129
Vales MI, Srivastava RK, Sultana R, Singh S, Singh I, Singh G, Patil SB, Saxena KB (2012) Development of new determinate and non-determinate super-early pigeonpea (*Cajanus cajan* L. Millspaugh) lines. Crop Sci 52:1–10
Varshney RK, Mohan MS, Gaur PM, Gangarao NVPR, Pandey MK, Bohra A, Sawargaonkar SL, Chitikineni A, Kimurto PK, Janila P, Saxena KB, Fikre A, Sharma M, Rathore A, Pratap A, Tripathi S, Datta S, Chaturvedi SK, Mallikarjuna N, Anuradha G, Babbar A, Choudhary AK, Mhase MB, Bharadwaj C, Mannur DM, Harer PN, Guo B, Liang X, Nadarajan N, Gowda CLL (2013) Achievements and prospects of genomics-assisted breeding in three legume crops of the semi arid tropics. Biotechnol Adv. http://dx.doi.org/10.1016/j.biotechadv.2013.01.0
Wang J, Gan YT, Clarke F, McDonald CL (2006) Response of chickpea yield to high temperature stress during reproductive development. Crop Sci 46:2171–2178
Willekens H, Chamnongpol S, Davey M, Schraudner M, Langebartels C, Van Montagu M, Inze D, Van Camp W (1997) Catalase is a sink for H_2O_2 and is indispensable for stress defence in C3 plants. EMBO J 16:4806–4816
Wu G, Wei ZK, Shao HB (2007) The mutual responses of higher plants to environment: physiological and microbiological aspects. Biointerfaces 59:113–119
Zagorchev L, Seal CE, Kranner I, Odjakova M (2013) A central role for thiols in plant tolerance to abiotic stress. Int J Mol Sci 14:7405–7432

Photosynthesis and Associated Aspects Under Abiotic Stresses Environment

Kiran P. Bhagat, R. Arun Kumar, Pasala Ratnakumar, Satish Kumar, S.K. Bal, and P.K. Agrawal

Abstract

Abiotic stresses are the prime reason of crop loss worldwide, reducing average yields for most of the major crop plants by more than 50 %. Plants as sessile organisms are persistently exposed to changes in environmental conditions. When these changes are swift and extreme, plants generally perceive them as stresses. However stresses are not necessarily a problem for plants because they have evolved effective mechanisms to avoid or reduce the possible damages. The response to changes in environment can be rapid, depending on the type of stress, and can involve adaptation mechanisms, which allow them to survive the adverse conditions. Extreme environmental conditions, such as high and low temperatures, waterlogging and deficits, salinity, and carbon dioxide (CO_2) and ozone (O_3) concentrations at the leaf surface strongly affect plant growth and development. Such abiotic stresses adversely affect on physiological mechanisms associated with plant responses, adaptation, and tolerance to stresses in terms of photosynthetic mechanisms, such as CO_2 diffusion through stomatal control, photosystem II repair, ribulose bisphosphate carboxylase/oxygenase (Rubisco) activity, and generation of reactive oxygen species (ROS), are susceptible to damage that causes great diminution in photosynthetic efficiency. Therefore, photosynthesis is one of the key processes to be affected by abiotic stresses, which results in decrease in CO_2 diffusion to the chloroplast and metabolic constraints. Although several structural and functional components of the photosynthetic apparatus are responsive to abiotic stresses, photosystem II (PS II) and Rubisco act as the major stress sensors. In addition, it is essential to systematize current knowledge on the complex network of interactions

K.P. Bhagat (✉) • P. Ratnakumar • S. Kumar • S.K. Bal
National Institute of Abiotic Stress Management (ICAR),
Malegaon, Baramati, Pune, Maharashtra 413115, India
e-mail: kbhagat@niam.res.in

R.A. Kumar (✉) • P.K. Agrawal
Vivekananda Parvatiya Krishi Anusandhan Sansthan
(I.C.A.R), Almora, Uttarakhand 263601, India
e-mail: agri_arun@rediffmail.com

R.K. Gaur and P. Sharma (eds.), *Approaches to Plant Stress and their Management*,
DOI 10.1007/978-81-322-1620-9_10, © Springer India 2014

and regulation of photosynthesis in plants exposed to abiotic stresses. In this chapter, we brought the update knowledge emphasizing on the regulation of photosynthesis and associated aspects that are affected by various abiotic stresses.

Keywords
Photosynthesis • Temperature • Water stress • Salinity • Elevated CO_2 and O_3 • Rubisco • Carbon metabolism

Introduction

Abiotic stress is an integral part of "climate change," a complex phenomenon with a wide range of unpredictable impacts on the environment. Abiotic stress causes changes in soil-plant-atmosphere continuum and is responsible for reduced yield in several major crops. Abiotic stress is already a major limiting factor in plant growth and will soon become even more severe as desertification covers more and more of the world's terrestrial area (Vinocur and Altman 2005). Therefore, the subject of abiotic stress response in plant metabolism, productivity, and sustainability is gaining considerable significance in the contemporary world. Prolonged exposure to these abiotic stresses results in altered metabolism and damage to biomolecules. Plants evolve defense mechanisms to tolerate these stresses by upregulation of osmolytes, osmo-protectants, enzymatic and nonenzymatic antioxidants, etc. Plants are subjected to various abiotic stresses because of unavoidable environmental conditions which adversely affect their growth and development and trigger a series of morphological, physiological, biochemical, and molecular changes. Plants show a range of responses and adaptations that help bring about abiotic stress tolerance. Some of these involve structural or chemical changes, while others involve restriction of the growing period according to conditions. To survive in several abiotic stresses, symbiotic relationships have developed by the plants as a response to stresses.

Plants have evolved to live in environments where they are often exposed to different stress factors in combination. Being sessile, they have developed specific mechanisms that allow them to detect precise environmental changes and respond to complex stress conditions, minimizing damage while conserving valuable resources for growth and reproduction. Plants activate a specific and unique stress response when subjected to a combination of multiple stresses (Rizhsky et al. 2004). Photosynthesis is the most important process in the world which involves a chain of events where light energy is converted into chemical energy by the plants through chemical reactions with water and carbon dioxide. As leaf cells absorb light, the electrons attached to water molecules become excited and move into a higher energy state. When this happens, water molecules are split. The splitting of water molecules releases oxygen into the atmosphere and also releases high-energy hydrogen atoms for use. From this point, cellular respiration processes manufacture ATP and NADPH. These two materials combine with hydrogen and carbon dioxide molecules to produce the sugar molecules that become food for the plant.

During the onset and development of abiotic stresses within a plant, all the major processes such as photosynthesis, protein synthesis, energy, and lipid metabolism are affected. Photosynthesis is the source of organic carbon and energy required by plants for their growth, biomass production, and yield. Therefore, we have discussed in this chapter under the separate subheadings regarding the available knowledge on regulation of photosynthesis under different abiotic stresses, such as changes in temperatures, water stress conditions, salinity, elevated CO_2 and O_3, and its associated aspects.

Photosynthesis Under Change in Temperatures

Photosynthesis and transpiration share a common pathway through the stomatal opening regulated by the guard cells. Although these are independent processes, huge amount of water is lost during photosynthetic periods. This high water loss rate also removes heat from leaves and keeps them relatively cool under full sunlight conditions. Since photosynthesis is a temperature-dependent process, therefore, it is important to remember this linkage between two processes influenced by the degree of stomatal opening. Temperature affects all biochemical reactions of photosynthesis as well as membrane integrity in chloroplasts, so it is not surprising that the responses to temperature are complex. Here, we will discuss the photosynthesis under high and low temperatures in separate subheadings.

I] Photosynthesis Under High Temperatures

There are several factors which are associated with the decline in photosynthesis due to high temperature:

(a) Optimal temperature
(b) Respiration rates
(c) Membrane-bound electron transport processes

Out of these, optimal temperature is considered as the primary reason for the sharp decrease in net photosynthesis at high temperatures (Taiz and Zeiger 2002). Optimal temperature is the point at which the capacities of the various steps of photosynthesis are optimally balanced, with some of the steps becoming limiting as the temperature decreases or increases. Optimal temperatures have strong genetic and environmental components. The chlorophyll fluorescence from leaves increases at high temperatures and causes irreversible damage to the photosynthetic apparatus as indicated by instability of the rate of net CO_2 exchange at supraoptimal temperatures (Taiz and Zeiger 2002).

High temperatures adversely affect plant growth and survival in a number of ways, but the impact of heat stress on the photosynthetic apparatus is considered to be of particular significance because photosynthesis is often inhibited before other cell functions are impaired (Berry and Björkman 1980). High temperatures affect photosynthesis by altering the excitation energy distribution by changing the structure of thylakoids (Weis and Berry 1988) and by changing the activity of the Calvin cycle and other metabolic processes such as photorespiration and product synthesis. Much of the attention has been focused on the former aspect, since thylakoids are highly sensitive to heat, whereas the restrictions imposed by high temperatures on carbon metabolism have been nearly exclusively interpreted as the effect on CO_2, availability. The diffusion of CO_2 and O_3 and the affinity for carboxylation of the Rubisco enzyme have been proven to be affected by increasing temperatures (Brooks and Farquhar 1985).

In general, under conditions of high solar irradiance, leaf temperature can increase several degrees centigrade above air temperature (Leakey et al. 2003), which further enhances the problem of high-temperature-dependent reduction of carbon assimilation. Components of the thylakoid membranes have for some time been recognized as being particularly sensitive to heat stress (Yordanov et al. 1986). For instance, heat stress results in the loss of grana stacking due to the dissociation of peripheral light-harvesting complexes from the core complex (Gounaris et al. 1984). Actually, photosystem II (PS II) has been identified as one of the most thermolabile components of the photosynthetic electron transport chain (Srivastava et al. 1997), whereas photosystem I (PS I) has been demonstrated to be comparatively heat resistant (Havaux 1996). Reduction of PS II activity by heat stress primarily results from an inactivation of the oxygen-evolving complex (Enami et al. 1994). A reversible increase in the permeability of the thylakoid membranes leading to proton leakage has been reported to occur at temperatures lower than those at which PS II activity gets inhibited (Bukhov et al. 1999). Besides affecting the membranes, heat stress has been shown to inhibit the export of photo-assimilates (Jiao and Grodzinski 1996). The heat-induced alterations described above, which in the light usually appear at rather high temperatures and are not rapidly reversible (Karim et al. 1999), are thought to arise because of changes in lipid-protein interactions that are associated with increased lipid fluidity of the thylakoid membranes at elevated temperatures (Yordanov et al. 1986). The activity of fully activated Rubisco is robust against high-temperature-dependent inhibition (Crafts-Brandner and Law 2000). Rubisco activase, a stromal enzyme, plays an essential role in the process of Rubisco activation (Portis 2003).

Inhibition of CO_2 assimilation by high temperature occurs under both photorespiratory and non-photorespiratory conditions, indicating that the phenomenon cannot be simply explained by the greater rate of photorespiration at high temperatures. It originates from changes related to the different solubility of CO_2 and O_2 and the kinetic properties of Rubisco (Sage and Sharkey 1987). Rubisco activase, as judged by measuring the thermal stability of its ATP hydrolysis activity, is particularly sensitive to inactivation by high temperature (Eckhardt and Portis 1997). Under moderate heat stress conditions, Rubisco activation is reduced because the activity of activase is insufficient to keep pace with the faster deactivation of Rubisco at high temperatures. Hence, the heat labile character of Rubisco activation has been attributed to the thermal sensitivity of activase. The variations induce due to changes in temperature in the activity of enzymes are not the same for all of the enzymes acting in a particular pathway. Changes that occur in the properties of enzymes caused by changes in temperature contribute to the regulation of carbohydrate metabolism (Stitt and Grosse 1988).

II] Photosynthesis Under Low Temperatures

Photosynthetic process is a thermosensitive function in higher plants (Havaux 1993). The primary effect of low temperatures on photosynthesis is apparent within some important processes, namely, inactivation of Calvin cycle enzymes through oxidation in the stroma (Wise 1995), delay in circadian rhythm of sucrose phosphate synthase activity in the cytosol (Jones et al. 1998), and oxidative damage to PS I and PS II reaction centers (Kudoh and Sonoike 2002). Low temperatures interfere with the photosynthetic process in several ways. They lower stomatal conductance, photochemical efficiency of the photosystem (PS) II, thylakoid electron transport rate, enzyme activity and carbon metabolism, as well as the photosynthetic pigment complex systems (Suzuki et al. 2008). The photosynthesis is strongly reduced below 18 °C (Ramalho et al. 2003), while temperatures around 4 °C dramatically depress photosynthetic performance and yield (Silva et al. 2004). However, after a period of exposure to low temperature, the photosynthetic assimilation acclimates to low temperatures which results in an increase in the capacity of photosynthesis (Berry and Björkman 1980).

At low temperatures, photosynthesis can also be limited by other factors such as phosphate availability at the chloroplast (Sage and Sharkey 1987). If the rate of triose phosphate utilization in the cytosol decreases, phosphate uptake into the chloroplast is inhibited and photosynthesis becomes phosphate limited (Geiger and Servaites 1994). Starch synthesis and sucrose synthesis decrease rapidly with temperature, reducing the demand for triose phosphates and causing the phosphate limitation at low temperatures.

Under the normal turnover conditions, oxidative damage to D1 (reaction center protein of PS II) comes along with the repair process. When the rate of repair keeps the pace with the rate of damage, then there is no inhibition of electron flow. Under such conditions there is a maximum potential for photosynthesis (light-saturated rates). Chilling can also decrease the D1 repair rates. Any decrease in CO_2 fixing or in the repair rate of D1 protein can shift the maximum light intensities needed for photosynthesis saturation to a much lower intensities (Kyle and Ohad 1986).

Photosynthesis Under Water Stress

I] Photosynthesis Under Waterlogging Conditions

Stomata closure, reduction of transpiration, and inhibition of photosynthesis are common responses that can occur in hours or days, depending on the tolerance to waterlogging of individual plants. When waterlogging is prolonged, waterlogging-susceptible plants drastically reduce their physiological activities and are often killed in a short time (Malik et al. 2001; Zaidi et al. 2004), whereas in waterlogging-tolerant plants, the same parameters could even be enhanced or have less effect due to the ability of roots to acclimate to waterlogging, such as by the ability to produce adventitious roots and aerenchyma formation (Li et al. 2007; Mollard et al. 2008). Stomatal conductance is the major factor affecting photosynthesis under waterlogging conditions in plants (Striker et al. 2005). In addition, factors regulating photosynthesis in plants grown in waterlogged soil may be reduced CO_2 transfer conductance from substomatal cavities to the site of carboxylation, or activity of photosynthetic enzymes at the carboxylation point.

A relevant limitation to photosynthesis under waterlogging conditions may be caused by a reduced capacity of Rubisco for CO_2 fixation, not translocation of CO_2 such as that occurring in drought-stressed plants (Massacci et al. 1996). Reduction of Rubisco is closely related to total soluble protein content and is logically related to total leaf nitrogen content (Irving et al. 2007). The remobilization of nitrogen from older plant parts to shoot for supporting shoot growth during waterlogging causes a reduction in photosynthetic rate (Mollard et al. 2008).

The leaf is also very receptive to waterlogging; respiration changes in the leaf, leaf chlorophyll content, and photosynthetic assimilation have been detected during a waterlogging period (Parolin 2000). In particular, chlorophyll fluorescence characteristics are usually influenced under waterlogging conditions. Under such conditions, reduction in the maximum quantum yield of photosystem II (Fv/Fm) after the onset of waterlogging has occurred in plants (Smethurst et al. 2005). This decline may be induced by the lowered stomatal conductance (Lawlor and Cornic 2002), altered hormonal status (Salisbury and Ross 2009), and disordered mineral nutrient uptake (Castonguay et al. 1993). Furthermore, these factors can also change other indices of chlorophyll fluorescence, such as electron transport, non-photochemical quenching (NPQ), and photochemical quenching (Smethurst et al. 2005) in waterlogging conditions.

II] Photosynthesis Under Water-Deficit Conditions

Water stress is one of the most limiting environmental factors to plant productivity worldwide and can be caused by both soil and atmospheric water deficits. In case of C_3 plants, the photosynthesis is negatively affected by water stress measured as changes in leaf water potential or relative water content. In the early phase of water stress, when leaf RWC is still greater than 70 %, the decline in CO_2 assimilation rates is largely the result of reduced intercellular CO_2 concentration due to decreased stomatal conductance. Under these conditions, maximal photosynthetic capacity and quantum yield remain unaffected when measured under saturating irradiance and CO_2 concentration. In addition, photosynthetic inhibition usually recovers relatively quickly when plants are rehydrated. If water stress persists and leaf RWC falls below 70 %, the loss of photosynthetic activity becomes increasingly less responsive to high CO_2, and assimilation rate fails to recover to prestress values following the removal of water stress. The exact mechanisms underlying this non-stomatal phase, also termed metabolic inhibition, are diverse and less well understood (Flexas et al. 2004). C_4 plants make a significant contribution to the global carbon budget, and C_4 crops, such as maize and sorghum, are pivotal to current and future global food security (Pingali 2001).

The limitation of plant growth imposed by low water availability is mainly due to reductions in plant carbon balance, which is dependent on the balance between photosynthesis and respiration. Since both the processes are intimately linked, therefore, it is imperative to increase our knowledge on the physiological bases of regulation of photosynthesis and respiration under water scarcity in order to be able to improve plant yield in semiarid regions and to face the climate change-driven increase in global aridity. Of the total carbon assimilated in photosynthesis, usually more than half is lost in respiratory processes necessary for growth and maintenance, but this balance may change under water stress. For instance, although photosynthesis may decrease up to 100 % becoming totally impaired under severe water shortage, the respiration rate may either increase or decrease under stress, but may never become totally impaired (Flexas et al. 2005).

If soil water becomes scarce, plant water status worsens, leading to cascading effects that can be severely adverse at both the leaf and plant scales (Porporato et al. 2001). To reduce these risks, plants under water deficit reduce transpirational water losses by reducing stomatal conductance. Moreover, water stress hinders leaf internal transport of CO_2, enzyme activity, and hence photosynthetic capacity. These so-called internal (metabolic and diffusive) limitations become predominant relative to stomatal limitation as water stress further increases (Ghannoum 2009; Lawlor and Tezara 2009). The complexity of these stomatal controls and feedbacks on leaf carbon assimilation and water loss poses a major challenge to modeling stomatal and internal limitations on photosynthesis under water stress.

Photosynthesis Under Saline Conditions

Salinity is one of the most widespread environmental threats to global crop production, especially in arid and semiarid climates, where land degradation, water shortage, and population growth are already a major concern (Geissler et al. 2010). Salinity has many direct and indirect effects on photosynthetic processes. Photosynthetic rates are usually lower in plants exposed to salinity and especially to sodium chloride. It is yet to clear whether these low photosynthesis rates are responsible for the reduced growth in salinized plants or if stunted plants control assimilation through a negative feedback of a reduced sink activity. Salt effects on photosynthetic processes fall into two major categories:

(i) Stomatal closure, the usual response of stomata to salinization of salt-sensitive plants
(ii) Effects on the capacity for CO_2 fixation apart from the altered diffusion limitations

On the other hand, the reduction in the photosynthesis rate of plants exposed to salinity usually depends on two aspects of salinization (Kalaji and Nalborczyk 1991):

(i) The total concentration of salts (osmotic effect)
(ii) Their ionic composition (specific ion effect)

In arid and semiarid regions, insufficient precipitation necessitates extensive irrigation, resulting in salinization problems. High salt concentrations accumulating in the soil solution create high osmotic potentials, which reduce the availability of water to the plants. Salinity causes chloroplasts to aggregate and leads to ultrastructural changes of the assimilating organs (Glagoleva et al. 1992). These include dilatation of thylakoid membranes, almost no sign of grana, and enlarged mesophyll cells (Mitsuya et al. 2000).

Turgor in salt-stressed plants is maintained mainly by means of the accumulation of organic and inorganic solutes in plant organs, usually leaves, through osmoregulation. The synthesis of these organic solutes, which are required for osmoregulation, requires sources of carbon and energy, derived mainly from photosynthesis. Sodium chloride increases CO_2 fixation into malate and decreases it into aspartate in plants (Murumkar and Chavan 1993).

The accumulation of compatible solutes (sucrose, proline, and glycinebetaine) contributes to decreases in leaf osmotic potential under sodium chloride stress, allowing plants to keep a positive cell turgor by continuing water uptake so that seedlings can grow under salt stress. Short-term adaptation to stress is primarily linked to stomatal regulation by reducing water loss by transpiration and maximizing CO_2 uptake; this tends to lead to a constant ratio between transpiration and photosynthesis. Long-term adaptation includes changes in biomass allocation, specific anatomical modifications, or sophisticated physiological mechanisms.

Photosynthesis Under Elevated CO_2 and O_3

CO_2 and tropospheric O_3 are two of the most abundant greenhouse gases in the atmosphere, and the concentrations of those gases are still increasing since the initial of industrial period as a consequence of anthropogenic activities and, principally, as a result of fossil fuel combustion, being nowadays major contributors to the greenhouse effect (IPCC 2007).

I] Photosynthesis Under Elevated CO_2 Concentrations

The concentration of CO_2 gas in the atmosphere is very trace, presently accounting for about 0.038 %, or 380 parts per million (ppm), of the air. The current atmospheric concentration of CO_2 is almost twice the concentration that has prevailed during most of the last 420,000 years, as measured from air bubbles trapped in glacial ice in Antarctica. Today's atmospheric CO_2 is likely higher than Earth has experienced in the last 2 million years (Taiz and Zeiger 2002). The current CO_2 concentration of the atmosphere is increasing by about 1–3 ppm each year, primarily because of the burning of fossil fuels such as coal oil and natural gas. CO_2 concentration in the atmosphere is changing in a regular manner. The major cause of this change is the emissions from fossil fuel combustion, despite there are other contributing factors as land-use change, and as these emissions are likely to continue, an increase in concentrations is expected with predicted values between 540 and 970 $\mu mol\ mol^{-1}$ by the year 2100 (IPCC 2007).

The short-term stimulation of photosynthesis to elevated CO_2 is well documented (Drake et al. 1997). There are two different reasons as to why short-term increases in CO_2 will have effects on Rubisco and thus photosynthesis:

(i) Since the atmospheric CO_2 of today is substrate limiting to Rubisco, the carboxylation rate will increase in elevated CO_2.
(ii) The net CO_2 uptake efficiency will increase as a result of inhibition of the oxygenation reaction.

The CO_2 lost in photorespiration will decrease, and a larger part of the energy achieved from the light reactions as ATP and NADPH will be used for assimilation instead of photorespiration (Long et al. 2004).

Photosynthetic acclimation can be defined as "a physiological change that occurs with growth at high CO_2," and the implication of this is that after medium- or long-term growth in elevated CO_2, the initial stimulation of net photosynthesis from elevated CO_2 might be reduced (Drake et al. 1997). There are uncertainties regarding both the carboxylation capacity and the electron transport capacity in future atmospheric CO_2. The elevated CO_2 concentrations in the boundary layer atmosphere of leaves may cause changes in stomatal aperture are water availability, light, and CO_2 concentration (Ainsworth and Rogers 2007). Stomata close in response to low water availability in the air or soil and in most species open in response to light and close in the dark. Stomatal conductance decreases in response to short-term increase in CO_2, while long-term effects on stomatal conductance from growth in elevated CO_2 are on the other hand more unclear and variable (Morison 2001).

Further, the short-term responses result from changes in stomatal aperture, and long-term responses could be physiological (acclimations) or anatomical/morphological (adjustments) (Morison 1998). Stomatal response to CO_2 and the way this response affects photosynthesis and transpiration will have effects on plant water regulation and growth, since virtually all of the CO_2 used by the plant passes through stomata. These responses might have influences on climate change through changes in the hydrological and carbon cycles (Morison 2001).

CO_2 enters in contact with plants through stomata, which respond to different internal and environmental stimuli, as light, humidity, or CO_2 (Paoletti and Grulke 2005). Guard cells sense CO_2 concentrations and react by changing their turgor pressure. These changes mediate the closure of stomata. Stomata respond to intercellular rather than to leaf surface CO_2 concentrations, but the biochemical and physiological mechanisms involved in this response are not well understood (Ainsworth and Rogers 2007; Assmann 1999). The degree to which the stomata are open determines stomatal conductance and, therefore, the resistance that atmospheric CO_2 or O_3 encounters to diffuse to sub-stomatal cavities. From there, CO_2 will diffuse to the active sites of Rubisco, and in this step, internal conductance plays a determinant role (Warren et al. 2003). At the active site of Rubisco, CO_2 is bound to ribulose-1, 5-bisphosphate (RubP), producing two molecules of 3-phosphoglyceric acid (Ainsworth and Rogers 2007). This reaction, in which Rubisco fixates CO_2, constitutes the first step of the Calvin cycle and is known as carboxylation. Rubisco is also able to bind O_2, in the reaction known as oxygenation which means that both substrates compete for the active site of the enzyme; approximately every third RubP molecule binds O_2 (Ainsworth and Rogers 2007). Rubisco is not saturated at the current atmospheric CO_2 concentrations, given its low affinity for CO_2 (Warren 2008). Hence, an increase in atmospheric CO_2 would result in higher photosynthetic efficiency derived from higher carboxylation rate and competitive inhibition of the oxygenation reaction (Long et al. 2004). Photosynthesis enhancement is a well-known effect of elevated carbon dioxide on plants, which leads to increased primary productivity and plant growth (IPCC 2001; Long et al. 2004). Photosynthesis enhancement results from a higher maximum carboxylation rate of Rubisco and lower photorespiration. The greater photosynthetic efficiency allows more growth and thus increased leaf area, which again potentiates greater productivity, and water use is also improved, supporting enhanced growth (Long et al. 2004).

II] Photosynthesis Under Elevated O_3 Concentrations

O_3 is present in the stratosphere, where it has a protective role absorbing ultraviolet radiation. In the troposphere, however, it behaves as a very reactive compound, resulting phytotoxic at certain levels (Finlayson-Pitts and Pitts 1997). Tropospheric O_3 is assumed to be the most important air pollutant in many parts of the world due to its potential threat to crops, (semi)natural grassland, and forest ecosystems (Mills et al. 2011). The concentration of O_3 has increased globally by about 36 % since preindustrial times (IPCC 2007) and is currently increasing at an annual rate of 0.5–2.0 % on a global scale (Vingarzan 2004). In terms of exposure, the global forest surface that experiences O_3 concentration over 60 ppb rose from 9 % in 1950 to around 25 % in 1990 and is expected to reach about 50 % by 2100. A 60 ppb concentration represents a level above the average background and can be considered damaging to vegetation (Fowler et al. 1999). O_3 has been shown to have phytotoxic properties (Percy et al. 2002), and nearly a quarter of the world's forests are currently at risk of damage (at O_3 more than 60 ppb) and reduced productivity, and by 2100 this will expand to half of the world's forests (Fowler et al. 1999). O_3 may cause significant damage to photosynthesis because of its strong oxidizing effects (Reich 1987) and decreases in photosynthesis as a consequence of loss of Rubisco activity (Farage et al. 1991). Ozone of today, approximately 40 ppb, suppresses net photosynthesis by on average 11 % compared with preindustrial O_3 levels (10 ppb), whereas stomatal conductance suppresses by on average 13 % compared to preindustrial O_3 levels (Wittig et al. 2007). Although the stomatal closure is a well-investigated response to O_3, there are uncertainties whether this is a direct response on stomata or an indirect response from the effects on photosynthesis (Robinson et al. 1997). The decrease in stomatal aperture is an indirect effect as a result of decreasing carboxylation capacity and hence declining intercellular CO_2 (Farage et al. 1991).

The formation of O_3 occurs naturally, but anthropogenic activities enhance the concentrations by the emission of precursors as volatile organic compounds, carbon monoxide, and nitrogen oxides (Finlayson-Pitts and Pitts 1997). The main source of nitrogen oxides is the combustion of fossil fuels, which are mostly released in the form of NO (Jenkin and Clemitshaw 2000). In the troposphere, nitric oxide reacts with O_3 to produce nitrogen dioxide and oxygen. By photolysis, nitrogen dioxide gives nitric oxide and nascent oxygen (1/2 O_2^-), able to bind to oxygen and thus yielding O_3. The O_3 that is produced in this way could react with nitric oxide to close the cycle, but in the presence of volatile organic compounds, radicals produced by photooxidation as the hydroperoxy and organic peroxy radicals (HO_2/RO_2) react with nitric oxide, impeding the consumption of O_3 and thus leading to increased concentrations in the troposphere (Jenkin and Clemitshaw 2000). O_3 is responsible for different injury symptoms and detrimental responses in plants, but they vary according to species, leaf and tree age, or growth conditions among other factors (Percy et al. 2002). In addition, environmental interactions modulate directly and indirectly these effects (Skärby et al. 1998). Therefore, the particular characteristics of the affected plant will determine the type and magnitude of the effects.

O_3 present at the ground level enters the leaf mainly through stomata. The uptake rate is determined by boundary layer and stomatal resistances. Once it has entered the leaf and it contacts a surface, it reacts by oxidizing another chemical compounds and producing free radicals (Percy et al. 2002). The principal consequences of O_3 exposure are stomatal closure and declined net photosynthesis. The decline in net photosynthesis is attributed to reduced rate of Rubisco and Rubisco content and stomatal closure (Percy et al. 2002). Stomatal closure is thought to be indirect effects of O_3, caused by increased internal CO_2 concentrations as photosynthesis is impaired (Reich 1987).

Chronic exposure to elevated O_3 significantly decreased net photosynthetic rates, accelerated leaf senescence, and increased dark respiration in many tree species, such as deciduous temperate tree species, evergreen Mediterranean trees (Paoletti 2009). O_3-induced reduction in photosynthesis rate has been attributed to two factors: stomatal and non-stomatal limitation. A reduction in the rate of Rubisco is considered to be the primary effect of O_3 on photosynthesis in plants, which induces stomatal closure via an increase in intercellular CO_2 concentration (Yamaguchi et al. 2007), although the stomatal regulation is considered as a very important factor in controlling the ozone sensitivity of plants (Novak et al. 2005) and O_3 uptake (Paoletti and Grulke 2005). O_3 can also alter photosynthetic process at the level of the electron transport, such as decreases in leaf chlorophyll content, reduction in the efficiency of excitation capture in plants, reduced numbers of intact or open photosystem II reaction centers, and increases in dissipation of energy through alternative means such as heat (Ryang et al. 2009).

Associated Factors with Photosynthesis

I] Rubisco

Protein synthesis and protein degradation are equally important for changes in the protein pattern and are of fundamental importance for the normal development, homeostasis, and final death of a plant cell (Vierstra 1996). Proteolysis in plants is a complex process involving many enzymes and multifarious proteolytic pathways in various cellular compartments (Grudkowska and Zagdańska 2004). Therefore, proteolysis is an important process in maintaining functional chloroplasts under optimal and stress conditions. Chloroplasts are a major site of protein degradation during senescence (Mae et al. 1984). Rubisco is the most abundant protein on earth and contributes a high percentage to the total leaf nitrogen in C_3 plants (Feller et al. 2008). During early stages of senescence, Rubisco accounts for about 90 % of the degraded proteins (Miller and Huffaker 1985).

The most abundant protein, Rubisco, catalyzes the assimilation of CO_2 by carboxylation of ribulose-1, 5-bisphosphate in photosynthetic carbon assimilation (Ellis 1979). However, the catalytic limitations of Rubisco compromise the efficiency of photosynthesis (Parry et al. 2007). Compared to other enzymes of the Calvin cycle, Rubisco has a low turnover number, meaning that relatively large amounts must be present to sustain sufficient rates of photosynthesis. Furthermore, Rubisco also catalyzes a competing and wasteful reaction with oxygen, initiating the process of photorespiration, which leads to a loss of fixed carbon and consumes energy. A net degradation of Rubisco and other chloroplast proteins endogenously initiated leaf senescence as well as during or after abiotic stress phases and allows the reutilization of the nitrogen in other organs after the transfer via the phloem (Thoenen et al. 2007). Not only the velocity of Rubisco degradation but also the mechanisms involved may depend on the environmental conditions (Feller et al. 2008).

Regulating Rubisco activity is essential to match the capacity for ribulose-1, 5-bisphosphate regeneration with the prevailing rate of RuBP utilization. This is not achieved solely by the availability of substrate since Rubisco in excess of that needed to sustain photosynthesis in the prevailing environment is deactivated (Sage et al. 1990). The continual process of Rubisco protein turnover (a function of synthesis, maintenance, and degradation) that represents a drain on cellular resources is uncertain, but on account of its abundance could be considerable, particularly in the presence of a relatively oxidizing environment induced by stress (Feller et al. 2008). In the light conditions, under which promote heat and cold stress, an increase in the active oxygen species in the chloroplast is likely to cause increased oxidative damage to thylakoid-bound and stromal proteins. Stress-induced oxidative modifications of specific residues on Rubisco mark the enzyme for degradation (Moreno et al. 2008).

II] Carbon Metabolism

Photosynthetic processes have direct impact on the carbon metabolisms. Abiotic stresses limit the photosynthetic capacity and growth of plants by disconcerting the carbon metabolism. Under such abiotic stresses, plants achieve a new balance between carbohydrate accumulations and export (Martindale et al. 1997). The carbon export is declined from the leaf (which itself may also be particularly sensitive to low temperature), resulting in an accumulation of soluble carbohydrate in leaves (Pollock 1984). One advantage of this accumulation of soluble sugars is that they exert a cryoprotective function in cells exposed to low temperature (Koster and Lynch 1992). What remains unclear is whether these changes in carbohydrate content result from imbalances between synthesis and export or whether they result from increases in carbohydrate synthesis. Similarly, the capacities of other processes may increase during photosynthesis acclimation. If photosynthetic acclimation is largely concerned with the removal of limitations in the ability to synthesize sucrose, then this acclimation should be more evident at higher CO_2 concentrations.

Feedback inhibition of photosynthesis as a result of decreased sink demand and the accumulation of sucrose or sugar phosphate intermediates in source leaves is a long-known phenomenon. Assimilates function as a link between source-sink tissues. Plants both produce and utilize carbohydrates and have developed mechanisms to regulate their sugar status and coordinate carbohydrate partitioning. High sugar levels result in a feedback inhibition of photosynthesis and an induction of storage processes. Thus, a negative correlation exists between photosynthesis and accumulation of soluble photosynthates in source leaves (Iglesias et al. 2002).

Regulation of Photosynthesis Under Abiotic Stress Conditions

A group of genes controlled by a certain type of transcription factors (TF) is known as a regulation, and the overexpression of CBF/DREB1 regulation in Arabidopsis plants showed with improved survival rates when exposed to salt, drought, and low temperatures (Jaglo-Ottosen et al. 1998; Kasuga et al. 1999). Along with this, improved stress tolerance by overexpression of CBF/DREB1 genes is associated with sustained photochemical efficiency and photosynthetic capacity as compared with wild-type plants (Hsieh et al. 2002; Savitch et al. 2005; Oh et al. 2007).

The overexpression of STZ represses several genes involved in photosynthesis and related metabolism in *Arabidopsis*. This means that STZ factor may be involved in growth retardation through repression of photosynthesis and carbohydrate metabolism genes observed in both the wild-type plants under abiotic stress and plants overexpressing CBF/DREB1 genes (e.g., CBF3/DREB1A). Analyzing all the promoters of the genes related to photosynthesis and carbohydrate metabolism and search for cis-elements where STZ may bind would be another thrust area in this regard (Sakamoto et al. 2004). Overexpression of CBF/DREB1 genes in transgenic plants under the control of RD2A stress-inducible promoter can be enhanced abiotic stress tolerance without totally compromising the yield (Kasuga et al. 2004; Pino et al. 2007).

Conclusion

Climate change is an ongoing phenomenon that invites various abiotic stress conditions to plants and affects their life cycle by changing the morphological, physiological, biochemical, and molecular processes. Plants get affected by their regular metabolism under either influence of single abiotic stress or combinations of two or more alterations in the photosynthetic activities in plants under abiotic stress conditions that hinder their growth, development, and yield. Stomatal closure plays by far the main role in the decline in leaf photosynthesis and photosynthetic machinery remains intact, thereby allowing the leaf to respond rapidly to changes in vapor pressure deficit. Increase in the concentrations of CO_2 and O_3 in the atmospheric air causes several effects on photosynthetic processes; altered photosynthetic capacity and stomatal conductance are likely to affect both the ability of plants to sequester carbon and plant water use, which in turn can affect whole carbon and hydrological cycles. The decrease in photosynthesis may be related to several factors:

- Reduction of the CO_2 supply because of hydro-active closure of the stomata
- Changes in the leaf temperature required for photosynthesis
- Dehydration of cell membranes, which reduce their permeability to CO_2
- Salt toxicity
- Enhanced senescence induced by salinity
- Changes in enzyme activity induced by changes in cytoplasmic structure
- Negative feedback by reduced sink activity

Rubisco is an utmost enzyme for photosynthetic processes, and its degradation depends on the intensity of abiotic stresses. Under abiotic stress conditions, stomatal closure imposed a diffusive limitation on photosynthesis by decreasing the availability of CO_2 for assimilation by Rubisco. In addition, the reduced capacity of evaporative cooling raised the leaf/canopy temperatures, imposing a metabolic limitation on photosynthesis through the inactivation of Rubisco. The various steps in Rubisco degradation, the proteolytic enzymes involved, and the subcellular compartmentation of these processes under various abiotic stress conditions remain to be elucidated in the future perspective. Other than Rubisco, different mechanisms might be relevant under various stress conditions. Although, combinations of multiple abiotic stresses hamper life cycle of plants, plants evolve themselves at least to survive in adverse abiotic stress conditions.

Future Research Perspectives to Improve Photosynthesis and Productivity of Crop Plants

Abiotic stresses in plants suppress them to yield at their optimum. Manipulation of the C_3 cycle offers an opportunity to increase photosynthesis and yield. It is now important that this knowledge is fully exploited in crops. Clearly, the reduction of the Rubisco oxygenase reaction remains a target for future improvement of photosynthesis. However, the current strategies that might be exploited to achieve this goal are conceptually straightforward, and all of these approaches will be technically demanding, requiring fundamental research to identify the genes involved. Improvement of the C_3 cycle is not just about increasing CO_2 fixation but should also aim to increase both nitrogen use efficiency and water use efficiency while maintaining high productivity. Therefore manipulation of the C_3 cycle to improve these parameters is also an important goal. The range of genetic and molecular techniques that are now available, together with the development and application of rapid in vivo techniques to allow infield analysis of a wider range of species in their natural environments, will facilitate the wider analysis of natural variation in photosynthetic carbon assimilation. This approach has enormous and unexplored potential for future exploitation to improve yield through manipulation of the C_3 cycle (Raines 2011). In considering opportunities to increase photosynthesis (Long et al. 2006), photosynthetic research can summarized some identified targets for hopefully changing the efficiency of photosynthetic light conversion:

- Rubisco with decreased oxygenase activity and with efficient or increased catalytic activity
- C_4 photosynthesis engineered into C_3 plants for C, N, and S assimilation
- Improved plant photosynthesis architecture deeper into crop canopies
- Increased photoprotection both in the amounts of protection and the recovery of photosynthesis
- Increased capacity to regenerate RuBP, e.g., by overexpression of SBPase or Rubisco 5-P kinase
- Engineering focused on the activities and role of light-sensitive PS enzymes: ribulose 5-P kinase, SBPase, FBPase, and Gly 3-P DH, NADPH dependent, in all green tissues
- Engineering of metabolic and PS activities to increase sink strength, especially in green, non-leaf, sources, and sinks, e.g., glumes, awns, pods, seeds, and fruits

References

Ainsworth EA, Rogers A (2007) The response of photosynthesis and stomatal conductance to rising CO_2: mechanisms and environmental interactions. Plant Cell Environ 30:258–270

Assmann SM (1999) The cellular basis of cell guard sensing of rising CO_2. Plant Cell Environ 22:629–637

Berry J, Björkman O (1980) Photosynthetic response and adaptation to temperature in higher plants. Annu Rev Plant Physiol 31:491–543

Brooks A, Farquhar GD (1985) Effect of temperature on the CO/O, specificity of ribulose-1,5-bisphosphate carboxylase/oxygenase and the rate of respiration in the light. Planta 165:397406

Bukhov NG, Wiese C, Neimanis S, Heber U (1999) Heat sensitivity of chloroplasts and leaves: leakage of protons from thylakoids and reversible activation of cyclic electron transport. Photosynth Res 59:81–93

Castonguay Y, Nadeau P, Simard RR (1993) Effects of flooding on carbohydrate and ABA levels in roots and shoots of alfalfa. Plant Cell Environ 16:695–702

Crafts-Brandner SJ, Law RD (2000) Effect of heat stress on the inhibition and recovery of the ribulose-1,5-bisphosphate carboxylase/oxygenase activation state. Planta 212:67–74

Drake BG, Gonzàlez-Meler MA, Long SP (1997) More efficient plants: a consequence of rising atmospheric CO_2. Annu Rev Plant Physiol Plant Mol Biol 48:609–639

Eckhardt NA, Portis AR Jr (1997) Heat denaturation profiles of ribulose-1,5-bisphosphate carboxylase/oxygenase (Rubisco) and Rubisco activase and the inability of Rubisco activase to restore activity of heat-denaturated Rubisco. Plant Physiol 113:243–248

Ellis RJ (1979) The most abundant protein in the world. Trends Biochem Sci 4:241–244

Enami I, Kitamura M, Tato T, Isokawa Y, Ohta H, Kato S (1994) Is the primary cause of thermal inactivation of oxygen evolution in spinach PS II membranes release of the 33 kDa protein or of Mn? Biochim Biophys Acta 186:52–58

Farage PK, Long SP, Lechner EG, Baker NR (1991) The sequence of change within the photosynthetic apparatus of wheat following short-term exposure to ozone. Plant Physiol 95:529–535

Feller U, Anders I, Demirevska K (2008) Degradation of rubisco and other chloroplast proteins under abiotic stress. Rubisco degradation under stress. Gen Appl Plant Physiol 34(1–2):5–18, Special Issue

Finlayson-Pitts BJ, Pitts JN Jr (1997) Tropospheric air pollution: ozone, airborne toxics, polycyclic aromatic hydrocarbons, and particles. Science 276:1044–1052

Flexas J, Bota J, Loreto F, Cornic G, Sharkey TD (2004) Diffusive and metabolic limitations to photosynthesis under drought and salinity in C_3 plants. Plant Biol 5:1–11

Flexas J, Galmés J, Ribas-Carbó M, Medrano H (2005) The effects of drought in plant respiration. In: Lambers H, Ribas-Carbó M (eds) Advances in photosynthesis and respiration: plant respiration, from cell to ecosystem. Kluwer Academic, Dordrecht, pp 85–94

Fowler D, Cape NJ, Coyle M, Flechard C, Kuylenstierna J (1999) The global exposure of forests to air pollutants. Water Air Soil Pollut 116:5–32

Geiger DR, Servaites JC (1994) Diurnal regulation of photosynthetic carbon metabolism in C_3 plants. Annu Rev Plant Physiol Plant Mol Biol 45:235–256

Geissler N, Hussin S, Koyro HW (2010) Elevated atmospheric CO_2 concentration enhances salinity tolerance in *Aster tripolium* L. Planta 231:583–594

Ghannoum O (2009) C_4 photosynthesis and water stress. Ann Bot 103:635–644

Glagoleva TA, Chulanovskaya MV, Pakhomova MV, Voznesenskaya EV, Gamalei YV (1992) Effect of salinity on the structure of assimilating organs and ^{14}C labelling patterns in C_3 and C_4 plants of Ararat plain. Photosynthetica 26:363–369

Gounaris K, Brain AP, Quinn PJ, Williams WP (1984) Structural reorganization of chloroplast thylakoid membranes in response to heat stress. Biochim Biophys Acta 766:198–208

Grudkowska M, Zagdańska B (2004) Multifunctional role of plant cysteine proteinases. Acta Biochim Pol 51:609–624

Havaux M (1993) Characterization of thermal damage to the photosynthetic electron transport system in potato leaves. Plant Sci 94:19–33

Havaux M (1996) Short-term responses of Photosystem I to heat stress. Photosynth Res 47:85–97

Hsieh TH, Lee JT, Yang PT (2002) Heterology expression of the *Arabidopsis* C-repeat/dehydration response element binding factor 1 gene confers elevated tolerance to chilling and oxidative stresses in transgenic tomato. Plant Physiol 129:1086–1094

Iglesias DJ, Lliso I, Tadeo FR, Talon M (2002) Regulation of photosynthesis through source: sink imbalance in citrus is mediated by carbohydrate content in leaves. Physiol Plant 116:563–572

IPCC (2001) The scientific basis: contributions of working group 1 to the fourth assessment report of the IPCC. Available at www.grida.no/publications/other/ipcc_tar/?src=/CLIMATE/IPCC_TAR/WG1/index.htm

IPCC (2007) Climate change 2007. Contribution of working group I to the fourth assessment report of the intergovernmental panel on climate change. Available at http://www.ipcc.ch/

Irving LJ, Sheng YB, Woolley D, Matthew C (2007) Physiological effects of waterlogging on two lucerne varieties grown under glasshouse conditions. J Agron Crop Sci 193:345–356

Jaglo-Ottosen KR, Gilmour SJ, Zarka DG, Schabenberger O, Thomashow MF (1998) *Arabidopsis* CBF1 overexpression induces COR genes and enhances freezing tolerance. Science 280:104–106

Jenkin ME, Clemitshaw KC (2000) Ozone and other secondary photochemical pollutants: chemical processes governing their formation in the planetary boundary layer. Atmos Environ 34:2499–2527

Jiao J, Grodzinski B (1996) The effect of leaf temperature and photo-respiratory conditions on export of sugars during steady-state photosynthesis in *Salvia splendens*. Plant Physiol 111:169–178

Jones TL, Tucker DE, Ort DR (1998) Chilling delays circadian pattern of sucrose phosphate synthase and nitrate reductase activity in tomato. Plant Physiol 118:149–158

Kalaji H, Nalborczyk E (1991) Gas-exchange of barley seedlings growing under salinity stress. Photosynthetica 25:197–202

Karim MA, Fracheboud Y, Stamp P (1999) Photosynthetic activity of developing leaves of *Zea mays* is less affected by heat stress than that of developed leaves. Physiol Plantarum 105:685–693

Kasuga M, Liu Q, Miura S, Yamaguchi-Shinozaki K, Shinozaki K (1999) Improving plant drought, salt, and freezing tolerance by gene transfer of a single stress-inducible transcription factor. Nat Biotechnol 17:287–291

Kasuga M, Miura S, Shinozaki K, Yamaguchi-Shinozaki K (2004) A combination of the Arabidopsis *DREB1A* gene and stress-inducible *rd29A* promoter improved drought- and low-temperature stress tolerance in tobacco by gene transfer. Plant Cell Physiol 45:346–350

Koster KL, Lynch DV (1992) Solute accumulation and compartmentation during the cold acclimation of Puma rye. Plant Physiol 98:108–113

Kudoh H, Sonoike K (2002) Irreversible damage to photosystem I by chilling in the light: cause of the degradation of chlorophyll after returning to normal growth temperature. Planta 215:541–548

Kyle DJ, Ohad I (1986) The mechanism of photoinhibition in higher plants and green algae. In: Staehelin LA, Arntzen CJ (eds) Encyclopedia of plant physiology. New Series, vol 19. Springer-Verlag, Berlin, pp 468–475

Lawlor DW, Cornic G (2002) Photosynthetic carbon assimilation and associated metabolism in relation to water deficits in higher plants. Plant Cell Environ 25:275–294

Lawlor DW, Tezara W (2009) Causes of decreased photosynthetic rate and metabolic capacity in water-deficient leaf cells: a critical evaluation of mechanisms and integration of processes. Ann Bot 103:561–579

Leakey ADB, Press MC, Scholes JD (2003) High-temperature inhibition of photosynthesis is greater under sun flecks than uniform irradiance in a tropical rain forest tree seedling. Plant Cell Environ 26:1681–1690

Li M, Yang D, Li W (2007) Leaf gas exchange characteristics and chlorophyll fluorescence of three wetland plants in response to long-term soil flooding. Photosynthetica 45:222–228

Long SP, Ainsworth EA, Rogers A, Ort DR (2004) Rising atmospheric carbon dioxide: plants FACE the future. Annu Rev Plant Physiol Plant Mol Biol 55:591–628

Long SP, Zhu XG, Naidu SL, Ort DR (2006) Can improvement in photosynthesis increase crop yields? Plant Cell Environ 29:315–330

Mae T, Kai N, Makino A, Ohira K (1984) Relation between ribulose bisphosphate carboxylase content and chloroplast number in naturally senescing primary leaves of wheat. Plant Cell Physiol 25:333–336

Malik AI, Colmer DTD, Lambers H, Schortemeyer M (2001) Changes in physiological and morphological traits of roots and shoots of wheat in response to different depth of waterlogging. Aust J Plant Physiol 28:1121–1131

Martindale W, Parry MAJ, Andralojc PJ, Keys AJ (1997) Synthesis of 2-carboxy-D-arabinitol-1-phosphate in French bean (*Phaseolus vulgaris* L.): a search for precursors. J Exp Bot 48:9–14

Massacci A, Battistelli A, Loreto F (1996) Effect of drought stress on photosynthetic characteristics, growth and sugar accumulation of field grown sweet sorghum. Aust J Plant Physiol 23:331–340

Miller BL, Huffaker RC (1985) Differential induction of endoproteinases during senescence of attached and detached barley leaves. Plant Physiol 78:442–446

Mills G, Hayes F, Simpson D, Emberson L, Norris D, Harmens H, Buker P (2011) Evidence of widespread effects of ozone on crops and (semi-) natural vegetation in Europe (1990–2006) in relation to AOT40- and flux-based risk maps. Global Change Biol 17:592–613

Mitsuya S, Takeoka Y, Miyake H (2000) Effects of sodium chloride on foliar ultrastructure of sweet potato (*Ipomoea batatas* Lam.) plantlets grown under light and dark conditions in vitro. J Plant Physiol 157:661–667

Mollard FOP, Striker GG, Polschuk EL, Vega AS, Insausti P (2008) Flooding tolerance of *Paspalum dilatatum* (*Poaceae: Paniceae*) from upland and lowland positions in a natural grassland. Flora 203:548–556

Moreno J, Garcâia-Murria MJ, Marâin-Navarro J (2008) Redox modulation of Rubisco conformation and activity through its cysteine residues. J Exp Bot 59 (7):1605–1614

Morison JIL (1998) Stomatal response to increased CO_2 concentration. J Exp Bot 49:443–452

Morison JIL (2001) Increasing atmospheric CO_2 and stomata. New Phytol 149:154–156

Murumkar CV, Chavan PD (1993) Alternations in photosynthetic carbon metabolism of chickpea (*Cicer arietinum* L.) due to imposed NaCl salinity. Agrochimica 37:26–32

Novak K, Schaub M, Fuhrer J, Skelly JM, Hug C, Landolt W, Bleuler P, Kräuchi N (2005) Seasonal trends in reduced leaf gas exchange and ozone-induced foliar injury in three ozone sensitive woody plant species. Environ Pollut 136:33–45

Oh SJ, Kwon CW, Choi DW, Song SI, Kim JK (2007) Expression of barley HvCBF4 enhances tolerance to abiotic stress in transgenic rice. Plant Biotechnol J 5:646–656

Paoletti E (2009) Ozone and mediterranean ecology: plants, people, problems. Environ Pollut 157:1397–1398

Paoletti E, Grulke NE (2005) Does living in elevated CO_2 ameliorate tree response to ozone? A review on stomatal responses. Environ Pollut 137:483–493

Parolin P (2000) Phenology and CO_2-assimilation of trees in central Amazonian floodplains. J Trop Ecol 16:465–473

Parry MAJ, Madgwick PJ, Carvahlo JFC, Andralojc PJ (2007) Prospects for increasing photosynthesis by overcoming the limitations of Rubisco. J Agric Sci 145:31–43

Percy KE, Awmach CS, Lindroth RL, Kubiske ME, Kopper BJ (2002) Altered performance on forest pest under atmospheres enriched by CO_2 and O_3. Nature 420:403–407

Pingali PI (2001) CIMMYT 1999–2000 facts and trends. Meeting the world maize needs: technological opportunities and priorities for the public sector. CIMMYT, Mexico City

Pino MT, Skinner JS, Park EJ (2007) Use of a stress inducible promoter to drive ectopic AtCBF expression improves potato freezing tolerance while minimizing negative effects on tuber yield. Plant Biotechnol J 5:591–604

Pollock CJ (1984) Sucrose accumulation and the initiation of fructan biosynthesis in *Lolium temulentum* L. New Phytol 96:527–532

Porporato A, Laio F, Ridolfi L, Rodriguez-Iturbe I (2001) Plants in water-controlled ecosystems: active role in hydrologic processes and response to water stress – III vegetation water stress. Adv Water Res 24:725–744

Portis AR Jr (2003) Rubisco activase – Rubisco's catalytic chaperone. Photosynth Res 75:11–27

Raines CA (2011) Increasing photosynthetic carbon assimilation in C_3 plants to improve crop yield: current and future strategies. Plant Physiol 155:36–42

Ramalho JC, Quartin VL, Leitao E, Campos PS, Carelli MLC, Fahl JI, Nunes MA (2003) Cold acclimation ability and photosynthesis among species of the tropical *Coffea* genus. Plant Biol 5:631–641

Reich PB (1987) Quantifying plant response to ozone: a unifying theory. Tree Physiol 3:63–91

Rizhsky L, Liang HJ, Shuman J, Shulaev V, Davletova S, Mittler R (2004) When defense pathways collide. The response of *Arabidopsis* to a combination of drought and heat stress. Plant Physiol 134:1683–1696

Robinson MF, Heath J, Mansfield TA (1997) Disturbances in stomatal behaviour caused by air pollutants. J Exp Bot 49:461–469

Ryang SZ, Woo SY, Kwon SY, Kim SH, Lee SH, Kim KN, Lee DK (2009) Changes of net photosynthesis, antioxidant enzyme activities, and antioxidant contents of *Liriodendron tulipifera* under elevated ozone. Photosynthetica 47:19–25

Sage RF, Sharkey TD (1987) The effect of temperature on the occurrence of O_2 and CO_2 insensitive photosynthesis in field grown plants. Plant Physiol 84:658–664

Sage RF, Sharkey TD, Seemann JR (1990) Regulation of ribulose-1,5-bisphosphate carboxylase activity in response to light intensity and CO_2 in the C3 annuals

Chenopodium album L. and *Phaseolus vulgaris* L. Plant Physiol 94:1735–1742

Sakamoto H, Maruyama K, Sakuma Y (2004) Arabidopsis Cys2/His2-type zinc-finger proteins function as transcription repressors under drought, cold, and high-salinity stress conditions. Plant Physiol 136:2734–2746

Salisbury FB, Ross CW (2009) Photosynthesis: environmental and agricultural aspects. In: Plant physiology. Wadsworth Publishing, Belmont, pp 249–265

Savitch LV, Allard G, Seki M (2005) The effect of overexpression of two Brassica CBF/DREB1-like transcription factors on photosynthetic capacity and freezing tolerance in *Brassica napus*. Plant Cell Physiol 46:1525–1539

Silva EA, Damatta FM, Ducatti C, Regazzi AJ, Barros RS (2004) Seasonal changes in vegetative growth and photosynthesis of *Arabica coffee* trees. Field Crops Res 89:349–357

Skärby L, Ropoulsen H, Wellburn FAM, Sheppard LJ (1998) Impacts of ozone on forests: a European perspective. New Phytol 139:109–122

Smethurst CF, Garnett T, Shabala S (2005) Nutritional and chlorophyll fluorescence responses of lucerne (*Medicago sativa*) to waterlogging and subsequent recovery. Plant Soil 270:31–45

Srivastava A, Guissé B, Greppin U, Strasser RJ (1997) Regulation of antenna structure and electron transport in photosystem II of *Pisum sativum* under elevated temperature probed by the fast polyphasic chlorophyll a fluorescence transient: OKJIP. Biochim Biophys Acta 1320:95–106

Stitt M, Grosse H (1988) Interactions between sucrose synthesis and CO_2 fixation. IV. Temperature-dependent adjustment of the relation between sucrose synthesis and CO_2 fixation. J Plant Physiol 133:392–400

Striker GG, Insausti P, Grimoldi AA, Ploschuk EL, Vasellati V (2005) Physiological and anatomical basis of differential tolerance to soil flooding of *Lotus corniculatus* L. and *Lotus glaber* Mill. Plant Soil 276:301–311

Suzuki K, Nagasuga K, Okada M (2008) The chilling injury induced by high root temperature in the leaves of rice seedlings. Plant Cell Physiol 49:433–442

Taiz L, Zeiger E (2002) Photosynthesis: physiological and ecological considerations. In: Plant physiology, 3rd edn. Sinauer Associates, Sunderland, pp 171–192

Thoenen M, Herrmann B, Feller U (2007) Senescence in wheat leaves: is a cysteine endopeptidase involved in the degradation of the large subunit of Rubisco? Acta Physiol Plant 29:339–350

Vierstra RD (1996) Proteolysis in plants: mechanisms and functions. Plant Mol Biol 32:275–302

Vingarzan R (2004) A review of surface ozone background levels and trends. Atmos Environ 38:3431–3442

Vinocur B, Altman A (2005) Recent advances in engineering plant tolerance to abiotic stress: achievements and limitations. Curr Opin Biotechnol 16(2):123–132

Warren CR, Dreyer E and Adams MA (2003) Photosynthesis– Rubisco relationships in foliage of *Pinus sylvestris* in response to nitrogen supply and the proposed role of Rubisco and amino acids as nitrogen stores. Trees 17:359–366

Warren CR (2008) Stand aside stomata, another actor deserves centre stage: the forgotten role of the internal conductance to CO_2 transfer. J Exp Bot 59:1475–1487

Weis E, Berry JA (1988) Plants and high temperature stress. In: Long SP, Woodward FI (eds) Symposia of the society for experimental biology number XLII. The Company of Biologist, Cambridge, pp 329–346

Wise RR (1995) Chilling-enhanced photooxidation: the production, action and study of reactive oxygen species produced during chilling in the light. Photosynth Res 45:79–97

Wittig VE, Ainsworth EA, Long SP (2007) To what extent do current and projected increases in surface ozone affect photosynthesis and stomatal conductance of trees? A meta-analytic review of the last 3 decades of experiments. Plant Cell Environ 30:1150–1162

Yamaguchi M, Watanabe M, Iwasaki M, Tabe C, Matsumura H, Kohno Y, Izuta T (2007) Growth and photosynthetic responses of *Fagus crenata* seedlings to O_3 under different nitrogen loads. Trees 21:707–718

Yordanov I, Dilova S, Petkova R, Pangelova T, Goltsev V, Süss KH (1986) Mechanisms of the temperature damage and acclimation of the photosynthetic apparatus. Photobiochem Photobiophys 12:147–155

Zaidi PH, Rafique S, Rai PK, Singh NN, Srinivasan G (2004) Tolerance to excess moisture in maize (*Zea mays* L.): susceptible crop growth stage and identification of tolerant genotypes. Field Crops Res 90:189–202

The Effects of Drought on Plant Communities in the Desert Rangelands of Tunisia

Mouldi Gamoun and Jamila Zammouri

Abstract

Plant communities of the Dahar plateau are characteristic of the Tunisian desert both in structure and in dynamics. Although a number of plant communities can be differentiated, four major vegetation types are often distinguished that differ in plant species cover and composition, as well as other factors, such as soil types and elevation. Among the abiotic factors which affect the vegetation structure and diversity of plants, climate is probably the most relevant. The desert vegetation in southern Tunisia is in a state of change and the most debilitating risk is that of drought in these desert areas. Under protection from grazing the dynamic nature of this vegetation is affected by such conditions such as drought. Precipitation probably explains part of the vegetation response to drought. The effects of drought stress on vegetation were tested for four plant communities who are *Stipagrostis pungens*, *Anthyllis sericea*, *Helianthemum kahiricum,* and *Hammada schmittiana*. Patterns of plant response to drought differed among the four vegetation types considered. Vegetation cover, species richness, and diversity were used for the characterization of the considered vegetation. Main results show that plant cover, richness, and diversity change with vegetation type and rainfall variations. Vegetation cover on *H. kahiricum* steppe is more affected by drought than the other steppes. Plant diversity is affected by drought in all plant community and mainly on the *A. sericea* and *H. kahiricum* steppes.

Keywords

Drought stress • Plant communities • Management • Desert rangeland

Introduction

Desert lands throughout the world are often degraded or increasingly at risk of degradation. These lands, including those at the border of arid

M. Gamoun (✉) • J. Zammouri
Laboratory of Range Ecology, Arid Regions Institute (IRA), Medenine 4119, Tunisia
+216 97 277 824
e-mail: gamoun.mouldi@yahoo.fr

R.K. Gaur and P. Sharma (eds.), *Approaches to Plant Stress and their Management*,
DOI 10.1007/978-81-322-1620-9_11, © Springer India 2014

regions, commonly exhibit accelerated soil erosion, losses of productivity, and impaired economic potential to support human populations. Desertification involves human and environmental drivers of change but is a regional symptom that emerges from degradation at finer spatial scales (Reynolds and Stafford 2002). Desertification does not describe cyclic phenomena, as when decadal variations of precipitation lead to periods of drought and to losses of vegetation cover that are fully restored when rains return (Tucker et al. 1994). Natural disturbances and often previously unappreciated human disturbances became a serious challenge to traditional succession beginning as early as the 1940s (e.g., Stearns 1949; Raup 1957), but really solidifying in 1970s. The increasing recognition of the pervasiveness of disturbances (White 1979) led to idea that ecological systems consisted of patches of different times since the last disturbance.

Dynamic changes in vegetation level are controlled by the balance between climatic and anthropogenic factor. Human activity has greatly accelerated the rate at which species have disappeared from the earth (May et al. 1995). The anthropogenic origin of phenomenon was underlined by many studies (Auclair et al. 1999): demographic pressure, overgrazing, and clearing of the ligneous family.

The primary causes of vegetation destruction on arid land and indirect causes of desertification are the clearing of land unsuitable for cultivation, prolonged overstocking and overgrazing, and the destruction of woody species by excessive firewood (Le Houérou 2009). In addition to the anthropogenic factors, climate influences a variety of ecological processes (Stenseth et al. 2002). These effects operate through local weather parameters such as temperature, wind, rain, snow, and ocean currents and the interactions occurring among them. There have been several recent studies on the impact of large-scale climatic force on ecological systems.

The relationships between climate, soil, vegetation, and other components of arid ecosystems have been described by many authors, including Shreve (1942), Hassib (1951), Tadros (1953), Vernet (1955), Kassas (1955), Chapman (1960), Zohary (1962), Batanouny (1973), Younes et al. (1983), Ayyad and El-Ghareeb (1984), Evenary et al. (1985), Zahran and Willis (2008), and Le Houérou (2009).

The Mediterranean bioclimates are characterized by winter rains and summer drought (Le Houérou 2005a, b). However, droughts can occur anywhere, and this simple statement tells us nothing about what constitutes a drought (Allaby 2003). In parts of North Africa, a drought occurs when no rain has fallen for at least 2 years. Perhaps, then, a drought is a period during which rainfall is insufficient to meet the needs of plants. Droughts meeting this definition have led to ongoing debates about desertification (Mainguet 1995; Thomas 1997).

Millions of people in rangelands depend directly on livestock for their livelihoods, but the management of these regions remains mired in controversy (Gillson and Hoffman 2007). The sustainable management of natural resources demands controls negating land degradation and desertification (Ganry and Campbell 1993). Although restoration models and practices have previously been applied to ecosystems, there is now a more recent focus on the "landscape perspective" of ecosystem restoration to improve nature conservation and management effectiveness (Moreira et al. 2006).

Among rare rehabilitation experiments, the one carried out by Le Floc'h et al. in 1999 in southern Tunisia enabled the reconstitution of a badly degraded steppe. In addition to this technique being generally beneficial to vegetation cover and species diversity (Gamoun et al. 2010a), it was found here that although sandy soil is more productive than limestone soil, the latter is more resistant to animal trampling (Gamoun et al. 2010b).

This chapter describes the effects of drought on the vegetation of the major plant community types of the desert rangelands in Tunisia from 2007 to 2009, with emphasis on cover, species richness, and diversity.

Study Area and Data Collection

Studies reported here were carried over 3 years on four protected rangelands, which are located in southern Tunisia, forming a collective steppe unit of the plate of Dahar to the south of Tataouine. The climate is characterized by hot, dry summers and cool, mild winters, so according to Emberger (1954), it has a Saharan superior Mediterranean bioclimate.

The studies reported here can be divided into four different types of plant communities designated since 2007. Soil type influences brush growth in the distribution and determines the type of plant community. These soils were grouped into sand accumulation, limestone, loam, or sandy soil. Consequently, soil type is the key determinant in defining plant community. These plant communities are distributed as follows:

- *Stipagrostis pungens* [(Desf.) de Winter] community on sand accumulation
- *Anthyllis sericea* [Lag. subsp. *henoniana* (Coss.) Maire] community on limestone soil
- *Helianthemum kahiricum* [(Desf.)] community on loamy soil
- *Hammada schmittiana* [(Pomel) Ilji] community on sandy soil

These plant communities are exploited by individual, and although overgrazing can occur, they are largely composed of all plants that have resisted or benefitted from grazing and drought (Gamoun 2005). On these rangelands each individual had an incentive to increase the number of animals, and no individual was entitled to prevent access to others (Hardin 1968).

Since each plant communities, three 20-m-long transects, were set up to measure Total Plant Cover using the quadrats point method (Jauffret and Visser 2003). Observations were made every 20 cm, providing a total of 100 points in each transect. The total plant cover is determined by the formula $TPC = (n/N) \times 100$, with n: number of points where the vegetation is present and N: numbers of total points in each transect (100 points in our case).

Shannon information index (H′; Shannon and Weaver 1948) is the most widely used measure of diversity in plant communities (Stirling and Wilsey 2001; Pélissier et al. 2003). Shannon-Wiener diversity index combines richness and relative abundance of plant species. It is calculated from centesimal frequencies of species (*fi*): $H' = -\Sigma\ ((fi/N) * \log2\ (fi/N))$, where *fi* is the number of *i*th species in the samples and N is the total number of species in the plant community.

A two-way ANOVA was used to test differences in the effects of vegetation types and years. The obtained data is subjected to several statistical analyses using SPSS for windows software v. 11.5 (SPSS Inc. 2002).

Results and Discussion

Vegetation cover, diversity, and species richness were assessed in March each year (2007, 2008 and 2009). The following classes of data were analyzed for each vegetation type: (1) cover, (2) species richness, and (3) diversity.

Weather Conditions

Rainfalls during the period from 1987 to 2009 averaged 75 mm/year (Fig. 1). The temporal rainfall distribution in Tataouine is eminently variable, with annual rainfall irregularly being either side of the norm. Various anomalies include considerable reduction in rainfall in 1987, 1998, 1998, 1998, 1999, 2001, and 2009 and contrasting periods of excessive humidity (1990, 1994, 1995, and 2002).

From 2003, there was a marked tendency in annual rainfall to constantly remain lower than the average precipitation established over one long period.

These rainy events in southern Tunisia are characterized by their great variability and very irregular distribution. During the past 20 years, several droughts have affected the south and caused significant losses, mainly in agricultural sectors, and these have endangered farms specializing in livestock. The research community perceives the apparent desertification as a transient response to reduced rainfall (Tucker et al. 1991, 1994). A number of studies have

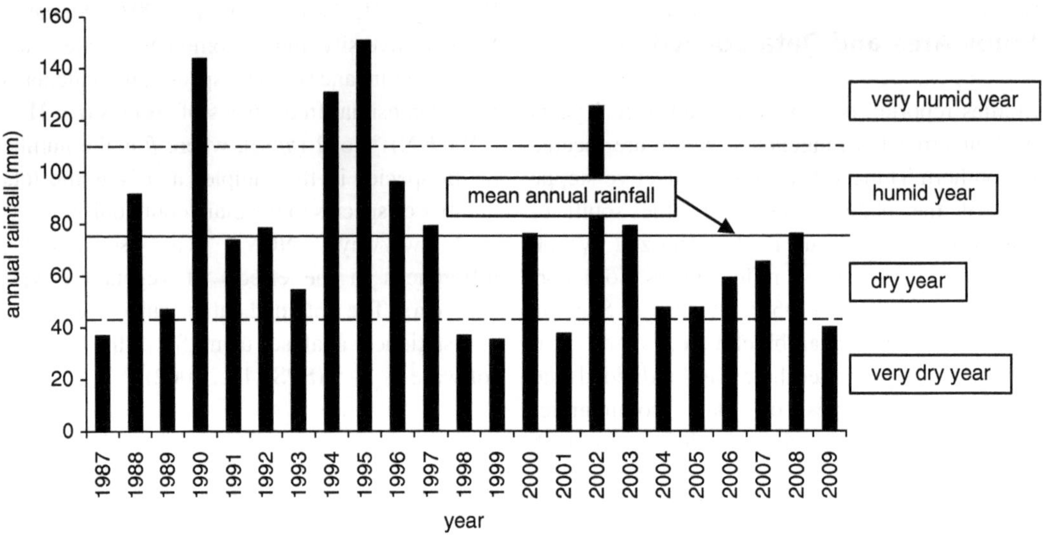

Fig. 1 Average annual rainfall in Remada region between 1987 and 2009

Fig. 2 Climatogram plans showing temperature and precipitation in Remada

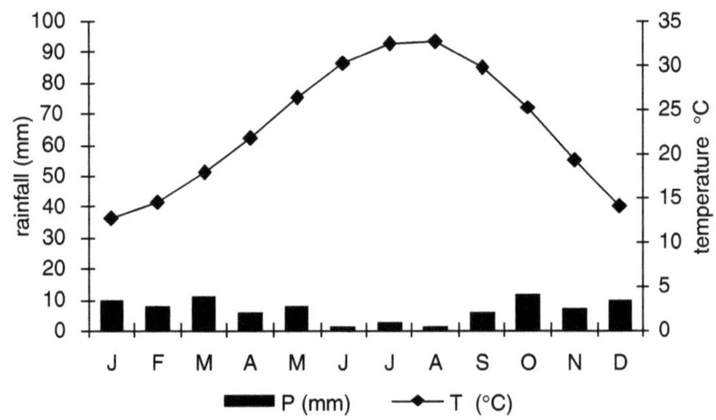

shown that degradation to this land is reversible: with vegetation restored when rain returns.

Temperatures exceeded 30 °C for 60 days, and the maximum exceeded 40 °C for 10 days. The average annual heat-under-shelter varied between 22 °C and 23 °C, while the winters were cold registering between 11 °C and 12 °C in January.

The rains are spread over part of the year, with a relative variability of rainfall above 95 % and a Martonne index of aridity of approximately 2–3. The most important rains generally fell in winter, spring, and autumn, with the heaviest at 11.57 mm in October. The minimum rainfall of 0 mm has occurred in summer, in July (Fig. 2).

During summer, temperatures are pleasant with 32 °C in July and August and 30 °C in June. Maximum temperatures are expected to increase with the growing season which is segmented into rain events separated by drought periods. If two successive years are dry, the probability of a third dry year in the south ranges from 14 % to 17 % (Benzarti and Habaieb 2001)

Trend in Vegetation Cover

Based on the ANOVA analysis, results showed significant vegetation type effect ($F = 9.14$; $p < 0.001$) and year effect ($F = 9.362$; $p = 0.001$) for vegetation cover. The vegetation type and year interaction was also significant ($F = 2.713$; $p = 0.037$). This interaction

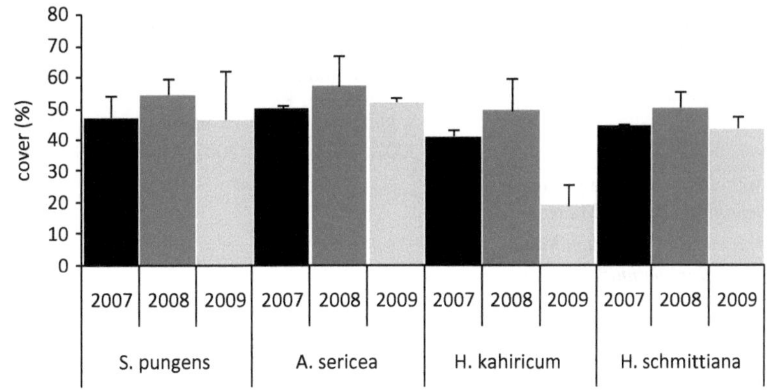

Fig. 3 Mean of vegetation cover of four plant community types in southern Tunisia. Cover is significantly different among vegetation type and year ($p < 0.05$). *Vertical bars* represent standard errors

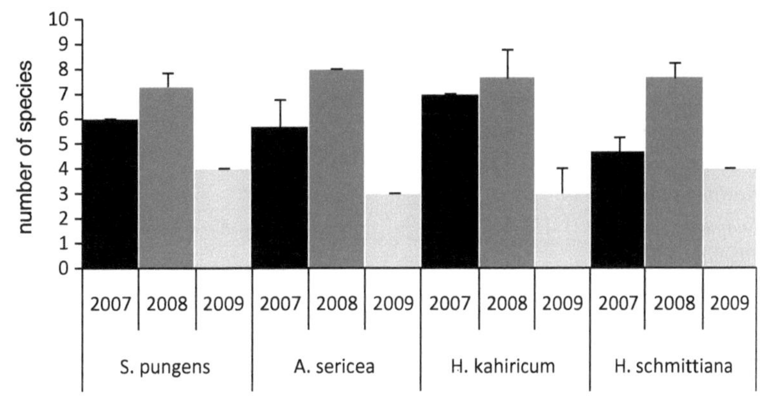

Fig. 4 Mean of plant species richness of four plant community types in southern Tunisia according to time after protection (2006). The effect of year is highly significant ($p < 0.05$), but the effect of vegetation type is not significant ($p > 0.05$). *Vertical bars* represent standard errors

suggests that available water content depends on the soil type.

Figure 3 gives the participation in vegetation composition of the four plant community types from year 2007 to year 2009 after protection. The vegetation covers measured on the *S. pungens* and *A. sericea* steppes are the highest followed by the *H. schmittiana* and the *H. kahiricum* steppes. The impact of drought on plant cover is important and varied with plant community type. For example, on the *H. kahiricum* steppe, the vegetation covers increase from 42 % to 50 % in wet year (2008) and then dropped to 19 % in dry year (2009), whereas on the other plant community types, the decrease does not exceed 9 %.

Trend in Species Richness

Community composition is further modified by the occurrence of rainfall through time and vegetation type. In this chapter, ANOVA results showed significant year effects ($F = 134.571$; $p < 0.001$) and no significant vegetation type effects ($F = 0.952$; $p = 0.431$), but showed significant interaction between the two factors tested year*vegetation type ($F = 4.667$; $p = 0.003$). During the first year of protection (2007) with low rainfall, plant species richness ranged from five species on the *H. schmittiana* and *S. pungens* steppes to seven species on the *H. kahiricum* steppe. During the rainy year (2008), species richness increased and reached eight species on all vegetation community. In 2009 species richness decreased with decreased rainfall, but *S. pungens* and *H. schmittiana* steppes remained richer than *H. kahiricum* and *A. sericea* steppes (Fig. 4). The distributions of species suggest that plant species respond to the variation of available water which in turn depends on vegetation type. Table 1 gives the total species richness in the four plant community types.

Table 1 Plant species occurrence in four plant community types during the 3 years of protection (Presence of each species in each sampling plant community during the three years of treatment is symbolized by (*), absent species is symbolized by (–))

	S. pungens			*A. sericea*			*H. kahiricum*			*H. schmittiana*		
	2007	2008	2009	2007	2008	2009	2007	2008	2009	2007	2008	2009
Anthyllis sericea Lag. subsp. *henoniana* (Coss.) Maire	–	–	–	*	*	*	–	–	–	–	–	–
Anabasis oropediorum Maire	–	–	–	–	–	–	*	*	–	–	–	–
Argyrolobium uniflorum (Decne.) Jaub. & Spach	–	–	–	–	*	–	*	*	–	–	–	–
Atractylis serratuloides Sieber ex Cass	–	–	–	*	*	*	*	*	*	–	–	–
Calligonum comosum L'Herit	–	*	–	*	*	*	–	–	–	–	*	
Cutandia dichotoma (Forssk.) Trab.	*	*	*	–	*	–	–	–	–	*	*	–
Daucus syrticus Murb.	–	–	–	*	*	*	–	–	–	–	–	–
Fagonia glutinosa Delile	–	–	–	–	–	–	–	*	–	–	–	–
Gymnocarpos decander Forssk.	–	–	–	–	–	–	*	*	*	–	–	–
Hammada schmittiana (Pomel) Ilji	*	*	*	*	*	*	*	*	*	*	*	*
Helianthemum kahiricum (Desf.)	–	–	–	–	–	–	*	*	*	–	–	–
Herniaria fontanesii J. Gay	–	–	–	–	–	–	*	*	*	–	–	–
Koelpinia linearis Pall.	–	–	–	*	*	–	–	*	–	–	–	–
Launaea resedifolia (L.) O. Kuntze	–	–	–	–	–	–	–	*	–	–	–	–
Plantago albicans L.	–	–	–	–	–	–	–	*	–	–	–	–
Polygonum equisetiforme Sibth. & Sm.	*	*	–	–	–	–	–	–	–	–	–	–
Retama raetam (Forssk.) Webb	*	*	*	–	–	–	–	–	–	–	–	–
Salsola vermiculata L.	–	–	–	–	–	–	–	–	–	*	*	*
Savignya parviflora (Del.) Webb	–	–	–	*	*	*	*	–	–	–	–	–
Schismus barbatus (L.) P. Beauv.	*	*	–	–	–	–	–	*	*	*	*	*
Stipa lagascae Roem. & Schult.	–	–	–	*	*	–	–	–	–	–	–	–
Stipagrostis pungens (Desf.) de Winter	*	*	*	–	–	–	–	–	–	*	*	*

Trend in Diversity Index

There are many different types of diversity indexes used by plant ecologists. Plant species diversity refers to number of species and their relative abundance in a defined area. Diversity measurements incorporate species richness, where S = the number of plant species in a community. The spatial scale strongly influences biodiversity (Crawley and Harral 2001; Symstad et al. 2003). Similarly, drought represents a determining component of arid ecosystem dynamics and biodiversity maintenance. Climate constitutes a crucial feature in desert rangelands, and drought often seriously affects the structure and diversity of vegetation communities, especially those on this steppe.

One key indicator of rangeland degradation is the diversity and its trend. In this chapter, the widely used Shannon-Weiner diversity index was used to assess the diversity of plant community in different years.

The diversity was significantly different between the four plant community types, and it was changing between years of study. In this chapter, ANOVA showed that diversity was significantly different between vegetation type ($F = 14.502$; $p < 0.001$) and year ($F = 122.932$; $p < 0.001$). Also vegetation type*year interaction is significant ($F = 3.215$; $p = 0.018$), indicating

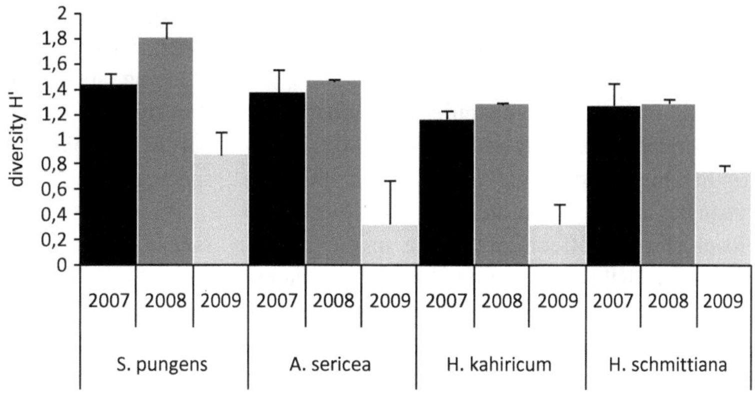

Fig. 5 Mean H′ diversity of four plant community types in southern Tunisia according to year. The effects of year and vegetation type are highly significant ($p < 0.05$). *Vertical bars* represent standard errors

that the effect of climate on plant diversity depended on steppe vegetation type.

The results obtained illustrate an increase in diversity in 2008 for *S. pungens*, *H. schmittiana*, *A. sericea,* and *H. kahiricum* steppes. Then it decreased in 2009 on all sites. In 2008, diversity reached 1.8 on the *S. pungens* steppe and decreased to 0.3 on the *A. sericea* and the *H. kahiricum* steppes in 2009 (Fig. 5). This may be explained by that the plant species on the *H. kahiricum* and *A. sericea* steppe would be more exposed to edaphic drought than on *H. schmittiana* steppe.

The effects of climate and vegetation type on plant have been studied by numerous researchers, and drought remains a major factor, if not the most difficult problem, in maintaining efficient, economical animal production enterprises on dryland pasture (Valentine 2001).

Grime (1979) reported that the environment can control species richness in two distinct ways, by regulating the expression of dominance and by affecting the potential richness (pool of suitable species). Previous studies in southern Tunisia have demonstrated relationships between vegetation type (Floret and Pontanier 1982) and rainfall (Le Houérou and Hoste 1977). Different resilience levels are intertwined with different vegetation type and intensity of natural disturbance. According to Floret and Pontanier (1984), climatic aridity can be increased or decreased if considering soil conditions and their utilization by man.

As was shown by Ludwig and Tongway (2000), once rangeland has been degraded, it is often possible to rehabilitate it and thus restore it to a level of utility, possibly not as good as its original state, but better than it was in its damaged state.

Desert rangelands have frequent droughty periods that have a marked effect on vegetation. Precipitation is extremely variable temporally. Variations of vegetation cover exhibited an overall ascending tendency at beginning then decreasing, but significant spatial contrast exists. Vegetation cover increased in 2008 where precipitation is strong. But in 2009 where it has relatively weak precipitation, vegetation cover deteriorated.

The dynamic variability of climatic parameters can also have significant implications for vegetation cover, species richness, and diversity. Changes in the vegetation diversity can be one of the most sensitive indicators of climatic variability, and they can be used to monitor the climate changes on different spatial scales. Hence, it is necessary to identify the different relationships existing between different climatic variables and vegetation dynamics. These relationships can have different patterns and magnitude on different spatial and temporal scales.

The soils in arid areas have characteristics that put limitations on the ecosystems (Lundholm 1976). As indicated by Whitford (2002), soil is the most important factor affecting vegetation structure in desert ecosystem. Sparsely distributed vegetation results in a heterogeneous horizontal pattern of vegetation patches alternating with areas of bare soil (Noy-Meir 1973). This spatial heterogeneity is now more broadly interpreted as cumulative outcome of the processes affecting the spatial and temporal distribution of vital resources such as water, topsoil organic matter, and propagules individually and collectively (Tongway and Ludwig 2005).

In this chapter we have shown that the changes in vegetation are affected more by the local climatic effects prevailing in the region. It is reasonable to assume that these changes will manifest themselves in the frequency of warmer months and warmer seasons. I show that the close relationship that exists among the vegetations and climatic variables is reflected by the impacts of the 2009 droughts. They also reveal that the dry season rainfall in the country will change by 25 % of vegetation cover on *H. kahiricum* steppe. These results would predict that these droughts would have resulted in substantial declines in vegetation.

The number of species is highest on *H. kahiricum* steppe, while *A. sericea* steppe reveals a slightly lower average. However, the number is lowest on *S. pungens* and *H. schmittiana* steppes.

The impact of species diversity on ecosystem functioning has generated considerable research and tremendous debate in view of accelerated depletion of biodiversity worldwide (Singh et al. 2005). In particular, many recent advances have indicated that diversity can be expected, on average, to give rise to ecosystem stability (Wolfe 2000; Chapin et al. 2000; Tilman 2000; McCann 2000). Although, for same period of drought stress, the real drought for plants in loamy soils is 60 % (*H. kahiricum* steppe) longer than in sandy soils (*H. schmittiana* steppe) (Floret and Pontanier 1984). These results show that *H. kahiricum and A. sericea* steppes are most affected by drought. This may be attributed to soil sealing, which plays a key role in crusts formation. In arid environments, sealing and crusting surfaces play major roles in ecosystem processes, particularly in water and soil flows, and are therefore critical for landscape structure and function (Eldridge et al. 1995; Zaady et al. 1997; Eldridge et al. 2000). They also play a major cause of decreased infiltration, increased erosion, and runoff, as well as low seedling emergence, on agricultural fields (Morin et al. 1989; Zhang and Miller 1996). Therefore, the reducing species richness and biodiversity in this study is said to crust, which is a physical barrier to plant germination, mainly annual species.

According to Noy-Meir (1998), the Mediterranean region includes a wide range of climatic and edaphic conditions so that it is difficult to generalize the results from one region or one ecosystem for all the Mediterranean areas. However, this chapter suggested that the stage of desert rangelands succession is an important factor which influences the evolution of vegetation composition and diversity after cessation of grazing in desert area.

Clearly, in these situations, the response to variation in species diversity cannot be separated from the response to environmental variation. This relationship is a central but contentious issue within ecology (Schmid 2002).

Today, the risks of desertification are substantial and clear. Under present scenarios of population growth, drought, and loss of ecosystem services, even within a stricto sensu classic sustainable development approach, the challenges posed by desertification are enormous and should therefore be easily comprehended. The desertification process embodies a strong disruptive potential in terms of rangelands stability. In desert rangelands, we suggested that there was an increase in soil vulnerability to wind erosion associated with fragile ecosystems resulting in desertification of the site (Fig. 6).

Conclusion

The desert area in southern Tunisia is characterized by average annual rainfall limits (isohyets) of 75 mm $years^{-1}$. It is clear that desertization is the consequence not only of mismanagement of the environment but also of droughts. The natural vegetation is in accordance with rainfall and vegetation type, mainly via the soil drought regime, since soils are enormously different from place to place. And that, in dry

Fig. 6 The desertification processes in protected desert rangelands; (**a**) wind erosion, (**b**) plant mortality, (**c–d**) sand accumulation

land, growing plants require particular soil textures and appropriate amounts of water.

We tested whether particular patterns of variation in cover, diversity, and richness could be applied generally to distinguish between mechanisms responsible for organizing vegetation.

The rangeland vegetation is highly dynamic due to climatic variability and the extensive ecosystem degradation resulting from increased population pressure brought about by both humans and animals. Spatial scale strongly influences vegetation, but, drought disturbance here was more evident. In the same way drought represents a determinant component of the arid ecosystem dynamics and for the maintenance of their biodiversity.

This chapter shows mainly that cover, species richness, and species diversity changes depending on spatial scale and rainfall. Vegetation cover on *H. kahiricum* steppe is very affected by drought than the other steppes, while diversity has been affected by drought on all steppes and mainly on *H. kahiricum* and *A. sericea* steppes.

Water condition is a limiting factor for vegetation, and precipitation takes a key role in the ecological distribution of vegetation. Variations of the annual precipitation from increasing, vegetation cover was improving, and conversely, variations of the annual precipitation from decreasing, vegetation cover was degraded. Moreover, the relation between precipitation and vegetation rapidly reached a significant level and steadily keeps at this level, indicating that vegetation change is very sensitive to precipitation change. Thus one can deduce that vegetation type affects the amount and structure of associated cover, and consequently the infiltration rate differs among vegetation types. The effects of cessation of grazing vary with vegetation composition, precipitation, and soil condition. The time needed for degraded overgrazed rangelands to recover and reach an excellent condition through natural succession differs depending on the climate and vegetation type.

In the light of our findings and also considering the short time of our experimentation, it is difficult to recommend optimal year duration of protection to regenerate the desert rangelands of southern Tunisia. Three years protection cannot increase more stability and biodiversity of desert rangelands.

References

Allaby M (2003) Droughts: dangerous weather, rev edn. Facts on File, New York, 224pp

Auclair L, Delaître E, Simonneaux V (1999) Mutations foncières et désertification dans le Sud Tunisien. In: Symposium "Jardin planétaire", Chambéry, 5 pp, 14–18 Mar 1999

Ayyad MA, El-Ghareeb RE (1984) Habitat and plant communities in the NE Desert of Egypt. Communication in Agrisciences and Development Research, College of Alexandria. University of Alexandriey 7(6):1–34

Batanouny KH (1973) Habitat features and vegetation of deserts and semi-deserts in Egypt. Vegetatio 27(4):181–199

Benzarti Z, Habaieb H (2001) Etude de la persistance de la sécheresse en Tunisie par utilisation des chaînes de Markov (1909–1996). Sécheresse 12(4):215–220

Chapin FS, Zavaleta ES, Eviner VT, Naylor RL, Vitousek PM, Reynolds HL, Hooper DU, Lavorel S, Sala OE, Hobbie SE, Mack MC, Diaz S (2000) Consequences of changing biodiversity. Nature 405:234–242

Chapman VI (1960) Salt marshes and salt deserts of the world. L. Hill, London, 392 pp

Crawley MJ, Harral JE (2001) Scale dependence in plant biodiversity. Science 291:864–868

Eldridge DJ, Chartres CJ, Greene RSB, Mott JJ (1995) Management of crusting and hardsetting soils under rangeland conditions. In: So HB, Smith GD, Raine SR, Schafer BM, Loch RJ (eds) Crusting, sealing and hardsetting soils. productivity and conservation. Australian Society of Soil Science, Brisbane, pp 381–399

Eldridge DJ, Zaady E, Shachak M (2000) Infiltration through three contrasting biological soil crusts in patterned landscapes in the Negev, Israel. Catena 40:323–336

Emberger L (1954) Une classification biogéographique des climats. Trav Lab Bot Zool Fac Sci Serv Montpellier 7:3–43

Evenary M, Noy-Meir I, Goodal DW (1985) Ecosystems of the world, vol 12A, Hot deserts and arid shrublands. Elsevier, Amsterdam, 365 pp

Floret C, Pontanier R (1982) L'aridité en Tunisie présaharienne. Travaux et Documents no 150. Orstom éditions, Paris, 544pp

Floret C, Pontanier R (1984) Aridité climatique et aridité édaphique. Bull Soc Bot Franc Actual Bot 131:265–276

Gamoun M (2005) Cartographie et dynamique de la végétation dans la région d'Ouled Chéhida, El Ouara Tataouine. Diplôme de mastère, faculté des sciences de Sfax, 95 pp

Gamoun M, Chaieb M, Ouled Belgacem A (2010a) Evolution de la diversité floristique des parcours en Tunisie aride. Ecol Mediterr 36(2):5–16

Gamoun M, Tarhouni M, Ouled Belgacem A, Hanchi B, Neffati M (2010b) Effects of grazing and trampling on primary production and soil surface in North African rangelands. Ekológia (Bratislava) 29(2):219–226

Ganry F, Campbell B (1993) Sustainable land management in African semi-arid and subhumid regions. In: Proceedings of the SCOPE workshop, Dakar, 15–19 Nov 1993. CIRADCA, Montpellier, 1995, 406p

Gillson L, Hoffman MT (2007) Rangeland ecology in a changing world. Science 315:53–54

Grime JP (1979) Plant strategies and vegetation processes. Wiley, Chichester, 222 pp

Hardin G (1968) The tragedy of the commons. Science 162:1243–1248

Hassib M (1951) Distribution of plant communities in Egypt. Bull Fac Sci Univ Fouad I Cairo Egypt 29:59–61

Jauffret S, Visser M (2003) Assigning life-history traits to plant species to better qualify arid land degradation in Presaharian Tunisia. J Arid Environ 55:1–28

Kassas M (1955) Rainfall and vegetation in arid NE Africa, Plant ecology proceedings of Montpellier symposium. UNESCO, Paris, pp 49–77

Le Floc'h E, Neffati M, Chaieb M, Floret C, Pontanier R (1999) Rehabilitation experiment at Menzel Habib, Southern Tunisia. Arid Soil Res Rehabil 13:357–368

Le Houérou HN (2005a) Atlas of climatic diagrams for the isoclimatic Mediterranean regions. Copymania, Montpellier, 220 pp

Le Houérou HN (2005b) The isoclimatic Mediterranean biomes, vol 2, Bioclimatology, diversity and phytogeography. Copymania, Montpellier, 765 pp

Le Houérou HN (2009) Bioclimatology and biogeography of Africa. Springer-Verlag, Heidelberg, 250 pp

Le Houérou HN, Hoste CH (1977) Rangeland production and annual rainfall relations in the Mediterranean basin and African Sahelo-Sudanian zone. J Range Manage 30:181–189

Ludwig JA, Tongway DJ (2000) Viewing rangelands as landscape systems. In: Arnalds O, Archer S (eds) Rangeland desertification. Kluwer Academic, Dordrecht, pp 39–52

Lundholm B (1976) Adaptations in arid ecosystems. Can desert encroachment be stopped? Ecol Bull (Stockholm) 24:19–27

Mainguet M (1995) L'homme et la sécheresse. Masson, Paris, 335 pp
May RM, Lawton JH, Stork NE (1995) Assessing extinction rates. In: Lawton JH, May RM (eds) Extinction rates. Oxford University Press, Oxford, pp 1–24
McCann KS (2000) The diversity-stability debate. Nature 405:228–233
Moreira F, Queiroz AI, Aronson J (2006) Restoration principles applied to cultural landscapes. J Nat Conserv 14:217–224
Morin J, Keren R, Benjamini Y, Ben-Hur M, Shainberg I (1989) Water infiltration as affected by soil crust and moisture profile. Soil Sci 148:53–59
Noy-Meir I (1973) Desert ecosystems: environment and producers. Annu Rev Ecol Syst 4:25–51
Noy-Meir I (1998) Effects of grazing on Mediterranean grasslands: the community level. In: Papanastasis VP, Peter D (eds) Proceedings of EU workshop, ecological basis of livestock grazing in Mediterranean ecosystems, Thessaloniki (Greece). Commission of European Communities, Brussels, pp 27–39, 23–25 Oct 1997
Pélissier R, Couteron P, Dray S, Sabatier D (2003) Consistency between ordination techniques and diversity measurements: two strategies for species occurrence data. Ecology 84:242–251
Raup HM (1957) Species versatility in shore habitats. J Arnold Arboretum 56:126–163
Reynolds JF, Stafford DMS (2002) Global desertification: do humans cause deserts? Dahlem workshop report 88. Dahlem University Press, Dahlem
Schmid B (2002) The species richness-productivity controversy. Trends Ecol Evol 17(3):113–114
Shannon CE, Weaver W (1948) The mathematical theory of communication. University of Illinois Press, Urbana, pp 3–91
Shreve F (1942) The desert vegetation of North America. Bot Rev 8:195–2046
Singh SP, Sah P, Tyagi V, Jina BS (2005) Species diversity contributes to productivity – evidence from natural grassland communities of the Himalaya. Curr Sci 89(3):548–552
Stearns FW (1949) Ninety years change in a northern hardwood forest in Wisconsin. Ecology 30:350–358
Stenseth NC, Mysterud A, Ottersen G, Hurrell JW, Chan KS, Lima M (2002) Ecological effects of climate fluctuations. Science 297(5585):1292–1296
Stirling G, Wilsey B (2001) Empirical relationships between species richness, evenness, and proportional diversity. Am Nat 158:286–297
Symstad AJ, Chapin FS, Wall DH, Gross LK, Huenneke FL, Mittelbach GG, Peters PCD, Tilman D (2003) Long-term and large-scale perspectives on the relationship between biodiversity and ecosystem functioning. BioScience 53:89–98
Tadros TM (1953) A phytosociological study of halophilous communities from Mareotis, Egypt. Vegetatio 4:102–124
Thomas D (1997) Science and the desertification debate. J Arid Environ 37:599–608
Tilman D (2000) Causes, consequences and ethics of biodiversity. Nature 405:208–211
Tongway DJ, Ludwig JA (2005) Heterogeneity in arid and semi arid lands. In: Lovett GM, Jones CG, Turner MG, Weathers KC (eds) Ecosystem function in heterogeneous landscapes, Xth Cary conference proceedings. Millbrook, New York
Tucker CJ, Dregne HE, Newcomb WW (1991) Expansion and contraction of the Sahara desert from 1980 to 1990. Science 253:299–301
Tucker CJ, Newcomb WW, Dregne HC (1994) AVHRR data sets for determination of desert spatial extent. Int J Remote Sens 15:3547–3566
Valentine JF (2001) Grazing management, 2nd edn. Academic Press, San Francisco, 659pp
Vernet A (1955) Climate and vegetation. In: Plant ecology proceedings of the Montpellier symposium, Montpellier, pp 75–101
White PS (1979) Pattern, process and natural disturbance in vegetation. Bot Rev 45:229–299
Whitford WG (2002) Ecology of desert system. Academic, London, 360 pp
Wolfe MS (2000) Crop strength through diversity. Nature 406:681–682
Younes HA, Zahran MA, El Qurashy ME (1983) Vegetation–soil relationships of a sea-landword transect, Red Sea coast, Saudi Arabia. J Arid Environ 6:349–356
Zaady E, Gutterman Y, Boeken B (1997) The germination of mucilaginous seeds of *Plantago coronopus*, *Reboudia pinnata*, and *Carrichtera annua* on cyanobacterial soil crust from the Negev Desert. Plant Soil 190:247–252
Zahran MA, Willis AJ (2008) The vegetation of Egypt. Springer, Dordrecht, 437 pp
Zhang XC, Miller WP (1996) Physical and chemical crusting processes affecting runoff and erosion in furrows. Soil Sci Soc Am J 60:860–865
Zohary M (1962) Plant life of Palestine, Israel and Jordan. Ronald Press, New York, 262 pp

Effect of Cold Stress on Photosynthesis of Plants and Possible Protection Mechanisms

S. Adam and S.D.S. Murthy

Abstract

Plants are subjected to various types of abiotic stresses; among them temperature stress is a common stress experienced by plants distributed all across the globe. Low temperature is one among the important environmental factors that limit the food productivity of agricultural fields around the world. To cope with cold stress, plant species have evolved several physiological and molecular adaptations to minimise damage from cold by adjusting their metabolic processes. The understanding of adaptations and protective regulations represents an additional mechanism of cold tolerance. As a consequence of these changes, plants undergo a process known as cold acclimation. In this chapter, a brief summary on the recent progress of research on how cold-sensitive plants perceive cold is mentioned. We have also explored how this perception is translated into protective mechanism within plants. Particular emphasis is placed on physiological parameters, and regulation of cold-induced photosynthetic processes that occur after exposure to low temperatures, leading to cold acclimation, is widely discussed. This chapter mainly emphasises on the various molecules and pigments synthesised to acclimatise during low-temperature exposure.

Keywords

Acclimatisation • Cold stress • Photosynthesis • Reactive oxygen species

Introduction

All along the life of plants, they adapt to variable environmental conditions such as changes in temperatures, starvation, drought, change in nutrient abundance, flooding, air pollution, soil pollution and osmotic alterations, which are various abiotic factors that cause stress (Apel and Hirt 2004). To constantly monitor the

S. Adam • S.D.S. Murthy (✉)
Department of Biochemistry, Sri Venkateswara University, Tirupati, AP 517502, India
e-mail: sdsmurthy@rediffmail.com

R.K. Gaur and P. Sharma (eds.), *Approaches to Plant Stress and their Management*,
DOI 10.1007/978-81-322-1620-9_12, © Springer India 2014

environmental changes and to adapt appropriately, plants develop enumerable number of mechanisms which are robust in their nature. The importance of ability to adapt to a changing environment was described in various research articles. In recent years there has been tremendous progress in understanding the mechanisms and processes involved in defence adaptations towards abiotic stress in numerable plant species (Hirayama and Shinozaki 2010; Jaspers and Kangasjärvi 2010).

Among various abiotic stresses, low temperature (LT) is one of the most important factors limiting the crop productivity and distribution of plants. Low temperatures are defined as low but not freezing temperatures from 0°C to −15°C; they are common in nature and can damage many plant species. Chilling temperatures range from 0°C to 15°C, whereas freezing temperature starts from 0°C and further. Chilling and freezing temperatures are different from each other and have different tolerance mechanisms. Low-temperature stress is a major factor that limits the agricultural productivity in hilly areas and high altitude areas especially (Xin and Browse 2001). Plants of tropical and subtropical origins are sensitive to chilling temperature (0–10°C) and are incapable of cold acclimation. Whereas plants belonging to temperate climatic regions are considered to be chilling tolerant with variable degrees of temperature and can increase their freezing tolerance by being exposed to chilling, nonfreezing temperatures, this is known as cold acclimation (Levitt 1980; Zhu et al. 2007). Brief exposures of plants to low-temperature stress may cause transitory changes in them, and they survive. However, prolonged exposure to low-temperature stress causes necrosis, wilting and chlorosis and finally leads to death of plants.

Various efforts to improve crop performance and productivity under low-temperature stress were not yet so successful, because of two main reasons: (1) incomplete understanding of underlying fundamental mechanisms of stress tolerance and (2) lack of knowledge about the interactions of different stressors. As a result of exposure to low temperatures, many physiological and morphological changes have been correlated with visible symptoms like wilting, chlorosis and necrosis that lead to death of plants.

Plants ultimately show marked changes in biochemical and cell functions such as the following:

1. Biomembrane lipid composition
2. Small molecule accumulation (Shinozaki and Yamaguchi-Shinozaki 1996; Thomashow 1998; Gilmour et al. 2000; Yamaguchi-Shinozaki and Shinozaki 2006)
3. Amino acids (Lalk and Dörfflung 1985; Wanner and Junttila 1999; Koster 1991; Shao et al. 2006)
4. Gene expression (Gulzar et al. 2011)
5. Cellular leakage of electrolytes
6. Redistribution of intracellular calcium ions (Knight et al. 1998)
7. Diversion of electron flow to alternate pathways (Seo et al. 2010)
8. Enzyme activities (Ruelland and Zachowski 2010)

Modifications to Plant Cell Membranes

A fall in temperature reduces the fluidity of membranes, and this appears to be an effective direct sensor of cold in cyanobacteria and plants. The lipid composition of plasma membrane and chloroplast envelopes in acclimated plants changes due to membrane damage during the low temperature, and this marks as relative difference with non-acclimated plants (Uemura and Steponkus 1999). The increase in fluidity of membranes is achieved with an increase in the number of unsaturated fatty acid content; this marks them to become more fluid as well as cold-adapted membranes (Vogg et al. 1998). Stabilisation of membranes during cold stress prevents damage of cell membranes and membrane-bound organelles leading to their death. Changes in chloroplast thylakoid structure, chlorophyll content, photosynthetic enzyme activity and electron transport are associated with winter hardening, although the relation among these physiological changes is unclear. It is understood from research that the changes in combination with stomatal closure are largely responsible for the declined net photosynthesis in winter (Paul et al. 1992). Sugars serve as osmoprotectants and play a role in protecting cellular membranes from damage caused by dehydration and freezing through interacting with the lipid bilayers (Anchordoguy et al. 1987; Shalaev and Steponkus 2001; Shao et al. 2006).

Cold Stress-Induced Alterations in Pigment Proteins and Photosynthetic Apparatus

Among the cell structure, the chloroplast is usually the only organelle that is more rapidly and deeply affected during cold stress. When plants are exposed to low-temperature stress, chlorophyll biosynthesis is affected. Therefore, an imbalance in PS II is created by an exposure to low temperature because of the alterations in the Chl antenna complexes (Ensminger et al. 2006). At the seedling stage the impact of chilling stress sharply increases the concentration of the accessory pigments (Chl *b* and carotenoids) when compared to Chl *a* probably in order to increase the photon capture. The bulk of the Chl and carotenoid present within the chloroplast thylakoid membrane is bound to be LHC b and LHC a families of light-harvesting polypeptides, associated with PS II and PS I, respectively. The xanthophylls cycle is the pathway by which violaxanthin is rapidly and reversibly de-epoxidised into zeaxanthin via the intermediate antheraxanthin. In response to low temperature, both the pools of xanthophyll (violaxanthin + antheraxanthin + zeaxanthin) and the level of zeaxanthin increase (Krol et al. 1988). The xanthophylls have intrinsic antioxidant capacity, i.e. independent of the non-photochemical quenching. In comparison, the antioxidant capacity of zeaxanthin is higher than violaxanthin. This antioxidant activity plays a key role during cold exposure.

In barely (*Hordeum vulgare L.*) under high light, cold or both in combination had only a small effect on the protein composition of the total xanthophyll cycle. Cold and high light applied synergistically induce the enhancement of the total xanthophylls. The fraction consisting of antheraxanthin plus zeaxanthin was up to four- to fivefold higher at 5°C than at 25°C. Carotenoid content increases linearly with cellular development, the highest amount being in the apical part of the leaf. In *Arabidopsis*, when exposed to high light at low temperature, wild-type plants exhibit symptoms of severe oxidative stress, lipid peroxidation, chlorophyll bleaching and photoinhibition. Similarly winter conifers exhibit long-term changes in the organisation of the photosynthetic apparatus that induces decrease in the number of functional PS II reaction centres, loss of light-harvesting Chl and the formation of a large thylakoid protein complex involved in LHC II, PS II and PS I (Savitch et al. 2002; Ensminger et al. 2006).

In contrast with the above, the low-light chilling treatment has a much more pronounced effect on PS I activity compared with PS II in *Arabidopsis* plants (Zhang and Scheller 2004). The major site of photo inhibition in PS II is the inactivation of D1 protein (Aro et al 1993; Greenberg et al. 1987). This inactivated protein must be newly synthesised to restore PS II activity. The actual extent of PS II photoinhibition in vivo depends on the balance between the inactivation of D1 and the recovery process which involves insertion of new D1 molecule into the thylakoid and their incorporation into the PS I complex. Recovery from PS II photoinhibition is strongly temperature dependent, i.e. low temperature will decrease the rate of repair (Gombos et al. 1994). Photo damage becomes apparent as low temperature interferes with the normal replacement rate of D1 in the turn over repair cycle. This has been attributed to changes in the expression of psb A, the plastid gene that encodes D1 and directs temperature effect on membranes (Damian et al. 2001).

Photo damage of PS II is very frequent in low-temperature stress. This is not primarily responsible for light chilled-induced inhibition of photosynthesis in thermophilic plants (Kee et al 1986; Kingston-Smith et al. 1987). The effects of the growth under chilling condition on PS I, PS II and CO_2 assimilation have been investigated (Harbinson and Foyer 1989). Therefore, a limited number of reports showed that PS I has a greater chilling sensitivity than PS II. It is also evidenced that PS I activity declines to a greater extent than PS II. It is not sufficient to identify PS I as a primary target of chilling. This is frequently assessed in isolated thylakoid membranes using artificial electron donors or acceptors. The use of absorption measurements at 820 nm facilitates assessment of the redox state of the PS I reaction centre, and hence, the quantum efficiency PS I electron transport in intact leaves can be assessed under cold stress (Kingston-Smith et al. 1997). Furthermore, polyamines bound to LHC II respond to environmental factors as if they were part of a regulatory mechanism and fine-tune the properties of PSII antenna, helping them to expand and capture as many electrons as possible (Navakoudis et al. 2007).

Effect of Cold Stress on CO_2 Fixation

Low temperature affects different aspects of photosynthesis such as sucrose synthesis in the cytosol, leading to accumulation of phosphorus-related intermediates. This results in the depletion of the available inorganic phosphate and in the decreased cycling of inorganic phosphate between the cytosol and chloroplast (Furbank et al. 1987; Hurry et al. 2000). This in turn impedes synthesis of the ATP necessary for the regeneration of ribulose 1,5 bis-phosphatase to support CO_2 fixation. Carbohydrate metabolism has been reported to have greater instantaneous low-temperature sensitivity than other component of photosynthesis (Leegood and Edwards 1996). Pollock and Lloyd (1987) have shown that starch synthesis is more sensitive to low temperature than sucrose and fructose synthesis in cold-tolerant species. Growth of winter rye at 5°C results in an increased capacity for CO_2-saturated O_2 evolution, dampening of the fluorescence induction transients and reductions in the lag time to reach steady-state photosynthesis. Low-temperature stress inhibits sucrose synthesis in the cytosol coursing decreased inorganic phosphate cycling between the cytosol and the chloroplast (Hurry et al. 2000). Concomitant with the upregulation of carbon metabolism, cold-acclimated winter wheat also exhibits stimulation of carbon export from the leaf compared to cold-stressed plants. Cold-induced tolerance includes increase in enzyme activities (e.g. from the Calvin cycle and sugar metabolism), reinforcement of energy dissipation mechanisms and antioxidative molecules and qualitative and quantitative changes in lipids and protein membrane composition and protects the cell structures by keeping metabolic pathway regulation mechanisms intact (Ramalho et al. 2003; Quartin et al. 2004).

In contrast, growth and development at low temperatures increased the expression and subsequent activity of Calvin cycle enzyme (Hurry et al. 2000). Decline in photosynthesis after chilling during dark and light photoperiods attributes loss in Rubisco activity. It has been suggested that chilling damages the Rubisco protein itself. In addition cold-tolerant, herbaceous plants grown at cold temperatures exhibit an increase in inorganic phosphorous availability in the chloroplast (Stitt and Hurry 2002) as well as in adenylates and phosphorylated intermediates and in the capacity for the regeneration of RUBP (Hurry et al. 1994).

Possible Protection Mechanism

Role of Sugars in Cold Sensation and Cold Defence

Accumulation of substantial amounts of solutes, such as soluble sugars (Wanner and Junttila 1999), amino acids and glycine betaine (Kishitani et al. 1994), are thought to play a key role in protection of cells from freezing injury. In response to cold stress, high concentrations of the sugars like fructose and its derivative sugars accumulate, as a response in grasses especially (Wang et al. 2008; Xiao et al. 2009). Sugars affect developmental programmes ranging from embryogenesis to senescence, which makes the research complex. During cold acclimation, soluble sugars accumulated theoretically cause adverse impact on photosynthesis and induce senescence. Sugar-sensing and sensing of the environment appear to be part of complex regulatory mechanisms (Gibson 2000). Cold-acclimated leaves do not show the typical repression by sugars on photosynthetic mechanism (Strand et al. 1997; Li and Huang 2008). It is also hypothesised that photosynthetic activity is regulated in a feedback manner by soluble sugars, and it is downregulated by them.

Cold girdling of petioles is another phenomenon that occurs to prevent sugar export out of the leaf. Genes related to chlorophyll-binding protein and Rubisco are repressed during cold (Smeekens 2000). Studies also show increased flux through the sucrose biosynthetic pathway; this reduces inhibition of photosynthesis in the cold (Strand et al. 2003). It was therefore conceivable that cold acclimation could interact with various metabolic regulations of senescence and inhibit the induction of senescence by sugar accumulation (Shao et al. 2008; Bressan et al. 2009). Another possibility is that, in light, photosynthetic cells may sense cold through an effect on photosystem II excitation pressure (Huner et al. 1998). Molecular studies have shown that cold acclimation in higher plants is complex involving multiple regulatory pathways (Fowler and Thomashow 2002).

Osmolytes

Accumulation of osmolytes in plants in response to various environmental stresses, namely, drought, salinity, extreme low temperatures, UV radiation and heavy metals, is well reported in literature (Serraj and Sinclair 2002; Ashraf and Foolad 2007). Osmolytes are low-molecular-weight and highly soluble compounds. The major role of osmolytes lies in protection of plants from abiotic stresses by different defence mechanisms such as adjustment of cellular osmoticum, detoxification of reactive oxygen species (ROS), maintenance of membrane integrity and stabilisation of enzymes and proteins. Additionally they protect cellular organelles from dehydration and injury (Sharma and Dietz 2006; Ashraf and Foolad 2007). The major osmolytes include sugars (glucose, fructose, sucrose, trehalose and raffinose), sugar alcohols and nitrogen-containing compounds such as proline and quaternary amino compounds (QACs) such as glycine betaine (GB), alanine betaine, proline betaine and polyamines (Mudgal et al. 2010). Soluble sugars those accumulate during freeze tolerance in higher plants are proposed to play a key role in the form of oligosaccharides to prevent crystallisation of sucrose and to facilitate glass formation within the cell and finally lead to protection of membrane phospholipids (Crowe et al. 1988). Generally, the accumulation of soluble sugars contributes to the increase in the cryostability of cellular membranes. Cryostability is a prerequisite for freezing tolerance because freeze-induced destabilisation of cellular membranes is the primary cause of injury in plants (Steponkus et al. 1993; Shao et al. 2008). Soluble sugars act as important hormones like primary messengers in signal transduction. Their role as signaling molecules is well illustrated by a variety of sugar-sensing and signaling mechanisms in free-living microorganisms such as bacteria and yeast and also along with plants (Rolland et al. 2001; Stulke and Hillen 1999; Wang et al. 2008; Xiao et al. 2009).

Glycine Betaine (GB)

Amino compounds like glycine betaine are known to accumulate in response to stress in many crop plants, including sugar beet (*Beta vulgaris*), spinach (*Spinacia oleracea*), barley (*Hordeum vulgare*), wheat (*Triticum aestivum*) and sorghum (*Sorghum bicolor*) (Weimberg et al. 1984; Fallon and Phillips 1989; Kent et al. 1990; Rhodes and Hanson 1993; Yang et al. 2003). Exogenous application of GB to low-accumulating or non-accumulating plants may help reduce adverse effects of environmental stresses (Somersalo et al. 1996; Makela et al. 1998; Yang and Lu 2005). Tolerance to low temperatures was improved in two potato cultivars by exogenous application of GB.

Antifreeze Proteins

Antifreeze activity has been found to increase in response to cold stress in winter rye due to the accumulation of antifreeze proteins (AFPs). The pathway for regulating AFPs production is independent of ABA (Griffith et al. 2005). Low-temperature-responsive genes are an important gene family synthesising AFPs. Dehydrins are another group of proteins which are shown to be cryoprotective in nature towards freeze-labile enzyme lactate dehydrogenase (Wisniewski et al. 1999). Late embryogenesis proteins (LEA) function as chaperones in dehydrated cells. Chaperonin expression and its activity are affected by low temperature (Anderson et al. 1993; Ukaji et al. 1999; Mendoza et al. 2000) as a response to the freezing environment of dehydrated cell proteins, which are likely to be denatured. Evidence also suggests the synthesis of cryoprotectins that bind to thylakoid membranes. Glucanases are enzymes that belong to the group of pathogenesis-related (PR) proteins. A class 1b-1,3-glucanase from tobacco has cryoprotective activity, and immunologically related proteins are accumulated in cabbage and spinach during frost hardening under natural conditions.

Freezing tolerance gene "regulon" in Arabidopsis and a family of transcriptional factor-controlling genes that control their expression have been recognised that help during defence. Genes involved in protection or repair mechanisms could be new targets for the improvement of plant plasticity and adaptive responses towards low-temperature stress (Hsieh et al. 2002). Finding new key genes responsible for abiotic stress tolerance phenotypes is of great importance not only for a better understanding of stress responses but also for promising future crop improvement.

References

Anchordoguy TJ, Rudolph AS, Carpenter JF, Crowe JH (1987) Modes of interaction of cryoprotectants with membrane phospholipids during freezing. Cryobiology 24:324–331

Anderson JM, Chow WS, Öquist G (1993) Dynamics of photosystem II: photoinhibition as a protective acclimation strategy. In: Yamamoto HY, Smith CM (eds) Photosynthetic responses to the environment. Amer Soc Plant Physiol. Rockville, pp 14–26

Apel K, Hirt H (2004) Reactive oxygen species: metabolism, oxidative stress and signal transduction. Annu Rev Plant Biol 55:373–399

Aro EM, Virgin I, Andersson B (1993) Photoinhibition of photosystem II. Inactivation, protein damage and turnover. Biochim Biophys Acta 1143(2):113–134

Ashraf M, Foolad MR (2007) Roles of glycine betaine and proline in improving plant abiotic stress resistance. Environ Exp Bot 59:206–216

Bressan R, Bohbert H, Zhu JK (2009) Abiotic stress tolerance: from gene discovery in model organisms to crop improvement. Mol Plant 2:1–2

Crowe JH, Crowe LM, Carpenter JF, Rudolph AS, Wistrom CA, Spargo BJ, Anchordoguy TJ (1988) Interactions of sugars with membranes. Biochem Biophys Acta 947:367–384

Damian, J. Allen, Ort DR (2001) Impacts of chilling temperatures on hotosynthesis in warm-climate plants. Trends Plant Sci 6(1):36–42

Ensminger I, Bosch F, Huner NPA (2006) Photo stasis and cold acclimation: sensing low temperature through photosynthesis. Physiol Plant 126(1):28–44

Fallon KM, Phillips R (1989) Responses to water stress in adapted carrot cell suspension cultures. J Exp Bot 40:681–687

Fowler S, Thomashow MF (2002) *Arabidopsis* transcriptome profiling indicates multiple regulatory pathways are activated during cold acclimation in addition to the CBF cold-response pathway. Plant Cell 14:1675–1690

Furbank RT, Foyer CH, Walker DA (1987) Regulation of photosynthesis in isolated spinach chloroplasts during orthophosphate limitation. Biochim Biophys Acta 894:552–561

Gibson SI (2000) Plant sugar-response pathways. Part of a complex regulatory web. Plant Physiol 124:1532–1539

Gilmour SJ, Sebolt AM, Salazar MP, Everard JD, Thomashow MF (2000) Over expression of the *Arabidopsis CBF3* transcriptional activator mimics multiple biochemical changes associated with cold acclimation. Plant Physiol 124:1854–1865

Gombos Z, Wada H, Hideg E, Murata N (1994) The unsaturation of membrane lipids stabilizes photosynthesis against heat stress. Plant Physiol 104:563–567

Greenberg BM, Gaba V, Matoo A, Edelman M (1987) Identification of a primary in vivo degradation product of the rapidly turning over 32kD protein of photosystem II. EMBO J 6:2865–2869

Griffith M, Lumb C, Wiseman SB, Wisniewski M, Johnson RW, Marangoni AG (2005) Antifreeze proteins modify the freezing process in planta. Plant Physiol 138:330–340

Gulzar SS, Wani SH, Hussain W, Singh NB (2011) Engineering cold stress tolerance in crop plants. Curr Genomics 12:30–43

Harbinson J, Foyer CH (1989) Relationship between the quantum efficiencies of photo systems I and II in pea leaves. Plant Physiol 90:1029–1034

Hirayama T, Shinozaki K (2010) Research on plant abiotic stress responses in the post-genome era: past present. Plant J 61:1041–1062

Hsieh TH, Lee JT, Yang PT, Chiu LH, Charng YY, Wang YC, Chan MT (2002) Heterology expression of the Arabidopsis C-repeat/dehydration response element binding factor 1 gene confers elevated tolerance to chilling and oxidative stresses in transgenic tomato. Plant Physiol 129:1086–1094

Huner NPA, Öquist G, Sarhan F (1998) Energy balance and acclimation to light and cold. Trends Plant Sci 3:224–230

Hurry VM, Malmberg G, Gardeström P, Öquist G (1994) Effects of a short-term shift to low temperature and of long-term cold hardening on photosynthesis and Ribulose-1,5-Bisphosphate carboxylase/oxygenase and sucrose phosphate synthase activity in leaves of winter rye (*Secale cereale* L.). Plant Physiol 106:983–990

Hurry V, Strand A, Furbank R, Stitt M (2000) The role of inorganic phosphate in the development of freezing tolerance and the acclimatization of photosynthesis to low temperature is revealed by the pho mutants of *Arabidopsis* thaliana. Plant J 24:383–396

Jaspers P, Kangasjärvi J (2010) Reactive oxygen species in abiotic stress signaling. Physiol Plant 138:405–413

Kee SC, Martin B, Ort DR (1986) The effects of chilling in the dark and in the light on photosynthesis of tomato: electron transfer reactions. Photosynth Res 8:41–51

Kent F, McCue RF, Hanson AD (1990) Drought and salt tolerance: towards understanding and application. Trends Biotech 8:358–362

Kingston-Smith AH, Harbinson J, Williams J, Foyer CH (1997) Effect of chilling on carbon assimilation, enzyme activation, and photosynthetic electron transport in the absence of photoinhibition in maize leaves. Plant Physiol 114:1039–1046

Kishitani S, Watanabe K, Yasuda S, Arakawa K, Takabe T (1994) Accumulation of glycine betaine during cold acclimation and freezing tolerance in leaves of winter and spring barley plants. Plant Cell Environ 17:89–95

Knight H, Brandt S, Knight MR (1998) A history of drought stress alters calcium signalling pathways in *Arabidopsis*. Plant J 16:681–687

Koster KL (1991) Glass formation and desiccation tolerance in seeds. Plant Physiol 96:302–304

Krol M, Huner NPA, Mac Intosh A (1988) Chloroplast biogenesis at cold hardening temperatures. Development of photosystem I and photosystem II activities in relation to pigment accumulation. Photosynth Res 14:97–112

Lalk I, Dörfflung K (1985) Hardening, abscisic acid, proline and freezing resistance in two winter wheat varieties. Physiol Plant 63:287–292

Leegood RC, Edwards GE (1996) Carbon metabolism and photorespiration: temperature dependence in relation to other environmental factors. In: Baker NR (ed) Photosynthesis and the environment. Kluwer, Dordrecht, pp 191–221

Levitt J (1980) Responses of plants to environmental stress. In: Chilling, freezing, and high temperature stress. Academic, New York

Li YS, Huang MB (2008) Pasture yield and soil water deletion of continuous growing alfalfa in the Loess Plateau of China. Agric Ecosyst Environ 124:24–32

Makela P, Jokinen K, Kontturi M, Peltonen-Sainio P, Pehu E, Somersalo S (1998) Foliar application of glycine bestaine – a novel product from Sugar beet – as an approach to increase tomato yield. Ind Crops Prod 7:139–148

Mendoza JA, Dulin P, Warren T (2000) The lower hydrolysis of ATP by the stress protein GroEL is a major factor responsible for the diminished chaperonin activity at low temperature. Cryobiology 41:319–323

Mudgal V, Madaan N, Mudgal A (2010) Biochemical mechanisms of salt tolerance in plants: a review. Int J Bot 6:136–143

Navakoudis E, Vrentzou K, Kotzabasis K (2007) A polyamine- and LHCII protease activity-based mechanism regulates the plasticity and adaptation status of the photosynthetic apparatus. Biochim Biophys Acta Bioenerg 1767(4):261–271

Paul MJ, Driscoll SP, Lawlor DW (1992) Sink-regulation of photosynthesis in relation to temperature in sunflower and rape. J Exp Bot 43:147–153

Pollock CJ, Lloyd EJ (1987) The effect of low growth temperature upon starch, sucrose and fructan synthesis in leaves. Ann Bot 60:231–235

Quartin VL, Ramalho JC, Campos PS, Nunes MA (2004) A importância da investigação na cafeícultura: o problema do frio. In: Tembo PJL, Neto JFC, e Pombal AM (eds) Livro de Actas do 1º Colóquio Angola - Agricultura, Sociedade e Desenvolvimento Rural, Instituto Superior de Agronomia Press, Lisboa, pp 64–72, ISBN 972-8669-09-7, 20–21 Junho 2002

Ramalho JC, Quartin V, Fahl JI, Carelli ML, Leitao AE, Nunes MA (2003) Cold acclimation ability of photosynthesis among species of the tropical *Coffea* genus. Plant Biol 5:631–641

Rhodes D, Hanson AD (1993) Quaternary ammonium and tertiary sulfonium compounds in higher-plants. Annu Rev Plant Physiol Plant Mol Biol 44:357–384

Rolland F, Winderickx J, Thevelein JM (2001) Glucose sensing mechanisms in eukaryotic cells. Trends Biochem Sci 26:310–317

Ruelland E, Zachowski A (2010) How plants sense temperature. Environ Exp Bot 69:225–232

Savitch LV, Leonardos ED, Krol M, Jansson S, Grodzinski B, Huner NPA, Oquist G (2002) Two different strategies for light utilization in photosynthesis in relation to growth and cold acclimation. Plant Cell Environ 25:761–771

Seo PJ, Kim MJ, Park JY, Kim SY, Jeon J, Lee YH, Kim J, Park CM (2010) Cold activation of a plasma membrane-tethered NAC transcription factor induces a pathogen resistance response in *Arabidopsis*. Plant J 61:661–671

Serraj R, Sinclair TR (2002) Osmolyte accumulation: can it really help increase crop yield under drought conditions. Plant Cell Environ 25:333–341

Shalaev EY, Steponkus PL (2001) Phase behavior and glass transition of 1,2-dioleoylphosphatidylethanolamine (DOPE) dehydrated in the presence of sucrose *Biochem.* Biophys Acta Biomembr 1514:100–116

Shao HB, Liang ZS, Shao MA (2006) Osmotic regulation of 10 wheat (*Triticum aestivum L.*) genotypes at soil water deficits. Biointerfaces 47:132–139

Shao HB, Chu LY, Shao MA, Zhao CX (2008) Advances in functional regulation mechanisms of plant aquaporins: their diversity, gene expression, localization, structure and roles in plant soil-water relations *(Review)*. Mol Membr Biol 25:1–12

Sharma SS, Dietz KJ (2006) The significance of amino acids and amino acid-derived molecules in plant responses and adaptation to heavy metal stress. J Exp Bot 57(4):711–726

Shinozaki K, Yamaguchi-Shinozaki K (1996) Molecular response to drought and cold stress. Curr Opin Plant Biol 7:161–167

Smeekens S (2000) Sugar-induced signal transduction in plants. Annu Rev Plant Physiol Plant Mol Biol 51:49–81

Somersalo S, Kyei-Boahen S, Pehu E (1996) Exogenous glycine betaine application as a possibility to increase low temperature tolerance of crop plants. Nordisk Jordbruksforskning 78:10

Steponkus PL, Uemura M, Webb MS (1993) A contrast of the cryostability of the plasma membrane of winter rye and spring oat-two species that widely differ in their freezing tolerance and plasma membrane lipid composition. In: Steponkus PL (ed) Adv Low-Temp Biol 2:211–312

Stitt M, Hurry V (2002) A plant for all seasons: alterations in photosynthetic carbon metabolism during cold acclimation in *Arabidopsis*. Curr Opin Plant Biol 5:199–206

Strand A, Hurry V, Gustafsson P, Gardestrom P (1997) Development of *Arabidopsis* thaliana leaves at low temperatures releases the suppression of photosynthesis and photosynthetic gene expression despite the accumulation of soluble carbohydrates. Plant J 12:605–614

Strand A, Foyer CH, Gustafsson P, Gardestrom P, Hurry V (2003) Altering flux through the Sucrose biosynthesis pathway in transgenic *Arabidopsis* thaliana modifies photosynthetic acclimation at low temperatures and the development of freezing tolerance. Plant Cell Environ 26:523–535

Stulke J, Hillen W (1999) Carbon catabolite repression in bacteria. Curr Opin Microbiol 2:195–201

Thomashow MF (1998) Role of cold-responsive genes in plant freezing tolerance. Plant Physiol 118:1–8

Uemura M, Steponkus PL (1999) Cold acclimation in plants: relationship between the lipid composition and the cryostability of the plasma membrane. J Plant Res 112:245–254

Ukaji N, Kuwabara C, Takezawa D, Arakawa K, Yoshida S, Fujikawa S (1999) Accumulation of small heat-shock protein homologs in the endoplasmic reticulum of cortical parenchyma cells in mulberry in association with seasonal cold acclimation. Plant Physiol 120:481–489

Vogg G, Heim R, Hansen J, Schäfer C, Beck E (1998) Frost hardening and photosynthetic performance of Scots pine (*Pinus sylvestris* L.) needles. I. Seasonal changes in the photosynthetic apparatus and its function. Planta 204:193–200

Wang ZY, Li FM, Xiong YC, Xu BC (2008) Soil-water threshold range of chemical signals and drought tolerance was mediated by ROS homeostasis in winter wheat during progressive soil drying. J Plant Growth Regul 27:309–319

Wanner LA, Junttila O (1999) Cold-induced freezing tolerance in *Arabidopsis*. Plant Physiol 120:391–400

Weimberg R, Lerner HR, Poljakoff-Mayber A (1984) Changes in growth and water soluble solute concentrations in *Sorghum bicolor* stressed with sodium and potassium. Physiol Plant 62:472–480

Wisniewski M, Webb R, Balsamo R, Close TJ, Yu XM (1999) Purification, immunolocalization, cryoprotective, and antifreeze activity of PCA60: a dehydrin from peach (*Prunus persica*). Physiol Plant 105:600–608

Xiao BZ, Chen X, Xiang CB, Tang N, Zhang QF, Xiong LZ (2009) Evaluation of seven function-known candidate genes for their effects on improving drought resistance of transgenic rice under field conditions. Mol Plant 2:73–83

Xin Z, Browse J (2001) Cold comfort farm: the acclimation of plants to freezing temperatures. Plant Cell Environ 23:893–902

Yamaguchi-Shinozaki K, Shinozaki K (2006) Transcriptional regulatory networks in cellular responses and tolerance to dehydration and cold stresses. Annu Rev Plant Biol 57:781–803

Yang X, Lu C (2005) Photosynthesis is improved by exogenous glycine betaine in salt-stressed maize plants. Physiol Plant 124:343–352

Yang WJ, Rich PJ, Axtell JD, Wood KV, Bonham CC, Ejeta G, Mickelbart MV, Rhodes D (2003) Genotypic variation for glycine betaine in sorghum. Crop Sci 43:162–169

Zhang S, Scheller HV (2004) Photoinhibition of photosystem I at chilling temperature and subsequent recovery in *Arabidopsis* thaliana. Plant Cell Physiol 45:1595–1602

Zhu XG, de Sturler E, Long SP (2007) Optimizing the distribution of resources between enzymes of carbon metabolism can dramatically increase photosynthetic rate: a numerical simulation using an evolutionary algorithm. Plant Physiol 145:513–526

Oxidative Stress in Plants and Its Management

Sachin Teotia and Deepali Singh

Abstract

We all live in an oxygen-rich environment which has to deal with the danger of oxidative stress. During normal cell metabolism, reactive oxygen species (ROS) are constantly produced, mainly by respiratory and photosynthetic components. These species mainly include superoxide radicals (O_2^-), singlet oxygen ($^1O_2$), hydrogen peroxide (H_2O_2), and hydroxyl radical (OH^-). The others are hydroperoxyl radical ($HO_2^{\cdot}$), alkoxy radical ($RO^{\cdot}$), peroxyl radicals ($ROO^{\cdot}$), and excited carbonyl (RO). But during stress conditions like salinity, drought, metal toxicity, herbicides, fungicides, air pollutants, hypoxia, and abnormal conditions of light, temperature, and topography, ROS are produced in excess amount. These highly reactive molecules can react with many cellular biomolecules and other components and damage DNA, proteins, and lipids. Thus, their concentration has to be tightly controlled. To counter the deleterious effects of ROS, aerobic organisms are equipped with antioxidant systems to scavenge ROS from the cells. Enzymatic antioxidants are mainly superoxide dismutase (SOD), catalase, ascorbate peroxidase, glutathione peroxidase, glutathione *S*-transferases, and peroxiredoxin, while the nonenzymatic antioxidants are mainly ascorbate, glutathione, proline, tocopherol, flavonoids, and carotenoids. These antioxidants protect against the oxidative damage by inhibiting or quenching free radicals and ROS. When the balance between the production of ROS and the quenching activities of antioxidants is disturbed, the cell faces the risk of oxidative stress and damage. These ROS creating stresses are numerous and often species or area specific. These stresses cause significant crop losses. There is a growing need to develop crops which can be resistant to the effects of various oxidative stresses. One such way is to develop transgenic plants overexpressing one or more

S. Teotia (✉) • D. Singh
School of Biotechnology, Gautam Buddha University,
Greater Noida 201312, India
e-mail: sachin25@gmail.co

R.K. Gaur and P. Sharma (eds.), *Approaches to Plant Stress and their Management*,
DOI 10.1007/978-81-322-1620-9_13, © Springer India 2014

antioxidants, which can confer resistance towards particular stress. Another way is to develop mutants which are resistant towards certain stresses.

Keywords
Reactive oxygen species • Oxidative stress tolerance • Enzymatic Antioxidants • Nonenzymatic antioxidants • Transgenic plants and mutants

Introduction

Oxygen is the primary source of life. In an oxygen atmosphere, the generation of ROS, especially under metabolic stress, is unavoidable. ROS are also produced continuously as by-products of various metabolic pathways localized in different cellular compartments such as chloroplast, mitochondria, peroxisomes, and apoplast (del Rio et al. 2006; Panieri et al. 2013). Many metabolic processes normally produce ROS, which comprise mainly of superoxide radical, hydrogen peroxide, hydroxyl radical, and singlet oxygen. ROS may also be produced by abiotic stress such as salinity, drought, heavy metals, nutrient deficiency, herbicides, fungicides, air pollutants, ozone, light, temperature, topography, and hypoxia and by biotic factors such as pathogen attacks. Oxidative stress is programmed to be a regulated process, and the equilibrium between ROS and its quenching determines the well-being of a plant. Oxidative damage of cells happens when the balance between the production of ROS and their quenching by antioxidants reaches the state of disequilibrium. The extent of oxidative stress depends on the type of ROS that is produced, the concentration and the site where it is released, their interaction with other molecules in the cell, and the developmental stage and potential of the cell (Moller et al. 2007). The enhanced and prolonged production of ROS can cause significant damage to cell structures in plants like lipid peroxidation, protein oxidation, damage to nucleic acids, enzyme inhibition, and programmed cell death (Mittler 2002; Perez-Perez et al. 2012) (Fig. 1). However, plants have evolved protective scavenging systems in response to these ROS. High levels of ROS are kept in check by dynamic and synergistic mechanisms of ROS-scavenging antioxidants that control the concentration of intracellular ROS (Apel and Hirt 2004). An antioxidant is described as a compound capable of scavenging ROS without itself undergoing conversion to a destructive radical (Noctor and Foyer 1998). These antioxidants can be conveniently divided into two groups: enzymatic and nonenzymatic antioxidants (Fig. 1).

Reactive Oxygen Species (ROS)

ROS are mostly free radicals produced as by-products of redox reactions. ROS is formed as a natural by-product of the normal metabolism in the presence of oxygen. In plant cells, ROS are continuously produced as a consequence of aerobic metabolism in all of the the intracellular organelles; as by-products in the electron transport chains of chloroplasts, mitochondria, and the plasma membrane (cytochrome b-mediated electron transfer); and in peroxisomes (Apel and Hirt 2004; Asada 1999). Several apoplastic enzymes may also lead to ROS production under normal and stress conditions. ROS include a number of molecules derived from oxygen, such as oxygen ions, free radicals, and peroxides. All of these molecules are highly reactive due to the presence of unpaired electrons at valence shell. ROS have important roles in cell signaling, cellular homeostasis, and oxidative stress (Kotchoni and Gachomo 2006; Neill et al. 2002). ROS play two main divergent roles in plants; when present in low concentrations, they act as signaling molecules mediating several responses in plant,

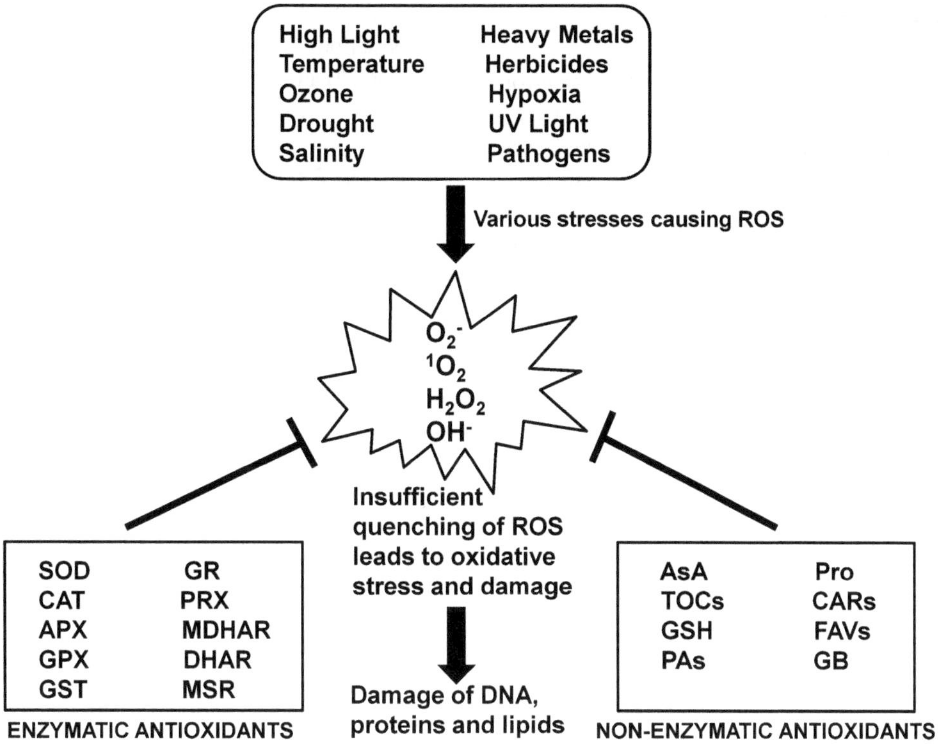

Fig. 1 Diagrammatic representation of various agents generating ROS in plants and various antioxidants scavenging those ROS molecules

including growth and development, and responses under stresses, whereas when present in high concentrations, they cause extensive damage to cellular components. However, as described above, during times of various environmental stresses, ROS levels can increase dramatically. Thus, ROS has multiple roles in cells, and therefore, it is not feasible to eliminate them completely, but at the same time, it is extremely necessary to control them tightly to avoid any oxidative damage. High levels of ROS can cause damage to cellular structures, nucleic acids, lipids, and proteins (Bergamini et al. 2004; Wiseman and Halliwell 1996). ROS is quenched by a number of enzymatic antioxidants and molecules (see next section) (Fig. 1). The problem arises in stressful conditions, when there is disequilibrium between ROS production and its quenching. Then, modulation of ROS by the antioxidants is necessary for the survival of the cell. When the level of ROS in a cell exceeds the antioxidative capacity of the antioxidants, a cell is said to be in a state of oxidative stress.

Superoxide Radicals (O_2^-)

In plants, photosynthesis takes place in chloroplasts, and oxygen is generated which can accept electrons passing through the photosystems, thus forming O_2^-. O_2^- is mainly produced in the thylakoid membrane-bound primary electron acceptor of photosystem I. Normally O_2 is converted to H_2O by transfer of four electrons, but occasionally O_2 can react with other electron transfer chain components and only one electron is transferred, producing the O_2^- (Giba et al. 1998). With one unpaired electron, O_2^- is a free radical (Forman et al. 2004). The generation of O_2^- may trigger the formation of more reactive ROS like OH^-, $^1O_2$, and also H_2O_2 (Halliwell 2006).

Singlet Oxygen ($^1O_2$)

Singlet oxygen, $^1O_2$, is the first excited electronic state of O_2 and is not related to electron transfer to O_2. During photosynthesis in plants, sometimes insufficient energy dissipation can lead to the formation of a chlorophyll triplet state that can transfer its excitation energy to ground-state O_2 to form $^1O_2$ (Halliwell 2006). This can oxidize chloroplast molecules and also trigger cell death. Additionally, $^1O_2$ is formed by low intercellular CO_2 concentration in the chloroplast resulting from the closed stomata because of various abiotic stresses such as salinity and drought. $^1O_2$ production has also been produced as a mechanism of resistance in plant–pathogen interactions through the production of phytoalexins (Flors and Nonell 2006; Flors et al. 2006). Formation of $^1O_2$ during photosynthesis has a substantial damaging effect on essential components of the whole photosynthetic machinery. $^1O_2$ activates a signaling cascade that can stimulate a specific gene expression response and can interact with signal cascades of other ROS, thereby activating several stress-response pathways (op den Camp et al. 2003). $^1O_2$ is highly diffusive and destructive, as it reacts with most biomolecules and rapidly oxidizes amino acids, lipids, pigments, and DNA (Agnez-Lima et al. 2012; Fischer et al. 2013). It reacts with nitric oxide (NO) to form peroxynitrite ($ONOO^-$).

Hydrogen Peroxide (H_2O_2)

H_2O_2 is mainly produced by dismutation of O_2^- by superoxide dismutase (SOD), NADPH oxidase, cell-wall peroxidase, amino oxidase, oxalate oxidase, and flavin-containing oxidase (Neill et al. 2002). Excess of H_2O_2 in the plant cells leads to the occurrence of oxidative stress. H_2O_2 is moderately reactive, relatively stable, and highly diffusible. At low concentrations, H_2O_2 acts as a signaling molecule involved in mediating the acquisition of tolerance to both biotic and abiotic stresses (Desikan et al. 2004), while at high concentrations, it leads to programmed cell death (Quan et al. 2008). H_2O_2 oxidizes the thiol groups of enzymes and thereby inactivates them. It plays a role as an intermediate in the formation of other ROS, including hypochlorous acid (HOCl) and •OH.

Hydroxyl Radicals (OH^-)

This radical is formed from H_2O_2 in a reaction catalyzed by metal ions (Fe^{2+} or Cu^+), often present in complex with different proteins or other molecules. This is known as the Fenton reaction: $Fe^{2+} + H_2O_2 \rightarrow Fe^{3+} + \bullet OH + OH^-$. OH^- can also be produced from O_2^- and H_2O_2 at neutral pH and ambient temperatures by the iron-catalyzed O_2^-. This is called the Haber–Weiss reaction: $O_2^- + H_2O_2 \rightarrow \bullet OH + OH^- + O_2$. OH^- is the most reactive among all ROS and dangerous as it can potentially react with all biomolecules like DNA, proteins, lipids, and almost any constituent of cells and ultimately leads to cell death (Halliwell 2006). It causes lipid peroxidation, protein damage, and membrane destruction.

Effects of ROS

On Metabolism

Lipid peroxidation: When higher ROS levels are reached, lipid peroxidation takes place in both cellular and organelle membranes, thereby not only directly affecting normal cellular functioning but also aggravating the oxidative stress through production of lipid-derived radicals (Catala 2010). Hydroperoxyl radical ($HO_2^{\cdot}$) which is formed from O_2^- by protonation in aqueous solutions can subtract hydrogen atoms from polyunsaturated fatty acids (PUFAs) and lipid hydroperoxides and cause lipid autooxidation. Because of the presence of double bonds, PUFAs are excellent targets for attack by free radicals (particularly $^1O_2$ and OH^-), forming mixture of lipid hydroperoxides (LOOH) (Moller et al. 2007). Extensive PUFA peroxidation decreases the membrane fluidity, increases leakiness, and causes secondary damage to membrane proteins (Moller et al. 2007).

Protein oxidation: It is defined as covalent modification of a protein which can be induced by ROS or components of oxidative stress. Reaction of proteins with ROS causes modifications in the form of the following: (i) side chains oxidation mainly at cysteine, methionine, and tryptophan; (ii) carbonylation; (iii) nitrosylation; and (iv) interaction with products of PUFA oxidation (Moller et al. 2007). Formation of a disulfide between two cysteine residues to form cystine and oxidation of methionine to form methionine sulfoxide are some of the common modifications. Other amino acids can also be target of redox modifications in proteins. Protein nitrosylation mainly involves the covalent attachment of a nitric oxide (NO) group to the thiol side chain of select cysteine residues. This attachment impacts the protein function. But among all modifications, protein carbonylation is the main feature of protein oxidation. Carbonylation of amino acid residues like arginine, lysine, threonine, or proline is one of the most commonly occurring oxidative modifications of proteins. This modification might lead to alteration in protein activity, its proteolytic breakdown or aggregate formation (Debska et al. 2012; Moller et al. 2007).

On Growth and Development

Oxidative stress has marked effect on growth and development of plants. The common stress-associated phenotypes seen under various abiotic stresses have been termed stress-induced morphogenetic response (SIMR; (Potters et al. 2007)). Typical SIMR responses include decreases in root length, stem height, and leaf area, altered xylem development, and redistribution of cell division and elongation (reviewed by (Potters et al. 2009)).

DNA Damage

Exposure to various biotic and abiotic stress factors might damage the DNA and exerts genotoxic stress (Balestrazzi et al. 2011; Tuteja et al. 2009). OH^- is most reactive and damages both purine and pyrimidine bases and also deoxyribose backbone of the DNA molecule (Cooke et al. 2003). This can induce cleavage of DNA, base deletion, formation of pyrimidine dimers, DNA–protein cross-links, and alkylation and oxidation of bases of DNA (Tuteja et al. 2001). $^1O_2$ primarily attacks guanine and converts it into eight-hydroxyguanine (Britt 1996). In addition to mutations, oxidative DNA modifications can lead to changes in the methylation of cytosines, which is important for regulating gene expression (Moller et al. 2007).

Cellular Protection Against Oxidative Stress by Scavenging of ROS

Under normal conditions, the ROS molecules are scavenged by various antioxidants (Foyer and Noctor 2005). In order to control ROS levels and protect the cells from oxidative damage, plants deploy a complex antioxidant defense system which scavenges the ROS. These antioxidant systems include various enzymes and nonenzymatic metabolites that may also play a significant role in ROS signaling in plants (Vranova et al. 2002). Many of these enzymes and molecules are overexpressed and accumulate in various stressed conditions.

Enzymatic Antioxidants

Superoxide Dismutase (SOD)

SOD is the most effective intracellular enzymatic antioxidant which is ubiquitously found in all cellular compartments of organisms (Table 1). SODs are metalloproteins that catalyze dismutation of superoxide radical (O_2^-) into O_2 and H_2O_2. SODs are classified by their metal cofactors into three known families: CuZnSODs, localized in cytosol or in plastids; MnSOD, mainly restricted to mitochondria; and FeSOD, localized in the plastid (Asensio et al. 2012). These SODs can be tissue specific such as CuZnSOD (Ogawa et al. 1997; Karlsson et al. 2005) or MnSOD (Corpas et al. 2006). SOD acts

Table 1 ROS-scavenging enzymes and molecules

Antioxidant enzymes	E. C. number	Subcellular localization	Reaction(s) catalyzed
Superoxide dismutase	1.15.1.1	Cyt, Chl, Mit, Per	$2O_2^- + 2H^+ \rightarrow 2H_2O_2 + O_2$
Catalase	1.11.1.6	Gly, Cyt, Mit, Per	$2H_2O_2 \rightarrow O_2 + 2H_2O$
Ascorbate peroxidase	1.11.1.11	Cyt, Per, Chl, Mit	$2AsA + H_2O_2 \rightarrow 2MDHA + 2H_2O$
Monodehydroascorbate reductase	1.6.5.4	Chl stroma, Mit, Cyt	$NADPH + 2MDHA \rightarrow NADP^+ + 2AsA$
Dehydroascorbate reductase	1.8.5.1	Cyt, Chl, Mit	$2GSH + DHA \rightarrow GSSG + AsA$
Guaiacol-type peroxidase	1.11.1.7	Vac, Chl, Cyt, Mit, ER	Donor + $H_2O_2 \rightarrow$ oxidized donor + $2H_2O$
Peroxiredoxin	1.11.1.15	Cyt, Chl, Mit, Nuc	$ROOH + 2RSH \rightarrow ROH + RSSR + H_2O$
Glutathione peroxidase	1.11.1.12	Cyt, Mit, Chl	$GSH + ROOH \rightarrow GSSG + ROH + 2H_2O$
	1.11.1.9		$GSH + H_2O_2 \rightarrow GSSG + 2H_2O$
Glutathione reductase	1.6.4.2	Cyt, Chl, Mit	$NADPH + GSSG \rightarrow NADP^+ + GSH$
Glutathione S-transferases	2.5.1.18	Cyt, Mit, Chl	$RX + GSH \rightarrow HX + R\text{-}S\text{-}GSH$
Methionine sulfoxide reductase	1.8.4.11	Cyt, Chl, Mit, ER, SP	Met. + thioredoxin disulfide $\rightarrow$ Met.(S)-S oxide + Thioredoxin
Thioredoxin	1.8.1.9	Cyt, Chl, Mit, Nuc	Thioredoxin + $NADP^+ \rightarrow$ thioredoxin disulfide + NADPH
Glutaredoxin	1.20.4.1	Cyt, Chl, Mit, Apo	Arsenate + glutaredoxin $\rightarrow$ arsenite + glutaredoxin disulfide + H_2O
Antioxidant molecules	**Subcellular localization**		**Reaction(s) catalyzed**
Ascorbate (vitamin C)	Chl, Apo, Cyt, Vac, Mit, Per		$O_2^- + AsA + H^+ \rightarrow A.^- + H_2O_2$
			$^1O_2 + AsA + H^+ \rightarrow H_2O_2 + DHA$
			$H_2O_2 + AsA \rightarrow MDHA + 2H_2O$
Glutathione (GSH)	Chl, Apo, Cyt, Vac, Mit, Per		$ROOH + 2GSH \rightarrow ROH + GSSG + H_2O$
			Scavenges H_2O_2, OH^-, and $^1O_2$
Proline	Cyt (normal conditions), Chl (stressed conditions)		Proline + $OH^-/^1O_2 \rightarrow$ proline nitroxide/proline peroxide
α-Tocopherol (vitamin E)	Cell and Chl memb		α-Tocopherol + $^1O_2 \rightarrow$ α-tocopherylquinone
			Also, scavenges OH, ROO,·and ROOH (lipid peroxyl) radicals in thylakoid membranes
Carotenoids (ß-Carotene and zeaxanthin)	Chl, chromoplast, elaioplast, and amyloplast		Quench $^1O_2$ by involving their own oxidation and forming β-carotene endoperoxide in high-light conditions, modulate formation of triplet chlorophyll, and prevent formation of $^1O_2$
Flavonoids	Vac, Nuc, Chl, Apo, Cyt, ER, Cwl		Scavenge H_2O_2 and OH^-
Polyamines	Nuc, Chl, Mit, Cyt, Vac, Cwl		Scavenge O_2^-, $^1O_2$, OH^-
Glycine betaine	Chl		Stabilizes PS II repair proteins and alleviate lipid peroxidation

$^1O_2$ singlet oxygen, *A.*$^-$, ascorbate free radical, *AsA* ascorbate, *Apo* apoplast, *APX* ascorbate peroxidase, *Cyt* cytosol, *Chl* chloroplast, *Cwl* cell wall, *DHA* dehydroascorbate, *DHAR* dehydroascorbate reductase, *ER* endoplasmic reticulum, *Gly* glyoxysomes, *GSH* glutathione, *GSSG* oxidized glutathione, *GSTs* glutathione *S*-transferases, *H^+* hydrogen ion, *H_2O* water, *H_2O_2* hydrogen peroxide, *Met* methionine, *MDHA* monodehydroascorbate, *MDHAR* monodehydroascorbate reductase, *Mit* mitochondria, *$NADP^+$* nicotinamide adenine dinucleotide phosphate, *NADPH* reduced $NADP^+$, *Nuc* nucleus, *O_2* oxygen, *O_2^-* superoxide radical, *OH*˙ hydroxyl radical, *Per* peroxisomes, *PSII* photosystem II, *R* may be an aliphatic, aromatic, or heterocyclic group, *ROO*$^-$ peroxy radical, *ROOH* organic peroxide, *S* sulfide, *SP* secreted pathway, *Vac* vacuole, *X* may be a sulfate, nitrite, or halide group

in the first line of defense against the toxic effects of elevated levels of ROS. The SODs remove superoxide radical and hence decrease the risk of OH^- formation. SODs are upregulated in response to many abiotic and biotic stresses and have a crucial role in the survival of plants under stressed conditions (Alscher et al. 2002; Raychaudhuri and Deng 2000). There have been many reports of the production of both biotic and abiotic stress-tolerant transgenic plants overexpressing different SODs (Table 2).

Catalase (CAT)

CAT is a heme-coordinated tetrameric protein encoded by nuclear genes that plays an important role in maintaining cellular concentration of hydrogen peroxide to a level, necessary for all aspects of normal plant growth and development. CATs are located mostly in peroxisomes and glyoxysomes, where they play a key role in the removal of H_2O_2 generated by various oxidases. CAT directly dismutates H_2O_2 into H_2O and O_2 and are required for ROS detoxification during stressed conditions (Mittler et al. 2004). CAT also reacts with some hydroperoxides.

Ascorbate Peroxidase (APX)

The APX are heme-containing enzymes. The APX family has many isoforms located at different subcellular locations like thylakoid (tAPX), glyoxisome membrane (gmAPX), chloroplast stroma (sAPX), and cytosol (cAPX) (Dąbrowska et al. 2007). APX is involved in scavenging of H_2O_2. APX catalyzes the reduction of H_2O_2 to water and uses ascorbate as a reductant for this reaction (Shigeoka et al. 2002; Asada 1999). The scavenging of H_2O_2 by APX is the first step of the ascorbate–glutathione (ASH-GSH) cycle. APX activity is enhanced in plants in response to different abiotic stresses. In *Arabidopsis thaliana* APX activity increased during exposure of plants to ozone, sulfur dioxide, chilling, and UV-B (Kubo et al. 1995; Rao et al. 1996).

Monodehydroascorbate Reductase (MDHAR)

MDHAR is a flavin adenine dinucleotide (FAD) enzyme, present in cytosol and chloroplast. MDHAR has a high specificity for monodehydroascorbate (MDHA) as the electron acceptor and NADPH as the electron donor. Oxidation of ascorbate (AsA) leads to the formation of MDHA. If MDHA is not reduced again to AsA by MDHAR, it will spontaneously convert into AsA and dehydroascorbate (DHA). Therefore, MDHAR rapidly reduce MDHA to AsA using NADPH. This rapid regeneration is necessary in order to maintain the antioxidative capacity of AsA.

Dehydroascorbate Reductase (DHAR)

DHAR is a monomeric thiol enzyme which regenerates ascorbate from dehydroascorbate (DHA) (Foyer and Mullineaux 1998). DHAR catalyzes the reduction of DHA to AsA using GSH as the reducing substrate. This is a key in conferring tolerance to various abiotic stresses which produce ROS. Stresses such as drought, chilling, ozone, and metal toxicity increase the activity of the DHAR in plants (Maheshwari and Dubey 2009; Yoshida et al. 2006). It has also been found that DHAR overexpression also enhances plant tolerance against various abiotic stresses (Lee et al. 2007) (Table 2).

Thus, MDHAR and DHAR are equally important in regulating the level of AsA and its redox state under oxidative stress (Eltayeb et al. 2006, 2007).

Guaiacol Peroxidase (GPOX)

GPOX is a heme-containing protein. GPOX has a role in the biosynthesis of lignin and defense against biotic stresses by consuming H_2O_2. GPOX in plants use guaiacol and pyrogallol as a reducing substrate to oxidize many substrates in the presence of H_2O_2 (Vianello et al. 1997). This means that GPOX oxidize certain substrates at the expense of H_2O_2 and rid the cell of excess peroxide produced, especially under stress conditions. They are also effective quenchers of reactive intermediary forms of O_2 and peroxyl radicals.

Peroxiredoxins (PRXs)

Peroxiredoxins are non-heme containing peroxidases, which have to rely on an external electron donor to reduce H_2O_2, alkyl hydroperoxide, and

Table 2 Transgenic plants accumulating various antioxidant enzymes and molecules

Gene and source	Transgenic organism	Response of transgenic against various stresses	Reference
Superoxide dismutase (SOD)			
FeSOD from *Arabidopsis thaliana*	*Nicotiana tabacum*	Enhanced tolerance to oxidative stress induced by methyl viologen (MV)	(Van Camp et al. 1996)
Cu/ZnSOD from *Hevea brasiliensis*	*Hevea brasiliensis*	Protection against ROS and tolerance to water deficit	(Leclercq et al. 2012)
Chloroplastic Cu/ZnSOD from *Pisum sativum*	*Nicotiana tabacum*	Resistance against MV-mediated oxidative stress	(Gupta et al. 1993a; Kwon et al. 2002)
MnSOD from *Triticum aestivum*	*Brassica napus*	Resistance to aluminum	(Basu et al. 2001)
MnSOD from *Nicotiana plumbaginifolia*	*Medicago sativa*	Resistance to cold stress	(McKersie et al. 1999)
MnSOD from Pisum sativum	*Oryza sativa*	Resistance to drought and MV and polyethylene glycol (PEG)-induced oxidative stress	(Wang et al. 2005)
Cu/Zn SOD from *Spinacia oleracea*	*Malus domestica*	Resistance to high and freezing temperatures	(Artlip et al. 2009)
Cu/Zn SOD from *Oryza sativa*	*Nicotiana tabacum*	Enhanced tolerance to salt, water, and PEG stresses	(Badawi et al. 2004)
Cu/Zn SOD from *Avicennia marina*	*Oryza sativa*	Tolerance to MV-mediated oxidative stress, salinity and drought stress	(Prashanth et al. 2008)
Cu/Zn SOD from *Solanum lycopersicum*	*Solanum tuberosum*	Elevated tolerance to MV	(Perl et al. 1993)
Cytosolic Cu/ZnSOD from *Solanum lycopersicum*	*Beta vulgaris*	Increased tolerance to MV and to leaf infection with the fungus *Cercospora beticola*	(Tertivanidis et al. 2004)
MnSOD from *Tamarix androssowii*	*Populus davidiana*	Enhanced salt tolerance	(Wang et al. 2010)
Catalase (CAT)			
Catalase from Broccoli	*Arabidopsis thaliana*	Tolerance against heat stress by removing H_2O_2	(Chiang et al. 2013)
Catalase gene, katE, from *E. coli*	*Oryza sativa*	Tolerance against salt stress by removing H_2O_2	(Nagamiya et al. 2007)
Catalase gene, *Cat2*, from *Zea mays*	*Nicotiana tabacum*	Tolerance to MV-mediated oxidative stress and resistance to bacterial pathogen	(Polidoros et al. 2001)
Catalase gene, *Cat2*, from *Nicotiana tabacum*	*Solanum tuberosum*	Enhanced resistance to fungal pathogen, *P. infestans*	(Yu et al. 1999)
Catalase gene from *Triticum aestivum*	*Oryza sativa*	Increased resistance to low-temperature stress by removing H_2O_2	(Matsumuraa et al. 2002)
Ascorbate peroxidase (APX)			
Ascorbate peroxidase-like 1 gene from *Capsicum annuum*	*Nicotiana tabacum*	Increased tolerance to MV-mediated oxidative stress and to the oomycete pathogen, *Phytophthora nicotianae*	(Sarowar et al. 2005)
tAPX from *Brassica napus*	*Brassica napus*	Increases resistance to salt stress and drought, and reduced accumulation of H_2O_2	(Wang et al. 2011)

(continued)

Table 2 (continued)

Gene and source	Transgenic organism	Response of transgenic against various stresses	Reference
cAPX from *Pisum sativum*	*Solanum lycopersicum*	Tolerance to chilling and salt stress	(Wang et al. 2005)
APX3 from *Arabidopsis thaliana*	*Nicotiana tabacum*	Increases protection against oxidative stress	(Wang et al. 1999)
Two cytosolic ascorbate peroxidases from *Oryza sativa*	*Arabidopsis thaliana*	Increases protection against salt stress	(Lu et al. 2007)
Glutathione reductase (GR)			
GR from bacteria	A poplar hybrid, *Populus tremula* x *Populus alba*	Tolerance to oxidative stress caused by MV	(Foyer et al. 1995)
GR from *E. coli*	*Nicotiana tabacum*	Tolerance to oxidative stress caused by MV, H_2O_2, heavy metal stress, and UV-B radiation	(Lederer and Boger 2003; Poage et al. 2011)
GR from *Pisum sativum*	*Nicotiana tabacum*	Tolerance to oxidative stress caused by MV	(Creissen et al. 1994)
Monodehydroascorbate reductase (MDHAR)			
cMDHAR from *Solanum lycopersicum*	*Solanum lycopersicum*	Resistant to salt- and PEG-induced osmotic stress, showing lower level of H_2O_2, higher APX activity. Enhanced tolerance to temperature and MV-mediated oxidative stresses	(Li et al. 2010a; Li et al. 2012b)
MDAR1 gene from *Arabidopsis thaliana*	*Nicotiana tabacum*	Enhanced tolerance against ozone, salt, and PEG stress	(Eltayeb et al. 2007)
Dehydroascorbate reductase (DHAR)			
DHAR from *Homo sapiens*	*Nicotiana tabacum*	Enhanced tolerance to MV, H_2O_2, low temperature, and salt	(Kwon et al. 2003)
DHAR from *Oryza sativa*	*Arabidopsis thaliana*	Enhanced resistance to salt stress	(Ushimaru et al. 2006)
cDHAR from *Arabidopsis thaliana*	*Nicotiana tabacum*	Enhanced tolerance to aluminum stress	(Yin et al. 2010)
cDHAR from *Arabidopsis thaliana*	*Nicotiana tabacum*	Enhanced tolerance to ozone, drought, salt, and PEG stresses	(Eltayeb et al. 2006)
Glutathione S-transferases (GSTs)			
GST and GPX from tobacco	*Nicotiana tabacum*	Enhanced tolerance to oxidative, chilling, and salt stress	(Roxas et al. 2000, 1997)
GST from cotton	*Nicotiana tabacum*	Enhanced tolerance to oxidative stress induced by a low concentration of MV	(Yu et al. 2003)
GST from maize	*Triticum aestivum*	Tolerant to herbicide	(Milligan et al. 2001)
GST from rice	*Oryza sativa*	Enhanced tolerance to low temperature	(Takesawa et al. 2002)
GST from *Trichoderma virens*	*Nicotiana tabacum*	Enhanced tolerance to cadmium	(Dixit et al. 2011)
GST, from wild soybean (*Glycine soja*)	*Nicotiana tabacum*	Enhanced tolerance to drought and salt	(Ji et al. 2010)
GST from maize	*Nicotiana tabacum*	Higher tolerance to alachlor	(Karavangeli et al. 2005)

(continued)

Table 2 (continued)

Gene and source	Transgenic organism	Response of transgenic against various stresses	Reference
GST from *Prosopis juliflora*	*Nicotiana tabacum*	Enhanced tolerance to drought	(George et al. 2010)
GST (GSTL1) from *Oryza sativa*	*Oryza sativa*	Enhanced tolerance to herbicides	(Hu et al. 2009)
GST (GSTL2) from *Oryza sativa*	*Arabidopsis thaliana*	Enhanced tolerance for heavy metals and other abiotic stresses like cold, osmotic stress and salt	(Kumar et al. 2013)
Peroxiredoxins (PRXs)			
2-Cys PRX from *Arabidopsis thaliana*	Tall fescue (*Festuca arundinacea*) and *Solanum tuberosum* L. cv. Atlantic	Increased tolerance against heat and MV	(Kim et al. 2010, 2011)
PRX-Q from *Gentiana triflora* cv. Yahaba Y514	*Nicotiana tabacum*	Improved resistance to MV and Botrytis infection	(Kiba et al. 2005)
PRX-Q from *Suaeda salsa*	*Arabidopsis thaliana*	Increased tolerance to salt and cold stress	(Jing et al. 2006)
1-Cys- PRX from *Oryza sativa*	*Nicotiana tabacum*	Increased tolerance to oxidative stress	(Lee et al. 2000)
Glutaredoxin (GRX)			
AtGRXS17 from *Arabidopsis thaliana*	*Solanum lycopersicum*	Enhanced tolerance to oxidative and heat stress	(Wu et al. 2012)
GRX5 from *Pteris vittata*	*Arabidopsis thaliana*	More tolerance to arsenic	(Sundaram et al. 2009)
Thioredoxin (TRX)			
TRX-hl from *Oryza sativa*	*Oryza sativa*	More tolerance to salt stress	(Zhang et al. 2011)
P-TRX from *Phalaris coerulescens*	*Hordeum vulgare*	Increased aluminum resistance	(Li et al. 2010b)
Methionine sulfoxide reductase (MSR)			
Peptide MSR from *Arabidopsis thaliana*	*Arabidopsis thaliana*	Enhanced tolerance to MV and high light in cold conditions	(Romero et al. 2004)
MSRB2 gene from *Capsicum annuum*	*Solanum lycopersicum*	Showed reduced production of H_2O_2 and resistance towards *Phytophthora capsici* and *Phytophthora infestans*	(Oh et al. 2010)
MSRB3 from *Arabidopsis thaliana*	*Arabidopsis thaliana*	Enhanced tolerance to MV and cold stress	(Kwon et al. 2007)
Cytosolic *MSRB7* and *MSRB8* from *Arabidopsis thaliana*	*Arabidopsis thaliana*	Enhanced tolerance to MV and H_2O_2 treatment	(Li et al. 2012a)
MSRA4.1 from *Oryza sativa*	*Oryza sativa*	Enhanced tolerance to salt stress	(Guo et al. 2009)
Glutathione Peroxidase (GPX)			
GPX from Chlamydomonas	*Nicotiana tabacum*	Increased tolerance to oxidative stress caused by MV, cold, and salt stress	(Yoshimura et al. 2004)
GST and GPX from *Nicotiana tabacum*	*Nicotiana tabacum*	Increased tolerance to cold and salt stress	(Roxas et al. 1997)
GPX2 from *Synechocystis*	*Arabidopsis thaliana*	Enhanced tolerance to oxidative damage caused by H_2O_2, MV, Fe ions, and other stresses such as chilling, high salinity, or drought	(Gaber et al. 2006)
Ascorbate (AsA)			
L-Gulono-gamma-lactone oxidase from rat	*Solanum tuberosum* L. cv. Taedong Valley	Increased levels of AsA in transgenic plants leading to enhanced tolerance to MV, salt, and mannitol	(Hemavathi et al. 2010)

(continued)

Table 2 (continued)

Gene and source	Transgenic organism	Response of transgenic against various stresses	Reference
D-Galacturonic acid reductase from strawberry (this enzyme converts D-galacturonic acid into L-Galactonic acid which is converted to L-galactono-1,4-lactone, the immediate precursor of AsA)	*Solanum tuberosum* L. cv. Taedong Valley	Increased levels of AsA in transgenic plants leading to enhanced tolerance to MV, salt, and mannitol	(Hemavathi et al. 2009)
Glutathione (GSH)			
Glutathione synthetase (GS) from *E. coli*	*Brassica juncea*	Enhanced tolerance to cadmium	(Liang Zhu et al. 1999)
Glutathione synthetase enzyme from *Streptococcus thermophilus*	*Nicotiana tabacum*	Enhancing tolerance to abiotic stress	(Liedschulte et al. 2010)
γ-Glutamylcysteine synthetase and Glutathione synthetase from *Brassica juncea*	*Brassica juncea*	Enhanced tolerance to atrazine; 1-chloro-2, 4-dinitrobenzene; phenanthrene; and metolachlor	(Flocco et al. 2004)
γ-Glutamylcysteine synthetase (γ-ECS) from *Oryza sativa*	*Oryza sativa*	Enhanced tolerance to salt and MV	(Choe et al. 2013)
γ-Glutamylcysteine synthetase (γ-ECS) from *E. coli*	Poplar hybrid (*Populus tremula* X *Populus alba*)	Enhanced tolerance to chloroacetanilide herbicides, acetochlor, and metolachlor	(Gullner et al. 2001)
Proline (Pro)			
P5CS from *Vigna aconitifolia* L.	*Nicotiana tabacum*	Increased tolerance to osmotic stress	(Kishor et al. 1995)
P5CS from *Vigna aconitifolia* L.	*Medicago truncatula*	Enhanced tolerance to osmotic stress	(Verdoy et al. 2006)
P5CS from *Vigna aconitifolia* L.	*Cicer arietinum*	Enhanced tolerance to salt stress	(Kiran Kumar Ghanti et al. 2011)
Osmotin gene	*Nicotiana tabacum*	Enhanced tolerance to salinity and drought	(Barthakur et al. 2001)
P5CS from *Vigna aconitifolia*	*Oryza sativa* L. ssp. *indica* cv. *ADT 43*	Enhanced tolerance to salt stress	(Karthikeyan et al. 2011)
P5CS from *Vigna aconitifolia*	*Triticum aestivum*	Enhanced tolerance to drought	(Vendruscolo et al. 2007)
P5CR from *Triticum aestivum*	*Arabidopsis thaliana*	Enhanced stress tolerance	(Ma et al. 2008)
P5CR from *Arabidopsis thaliana*	*Glycine max*	Enhanced tolerance to heat and drought stress	(De Ronde et al. 2004)
Ornithine-δ-aminotransferase (δ-OAT) from *Arabidopsis thaliana*	*Oryza sativa* L. ssp. *japonica* cv. Zhongzuo 321	Accumulation of proline and enhanced tolerance to drought and salt stress	(Liangqi et al. 2003)
Antisense of proline dehydrogenase (ProDH) from *Arabidopsis thaliana*	*Arabidopsis thaliana*	Proline accumulation and more tolerance to freezing and high salinity	(Nanjo et al. 1999)
Tocopherols (TOCs)			
Tocopherol cyclase (VTE1) from *Arabidopsis thaliana*	*Nicotiana tabacum*	Enhanced tolerance to drought induced oxidative stress	(Liu et al. 2008)
VTE2.1 from *Solanum chilense*	*Nicotiana tabacum*	Increased tolerance to oxidative stress damage as evidenced by reduced lipid peroxidation and delayed leaf senescence	(Espinoza et al. 2013)
Carotenoids (CARs)			
Beta-carotene ketolase gene (bkt) from green algae	*Daucus carota*	Plants accumulate ketocarotenoids and show enhanced tolerance to UV-B radiation, H_2O_2, and MV	(Jayaraj and Punja 2008)

(continued)

Table 2 (continued)

Gene and source	Transgenic organism	Response of transgenic against various stresses	Reference
chyB gene that encodes beta-carotene hydroxylase to make zeaxanthin from *Arabidopsis thaliana*	*Arabidopsis thaliana*	More tolerance to high light, high temperature, and lipid peroxidation	(Davison et al. 2002)
chyB gene that encodes beta-carotene hydroxylase from *Arabidopsis thaliana*	*Nicotiana tabacum*	More tolerance to drought stress, reduced lipid peroxidation	(Zhao et al. 2013)
DSM2 gene which encodes β-carotene hydroxylase (BCH) from *Oryza sativa*	*Oryza sativa*	Increased resistance to drought and oxidative stresses and increase of the xanthophylls and non-photochemical quenching	(Du et al. 2010)
Flavonoids (FLVs)			
Anthocyanidin synthase (ANS) from *Oryza sativa*	*Oryza sativa* mutant Nootripathu (NP)	Accumulation of a mixture of flavonoids and anthocyanins, with increased antioxidant potential	(Reddy et al. 2007)
Flavonol synthase 1 (FLS1) from *Zea mays*	*Arabidopsis thaliana*	Increased resistance to UV-B radiation	(Emiliani et al. 2013)
Polyamines (PAs)			
S-Adenosyl-l-methionine decarboxylase (SAMDC) from yeast	*Solanum lycopersicum*	Enhanced tolerance to high-temperature stress	(Cheng et al. 2009)
Arginine decarboxylase (ADC) from *Datura stramonium*	*Oryza sativa*	Enhanced tolerance to drought	(Capell et al. 2004)
Spermidine synthase (SPDS) from apple (*Malus domestica*)	European pear (*Pyrus communis* L. 'Ballad')	Enhanced tolerance to salinity, osmotic, and heavy metal stress	(Wen et al. 2008)
Spermidine synthase (SPDS) from *Cucurbita ficifolia*	*Arabidopsis thaliana*	Enhanced tolerance to chilling, freezing, salinity, hyperosmosis, drought, and MV	(Kasukabe et al. 2004)
ADC from oat (*Avena sativa*)	*Oryza sativa*	Enhanced tolerance to salt stress	(Roy and Wu 2001)
SAMDC from human	*Nicotiana tabacum* var. *Xanthi*	Enhanced tolerance to salinity, drought, and fungal pathogens (*V. dahliae* and *F. oxysporum)*	(Waie and Rajam 2003)
SAMDC from carnation (*Dianthus caryophyllus* L.) flower	*Nicotiana tabacum*	Enhanced tolerance to salt, cold, acidic, and abscisic acid stress	(Wi et al. 2006)
SPDS from *Cucurbita ficifolia*	Ipomoea batatas, cv. Kokei 14	Increased tolerance to chilling, heat, and MV	(Kasukabe et al. 2006)
Glycine betaine (GB)			
Choline oxidase (*codA*) from *Arthrobacter globiformis*	*Solanum lycopersicum* Mill. cv. Moneymaker	Enhanced tolerance to chilling, high salt, and oxidative stresses	(Park et al. 2007)
Choline oxidase (*codA*) from *Arthrobacter globiformis*	*Solanum tuberosum* L. cv. Superior	Enhanced tolerance to salt, drought, and MV	(Ahmad et al. 2008b)
Choline monooxygenase *(CMO)* from *Atriplex hortensis*	*Gossypium hirsutum*	Increased tolerance to salt stress	(Zhang et al. 2009)
Choline dehydrogenase (*CDH*), encoded by *betA* from *E. coli*	*Nicotiana tabacum*	Improved tolerance to salinity and cold stress	(Holmstrom et al. 2000)
Choline monooxygenase *(CMO)* from *Spinacia oleracea*	*Oryza sativa*	Enhanced tolerance to salt stress and temperature stress	(Shirasawa et al. 2006)
Betaine aldehyde dehydrogenase (*BADH*) from *Spinacia oleracea*	*Nicotiana tabacum*	Protection of Rubisco activity in high-temperature stress	(Yang et al. 2005)
A chloroplastic *BADH* from *Spinacia oleracea*	*Ipomoea batatas*, cv.-Sushu-2	Improved tolerance to salt, oxidative stress, and low temperature	(Fan et al. 2012)

(continued)

Table 2 (continued)

Gene and source	Transgenic organism	Response of transgenic against various stresses	Reference
Others			
Aldose/aldehyde reductase from *Medicago sativa*	*Nicotiana tabacum*	Tolerance against oxidative damage caused by paraquat, heavy metal treatment, and drought	(Oberschall et al. 2000)
Oxalate oxidase (OxO) from wheat	*Nicotiana tabacum*	Increased tolerance to MV or high light-induced oxidative stress	(Wan et al. 2009)
Apolipoprotein D ortholog (*AtTIL*) from *Arabidopsis thaliana*	*Arabidopsis thaliana*	Enhances tolerance to freezing and oxidative stress	(Charron et al. 2008)
Aldehyde dehydrogenase ALDH3I1 and ALDH7B4 from *Arabidopsis thaliana*	*Arabidopsis thaliana*	Enhances tolerance to osmotic and oxidative stress	(Kotchoni et al. 2006)
BcZAT12 from *Brassica carinata*	Solanum lycopersicum, cv. H-86	Tolerance to heat-shock (HS)-induced oxidative stress	(Shah et al. 2013)
Serotonin N-acetyltransferase (NAT) from sheep, producing more melatonin	*Oryza sativa*	Increased resistance to the singlet-oxygen-generating peroxidizing herbicide butafenacil and increased SOD and CAT activity	(Park et al. 2013)
Isoprene synthase from Populus alba, producing more isoprene	*Nicotiana tabacum*	Increased resistance to ozone-induced oxidative damage and high temperature	(Vickers et al. 2009)

other peroxides. This electron donor is often reduced thioredoxin; hence, PRXs are often called thioredoxin peroxidases. The PRX family in plants can be divided into four groups (A to D) (Dietz 2011). A-type PRX are 2-Cys peroxiredoxin (2-CysPRX), the B-type PRX are 1-Cys peroxiredoxin (1-CysPRX), the C-type PRX are peroxiredoxin Q (PRX-Q), and the D-type PRXs are type II peroxiredoxins (PRXII). These four groups can be further divided depending on their subcellular locations. In plants, 2-Cys-PRXs are the most abundant PRXs and are located in chloroplasts. The 2-Cys-PRXs reduce peroxides through a thiol-based mechanism. During catalysis, these enzymes are sometimes inactivated by the substrate-dependent oxidation of the catalytic cysteine to the sulfinic acid ($-SO_2H$) form and are reactivated by reduction carried by sulfiredoxin (SRX) (Jonsson et al. 2008).

Glutathione Peroxidase (GPX)

GPXs are considered as a fifth class of PRX, but evolutionary PRX and GPX are considered two different protein families. GPXs are a large family of diverse isozymes (Rodriguez Milla et al. 2003). GPXs help plants alleviate oxidative stress by reducing a broad range of hydroperoxides, including H_2O_2 and organic and lipid hydroperoxides (LOOH) (Arthur 2000). They use GSH to do this function. They help prevent lipid peroxidation of cellular membranes by removing free peroxide in the cell. GPX also functions as an oxidative signal transducer (Miao et al. 2006). GPXs are reduced by thioredoxins.

Glutathione Reductase (GR)

GR is a flavoprotein oxidoreductase and is a key player against ROS defense by maintaining the reduced status of glutathione (GSH). It is localized mainly in chloroplasts but also in small amounts in mitochondria and cytosol (Edwards et al. 1990). GR catalyzes the conversion of GSSG into GSH (Meister and Anderson 1983). GR transfers electrons from NADPH to GSSG to generate GSH. Thus, GR maintains a high ratio of GSH/GSSG in plant cells which is important for scavenging H_2O_2. GR and GSH play a crucial role in determining the tolerance of a plant under various stresses (Foyer et al. 1997). In rice, expression of GR was found to be induced by ABA and chilling, drought, and salinity (Kaminaka et al. 1998).

Glutathione *S*-Transferases (GSTs)

Plant GSTs can be divided into eight classes of: phi, tau, theta, zeta, lambda, dehydroascorbate reductase

(DHAR), EF1Bγ and tetrachlorohydroquinone dehalogenase (TCHQD). Arabidopsis encodes about 55 GSTs (Dixon and Edwards 2010). GSTs catalyze the transfer of the tripeptide glutathione (γ-glutamyl-cysteinyl-glycine; GSH) to a cosubstrate (R-X) containing a reactive electrophilic center to form a polar *S*-glutathionylated reaction product (R-SG). GSTs detoxify electrophilic herbicides and other xenobiotics by catalyzing their conjugation with GS, to produce less toxic and more water-soluble conjugates. Apart from herbicide detoxification, plant GSTs are known to function in hormone homeostasis, tyrosine metabolism, hydroxyperoxide detoxification, and plant responses to various stresses. GST activity in plants is induced in response to many abiotic stresses (Dixon et al. 2010). GSTs safeguard proteins from oxidative damage and maintain redox homeostasis by regenerating AsA from DHA (Dixon and Edwards 2010).

Methionine Sulfoxide Reductase (MSR)

In proteins, methionine (Met) residues are especially sensitive to oxidation, as ROS can oxidize them to form S and R methionine sulfoxide (MetSO) diastereoisomers. Thus, Met residues form Met-*S*-sulfoxide or Met-*R*-sulfoxide, causing inactivation or malfunction of the proteins. To rescue the proteins, the oxidized forms of methionine, S-MetSO and R-MetSO, are reduced back to Met by the MetSO reductases, MsrA and MsrB, respectively (Sharov and Schoneich 2000). These proteins catalyze the thioredoxin-dependent reduction of MetSO back to Met (Brot et al. 1981), thereby repairing proteins. MSRs are proposed to act as a last-chance antioxidants and importantly repair proteins damaged from oxidative stress (Cabreiro et al. 2006).

Glutaredoxin (GRX) and Thioredoxin (TRX)

Thioredoxins (TRX) and glutaredoxins (GRX) constitute families of thiol oxidoreductases, furnishing reducing power to PRX, MSR, and arsenate reductases, which are key players for the plant response to the oxidative environment.

TRXs are small redox proteins, widely distributed, and function in redox regulation in a broad spectrum of cellular reactions. Plant TRXs are composed of six well-defined types (TRXs f, m, x, y, h, and o) that reside in different cell compartments and function in different processes (Meyer et al. 2005). TRX can exist either in reduced (dithiol) or in oxidized (disulfide) form. Reduced TRX acts to directly reduce protein disulfides and cysteine sulfenic acid (Meyer et al. 2012).

Land plants contain a large GRX family. GRXs are small, nearly ubiquitous, oxidoreductases constituting an alternative reducing system to TRXs. GRXs catalyze the reduction of disulfide bonds of their substrate proteins in the presence of glutathione (GSH) and help binding of iron–sulfur clusters (Rouhier 2010). GRXs are important proteins for the response of plants to oxidative stress because they help in regenerating antioxidant enzymes. GRXs are directly reduced by GSH to produce GSSG.

Nonenzymatic Antioxidants

Ascorbate (AsA)

AsA is the most potent and abundant antioxidant that protects the cell from the damage caused by ROS in plants (Foyer and Noctor 2011; Smirnoff 2007). Exogenous application of AsA renders the plants to be resistant to salt stress as shown in case of durum wheat (Azzedine et al. 2011). AsA can directly scavenge O_2^-, OH^-, and $^1O_2$ and reduce H_2O_2 to water via ascorbate peroxidase reaction (Noctor and Foyer 1998). APX requires a reducing substrate, ascorbate, which is then oxidized to monodehydroascorbate (MDHA). Transgenic plants overexpressing genes leading to increased ascorbate content confer resistance to oxidative and other stresses (Table 2). AsA is a reduced form, while its oxidized forms are MDHA and DHA. Regeneration of AsA is catalyzed by either MDHAR (from MDHA) or DHAR (from DHA) by using NADPH or reduced glutathione (GSH), respectively. AsA also maintains α-tocopherol in a reduced state. AsA regenerate tocopherol from tocopheroxyl radicals, thus providing protection to the membranes. Exogenous application of AsA positively influences the activity of many antioxidative enzymes and minimizes the oxidative damage (Shalata and Neumann 2001).

Glutathione (GSH)

Reduced glutathione (GSH) is a major water-soluble antioxidant in plant cells, localized in all cell compartments (Table 1) (Moran et al. 2000). GSH is a tripeptide (γ-glutamyl-cysteinyl-glycine), which is synthesized from Cys and that exists interchangeably with the oxidized form, GSSG, and is vital for normal cellular function. Two sequential ATP-dependent reactions allow the synthesis of γ-glutamylcysteine (γ-EC) from L-glutamate and L-cysteine, followed by the formation of GSH by addition of glycine to the C-terminal end of γ-EC (Meister 1988). These reactions are catalyzed by γ-glutamylcysteine synthetase (γ-ECS) and glutathione synthetase (GS). Glutathione plays important roles in protecting cells from biotic and abiotic stress. In a cell, it is the major antioxidant and major cellular redox buffer which directly scavenges most free radicals and reactive oxygen species (Noctor and Foyer 1998). GSH is a key ROS scavenger and can protect macromolecules like proteins, lipids, and DNA by acting as a proton donor forming GSSG. The reduced state of cells brought by GSH counteracts the effects of oxidative stress (Meyer 2008) by scavenging $^1O_2$, H_2O_2, and OH^- (Alscher 1989). Additionally, GSH is critical in regenerating another antioxidant like ascorbate (AsA), via the ASH-GSH cycle (Foyer and Halliwell 1976). Biosynthesis of glutathione is stimulated under oxidative stress conditions, as GSH gets converted to GSSG. In oxidative stress, GSH prevents the denaturation of proteins caused by the oxidation of protein thiol groups. Moreover, GSH acts as a substrate for GPX and GST, which are also involved in the removal of ROS (Noctor et al. 2002).

The AsA-GSH cycle constitutes one of the most important antioxidant systems in plants. In this cycle, the ascorbate and the glutathione are utilized as reducers which are recycled through consuming the ATP and NADPH by the action of four enzymes: APX, MDHAR, DHAR, and GR.

Proline (Pro)

Pro is also considered as a potent antioxidant and potential inhibitor of adverse effects of ROS (Krishnan et al. 2008; Matysik et al. 2002; Szabados and Savoure 2010). In plants the synthesis of L-Pro takes place from L-glutamic acid by the action of enzymes D1-pyrroline-5-carboxylate synthetase (P5CS) and D1-pyrroline-5-carboxylate reductase (P5CR) (Verbruggen and Hermans 2008). Following salt, drought, and metal stress, there is a dramatic accumulation of Pro. Free Pro has been proposed to act as an osmoprotectant, a protein stabilizer, a metal chelator, maintainer of redox homeostasis, and OH^- and $^1O_2$ scavenger (Ashraf and Foolad 2007; Matysik et al. 2002). Pro appeared as an effective scavenger of OH^- (Smirnoff and Cumbes 1989). The constitutive or stress-inducible expression of P5CS cDNA in plants leads to Pro accumulation and confers tolerance to various abiotic stresses (Hmida-Sayari et al. 2005; Su and Wu 2004) (Table 2).

Tocopherols (TOCs)

TOCs are lipid-soluble antioxidant and are potential scavengers of ROS (Shao et al. 2007). TOCs are considered general antioxidants for protection of membrane stability, including quenching or scavenging ROS like $^1O_2$ and OH^-(Krieger-Liszkay and Trebst 2006). TOCs are localized in plants in the thylakoid membrane of chloroplasts. Out of the four isomers of TOCs (α, β, γ, δ) found in plants, *α*- and *γ*-tocopherol are predominant. *α*-tocopherol has the highest antioxidant activity, which together with the hydrophilic antioxidants, glutathione and ascorbate participates in the detoxification of ROS (Kamal-Eldin and Appelqvist 1996). TOCs also reduce lipid peroxyl radicals (LOO^-) to their corresponding hydroperoxides (Maeda et al. 2005). TOCs also participate in cell signaling. Oxidative stress activates the expression of genes responsible for the synthesis of TOCs in higher plants (Ahmad et al. 2008a). Tocopherol cyclase (VTE1) catalyzes the penultimate step of TOC synthesis Porfirova et al. 2002.

Carotenoids (CARs)

CARs are the most abundant pigmented plant-derived compounds. CARs are considered to be the first line of defense of plants against toxicity. Like TOCs, CARs are lipid-soluble antioxidants that play a role in oxidative stress tolerance (Edge et al. 1997). Oxygenated CARs are known as xanthophylls. Examples of these compounds are zeaxanthin and lutein. In all photosynthetic organisms, the carotenoids β-carotene and zeaxanthin, together with TOCs play a photoprotective role, either by dissipating excess excitation energy as heat or by scavenging ROS and suppressing lipid peroxidation. They play a role of an antioxidant by preventing the formation of singlet oxygen by quenching the triplet chlorophylls (Chl3) and other harmful free radicals which are naturally formed during photosynthesis (Ramel et al. 2012).

Flavonoids (FLVs)

FLVs are a group of polyphenolic compounds produced as secondary metabolites by plants. FLVs occur widely in the plant kingdom and accumulate in the plant vacuole as glycosides and also as exudates on the surface of leaves and other aerial plant parts. FLVs serve as ROS scavengers by neutralizing harmful radicals under adverse environmental conditions (Agati et al. 2007). FLVs absorb UV light, and plants able to synthesize these compounds were more tolerant to high UV irradiation than mutants impaired in the flavonoid pathway (Emiliani et al. 2013). FLVs play a key role in quenching free radicals, $^1O_2$, and decomposing peroxides (Vieyra et al. 2009). Many flavonoid biosynthetic genes are induced under stress conditions (Kim et al. 2012).

Polyamines (PAs)

PAs are a group of natural compounds with aliphatic nitrogen structure and present in almost all living organisms. Putrescine, spermidine, and spermine are the most commonly found PAs in higher plants and could be present in free, soluble-conjugated, and insoluble-bound forms. PAs play important roles in plant growth and development (Kusano et al. 2008). They are also potent ROS scavengers and inhibitors of lipid peroxidation (Belle et al. 2004). The accumulation of conjugated and free polyamines in plants is very important for their protection against oxidative stress induced by abiotic factors (Jang et al. 2012; Nayyar and Chander 2004). Among the common polyamines, putrescine appears to be the most sensitive for external stress. Plant PAs are involved in imparting tolerance to such stresses such as cold, heat, salinity, hyperosmosis, hypoxia, and atmospheric pollutants (Liu et al. 2007). An exogenous supply of polyamines can protect plant against ozone damage (Bors et al. 1989). PAs are powerful OH^- scavengers and can also quench O_2^- at a higher concentrations (Drolet et al. 1986). Spermine or spermidine also can quench $^1O_2$ at higher concentrations (Das and Misra 2004). Exogenously applied PAs counteracted the toxic effects of paraquat in Arabidopsis (Kurepa et al. 1998a).

Glycine Betaine (GB)

GB is a nitrogenous compound, a quaternary amine. GB is synthesized by either oxidation of choline or N-methylation of glycine (Chen and Murata 2002). In plants, the enzyme choline monooxygenase (CMO) first converts choline into betaine aldehyde, followed by the action of betaine aldehyde dehydrogenase (BADH) (a NAD^+, dependent enzyme), to produce glycine betaine. These enzymes are mainly found in chloroplast stroma. GB biosynthetic genes have been widely used to improve abiotic stress in transgenic plants (Chen and Murata 2011). GB biosynthetic gene, choline oxidase (*codA*) from *Arthrobacter globiformis*, has been widely used for GB production in transgenic plants (Ahmad et al. 2008b; Park et al. 2004). *codA* converts choline into GB in a single step. GB has been implicated in inhibiting ROS accumulation and activation of some stress-related genes. It helps in controlling water balance but can also help to maintain protein and membrane structure. During salt or drought stress, synthesis of proteins involved in PSII repair is affected, leading to photoinhibition. GB stabilizes those PSII repair proteins and thus helps in the repair of PSII, which eventually increases stress tolerance (Li et al. 2013). There have been reports of GB alleviating lipid peroxidation (Li et al. 2013; Cruz et al. 2013).

Genetic Engineering of Oxidative Stress Resistance in Plants

Development of Transgenic Plants

A number of transgenic plants with improved tolerance to various abiotic stresses have been achieved through development of plants overexpressing enzymes involved in oxidative protection, such as GPX, SOD, APX, GST, and GR (Gupta et al. 1993a, b; Lee et al. 2007; Lu et al. 2007; Miao et al. 2006; Roxas et al. 1997) (Table 2). Sometimes, overexpression of one gene may not be enough to confer desired stress resistance upon transgenic plants (Lee et al. 2009). In those cases combinations of two or more antioxidants in transgenic plants have shown to have synergistic effects on stress tolerance (Tseng et al. 2008). Therefore, there has been increased emphasis on production of such transgenic plants.

Development of Mutants

Many plant mutants show reduced tolerance to oxidative stress (Charron et al. 2008; Filkowski et al. 2004; Li et al. 2011; Ning et al. 2010; Shin et al. 2009). But there are several examples of plant mutants which show enhanced tolerance to oxidative stress as summarized in Table 3.

Conclusions and Future Perspectives

ROS are unavoidable part of cell metabolism. They are generated by electron transport activities of chloroplast, mitochondria, and plasma membrane or as a by-product of various metabolic pathways in different cellular compartments. ROS can also be produced as result of various prolonged abiotic stresses. Under normal environmental conditions, ROS production in various cell compartments is low. These ROS are highly reactive and toxic and ultimately result in oxidative stress and damage to the cell. In oxidative stress ROS or free radicals are generated which can damage the biomolecules and cell structures and homeostasis, including oxidative damage to nucleic acids, lipids, and proteins. This leads to altered membrane properties like fluidity, loss of enzyme activity, protein structures, folding and cross-linking, inhibition of protein synthesis, DNA damage, impaired ion transport, and apoptosis. The free radicals of ROS interact with each other and also with antioxidant systems. If ROS has to play a role of signaling molecules or preventing the spread of pathogens, their localization and concentration needs to be controlled. For this purpose, plant cell and its compartments like chloroplast, mitochondria, and peroxisomes deploy antioxidant defense systems to protect themselves against the oxidative damage caused by ROS. This repertoire of antioxidants comprises of enzymatic and nonenzymatic components. When ROS is produced in excess or when the antioxidant defense system is not properly functioning, the cell faces the danger of oxidative damage.

To evaluate the negative role of ROS, it is important to understand mechanisms of its resistance and tolerance in plants. In the recent years, a lot of progress has been done in the field of oxidative stress, but still a lot of gaps in our knowledge of ROS metabolism and their effects on plants are left. Further progress in the fields of genomics, proteomics, and metabolomics will help in untying the knots of hidden biochemical networks involved in establishing oxidative stress in the cell. Knowledge of improved understanding of these, together with the biotechnological tools, will be helpful in producing plants with enhanced levels of tolerance to ROS. Past and ongoing research has already proven that induced expression of various antioxidant enzymes and accumulation of various antioxidant molecules have key roles in detoxification of ROS. Overexpression of ROS-scavenging enzymes like SODs, CAT, APX, GPX, PRX, GR, MDHAR, DHAR, and GST results in abiotic stress tolerance in various crop plants due to increased ROS-scavenging capacity. Significant loss to the yield of crops is done because of the cumulative effect of abiotic stress factors. Therefore, steps for better understanding of the mechanisms of abiotic stress and finding the ways that would increase stress tolerance in plants are crucial for nations' economy and worldwide agriculture. ROS detoxification system is very complex and controlled at multiple levels in various subcellular locations, and modulating one component of the antioxidative defense system

Table 3 Mutants showing tolerance to oxidative stress

Mutated gene	Species	Function(s)	Response against oxidative stresses	Reference
AAL-TOXIN RESISTANT (ATR) 1, 2, 7, 9	*Arabidopsis thaliana*	Not known	Resistance to ROS generating herbicides aminotriazole (AT) and MV	(Gechev et al. 2008; Qureshi et al. 2011)
ORE9 (AT2G42620)	*Arabidopsis thaliana*	Involved in karrikin and strigolactone signaling	Increased tolerance to MV and H_2O_2	(Woo et al. 2004)
ORE1 (AT5G39610)	*Arabidopsis thaliana*	NAC-domain transcription factor regulates senescence in leaves	Increased tolerance to MV and H_2O_2	(Woo et al. 2004)
ORE3 (AT5G39610)	*Arabidopsis thaliana*	Involved in ethylene signal transduction	Increased tolerance to MV and H_2O_2	(Woo et al. 2004)
ELONGATOR subuntis, ELP2 and ELP6	*Arabidopsis thaliana*	Forms subunits of histone acetyltransferase complex	Increased tolerance to MV and CsCl	(Zhou et al. 2009)
HYS1/CPR5	*Arabidopsis thaliana*	Involved in signal transduction of plant defense and trichome development	Increased tolerance to H_2O_2	(Hong-Ying et al. 2010)
PHOTOAUTOTROPHIC SALT TOLERATE 1 (PST1)	*Arabidopsis thaliana*	Involved in photoautotrophic salt tolerance	Increased tolerance to MV, high-light intensity	(Tsugane et al. 1999)
RADICAL-INDUCED CELL DEATH1 (RCD1)	*Arabidopsis thaliana*	A putative PARP protein, interacts with many stress-related transcription factors	Increased tolerance to MV, UV-B, freezing, and osmotic stress	(Ahlfors et al. 2004; Overmyer et al. 2000; Fujibe et al. 2004)
SIMILAR TO RCD ONE 1 (SRO1)	*Arabidopsis thaliana*	A putative PARP protein has redundant roles with RCD1	Increased tolerance to H_2O_2, salt, and osmotic stress	(Teotia and Lamb 2009)
GIGANTEA	*Arabidopsis thaliana*	Promotes flowering under long days in a circadian clock-controlled flowering	Increased tolerance to MV	(Kurepa et al. 1998b)
PARAQUAT RESISTANT2	*Arabidopsis thaliana*	Encodes an S-nitrosoglutathione reductase that is a key regulator of cell death	Increased tolerance to MV	(Chen et al. 2009)
UDP-GLUCOSYL TRANSFERASE 71C1 (UGT71C1)	*Arabidopsis thaliana*	Glycosylate the 3-OH of hydroxycinnamates and flavonoids	Increased tolerance to MV	(Lim et al. 2008)
POLY(ADP-RIBOSE) POLYMERASE 1 AND 2 (PARP1 AND PARP2)	*Arabidopsis thaliana*	Poly (ADP-ribosylation) of target proteins	Increased tolerance to MV, high light, drought, and heat	(De Block et al. 2005)
DET2	*Arabidopsis thaliana*	Involved in the brassinolide biosynthetic pathway	Increased tolerance to oxidative stress	(Cao et al. 2005)
Silencing of γ-tocopherol methyltransferase (γ-TMT)	*Nicotiana tabacum*	Reduction of α-tocopherol and increase in γ-tocopherol	Increased tolerance to MV and sorbitol	(Abbasi et al. 2007)
Lysine decarboxylase-like 1	*Oryza sativa*	Accumulation of the polyamines, putrescine, spermidine, and spermine under conditions of oxidative stress	Reduced accumulation of ROS after exposure to oxidative, high salt, and acid stresses	(Jang et al. 2012)
GLUTATHIONE S-TRANSFERASE U17 (GSTU17)	*Arabidopsis thaliana*	Participates in light signaling and affects GSH and ABA accumulation	Plants were more tolerant to drought and salt stresses	(Chen et al. 2012)
Peroxiredoxin Q (PRX-Q)	*Arabidopsis thaliana*	Peroxiredoxin Q decomposes peroxides using thioredoxin as an electron donor	Decreased oxidative stress sensitivity	(Lamkemeyer et al. 2006)

might not be enough to confer resistance to the whole ROS pathway which may be emanating from multiple stressors. Genetic engineering to develop transgenic crops with gene stacking of different classes of ROS-scavenging enzymes and their isoforms may also be used to obtain synergistic and diversified tolerance to multiple environmental stresses. Similarly, mutants with enhanced tolerance to various stresses are also the answer to the growing demand of developing crops to withstand harsh environmental conditions. Therefore, plants with the ability to control or alleviate ROS levels are the need of the hour and answer to the future to enhance food production.

References

Abbasi AR, Hajirezaei M, Hofius D, Sonnewald U, Voll LM (2007) Specific roles of alpha- and gamma-tocopherol in abiotic stress responses of transgenic tobacco. Plant Physiol 143:1720–1738

Agati G, Matteini P, Goti A, Tattini M (2007) Chloroplast-located flavonoids can scavenge singlet oxygen. New Phytol 174:77–89

Agnez-Lima LF, Melo JT, Silva AE, Oliveira AH, Timoteo AR, Lima-Bessa KM, Martinez GR, Medeiros MH, Di Mascio P, Galhardo RS, Menck CF (2012) DNA damage by singlet oxygen and cellular protective mechanisms. Mutat Res, 751(1):15–28

Ahlfors R, Lang S, Overmyer K, Jaspers P, Brosche M, Tauriainen A, Kollist H, Tuominen H, Belles-Boix E, Piippo M, Inze D, Palva ET, Kangasjarvi J (2004) Arabidopsis *RADICAL-INDUCED CELL DEATH1* belongs to the WWE protein-protein interaction domain protein family and modulates abscisic acid, ethylene, and methyl jasmonate responses. Plant Cell 16:1925–1937

Ahmad P, Sarwat M, Sharma S (2008a) Reactive oxygen species, antioxidants and signaling in plants. J Plant Biol 51:167–173

Ahmad R, Kim MD, Back KH, Kim HS, Lee HS, Kwon SY, Murata N, Chung WI, Kwak SS (2008b) Stress-induced expression of choline oxidase in potato plant chloroplasts confers enhanced tolerance to oxidative, salt, and drought stresses. Plant Cell Rep 27:687–698

Alscher RG (1989) Biosynthesis and antioxidant function of glutathione in plants. Physiol Plant 77:457–464

Alscher RG, Erturk N, Heath LS (2002) Role of superoxide dismutases (SODs) in controlling oxidative stress in plants. J Exp Bot 53:1331–1341

Apel K, Hirt H (2004) Reactive oxygen species: metabolism, oxidative stress, and signal transduction. Annu Rev Plant Biol 55:373–399

Arthur JR (2000) The glutathione peroxidases. Cell Mol Life Sci 57:1825–1835

Artlip TS, Wisniewski ME, Macarisin D, Norelli JL (2009) Ectopic expression of a Spinach SOD gene in young apple trees enhances abiotic stress resistance. Acta Hortic 839:645–650

Asada K (1999) The water-water cycle in chloroplasts: scavenging of active oxygen and dissipation of excess photons. Annu Rev Plant Physiol Plant Mol Biol 50:601–639

Asensio AC, Gil-Monreal M, Pires L, Gogorcena Y, Aparicio-Tejo PM, Moran JF (2012) Two Fe-superoxide dismutase families respond differently to stress and senescence in legumes. J Plant Physiol 169:1253–1260

Ashraf M, Foolad MR (2007) Roles of glycine betaine and proline in improving plant abiotic stress resistance. Environ Exp Bot 59:206–216

Azzedine F, Hocine G, Mebarek B (2011) Improvement of salt tolerance in durum wheat by ascorbic acid application. Plant Physiol Biochem 7:27–37

Badawi GH, Yamauchi Y, Shimada E, Sasaki R, Kawano N, Tanaka K, Tanaka K (2004) Enhanced tolerance to salt stress and water deficit by overexpressing superoxide dismutase in tobacco (Nicotiana tabacum) chloroplasts. Plant Sci 166: 919–928

Balestrazzi A, Confalonieri M, Macovei A, Dona M, Carbonera D (2011) Genotoxic stress and DNA repair in plants: emerging functions and tools for improving crop productivity. Plant Cell Rep 30:287–295

Barthakur S, Babu V, Bansa KC (2001) Over-expression of osmotin induces proline accumulation and confers tolerance to osmotic stress in transgenic tobacco. J Plant Biochem Biotechnol 10:31–37

Basu U, Good AG, Taylor GJ (2001) Transgenic *Brassica napus* plants overexpressing aluminium-induced mitochondrial manganese superoxide dismutase cDNA are resistant to aluminium. Plant Cell Environ 24:1278–1269

Belle NAV, Dalmolin GD, Fonini G, Rubin MA, Rocha JBT (2004) Polyamines reduces lipid peroxidation induced by different pro-oxidant agents. Brain Res 1008:245–251

Bergamini CM, Gambetti S, Dondi A, Cervellati C (2004) Oxygen, reactive oxygen species and tissue damage. Curr Pharm Des 10:1611–1626

Bors W, Langebartels C, Michel C, Sandermann H Jr (1989) Polyamines as radical scavengers and protectants against ozone damage. Phytochemistry 28:1589–1595

Britt AB (1996) DNA damage and repair in plants. Annu Rev Plant Physiol Plant Mol Biol 47:75–100

Brot N, Weissbach L, Werth J, Weissbach H (1981) Enzymatic reduction of protein-bound methionine sulfoxide. Proc Natl Acad Sci USA 78:2155–2158

Cabreiro F, Picot CR, Friguet B, Petropoulos I (2006) Methionine sulfoxide reductases: relevance to aging and protection against oxidative stress. Ann N Y Acad Sci 1067:37–44

Cao S, Xu Q, Cao Y, Qian K, An K, Zhu Y, Binzeng H, Zhao H, Kuai B (2005) Loss-of-function mutations in DET2 gene lead to an enhanced resistance to oxidative stress in *Arabidopsis*. Physiol Plant 123:57–66

Capell T, Bassie L, Christou P (2004) Modulation of the polyamine biosynthetic pathway in transgenic rice confers tolerance to drought stress. Proc Natl Acad Sci USA 101:9909–9914

Catala AA (2010) Synopsis of the process of lipid peroxidation since the discovery of the essential fatty acids. Biochem Biophys Res Commun 399:318–323

Charron JB, Ouellet F, Houde M, Sarhan F (2008) The plant Apolipoprotein D ortholog protects Arabidopsis against oxidative stress. BMC Plant Biol 8:86

Chen TH, Murata N (2002) Enhancement of tolerance of abiotic stress by metabolic engineering of betaines and other compatible solutes. Curr Opin Plant Biol 5:250–257

Chen TH, Murata N (2011) Glycinebetaine protects plants against abiotic stress: mechanisms and biotechnological applications. Plant Cell Environ 34:1–20

Chen R, Sun S, Wang C, Li Y, Liang Y, An F, Li C, Dong H, Yang X, Zhang J, Zuo J (2009) The Arabidopsis *PARAQUAT RESISTANT2* gene encodes an S-nitrosoglutathione reductase that is a key regulator of cell death. Cell Res 19:1377–1387

Chen JH, Jiang HW, Hsieh EJ, Chen HY, Chien CT, Hsieh HL, Lin TP (2012) Drought and salt stress tolerance of an Arabidopsis glutathione S-transferase U17 knockout mutant are attributed to the combined effect of glutathione and abscisic acid. Plant Physiol 158:340–351

Cheng L, Zou Y, Ding S, Zhang J, Yu X, Cao J, Lu G (2009) Polyamine accumulation in transgenic tomato enhances the tolerance to high temperature stress. J Integr Plant Biol 51:489–499

Chiang C, Chen S, Chen L, Chiang M, Chien H, Lin K (2013) Expression of the broccoli catalase gene (*BoCAT*) enhances heat tolerance in transgenic Arabidopsis. J Plant Biochem Biotechnol. DOI:10.1007/s13562-013-0210-1

Choe YH, Kim YS, Kim IS, Bae MJ, Lee EJ, Kim YH, Park HM, Yoon HS (2013) Homologous expression of gamma-glutamylcysteine synthetase increases grain yield and tolerance of transgenic rice plants to environmental stresses. J Plant Physiol 170:610–618

Cooke MS, Evans MD, Dizdaroglu M, Lunec J (2003) Oxidative DNA damage: mechanisms, mutation, and disease. FASEB J 17:1195–1214

Corpas FJ, Fernandez-Ocana A, Carreras A, Valderrama R, Luque F, Esteban FJ, Rodriguez-Serrano M, Chaki M, Pedrajas JR, Sandalio LM, del Rio LA, Barroso JB (2006) The expression of different superoxide dismutase forms is cell-type dependent in olive (*Olea europaea* L.) leaves. Plant Cell Physiol 47:984–994

Creissen GP, Broadbent P, Kular B, Reynolds H, Wellburn AR, Mullineaux PM (1994) Manipulation of glutathione reductase in transgenic plants: implications for plants' responses to environmental stress. Proc R Soc Edinb Sect B Biol Sci 102:167–175

Cruz FJR, Castro GLS, Silva JÃºnior DD, Festucci-Buselli RA, Pinheiro HA (2013) Exogenous glycine betaine modulates ascorbate peroxidase and catalase activities and prevent lipid peroxidation in mild water-stressed *Carapa guianensis* plants. Photosynthetica 51:102–108

Dąbrowska G, Kata K, Goc A, Szechyńska-Hebda M, Skrzypek E (2007) Characteristics of the plant ascorbate peroxidase family. Acta Biol Cracov Bot 49:7–17

Das K, Misra H (2004) Hydroxyl radical scavenging and singlet oxygen quenching properties of polyamines. Mol Cell Biochem 262:127–133

Davison PA, Hunter CN, Horton P (2002) Overexpression of beta-carotene hydroxylase enhances stress tolerance in Arabidopsis. Nature 418:203–206

De Block M, Verduyn C, De Brouwer D, Cornelissen M (2005) Poly(ADP-ribose) polymerase in plants affects energy homeostasis, cell death and stress tolerance. Plant J 41:95–106

De Ronde JA, Cress WA, Kruger GH, Strasser RJ, Van Staden J (2004) Photosynthetic response of transgenic soybean plants, containing an Arabidopsis P5CR gene, during heat and drought stress. J Plant Physiol 161:1211–1224

Debska K, Bogatek R, Gniazdowska A (2012) Protein carbonylation and its role in physiological processes in plants. Postepy Biochem 58:34–43

del Rio LA, Sandalio LM, Corpas FJ, Palma JM, Barroso JB (2006) Reactive oxygen species and reactive nitrogen species in peroxisomes. Production, scavenging, and role in cell signaling. Plant Physiol 141:330–335

Desikan R, Cheung MK, Bright J, Henson D, Hancock JT, Neill SJ (2004) ABA, hydrogen peroxide and nitric oxide signalling in stomatal guard cells. J Exp Bot 55:205–212

Dietz KJ (2011) Peroxiredoxins in plants and cyanobacteria. Antioxid Redox Signal 15:1129–1159

Dixit P, Mukherjee PK, Ramachandran V, Eapen S (2011) Glutathione transferase from *Trichoderma virens* enhances cadmium tolerance without enhancing its accumulation in transgenic *Nicotiana tabacum*. PLoS One 6:e16360

Dixon DP, Edwards R (2010) Glutathione transferases. Arabidopsis Book 8:e0131

Dixon DP, Skipsey M, Edwards R (2010) Roles for glutathione transferases in plant secondary metabolism. Phytochemistry 71:338–350

Drolet G, Dumbroff EB, Legge RL, Thompson JE (1986) Radical scavenging properties of polyamines. Phytochemistry 25:367–371

Du H, Wang N, Cui F, Li X, Xiao J, Xiong L (2010) Characterization of the beta-carotene hydroxylase gene DSM2 conferring drought and oxidative stress resistance by increasing xanthophylls and abscisic acid synthesis in rice. Plant Physiol 154:1304–1318

Edge R, McGarvey DJ, Truscott TG (1997) The carotenoids as anti-oxidants — a review. J Photochem Photobiol B Biol 41:189–200

Edwards EA, Rawsthorne S, Mullineaux PM (1990) Subcellular distribution of multiple forms of glutathione reductase in leaves of pea (*Pisum sativum* L.). Planta 180:278–284

Eltayeb AE, Kawano N, Badawi GH, Kaminaka H, Sanekata T, Morishima I, Shibahara T, Inanaga S, Tanaka K (2006) Enhanced tolerance to ozone and drought stresses in transgenic tobacco overexpressing dehydroascorbate reductase in cytosol. Physiol Plant 127:57–65

Eltayeb AE, Kawano N, Badawi GH, Kaminaka H, Sanekata T, Shibahara T, Inanaga S, Tanaka K (2007) Overexpression of monodehydroascorbate reductase in transgenic tobacco confers enhanced tolerance to ozone, salt and polyethylene glycol stresses. Planta 225:1255–1264

Emiliani J, Grotewold E, Falcone Ferreyra ML, Casati P (2013) Flavonols protect Arabidopsis plants against UV-B deleterious effects. Mol Plant, 6 (4): 1376–1379

Espinoza A, San Martin A, López-Climent M, Ruiz-Lara S, Gómez-Cadenas A, Casaretto JA (2013) Engineered drought-induced biosynthesis of α-tocopherol alleviates stress-induced leaf damage in tobacco. J Plant Physiol 170:1285–1294

Fan W, Zhang M, Zhang H, Zhang P (2012) Improved tolerance to various abiotic stresses in transgenic sweet potato (Ipomoea batatas) expressing spinach betaine aldehyde dehydrogenase. PLoS One 7:e37344

Filkowski J, Kovalchuk O, Kovalchuk I (2004) Genome stability of vtc1, tt4, and tt5 *Arabidopsis thaliana* mutants impaired in protection against oxidative stress. Plant J 38:60–69

Fischer BB, Hideg E, Krieger-Liszkay A (2013) Production, detection, and signaling of singlet oxygen in photosynthetic organisms. Antioxid Redox Signal, 18: 2145–2162

Flocco CG, Lindblom SD, Smits EA (2004) Overexpression of enzymes involved in glutathione synthesis enhances tolerance to organic pollutants in *Brassica juncea*. Int J Phytoremediation 6:289–304

Flors C, Nonell S (2006) Light and singlet oxygen in plant defense against pathogens: phototoxic phenalenone phytoalexins. Acc Chem Res 39:293–300

Flors C, Ogilby PR, Luis JG, Grillo TA, Izquierdo LR, Gentili PL, Bussotti L, Nonell S (2006) Phototoxic phytoalexins. Processes that compete with the photosensitized production of singlet oxygen by 9-phenylphenalenones. Photochem Photobiol 82:95–103

Forman HJ, Fukuto JM, Torres M (2004) Redox signaling: thiol chemistry defines which reactive oxygen and nitrogen species can act as second messengers. Am J Physiol Cell Physiol 287:C246–C256

Foyer C, Halliwell B (1976) The presence of glutathione and glutathione reductase in chloroplasts: a proposed role in ascorbic acid metabolism. Planta 133:21–25

Foyer CH, Mullineaux PM (1998) The presence of dehydroascorbate and dehydroascorbate reductase in plant tissues. FEBS Lett 425:528–529

Foyer CH, Noctor G (2005) Redox homeostasis and antioxidant signaling: a metabolic interface between stress perception and physiological responses. Plant Cell 17:1866–1875

Foyer CH, Noctor G (2011) Ascorbate and glutathione: the heart of the redox hub. Plant Physiol 155:2–18

Foyer CH, Souriau N, Perret S, Lelandais M, Kunert KJ, Pruvost C, Jouanin L (1995) Overexpression of glutathione reductase but not glutathione synthetase leads to increases in antioxidant capacity and resistance to photoinhibition in poplar trees. Plant Physiol 109:1047–1057

Foyer CH, Lopez-Delgado H, Dat JF, Scott I (1997) Hydrogen peroxide- and glutathione-associated mechanisms of acclimatory stress tolerance and signalling. Physiol Plant 100:241–254

Fujibe T, Saji H, Arakawa K, Yabe N, Takeuchi Y, Yamamoto KT (2004) A methyl viologen-resistant mutant of Arabidopsis, which is allelic to ozone-sensitive rcd1, is tolerant to supplemental ultraviolet-B irradiation. Plant Physiol 134:275–285

Gaber A, Yoshimura K, Yamamoto T, Yabuta Y, Takeda T, Miyasaka H, Nakano Y, Shigeoka S (2006) Glutathione peroxidase-like protein of Synechocystis PCC 6803 confers tolerance to oxidative and environmental stresses in transgenic Arabidopsis. Physiol Plant 128:251–262

Gang W, Zhen-Kuan W, Yong-Xiang W, Li-Ye C, Hong-Bo S (2007) The mutual responses of higher plants to environment: physiological and microbiological aspects. Colloids Surf B Biointerfaces 59:113–119

Gechev TS, Ferwerda MA, Mehterov N, Laloi C, Qureshi MK, Hille J (2008) Arabidopsis AAL-toxin-resistant mutant *atr1* shows enhanced tolerance to programmed cell death induced by reactive oxygen species. Biochem Biophys Res Commun 375:639–644

George S, Venkataraman G, Parida A (2010) A chloroplast-localized and auxin-induced glutathione S-transferase from phreatophyte *Prosopis juliflora* confer drought tolerance on tobacco. J Plant Physiol 167:311–318

Giba Z, Todorivic S, Grubisic D, Konjevic R (1998) Occurrence and regulatory roles of superoxide anion radical and nitric oxide in plants. Iugoslav Physiol Pharmacol Acta 34:447–461

Gullner G, Komives T, Rennenberg H (2001) Enhanced tolerance of transgenic poplar plants overexpressing gamma-glutamylcysteine synthetase towards chloroacetanilide herbicides. J Exp Bot 52:971–979

Guo X, Wu Y, Wang Y, Chen Y, Chu C (2009) OsMSRA4.1 and OsMSRB1.1, two rice plastidial methionine sulfoxide reductases, are involved in abiotic stress responses. Planta 230:227–238

Gupta AS, Heinen JL, Holaday AS, Burke JJ, Allen RD (1993a) Increased resistance to oxidative stress in transgenic plants that overexpress chloroplastic Cu/Zn superoxide dismutase. Proc Natl Acad Sci USA 90:1629–1633

Gupta AS, Webb RP, Holaday AS, Allen RD (1993b) Overexpression of superoxide dismutase protects plants from oxidative stress (induction of ascorbate peroxidase in superoxide dismutase-overexpressing plants). Plant Physiol 103:1067–1073

Hagar H, Ueda N, Shah SV (1996) Role of reactive oxygen metabolites in DNA damage and cell death in chemical hypoxic injury to LLC-PK1 cells. Am J Physiol 271:209–215

Halliwell B (2006) Reactive species and antioxidants. Redox biology is a fundamental theme of aerobic life. Plant Physiol 141:312–322

Hemavathi, Upadhyaya CP, Akula N, Young KE, Chun SC, Kim DH, Park SW (2010) Enhanced ascorbic acid accumulation in transgenic potato confers tolerance to various abiotic stresses. Biotechnol Lett 32:321–330

Hemavathi, Upadhyaya CP, Younga KE, Akulaa N, Kimb HS, Heungb JJ, Ohc OM, Aswathd CR, Chuna SC, Kima DH, Parka SW (2009) Over-expression of strawberry D-galacturonic acid reductase in potato leads to accumulation of vitamin C with enhanced abiotic stress tolerance. Plant Sci 177:659–667

Hmida-Sayari A, Gargouri-Bouzid R, Bidani A, Jaoua L, SavourÃ A, Jaoua S (2005) Overexpression of ÎΔ^1-pyrroline-5-carboxylate synthetase increases proline production and confers salt tolerance in transgenic potato plants. Plant Sci 169:746–752

Holmstrom K-O, Somersalo S, Mandal A, Palva TE, Welin B (2000) Improved tolerance to salinity and low temperature in transgenic tobacco producing glycine betaine. J Exp Bot 51:177–185

Hong-Ying H, Wen-Ling L, Li-Ying S, Zhan S, Cheng-Wei Y, Chang-Lian P (2010) Increased resistance of Arabidopsis cpr5 mutant to H_2O_2- induced photooxidation. Pak J Bot 42:1041–1049

Hu T, Qv X, Xiao G, Huang X (2009) Enhanced tolerance to herbicide of rice plants by over-expression of a glutathione S-transferase. Mol Breed 24:409–418

Jang S, Wi S, Choi Y, An G, Park K (2012) Increased polyamine biosynthesis enhances stress tolerance by preventing the accumulation of reactive oxygen species: T-DNA mutational analysis of *Oryza sativa* lysine decarboxylase-like protein 1. Mol Cells 34:251–262

Jayaraj J, Punja ZK (2008) Transgenic carrot plants accumulating ketocarotenoids show tolerance to UV and oxidative stresses. Plant Physiol Biochem 46:875–883

Ji W, Zhu Y, Li Y, Yang L, Zhao X, Cai H, Bai X (2010) Over-expression of a glutathione S-transferase gene, GsGST, from wild soybean (*Glycine soja*) enhances drought and salt tolerance in transgenic tobacco. Biotechnol Lett 32:1173–1179

Jing L-W, Chen S-H, Guo X-L, Zhang H, Zhao Y X (2006) Overexpression of a chloroplast-located peroxiredoxin Q gene, SsPrxQ, increases the salt and low-temperature tolerance of Arabidopsis. J Integr Plant Biol 48:1244–1249

Jonsson TJ, Murray MS, Johnson LC, Lowther WT (2008) Reduction of cysteine sulfinic acid in peroxiredoxin by sulfiredoxin proceeds directly through a sulfinic phosphoryl ester intermediate. J Biol Chem 283:23846–23851

Kamal-Eldin A, Appelqvist LA (1996) The chemistry and antioxidant properties of tocopherols and tocotrienols. Lipids 31:671–701

Kaminaka H, Morita S, Nakajima M, Masumura T, Tanaka K (1998) Gene cloning and expression of cytosolic glutathione reductase in rice (*Oryza sativa* L.). Plant Cell Physiol 39:1269–1280

Karavangeli M, Labrou NE, Clonis YD, Tsaftaris A (2005) Development of transgenic tobacco plants overexpressing maize glutathione S-transferase I for chloroacetanilide herbicides phytoremediation. Biomol Eng 22:121–128

Karlsson M, Melzer M, Prokhorenko I, Johansson T, Wingsle G (2005) Hydrogen peroxide and expression of hipI-superoxide dismutase are associated with the development of secondary cell walls in *Zinnia elegans*. J Exp Bot 56:2085–2093

Karthikeyan A, Pandian S, Ramesh M (2011) Transgenic indica rice cv. ADT 43 expressing a Δ^1-pyrroline-5-carboxylate synthetase (P5CS) gene from *Vigna aconitifolia* demonstrates salt tolerance. Plant Cell Tissue Organ Cult (PCTOC) 107:383–395

Kasukabe Y, He L, Nada K, Misawa S, Ihara I, Tachibana S (2004) Overexpression of spermidine synthase enhances tolerance to multiple environmental stresses and up-regulates the expression of various stress-regulated genes in transgenic *Arabidopsis thaliana*. Plant Cell Physiol 45:712–722

Kasukabe Y, He L, Watakabe Y, Otani M, Shimada T, Tachibana S (2006) Improvement of environmental stress tolerance of sweet potato by introduction of genes for spermidine synthase. Plant Biotechnol 23:75–83

Kiba A, Nishihara M, Tsukatani N, Nakatsuka T, Kato Y, Yamamura S (2005) A peroxiredoxin Q homolog from gentians is involved in both resistance against fungal disease and oxidative stress. Plant Cell Physiol 46:1007–1015

Kim KH, Alam I, Lee KW, Sharmin SA, Kwak SS, Lee SY, Lee BH (2010) Enhanced tolerance of transgenic tall fescue plants overexpressing 2-Cys peroxiredoxin against methyl viologen and heat stresses. Biotechnol Lett 32:571–576

Kim MD, Kim YH, Kwon SY, Jang BY, Lee SY, Yun DJ, Cho JH, Kwak SS, Lee HS (2011) Overexpression of 2-cysteine peroxiredoxin enhances tolerance to methyl viologen-mediated oxidative stress and high temperature in potato plants. Plant Physiol Biochem 49:891–897

Kim B G, Lee E R, Ahn J-H (2012) Analysis of flavonoid contents and expression of flavonoid biosynthetic genes in *Populus euramericana* Guinier in response to abiotic stress. J Korean Soc Appl Biol Chem 55:141–145

Kiran Kumar Ghanti S, Sujata KG, Vijay Kumar BM, Nataraja Karba N, Janardhan Reddy K, Srinath Rao M, Kavi Kishor PB (2011) Heterologous expression of P5CS gene in chickpea enhances salt tolerance without affecting yield. Biol Plant 55:634–640

Kishor PBK, Hong Z, Miao GH, Hu CAA, Verma DPS (1995) Overexpression of [delta]-pyrroline-5-carboxylate synthetase increases proline production and confers osmotolerance in transgenic plants. Plant Physiol 108:1387–1394

Kotchoni SO, Gachomo EW (2006) The reactive oxygen species network pathways: an essential prerequisite for perception of pathogen attack and the acquired disease resistance in plants. J Biosci 31:389–404

Kotchoni SO, Kuhns C, Ditzer A, Kirch HH, Bartels D (2006) Over-expression of different aldehyde dehydrogenase genes in *Arabidopsis thaliana* confers tolerance to abiotic stress and protects plants against lipid peroxidation and oxidative stress. Plant Cell Environ 29:1033–1048

Krieger-Liszkay A, Trebst A (2006) Tocopherol is the scavenger of singlet oxygen produced by the triplet states of chlorophyll in the PSII reaction centre. J Exp Bot 57:1677–1684

Krishnan N, Dickman MB, Becker DF (2008) Proline modulates the intracellular redox environment and protects mammalian cells against oxidative stress. Free Radic Biol Med 44:671–681

Kubo A, Saji H, Tanaka K, Kondo N (1995) Expression of Arabidopsis cytosolic ascorbate peroxidase gene in response to ozone or sulfur dioxide. Plant Mol Biol 29:479–489

Kumar S, Asif MH, Chakrabarty D, Tripathi RD, Dubey RS, Trivedi PK (2013) Expression of a rice Lambda class of glutathione S-transferase, OsGSTL2, in Arabidopsis provides tolerance to heavy metal and other abiotic stresses. J Hazard Mater 248–249:228–237

Kurepa J, Smalle J, Montagu MV, Inzé D (1998a) Polyamines and paraquat toxicity in *Arabidopsis thaliana*. Plant Cell Physiol 39:987–992

Kurepa J, Smalle J, Van Montagu M, Inzé D (1998b) Oxidative stress tolerance and longevity in Arabidopsis: the late-flowering mutant *gigantea* is tolerant to paraquat. Plant J 14:759–764

Kusano T, Berberich T, Tateda C, Takahashi Y (2008) Polyamines: essential factors for growth and survival. Planta 228:367–381

Kwon SY, Jeong YJ, Lee HS, Kim JS, Cho KY, Allen RD, Kwak SS (2002) Enhanced tolerances of transgenic tobacco plants expressing both superoxide dismutase and ascorbate peroxidase in chloroplasts against methyl viologen-mediated oxidative stress. Plant Cell Environ 25:873–882

Kwon SY, Choi SM, Ahn YO, Lee HS, Lee HB, Park YM, Kwak SS (2003) Enhanced stress-tolerance of transgenic tobacco plants expressing a human dehydroascorbate reductase gene. J Plant Physiol 160:347–353

Kwon SJ, Kwon SI, Bae MS, Cho EJ, Park OK (2007) Role of the methionine sulfoxide reductase MsrB3 in cold acclimation in Arabidopsis. Plant Cell Physiol 48:1713–1723

Lamkemeyer P, Laxa M, Collin V, Li W, Finkemeier I, Schottler MA, Holtkamp V, Tognetti VB, Issakidis-Bourguet E, Kandlbinder A, Weis E, Miginiac-Maslow M, Dietz KJ (2006) Peroxiredoxin Q of *Arabidopsis thaliana* is attached to the thylakoids and functions in context of photosynthesis. Plant J 45:968–981

Leclercq J, Martin F, Sanier C, Clement-Vidal A, Fabre D, Oliver G, Lardet L, Ayar A, Peyramard M, Montoro P (2012) Over-expression of a cytosolic isoform of the HbCuZnSOD gene in *Hevea brasiliensis* changes its response to a water deficit. Plant Mol Biol 80:255–272

Lederer B, Boger P (2003) Antioxidative responses of tobacco expressing a bacterial glutathione reductase. Z Naturforsch C 58:843–849

Lee KO, Jang HH, Jung BG, Chi YH, Lee JY, Choi YO, Lee JR, Lim CO, Cho MJ, Lee SY (2000) Rice 1Cys-peroxiredoxin over-expressed in transgenic tobacco does not maintain dormancy but enhances antioxidant activity. FEBS Lett 486:103–106

Lee YP, Kim SH, Bang JW, Lee HS, Kwak SS, Kwon SY (2007) Enhanced tolerance to oxidative stress in transgenic tobacco plants expressing three antioxidant enzymes in chloroplasts. Plant Cell Rep 26:591–598

Lee S-C, Kwon S-Y, Kim S-R (2009) Ectopic expression of a cold-responsive CuZn superoxide dismutase gene, SodCc1, in transgenic rice (*Oryza sativa* L.). J Plant Biol 52:154–160

Li F, Wu QY, Sun YL, Wang LY, Yang XH, Meng QW (2010a) Overexpression of chloroplastic monodehydroascorbate reductase enhanced tolerance to temperature and methyl viologen-mediated oxidative stresses. Physiol Plant 139:421–434

Li QY, Niu HB, Yin J, Shao HB, Niu JS, Ren JP, Li YC, Wang X (2010b) Transgenic barley with overexpressed PTrx increases aluminum resistance in roots during germination. J Zhejiang Univ Sci B 11:862–870

Li G, Nasar V, Yang Y, Li W, Liu B, Sun L, Li D, Song F (2011) Arabidopsis poly(ADP-ribose) glycohydrolase 1 is required for drought, osmotic and oxidative stress responses. Plant Sci 180:283–291

Li CW, Lee SH, Chieh PS, Lin CS, Wang YC, Chan MT (2012a) Arabidopsis root-abundant cytosolic methionine sulfoxide reductase B genes MsrB7 and MsrB8 are involved in tolerance to oxidative stress. Plant Cell Physiol 53:1707–1719

Li F, Wu QY, Duan M, Dong XC, Li B, Meng QW (2012b) Transgenic tomato plants overexpressing chloroplastic monodehydroascorbate reductase are resistant to salt- and PEG-induced osmotic stress. Photosynthetica 50:120–128

Li M, Li Z, Li S, Guo S, Meng Q, Li G, Yang X (2013) Genetic engineering of glycine betaine biosynthesis reduces heat-enhanced photoinhibition by enhancing antioxidative defense and alleviating lipid peroxidation in tomato. Plant Mol Biol Rep 1–10. DOI:10.1007/s11105-013-0594-z

Liang Zhu Y, Pilon-Smits EA, Jouanin L, Terry N (1999) Overexpression of glutathione synthetase in Indian mustard enhances cadmium accumulation and tolerance. Plant Physiol 119:73–80

Liangqi W, Zhanmin F, Lei G, Yongqing L, Wenjing Z, Li-Jia Q, Zhangliang C (2003) Over-expression of an Arabidopsis δ-OAT gene enhances salt and drought tolerance in transgenic rice. Chin Sci Bull 48:2594–2600

Liedschulte V, Wachter A, Zhigang A, Rausch T (2010) Exploiting plants for glutathione (GSH) production: uncoupling GSH synthesis from cellular controls results in unprecedented GSH accumulation. Plant Biotechnol J 8:807–820

Lim CE, Choi JN, Kim IA, Lee SA, Hwang YS, Lee CH, Lim J (2008) Improved resistance to oxidative stress by a loss-of-function mutation in the Arabidopsis UGT71C1 gene. Mol Cells 25:368–375

Liu JH, Kitashiba H, Wang J, Ban Y, Moriguchi T (2007) Polyamines and their ability to provide environmental stress tolerance to plants. Plant Biotechnol 24:117–126

Liu X, Hua X, Guo J, Qi D, Wang L, Liu Z, Jin Z, Chen S, Liu G (2008) Enhanced tolerance to drought stress in transgenic tobacco plants overexpressing VTE1 for increased tocopherol production from *Arabidopsis thaliana*. Biotechnol Lett 30:1275–1280

Lu Z, Liu D, Liu S (2007) Two rice cytosolic ascorbate peroxidases differentially improve salt tolerance in transgenic Arabidopsis. Plant Cell Rep 26:1909–1917

Ma L, Zhou E, Gao L, Mao X, Zhou R, Jia J (2008) Isolation, expression analysis and chromosomal location of P5CR gene in common wheat (*Triticum aestivum* L.). S Afr J Bot 74:705–712

Maeda H, Sakuragi Y, Bryant DA, Dellapenna D (2005) Tocopherols protect Synechocystis sp. strain PCC 6803 from lipid peroxidation. Plant Physiol 138:1422–1435

Maheshwari R, Dubey RS (2009) Nickel-induced oxidative stress and the role of antioxidant defence in rice seedlings. Plant Growth Regul 59:37–49

Matsumuraa T, Tabayashib N, Kamagata Y, Souma C, Saruyama H (2002) Wheat catalase expressed in transgenic rice plants can improve tolerance against low temperature injury. Physiol Plant 116:317–327

Matysik J, Alia BB, Mohanty P (2002) Molecular mechanisms of quenching of reactive oxygen species by proline under stress in plants. Curr Sci 82:525–532

McKersie BD, Bowley SR, Jones KS (1999) Winter survival of transgenic alfalfa overexpressing superoxide dismutase. Plant Physiol 119:839–848

Meister A (1988) Glutathione metabolism and its selective modification. J Biol Chem 263:17205–17208

Meister A, Anderson ME (1983) Glutathione. Annu Rev Biochem 52:711–760

Meyer AJ (2008) The integration of glutathione homeostasis and redox signaling. J Plant Physiol 165:1390–1403

Meyer Y, Reichheld JP, Vignols F (2005) Thioredoxins in Arabidopsis and other plants. Photosynth Res 86:419–433

Meyer Y, Belin C, Delorme-Hinoux V, Reichheld JP, Riondet C (2012) Thioredoxin and glutaredoxin systems in plants: molecular mechanisms, crosstalks, and functional significance. Antioxid Redox Signal 17:1124–1160

Miao Y, Lv D, Wang P, Wang XC, Chen J, Miao C, Song CP (2006) An Arabidopsis glutathione peroxidase functions as both a redox transducer and a scavenger in abscisic acid and drought stress responses. Plant Cell 18:2749–2766

Milligan AS, Daly A, Parry MAJ, Lazzeri PA, Jepson I (2001) The expression of a maize glutathione S-transferase gene in transgenic wheat confers herbicide tolerance, both *in planta and in vitro*. Mol Breed 7:301–315

Mittler R (2002) Oxidative stress, antioxidants and stress tolerance. Trends Plant Sci 7:405–410

Mittler R, Vanderauwera S, Gollery M, Van Breusegem F (2004) Reactive oxygen gene network of plants. Trends Plant Sci 9:490–498

Moller IM, Jensen PE, Hansson A (2007) Oxidative modifications to cellular components in plants. Annu Rev Plant Biol 58:459–481

Moran JF, Iturbe-Ormaetxe I, Matamoros MA, Rubio MC, Clemente MR, Brewin NJ, Becana M (2000) Glutathione and homoglutathione synthetases of legume nodules. Cloning, expression, and subcellular localization. Plant Physiol 124:1381–1392

Nagamiya K, Motohashi T, Nakao K, Prodhan SH, Hattori E, Hirose S, Ozawa K, Ohkawa Y, Takabe T, Takab T, Komamine A (2007) Enhancement of salt tolerance in transgenic rice expressing an *Escherichia coli* catalase gene, katE. Plant Biotechnol Rep 1:49–55

Nanjo T, Kobayashi M, Yoshiba Y, Kakubari Y, Yamaguchi-Shinozaki K, Shinozaki K (1999) Antisense suppression of proline degradation improves tolerance to freezing and salinity in *Arabidopsis thaliana*. FEBS Lett 461:205–210

Nayyar H, Chander S (2004) Protective effects of polyamines against oxidative stress induced by water and cold stress in Chickpea. J Agron Crop Sci 190:355–365

Neill SJ, Desikan R, Clarke A, Hurst RD, Hancock JT (2002) Hydrogen peroxide and nitric oxide as signalling molecules in plants. J Exp Bot 53:1237–1247

Ning J, Li X, Hicks LM, Xiong L (2010) A Raf-like MAPKKK gene DSM1 mediates drought resistance through reactive oxygen species scavenging in rice. Plant Physiol 152:876–890

Noctor G, Foyer CH (1998) Ascorbate and glutathione: keeping active oxygen under control. Annu Rev Plant Physiol Plant Mol Biol 49:249–279

Noctor G, Gomez L, Vanacker H, Foyer CH (2002) Interactions between biosynthesis, compartmentation and transport in the control of glutathione homeostasis and signalling. J Exp Bot 53:1283–1304

Oberschall A, Deak M, Torok K, Sass L, Vass I, Kovacs I, Feher A, Dudits D, Horvath GV (2000) A novel aldose/aldehyde reductase protects transgenic plants against lipid peroxidation under chemical and drought stresses. Plant J 24:437–446

Ogawa K, Kanematsu S, Asada K (1997) Generation of superoxide anion and localization of CuZn-superoxide dismutase in the vascular tissue of spinach hypocotyls: their association with lignification. Plant Cell Physiol 38:1118–1126

Oh SK, Baek KH, Seong ES, Joung YH, Choi GJ, Park JM, Cho HS, Kim EA, Lee S, Choi D (2010) CaMsrB2, pepper methionine sulfoxide reductase B2, is a novel defense regulator against oxidative stress and pathogen attack. Plant Physiol 154:245–261

op den Camp RG, Przybyla D, Ochsenbein C, Laloi C, Kim C, Danon A, Wagner D, Hideg E, Gobel C, Feussner I, Nater M, Apel K (2003) Rapid induction of distinct stress responses after the release of singlet oxygen in Arabidopsis. Plant Cell 15:2320–2332

Overmyer K, Tuominen H, Kettunen R, Betz C, Langebartels C, Sandermann H Jr, Kangasjarvi J (2000) Ozone-sensitive arabidopsis *rcd1* mutant reveals opposite roles for ethylene and jasmonate signaling pathways in regulating superoxide-dependent cell death. Plant Cell 12:1849–1862

Panieri E, Gogvadze V, Norberg E, Venkatesh R, Orrenius S, Zhivotovsky B (2013) Reactive oxygen species generated in different compartments induce cell death, survival, or senescence. Free Radic Biol Med 57:176–187

Park J, Gu Y, Lee Y, Yang Z (2004) Phosphatidic acid induces leaf cell death in Arabidopsis by activating the Rho-related small G protein GTPase-mediated pathway of reactive oxygen species generation. Plant Physiol 134:129–136

Park EJ, Jeknic Z, Pino MT, Murata N, Chen TH (2007) Glycinebetaine accumulation is more effective in chloroplasts than in the cytosol for protecting transgenic tomato plants against abiotic stress. Plant Cell Environ 30:994–1005

Park S, Lee D-E, Jang H, Byeon Y, Kim Y-S, Back K (2013) Melatonin-rich transgenic rice plants exhibit resistance to herbicide-induced oxidative stress. J Pineal Res 54:258–263

Perez-Perez ME, Lemaire SD, Crespo JL (2012) Reactive oxygen species and autophagy in plants and algae. Plant Physiol 160:156–164

Perl A, Perl-Treves R, Galili S, Aviv D, Shalgi E, Malkin S, Galun E (1993) Enhanced oxidative-stress defense in transgenic potato expressing tomato Cu, Zn superoxide dismutases. Theor Appl Genet 85:568–576

Poage M, Le Martret B, Jansen MA, Nugent GD, Dix PJ (2011) Modification of reactive oxygen species scavenging capacity of chloroplasts through plastid transformation. Plant Mol Biol 76:371–384

Polidoros AN, Mylona PV, Scandalios JG (2001) Transgenic tobacco plants expressing the maize Cat2 gene have altered catalase levels that affect plant-pathogen interactions and resistance to oxidative stress. Transgenic Res 10:555–569

Potters G, Pasternak TP, Guisez Y, Palme KJ, Jansen MA (2007) Stress-induced morphogenic responses: growing out of trouble? Trends Plant Sci 12:98–105

Potters G, Pasternak TP, Guisez Y, Jansen MA (2009) Different stresses, similar morphogenic responses: integrating a plethora of pathways. Plant Cell Environ 32:158–169

Porfirova S, Bergmüller E, Tropf S, Lemke R, Dörmann P (2002) Isolation of an Arabidopsis mutant lacking vitamin E and identification of a cyclase essential for all tocopherol biosynthesis. Proc Natl Acad Sci USA 99:12495–12500

Prashanth SR, Sadhasivam V, Parida A (2008) Over expression of cytosolic copper/zinc superoxide dismutase from a mangrove plant *Avicennia marina* in indica rice var Pusa Basmati-1 confers abiotic stress tolerance. Transgenic Res 17:281–291

Quan LJ, Zhang B, Shi WW, Li HY (2008) Hydrogen peroxide in plants: a versatile molecule of the reactive oxygen species network. J Integr Plant Biol 50:2–18

Qureshi M, Radeva V, Genkov T, Minkov I, Hille J, Gechev T (2011) Isolation and characterization of Arabidopsis mutants with enhanced tolerance to oxidative stress. Acta Physiol Plant 33:375–382

Ramel F, Birtic S, Cuine S, Triantaphylides C, Ravanat JL, Havaux M (2012) Chemical quenching of singlet oxygen by carotenoids in plants. Plant Physiol 158:1267–1278

Rao MV, Paliyath G, Ormrod DP (1996) Ultraviolet-B- and ozone-induced biochemical changes in antioxidant enzymes of *Arabidopsis thaliana*. Plant Physiol 110:125–136

Raychaudhuri SS, Deng XW (2000) The role of superoxide dismutase in combating oxidative stress in higher plants. Bot Rev 66:89–98

Reddy AM, Reddy VS, Scheffler BE, Wienand U, Reddy AR (2007) Novel transgenic rice overexpressing anthocyanidin synthase accumulates a mixture of flavonoids leading to an increased antioxidant potential. Metab Eng 9:95–111

Rodriguez Milla MA, Maurer A, Rodriguez Huete A, Gustafson JP (2003) Glutathione peroxidase genes in *Arabidopsis* are ubiquitous and regulated by abiotic stresses through diverse signaling pathways. Plant J 36:602–615

Romero HM, Berlett BS, Jensen PJ, Pell EJ, Tien M (2004) Investigations into the role of the plastidial peptide methionine sulfoxide reductase in response to oxidative stress in Arabidopsis. Plant Physiol 136:3784–3794

Rouhier N (2010) Plant glutaredoxins: pivotal players in redox biology and iron-sulphur centre assembly. New Phytol 186:365–372

Roxas VP, Smith RK Jr, Allen ER, Allen RD (1997) Overexpression of glutathione S-transferase/glutathione peroxidase enhances the growth of transgenic tobacco seedlings during stress. Nat Biotechnol 15:988–991

Roxas VP, Lodhi SA, Garrett DK, Mahan JR, Allen RD (2000) Stress tolerance in transgenic tobacco seedlings that overexpress glutathione S-transferase/glutathione peroxidase. Plant Cell Physiol 41:1229–1234

Roy M, Wu R (2001) Arginine decarboxylase transgene expression and analysis of environmental stress tolerance in transgenic rice. Plant Sci 160:869–875

Sarowar S, Kim EN, Kim YJ, Ok SH, Kim KD, Hwang BK, Shin JS (2005) Overexpression of a pepper ascorbate peroxidase-like 1 gene in tobacco plants enhances tolerance to oxidative stress and pathogens. Plant Sci 169:55–63

Shalata A, Neumann PM (2001) Exogenous ascorbic acid (vitamin C) increases resistance to salt stress and reduces lipid peroxidation. J Exp Bot 52:2207–2211

Shao H, Chu LY, Lu ZH, Kang CM (2007) Primary antioxidant free radical scavenging and redox signaling pathways in higher plant cells. Int J Biol Sci 4:8–14

Shah K, Singh M, Rai AC (2013) Effect of heat-shock induced oxidative stress is suppressed in BcZAT12 expressing drought tolerant tomato. Phytochemistry 95:109–117

Sharov VS, Schoneich C (2000) Diastereoselective protein methionine oxidation by reactive oxygen species and diastereoselective repair by methionine sulfoxide reductase. Free Radic Biol Med 29:986–994

Shigeoka S, Ishikawa T, Tamoi M, Miyagawa Y, Takeda T, Yabuta Y, Yoshimura K (2002) Regulation and function of ascorbate peroxidase isoenzymes. J Exp Bot 53:1305–1319

Shin JH, Yoshimoto K, Ohsumi Y, Jeon JS, An G (2009) OsATG10b, an autophagosome component, is needed for cell survival against oxidative stresses in rice. Mol Cells 27:67–74

Shirasawa K, Takabe T, Kishitani S (2006) Accumulation of glycinebetaine in rice plants that overexpress choline monooxygenase from spinach and evaluation of their tolerance to abiotic stress. Ann Bot 98:565–571

Smirnoff N (2007) Ascorbate, tocopherol and carotenoids: metabolism, pathway engineering and functions. In: Smirnoff N (ed) Antioxidants and reactive oxygen species in plants. Blackwell Publishing Ltd, Oxford, UK, pp 53–86

Smirnoff N, Cumbes QJ (1989) Hydroxyl radical scavenging activity of compatible solutes. Phytochemistry 28:1057–1060

Su J, Wu R (2004) Stress-inducible synthesis of proline in transgenic rice confers faster growth under stress conditions than that with constitutive synthesis. Plant Sci 166:941–948

Sundaram S, Wu S, Ma LQ, Rathinasabapathi B (2009) Expression of a *Pteris vittata* glutaredoxin PvGRX5 in transgenic *Arabidopsis thaliana* increases plant arsenic tolerance and decreases arsenic accumulation in the leaves. Plant Cell Environ 32:851–858

Szabados L, Savoure A (2010) Proline: a multifunctional amino acid. Trends Plant Sci 15:89–97

Takesawa T, Ito M, Hiroyuki Kanzaki H, Kameya N, Nakamura I (2002) Over-expression of ζ glutathione S-transferase in transgenic rice enhances germination and growth at low temperature. Mol Breed 9:93–101

Teotia S, Lamb RS (2009) The paralogous genes *RADICAL-INDUCED CELL DEATH1 and SIMILAR TO RCD ONE1* have partially redundant functions during Arabidopsis development. Plant Physiol 151:180–198

Tertivanidis K, Goudoula C, Vasilikiotis C, Hassiotou E, Perl-Treves R, Tsaftaris A (2004) Superoxide dismutase transgenes in sugarbeets confer resistance to oxidative agents and the fungus *C. beticola*. Transgenic Res 13:225–233

Tseng M-J, Liu C-W, Yiu J-C (2008) Tolerance to sulfur dioxide in transgenic Chinese cabbage transformed with both the superoxide dismutase containing manganese and catalase genes of *Escherichia coli*. Sci Hortic 115:101–110

Tsugane K, Kobayashi K, Niwa Y, Ohba Y, Wada K, Kobayashi H (1999) A recessive Arabidopsis mutant that grows photoautotrophically under salt stress shows enhanced active oxygen detoxification. Plant Cell 11:1195–1206

Tuteja N, Singh MB, Misra MK, Bhalla PL, Tuteja R (2001) Molecular mechanisms of DNA damage and repair: progress in plants. Crit Rev Biochem Mol Biol 36:337–397

Tuteja N, Ahmad P, Panda BB, Tuteja R (2009) Genotoxic stress in plants: shedding light on DNA damage, repair and DNA repair helicases. Mutat Res 681:134–149

Ushimaru T, Nakagawa T, Fujioka Y, Daicho K, Naito M, Yamauchi Y, Nonaka H, Amako K, Yamawaki K, Murata N (2006) Transgenic Arabidopsis plants expressing the rice dehydroascorbate reductase gene are resistant to salt stress. J Plant Physiol 163:1179–1184

Van Camp W, Capiau K, Van Montagu M, Inze D, Slooten L (1996) Enhancement of oxidative stress tolerance in transgenic tobacco plants overproducing Fe-superoxide dismutase in chloroplasts. Plant Physiol 112:1703–1714

Vendruscolo EC, Schuster I, Pileggi M, Scapim CA, Molinari HB, Marur CJ, Vieira LG (2007) Stress-induced synthesis of proline confers tolerance to water deficit in transgenic wheat. J Plant Physiol 164:1367–1376

Verbruggen N, Hermans C (2008) Proline accumulation in plants: a review. Amino Acids 35:753–759

Verdoy D, Coba De La Peña T, Redondo FJ, Lucas MM, Pueyo JJ (2006) Transgenic *Medicago truncatula* plants that accumulate proline display nitrogen-fixing activity with enhanced tolerance to osmotic stress. Plant Cell Environ 29:1913–1923

Vianello A, Zancani M, Nagy G, Macri F (1997) Guaiacol peroxidase associated to soybean root plasma membranes oxidizes ascorbate. J Plant Physiol 150:5–5

Vieyra FE, Boggetti HJ, Zampini IC, Ordonez RM, Isla MI, Alvarez RM, De Rosso V, Mercadante AZ,

Borsarelli CD (2009) Singlet oxygen quenching and radical scavenging capacities of structurally-related flavonoids present in *Zuccagnia punctata* Cav. Free Radic Res 43:553–564

Vickers CE, Possell M, Cojocariu CI, Velikova VB, Laothawornkitkul J, Ryan A, Mullineaux PM, Nicholas Hewitt C (2009) Isoprene synthesis protects transgenic tobacco plants from oxidative stress. Plant Cell Environ 32:520–531

Vranova E, Inze D, Van Breusegem F (2002) Signal transduction during oxidative stress. J Exp Bot 53:1227–1236

Waie B, Rajam MV (2003) Effect of increased polyamine biosynthesis on stress responses in transgenic tobacco by introduction of human S-adenosylmethionine gene. Plant Sci 164:727–734

Wan X, Tan J, Lu S, Lin C, Hu Y, Guo Z (2009) Increased tolerance to oxidative stress in transgenic tobacco expressing a wheat oxalate oxidase gene via induction of antioxidant enzymes is mediated by H_2O_2. Physiol Plant 136:30–44

Wang J, Zhang H, Allen RD (1999) Overexpression of an Arabidopsis peroxisomal ascorbate peroxidase gene in tobacco increases protection against oxidative stress. Plant Cell Physiol 40:725–732

Wang FZ, Wang QB, Kwon SY, Kwak SS, Su WA (2005) Enhanced drought tolerance of transgenic rice plants expressing a pea manganese superoxide dismutase. J Plant Physiol 162:465–472

Wang Y, Wisniewski M, Meilan R, Cui M, Webb R, Fuchigami L (2005) Overexpression of cytosolic ascorbate peroxidase in tomato confers tolerance to chilling and salt stress. J Am Soc Hortic Sci 130:167–173

Wang YC, Qu GZ, Li HY, Wu YJ, Wang C, Liu GF, Yang CP (2010) Enhanced salt tolerance of transgenic poplar plants expressing a manganese superoxide dismutase from *Tamarix androssowii*. Mol Biol Rep 37:1119–1124

Wang J, Fan Z, Liu Z, Xiang J, Chai L, Li X, Yang Y (2011) Thylakoid-bound ascorbate peroxidase increases resistance to salt stress and drought in *Brassica napus*. Afr J Biotechnol 10:8039–8045

Wen XP, Pang XM, Matsuda N, Kita M, Inoue H, Hao YJ, Honda C, Moriguchi T (2008) Over-expression of the apple spermidine synthase gene in pear confers multiple abiotic stress tolerance by altering polyamine titers. Transgenic Res 17:251–263

Wi S, Kim W, Park K (2006) Overexpression of carnation S-adenosylmethionine decarboxylase gene generates a broad-spectrum tolerance to abiotic stresses in transgenic tobacco plants. Plant Cell Rep 25:1111–1121

Wiseman H, Halliwell B (1996) Damage to DNA by reactive oxygen and nitrogen species: role in inflammatory disease and progression to cancer. Biochem J 313(Pt 1):17–29

Woo HR, Kim JH, Nam HG, Lim PO (2004) The delayed leaf senescence mutants of Arabidopsis, *ore1, ore3,* and *ore9* are tolerant to oxidative stress. Plant Cell Physiol 45:923–932

Wu Q, Lin J, Liu JZ, Wang X, Lim W, Oh M, Park J, Rajashekar CB, Whitham SA, Cheng NH, Hirschi KD, Park S (2012) Ectopic expression of Arabidopsis glutaredoxin *AtGRXS17* enhances thermotolerance in tomato. Plant Biotechnol J 10:945–955

Yang X, Liang Z, Lu C (2005) Genetic engineering of the biosynthesis of glycinebetaine enhances photosynthesis against high temperature stress in transgenic tobacco plants. Plant Physiol 138:2299–2309

Yin L, Wang S, Eltayeb AE, Uddin MI, Yamamoto Y, Tsuji W, Takeuchi Y, Tanaka K (2010) Overexpression of dehydroascorbate reductase, but not monodehydroascorbate reductase, confers tolerance to aluminum stress in transgenic tobacco. Planta 231:609–621

Yoshida S, Tamaoki M, Shikano T, Nakajima N, Ogawa D, Ioki M, Aono M, Kubo A, Kamada H, Inoue Y, Saji H (2006) Cytosolic dehydroascorbate reductase is important for ozone tolerance in *Arabidopsis thaliana.* Plant Cell Physiol 47:304–308

Yoshimura K, Miyao K, Gaber A, Takeda T, Kanaboshi H, Miyasaka H, Shigeoka S (2004) Enhancement of stress tolerance in transgenic tobacco plants overexpressing Chlamydomonas glutathione peroxidase in chloroplasts or cytosol. Plant J 37:21–33

Yu D, Xie Z, Chen C, Fan B, Chen Z (1999) Expression of tobacco class II catalase gene activates the endogenous homologous gene and is associated with disease resistance in transgenic potato plants. Plant Mol Biol 39:477–488

Yu T, Li YS, Chen XF, Hu J, Chang X, Zhu YG (2003) Transgenic tobacco plants overexpressing cotton glutathione S-transferase (GST) show enhanced resistance to methyl viologen. J Plant Physiol 160:1305–1311

Zhang H, Dong H, Li W, Sun Y, Chen S, Kong X (2009) Increased glycine betaine synthesis and salinity tolerance in AhCMO transgenic cotton lines. Mol Breed 23:289–298

Zhang C-J, Zhao B-C, Ge W-N, Zhang Y-F, Song Y, Sun D-Y, Guo Y (2011) An apoplastic H-type thioredoxin is involved in the stress response through regulation of the apoplastic reactive oxygen species in rice. Plant Physiol 157:1884–1899

Zhao Q, Wang G, Ji J, Jin C, Wu W, Zhao J (2013) Over-expression of Arabidopsis thaliana β-carotene hydroxylase (chyB) gene enhances drought tolerance in transgenic tobacco. J Plant Biochem Biotechnol 1–9. DOI: 10.1007/s13562-013-0201-2

Zhou X, Hua D, Chen Z, Zhou Z, Gong Z (2009) Elongator mediates ABA responses, oxidative stress resistance and anthocyanin biosynthesis in Arabidopsis. Plant J 60:79–90

Weed Stress in Plants

Meenal Rathore, Raghwendra Singh,
Partha Pratim Choudhary, and Bhumesh Kumar

Abstract

Few plant species utilize available natural resources more efficiently as compared to other species in order to establish themselves in prevailing environmental conditions. Once established in early phase of growth, they can be sustained throughout the growing season by virtue of better stand. And this is the strategy behind the success of most of the notorious weeds in any cropping system. They, hence, continue to cause huge yield losses despite every effort made by farmers to manage them. Herbicides are largely used to manage weeds globally, but its application is also known to cause stress, though minimal, in crop plants. Another factor which contributes to the success of weeds is their hardiness and resilience to abiotic and biotic stress factors. Molecular mechanism(s) responsible for traits like competitiveness and invasiveness of weeds is poorly understood till date. However, development and availability of sophisticated molecular tools pave the way to dissect the mechanism of weed dominance. Competitiveness and tolerance to stress factors are important traits observed among different weed species that can be exploited in attempts to develop crop plants tolerant to abiotic/biotic stress(es). The need of the hour is to understand the molecular mechanisms underlying weed competitiveness over crop plants in field and to utilize the responsible gene(s) by transferring them into crop plants. However, success of such approaches depends upon integration and collaboration to bring expertise together from weed science, molecular biology, and plant physiology. An effort has been made to review the traits available among weed species that make them competitive and hardy.

Keywords

Crop–weed interaction • Nutrient stress • Light stress • Water stress • Herbicide stress

M. Rathore (✉) • R. Singh • P.P. Choudhary • B. Kumar
Directorate of Weed Science Research, Jabalpur, India
e-mail: mnl.rthr@gmail.com

R.K. Gaur and P. Sharma (eds.), *Approaches to Plant Stress and their Management*,
DOI 10.1007/978-81-322-1620-9_14, © Springer India 2014

Introduction

Agricultural produce is nowadays subjected to a lot of stress factors (abiotic and/or biotic) that affect quality and quantity of crop and thereby yield. Biotic stresses include viruses, fungi, bacteria, insects, pests, and pathogens and also weeds. Out of the total plant species, 10 % are weeds and account approximately 30,000 in terms of number. Of these, 1,800 pose very serious threat not only to crop production but also to environment and health hazards. Some 300 weed species plague cultivated crops throughout the world and have been reckoned to have a direct negative impact on crop production and crop protection (Ware and Whitacre 2004).

Weeds are plants out of place and out of time or growing where/when they are not wanted and interfering with the activities or welfare of mankind. These weeds thrive in disturbed habitats and produce seed in abundance that is not useful to humans (Manning 2004). Agriculture too is based on few crop plants that grow in disturbed (or cropped) habitat and produce abundance of seeds, but these are useful to mankind in one or other ways. There are other characteristics of weeds which make them different from crop plants and enable to compete with crop plants. These include rapid growth at early stage, shorter vegetative phase, early maturity, dual modes of reproduction in few cases, high fecundity, environmental plasticity and tolerance to stresses, seed dormancy, seed longevity, self-compatibility, extensive root system, and competitiveness for nutrients, water, and light. Losses in agriculture due to weeds are already established. Apart from competing with crop plants, weeds also add up to the protection costs because they harbor other pests, reduce farm produce quality, affect health of animals and humans, increase product processing costs, and interfere with water management in irrigated agriculture. This chapter will however limit to stress on crop plants due to weeds with emphasis on crop–weed competition.

The competition between crop and weeds is real and for survival of the fittest. In weed science, huge competition is among the plants (Donald 1963) for acquiring of natural resources. Plant competition also means a reduction in performance of a given plant species of importance due to shared use of a limited available resources (Gurevitch et al. 2009). Competition between plants is basically passive, not visible in the beginning of development (Floss 2008). Crop plants, in the process of domestication, have acquired desired productive features but also lost few characteristics which, apparently, weeds have retained. Some of such characteristics can be mentioned as to have an idea about the unique traits of weeds which make them weed. In addition to competitiveness, robustness, and invasiveness, their ability to cope with adverse environments, resistance against diseases and insects, shorter life cycle, variation in maturity time, variable degree of shattering, and dormancy of seeds (even within the species) are worth mentioning. All these factors together make weeds more competitive and hardier as compared to crop plants. How weeds succeed in over-competing crop plants can be understood by a very simple example. Weedy rice (which is dominant problem in direct seeded rice) has vigorous growth and by virtue of that competes with cultivated rice and reduces the crop yield. Grains of weedy rice cannot be harvested as it matures earlier and shatters the grains into soil for the next generation and become dormant till next season, thus establishing a potential seed bank. All these characters together seem to be a strategy which provides an edge over cultivated rice. Efforts have been made and being made to find out what is responsible for success of weedy rice, and it was suggested that nitrogen uptake and nitrogen use efficiency for biomass production was higher in weedy rice as compared to that in cultivated rice which gave weedy rice an edge in terms of vigorous growth and biomass accumulation (Burgos et al. 2006). In initial stage of establishment of seedling, the fast and vigorous growth can be seen in most of the weed species in agricultural fields, i.e., *Phalaris minor* in wheat crop and *Echinochloa* species in rice crop. The strategy behind the success of weeds seems to be simple but effective. During the early growth, weeds use

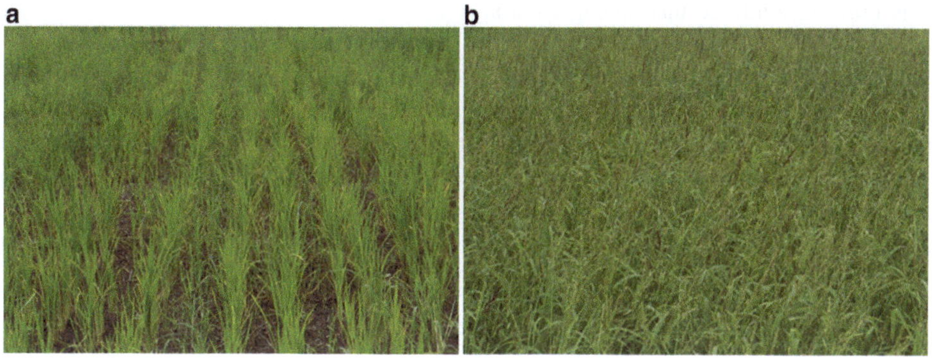

Fig. 1 (**a**) Establishment of weeds between rows in a rice field. (**b**) Weeds out competing rice in infested fields

available soil resources, i.e., soil nitrogen (even from depleted soil which cannot be utilized by crop plants), more effectively than crop plants in order to achieve a faster and vigorous growth resulting in greater biomass production. Once they succeed in doing so, they can easily capture the other resources like water, space, and more importantly photosynthetically active radiation and thus provide a tough competition to the crop plants.

In simple term, weed competition poses stress on the plants. This competition is critical for the crop when weed establishes together or before the crop. An early establishment of weeds covers the soil surface preventing access of plants to light – a very detrimental factor during canopy development (Floss 2008; Gurevitch et al. 2009). Figure 1 reveals establishment of weeds with crop (a) and how these weeds out compete the crop (b) so much so that even the rows are not visible. Early establishment of weeds in the field gives a better weed stand and helps them retain an upper hand in utilization of nutrients, light, and water.

In addition to the direct competition from weed species, several other indirect effects also affect the crop performance.

Nutrient Stress

Weeds are known to absorb mineral nutrients faster than many of our crop plants and accumulate them in tissues in relatively large amounts. Success in absorbing nutrients leads to rapid growth and successful competition for light and water. Weeds have a large nutrient demand, and to fulfill that demand, they keep absorbing nutrients more than that in crop plants. For example, *Cyperus rotundus* (a weed of rice crop) can accumulate about 2.17 % nitrogen and 2.73 % potassium. Similarly, *Echinochloa colonum* can accumulate 2.98 % nitrogen, but rice (*Oryza sativa*) can accumulate only 1.13 % nitrogen and 1.1 % potassium on dry weight basis. *Digitaria sanguinalis* can accumulate about 3.36 % phosphorus, while wheat can accumulate only 0.59 % on dry weight basis. Scientists found that from a weedy sugarcane field at Delhi, weeds remove approximately 162 kg N, 24 kg P_2O_5, and 203 kg K_2O per ha against 28 kg N, 17 kg P_2O_5, and 64 kg K_2O per ha by the crop (Gupta 2000). Rajan and Sankaran (1974) estimated nutrient removal by weeds during the first 30 days of maize crop to be seven to ten times higher than nutrient removal by the crop. Ronchi et al. (2003) observed that *Bidens pilosa* accumulates 5.53, 11.19, and 5.32 times more N, P, and K, respectively, in comparison to coffee crop with more accumulation of dry matter at pre-flowering stage.

Dunbabin (2006) found that the most important parameters for plant competition with weeds were those controlling the intensity of soil foraging (spatial efficiency and intensity) and rate of root system establishment. This directly highlights the need of plant to occupy soil volume during its establishment phase to out compete the weeds.

There is growing evidence that among environmental cues light quality in terms of far-red/red light ratio perceived by a plant can play an important role in affecting interaction among neighbors (Ballaré and Casal 2000; Ballare' et al. 1990). It has been suggested that weeds growing with maize modify its ability to compete for soil resources by stimulating reduction in its root to shoot ratio (Rajcan and Swanton 2001; Rajcan et al. 2004). It was also speculated that weed infestation could alter maize and weed root system's size and morphology, but no evidence was then available. Further, Afifi and Swanton (2011) tested the hypothesis that light signals of red to far-red ratio(R/FR) reflected from stem and leaf surfaces of neighboring weeds can trigger a shade avoidance response in maize seedlings resulting in reduction in root biomass. They first reported a reduction in total biomass and volume of roots originating from seeds or stem tissue in response to proximity of weeds.

Most studies on abiotic and biotic stresses have been done on weed *Arabidopsis* (now a model plant), and in context of nutrient uptake, studies reveal that in low phosphorus soils, *Arabidopsis* favors lateral growth (root density and length) over primary root growth. The ability of roots to respond to phosphate availability was independent of sucrose supply and auxin signaling (Williamson et al. 2001). It has also been shown that when *Arabidopsis* roots are subjected to low P, there is early differentiation of cells that actively transcribed genes involved in P uptake and scavenging leading to formation of roots specialized for P extraction from the soil. The Arabidopsis *ANR1* has been identified to encode a member of the MADs box family of transcription factors and is a key determinant of the developmental plasticity in *Arabidopsis* (Zhang and Forde 1998).

Bocrboom (2007) found that nitrogen uptake by weeds in a corn field was sufficient enough to limit the available nitrogen for corn resulting tremendous yield losses. Studies conducted by Hansen et al. (2010) suggest that corn growth is affected during early growth by weeds in terms of delayed development and decrease in leaf area and biomass even before measurable water and nitrogen competition and further yield reduction occurred if weed competition remained till later stages.

Light Stress

Light is the environmental resource that has to be utilized as and when it is received with a chance of storage for later use, or else it is lost forever (Donald 1963). Plant height and vertical leaf area distribution also determine the efficient utilization of light and affect crop–weed interactions (Graf and Hill 1992). Weeds may affect crop development if they establish early and cover more space than crop, thus inhibiting proper transmittance of sunlight. Wild oats reduce light penetration and growth of wheat by growing taller. And if the oat is clipped to the height of wheat, the light penetration in the mixed weed–crop canopy is similar to that of a monocropped wheat (Cudney et al. 1991). In velvetleaf–soybean competition, the greater height and dry weight allocation to upper branches of velvetleaf allowed more interception of light by the weed (Akey et al. 1990). Moriles et al. (2012) while studying effects of weed interference, shade, and water on crop growth found that shade had less effect than weeds on plants, and hence the response to weeds is not totally due to reduction in intercepted light. They also found during microarray analysis that gene expression in shade and weed stress clustered more tightly together, but shared only three ontologies, namely, O-methyl transferase activity (lignification process), poly-U binding activity (posttranscriptional gene regulation), and stomatal movement. Based on morphologic and genomic observations, weed stress to corn was not explained by individual effects of nitrogen or light stress. Therefore, we hypothesize that these stresses share limited signaling mechanisms. Principle component analysis of the differentially expressed genes revealed that shade/light and weed stress had more similar gene expression patterns (252 genes) to each other. Genes related to photosynthesis, energy conversion, and signaling were downregulated in response to weed stress (Moriles et al. 2012).

Light competition becomes important when there is adequate soil moisture and high fertile soils as in such cases plants have broader leaves. Leaf area index is the one-sided green leaf area per unit ground area in broadleaf canopies, or the projected needle leaf area per unit ground area in needle canopies. It is directly correlated with potential light interception. Successful competitors for shade have a high LAI, but not necessarily high foliage. Architecture of the plant leaf, angle of inclination, and its arrangement on shoot are important here so that it is in the most advantageous position for light interception. Plants with leaves horizontal to the ground are more competitive for light than plants with leaves perpendicular to ground or at an angle. Also leaves arranged alternately and spirally provide space for light to reach lower leaves than those arranged one above the other.

Even if light is intercepted, the plant should be able to utilize it efficiently, and this is measured in terms of light use efficiency (LUE). Santos et al. (2003) evaluated the LUE of bean, soybean, and weeds *Euphorbia heterophylla, Bidens pilosa,* and *Desmodium tortuosum.* It was concluded that LUE of these weeds is lesser then crop plants. Although weeds are less efficient than crops in utilizing light, they are more competitive for extracting the resources of nutrition and water (Concenco et al. 2012). This light use efficiency is "in turn" dependent on the specific photoreceptors that perceive light, namely, phytochromes, cryptochromes, phototropin, and the UV-B photoreceptors UVR8, that signal photo morphogenic responses and affect usage of light resource and ability of plants to capture additional features (Ballaré and Casal 2000). These signals include the low irradiance and low red/far-red ratio typical of dense canopies. The response they affect includes phyllotaxy, changes in stem height, photosynthetic capacity of green tissue, and also plant defenses (Casal 2013). An example is nitrate reductase activity that is known to be stimulated by light intensity and duration, and it may also respond to different water contents in soil (Sung 1993). The activity is also influenced by carbohydrate content.

Figure 2 shows plants of *Sesamum indicum* in plots that are weed-free and full of weeds. Both the plots have received identical treatment in terms of irrigation and fertilizer dose. The stress generated due to presence of weeds is clearly visible in the crop plants as they are smaller and weaker than those in the weed-free plot. Sesame plants in plot free of weeds are taller with broader leaves and visibly healthier.

Fig. 2 Development of sesame plants in a plot free of weeds maintained by hand weeding and in a plot full of weeds

The message is that physiology of plants affects competitive ability between crops and weeds, and light perception affects plant physiology, obviously. And hence shade stress caused by weeds has ability to affect the crop severely.

Water Stress

Water is an essential natural resource required by both weeds and crop plants and whose deficit becomes a significant limiting constraint in their development (King 1996). In a weed-infested field, the weeds extract soil moisture and nutrients, reducing its availability to crop plant and thus imposing stress on them. This stress reduces growth and also yield of crop. Competition for water also causes competition of nutrients and light too, though indirectly (Silva et al. 2007). The competitive ability of a crop for water uptake is related to its ability to explore soil for moisture and the volume of soil explored for the same. This in turn relates to the root system, physiological characteristics of plant, stomatal regulation, osmotic adjustment in roots, and hydraulic conductivity capacity of roots (Floss 2008).

In dry farming areas, in general, for producing equal dry matter, weeds transpire more water than most of our crop plants. Hence, under such areas the evapotranspiration from weedy fields is more than from weed-free field. Shahi (1978) observed that *Chenopodium* had higher water consumptive use and removed water evenly from up to 90 cm soil depth, while wheat largely up took moisture from up to 15 cm depth. The root system volume and efficacy seems to be important (Gupta 2000).

Water use efficiency is an important parameter assessing amount of dry matter accumulated as a function of water used at the same period. Plants with higher WUE use less water to produce per unit dry mass, and it is possible that plants with higher WUE adapt better when subjected to moisture stress over those with less WUE (Radosevich et al. 2007). But few weeds may have distinct WUE throughout the cycle but be more competitive for water in certain developmental stages (Silva et al. 2007).

As roots grow and spread early in plant life, the competition for water and nutrition begins before competition for light. Less competition will be there if the roots of weeds and plants acquire different soil zones. The most competitive, vigorously growing plants having a larger root system will be able to explore larger soil areas. If plants have similar root lengths, then those with widely spreading and less branched root systems will be more competitive for moisture extraction. The weed *Bidens pilosa* is not highly efficient in using water but is very efficient in extracting water from soil. It is capable of maintaining high growth rates under poor soil water potentials wherein most crops and weed species reach the point of permanent wilting (Procópio et al. 2004b). Similarly, *B. plantaginea* is also highly efficient in using the water extracted or absorbed from soil.

CO_2 Competition

Competition for CO_2 is actually competitive ability in terms of photosynthesis that also affects other parameters like water use efficiency. Photosynthesis and respiration depend upon a constant flux of CO_2 and O_2 within the cell which is related to concentration of CO_2 and O_2 in intercellular spaces. Concentration of these two gases, in turn, is dependent on gas flux through stomata and therefore is dependent on stomatal opening and closure which is controlled by turgor of stomatal guard cells and epidermal cells of stomata (Humble and Hsiao 1970; Messinger et al. 2006; Concenço et al. 2012). Low water potential will reduce stomatal opening and leaf conductance and inhibit photosynthesis (Attridge 1990). Water use efficiency is a relation between CO_2 incorporated within cells and water lost via transpiration during the same period (Gurevitch et al. 2009). Hence, presence of weeds with high photosynthetic ability adds up to the stress faced by crop plants. Plants efficient in photosynthetic ability are generally more competitive, and weeds, in general, are more efficient photosynthetically (Black et al. 1969).

Molecular Biology in Crop–Weed Competition

Presence of weeds in a field of crop is always stressful for the crop plant as they now have to compete for nutrients, light, and moisture. There is nutrient and water stress beneath the soil and shade stress above the soil. Weeds, when more competitive than crop in the prevailing environmental conditions, are so as they are able to extract the limiting nutrients and water in a more efficient manner. And this efficiency is either inherent in terms of the genes regulating root system and its architecture or adaptive developed through phenotypic plasticity of weed on exposure to competitive regimes. Weeds are supposed to be more efficient photosynthetically probably because of a more efficient system to capture available CO_2 and utilize it. Studies on *Arabidopsis* have helped us to understand the underlying reasons for competitiveness in weeds to an extent, some of which will be listed here.

Phosphate is one of the difficult nutrient elements to be extracted from soil and is also a limiting factor in crop yield. It is known that changes in root architecture occur as part of adaptive strategies to compensate for local phosphate deficiencies. The AINTEGUMENTA-like gene known as PRD gene (phosphate root development) rapidly decreases in roots under low P_i conditions. Complementation studies of prd mutants in *Arabidopsis* confirmed the role of PRD in regulation of root architectural responses to P_i starvation by controlling primary and lateral root elongation (Camacho-Cristobal et al. 2008). Hamburger et al. (2002) identified the *Arabidopsis* pho1 gene to be required for transferring the phosphate from root epidermal and cortical cells into the xylem by a map-based cloning strategy. The cysteine-2/histidine-2 zinc finger transcription factor ZAT6 is a repressor of primary root growth and regulates Pi homeostasis by controlling root architecture (Devaiah et al. 2007)

The ion exchangers NHX1 and NHX2 mediate active potassium uptake into vacuoles to regulate cell turgor and stomatal function in *Arabidopsis* (Barragan et al. 2012). Transcriptomics studies in rice and *Arabidopsis* revealed many similar differences in response to K ion deficiency. However, there were more genes related to stress responses and development in *Arabidopsis* than rice. Auxin-related gene responded to K ion deficiency in rice, while jasmonic acid-related genes had a more important role in K ion signaling in *Arabidopsis* (Ma et al. 2012). This indicates differences in transcriptional regulation between monocots and dicots. It is expected there will be similar or maybe even more differences if studies on transcriptomics in crop and related weeds are done for stress(es) caused due to presence of weeds.

While working with accessions of *Arabidopsis thaliana,* Chardon et al. (2010) showed that nitrogen use efficiency was exclusively genetically determined. In a study to assess differential response of weeds, the shoot and root growth response of 23 agricultural weeds to N fertilizer with wheat and canola was done. All crop and weed species extracted >80 % of available nitrogen at low soil levels. Seventeen of 23 weed species took up similar or greater amounts of soil nitrogen than did wheat, and 6 weed species took up nitrogen in similar amounts to that of canola (Blackshaw et al. 2003). It is understood that infestation of weeds during the critical period definitely results in yield loss. A few genes have recently been discovered in *Arabidopsis* and discussed by Chen et al. (2012) for their functions related to light-mediated signal cascades and possible involvement in circadian clock.

Overcoming Competition to Manage Stress

There are basically two ways to overcome any stress (weeds in present case) – remove or manage the factor causing stress, i.e., weeds, or use crops that are more competitive than weeds.

Development of hardier crops that yield more and are tolerant/resistant to pests has been largely taken care of plant breeders, biotechnologists, and geneticists. But development of crop cultivars that are actually competitive to weeds and stress caused due to them needs to be seriously looked into. Herbicide-tolerant crops are developed based on natural selection on plants with inherent ability to tolerate a herbicide which are then crossed with crop plants through traditional plant breeding methods, or they are genetically modified to be tolerant/resistant to the herbicide. For example, Clearfield crops are resistant to imidazolinone herbicides, roundup-ready crops are resistant to glyphosate, and STax crops have multiple genes making them tolerant to herbicides as well as pests.

The most widely used method to overcome competition is managing the weeds. Weed management is the combination of techniques of prevention, eradication, and control to manage weeds in a crop, cropping system, or environment (Zimdahl 2007). This includes agronomic practices like crop rotation, mulching, soil solarization, companion cropping, use of fertilizers, and use of herbicide-resistant crops. Mechanical methods of control rely on agricultural implements like tilling, hand hoeing, mowing, and hand pulling. But the most widely accepted method is the chemical mean of managing weeds by use of herbicides.

Herbicides: Causing Stress While Managing Weeds

A herbicide is designed chemically so that it kills only the unwanted plants in a crop field and not the crop plants. This selectivity of herbicides in killing weeds is based on certain biochemical differences among plants, namely, interception and uptake of herbicides, metabolic pathways and rates, target-enzyme sensitivity, and tolerating product phytotoxicity. Factors other than biotic ones, such as product characteristics, mode of application, time of application, use of antidotes or safeners, and genetic engineering (Oliveira Júnior 2001), may be responsible for selectivity by influencing the interaction between herbicides and enzyme systems. In many cases, herbicides applied in a crop to control associated weeds result in ill growth of the crop too, and this adverse effect is expressed in many forms, namely, stunted growth, chlorosis, and leaf-tip burn. This adverse effect of any herbicide on a crop is called phytotoxicity, albeit in strict sense it is associated with any plants. Phytotoxicity is, thus, the capacity of a pesticide to cause temporary or long-lasting damage to plants.

The phenomenon of phytotoxicity is expressed in a crop plant only when the herbicide gets access to the enzyme it acts upon. For crops the biochemical basis of selectivity must be perfect, not allowing interaction between herbicide and enzyme to avoid phytotoxicity. In tolerant plants, a part of the herbicide is taken up, quickly metabolized, and inactivated before exerting its phytotoxic effect (de Carvalho et al. 2009). Such plant species contains mechanisms that can alter or degrade the herbicide molecule by biochemical reaction producing nontoxic products. The efficiency in metabolizing the herbicide makes the difference between tolerant crop and susceptible weed. Herbicide metabolism takes place through four phases (Yuan et al. 2007): Phase I, also known as conversion, Phase II or conjugation, Phase III with secondary conversion and transport into vacuole, and Phase IV with deposition of final metabolite (Devine et al. 1993; Cole 1994; van Eerd et al. 2003; Yuan et al. 2007).

During Phase I of herbicide metabolism, the active ingredient molecules suffer chemical modifications, such as oxidation, reduction, hydrolysis, oxygenation, or hydroxylation, when functional groups (OH, NH_2, COOH) are introduced or revealed, making the molecule more hydrophilic and therefore less phytotoxic (Devine et al. 1993; van Eerd et al. 2003; Edwards et al. 2005; Yuan et al. 2007). During Phase II, the herbicide molecule or the metabolite derived from Phase I is conjugated with sugars, amino acids, or with the tripeptide glutathione, increasing the solubility in water, while reducing phytotoxicity. During Phase III, the metabolites derived from Phase II are actively transported to the vacuole by mostly ABC (ATP-binding cassette) transporters (Yuan et al. 2007). Secondary conjugations might also occur during Phase III, giving origin to non-phytotoxic compounds (Hatzios 1991). Later, in Phase IV, the metabolites of the detoxification process, compartmentalized in vacuoles, may be associated with components of the cell wall (pectin, lignin, polysaccharides, and protein fractions) forming insoluble residues (Pillmoor et al. 1984; Langebartels and Harms 1985; Cole 1994; Edwards et al. 2005).

Phytotoxicity is caused mainly through direct application of selective herbicides that may enable it to be absorbed through roots and foliages of crop plants. Soil-applied herbicides, which persist for a longer duration, are carried over to the next crop season. The successive crop absorbs the herbicides and may be injured showing phytotoxicity. The drift of herbicide formulation may also cause phytotoxicity to neighboring crops.

In the crop plant most of the applied herbicide is detoxified by metabolic processes (Shimabukuro RH 1985). A part of the herbicide reaches the target site of action, leading the plant to express the symptoms. Phytotoxicity symptoms can be divided into structural damage (chlorosis, necrosis, albinism, wilt, epinasty, and leaf shriveling or rolling) or physiological damage (cycle reduction and growth rate reduction) (de Carvalho et al. 2009). These symptoms are the results of the interaction between herbicides and their sites of action. Therefore, all herbicides may not necessarily result the same symptoms.

Phenoxys: These synthetic hormones upset the hormonal balance in plants. Due to excessive cellular proliferation, abnormal growth of affected plants takes place. The first symptom of injury is usually stem twisting followed by deformities in terminal tissue which may lead to cupping or strapping of the leaves and total bending and twisting of the stems. Corn injury occurs in the form of onion leafing, proliferation of roots, or abnormal brace root formation.

Dinitroanilines: Members of this group, namely, pendimethalin, oryzalin, and trifluralin, are microtubules inhibitor. Dinitroanilines affect root development showing the symptoms of swollen root tips and stunted lateral and secondary roots. Soybean injury from dinitroanilines is characterized by root pruning.

Triazines: Herbicides of this group bind to a protein involved in electron transport in photosystem II inhibiting photosynthesis. This results in chlorosis of plant leaves followed by necrosis of leaf tissue. Injury to corn occurs as yellowing of leaf margins and tips followed by browning. In soybean yellowing of outer leaf margins is observed. The entire leaf turns yellow, but veins usually remain somewhat green. Lower leaves are most affected and new leaves remain unaffected.

Substituted urea: Plants show chlorosis and drying of green tissue and yellowing of leaf tips and margins. Lower leaves may be affected first, and newer leaves may be unaffected.

Glyphosate: This nonselective herbicide interferes with amino acid biosynthesis inhibiting the EPSPS enzyme. Leaves of affected plants slowly wilt and turn brown. Injury symptoms appear slowly.

Bipyridyliums: Paraquat and diquat, the important members of this family, exert their weed killing action through the inhibition of photosystem I. Affected plants look burned and wilted, with mottled yellowing of leaves followed by wilting and rapid drying. Drift of paraquat may cause spots of dead leaf tissue wherever spray droplets contact the leaves.

Sulfonylureas and imidazolinones: Herbicides of these classes are ALS inhibitors. Affected plants exhibit chlorosis, necrosis, stunting, epinasty, and shortening of internodes. Sulfonylureas kill germinating seedlings. They also exhibit foliar activity. Herbicide symptoms develop slowly. Some deformation of new growth can occur. Soybeans often develop reddish leaves. In case of corn, secondary root formation decreases and roots become stunted.

Aryloxyphenoxypropionate: Herbicides of this class, namely, quizalofop, propaquizafop, and fluazifop, are ACCase inhibitors. Irregular bleaching of leaves or bands of chlorotic tissue may appear as symptoms on affected leaves (Turner and Pernich 2002).

In general, phytotoxicity or injury to crop does not cause yield loss if recommended application of herbicides is followed. Crop yield loss occurs only when the stress due to herbicide phytotoxicity reaches the tolerance threshold of the crop. But plants are to spend a substantial amount of energy during the metabolic detoxification of the absorbed herbicides (de Carvalho et al. 2009 and Dragicevic et al. 2012). This energy may be available naturally and it does not cause any yield loss to crop. Nevertheless, the energy required for the repairing of damaged structures in plant recovery process can result yield loss.

Herbicides are mostly degraded in the environment either by chemical processes or by microbial transformation. Most herbicides are breakdown in a few weeks. But some herbicides often remain in the soil for several months because that is their intended purpose, and a few may last for more than a year. It is possible to manage the herbicide residues by enhancing organic matter status of soil. Selection of herbicide for a particular crop variety is also crucial for the management of crop phytotoxicity.

Conclusion

Weeds affect crop production more than known pests, yet they are generally overlooked as potential threats or even stress-causing plants. But the truth is that weeds are real examples of Darwin's theory of survival of the fittest in natural conditions. Still undomesticated, they are robust enough to compete with crop for light, nutrition, water, and air and cause a deficit of these natural reserves in soil imposing abiotic stress to the crop. Detailed studies reflect these stresses to be more or less interdependent and also to communicate at the molecular level. Stress caused by weeds is largely taken care of by removing weeds or by use of herbicides. But these chemicals too impose a stress on the plants known as phytotoxicity. And so, in totality, presence of weeds should be considered as important and alarming as presence of pest because the stress weeds impose is equally harmful. Also, there is a need to emphasize in-depth studies on weeds other than *Arabidopsis* to actually understand the phenomenon of weed–crop competitiveness.

References

Afifi M, Swanton C (2011) Maize seed and stem roots differ in response to neighboring weeds. Weed Res 51:442–450

Akey WC, Jurik TW, Dekker J (1990) Competition for light between velvetleaf (*Abutilon theophrasti*) and soybean (*Glycine max*). Wed Res 31:63–72

Attridge TH (1990) The natural environment. In: Attridge TH (ed) Light and plant responses. Edward Arnold, London, pp 1–5. ISBN 978-052-1427-48-7

Ballaré CL, Casal JJ (2000) Light signals perceived by crop and weed plants. Field Crops Res 67(2):149–160, 0378–4290

Ballare' CL, Scopel AL, Sa'nchez RA (1990) Far-red radiation reflected from adjacent leaves: an early signal of competition in plant canopies. Science 247:329–332

Barragan V, Leidi EO, Andrez Z, Rubio L, DeLuca A, Fernandez JA, Cubero B, Pardo JM (2012) Ion exchangers NHX1 and NHX2 mediate active potassium uptake into vacuoles to regulate cell turgor and stomatal function in Arabidopsis. Plant Cell 24(3):1127–1142

Black CC, Chen TM, Brown RH (1969) Biochemical basis for plant competition. Weed Sci 17:338–344

Blackshaw RE, Brandt RN, Janzen HH, Entz T, Grant CA, Derksen DA (2003) Differential response of weed species to added nitrogen. Weed Sci 51:532–539

Boerboom C (2007) Are weeds in corn risky business? In: Indiana CCA conference proceedings. http://www.agry.purdue.edu/CCA/2007/2007/proceedingss/Chris Boerboom-CCAProceedings_KLS.doc

Burgos NR, Norman RJ, Gealy DR, Black H (2006) Competitive uptake between rice and weedy rice. Fields Crop Res 99:9–105

Camacho-Cristobal JJ, Rexach J, Conejero G, Al-Ghazi Y, Nacry P, Doumas P (2008) PRD. An Arabidopsis AINTEGUMENTA-like gene, is involved in root architectural changes in response to phosphate starvation. Planta 22(8):511–522

Casal JJ (2013) Canopy light signals and crop yield in sickness and in health. ISRN agronomy. Article ID. 650439. Hindawi Publishing Corporation. http://dx.doi.org/10.1155/2013/650439

Chardon F, Bartheyemy J, Daniel-vedele F, Masclaux-Daubresse C (2010) Natural variation of nitrate uptake and nitrogen use efficiency in *Arabidopsis thaliana* cultivated with limiting and ample nitrogen supply. J Exp Bot 61(9):2293–2302. doi:10.1093/jxb/erq059

Chen C, Xiao YG, Ni M (2012) Light regulated stomatal aperture in Arabidopsis. Mol Plant 5(3):566–572

Cole DJ (1994) Detoxification and activation of agrochemicals in plants. Pesticide Sci 42:209–222

Concenço G, Aspiazu I, Ferreira EA, Galon L, da Silva AF (2012) Physiology of crops and weeds under biotic and abiotic stresses. In: Najafpour M (ed) Applied photosynthesis. Intechopen, pp 257–280. ISBN 978-953-0061-4. doi.10.5772/30691. http://www.intechopen.com/books/applied-photosynthesis/physiology-of-crops-and-weeds-under-biotic-and-abiotic-stressespublisher.

Cudney DW, Jordan LS, Hall AE (1991) Effect of wild oat (*Avena fatua*) infestations on light interception and growth rate of wheat (*Triticm aestivum*). Weed Sci 39:175–179

Devaiah BN, Nagarajan VK, Raghothama KG (2007) Phosphate homeostasis and root development in Arabidopsis are synchronized by the zinc finger, transcription factor ZAT1. Plant Physiol 145:147–159

de Carvalho SJP, Nicolai M, Ferreira RR, de Figueira OAV, Christoffoleti PJ (2009) Herbicide selectivity by differential metabolism: considerations for reducing crop damages. Sciencia Agricola (Piracicaba, Braz) 66:136–142

Devine MD, Duke SO, Fedtke C (1993) Herbicide translocation. In: Physiology of herbicide action. Prentice Hall, Englewood Cliffs, NJ: pp 67–94

Donald CM (1963) Competition among crop and pasture plants. Adv Agron 15:1–118

Dragicevic VD, Simic M, Brankov M, Spasojevic I, Secanski M, Kresovic B (2012) Thermodynamic characterization of early phytotoxic effects of sulfonyl urea herbicides to maize lines. Pestic Phytomed 27 (3):231–237

Dunbabin V (2006) Using the ROOTMAP model of crop root growth to investigate root-soil interactions. "Ground breaking stuff" proceedings of the 13th ASA conference, 10–14 Sept 2006, Perth, www.agronomy.org.au

Edwards R, Brazier-Hicks M, Dixon DP, Cummins I (2005) Chemical manipulation of antioxidant defenses in plants. Adv Bot Res 42:1–32

Floss EL (2008) Fisiologia das plantas cultivadas, 4th edn. Universidade de Passo Fundo, Passo Fundo. ISBN 978-857-5156-41-4

Graf B, Hill JE (1992) Modelling the competition for light and nitrogen between rice and *Echinochloa crusgalli*. Agri Syst 40(94):345–359

Gupta OP (2000) The nature of weed–crop competition In: Modern weed management. Agrobios, Inida, pp 18–37

Gurevitch J, Scheiner SM, Fox GA (2009) Ecologia vegetal, 2nd edn. Artmed, Porto Alegre. ISBN 978-853-6319-18-6

Hamburger D, Rezzonico E, Petetot CMC, Somerville C, Poirera Y (2002) Identification and characterization of the Arabidopsis PHO1 gene involved in phosphate loading to the xylem. Plant Cell 14:889–902

Hansen SA, Clay SA, Moriles J, Clay D, Horvath D (2010) Investigating early growth and development response of corn to weed competition using transcriptome analysis. WSSa abstract, February, Denver

Hatzios KK (1991) Biotransformations of herbicides in higher plants. In: Grover R, Cessna AJ (eds) Environmental chemistry of herbicides. CRC Press, Boca Raton, pp 141–185

Humble GD, Hsiao TC (1970) Light-dependent influx and efflux of potassium of guard cells during stomatal opening and closing. Plant Physiol 46(3):483–487, ISSN 0032–0889

King LJ (1996) Weeds of the world-biology and control. Interscience, New York, p 270

Langebartels C, Harms H (1985) Analysis for nonextractable (bound) residues of pentachlorophenol in plant cells using a cell wall fractionation procedure. Ecotoxicol Environ Saf 10:268–279

Ma T, Wu W, Wang Y (2012) Transcriptomic analysis of rice root responses to potassium deficiency. BMC Plant Biol 12:161. doi:10.1186/1147-2229-12-161

Manning R (2004) Against the grain: how agriculture has hijacked civilization. North Point Press, New York, 232

Messinger SM, Buckley TN, Mott KA (2006) Evidence for involvement of photosynthetic processes in the stomatal response to CO_2. Plant Physiol 140(2):771–778, ISSN 0032–0889

Moriles J, Hansen S, Horvath DP, Reicks G, Clay DE, Clay SA (2012) Microarray and growth analysis identify differences and similarities of early season corn response to weeds, shade and nitrogen stress. Weed Sci 60(2):158–166

Oliveira Júnior RS (2001) Seletividade de herbicidas para culturas e plantas daninhas. In: Oliveira Júnior RS, Constantin JP (eds) Plantas daninhas e seu manejo. Guaíba, Agropecuária, pp 291–313

Pillmoor JB, Gaunt JK, Roberts TR (1984) Examination of bound (non-extractable) residues of MCPA and flamprop in wheat straw. Pesticide Sci 15:375–381

Procópio, SO, Santos, JB, Silva, AA, Donagemma, GK and Mendonça, ESV (2004b) Ponto de murcha permanente de soja, feijão e plantas daninhas. Planta Daninha, 22(1):35–41

Radosevich SR, Holt JS, Ghersa CM (2007) Ecology of weeds and invasive plants: relationship to agriculture and natural resource management, 3rd edn. Wiley, Hoboken. ISBN 978-047-1767-79-4

Rajan AV, Sankaran S (1974) Studies on crop-weed competition for nutrients and its effect on grain yield of maize (vr. Ganga-5). Madras Agr J 61(4):413–416

Rajcan I, Swanton CJ (2001) Understanding maize-weed competition: resource competition, light quality and the whole plant. Field Crops Res 71(2):139–150, ISSN 0378–4290

Rajcan I, Chandler J, Swanton CJ (2004) Red-far-red ratio of reflected light: a hypothesis of why early season weed control is important in corn. Weed Sci 52(5):774–778

Ronchi CP, Terra AA, Silva AA, Ferreira LR (2003) Nutrient contents of coffee plants under weed interference. Planta Daninha 21(2):217–227, ISSN 1806–9681

Santos JB, Procópio SO, Silva AA, Costa LC (2003) Captação e aproveitamento da radiação solar pelas culturas da soja e do feijão e por plantas daninhas. Bragantia 62(1):147–153, ISSN 0006–8705

Shahi HN (1978) Competitive effects of *Chenopodium album* for soil moisture and nutrients in wheat. Ind Pest Control 20(4):14–16

Shimabukuro RH (1985) Detoxification of herbicides. In: Duke SO (ed) Weed physiology, vol 2. CRC Press, Boca Raton, pp 215–240

Silva AA, Ferreira FA, Ferreira LR, Santos JB (2007) Biologia de plantas daninhas. In: Silva A, Silva JF (eds) Tópicos em Manejo de Plantas Daninhas. Universidade Federal de Viçosa, Viçosa, pp 17–61. ISBN 978-857-2692-75-5

Sung FJM (1993) Waterlogging effect on nodule nitrogenase and leaf nitrate reductase activities in soybeans. Field Crops Res 35:183–189, ISSN 0378–4290

Turner JA, Pernich DJ (2002) Origin of enantiomeric selectivity in the aryloxyphenoxypropionic acid class of herbicidal acetyl coenzyme a carboxylase (ACCase) inhibitors. J Agric Food Chem 50:4554–4566

van Eerd LL, Hoagland RE, Zablotowicz RM, Hall JC (2003) Pesticide metabolism in plants and microorganisms. Weed Sci 51:472–495

Ware GW, Whitacre DM (2004) An introduction to herbicides, 2nd edn. In: The pesticide book 6th edn. (http://ipmworld.umn.edu/chapters/whitacreherb.htm)

Williamson LC, Ribrioux SPCP, Fitter AH, Leyser HMO (2001) Phosphate availability regulates root system architecture in Arabidopsis. Plant Physiol 126(2):875–882

Yuan JS, Tranel PJ, Stewart CN Jr (2007) Non-target-site herbicide resistance: a family business. Trends Plant Sci 12:6–13

Zhang H, Forde BG (1998) An Arabidopsis MADS box genet hat controls nutrient induced changes in root architecture. Science 279(5349):407–409. doi:10.1126/science.279.5349.407

Zimdahl RL (2007) Weed ecology. In: Fundamentals of weed science, 3rd edn. Elsevier, UK, pp 123–180

Heavy Metals Stress on Poplar: Molecular and Anatomical Modifications

Luca Sebastiani, Alessandra Francini, Stefania Romeo, Andrea Ariani, and Antonio Minnocci

Abstract

Heavy metal stress responses vary from plant to plant depending on the type of heavy metals and require a coordinated interplay of complex physiological and biochemical processes, gene expression, protein modification and changes in metabolites compositions leading to proper stress signal and tolerance.

Fast-growing tree species, such as poplar, have been studied as possible candidate in phytoremediation approaches to clean up soil or water polluted by organic and inorganic compounds. In particular poplar is known both for the ability to uptake (i.e. phytoextraction) and to stabilise heavy metals (i.e. phytostabilisation) into their tissues, thus reducing the mobility of these contaminants in the soil profile. Compared to other plant species, poplar trees have several advantageous characteristics, such as deeper root system, higher transpiration activity, and productivity. Moreover, they produce economically valuable nonfood biomass exploitable both for wood and bioenergy production. Since the availability of the genome sequence of *Populus trichocarpa* and the development of high-throughput technologies, poplar has also emerged as the model system for tree biology studies.

In this chapter, we examine the effects of heavy metals on anatomical traits and molecular machinery that are responsible for their accumulation and tolerance in poplar. Starting with this deeper molecular information, this chapter could provide new ideas for improving poplar trees with traits conferring heavy metals tolerance.

L. Sebastiani (✉) • A. Francini • S. Romeo
A. Ariani • A. Minnocci
BioLabs, Institute of Life Sciences, Scuola Superiore Sant'Anna, Piazza Martiri della Libertà, 33, 56127 Pisa, Italy
e-mail: luca.sebastiani@sssup.it

R.K. Gaur and P. Sharma (eds.), *Approaches to Plant Stress and their Management*,
DOI 10.1007/978-81-322-1620-9_15, © Springer India 2014

Keywords
Cadmium • Copper • Model system • Organs anatomy • Oxidative stress • Phytoremediation • Proteomic • Transcriptomic • Zinc

Introduction

The term "heavy metals" (HMs) has been widely used in scientific literature for many years, even if the International Union of Pure and Applied Chemistry consider this term meaningless and misleading (Duffus 2002). Authoritative definition cannot be found in the relevant literature, but usually the name identifies both metals and metalloids associated with toxicity and environmental contamination. In this chapter, it is not possible to discuss in detail these definition problems and at the same time we cannot ignore the scientific publications using these terms. For such reasons we maintain the general term "heavy metals" and will use it more specifically to identify a group of metals classically associated with the scientific literature on poplar tree: Cu, Cd, Cr, Fe, Mn, Ni, Pb and Zn.

Heavy metals can be found as natural constituents of the Earth's crust and geological processes, such as alteration and erosion of the geological underground materials can release these elements in soils, ground- and surface water, and atmosphere. In addition to the chemical composition of the parent material, emissions from volcanoes, degassing processes in the Earth's crust and forest fires can be also important sources of HMs in the environment (Palumbo et al. 2000; Lado et al. 2008). Besides the geological processes of the parental material, human activities can drastically alter the HM geochemical cycles and balance, producing the well-known agricultural, industrial and urban pollution problems. Actually, large areas of land result contaminated with HMs deriving from urban activities (municipal sewage, waste incinerators, traffic emissions), agricultural operations (fertilisers and pesticides) and industrial processing (metalliferous mining, smelting industry, paint factory and tannery) (Xiangdong et al. 2001; Lado et al. 2008).

Heavy metal pollution is responsible for several environmental problems and risks to human health, including decreased soil microbial activity and fertility and yield losses or contamination of plants (McGrath et al. 1995). Agricultural land contaminated by HMs pose very complex scientific issue than those related to yield losses and contamination. These soils can be excavated and disposed in landfills or treated with chemical or physical processes (immobilisation), but these solutions are very expensive and in long-term period negatively affect human health, wild flora and fauna. So, it is expected that in the next future, the management of heavy metal-contaminated soil, water and biosolids will become more challenging as stricter regulations to improve water quality and soil fertility are expected to be imposed.

Due to the scarcity of arable land in the world, especially in industrialised and fast economically growing country such as China and India, contaminated soils must be used by farmers. For such reasons it would be compulsory to have crops or new cultivars that limit the absorption and accumulation of HMs in the edible organs. On the other side of the coin, it would be interesting to have not edible (bioenergy or wood) crops that are able to accumulate high amounts of metal contaminants as natural decontamination systems for HM-polluted soils and as an additional income for farmers.

Fast-growing tree species, such as poplar, have been studied as possible candidate in phytoremediation approaches to clean up soil or water polluted by atrazine- (Burken and Schnoor 1998), trichloroethylene- (Newman et al. 1997), chloroacetanilide herbicides- (Gullner et al. 2001), Cd- (Robinson et al. 2000), Se- (Pilon-Smits et al. 1998), and Zn-polluted (Di Baccio et al. 2003) soil or water. Regarding the removal of HMs, phytoremediation approaches are based on plant ability to take up HMs into their tissues,

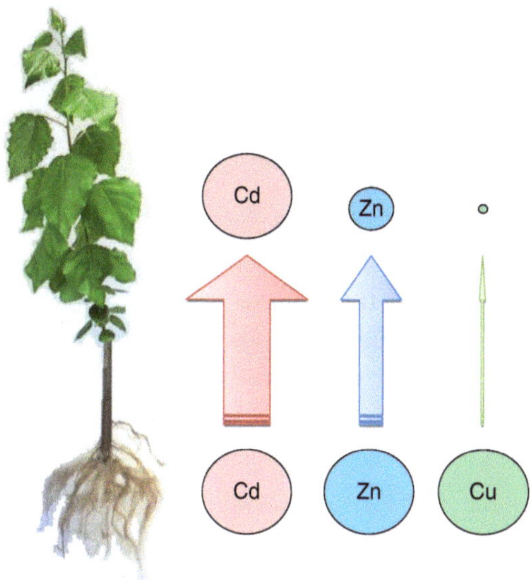

Fig. 1 Schematic representation of HMs translocation from soil to mature leaves in poplar. Arrow and circle size are indicative of the difference in HMs translocation. Scheme is a graphical elaboration of data from Laureysens et al. (2004)

i.e. phytoextraction. Alternatively, contaminated sites and sediments can be stabilised using vegetation, which mitigates the migration of HM contaminants in the soil profile, i.e. phytostabilisation. Poplars can provide promising examples for both approaches (Schnoor 2000; Di Baccio et al. 2003; Sebastiani et al. 2004) depending on the HM type (Fig. 1) and concentrations in soil. Moreover, compared to herbaceous species, poplar trees have several advantageous characteristics, such as a deeper root system, a higher transpiration activity and productivity, and produce economically valuable nonfood biomass exploitable both for wood and bioenergy production.

Heavy metal stress responses vary from plant to plant depending on the type of HM and require a coordinated interplay of complex physiological and biochemical processes, gene expression, protein modification and changes in metabolites compositions leading to proper stress signal and tolerance (Fig. 2). To cope HMs stress plants have evolved complex networks of molecular processes such as regulation of metal homeostasis, detoxification, repair capabilities and signalling molecules (Thapa et al. 2012). However, these processes have been explored mostly in herbaceous plants while there is less knowledge of these mechanisms in perennial woody plants.

In this chapter, we examine the effects of HMs on anatomical traits and molecular machinery that are responsible for HMs accumulation and stress tolerance in poplar. This deeper molecular information is now necessary to identify the key mechanisms of poplar response to HMs and provide the guidelines for producing or selecting genetically modified or not genetically modified poplar germ plasm more adapt for HMs specific phytoremediation applications.

Transcriptomic and Proteomic Analysis

Since the availability of the genome sequence of *Populus trichocarpa* (Tuskan et al. 2006) and the development of high-throughput technologies, poplar has emerged as the model system for plant biology studies of tree-specific traits (Jansson and Douglas 2007). However, the genus *Populus* is characterised by a high inter- and intraspecific variability that could affect plant growth and adaption to different environment (Lojewski et al. 2009). Moreover several studies suggest that this genetic variability could affect heavy metal accumulation and the partitioning of elements within the tree organs (Laureysens et al. 2004; Sebastiani et al. 2004; Pietrini et al. 2010). In the last years, several transcriptomic analyses have been performed in different poplar clones, focusing on the study of different tissues and metal stresses (Di Baccio et al. 2011; Guerra et al. 2009; He et al. 2013).

Regarding proteomics, that is, the systematic analysis of the proteins encoded by a genome, it is a powerful tool to describe complete proteomes at the organelle, cell, organ or tissue levels and also to compare proteomes affected by different physiological conditions. By proteomics it is possible to identify individual proteins or group of proteins associated with HM stress and gain insight the mechanisms of metal

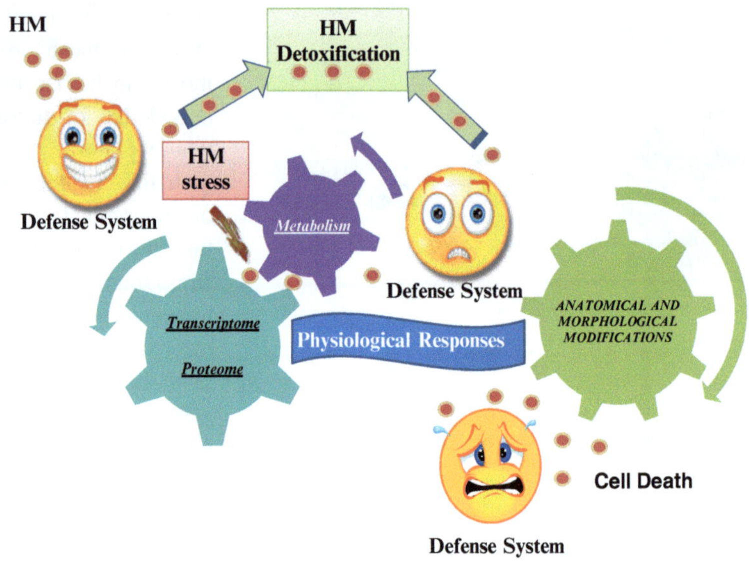

Fig. 2 Heavy metal stress responses require a coordinated interplay of complex physiological and biochemical processes involving gene expression, protein modification and changes in metabolites compositions. These changes lead to proper stress signal and defence mechanisms activation as well as to anatomical and morphological modifications. If the HM stress overcame the defence system capabilities plant cells can death

toxicity. In poplar, for example, several studies have been done on Cd toxicity (Kieffer et al. 2008, 2009; Durand et al. 2010). The metabolic impact of Cd was different in roots when compared to leaf tissues, maybe because roots are the first organ exposed. However, both in leaves and roots, Cd had a strong negative impact on proteins related to primary carbon metabolism. Regarding glutathione metabolism, it plays an important role in Cd toxicity, and significant changes for glutathione-S-transferases were observed in leaves and roots; also some typical stress-related proteins (HSP, proteinases and PR) showed an increase in their abundance in both tissues. Increase in abundance of proteins related to glycolysis and TCA cycle in Cd-exposed plants pointed out an increased in energy supply, by activation of mitochondrial respiration.

Metallothioneins (MTs) and phytochelatins are Cys-rich metal chelators that represent the two principal groups of metal-binding proteins and have been involved in heavy metal homeostasis and tolerance (Cobbett and Goldsbrough 2002). However the role of MTs in this process has not been conclusively shown in plants; under Cu-stress condition, for example, MTs were highly downregulated. This downregulation has been observed previously in poplar under pathogens infection (Smith et al. 2004) and has been interpreted as a mechanism to inducing a transient reactive oxygen species (ROS) accumulation necessary to trigger a general resistance response in plants.

All these transcriptomic and proteomic studies highlight some common responses among different tissues and genotypes but also some tissue- and metal-specific processes that were activated in response to metal excess. In the following subsections some case study are reported and analysed more in detail:

Copper stress in *P. deltoides* roots: Guerra et al. (2009) analysed the transcriptomic changes of Cu-tolerant *P. deltoides* clone in response to two different Cu concentration (30–60 μM $CuSO_4$) and two time of exposure (12–24 h) and identify a series of genes encoding defence proteins, such as pathogenesis-related (PR) proteins and trypsin inhibitors, which are significantly upregulated in all stress treatments. This kind of proteins were upregulated also in response to different biotic and abiotic stresses including insects (Major and Constabel 2008), pathogenic fungus (Rinaldi et al. 2007), wounding (Christopher et al. 2004), and ozone (Gupta et al. 2005), suggesting that this upregulation depends on the crosstalk among the stress-related signalling pathways, such as those induced by salicylate (SA) and nitric oxide (NO) (Poschenrieder et al. 2006). As confirmation of this hypothesis, there is the differential expression in all Cu treatments of protein related to signal transduction, like Ca^{2+}-signalling proteins, MAP kinases and Rab small G proteins (Rab GTP-binding protein), and accumulation of transcript coding for ethylene responsive elements suggests the participation of SA, jasmonic acid (JA) and ethylene in the Cu-mediated response in root tissues of *P. deltoides* (Guerra et al. 2009). Defence genes differentially expressed by Cu stress also included several enzymes involved in ROS scavenging such as peroxidases, Cu/Zn superoxide dismutases and catalases that have shown a downregulation under this condition. This phenomenon has been observed also in Cd-stressed roots of *P. x canescens* (Schützendübel et al. 2002) suggesting conserved responses of these enzymes in different poplar clones subjected to heavy metal stress.

Cadmium stress in *P. x canescens* bark: He et al. (2013) analysed the transcriptional modification of bark of *P. x canescens* exposed to 200 μM $CdSO_4$ identifying a transcriptomic network that could underlie the physiological and microstructural response of this tissue towards Cd stress. The main categories of identified differentially expressed genes are related to transport, energy metabolism, photosynthesis, isopentenyl diphosphate biosynthesis and stress response. Among the genes related to transport function, two ABC-like transporter protein, a NRAMP1 and a heavy metal transport/detoxification protein have been identified as differentially expressed in response to Cd excess in bark tissue. This gene could be directly linked with Cd homeostasis in this tissue since ABC-like protein are implicated in vacuolar detoxification of glutathione- or PC-conjugated Cd in *Arabidopsis* and yeast (Li et al. 1997; Park et al. 2012), while NRAMP1-like and heavy metal transport/detoxification protein are directly involved in Cd transport in *Arabidopsis* (Thomine et al. 2000; Wintz and Vulpe 2002).

This study also reports the downregulation of genes involved in photosynthetic apparatus like PSII subunit P−1 and NADH-ubiquinone oxidoreductase-related genes that were probably involved in thylakoid membrane formation and respiration, respectively (Meyer et al. 2008; Yi et al. 2009). This downregulation could be linked with the degeneration of cellular, chloroplast and mitochondrial membranes observed in bark cells of Cd-treated plants, and a similar phenomenon has been previously reported and poplar leaves exposed to Cd (Kieffer et al. 2008).

Genes involved in nutrient homeostasis and carbohydrate metabolisms are differentially expressed in response to Cd excess in bark tissue of *P. x canescens*. Among them the homologous of AtNRT2.4 and AtPHT1.4, a nitrate and a phosphate transporter, respectively (Lei et al. 2011; Kiba et al. 2012), are both downregulated in bark tissue like genes involved in inositol and mannitol biosynthesis, such as galactinol synthase1 and raffinose synthase 5, respectively (Zuther et al. 2004; Nishizawa et al. 2008).

Zinc stress in *P. x euramericana* leaves: Di Baccio et al. (2011) analysed the trancriptomic response in leaves of *P. x euramericana* clone I-214 in response sublethal dose of Zn (1 mM) and highlight the differential expression of genes mainly involved in response to both biotic and abiotic stresses, redox states, photosynthesis and cell division, cycle and development. The genes responsive to biotic stress mainly coded for pathogenesis-related and hormone-signalling protein, such as ethylene and SA, suggesting a crosstalks within this pathway or a common response against different stressors in plants (Poschenrieder et al. 2006) that could constitute a constitutive and conserved response to different HM stress in several tissue. Indeed the activation of similar proteins has been reported in Cu-stressed roots of *P. deltoides* by Guerra et al. (2009). The majority of genes involved in regulation of redox states activated in response to Zn excess belong to the ascorbate/GSH cycle, thioredoxin system and GSH metabolism, suggesting a GSH-mediated detoxification of the oxidative burst caused by Zn excess, as also previously described in Di Baccio et al. (2005). Indeed all the key enzymes of GSH metabolism, like the γ-ECS that catalyse GSH synthesis, the GR and GPXs that control redox state and GST that conjugates GSH with xenobiotics (Rennenberg and Brunold 1994; Xiangdong et al. 2001), were upregulated in response to Zn excess. Together with genes involved in GSH metabolism also the key enzyme of Cys biosynthesis were upregulated, suggesting a link between the biosynthetic pathway of GSH and its precursor (Rennenberg and Brunold 1994). The activation of this mechanism is probably involved in the protection of photosynthetic apparatus, which are the most impaired in response to metal stress in I-214 clone (Di Baccio et al. 2003, 2009). Indeed the genes involved in light reaction, such as PS subunits, cytochromes and electron chain's proteins, are mainly downregulated under this condition, like the genes involved in carbon fixation, such as the large subunit of ribulose-1,5-bisphosphate-carboxylase/oxygenase (RuBisCO).

Biochemical Aspects

One of the main biochemical effects of HMs in plants involves oxidative stresses. Oxidative stress, defined as a shift of the balance between

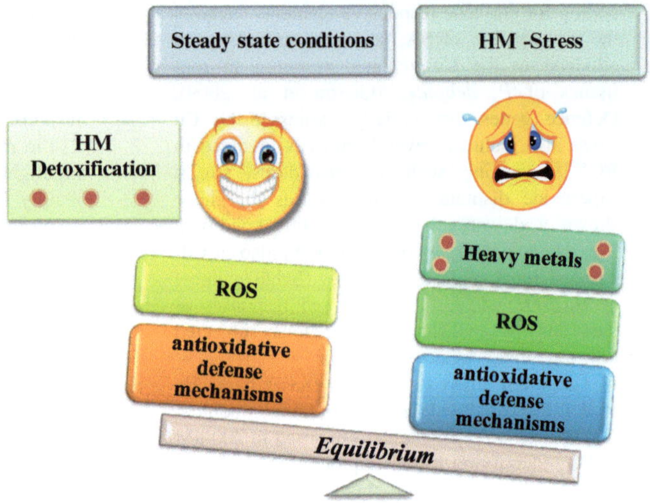

Fig. 3 One of the main biochemical effects of HMs in plants involves oxidative stress. The equilibrium between the production and the scavenging of ROS may be perturbed when HMs exceed the defence mechanism capabilities. Equilibrium disturbances lead to sudden increase in intracellular levels of ROS which can cause significant damage to cell structures

prooxidative and antioxidative reactions in favour of the former, it seems to be a common denominator of the action of various agents on living organisms (Bartosz 1997). During growth plants are exposed to oxidative stresses and ROS production. However when they are exposed to stressful conditions (extreme temperatures, high light intensity, drought, salinity, HMs and herbicides) ROS such as singlet oxygen ($^1O_2$), superoxide (O_2), hydrogen peroxide (H_2O_2) and hydroxyl radical (OH^-) are produced continuously and in higher number in different cellular compartments. Then, the equilibrium between the production and the scavenging of ROS may be perturbed, and these equilibrium disturbances lead to sudden increase in intracellular levels of ROS which can cause significant damage to cell structures (Fig. 3).

In the literature, the role of ROS in causing cellular damage during stress is well established and the necessity of efficient cellular systems to scavenge ROS is now generally accepted. Essentially, antioxidants defence fall into three general classes comprising: (i) water-soluble reductants, i. e. glutathione and ascorbate; (ii) liposoluble vitamins, i.e. lutein and α-tocopherol; and (iii) enzymatic antioxidants, i.e. superoxide dismutase (SOD), catalase (CAT) and peroxidases. Enzymatic antioxidants include SOD, CAT, APX, MDHAR, DHAR and GR, and nonenzymatic antioxidants are GSH, ascorbic acid (AsA) – both water soluble – carotenoids and tocopherols (lipid soluble). In plant cells, one of the most efficient detoxification mechanism is the ascorbate/glutathione cycle (Asada–Halliwell cycle) by which hydrogen peroxide (H_2O_2) is scavenged. This cycle appears essential in photosynthetic tissues but also in roots.

The role of oxidative stress in HMs toxicity has been assessed by measuring alterations in the redox metabolic components of stressed plants (Dietz et al. 1999). Over the past years major progress has been achieved, particularly by comparing HM-tolerant and/or HMs hyperaccumulator genotypes with their non-tolerant relatives and by using transgenic plants that overexpress or lack specific redox elements. These approaches provided novel insight into the relationship between metal sensitivity and cellular redox imbalance. In genetically modified poplar plants, overexpressing GR evidence to support the role of GSH and GR in maintaining the foliar ascorbate pool has been observed (Foyer et al. 1995).

Several authors tried to explain the relationship between antioxidant defence and HMs response, but the response of antioxidant systems to heavy metals depends on plant species, tissue analysed, HM used and the intensity of the stress, making comparison between experiments difficult. Moreover some HMs (Zn^{2+} and Cd^{2+}) are physiologically non-redox-active, while others (Fe, Cu, Cr, V and Co) are redox-active enabling redox reactions in the cell (Valko et al. 2005). They are involved in the formation of OH from H_2O_2 via Haber–Weiss and Fenton reactions and initiate nonspecific lipid peroxidation (Dietz et al. 1999).

The antioxidant protection is especially important in fine roots (defined as all roots up to 2 mm in diameter), which are the most active part of roots in absorbing water and nutrients. On the other hand, fine roots are also the gate for HMs entering the plants. For this reason, they are the most endangered of the plant's organs and the first line of defence against soil toxicity. Below we report, as example, some published paper related to the answer of antioxidant defences on poplar under metal stress:

Cadmium and oxidative stress in *Populus* × *canescens* roots: Clonal, hydroponically grown poplar plants (*Populus tremula* × *Populus alba*) were exposed to Cd in order to find out whether Cd-induced injury was related to the disturbance of the cellular redox control in root tips (Schützendübel et al. 2002).

Cd exposure resulted in an inhibition of antioxidative enzymes (SOD, CAT, APX, MDHAR, GR) but had fewer effects on DHAR activities. In *Populus x canescens* glutathione concentrations decreased, whereas ascorbate remained unaffected by Cd. Fifty micromoles of Cd-retarded shoot growth faster than root growth caused a more severe loss in antioxidative capacity than 5 μM Cd and resulted in an accumulation of H_2O_2 in roots. One possible explanation for the observed decreases in antioxidative enzymes activities was discussed by authors as the likelihood of enzyme inactivation increases with extended exposure to Cd, and that this gradual inactivation did not induce compensatory gene activation. As discuss in the paper, the functioning of the ascorbate-glutathione pathway requires supply with reductant NAD(P)H. It is possible that under highly oxidising conditions as generated by Cd exposure, the energy supply to roots is mainly consumed by the attempt to rescue the cellular redox balance. Consequently, growth would be significantly inhibited.

Heavy metal toxicity in cuttings *Populus nigra* L. fine roots: The effects of increasing concentrations of Cu and Pb in polluted soils collected from direct neighbourhood of a copper smelter were studied by analysing the activity of lipid peroxidation and antioxidant enzymes in the fine roots of cuttings of black poplar by Stobrawa and Lorenc-Plucińska (2008). The experiments involved two series, one consisting of soil from Głogów (Poland) and the other of that from Bogomice (Poland). Each experimental series consisted of five treatments, control (unpolluted soil), unpolluted soil mixed 25 %, 50 %, and 75 % by weight with polluted soil, and polluted soil alone. Copper and Pb were the major pollutants found. Despite clear differences in concentrations of HMs in the soil, only two treatments, namely, those comprising 75 % and 100 % of Głogów soil, can be regarded as definitively toxic on the basis of growth parameters of the shoots, whereas only Cu can be regarded as present at toxic concentrations (20–100 ppm) in the fine roots. A closer look into root disorders showed a clear increase activity of CAT and APX, suggesting intensive action of plants against H_2O_2. The authors conclude that threshold of breakdown of HM tolerance for black poplar is significantly higher than in other plants, indicating this tree as a good candidate for phytoextraction applications.

Zn excess and glutathione metabolism on *P. x euramericana*: Di Baccio et al. (2005) studying *P. x euramericana* clone I-214 reported that oxidative injuries have been observed in plants exposed to heavy metals without redox properties, such as Zn and Cd. In this study treatment with Zn decreased the total GSH content in the young leaves with a simultaneous increase of GSSG. The elevated GSSG content and the modification of GR activity in leaves exposed to high Zn concentrations indicate the induction of oxidative stress. From the results of this study, authors suggest that the poplar response to excess Zn involves a complex combination of metabolic processes in which GSH is involved. A progressive increase of the Zn concentration in the nutrient solution induces a progressive increase of the metal concentration in young leaves of poplar clone I-214, suggesting a direct correlation between Zn availability and Zn uptake. Modifications of glutathione pathway confirm that heavy metals cause oxidative stress, and that Zn is intimately linked to oxygen metabolism in redox processes of cells. The key role of GSH as scavenger of HM Cd was also demonstrated on transgenic poplar accumulating GSH to a level 3.5 fold higher than WT (Arisi et al. 2000).

Anatomical and Morphological Modifications

The effects of HMs on the anatomy and morphology of plants organs are complex, considering the differences between organs, their accumulation capacity and dynamic of exposure.

In the following subsections some case study are reported and analysed more in detail:

Root: Plant root is usually the first organ exposed to HM stress. Most studies have been done on Cd, and they showed that these HMs enter plants from the soil solution, traverses the root through symplastic or apoplasmic pathways before entering the xylem and being translocated to the shoot. Apoplasmic movement of Cd to the xylem can be restricted by the development of extracellular barriers (exodermis and endodermis). Moreover, the presence of Cd in the rhizosphere inhibits root elongation and influences root anatomy. Tissue Cd concentrations decrease from peripheral to inner root tissues. The development of apoplasmic barriers to Cd movement to the xylem indicates that their maturation is accelerated by high Cd concentrations in their immediate locality. Accelerated maturation of the endodermis in response to local Cd availability is of functional significance in protecting the shoot from excessive Cd loads (Lux et al. 2011).

To discriminate *Populus alba* clones (6 K3 and 14P11) in their tolerance to high Cd concentrations (50 μM), the localisation of Cd in root was investigated by scanning electron microscope coupled with energy-dispersive X-ray microanalysis (Cocozza et al. 2008). Data proved a differential pattern between the two clones in accumulating Cd within the root profile, although overall concentration and content of Cd in the root system did not differ between clones. Cd was found in different root tissues (parenchyma cells of cortex, endodermis, central cylinder and xylem vessels). However, compared with other ions, the distribution of Cd within the root was preferentially localised in the surroundings of central cylinder. In other experiments with the same Cd concentrations (50 μM) but different poplar genotypes (A4A and I-214 *Populus x canadensis* Mönch., Poli and 58–861 *Populus nigra* L.), the same authors (Cocozza et al. 2011) showed that I-214 plants were characterised by high Cd contents in the epidermis, whereas 58–861, A4A, and Poli showed high Cd contents in the cortex.

In *Populus* × *euramericana* cv. Robusta cuttings grown in 10 μM $Cd(NO3)_2$, the root apices, rhizodermis and cortex were the most seriously damaged root parts. Cd-treated poplar roots exhibited unusual defence activity of root apical meristem and accumulation of darkly stained material around central cylinder (Lunáčková et al. 2003). *Populus* × *canescens* expressing a heterologous GH3::GUS reporter gene exposed to 50 μM Cd in hydroponic solutions caused strong increases in the vascular system of roots as well as in parenchymatic cells in the xylem (Elobeid et al. 2012).

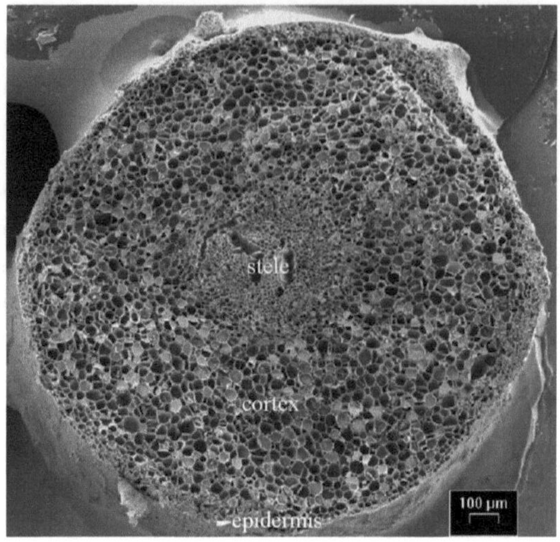

Fig. 4 Cryo-SEM. Transversal fracture of a frozen-hydrated root of poplar (clone I-214), which shows the different tissues from the outside inwards (epidermis, cortex and stele). Stolarikova et al. (2012) proved that high Zn concentrations (1 mM) determine the development of apoplasmic barriers, and by X-ray microanalysis and cryo-SEM, it was possible to prove that most of the Zn was localised in the cortical tissues. Similar data were obtained by Di Baccio et al. (2009) with higher Zn concentrations (10 mM); also in this experiment Zn was preferentially localised in epidermis and cortex tissues of roots

Regarding other HMs such as Zn, in hydroponically grown *Populus* × *euramericana* clone I-214, Stolarikova et al. (2012) proved that high Zn concentrations (1 mM) determine the development of apoplasmic barriers (Casparian bands and suberin lamellae in endodermis) closer to the root apex than in control root. Using energy-dispersive X-ray microanalysis and cryo-scanning electron microscopy (cryo-SEM) freeze-fractured and frozen-hydrated transversal root samples, it was possible to prove that most of the Zn was localised in the cortical tissues, and four-time less Zn was found in the inner part of the root below the endodermis. This indicates that endodermis serves as efficient barrier of apoplasmic Zn transport across the poplar root. Similar data were obtained by Di Baccio et al. (2009) with higher Zn concentrations (10 mM); also in this experiment Zn was preferentially localised in epidermis and cortex tissues of roots (Fig. 4).

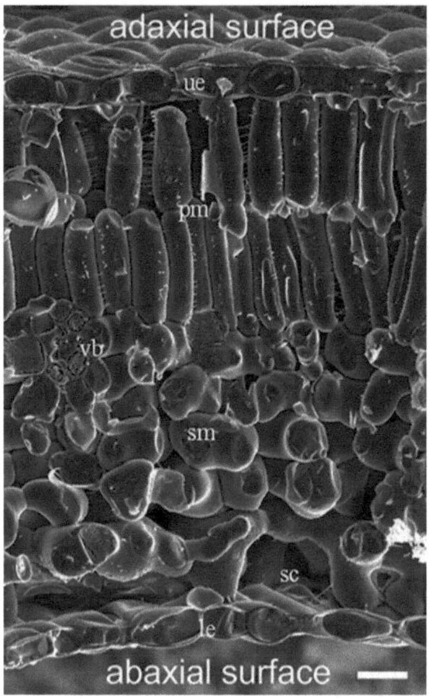

Fig. 5 Cryo-SEM. Transversal freeze-fracture plane of a frozen-hydrated poplar leaf (clone Eridano), showing from above: upper epidermis (ue), palisade (pm) and spongy mesophyll (sm), vascular bundle (vb), substomatal chamber (sc) and lower epidermis (le) (bar = 10 μm). HM such as Zn induces variations in leaf dry mass and area determined by different mesophyll thickness and intercellular spaces distributions. These changes were consistent with physiological impairments (i.e. photosynthesis) and may be explained as adaptive mechanisms to the stress caused by Zn excess (Tognetti et al. 2004; Di Baccio et al. 2010)

Leaf: Leaf modifications under HM stress were extensively investigated in poplar clones Eridano (*Populus deltoides* × *maximowiczii*), and I-214 (*P.* × *euramericana*) exposed for a growing season to soil amended with tanneries waste, which contains toxic amounts of several HMs (Tognetti et al. 2004). Stomatal density and leaf layer thickness were studied by cryo-SEM (Fig. 5) showing that leaf thickness was significantly decreased in Eridano but slightly increased in I-214 after soil treatment with HM-enriched waste. This difference was a result of differences in palisade mesophyll thickness. Hence foliage structures of these clones were either unaffected or improved by uptake of HMs in soil amended with this type of industrial biosolids, containing HMs.

In other study with Zn concentration gradient (from 0.001 to 10 mM) on *Populus* × *euramericana* clone I-214 significant modifications in foliage area, stomatal density and leaf layer thickness were found (Di Baccio et al. 2009). In young leaves exposed to high Zn, the mean area decreases of 61 and 70 % at 5 and 10 mM Zn, and stomatal density increases on abaxial surfaces. In old leaves, stomatal density on abaxial surfaces increased by 93, 77 and 62 % with 1, 5 and 10 mM treatments, respectively, while leaf area decreased only at 5 and 10 mM. Also the total leaf thickness was significantly increased by Zn treatments, 14 % at 1 mM Zn, 15 % at 5 mM Zn and 23 % at 10 mM Zn, while in old leaves total leaf thickness was enhanced (8 %) only at 10 mM. At the highest concentration (10 mM), the localisation of Zn by cryo-SEM and energy-dispersive X-ray microanalysis showed that Zn was preferentially present in the photosynthetic tissues of shoots.

In other studies on *P. x euramericana* (clone I-214) subjected to elevated, but sub-symptomatic concentrations of Zn (0.1 and 1 mM) (Di Baccio et al. 2010) impairments in leaf morphological and anatomical traits was found. Zn excess induced variations in leaf dry mass, area, mesophyll thickness, intercellular spaces, stomatal density and size but no visual symptoms of injury. Stronger modifications, especially concerning stomata characteristics induced by 1 mM Zn, were consistent with physiological impairments, while those induced by 0.1 mM Zn suggested a compensatory strategy for maintaining functional integrity. Differences in stomata number, density and size as well as leaf area and mesophyll modifications may be explained as adaptive mechanisms to the stress caused by Zn excess.

In *Populus alba* cv. Villafranca plants, supplemented with 300 mg kg^{-1} of soil Zn leaf morphology and ultrastructure were significantly modified and determine a significant lamina thickening (Todeschini et al. 2011). Zinc was mostly detected in the cell walls of the xylem and of the parenchyma cells surrounding the bundles of Zn-treated plants. A significant increase in the number of calcium oxalate crystals was observed in the leaves of Zn-treated plants, but the latter element was not incorporated in crystals, suggesting an increase of free calcium following Zn accumulation in the cell walls.

Leaf of *Populus tremula* subjected to Cd and Zn (Hermle et al. 2007) showed palisade cells with thickened cell walls, vacuolated cytoplasm, chloroplasts with size reduction, and an increased frequency of starch grains and osmiophilic plastoglobuli. In Populus tremula leaves, Zn and Cd preferentially accumulated in older foliage.

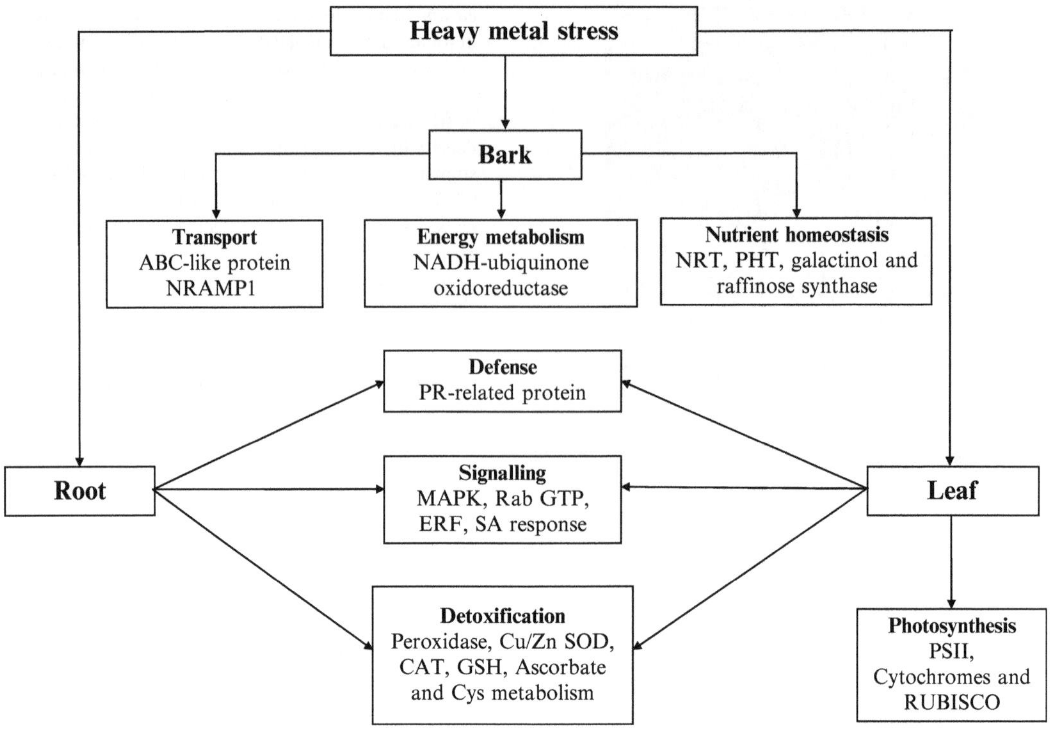

Fig. 6 Schematic representation of some relevant molecular interactions and targets in poplar trees under heavy metal stress

Moreover, excess Zn irregularly accumulated inside leaf tissues tended to saturate the veins and stored in cell symplast than apoplast (Vollenweider et al. 2011).

Conclusions

In this chapter, we have examined the effects of HMs on anatomical traits and molecular machinery that are responsible for HMs accumulation and stress tolerance in poplar. The potential of poplars to cope with HM stress has been studied under several experiment conditions, and results proved that the degree of tolerance is different among species and varieties depending also by the HMs type and concentrations. Poplar is becoming an interesting biotechnological platform, and relevant advances have been done to characterise fundamental aspects of poplar response to HMs: tolerance thresholds, metal distribution patterns, physiological effects and adaptation mechanisms. Moreover, the recent advances in genomics and proteomics are useful to understand the molecular basis of HMs tolerance and accumulation (Fig. 6). Molecular data can provide new guidelines for the genetic improvement of poplars by traditional and biotechnological approaches helping breeders in selecting poplar germ plasm more adapted to HM-polluted soils. Depending on the selection strategies (Fig. 7), new genotypes can be used for HM phytoremediation applications (improving uptake and tolerance) or for profitable cultivation (limiting uptake and improving tolerance) of poplar in highly HM-polluted soils.

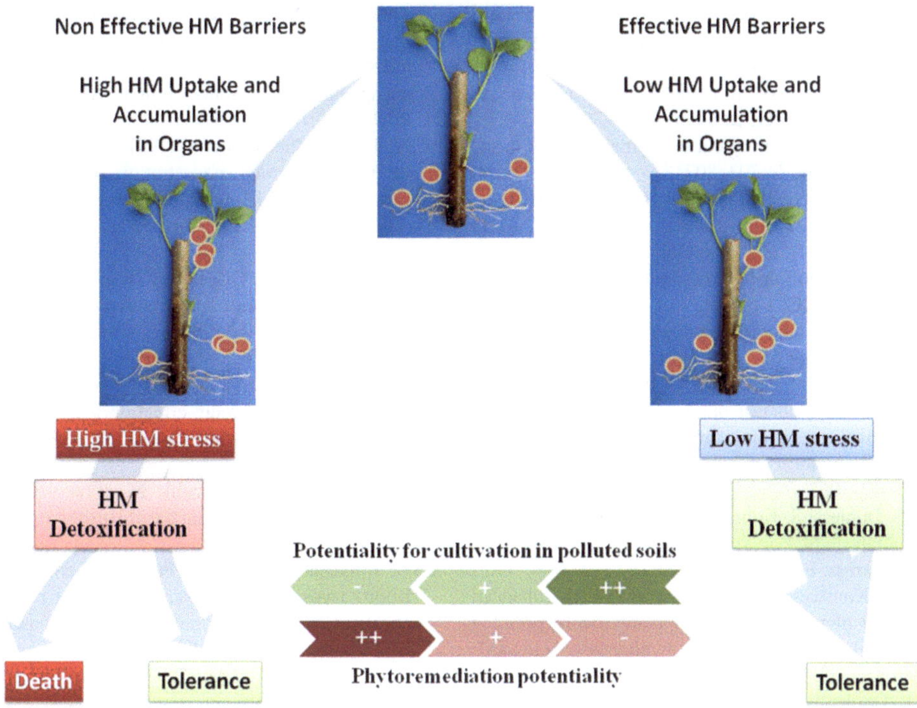

Fig. 7 Two possible selection strategies for poplar. New genotypes can be selected searching for molecular and anatomical modification that allow low HMs uptake and accumulation in organs maintaining or improving the detoxification system. These genotypes will be given more profitable cultivation in highly HM-polluted soils but should have a lower phytoremediation potentiality. Otherwise selection can search for high HMs uptake and accumulation in organs; in this scheme a very efficient HMs detoxification system should be selected too. These genotypes will be more useful for HMs phytoremediation applications

References

Arisi ACM, Mocquot B, Lagriffoul A et al (2000) Responses to cadmium in leaves of transformed poplars overexpressing g-glutamylcysteine synthetase. Physiol Plant 109:143–149

Bartosz G (1997) Oxidative stress in plants. Acta Phys Plant 19:47–64

Burken JG, Schnoor JL (1998) Predictive relationships for uptake of organic contaminants by hybrid poplar trees. Environ Sci Technol 32:3379–3385

Christopher M, Miranda M, Major I et al (2004) Gene expression profiling of systemically wound-induced defenses in hybrid poplar. Planta 219:936–947

Cobbett C, Goldsbrough P (2002) Phytochelatins and metallothioneins: roles in heavy metal detoxification and homeostasis. Annu Rev Plant Biol 53:159–182

Cocozza C, Minnocci A, Tognetti R et al (2008) Distribution and concentration of cadmium in root tissue of *Populus alba* determined by scanning electron microscopy and energy-dispersive x-ray microanalysis. I For 1:96–103

Cocozza C, Maiuro L, Tognetti R (2011) Mapping cadmium distribution in roots of *Salicaceae* through scanning electron microscopy with x-ray microanalysis. I For 4:113–120

Di Baccio D, Tognetti R, Sebastiani L et al (2003) Responses of *Populus deltoides x Populus nigra* (*Populus x euramericana* clone I-214) to high zinc concentration. New Phytol 159:443–452

Di Baccio D, Kopriva S, Sebastiani L et al (2005) Does glutathione metabolism have a role in the defense of poplar against zinc excess? New Phytol 167:73–80

Di Baccio D, Tognetti R, Minnocci A et al (2009) Responses of *Populus x euramericana* clone I-214 to zinc excess: carbon assimilation, structural modification, metal distribution and cellular localization. Environ Exp Bot 67:153–163

Di Baccio D, Minnocci A, Sebastiani L (2010) Leaf structural modifications in *Populus x euramericana* subjected to Zn excess. Biol Plant 54(3):502–508

Di Baccio D, Galla G, Bracci T et al (2011) Transcriptome analysis of *Populus x euramericana* clone I-214 leaves exposed to excess zinc. Tree Physiol 31:1293–1308

Dietz KJ, Krämer U, Baier M (1999) Free radicals and reactive oxygen species as mediators of heavy metal toxicity in plants. In: Prasad MNV, Hagemeyer J (eds) Heavy metal stress in plants: from molecules to ecosystems. Springer, Berlin, pp 73–97

Duffus JH (2002) Heavy metals – a meaningless term? Pure Appl Chem 74:793–807

Durand TC, Sergeant K, Planchon S et al (2010) Acute metal stress in *Populus tremula x P. alba* (717−1B4 genotype): leaf and cambial proteome changes induced by cadmium^{2+}. Proteomics 10:349–368

Elobeid M, Göbel C, Feussner I et al (2012) Cadmium interferes with auxin physiology and lignification in poplar. J Exp Bot 63:1413–1421

Foyer CH, Souriau N, Perret S et al (1995) Overexpression of glutathione reductase but not glutathione synthetase leads to increases in antioxidant capacity and resistance to photoinhibition in poplar trees. Plant Physiol 109:1047–1057

Guerra F, Duplessi S, Kohler A et al (2009) Gene expression analysis of *Populus deltoides* roots subjected to copper stress. Environ Exp Bot 67:335–344

Gullner G, Komives T, Rennenberg H (2001) Enhanced tolerance of transgenic poplar plants overexpressing gamma-glutamylcysteine synthetase towards chloroacetanilide herbicides. J Exp Bot 52:971–979

Gupta P, Duplessis S, White H et al (2005) Gene expression patterns of trembling aspen trees following long-term exposure to interacted elevated CO_2 and tropospheric O_3. New Phytol 167:129–142

He J, Li H, Luo J et al (2013) A transcriptomic network underlies microstructural and physiological responses to cadmium in *Populus x canescens*. Plant Physiol 162:424–439

Hermle S, Vollenweider P, Günthardt-Goerg MS et al (2007) Leaf responsiveness of *Populus tremula* and *Salix viminalis* to soil contaminated with heavy metals and acidic rainwater. Tree Physiol 27:1517–1531

Jansson S, Douglas CJ (2007) *Populus*: a model system for plant biology. Annu Rev Plant Biol 58:435–458

Kiba T, Feria-Bourrellier AB, Lafouge F et al (2012) The *Arabidopsis* nitrate transporter NRT2.4 plays a double role in roots and shoot of nitrogen-starved plants. Plant Cell 24:245–258

Kieffer P, Domenes J, Hoffmann L et al (2008) Quantitative changes in protein expression of cadmium-exposed poplar plants. Proteomics 8:2514–2530

Kieffer P, Planchon S, Oufir M et al (2009) Combining proteomics and metabolite analyses to unravel cadmium stress-response in poplar leaves. J Proteome Res 8:400–417

Lado LR, Hengl T, Reuter HI (2008) Heavy metals in European soils: a geostatistical analysis of the FOREGS geochemical database. Geoderma 148:189–199

Laureysens I, Blust R, Temmerman L et al (2004) Clonal variation and heavy metal accumulation and biomass production in a poplar coppice culture: I. seasonal variation in leaf, wood and bark concentration. Environ Pollut 131:484–494

Lei MG, Zhu CM, Liu YD et al (2011) Ethylene signalling is involved in regulation of phosphate starvation-induced gene expression and production of acid phosphatases and anthocyanin in *Arabidopsis*. New Phytol 189:1084–1095

Li SZ, Lu YP, Zhen RG et al (1997) A new pathway for vacuolar cadmium sequestration in *Saccharomyces cerevisiae* YCF1-catalyzed transport of bis (gluthionato)cadmium. Proc Natl Acad Sci USA 94:42–47

Lojewski NR, Fischer DG, Bailey JK et al (2009) Genetic basis of the aboveground productivity in two native *Populus* species and their hybrids. Tree Physiol 29:1133–1142

Lunáčková L, Šottníková A, Masarovičová E et al (2003) Comparison of cadmium effect on willow and poplar in response to different cultivation conditions. Biol Plant 47:403–411

Lux A, Martinka M, Vaculík M et al (2011) Root responses to cadmium in the rhizosphere: a review. J Exp Bot 62:21–37

Major I, Constabel P (2008) Functional analysis of the Kunitz trypsin inhibitor family in poplar reveals biochemical diversity and multiplicity in defense against herbivores. Plant Physiol 146:883–903

McGrath SP, Chaudri AM, Giller KE (1995) Long-term effects of metals in sewage sludge on soils, microorganisms and plants. J Ind Microbiol 14:94–104

Meyer EH, Taylor NL, Millar AH (2008) Resolving and identifying protein components of plant mitochondrial respiratory complexes using three dimensions of gel electrophoresis. J Proteome Res 7:786–794

Newman LA, Strand SE, Choe N et al (1997) Uptake and biotransformation of trichloroethylene by hybrid poplars. Environ Sci Technol 31:1062–1067

Nishizawa A, Yabuta Y, Shigeoka S (2008) Galactinol and raffinose constitute a novel function to protect plants from oxidative damage. Plant Physiol 147:1251–1263

Palumbo B, Angelone M, Bellanca A et al (2000) Influence of inheritance and pedogenesis on heavy metal distribution in soils of Sicily, Italy. Geoderma 95(3–4):247–266

Park J, Song WY, Ko D et al (2012) The phytochelatin transporters AtABCC1 and AtABCC2 mediate tolerance to cadmium and mercury. Plant J 69:278–288

Pietrini F, Zacchini M, Iori V et al (2010) Screening of poplar clones for cadmium phytoremediation using photosynthesis, biomass and cadmium content analyses. Int J Phytoremediation 12:105–120

Pilon-Smits EAH, De Souza MP, Lytle CM et al (1998) Selenium volatilization and assimilation by hybrid poplar (*Populus tremula* × *alba*). J Exp Bot 49:1889–1892

Poschenrieder C, Tolra R, Barcelo J (2006) Can metals defend plants against biotic stress? Trends Plant Sci 11:288–295

Rennenberg H, Brunold C (1994) Significance of glutathione metabolism in plants under stress. Prog Bot 55:142–155

Rinaldi C, Kohler A, Frey P et al (2007) Transcript profiling of poplar leaves upon infection with compatible and incompatible strains of the foliar rust *Melampsora larici-populina*. Plant Physiol 144:347–366

Robinson BH, Mills TM, Petit D et al (2000) Natural and induced cadmium-accumulation in poplar and willow: implications for phytoremediation. Plant Soil 227:301–306

Schnoor JL (2000) Phytostabilization of metals using hybrid poplar trees. In: Raskin I, Ensley BD (eds) Phytoremediation of toxic metals: using plants to clean up the environment. Wiley, New York, pp 133–150

Schützendübel A, Nikolova P, Rudolf C, Polle A (2002) Cadmium and H_2O_2-induced oxidative stress in *Populus x canescens* roots. Plant Physiol Biochem 40:577–584

Sebastiani L, Scebba F, Tognetti R (2004) Heavy metal accumulation and growth responses in poplar clones Eridano (*Populus deltoides* × *maximowiczii*) and I-214 (*P.* × *euramericana*) exposed to industrial waste. Environ Exp Bot 52:79–88

Smith C, Rodriguez-Buey M, Karlsson J et al (2004) The response of the poplar transcriptome to wounding and subsequent infection by a viral pathogen. New Phytol 164:123–136

Stobrawa K, Lorenc-Plucińska G (2008) Thresholds of heavy-metal toxicity in cuttings of European black poplar (*Populus nigra* L.) determined according to antioxidant status of fine roots and morphometrical disorders. Sci Total Environ 390:86–96

Stoláriková M, Vaculík M, Lux A et al (2012) Anatomical differences of poplar (*Populus x euramericana* clone I-214) roots exposed to zinc excess. Biologia 67(3):483–489

Thapa G, Sadhukan A, Panda SK et al (2012) Molecular mechanistic model of plant heavy metal tolerance. Biometals 25:489–505

Thomine S, Wang RC, Ward JM et al (2000) Cadmium and iron transport by members of a plant metal transporter family in *Arabidopsis* with homology to Nramp genes. Proc Natl Acad Sci USA 97:4991–4996

Todeschini V, Lingua G, D'Agostino G et al (2011) Effects of high zinc concentration on poplar leaves: a morphological and biochemical study. Environ Exp Bot 71(1):50–56

Tognetti R, Sebastiani L, Minnocci A (2004) Gas exchange and foliage characteristics of two poplar clones grown in soil amended with industrial waste. Tree Phys 24:75–82

Tuskan GA, Di Fazio S, Jansson S et al (2006) The genome of black cottonwood, *Populus trichocarpa* (Torr. & Gray). Science 313:1596–1604

Valko M, Klement R, Pelikan P et al (2005) Metals, toxicity and oxidative stress. Curr Med Chem 12:1161–1208

Vollenweider P, Menard T, Günthardt-Goerg MS (2011) Compartmentation of metals in foliage of *Populus tremula* grown on soils with mixed contamination. I. From the tree crown to leaf cell level. Environ Pollut 159:324–336

Wintz H, Vulpe C (2002) Plant copper chaperones. Biochem Soc Trans 30:732–735

Xiangdong L, Chi-sun P, Pui Sum L (2001) Heavy metal contamination of urban soils and street dusts in Hong Kong. Appl Geochem 16:1361–1368

Yi XP, Hargett SR, Frankel LK et al (2009) The PsbP protein, but not the PsbQ protein, is required for normal thylakoid architecture in *Arabidopsis thaliana*. FEBS Lett 583:2142–2147

Zuther E, Büchel K, Hundertmark M et al (2004) The role of raffinose in the cold acclimation response of *Arabidopsis thaliana*. FEBS Lett 576:169–173

Molecular Mechanism of Benign Microbe-Elicited Alleviation of Biotic and Abiotic Stresses for Plants

Anukool Vaishnav, Shekhar Jain, Amrita Kasotia, Sarita Kumari, Rajarshi Kumar Gaur, and Devendra Kumar Choudhary

Abstract

In continuous agricultural systems, crop yields are directly dependent on the inherent soil fertility with microbial processes that governs the mineralization and mobilization of nutrients required for plant growth. The impact of different crop species that are used in various combinations is likely to be an important factor in determining the structure of plant benign microbial communities that function in nutrient cycling, the production of plant growth hormones, and suppression of root diseases. In the present scenario, a perceived role of biotechnology is to introduce multiple choreographed genes into plants that would elicit multiple benefits to the plants such as resistance to stress, productivity, and quality. Microbial genomes that have coevolved with native plant species may already be choreographed and compatible with a wide range of plant genomes and available in this vast unexplored genetic reservoir. Understanding of microbial genome and how it communicates with plant genome for their mutual welfare could lead to innovative methods of plant improvement. Increased adverse effects of abiotic and biotic stresses impacting productivity in principal crops are being witnessed all over the world. Extreme events like prolonged droughts, intense rains and flooding, heat waves, and frost damages are likely to further increase in future due to climate change. A wide range of adaptations and mitigation strategies are required to cope with such impacts. Efficient resource management and crop improvement for evolving better breeds can help to overcome abiotic stresses to some extent. However, such strategies being long drawn and

A. Vaishnav • S. Jain • A. Kasotia • S. Kumari • R.K. Gaur
D.K. Choudhary (✉)
Department of Science, Faculty of Arts, Science & Commerce (FASC), Mody Institute of Technology & Science (MITS), Laxmangarh, Sikar 332311, Rajasthan, India
e-mail: dkchoudhary.fasc@mitsuniversity.ac.in; devmicro@rediffmail.com

R.K. Gaur and P. Sharma (eds.), *Approaches to Plant Stress and their Management*,
DOI 10.1007/978-81-322-1620-9_16, © Springer India 2014

cost intensive, there is a need to develop simple and low cost-effective biological methods for the management of abiotic stress, which can be used on long-term basis. Therefore, studies are needed to elucidate the molecular mechanisms that result from treatment of plants with benign microbes under stress conditions and only then will the full benefits of plant-microbe interaction be understood.

Keywords
Agroecosystem • Induced systemic resistance • Plant growth-promoting bacteria • Volatiles • Stresses

Introduction

Sustainable agriculture involves designing farm system, employing nature as a model. In most natural ecosystem, the greater the diversity the more resistant an ecosystem to change and is better able to recover from disturbances. In an agricultural ecosystem or the so-called agroecosystems (AESs), disturbance is much more frequent, regular, and intense. The ecological concepts of disturbance and their recovery through succession play an important role in AESs management. AESs are undergoing disturbances in the form of cultivation, soil preparation, sowing, planting, irrigation, fertilizer application, pest management, pruning, harvesting, and burning. AESs are limited in earliest stages of succession when disturbance is frequent, widespread, and intense as it is in conventional agriculture (Choudhary et al. 2011, 2012). The diversity and intensity of AESs in developing (Kassam et al. 2009) and developed (Izaurralde et al. 2003) countries have been changing over time in response to a number of interacting biophysical and social factors at the local, regional, and global levels. The impact of increased spatiotemporal climate variability on AESs is likely to be intensified by climate change, which will disrupt many ecosystem functions, altering their capacity to provide goods and services and rendering them more susceptible to degradation (IPCC 2007; Friend 2010). In addition, the security of food supply to an increasing world population has turned into a pressing issue worldwide. Sustainable food production can be achieved by avoiding excessive disturbance and allowing successional processes to generate greater AESs stability. One can enhance the ability of AESs to maintain both fertility and productivity through appropriate management of disturbance and recovery (Friend 2010).

In AESs plant productivity is often limited by soil nutrient availability and relies on the interface, the rhizosphere, between living roots and soils. This is the central area of exchange involving the organic carbon flux from root fuels and microbial decomposers that make nutrients available to roots. It is virtually impossible to investigate the intricacies of potential rhizosphere interactions in every environmental condition because of the tremendous diversity of soil microbes, soil fauna, and plants. An understanding of controls over the belowground function constitutes an important challenge as natural and agroecosystem around the globe are exposed to anthropogenic pressures (Pregitzer et al. 2006). The physicochemical and structural properties of soils, including their development, have been greatly affected by the action of the rhizosphere over consecutive evolutionary time frames, and the evolution of true plant roots, along with their extension deep into substrate, is considered to have led to a revolution in planetary carbon and water cycling that affects the biogeochemical functions of the rhizosphere on Earth today (Beerling and Berner 2005; Richter et al. 2006). It has long been recognized that the activities of soil microorganisms play an intrinsic role in residue decomposition, nutrient cycling, and crop production. Any shift in microbial community structure can be caused by implementation of various land usage and management systems that lead to development of best management practices for the AESs (Peacock et al. 2001).

Farmers use inputs and materials that replace what is removed at harvest or altered through cultivation, and therefore, constant disturbances keep the agroecosystem at an early stage of succession, where a greater proportion of gross productivity is available as net productivity or harvestable biomass. In subsistence agricultural systems, crop yields are directly dependent on the inherent soil fertility and on microbial processes that govern the mineralization and mobilization of nutrients required for plant growth. In addition, the impact of different crop species that are used in various combinations is likely to be an important factor in determining the structure of plant beneficial microbial communities that function in nutrient cycling, the production of plant growth hormones, and suppression of root diseases (Alvey et al. 2003). In AESs, sustainability is dependent on the biological balance in the soils that is governed by the activity of microbial communities. Soil microbial populations are involved in various interactions known to affect plant fitness and soil quality, thereby the stability and productivity of both the agroecosystem and natural ecosystem (Barea et al. 2005). The global necessity to increase agricultural productivity from steadily decreasing land resources base has placed significant strain on the fragile agroecosystems. Therefore, it has become necessary to adopt strategies to maintain and improve agricultural productivity through the employment of high input practices. Improvement in agricultural sustainability requires optimal use and management of soil fertility and soil physical properties and relies on soil biological processes and soil biodiversity. Hence, it is necessary to understand perspectives of microbial diversity in the agricultural context in order to arrive at measures, which can act as indicators of soil quality and plant productivity (Tilak et al. 2005).

Microbial Diversity in the Rhizosphere and Soil Functions

The resilience of the soil is associated with biodiversity, i.e., microbial diversity that increases its resilience capacity. The functional diversity of microbial population in soils is determined by measuring the expression of different enzymes with respect to carbon utilization pattern. The most important aspect of soil biodiversity is soil suppressiveness, i.e., an indicator of the capacity of soils to suppress specific plant pathogens with the help of inherent biotic and abiotic factors (Dorninguez et al. 2001). The link between microbial diversity and the function of soil ecosystem has been described and its importance studied mainly in the above-ground system. Loreau et al. (2001) reported the well-known relation between microbial diversity and function by describing the formation of hump-shaped curve wherein there is an increase in plant production, i.e., the function, concurrent with increasing microbial diversity until a certain point is reached when a sharp reduction occurs in plant production because of further increase in microbial biodiversity. It is difficult to measure stability (resistance and resilience) in soil. The capacity of soil to recover from perturbation can be assessed by monitoring microbial activities. It is quite tedious to measure microbial diversity as a whole, and it does not reflect the link between microbial diversity and function as such. The multifunctionality of soil measured by determining the rates of microbial processes without knowing the microbial species effectively involved in the measured process does not do justice to stability within the system. Griffith et al. (2000) have described two approaches, i.e., destructive and constructive to describe the link between microbial diversity and soil function. In the destructive approach, repeated $CHCl_3$ fumigation of soils decreases microbial diversity, and specific biocides are responsible for killing specific soil microorganisms, whereas constructive approach includes sterile soils inoculated with soil microorganisms. Degens et al. (2001) investigated the relation between microbial diversity and soil functioning by measuring catabolic evenness in two silt clay loam soils subjected to three different stresses (decline in pH, increase in EC, and increasing Cu concentration) and two cyclic disturbances, namely, wetting and drying and freezing and thawing. These workers reported that pasture soil was more resistant to cell stresses than all other perturbations. Finally, the effect of microbial diversity on microbial function in soils depends on the measured function (Griffith et al. 2001a, b). The central problem with the link between microbial diversity and soil function is to understand the relation between genetic diversity and community structure and between community structure and function (O'Donnell et al. 2001). At present various molecular techniques have been employed to determine the composition of soil microflora wherein recent advances include bacterial artificial chromosome (BAC) cloning libraries that allow the functional and taxonomic analysis of soil DNA, i.e., metagenome having insights into several other microbial processes, namely, pathogenicity, competitiveness, substrate range, and bioactive molecules produced by soil microorganisms.

Microbial Interactions in the Rhizosphere and Soil Health

Physicochemical properties of soil are fundamental to soil health including soil texture which is one of the most influential factors. Soil particles held together cohesively influence the precise pore structure of the soil. Soil texture stability reflects the prevention of soil erosion when the soil is exposed to climatic stresses. A well-aggregated soil structure ensures soil tilth, soil-plant water relation, water infiltration rates, soil aeration, root penetrability, and organic matter accumulation which all contribute to soil health (Miller and Jastrow 2000; Buscot 2005). It has been demonstrated that microbial cooperation in the rhizosphere reflects the formation and stabilization of soil aggregates wherein soil particles are held together by bacterial products followed by hyphae of saprophytic and arbuscular mycorrhizae (AMF) which form stable microaggregates of size 2–20 μm in diameter. These microaggregates are bound by the microbial products again into quite large microaggregates (20–250 μm in diameter) with bacterial polysaccharides acting as the binding agents. Finally, microaggregates are then bound into macroaggregates of size >250 μm in diameter with bacterial polysaccharides and AM mycelia that increase the size of microaggregates. The branching habit and three-dimensional structure of the external mycelium of AM that colonizes the soil surrounding the roots allows persistence up to 22 weeks after the plant has died (Miller and Jastrow 2000).

The formation of water-stable soil aggregates is evident in different ecological situations as a result of the effect of AM fungi in cooperation with other microbes and the involvement of glomalin, a glycoprotein produced by the external hyphae of AM fungi. Glomalin participates in the initiation and stabilization of soil aggregates because of its glue-like hydrophobic nature (Requena et al. 2001). Distribution of natural plant communities is accompanied by loss of physicochemical and biological properties of soil, e.g., soil texture, plant nutrient availability, OM content, and microbial activity which is the ultimate result of degradation/desertification processes. It has been investigated frequently that management of AM fungi together with rhizobacteria can restore soil traits (Jeffries and Barea 2001; Requena et al. 2001). The increase in N content in the rhizosphere of the legumes considerably accounts for improvement in nodulation and N-fixing capacity, resulting from cooperative interaction of the symbionts, e.g., *Rhizobium* and AM fungi (Barea et al. 2005). There is considerable experimental evidence to show that several bacteria and fungi can colonize the root-soil environment where they carry out a variety of interactive activities known to benefit plant growth and health, as also soil quality.

The varied genetic and functional activities of the microbial populations impart critical impact on soil functions based on the fact that microbes are driving forces for fundamental metabolic processes which involve specific enzyme activities. Most of the microbial interactions in rhizosphere are responsible for key environmental processes, namely, the biogeochemical cycling of nutrients and matter and the maintenance of plant and soil health (Nannipieri et al. 2003; Barea et al. 2004). Several investigators of investigations have reported that soil-borne microbes interact with plant roots and soil constituents at the root-soil interface wherein C fluxes are crucial determinants of rhizosphere function (Toal et al. 2000). The release of root exudates provides sources of C compounds for the heterotrophic soil biota whereby microbial activity in the rhizosphere affects rooting pattern and the supply of available nutrients to plants (Gryndler 2000).

Reciprocal Relation Between Soil Health and Plant Growth

Soil microorganisms play an important role in soil processes that determine plant and soil productivity. Exhaustive efforts have been made to explore soil microbial diversity of indigenous community, their distribution and behavior in soil habitats to understand the successful functioning of introduced microbial bio-inoculants, and their influence on soil health. Improvement in agricultural sustainability requires optimal use and management of soil fertility and soil physical property and relies on soil biological processes and soil biodiversity (Tilak et al. 2005; Roesti et al. 2006). Plants play an important role in selecting and enriching the type of bacteria by the constituents of their root exudates. The bacterial community develops in the rhizosphere which is a result of diverse nature and concentration of organic constituents of exudates and the corresponding ability of the bacteria to utilize these as sources of energy. Therefore, rhizosphere bacterial community has an efficient system for uptake and catabolism of organic compounds present in root exudates (Barraquio et al. 2000).

It has been described frequently that a plant obtains almost everything directly from the soil to support growth. The soil must have a structure that is physically capable of supporting the above-ground half of the plant through its developing root system as it grows. The soil needs to be maintained at an appropriate pH, provide protection from toxic substances and pathogens, and contain suitable water table. In addition to this all the essential mineral elements that a plant requires are obtained from the soil. Most of these elements are taken from the soil solution in their ionic form (White 2003). The interaction between plant roots and organisms within rhizosphere assists in acquiring essential mineral nutrients and prevents the accumulation of toxic elements. The essential mineral element that most frequently limit plant growth is P; it is taken up in the form of inorganic phosphate (P_i, $H_2PO_4^-$) from the soil solution. The concentration of P_i in the soil solution (2–10 μm) is very low which limits P_i diffusion to the root system with the resultant P_i depletion in the rhizosphere. Plants have evolved several strategies against limiting nature of P_i, to release and acquire P_i from the soil wherein plant increases its carbohydrate allocation to the roots which results in an increased root-shoot ratio and alters the morphology of root system by accelerating lateral root growth and produces long root hairs to increase the volume of soil explored. In addition, P deficiency increases the abundance of P_i transporter proteins and promotes the exudation of organic acids, RNases, and phosphatases to mobilize P from organic/insoluble compounds (Raghothama 2005). It is not surprising that a series of generalized and specific plant-microbe associations in rhizosphere exist that allow efficient solubilization of all the minerals that a plant requires.

Microbial-Elicited Induced Systemic Resistance

Studies on mechanisms of induced systemic resistance (ISR) are suggested to be valuable in extension of microbial-elicited ISR to practical agriculture. Choudhary et al. (2007, 2008) elaborately described induced resistance and its mechanism of action in plants. Plants have the ability to acquire enhanced level of resistance to pathogens after exposure to biotic stimuli provided by many different plant growth-promoting rhizobacteria (PGPR). These in association with plant roots elicit a steady state of defense or ISR in plants. This is often referred to as rhizobacteria-mediated ISR. PGPR-elicited ISR was initially observed in carnation, common bean, and in cucumber with reduced susceptibility to *Fusarium* wilt, halo blight, and *Colletotrichum orbiculare,* respectively. Several PGPR that colonize root systems with seed applications protect plant against foliar disease including *Pseudomonas fluorescens, P. putida, Bacillus pumilus,* and *Serratia marcescens* (Thomma et al. 2001). Induced resistance is a physiological "state of enhanced defensive capacity" elicited by specific environmental stimuli, whereby the plant's innate defenses are potentiated against subsequent biotic challenges. This enhanced state of resistance is effective against a broad range of pathogens and parasites (VanLoon 2000). Unlike systemic acquired resistance (SAR), ISR does not involve the accumulation of pathogenesis-related proteins or salicylic acid, but instead, relies on pathways regulated by jasmonate and ethylene (Yan et al. 2002). A network of interconnected signaling pathways regulates induced defenses of plants against pathogens. The primary components of the network are plant signal molecules – salicylic acid (SA), jasmonic acid (JA), ethylene (ET), and probably nitric oxide (NO).

PGPR can suppress diseases through antagonism between bacteria and soil-borne pathogens, as well as by inducing a systemic resistance in the plant against both root and foliar pathogens (Choudhary 2011). The induced resistance constitutes an increase in the level of basal resistance to several pathogens simultaneously, which is of benefit under natural conditions where multiple pathogens exist (Van Loon and Glick 2004). Plants possess a range of active defense apparatuses that can be actively expressed in response to biotic stresses (pathogens and parasites) of various scales (ranging from microscopic viruses to phytophagous insect). The timing of this defense response is critical and reflects on the difference between coping and succumbing to such biotic challenge of necrotizing pathogens/parasites (Choudhary et al. 2007). Pathogenic microorganisms affecting plant health are major and chronic threats to food production and ecosystem stability worldwide. Despite inconsistency in field performance, biological control is considered as an alternative or a supplemental way of reducing root diseases in agroecosystem (Sharma and Johri 2003). The widely recognized mechanism of biocontrol mediated by PGPR is competition for an ecological niche/substrate, production of inhibitory allelochemicals, and ISR in host plants to a broad spectrum of pathogens. Earlier attempts to commercialize products containing fluorescent pseudomonad strains of PGPR generally failed due to lack of long-term viability of these asporogenous bacteria. Although commercialization of PGPR is mainly proceeding with *Bacillus* spp. rather than pseudomonads, the preponderance of research on PGPR as elicitors of growth promotion or ISR employs PGPR strains that are fluorescent pseudomonads. Compared to plant growth-promoting *Pseudomonas* rhizobacteria, relatively little is known about the lifestyle of plant-associated *Bacillus* spp., which was originally considered as typical soil bacteria, despite their well-established advantages for beneficial action on plant growth and biocontrol (Kloepper et al. 2004a).

In the literature on elicitation of ISR by pseudomonads, the most often investigated component of mechanisms accounting for ISR is the study of signaling pathways in the plant *Arabidopsis thaliana* (Van Loon and Glick 2004). Fewer published accounts of ISR by *Bacillus* spp. are available which showed that specific strains of the species *B. amyloliquefaciens*, *B. subtilis*, *B. pasteurii*, *B. cereus*, *B. pumilus*, *B. mycoides*, and *B. sphaericus* elicit significant reductions in the incidence or severity of various diseases on a diversity of hosts. One aspect of mechanisms is to determine what compounds associated with plant defense against pathogens are produced during PGPR-elicited ISR. Elicitation of ISR in sugar beet was associated with enhanced peroxidase activity and increased production of one chitinase isozyme and two isozymes of β-1,3-glucanase produced by *B. mycoides* strain Bac J

(Bargabus et al. 2002) and *B. pumilus* strains 203–206 and 203–207 (Bargabus et al. 2004), respectively. In the tobacco blue mold system, Zhang et al. (2002) reported that plants treated with *B. pumilus* strain SE34 had greatly increased levels of salicylic acid, compared with that of nontreated plants or plants treated with two gram-negative bacteria, 1 day after challenge inoculation with the pathogen. In the tomato late blight system reported by Yan et al. (2002), elicitation of ISR by *B. pumilus* SE34 on tomato lines with various mutations in signaling pathways was tested. ISR was elicited on *nah*G lines, which breakdown endogenous salicylic acid, but not in the ethylene-insensitive NR/NR line or in the jasmonic acid-insensitive *dfl*/*dfl* line. These results are consistent with studies on several strains of *Pseudomonas* spp. that elicit ISR in *Arabidopsis thaliana* (Van Loon and Glick 2004) where ISR is typically independent of salicylic acid and does not result in activation of the PR1a gene that encodes production of the pathogenesis-related (PR) protein PR1a. Similar results were reported by Zhang et al. (2002). In the tobacco blue mold system, SE34 and two strains of gram-negative bacteria elicited ISR on both wild-type and *nah*G transgenic tobacco lines as evidenced by significant reductions in the severity of blue mold on bacterized plants compared with that on nonbacterized plants. The conclusion that SE34 elicits ISR via salicylic acid-independent pathways confirms to the model with ISR elicited by *Pseudomonas* spp. in *A. thaliana* (van Loon and Glick 2004). De Vleesschauwer and Höfte (2009) elaborately described pathogen- and effector-triggered immunity in plant (Fig. 1).

Various results were found with *B. pumilus* strain T4 (Park and Kloepper 2000) that elicited ISR in tobacco against wildfire, caused by *Pseudomonas syringae* pv. *tabaci*. In this system, a bacterized transgenic line of tobacco (*Nicotianatabacum* cv. Xanthi nc) with a GUS reporter gene fused to the PR1a promoter had significantly reduced severity of wildfire compared with nonbacterized controls. Elicitation of ISR by strain T4 was associated with a significant increase in GUS activity in microtiter plate and whole-plant bioassays. Hence, with strain T4, elicitation of ISR results in activation of PR1a, which is activated during salicylic acid-dependent signaling pathway (Van Loon and Glick 2004). In another study of signaling pathways, Ryu et al. (2003) found different results with strain T4 in *Arabidopsis* spp. In this study, *B. pumilus* T4 and SE34, *B. amyloliquefaciens* IN937a, and *B. subtilis* GB03 were evaluated for elicitation of ISR against two different pathovars of *Pseudomonas syringae* (pvs. *tomato* and *maculicola*). Strains T4 and SE34 elicited ISR against both pathogens whereas strains IN937a and GB03 did not elicit protection against either pathovar. When tested on nahG plants, both T4 and SE34 elicited ISR against *Pseudomonas syringae* pv. *maculicola*. However, against *Pseudomonas syringae* pv. *tomato*, ISR was elicited by T4 but not by SE34. Hence, while a salicylic acid-independent pathway was dominant in the tests, a salicylic acid-dependent pathway appeared to be activated during ISR elicited by strain SE34 against one pathovar. Additional tests of T4 and SE34 on various mutant lines of *Arabidopsis* spp. (Ryu et al. 2003) revealed that in agreement with results on signaling during ISR elicited by *Pseudomonas* spp., ISR elicited by strain SE34 was dependent on NPR1, jasmonic acid, and ethylene, while ISR elicited by strain T4 was dependent on ethylene. In contrast to results on signaling during ISR elicited by *Pseudomonas* spp., ISR elicited by strain T4 was independent of NPR1 and jasmonic acid.

The role of volatiles of microbial origin as signal molecules for plant defense has come to light recently. A comparison has been drawn between herbivores-induced plant volatiles (HIPVs) as an elicitor of plant defenses and two other classes of signaling molecules, C6 green leaf volatiles (GLVs) and C4 bacterial volatiles, which appear to prime plant defenses thereby enhancing the capacity to mobilize cellular defense responses when plants are faced with herbivore/pathogen attacks (Choudhary et al. 2008). Volatile signals generated by certain non-pathogenic bacteria have also been shown to

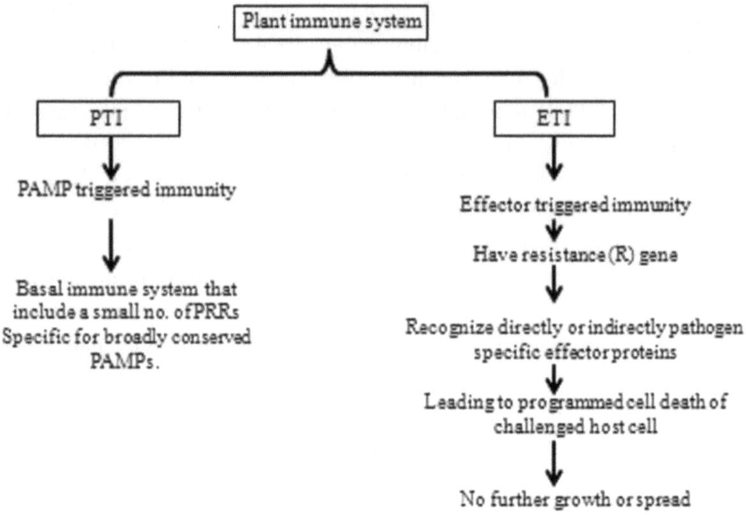

Fig. 1 Role of PTI and ETI in induced resistance

trigger defense responses in *Arabidopsis* (Ryu et al. 2003). Ryu et al. (2004) examined the role of airborne bacterial metabolites in triggering ISR by growing PGPR and *Arabidopsis* seedlings on separate sides of divided Petri dishes. ISR was activated by exposure of *Arabidopsis* seedlings to volatile organic compounds (VOCs) from the *Bacillus* sp. on continuous exposure for as short as 4 days by a significant reduction in symptomatic leaves inoculated with the soft rot-causing pathogen *Erwinia carotovora*. VOCs collected from growth-promoting bacteria *B. subtilis* GB03 and *B. amyloliquefaciens* IN937a showed consistent difference in the composition of volatile blends compared to VOCs which were recovered from non-growth-promoting bacterial strain DH5-α. Strains GB03 and IN937a consistently released two most abundant compounds, 2,3-butanediol and 3-hydroxy-2-butanone (acetoin), while these metabolites were not released from DH5-α or water-treated MS media (Ryu et al. 2003). Several other VOCs were also observed which included dodecane, 2-undecanone, 2-tridecanone, 2-tridecanol, and tetramethylpyrazine from complex bacterial bouquet that did not exhibit ISR priming activity. Bacteria employ different mechanisms to produce VOCs, e.g., in *Bacillus* sp. strain GB03 and IN937a, 2,3-butanediol and acetoin were produced under low atmospheric O_2 partial pressure to provide an alternative electron sink for the regeneration of NAD^+ when usual respiration was not possible (Ryu et al. 2004).

No disease protection was observed when *Bacillus* strains were genetically blocked for the production of 2,3-butanediol; this confirmed the priming activity of the compound to induce resistance against disease. The involvement of known signaling pathways in *Arabidopsis* was screened by exposing defined mutants and transgenic plant lines to bacterial emissions containing VOCs especially 2,3-butanediol. ISR triggered by GB03 VOC was independent of SA, NPR1, and JA signaling pathway but was more or less mediated by ethylene. Interestingly, ISR activation by strain IN937a was independent of all the signaling pathways and this opens up the possibility of involvement of additional VOCs which utilize alternative pathways to trigger ISR. From a more general perspective, the diversity within populations of antagonistic microorganisms with a common biocontrol trait is a means to improving biocontrol. This approach builds on existing knowledge of mechanisms while exploiting genetic differences that have evolved to enable microbial populations to compete successfully in diverse soil and rhizosphere environments. Understanding the diversity within populations of biocontrol agents holds the promise of pairing specific genotypes with their most supportive plant hosts or soil environments to maximize root colonization and disease suppression.

In addition, agricultural management practices as well as the "history" of cultivation in a crop rotation cycle may be supportive or contra-productive for the successful establishment of biocontrol active Bacilli in a given crop.

Microbial-Elicited Induced Systemic Tolerance

Environmental stresses such as drought, temperature, salinity, air pollution, heavy metals, pesticides, and soil pH are major limiting factors in crop production because they affect almost all plant functions. Habitat-imposed abiotic and biotic stress is a serious condition and also land degradation problem in arid and semiarid regions, causing major problem for crop productivity. About 20 % of cultivable and a least half of irrigated lands around the world are severely affected by environmental stresses. However, in these conditions, there are plant populations successfully adapted and evolutionarily different in their strategy of stresses tolerance. Vascular plants do not function as autonomous individuals, but house diverse communities of symbiotic microbes. The role of these microbes can no longer be ignored. Microbial interactions are critical not only for host but also for fungal survival in stressed environments. To date, improvements in plant quality, production, abiotic and biotic stress resistance, and nutrient and water use have relied largely on manipulating plant genomes by breeding and genetic modification. Increasing evidence indicates that the function of symbiotic microbes seem to parallel more than one of these characteristics (Choudhary 2012).

Besides developing mechanisms for stress tolerance, microorganisms can also impart some degree of tolerance to plants towards abiotic stresses like drought, chilling injury, salinity, metal toxicity, and high temperature. In the last decade, bacteria belonging to different genera including *Rhizobium, Bacillus, Pseudomonas, Pantoea, Paenibacillus, Burkholderia, Achromobacter, Azospirillum, Microbacterium, Methylobacterium, Variovorax, and Enterobacter* have been reported to provide tolerance to host plants under different abiotic stress environments. Use of these microorganisms per se can alleviate stresses in agriculture thus opening a new and emerging application of microorganisms. Production of indole acetic acid, gibberellins, and some unknown determinants by microbes results in increased root length, root surface area, and number of root tips, leading to enhanced uptake of nutrients, thereby improving plant health under stress conditions (Egamberdieva and Kucharova 2009). Plant growth-promoting bacteria have been found to improve growth of tomato, pepper, canola, bean, and lettuce under saline conditions (Barassi et al. 2006; Yildirim and Taylor 2005). Some microbial strains produce cytokinin and antioxidants, which result in abscisic acid (ABA) accumulation and degradation of reactive oxygen species. Inoculation of *Azospirillum brasilense* Sp245 in wheat (*Triticumaestivum*) under drought stress resulted in a better water status and an additional "elastic adjustment" resulting in better grain yield and mineral quality (Mg, K, and Ca) at harvest. Another microbial strain, *Achromobacter piechaudii* ARV8 which produced 1-aminocyclopropane-1-carboxylate (ACC) deaminase, conferred induced systemic tolerance (IST) against drought and salt in pepper and tomato (Mayak et al. 2004).

Many aspects of plant life are regulated by ethylene levels, and the biosynthesis of ethylene is subjected to tight regulation, involving transcriptional and posttranscriptional factors regulated by environmental cues, including biotic and abiotic stresses (Hardoim et al. 2008). In the biosynthetic pathway of ethylene, S-adenosylmethionine (S-AdoMet) is converted by 1-aminocyclopropane-1-carboxylate synthase (ACS) to 1-aminocyclopropane-1-carboxylate (ACC), the immediate precursor of ethylene. Under stress conditions, the plant hormone ethylene endogenously regulates plant homoeostasis and results in reduced root and shoot growth. In the presence of ACC deaminase-producing bacteria, plant ACC is sequestered and degraded by bacterial cells to supply nitrogen and energy. Furthermore, by removing ACC, the bacteria reduce the deleterious effect of ethylene, ameliorating plant stress and promoting plant growth (Glick 2007). Saleem et al. (2007) have reviewed the role of microbes containing ACC deaminase, in stress agriculture. Inoculation with ACC deaminase-containing bacteria induces longer roots which might be helpful in the uptake of relatively more water from deep soil under drought stress conditions, thus increasing water use efficiency of the plants under drought conditions (Zahir et al. 2008). Microbial polysaccharides can bind soil particles to form microaggregates and macroaggregates. Plant roots and fungal hyphae fit in the pores between microaggregates and thus stabilize macroaggregates. Plants treated with exopolysaccharide (EPS)-producing bacteria display increased resistance to water stress due to improved soil structure (Sandhya et al. 2009b). EPS can also bind to cations including Na^+, thus making it unavailable to plants under saline conditions. Chen et al. (2007) correlated proline accumulation with drought and salt tolerance in plants. Introduction of proBA genes derived from *Bacillus subtilis* into *A. thaliana* resulted in production of higher levels of free proline resulting in increased tolerance to osmotic stress in the transgenic plants. Increased production of proline along with decreased electrolyte leakage, maintenance of relative water content of leaves, and selective uptake of K^+ ions resulted in salt tolerance in *Zea mays* coinoculated with *Rhizobium* and *Pseudomonas* (Bano and Fatima 2009). Accumulation of proline buffers cellular redox potential under environmental stresses (Wahid and Close 2007). Trehalose metabolism in rhizobia is key for signaling plant growth, yield, and adaptation to abiotic stress, and its manipulation had a major agronomical impact on leguminous plants (Suarez et al. 2008). Figueiredo et al. (2008) reported increased plant growth, N content, and nodulation of *Phaseolus vulgaris* L. under drought stress due to coinoculation of *Rhizobium tropici* and *P. polymyxa*. *Phaseolus vulgaris* (common bean) plants inoculated with *Rhizobium etli* overexpressing trehalose-6-phosphate synthase gene had more nodules with increased nitrogenase activity and high biomass compared with plants inoculated with wild-type *R. etli*. Three-week-old plants subjected to drought stress fully recovered, whereas plants inoculated with a wild-type *Rhizobium etli* died. Microarray analysis of 7,200 expressed sequence tags from nodules of plants inoculated with strain over expressing trehalose-6-phosphate synthase gene revealed upregulation of genes involved in stress tolerance, suggesting a signaling mechanism for trehalose (Figueiredo et al. 2008).

Some of the volatile organic compounds (VOCs) emitted from *Bacillus* (Ryu et al. 2004) are bacterial determinants involved in IST. The volatiles emitted by PGPR downregulate hkt1 (high affinity k^+ transporter 1) expression in roots but upregulates it in shoots, orchestring lower Na^+ levels and recirculation of Na^+ in the whole plant under salt conditions (Zhang et al. 2008). Root colonization of *A. thaliana* by *Pseudomonas chlororaphis* O6 induced tolerance in the plants against biotic and abiotic stresses due to the production of a volatile metabolite, 2R, 3R-butanediol. Studies with Arabidopsis mutant lines indicated that induced drought tolerance requires salicylic acid (SA), ethylene, and jasmonic acid-signaling pathways (Cho et al. 2008). Higher temperatures influence photosynthetic rate, plant water relations, flowering, and fruit set in tropical and temperate crops. Similarly low temperature is a major factor limiting the productivity and geographical distribution of many species, including important agricultural crops. Srivastava et al. (2008) isolated a thermotolerant *Pseudomonas putida* NBRI0987 from drought-affected rhizosphere of chickpea. Overproduction of stress sigma (S) (RpoS) was observed by this microorganism when grown under high-temperature stress at 40 C compared with 30° C. A thermotolerant *Pseudomonas* sp. strain AMK-P6 induced thermotolerance in sorghum seedlings due to synthesis of high molecular weight protein in leaves and improved plant biomass as well as biochemical status in terms of proline, sugar, amino acid, and chlorophyll content (Ali et al. 2009). A plant growth-promoting bacterium *Burkholderia phytofirmans* PsJN capable of epiphytic and endophytic colonization of grapevine tissue and organs (Compant et al. 2005) could protect the plants against heat as well as chilling stress (AitBakra et al. 2006). The bacterized plantlets showed significantly increased levels of starch, proline, and phenolics.

The role of abscisic acid (ABA) had been suggested behind AM-mediated stress response of plants (Aroca et al. 2008). The addition of exogenous ABA considerably enhanced the ABA content in shoots of non-AM plants, concomitant with the expression of the stress marker genes *Lsp5cs* and *Ls1ea* and the gene *Lsnced*. By contrast, the addition of exogenous ABA decreased the content of ABA in shoots of AM plants and did not produce any further enhancement of the expression. Coinoculation of lettuce with PGPR *Pseudomonas mendocina* and *G. intraradices* or *G. mosseae* augmented an antioxidative catalase under severe drought conditions, suggesting that they could be used in inoculants to alleviate the oxidative damage (Kohler et al. 2008).

A 14-3-3 protein encoding gene from *Glomus intraradices* growing in vitro and subjected to drought stress was identified (Porcel et al. 2006). Role of these proteins that regulate both signaling pathways and also effector proteins was suggested in imparting protection to the host plants against drought stress. Glutathione and ascorbate have an important role in conferring protection and maintain metabolic function of plants under water deficit conditions. Low accumulation of these compounds in the lavender plants colonized by autochthonous drought-tolerant *Glomus intraradices* and *Glomus* sp. strain indicated high drought tolerance in plants (Marulanda et al. 2007). Mycorrhized lavender plants showed improved water content, root biomass, and N and K content. AM symbiosis has frequently increased resilience of host plants to salinity stress, perhaps with greater consistency than to drought stress. Growth in saline soils was increased by inoculation with *Glomus* spp., with AM plants having increased phosphate and decreased Na^+ concentrations in shoots compared to uninoculated controls (Giri and Mukerji 2004). Salt resistance was improved by AM colonization in maize (Feng et al. 2002) and clover (Ben Khaled et al. 2003) with AM effect correlated with improved osmoregulation or proline accumulation. AM inoculation also improved NaCl resistance in tomato, with extent of improvement related to salt sensitivity of the cultivar (Al-Karaki et al. 2001).

Future Perspectives

Plant-associated microorganisms can play an important role in alleviation of biotic and abiotic stresses. These organisms could include rhizoplane and endophytic bacteria and symbiotic fungi and operate through a variety of mechanisms like triggering osmotic response and induction of novel genes in plants. The development of stress-tolerant crop varieties through genetic engineering and plant breeding is essential but a long-drawn process, whereas microbial inoculation to alleviate stresses in plants could be a more cost-effective environmental friendly option which could be available in a shorter time frame. Taking the current leads available, concerted future research is needed in this area, particularly on field evaluation and application of potential organisms. It is our contention that native plants survive and flourish in stressed ecosystems because of endosymbiotic organisms that have coevolved and were essential for their adaptation to changing environments. Some of these microbial components are non-cultivable and vertically transmitted from generation to generation. They represent a vast reservoir of heritable DNA that can enhance plant performance in changing environments and add genetic flexibility to adaptation of long-lived plants. If such endophytes can be identified that not only persist in progeny of novel hosts but can confer benefits in mechanized, agricultural systems, they would be increasingly important in agricultural production and lead to a rapid and economical method of providing novel germplasms of native and crop plants. Many questions must be answered before systemic endophyte transfer to crop or native plants can become a standard practice. Better methods for identifying what are being transferred and for monitoring and how long these associations persist in field settings are required. The answer to these questions and others will require novel approaches of molecular technology.

The growing consensus is that microbial associations with higher plants are universal. Plant growth and development cannot be adequately described without acknowledging microbial interactions. We need to determine the extent of microbial associations in the plant kingdom. This question will only be answered as technology is developed to detect their presence in plant tissues. What we have learned is that there is a need to understand how plant microbes communicate in these endosymbiotic relationships and how they regulate basic genetic and physiological functions. There is an increasing interest in understanding the cooperative activities among microbial populations because of current public concerns about the adverse effect of agrochemicals and their effect on the agroecosystem. Two main types of interactions in the rhizosphere are recognized, one of which is based on dead plant material (the detritus-based interactions) and the other involves living plant roots. Both types of interactions are relevant to agronomy and ecology. Microbial activity in the rhizosphere affects rooting pattern and the supply of available nutrients to plants thereby modifying the quality and quantity of root exudates. The specific structure and diversity of the rhizosphere bacterial community varies between plant species and over time, and the different root zones present on the same plant can support distinct bacterial communities that reflect qualitative and quantitative differences in root exudation. In addition, the functioning of bacterial communities in an agroecosystem is affected by soil type which plays a key role in determining the specific dominant bacteria colonizing the rhizosphere. Comprehension of the dynamics of the microbial populations could shed light on the process of selecting successful strains that promote plant growth and/or suppressing diseases. Recent advances in the study of the intra- and interspecies signaling are providing an important area for scientific research as well as relevant application. Understanding quorum-sensing systems in antifungal metabolite production and identification of promoters that can be induced or increased in the rhizosphere provide new approaches for the development of new biological control agents.

Acknowledgment Some of the research work in the chapter has partially been supported by DBT and DST-SERB grant no. BT/PR1231/AGR/21/340/2011 and SR/FT/LS-129/2012 respectively to DKC for financial support.

References

Aitbakra E, Nowak J, Clement C (2006) Enhancement of chilling resistance of inoculated grapevine plantlets with a plant growth promoting rhizobacterium, *Burkholderia phytofirmans* strain PsJN. Appl Environ Microbiol 72(11):7246–7252

Ali Sk Z, Sandhya V, Grover M, Kishore N, Rao LV, Venkateswarlu B (2009) *Pseudomonas* sp. strain AKM-P6 enhances tolerance of sorghum seedlings to elevated temperatures. Biol Fertil Soils 46:45–55

Al-Karaki GN, Ammad R, Rusan M (2001) Response of two tomato cultivars differing in salt tolerance to inoculation with mycorrhizal fungi under salt stress. Mycorrhiza 11:43–47

Alvey S, Yang C-H, Buerkert A, Crowley DE (2003) Cereal/legume rotation effects on rhizosphere bacterial community structure in West African soils. Biol Fertil Soils 37:73–82

Aroca R, Vernieri P, Ruiz-Lozano JM (2008) Mycorrhizal and non-mycorrhizal *Lactuca sativa* plants exhibit contrasting responses to exogenous ABA during drought stress and recovery. J Exp Bot 59 (8):2029–2041

Bano A, Fatima M (2009) Salt tolerance in *Zea mays* (L.) following inoculation with *Rhizobium* and *Pseudomonas*. Biol Fertil Soils 45:405–413

Barassi CA, Ayrault G, Creus CM, Sueldo RJ, Sobero MT (2006) Seed inoculation with *Azospirillum mitigates* NaCl effects on lettuce. Sci Hortic 109:8–14

Barea JM, Azcon R, Azcón-Aguilar C (2004) Mycorrhizal fungi and plant growth promoting rhizobacteria. In: Varma A, Abbott L, Werner D, Hampp R (eds) Plant surface microbiology. Springer-Verlag, Heidelberg, pp 351–371

Barea J-M, Pozo MJ, Azcón R, Azcón-Aguilar C (2005) Microbial co-operation in the rhizosphere. J Exp Bot 56:1761–1778

Bargabus RL, Zidack NK, Sherwood JW, Jacobsen BJ (2002) Characterization of systemic resistance in sugar beet elicited by a non-pathogenic, phyllosphere-colonizing Bacillus mycoides, biological control agent. Physiol Mol Plant Pathol 61:289–298

Bargabus RL, Zidack NK, Sherwood JW, Jacobsen BJ (2004) Screening for the identification of potential biological control agents that induce systemic acquired resistance in sugar beet. Biol Control 30:342–350

Barraquio WL, Segubre EM, Gonzalez MAS, Verma SC, James EK, Ladha JK, Tripathi AK (2000) Diazotrophic enterobacteria: what is their role in the rhizosphere of rice? In: Ladha JK, Reddy PM (eds) The quest for nitrogen fixation in rice. IRRI, Los Banos, pp 93–118

Beerling DJ, Berner RA (2005) Feedbacks and the coevolution of plants and atmospheric CO_2. Proc Natl Acad Sci U S A 102:1302–1305

Ben Khaled L, Gomez AM, Ourraqi EM, Oihabi A (2003) Physiological and biochemical responses to salt stress of mycorrhized and/or nodulated clover seedlings (*Trifolium alexandrinum* L.). Agronomie 23:571–580

Buscot F (2005) What are soils? In: Buscot F, Varma S (eds) Micro-organisms in soils: roles in genesis and functions. Springer-Verlag, Heidelberg, pp 3–18

Chen M, Wei H, Cao J, Liu R, Wang Y, Zheng C (2007) Expression of *Bacillus subtilis* proAB genes and reduction of feedback inhibition of proline synthesis increases proline production and confers osmotolerance in transgenic *Arabidopsis*. J Biochem Mol Biol 40:396–403

Cho S, Kang BR, Han SH, Anderson AJ, Park JY, Lee YH, Cho BH, Yang KY, Ryu CM, Kim YC (2008) 2R, 3r-butanediol, a bacterial volatile produced by *Pseudomonas chlororaphis* O6, is involved in induction of systemic tolerance to drought in *Arabidopsis thaliana*. Mol Plant Microbe Interact 21(8):1067–1075

Choudhary DK (2011) Plant growth-promotion (PGP) activities and molecular characterization of rhizobacterial strains isolated from soybean (*Glycine* max) plants against charcoal rot pathogen, *Macrophomina phaseolina*. Biotechnol Lett 33:2287–2295

Choudhary DK (2012) Microbial rescue to plant under habitat-imposed abiotic and biotic stresses. Appl Microbiol Biotechnol 96:1137–1155

Choudhary DK, Prakash A, Johri BN (2007) Induced systemic resistance (ISR) in plants: mechanism of action. Indian J Microbiol 47:289–297

Choudhary DK, Johri BN, Prakash A (2008) Volatiles as priming agent that initiate plant growth and defense responses. Curr Sci 94:595–604

Choudhary DK, Sharma KP, Gaur RK (2011) Biotechnological perspectives of microbes in agro-ecosystems. Biotechnol Lett 33:1905–1910

Compant S, Reiter B, Sessitsch A, Nowak J, Clement C, AitBakra E (2005) Endophytic colonization of *Vitis vinifera* L. by plant growth promoting bacterium *Burkholderia* sp. strain PsJN. Appl Environ Microbiol 71:1685–1693

De Vleesschauwer D, Höfte M (2009) Rhizobacteria-induced systemic resistance. Adv Bot Res 51:223–281

Degens BP, Schipper LA, Sparling GP, Duncan LC (2001) Is the microbial community in a soil with reduced catabolic diversity less resistant to stress or disturbance? Soil Biol Biochem 33:1143–1153

Dorninguez J, Negrin MA, Rodriguez CM (2001) Aggregate water stability, particle-size and soil solution properties in conductive and suppressive soil to *Fusarium* wilt of banana from Canary Islands (Spain). Siol Biol Biochem 33:499–455

Egamberdieva D, Kucharova Z (2009) Selection for root colonizing bacteria stimulating wheat growth in saline soils. Biol Fertil Soils 45:563–571

Feng G, Zhang FS, Li XL, Tian CY, Tang C, Renegal Z (2002) Improved tolerance of maize plants to salt stress by arbuscular mycorrhiza is related to higher accumulation of leaf P-concentration of soluble sugars in roots. Mycorrhiza 12:185–190

Figueiredo MVB, Burity HA, Martinez CR, Chanway CP (2008) Alleviation of drought stress in common bean (*Phaseolus vulgaris* L.) by co-inoculation with *Paenibacillus polymyxa* and *Rhizobium tropici*. Appl Soil Ecol 40:182–188

Friend AD (2010) Terrestrial plant production and climate change. J Exp Bot 61:1293–1309

Giri B, Mukerji KG (2004) Mycorrhizal inoculant alleviate salt stress in *Sesbania aegyptiaca* and *Sesbania grandiflora* under field conditions: evidence for reduced sodium and improved magnesium uptake. Mycorrhiza 14:307–312

Glick BR (2007) Promotion of plant growth by bacterial ACC deaminase. Crit Rev Plant Sci 26:227–242

Griffiths BS, Ritz K, Bardgett RD, Cook R, Christensen S, Ekelund F et al (2000) Ecosystem response of pasture soil communities to fumigation-induced microbial diversity reductions: an examination of the biodiversity–ecosystem function relationship. Oikos 90:279–294

Griffiths BS, Bonkowski M, Riy J, Ritz K (2001a) Functional stability, substrate utilization and biological indicators of soils following environmental impacts. Appl Soil Ecol 16:49–61

Griffiths BS, Ritz K, Wheatley RE, Kuan HL, Boag B, Christensen S et al (2001b) An examination of the biodiversity–ecosystem function relationship in arable soil microbial communities. Soil Biol Biochem 33:1713–1722

Gryndler M (2000) Interactions of arbuscular mycorrhizal fungi with other soil organisms. In: Kapulnik Y, Douds DD Jr (eds) Arbuscular mycorrhizas: physiology and functions. Kluwer Academic, Dordrecht, pp 239–262

Hardoim PR, van Overbeek SV, van Elsas JD (2008) Properties of bacterial endophytes and their proposed role in plant growth. Trends Microbiol 16:463–471

IPCC, Climate Change (2007) Impacts, adaptation and vulnerability. In: Parry ML, Canziani OF, Palutikof JP, van der Linden PJ, Hanson CE (eds) Contribution of working group II to the fourth assessment report of the intergovernmental panel on climate change. Cambridge University Press, UK, pp 976

Izaurralde R, Rosenberg N, Brown R, Thomson A (2003) Integrated assessment of Hadley Center (HadCM2) climate-change impacts on agricultural productivity and irrigation water supply in the conterminous United States part II. Regional agricultural production in 2030 and 2095. Agric Forest Meteor 117:97–122

Jeffries P, Barea JM (2001) Arbuscular mycorrhiza: a key component of sustainable plant–soil ecosystems.

In: Hock B (ed) The Mycota: fungal associations, vol IX. Springer, Berlin Heidelberg, pp 95–113

Kassam A, Friedrich T, Shaxson F, Pretty J (2009) The spread of conservation agriculture: justification, sustainability and uptake. Internl J Agric Sustain 7:292–320

Kloepper JW, Ryu C-M, Zhang S (2004) Induced systemic resistance and promotion of plant growth by *Bacillus* spp. Phytopathology 94:1259–1266

Kohler J et al (2008) Plant growth promoting rhizobacteria and arbuscular mycorrhizal fungi modify alleviation biochemical mechanisms in water stressed plants. Funct Plant Biol 35:141–151

Loreau M, Naeem S, Inchausti P, Bengtsson J, Grime JP, Hector A, Hooper DU, Huston MA, Raffaelli D, Schmid B, Tilman D, Wardle DA (2001) Biodiversity and ecosystem functioning: current knowledge and future challenges. Science 294:804–808

Marulanda A, Porcel R, Barea JM, Azcon R (2007) Drought tolerance and antioxidant activities in lavender plants colonized by native drought tolerant or drought sensitive *Glomus* species. Microb Ecol 54 (3):543–552

Mayak S et al (2004) Plant growth-promoting bacteria confer resistance in tomato plants to salt stress. Plant Physiol Biochem 42:565–572

Miller RM, Jastrow JD (2000) Mycorrhizal fungi influence soil structure. In: Kapulnik Y, Douds D (eds) Arbuscular mycorrhizas: physiology and function. Kluwer Academic, Dordrecht, pp 4–18

Nannipieri P, Ascher J, Ceccherini MT, Landi L, Pietramellara G, Renella G (2003) Microbial diversity and soil functions. Eur J Soil Sci 54:655–670

O'Donnell AG, Seasman M, Macrae A, Waite I, Davies JT (2001) Plants and fertilizers as drivers of change in microbial com-munity structure and function in soils. Plant Soil 232:135–145

Park KS, Kloepper JW (2000) Activation of PR-1a promoter by rhizobacteria that induce systemic resistance in tobacco against Pseudomonas syringae pv. tabaci. Biol Control 18:2–9

Peacock AD, Muller MD, Ringelberg DB, Tyler DD, Hedrick DB, Gale PM, White DC (2001) Soil microbial community resources to dairy manure or ammonium nitrate applications. Soil Biol Biochem 33:1011–1019

Porcel R, Aroca R, Cano C, Bago A, Ruiz-Lozano JM (2006) Identification of a gene from the arbuscular mycorrhizal fungus *Glomus intraradices* encoding for a 14-3-3 protein that is up-regulated by drought stress during the AM symbiosis. Micro Ecol 52 (3):575–582

Pregitzer KS, Zak DR, Loya WM, Karberg NJ, King JS, Burton AJ (2006) The contribution of root systems to biogeochemical cycles in a changing world. In: Cardon ZG, Whitebeck JL (eds) The rhizosphere: an ecological perspective. Elsevier, Oxford, pp 155–178

Raghothama KG (2005) Phosphorus. In: Broadly MR, White PJ (eds) Plant nutritional genomics. Blackwell, Oxford, pp 112–126

Requena N, Perez-Solis E, Azcon-Aguilar C, Jeffries P, Barea JM (2001) Management of indigenous plant-microbe symbioses aids restoration of desertified. Appl Environ Microbiol 67:495–498

Richter D de B, Oh NH, Fimmen R, Jackson J (2006) The rhizosphere and soil formation. In: Cardon ZG, Whitebeck JL (eds) The rhizosphere: an ecological perspective. Elsevier, Oxford, pp 179–200

Roesti D, Gaur R, Johri BN, Imfeld G, Sharma S, Kawaljeet K, Aragno M (2006) Plant growth stage, fertilizer management and bio-inoculation of *Arbuscular mycorrhizal* fungi and plant growth promoting rhizobacteria affect the rhizobacterial community structure in rain-fed wheat fields. Soil Biol Biochem 38:1111–1120

Ryu C-M, Hu C-H, Reddy MS, Kloepper JW (2003) Different signaling pathways of induced resistance by rhizobacteria in *Arabidopsis thaliana* against two pathovars of Pseudomonas syringae. New Phytol 160:413–420

Ryu CM, Farag MA, Hu CH, Reddy MS, Wei HX, Kloepper JW, Paré PW (2004) Bacterial volatiles induce systemic resistance in *Arabidopsis*. Plant Physiol 134:1017–1026

Saleem M, Arshad M, Hussain S, Bhatti AS (2007) Perspective of plant growth promoting *Rhizobacteria* (PGPR) containing ACC deaminase in stress agriculture. J Ind Microbiol Biotechnol 34:635–648

Sandhya V, Ali SZ, Grover M, Reddy G, Venkateswarlu B (2009) Alleviation of drought stress effects in sunflower seedlings by exopolysaccharides producing *Pseudomonas putida* strain P45. Biol Fertil Soil 46:17–26

Sharma A, Johri BN (2003) Combat of iron-deprivation through a plant growth promoting fluorescent *Pseudomonas* strain GRP3A in mung bean. Microbiol Res 158:77–81

Srivastava S, Yadav A, Seem K, Mishra S, Chaudhary V, Srivastava CS (2008) Effect of high temperature on *Pseudomonas putida* NBRI0987 biofilm formation and expression of stress sigma factor RpoS. Curr Microbiol 56(4):453–457

Suarez R, Wong A, Ramirez M, Barraza A, OrozcoMdel C, Cevallos MA, Lara M, Hernandez G, Iturriaga G (2008) Improvement of drought tolerance and grain yield in common bean by over expressing trehalose-6-phosphate synthase in rhizobia. Mol Plant Microbe Interact 21:958–966

Thomma BPHJ, Tierens KFM, Penninckx IAMA, Mauch Mani B, Broekaert WF, Cammue BPA (2001) Different micro-organisms differentially induce *Arabidopsis* disease response pathways. Plant Physiol Biochem 39:673–680

Tilak KVBR, Ranganayaji N, Pal KK, De R, Saxena AK, Nautiyal CS, Mittal S, Tripathi AK, Johri BN (2005) Diversity of plant growth and soil health supporting bacteria. Curr Sci 89:136–150

Toal ME, Yeomans C, Killham K, Meharg AA (2000) A review of rhizosphere carbon flow modelling. Plant Soil 222:263–281

Van Loon LC, Glick BR (2004) Increased plant fitness by rhizobacteria. In: Sandermann H (ed) Molecular ecotoxicology of plants. Springer, Berlin, pp 177–205

VanLoon LC (2000) Systemic induced resistance. In: Slusarenko AJ, Fraser RSS, van Loon LC (eds) Mechanisms of resistance to plant diseases. Kluwer, Dordrecht, pp 521–574

Wahid A, Close TJ (2007) Expression of dehydrins under heat stress and their relationship with water relations of sugarcane leaves. Biol Plant 51:104–109

White PJ (2003) Ion transport. In: Thomas B, Murphy DJ, Murray BG (eds) Encyclopaedia of applied plant sciences. Academic, London, pp 625–634

Yan Z, Reddy MS, Ryu C-M, McInroy JA, Wilson M, Kloepper JW (2002) Induced systemic resistance against tomato late blight elicited by plant growth-promoting rhizobacteria. Phytopathology 92:1329–1333

Yildirim E, Taylor AG (2005) Effect of biological treatments on growth of bean plans under salt stress. Ann Rep Bean Improv Cooper 48:176–177

Zahir ZA, Munir A, Asghar HN, Shaharoona B, Arshad M (2008) Effectiveness of rhizobacteria containing ACC deaminase for growth promotion of peas (*Pisum sativum*) under drought conditions. J Microbiol Biotechnol 18(5):958–963

Zhang S, Moyne A-L, Reddy MS, Kloepper JW (2002) The role of salicylic acid in induced systemic resistance elicited by plant growth-promoting rhizobacteria against blue mold of tobacco. Biol Control 25:288–296

Zhang H, Kim MS, Sun Y, Dowd SE, Shi H, Paré PW (2008) Soil bacteria confer plant salt tolerance by tissue-specific regulation of the sodium transporter HKT1. Mol Plant Microbe Interact 21:737–44

MAPK Signaling Cascades and Transcriptional Reprogramming in Plant–Pathogen Interactions

Gohar Taj, Priyanka Giri, Mohd Tasleem, and Anil Kumar

Abstract

In plants, innate immunity is triggered through pattern recognition receptors (PRRs) in response to microbe-associated molecular patterns (MAMPs) to provide the first line of inducible defense. Plant receptor protein kinases (RPKs) represent the main plasma membrane PRRs perceiving diverse MAMPs. RPKs trigger mitogen-activated protein kinase (MAPK) module which is one of the earliest signaling events after plant sensing of the invading pathogen as they link the perception of external stimuli to cellular responses. MAPK signaling networks serve specific and overlapping roles in controlling the activities and synthesis of a plethora of transcription factors (TFs), enzymes, hormones, peptides, and antimicrobial chemicals, contributing to resistance against bacteria, oomycetes, and fungi. Transcriptional reprogramming has been carried out by one of the most studied WRKY family of transcription factors. Recently, genetic evidence directly proved its significance as positive and negative regulators of disease resistance. WRKY genes were shown to be functionally connected forming a transcriptional network composed of positive and negative feedback loops and feed-forward modules. Within a web of partially redundant elements, some WRKY factors hold central positions mediating fast and efficient activation of defense programs.

Keywords

Mitogen-activated protein kinase signaling • Signaling networks • MAMPs • Defense

Introduction

Plants face constant threat from a wide range of microorganisms in their natural habitat. The attack by pathogen is countered by sophisticated defense system possessed by plants. It is suggested that plant initially responded in

G. Taj (✉) • P. Giri • M. Tasleem • A. Kumar
Molecular Biology and Genetic Engineering, College of Basic Science and Humanities, G.B. Pant University of Agriculture & Technology, Dist. US Nagar, Pantnagar 263145, Uttarakhand, India
e-mail: gohartajkhan@rediffmail.com

R.K. Gaur and P. Sharma (eds.), *Approaches to Plant Stress and their Management*,
DOI 10.1007/978-81-322-1620-9_17,

similar ways to host and nonhost pathogen. In plant–pathogen interaction, the first step is the recognition of elicitors by plant receptors found in the plasma membrane or cytoplasm of plant cells. From here, signaling pathways are initiated, leading to number of events like reversible phosphorylation/dephosphorylation of plasma membrane/cytosolic proteins; cytosolic calcium level spike; plasma membrane depolarization; Cl^- and K^+ efflux/H^+ influx; extracellular alkalinization and cytoplasmic acidification; mitogen-activated protein kinase (MAPK) activation; NADPH oxidase activation; ROS production; early defense gene expression; ethylene, salicylic acid, and jasmonate production; late defense gene expression; and secondary metabolite accumulation (Zhao et al. 2005). In the classical concept, plant defenses can be divided into nonhost resistance and host resistance, whereas in the modern concept, defense governs by two layers. The first layer relies on the perception of pathogen-associated molecular patterns (PAMPs)/microbe-associated molecular patterns (MAMPs) (Zipfel and Robatzek 2010). The responses to PAMPs are called PAMP-triggered immunity (PTI). PTI effectively prevents colonization of plant tissues by potential pathogens (Chisholm et al. 2006; Boller and He 2009; Jones and Dangl 2006). The second layer is the recognition of pathogen effectors, which can promote pathogen fitness through host resistance (R) proteins. Effector recognition by R protein leads to effector-triggered immunity (ETI) which acts mostly inside the cell. ETI is an accelerated and magnified defense response compared to PTI and is often accompanied by localized cell death termed a hypersensitive response (HR) (Jones and Dangl 2006). One of the earliest signaling events after plant sensing of invading pathogens is the activation of mitogen-activated protein kinases (MAPKs) signaling cascade (Tena et al. 2001; Zhang and Klessig 2001; Ichimura et al. 2002; Nakagami et al. 2005). Although there is a growing body of literature regarding role of individual MAPK in the plant–pathogen interaction, very little is known about the complete pathway or cascade component involved in the recognition of pathogen up to the response of the plant. In this chapter, we tried to cover all the components from the receptor to transcription factor. Therefore in order to better understand the complex picture of plant and pathogen interaction, an attempt has been taken to compile the recent information of MAPK signaling along with the transcriptional reprogramming machinery.

Mitogen-Activated Protein Kinases (MAPKs): A Key Molecule of Signal Transduction Pathway Involved in Defense Against Pathogen Attack

Mitogen-activated protein kinases (MAPKs) are encoded by a large family of serine/threonine protein kinases that are found in all eukaryotes. Activation of MAPKs is brought about by upstream MAPK kinases (denoted as MAPKKs) through phosphorylation of the conserved threonine and tyrosine residues that are located close to kinase domain VIII in all MAPKs. A given dual specificity MAPKK can only activate a specific MAPK and cannot functionally substitute other MAPKKs. MAPKKs are themselves activated by phosphorylation through upstream kinases that belong to the class of MAPKK kinases (MAPKKKs), raf and mos proteins. A specific set of three functionally interlinked protein kinases (MAPKKK-MAPKK-MAPK) forms the basic module of a MAPK pathway (Hirt 2009). An overall schematic representation of MAPK signaling was given in Fig. 1.

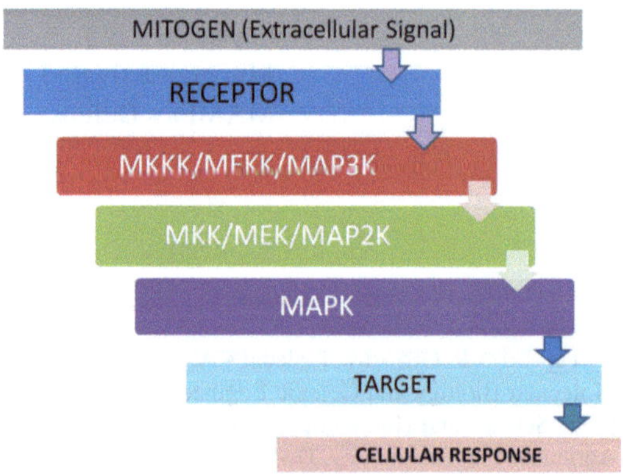

Fig. 1 MAPK module in the plant–pathogen interaction

MAP kinase first comes to light when Strungill and Ray (1986) identified a protein kinase from insulin-treated 3T3 l-1 cell extracts that would phosphorylate microtubule-associated protein-2 on both serine and threonine. In plant, MAPKs were firstly identified in 1993 in alfalfa as MsERK1 and D5 kinase in pea (Rodriguez et al. 2010). Despite the general name mitogen-activated protein kinase, this family of kinases is not only activated by mitogens such as epidermal growth factors (EGF) or insulin; there are many other elicitors such as hyper- and hypo-osmolarity, UV light, genotoxic agents, inflammatory mediators, thrombin, heat shock, or mechanical stretching (Morris 2001). By responding to external stimuli, the MAPK cascade plays a critical role in gene expression, metabolism, cell death, proliferation, and differentiation in animals, plants, and yeast (Widmann et al. 1999; Chen et al. 2001). Molecular and biochemical studies using specific antibodies to particular MAPKs have revealed that MAPK activation correlates with stimulatory treatments such as pathogen infection, wounding, low temperature, drought, hyper- and hypo-osmolarity, high salinity, touch, and reactive oxygen species (Romeis 2001; Morris 2001; Zhang and Klessig 2001).

MAMP Perception Through Cell-Surface LRR-RLK Receptors

A key aspect of active defense mechanisms is a prompt and efficient detection of microbial invaders. Plants respond to a wide array of MAMPs from both nonpathogenic and pathogenic microbes, including bacterial flagellin, lipopolysaccharide (LPS), elongation factor EF-Tu, and harpin (HrpZ), oomycete-derived Pep13, NEP1-like proteins (NLPs), and β-glucan, as well as fungal chitin and ergosterol (Boller 1995; Ligterink et al. 1997; Zhang et al. 1998; Cardinale et al. 2000; Desikan et al. 2001; Lee et al. 2001; Asai et al. 2002; Fellbrich et al. 2002; Wan et al. 2004; Daxberger et al. 2006; Qutob et al. 2006; Zipfel et al. 2006). The perception of several MAMPs is mediated by cell-surface receptors. In *Arabidopsis*, 610 receptor-like kinases (RLKs) and 56 receptor-like proteins (RLPs) have been identified (Shiu and Bleecker 2001; Fritz-Laylin et al. 2005). Schematic representation of MAMPs and receptors during plant–pathogen interaction is shown in Fig. 2.

It is not known to what extent RLKs and RLPs are involved in plant immunity. A large number of genes encoding RLKs and RLPs are transcriptionally induced upon MAMP treatment, which suggests a potential role in defense (Zipfel et al. 2004; Zipfel et al. 2006). In plants, the first identified and best studied pattern recognition receptor (PRR) is FLS2, the flagellin receptor (Gomez-Gomez and Boller 2000). It consists of an N-terminal signal peptide, 28 leucine-rich repeats (LRRs), a transmembrane domain, and a cytoplasmic kinase domain. The 22-amino-acid peptide (flg22) corresponding to the highly conserved amino terminus of flagellin is sufficient to trigger immune responses in *Arabidopsis,* tomato, tobacco, and barley but not rice (Felix et al. 1999; Peck et al. 2001; Taguchi et al. 2003; Chinchilla et al. 2006; Hann and Rathjen 2007; Shen et al. 2007). FLS2 and flg22 were found to coprecipitate by affinity cross-linking and immunoprecipitation, which suggests physical interaction between them (Chinchilla et al. 2006). EF-Tu is another most abundant and conserved bacterial proteins, which has been found to function as an MAMP in *Arabidopsis* (Zipfel et al. 2006). EF-Tu receptor (EFR) is another LRR RLK that perceives the 18-amino-acid minimal PAMP motif elf18 from bacterial elongation factor Tu. It is a close homolog of FLS2 and belongs to the same subclade XII of LRR RLKs (Zipfel et al. 2006). Expression of EFR in *Nicotiana benthamiana*, which normally lacks a perception system for EF-Tu, conferred elf18 responsiveness. This suggests physical interaction of elf18 and EFR. Interestingly, EFR contains a typical endocytic motif indicating possible intracellular trafficking. Another LRR RLK involved in perception of pathogens is rice Xa21, which recognizes the effector-type molecule AvrXa21 (Lee et al. 2006; Xu et al. 2006b). The wall-associated kinase (WAK) gene family is another unique subfamily of receptor-like kinases (RLKs) in plants. Each

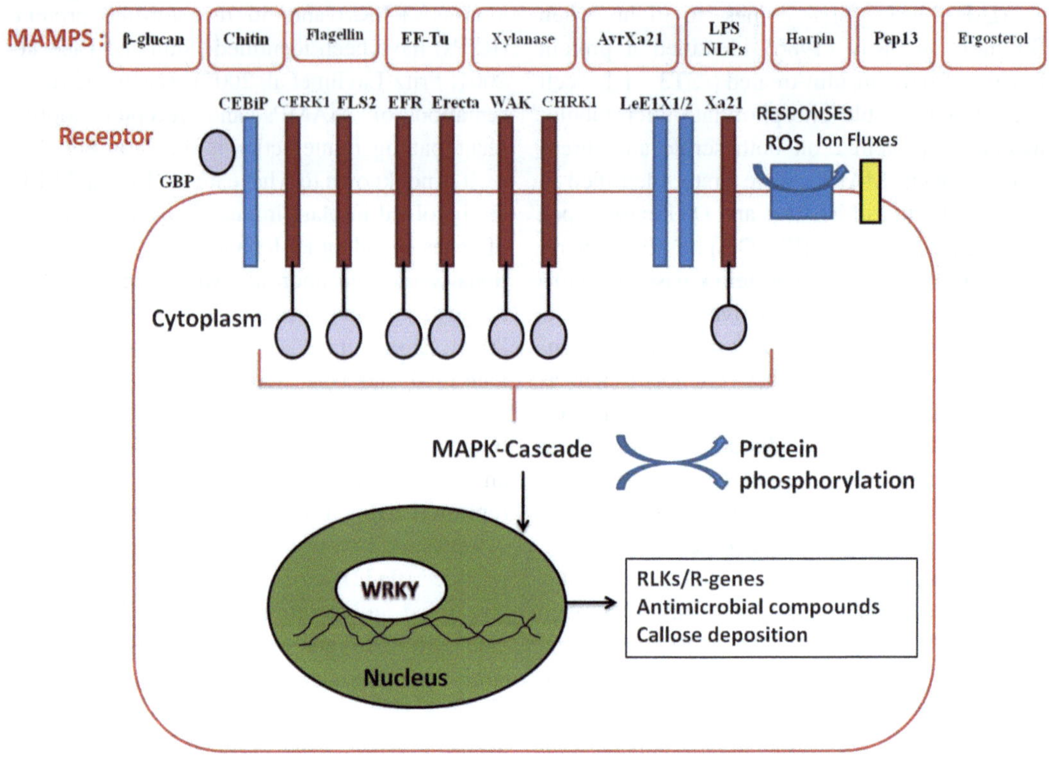

Fig. 2 Schematic representation of currently known MAMPs and cognate receptors mediating plant immunity. MAMPs are recognized by soluble-binding proteins (*GBP*), transmembrane LysM-containing RLPs (*CEBiP*), LRR-type RLPs (*LeEix*), and LRR RLKs (*FLS2*, *EFR*, *Xa21*, *Erecta*, *WAK*, *CERK1*, *CHRK1*). MAMP-mediated receptor signaling triggers activation of at least two MAP kinase cascades positively and negatively regulating transcriptional changes possibly by targeting WRKY transcription factors. Transcriptional changes include upregulation of RLKs and R-genes, production of antimicrobial compounds, and callose deposition. Furthermore, MAMP perception stimulates ion fluxes across the plasma membrane and generation of reactive oxygen species (*ROS*), NO, and ethylene and elicits differential protein phosphorylation

encodes a transmembrane protein with a cytoplasmic Ser/Thr kinase (STK) domain and an extracellular region with similarity to vertebrate epidermal growth factor (EGF)-like domains. WAKs are thought to physically link the extracellular matrix and the cytoplasm and to serve a signaling function between them (He et al. 1996; Kohorn 2000; Wagner and Kohorn 2001). WAK-RLKs play roles in cell expansion, pathogen resistance, and metal tolerance in *Arabidopsis* (Wang et al. 2012; He et al. 1998). *WAK1* is induced by the exposure of both bacterial and fungal pathogens; moreover, it is also induced by defense-related signaling molecules like salicylic acid, methyl jasmonate, and ethylene. It is also upregulated during systemic-acquired resistance (Maleck et al. 2000; Schenk et al. 2000). Erecta (ER) is another RLK that regulates both plant development and immunity. Erecta gene comprises an extracellular LRR domain, a transmembrane region, and an intracellular kinase domain (Torii et al. 1996). Erecta (ER) LRR RLK is required for resistance in *Arabidopsis* to the soilborne bacterium *Ralstonia solanacearum*, the necrotrophic fungus *Plectosphaerella cucumerina*, and the damping-off oomycete *Pythium irregulare* (Rodríguez et al. 2009). It has also been proved that mutation in this gene enhances the susceptibility to *P. cucumerina*. Chitinase-related receptor-like kinase, designated *CHRK1*, is a new type of receptor-like kinase containing a chitinase-like sequence in the putative extracellular domain. The C-terminal kinase domain (KD) of CHRK1 contained all of the conserved amino acids of serine/threonine protein kinases. Findings suggest that accumulation of chitinase-related RLK1 (*CHRK1*) mRNA was strongly stimulated by fungal pathogen and tobacco mosaic virus (TMV) infection. CHRK1 appears to be localized in membranes in plant cells (Kim et al. 2000). One more leucine-rich repeat (LRR) receptor-like protein kinase (LRPKm1) of *Malus* × *domestica* cv. Florina has been isolated using as a heterologous probe a cloned gene encoding a polygalacturonase-inhibiting protein (PGIP) of *Phaseolus vulgaris* L. comprising a 21-amino-acid signal peptide for secretion, a leucine zipper, 23 LRRs, a putative membrane-spanning region, and a serine/threonine protein kinase domain. LRPKm1 shows homology to the *A. thaliana* receptor-like protein kinase RLK5 (Komjanc et al. 1999). LysM motif proteins are another class of receptors which belongs to non-LRR receptors. LysM receptor-like protein (LysM RLK1/CERK1) is required for the perception of chitin oligomers that are structurally related to peptidoglycans. A mutation in this gene blocked the induction of almost all chitooligosaccharide-responsive genes and led to more susceptibility to fungal pathogen (*Alternaria brassicicola*) but had no effect on infection by a bacterial pathogen (Postel and Kemmerling 2009). The LysM RLK1-mediated chitin signaling pathway is unique, but it may share a conserved downstream pathway with the FLS2/flagellin- and EFR/EF-Tu-mediated signaling pathways. Interestingly, two RLKs with LysM motif have been reported as the receptors of Nod factor, a lipochitin oligosaccharide produced by rhizobial bacteria to establish symbiosis with legume plants (Madsen et al. 2003; Radutoiu et al. 2003). There is a possible evolutionary relationship between the chitin and Nod factor perception mechanisms due to the similarities between their potential receptors and between the signal molecules perceived by them (Wan et al. 2008; Hamel and Beaudoin 2010). Endogenous plant damage-derived signals (DAMPs) are also perceived by PRRs in plants. PEPR1 is a receptor for a wound-released peptide AtPEP1 that triggers weak antifungal activity in plants (Yamaguchi et al. 2006). Recently, Yamaguchi et al. (2010) reported that PEPR2 which is a second receptor for the Pep1 and Pep2 peptides contributes defense responses in *Arabidopsis*. The second class of surface receptors, the RLPs, has been described mainly as mediators of effector recognition (Fritz-Laylin et al. 2005). Plasma membrane RLPs consist of an extracellular (LRR or LysM) domain, a transmembrane domain, but lack a cognate signaling domain in the cytoplasm which is an additional candidate to form receptor complexes. RLPs are also involved in MAMP detection. CEBiP, the high affinity-binding site for fungal chitin, has been identified in rice. It carries two LysM motifs in its extracellular domain but lacks the kinase domain. CEBiP-specific mutant resulted in strong suppression of the chitin-induced generation of reactive oxygen species (ROS) (Kaku et al. 2006). The 22 kDa fungal protein ethylene-inducing xylanase (EIX) activates defense responses independent of its enzymatic activity in many plant species (Furman-Matarasso et al. 1999). Two tomato genes, *LeEix1* and *LeEix2*, have been identified by map-based cloning. They encode proteins with EIX-binding activities when ectopically expressed in plant and mammalian cells (Ron and Avni 2004). The EIX receptors contain a leucine zipper, an extracellular LRR domain, a transmembrane domain, and a C-terminal domain with a mammalian endocytosis signal. The structure of EIX receptors is similar to a family of RLPs, such as tomato Cf2 and Cf9, the plasma membrane R proteins that respond to specific extracellular fungal avirulence (Avr) proteins for effector-triggered immunity. A soluble extracellular protein lacking a transmembrane domain has been identified as the binding site for β-glucans (Fliegmann et al. 2004, 2005). The glucan-binding protein (GBP) binds the heptaglucoside elicitor from oomycetes and has intrinsic endo-β-glucanase activity. It is proposed to act firstly as a glucan hydrolase on heptaglucosides, releasing β-glucans, which are subsequently perceived by a different domain of

GBP. The receptor component that is involved in signal transduction remains to be identified. GBP is predominantly localized to the cytoplasmatic side of the cell wall but also to intracellular vesicles.

MAPK Cascade

MAPKKKs

It is not surprising to find that MAPKKK gene families are frequently the largest of the three having 60 MAPKKKs, which may allow for a diversity of incoming signals from different stimuli to feed into specific MAPK cascade modules. *Arabidopsis* MAPKKKs fall into three main classes: MEKKs, RAF-like, and ZIK-like. The ZIK groups of plant MAPK kinase kinases were originally named after ZIK1, a putative MAPKKK that is associated with the MAPK regulator ZR1 (Wrzaczek and Hirt 2001). However, it is important to note that for only a couple of MAPKKKs and none of the RAF- and ZIK-related kinases, it has been demonstrated that they actually function as MAPKKKs. Therefore, only the MEKK-like genes are referred as MAPKKKs (MAPKKK1–20) and use the nomenclature of RAF (RAF1–48) and ZIK (ZIK1–10) for the two other groups (Jonak et al. 2002). They contain different potential regulatory domains outside the catalytic domain, which means they can be regulated by a variety of upstream signals and then selectively activate MKKs (Menges et al. 2008). It has largest number of members but very few members of this family have assigned biological functions. For example, none of the MAPKKKs in *Arabidopsis* have been shown to function as MAPKK activators in the strict sense which open a possibility that not all of them are true MAPKKKs but are only similar in their amino acid sequence to those having this function (Jonak et al. 2002). Nakagami et al. (2004) reported that *Medicago* OMTK1, which is a KKK1, is involved in oxidative stress-induced cell death, and also some MAPKKKs seem able to serve as scaffold proteins, assembling specific MAPK pathway components into particular modules to prevent cell death and SA/H_2O_2 accumulation (Nakagami et al. 2004). Moreover, silencing of the genes encoding *Nicotiana tabacum* NPK1 (NtNPK1) and *Nicotiana benthamiana* MAPKKKα (NbMAPKKKα) suppresses the N gene-mediated hypersensitive response (HR) induced by the helicase domain of tobacco mosaic virus (TMV) replicase and Pto-mediated HR induced by *Pseudomonas syringae* pv. Tomato (Pst) effector avrPto, respectively (Jin et al. 2002; del Pozo et al. 2004). Recently, Hashimoto et al. (2012) identified that NbMAPKKKα, NbMAPKKKβ, and NbMAPKKKγ form a linear signaling pathway which leads to programmed cell death in *Nicotiana benthamiana*. Mizoguchi et al. (1996) cloned and characterized a cDNA from *Arabidopsis* with high sequence homology to known mammalian MAPKKKs. This kinase was named AtMEKK1 (*Arabidopsis thaliana* ERK kinase kinase 1) and was found to share 46 % sequence similarity to NDR1 from tobacco plants, 42 % to Byr2 from *Schizosaccharomyces pombe,* and 42 % similarity to Bck1 from *Saccharomyces cerevisiae,* all of which are involved in pathogen-induced responses. This was taken as a strong indication that this particular MAPKKK is in fact a part of a signal transduction cascade involved in pathogen resistance in *Arabidopsis*. Oh et al. (2010) reported that tomato 14-3-3 Protein 7 positively regulates immunity-associated programmed cell death by enhancing protein abundance and signaling ability of MAPKKKa, and Frye et al. (2001) reported that MAPKKK gene EDR1 negatively regulates the plant defense responses, including programmed cell death. Melech-Bonfil and Sessa (2010) reported that SlMAPKKKe is a signaling molecule that positively regulates cell death networks associated with plant immunity. AtMEKK1 can interact and activate four different MAPKKs, out of which MKK4 and MKK5 function in pathogen defense (Asai et al. 2002) and MKK2 gets activated during abiotic stress (Teige 2004). MEKK1 can also activate MPK4 in response to signals, such as flg22 or H_2O_2, by interacting with MPK4 shown by yeast two-hybrid assay with its regulatory domain instead of the kinase domain. The functional significance of MPK4 activated by external signals through MEKK1 remains to be elucidated (Teige et al. 2004; Nakagami et al. 2006; Ichimura et al. 1998). Many groups have reported the isolation and characterization of *mekkl* knockout plants (Ichimura et al. 2006; Nakagami et al. 2006; Rodriguez et al. 2007). The *mekk1*-deficient plants display constitutive cell death during the emergence of first pair of true leaves accompanied by the production of H_2O_2, deposition of callose, and activation of pathogenesis-related (PR) genes. Surprisingly, the lethal defect of *mekkl* knockout plants is rescued by the MEKK1 protein without the kinase activity (Suarez-Rodriguez et al. 2007). The results suggest that MEKK1 has both protein kinase (PK)-dependent and PK-independent functions. In another report, Krysan et al. (2002) isolated and analyzed the knockout mutants of the *ANP* family of MAPKKK genes in *Arabidopsis* and showed that the ANP kinases likely are involved in the control of cell division. Similarly, Kovtun et al. (2000) found that the H_2O_2-mediated MAPK cascade initiated by ANP1 in *Arabidopsis* protoplasts likely involves the activation of MPK3 and MPK6. In another study, Eckardt (2002) reported the evidence of *ANP/NPK1* gene to function against various stress responses. Menges et al. (2008) reported that MAPKKK1, MAPKKK2, and MAPKKK12 play a role in mitosis.

MAPKKs

The ten MAPKKs are divided into four groups based on their structures. A common characteristic for all MAPKKs is that they have a putative MAPK docking domain at their N-terminus (Cvetkovska et al. 2005). This domain has the general structure K/R-K/R-K/R-X (1–6)-L-X-L/V/I, and its function is to assist in the binding of the MAPK to the MAPKK (Jonak et al. 2002). The Ser/Thr active site is crucial for the activity of the kinase and comprises an activation loop, which has the general structure S/TXXXXXS/T. MAPKK is activated by MAPKKK through phosphorylation at the Ser/Thr residues (Hirt 2000; Xing et al. 2002). Zhang et al. (2000) have shown that a single MAPKK can interact with and activate more than one MAPK and thus acts as another divergent factor in the module.

AtMKK1 and AtMKK2 have been implicated in biotic and abiotic stress responses as part of a signaling cascade including MEKK1 and MPK4 (Qui et al. 2008). Other studies show that AtMKK1 (subgroup A) seems to mediate cold, drought, and wound signaling (Matsuoka et al. 2002) whereas AtMKK2 mediates cold and salt signaling (Hadiarto et al. 2006). MKK1 also involves in the activation of MPK4 and MPK6 during bacterial or fungal elicitors (Teige et al. 2004). In a report given by Mészáros et al. (2006), MKK1 participates in defense responses to the bacterial elicitor flagellin. Brader et al. (2007) reported that MKK2 plays a role in plant disease resistance as *MKK2-EE* lines are more resistant to *P. syringae* pv. *Tomato* DC3000 and *E. carotovora* subsp. *carotovora* SCC1, but more sensitive to *A. brassicicola.* Recently, You et al. (2007) identified a putative OmMKK1 and OsMKK1 responsive to biotic stresses. The analysis of one *mkkl* mutant has also suggested very complex roles of MKK1 as both a positive regulator in MPK3/4/6 activation by flg22 and a negative regulator in flg22 activation of gene expression (Mészáros et al. 2006). Other MKKs like MKK4, MKK5, MKK7, and MKK9 have been reported to function in hormone, stress, and pathogen signaling networks (Zhang et al. 2007; Xu et al. 2008). Wan et al. (2004) reported a MAPK cascade consisting of MKK4/5-MPK3/6 which might be involved in chitin signaling. Recently, Kaboshi et al. (2010) characterized a novel MAPK cascade in rice (OsMKK4–OsMPK3/OsMPK6) that induces production of diterpenoid phytoalexins. These components get activated by chitin elicitor. Doczi et al. (2007) reported the role of AtMKK3 during pathogenesis. In another study, Takahashi et al. (2007) provide evidence that MKK3, the only member of group B MAPKKs, plays a role in jasmonate (JA)-mediated developmental signaling. Dai et al. (2006) reported that

MKK7 was an inhibitor of polar auxin transport. It is also being involved in activities like cytokinesis and in generating the mobile signal for SAR (Zhang et al. 2007). Recently, Zhang et al. (2012) reported that cotton GhMKK5 affects disease resistance and induces HR-like cell death. There is no information on the function of group D MAPKKs. MKK9, an MAPKK, is an upstream activator of the MPK3 and MPK6 (Xu et al. 2008). Expression of active MKK9 protein in transgenic plants induces the synthesis of ethylene and camalexin through the activation of the endogenous MPK3 and MPK6 kinases. It inhibits hypocotyl elongation in the etiolated seedlings and enhances the sensitivity of transgenic seedlings to salt stress, whereas loss of MKK9 activity reduces salt sensitivity indicating a role for MKK9 in the salt stress response (Zhou et al. 2009). It was hypothesized that MKK7/MKK9 regulates cell death during pathogen defense mechanism (Popescu et al. 2009).

MAPK

In the past decade, gene encoding MAP kinase and other components of MAPK cascades in the signal transduction pathway in plants have been identified. MAPKs are serine/threonine kinases that phosphorylate a variety of substrates including transcription factors, protein kinases, and cytoskeleton-associated proteins (Nakagami et al. 2005) which affect gene expression (Tim and Jordan 2000). Moreover, MAPKs also affect cellular and physiological activities as well as stress responses (Nadarajah et al. 2009). The 23 MAP kinases are grouped into four subfamilies (A–D). Most of the MAPKs characterized so far belong to groups A and B, which include AtMPK3/4/6 as the most studied members (Teige et al. 2004; Asai et al. 2002; Ecker 2004; Lampard et al. 2008; Petersen et al. 2000; Hord et al. 2008; Wang et al. 2008; Qiu et al. 2008b; Bethke et al. 2009). MPK3 and MPK6 function together in a single MAPK cascade because they share common upstream kinases. They are coactivated together and are functionally redundant (Asai et al. 2002; Ren et al. 2002; Wang et al. 2008). MPK3 and MPK6 are orthologous to tobacco (*Nicotiana tabacum*) WIPK and SIPK, respectively (Zhang and Klessig 2001; Ichimura et al. 2002; Ren et al. 2002). Loss- and gain-of-function studies provide genetic evidence supporting a positive role of the MPK3/MPK6 cascade in signaling plant disease resistance (Yang et al. 2001; Asai et al. 2002; Jin et al. 2003; Kroj et al. 2003; Del Pozo et al. 2004; Menke et al. 2004; Beckers et al. 2009). Moreover, many researchers proved that MPK3/MPK6 regulates ethylene production by phosphorylating a subset of ACC synthase (ACS) isoforms (Liu and Zhang 2004; Joo et al. 2008; Han et al. 2010). Ethylene plays important roles in plant defense (Broekaert et al. 2006; Van Loon et al. 2006). In another study, ERF104 (an ethylene response factor) has been reported to be a MPK6 substrate that plays important roles in plant resistance to a nonadapted bacterial pathogen (Bethke et al. 2009). Not only this MPK3/MPK6 cascade is also involved in defense gene activation, reactive oxygen species generation, and hypersensitive response-like cell death (Ren et al. 2002; Kroj et al. 2003; Kim and Zhang 2004; Liu et al. 2007). It was also studied that MPK3/MPK6 cascade plays a positive role in regulating the biosynthesis of camalexin (3-thiazol-2'-yl-indole) (Tsuji et al. 1992). Major phytoalexin in *Arabidopsis* (Ren et al. 2008). Moreover, *Arabidopsis* MPK6 was also shown to be involved in pathogen-induced signaling (Mizoguchi et al. 1993). In *Oryza sativa*, OsMPK3 (previously named as OsMAPK5 in Xiong and Yang 2003), OsMPK4, and OsMPK6 (previously named as OsMPK2 in Kurusu et al. 2005 or OsMAPK6 in Lieberherr et al. 2005) are closely related to AtMPK3, AtMPK4, and AtMPK6, respectively. It has been reported that OsMPK6 is activated by several MAMPs (Kurusu et al. 2005; Lieberherr et al. 2005) and OsMPK3 by blast infection (Xiong and Yang 2003). In the case of wheat, TaMPK3 and TaMPK6 are differentially regulated at multiple levels during compatible disease interactions with necrotrophic fungal pathogen *Mycosphaerella graminicola* (Rudd et al. 2008). Recently, *Arabidopsis* group C MAPKs, including MPK1, MPK2, MPK7, and MPK14, were reported to be activated by MKK3 (Doczi et al. 2007). In another study, Shi et al. (2010) analyzed that cotton GhMPK7 might play a role in pathogen resistance and plant growth and development. On the basis of the phylogenetic analysis and pairwise comparison, it was proposed that the rice genome contains more MAPKs with a TDY phosphorylation motif (11 members) than with a TEY motif (six members). In contrast, the *Arabidopsis* genome contains more MAPKs with a TEY motif (12 members) than with a TDY motif (eight members) (Reyna and Yang 2006). In another study, Cheong et al. (2003) reported that overexpression of OsBWMK1 (also known as OsMPK12) in tobacco resulted in constitutive PR gene expression and enhanced resistance to fungal and bacterial infections. Lalle et al. (2005) shed exciting new insight in maize that ZmMPK6 is able to interact with a 14-3-3 protein and suggest that these data represent the first evidence of the possible involvement of 14-3-3 proteins in the regulation of MAPK cascades in plants. In another study, *Arabidopsis* MPK9 (a group D MAPK) and MPK12 (a group B MAPK) were found to be preferentially expressed in guard cells, share functional redundancy, and function as positive regulators downstream of reactive oxygen species (ROS) in guard cell abscisic acid (ABA) signaling (Jammes et al. 2009). Walia et al. (2009) reported that *Arabidopsis* MPK18 helps in mediating cortical microtubule functions in plant cells. Recently, Shi et al. (2011) proved that ectopic expression of GhMPK16 in *Arabidopsis* results in enhanced disease resistance against bacteria (*P. solanacearum*) and fungi (*C. nicotianae* and *A. alternata*). *Arabidopsis* MPK4 forms another independent MAPK cascade with upstream MKK1/MKK2 and MEKK1 (Petersen et al. 2000; Suarez-Rodriguez et al. 2007; Qiu et al. 2008a). MPK4 is found to be an important negative regulator of systemic-acquired resistance (SAR). They are involved mainly in stress responses and can be activated by a diverse set of stresses, including pathogens, osmotic stress, cold stress, and oxidative stress (Yuasa et al. 2001; Mishra et al. 2006).

Transcriptional Regulatory Networks Governing the Plant Immune Response

The PTI and ETI initiate massive transcriptional reprogramming. The major differences between PTI and ETI appear more quantitative rather than qualitative, suggesting that most pathogens trigger a common/interconnected plant signaling network. The graded transcriptional responses associated with immunity clearly indicate the existence of a complex regulatory network comprised of transcriptional activators and repressors fine-tuning the expression of defense genes. Schematic representation of MAPK cascade components and WRKY transcriptional network in plant and pathogen interaction is shown in Fig. 3.

The nuclear end of the signaling cascades remains poorly studied. Very little is known about the biochemical signals and how they are relayed and linked to nuclear components and specific transcription factors. Equally fragmentary are our insights into the intricate transcriptional machinery responsible for executing proper temporal and spatial control of immune response genes. Although five major families of plant transcription factors, including bZIP, WRKY, MYB, EREBF, and homeodomain protein, have been shown to play roles in the regulation of the plant defense response (Rushton and Somssich 1998), little is known about the exact function or mechanism of individual transcription factors. Generally, it is believed that the WRKY family of transcription factors plays major roles in plant responses to biotic and abiotic stresses and during development (Zhou et al. 2008; Wang et al. 2010).

WRKY proteins are characterized by a stretch of the amino acids tryptophan (W), arginine (R), lysine (K), and tyrosine (Y), followed by a typical zinc-finger domain, and constitute a large class of DNA-binding proteins in plants (Zhang and Wang 2005). The WRKY transcription factor (TF) superfamily consists of 74 and 109 members in *Arabidopsis* and rice *(Oryza sativa),* respectively (Eulgem and Somssich 2007; Ross

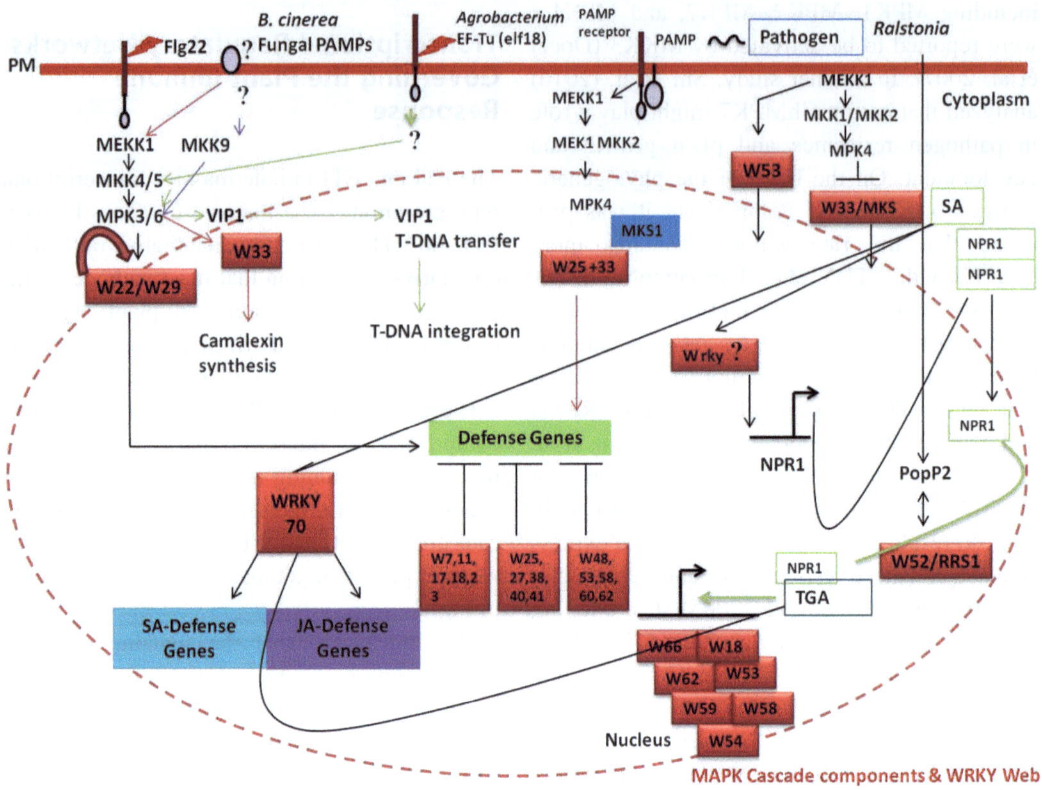

Fig. 3 Overview of MAPK cascade components and WRKY transcriptional network in plant and pathogen interaction

et al. 2007). The domain generally binds to the DNA element termed the W box (C/TTGACT/C), although alternative binding sites have been identified (Sun et al. 2003; Cai et al. 2008; Ciolkowski et al. 2008; Van Verk et al. 2008). In parsley, it was shown that clustering of W boxes is important for a strong transcriptional response (Eulgem et al. 1999; Rushton et al. 1996). The first WRKY-cDNA clone was characterized in 1994 from sweet potato (Ishiguro and Nakamura 1994), and their description as a class of transcription factors followed soon afterwards (Eulgem et al. 2000). WRKY family members are divided into three groups based on the number of WRKY domains and certain features of the zinc-finger-like motifs (Eulgem et al. 2000; Ishihama and Yoshioka 2012). The direct evidence for the involvement of WRKY proteins in disease resistance is limited (Dong et al. 2003; Pandey and Somssich 2009; Rushton et al. 2010). However, WRKY proteins have been found as transcriptional activators at the end of the PAMP signaling cascade involved in the response of *Arabidopsis* to the flagellin fragment flg22. In this case, signal transduction via the MAPK cascade MEKK1–MKK4/MKK5–MPK3/MPK6 leads to the activation of downstream WRKY22 and WRKY29. These WRKY factors are suggested to amplify their expression levels via multiple WRKY-binding sites in their own promoters, thereby creating a positive feedback loop. The induced expression of these WRKY factors would then allow induction of resistance to both bacterial and fungal pathogens (Asai et al. 2002). Recently, Abbruscato et al. (2012) reported that OsWRKY22 plays a role in the resistance response to blast. Soft-rot symptoms caused by infection with the fungal pathogen *Botrytis cinerea* were also effectively suppressed when DMEKK1, MKK4a, or WRKY29 was transiently expressed in leaves (Asai et al. 2002). In another study in *Arabidopsis* against *Botrytis cinerea* infection, WRKY33 gets phosphorylated and it functions downstream of MPK3/MPK6 cascade in reprogramming the expression of camalexin biosynthetic genes, which drives the metabolic flow to camalexin production (Mao et al. 2011). AtWRKY33 functions as a positive regulator of resistance toward the necrotrophic fungi *Alternaria brassicicola* and *Botrytis cinerea* (Birkenbihl et al. 2012) as wrky33 mutants are more susceptible to these pathogens (Zheng et al. 2006; Lippok et al. 2007). In the same studies, it was also demonstrated that WRKY33 is nuclear localized and that it binds to the W box cis-element. Qiu et al. (2008b) also showed that WRKY33 was essential for the induction of camalexin biosynthesis in *Arabidopsis* infected with *Pseudomonas syringae*, as it directly binds to the PAD3 promoter. Recently, Kaboshi et al. (2010) identified OsTGAP1 as transcriptional activators of phytoalexin biosynthesis in *Arabidopsis* and rice. Two closely related WRKY TFs, AtWRKY3 and AtWRKY4, play a positive role in plant resistance toward necrotrophic pathogens, as *Atwrky4, Atwrky3*, and *Atwrky3 wrky4* mutants showed increasing susceptibility toward the fungus *B. cinerea* whereas overexpression of AtWRKY4 enhanced susceptibility toward the biotrophic bacterium *Pseudomonas syringae* (Lai et al. 2008).

Overexpression of GhWRKY15 is involved in defense resistance to both viral and fungal infections, probably through regulating the reactive oxygen species (ROS) system via multiple signaling pathways in tobacco (Yu et al. 2012). It has also been reported that overexpression of grapevine (*Vitis vinifera*) VvWRKY1 resulted in enhanced resistance to the necrotrophic fungi *Alternaria tenuis*, *B. cinerea*, and *Pythium* (Mzid et al. 2007). Recently, Marchive et al. (2013) reported that overexpression of VvWRKY1 in grapevines induces expression of jasmonic acid pathway-related genes and confers higher tolerance to the downy mildew. Similarly, a novel *Gossypium barbadense* WRKY gene, GbWRKY1, was also induced by the infection with *Verticillium dahliae* (Shu-Ling et al. 2012). In other studies, overexpression of OsWRKY13 enhances resistance to the bacterial blight *Xanthomonas oryzae* pv. oryzae (Xoo) and the fungal blast *Magnaporthe grisea*. It exerts its function by activating SA-biosynthesis and SA-response genes while suppressing JA signaling (Qiu et al. 2007, 2008a). Similarly, OsWRKY53 overexpressor lines are more resistant to *M. grisea* and may act as a positive regulator of basal defense (Chujo et al. 2007). Enhanced resistance to *M. grisea* was observed with OsWRKY45 overexpressor lines but not with plants overexpressing OsWRKY19, −62, and −76 (Shimono et al. 2007; Shimono 2012). OsWRKY89 overexpression seems to positively contribute to resistance against fungal blast. Likewise, WRKY41 overexpression in *Arabidopsis* exhibited enhanced resistance to *Pseudomonas syringae* but increased susceptibility to *Erwinia carotovora* (Higashi et al. 2008). In another study, overexpression of WRKY28 in *Arabidopsis* protoplasts leads to induction of a *β-GLUCURONIDASE* (*GUS*) reporter gene under control of the 1 kb *ICS1* upstream promoter region as well as elevated levels of endogenous *ICS1* mRNA. This may indicate a link between PAMP signaling and the biosynthesis of SA (Navarro et al. 2004).

Transcripts of GhWRKY3 are enhanced after infection with *Rhizoctonia solani*, *Colletotrichum gossypii,* and *Fusarium oxysporum* f. sp. vasinfectum, and it might play an important role in plant defense responses (Guo et al. 2011). Constitutive expression of rice WRKY30 gene increases the endogenous jasmonic acid accumulation, PR gene expression, and resistance to fungal pathogens in rice (Peng et al. 2012). WRKY62 also acts in the cross talk between SA and JA signaling by repressing downstream JA targets such as *LOX2* and *VSP2* (Kim et al. 2008; Mao et al. 2007). Both WRKY18 and WRKY53 are positive regulators of *PR* gene expression and SAR. Functional WRKY18 is required for full induction of SAR and is linked to the activation of *PR-1* (Wang et al. 2006). Together with WRKY18, WRKY40 and WRKY60 play partly redundant roles in regulating disease resistance. These three

WRKY proteins can interact physically and functionally in their responses to different microbial pathogens. While WRKY18 enhances resistance against *Ps. syringae*, co-expression of WRKY40 or WRKY60 renders plants more susceptible to this pathogen (Xu et al. 2006a). Recently, Dang et al. (2012) reported that CaWRKY40, a WRKY protein of pepper, plays an important role in the regulation of tolerance to heat stress and resistance to *Ralstonia solanacearum* infection, whereas Wang et al. (2013) reported that CaWRKY58 negatively regulates resistance to *Ralstonia solanacearum* infection. In another study, *Capsicum annuum* WRKY transcription factor d (CaWRKYd) regulates hypersensitive response and defense response upon tobacco mosaic virus infection (Huh et al. 2012). WRKY46 is a transcription factor that is rapidly induced downstream of avirulence effectors. These results suggest an involvement of WKRY46 in the signaling cascade of avirulence effector recognition and the subsequent accumulation of SA (He et al. 2006). AtWRKY52 (also designated RRS1) is a novel protein comprising structural features of NB-LRR-type R gene products and a WRKY domain. It is present in the nucleus and interacts with PopP2 effector of *Ralstonia solanacearum.* If plants are challenged with strains of *R. solanacearum* that lack the *popP2* gene, they are highly susceptible to this pathogen. This clearly indicates the importance of WRKY52 in resistance against this pathogen (Deslandes et al. 2002; Caplan et al. 2008; Liu and Coaker 2008).

AtWRKY70 acts at a convergence point determining the balance between SA- and JA-dependent defense pathways as well as being required for R gene-mediated resistance (Li et al. 2006; Knoth et al. 2007). WRKY70 and the functional homolog WRKY54 have dual roles in SA-mediated gene expression and resistance. On high accumulation of SA, WRKY54 and WRKY70 act as negative regulators of SA biosynthesis, probably by direct negative regulation of *ICS1*. Besides this negative role, they activate other SA-regulated genes (Kalde et al. 2003; Wang et al. 2006). Loss of AtWRKY70 function rendered plants susceptible to the bacteria *Erwinia carotovora* and *Pseudomonas syringae* as well as the fungi *Erysiphe cichoracearum* and *Botrytis cinerea* (Li et al. 2004, 2006;,Wang et al. 2006). Recently, Shim et al. (2013) showed that AtMYB44 regulates WRKY70 expression and modulates antagonistic interaction between salicylic acid and jasmonic acid signaling. Sometimes, WRKY transcription factor gets activated via targeted degradation of bound suppressors, as it was seen in the MAPK cascade (MEKK1–MEK1/MKK2–MAPK4), which was induced by challenge inoculation with *Ps. syringae* or treatment with flg22. After phosphorylation of MKS1, WRKY33 gets released into the nucleus to initiate positive regulation of JA-induced defense genes and negative regulation of SA-related defense genes. The barley R protein Mildew a (MLA) appears to interfere with the PAMP-inducible repressors of basal resistance HvWRKY1 and HvWRKY2. In this manner, the repressor effect of the PAMP-induced WRKY genes is derepressed, thereby triggering basal defense responses (Shen et al. 2007). Many WRKY TFs act as negative regulators of defense signaling, including AtWRKY7, −11, −17, −18, −23, −25, −27, −38, −40, −41, −48, −53, −58, −60, and −62 (Andreasson et al. 2005; Brodersen et al. 2006; Higashi et al. 2008; Journot-Catalino et al. 2006; Qiu et al. 2008a; Kim et al. 2008).

Conclusion

In this chapter, we summarize the recent information regarding the receptors of MAMPs and components of MAPK cascade along with the transcriptional reprogramming in plant–pathogen interaction. Plants have many proteins that act as pattern recognition receptors (PRRs) at the cell surface or within the cytoplasm. They have a crucial role in monitoring the microbial communities according to their molecular patterns and thereby control pathogen infection. In recent years, a number of exciting studies have revived attention to MAMP perception and have provided a further understanding of plant immunity. Distinct MAMPs seem to elicit largely overlapping immune responses, and MAMP signaling employs at least two MAP kinase cascades. MAP kinase components reflect their central role in plant defense. Therefore it is essential to understand the molecular processes connecting transient and short-term MAPK cascade activation within 1–60 min to long-term responses observed in mutants or after 3–24 h, or even a few days of treatment, involving indirect, divergent, and peripheral pathways leading to immunity against broad-spectrum microbes, herbivores, and pathogens, it is also essential to determine the upstream receptors that monitor the stimuli as well as the downstream effectors that regulate the responses. A lot of effort is still required to uncover in detail each MAP kinase module to understand the complexity of the signal transduction pathways. Network modeling has great potential power to predict and manipulate plant protections against diverse pathogens in a variety of environments. To reach such practical goals in agricultural improvement and environmental protection, comprehensive and accurate data sets are a prerequisite.

References

Abbruscato P, Nepusz T, Mizzi L, Del Corvo M, Morandini P, Fumasoni I, Michel C, Paccanaro A, Guiderdoni E, Schaffrath U, Morel JB, Piffanelli P, Faivre-Rampant O (2012) OsWRKY22, a monocot WRKY gene, plays a role in the resistance response to blast. Mol Plant Pathol 13:828–841

Andreasson E, Jenkins T, Brodersen P, Thorgrimsen S, Petersen NH, Zhu S, Qiu JL, Micheelsen P, Rocher A, Petersen M (2005) The MAP kinase substrate MKS1 is a regulator of plant defense responses. EMBO J 24:2579–2589

Asai T, Tena G, Plotnikova J, Willmann MR, Chiu WL, Gomez-Gomez L, Boller T, Ausubel FM, Sheen J (2002) MAP kinase signalling cascade in *Arabidopsis* innate immunity. Nature 415:977–983

Beckers GJM, MI J, Liu Y, Underwood WR, He SY, Zhang S, Conrath U (2009) Mitogen-activated protein kinases 3 and 6 are required for full priming of stress responses in *Arabidopsis thaliana*. Plant Cell 21:944–953

Bethke G, Unthan T, Uhrig JF, Poschi Y, Gust AA, Scheel D, Lee J (2009) Flg22 regulates the release of an ethylene response factor substrate from MAP kinase 6 in *Arabidopsis* via ethylene signaling. Proc Natl Acad Sci USA 106:8067–8072

Birkenbihl RP, Diezel C, Somssich IE (2012) *Arabidopsis* WRKY33 is a key transcriptional regulator of hormone and metabolic responses towards *Botrytis cinerea* infection. Plant Physiol 159:266–285

Boller T (1995) Chemoperception of microbial signals in plant cells. Annu Rev Plant Physiol Plant Mol Biol 46:189–214

Boller T, He SY (2009) Innate immunity in plants: an arm race between pattern recognition in plants and effectors in microbial pathogens. Science 324:742–744

Brader G, Djamei A, Teige M, Palva ET, Hirt H (2007) The MAP kinase kinase MKK2 affects disease resistance in *Arabidopsis*. Mol Plant Microbe Interact 20:589–596

Brodersen P, Petersen M, Bjørn Nielsen H, Zhu S, Newman MA, Shokat KM, Rietz S, Parker J, Mundy J (2006) *Arabidopsis* MAP kinase 4 regulates salicylic acid- and jasmonic acid/ethylene-dependent responses via EDS1 and PAD4. Plant J 47:532–546

Broekaert WF, Delaure SL, De Bolle MFC, Cammue BPA (2006) The role of ethylene in host-pathogen interactions. Annu Rev Phytopathol 44:393–416

Cai M, Qiu D, Yuan T, Ding X, Li H, Duan L, Xu C, Li X, Wang S (2008) Identification of novel pathogen-responsive cis-elements and their binding proteins in the promoter of OsWRKY13, a gene regulating rice disease resistance. Plant Cell Environ 31:86–96

Caplan J, Padmanabhan M, Dinesh-Kumar SP (2008) Plant NB-LRR immune receptors: from recognition to transcriptional reprogramming. Cell Host Microbe 3:126–135

Cardinale F, Jonak C, Ligterink W, Niehaus K, Boller T, Hirt H (2000) Differential activation of four specific MAPK pathways by distinct elicitors. J Biol Chem 275:36734–36740

Chen Z, Gibson T, Robinson B, Silvestrol R, Pearson G, Xu B, Wright A, Vandrabilt C, Cobb MH (2001) MAP kinases. Chem Rev 101:2449–2476

Cheong YH, Moon BC, Kim JK, Kim CY, Kim MC, Kim IH, Park CY, Kim JC, Park BO, Koo SC et al (2003) BWMK1, a rice mitogen-activated protein kinase, locates in the nucleus and mediates pathogenesis-related gene expression by activation of a transcription factor. Plant Physiol 132:1961–1972

Chinchilla D, Bauer Z, Regenass M, Boller T, Felix G (2006) The *Arabidopsis* receptor kinase FLS2 binds flg22 and determines the specificity of flagellin perception. Plant Cell 18:465–476

Chisholm ST, Coaker G, Day B, Staskawicz BJ (2006) Host- microbe interactions: shaping the evolution of the plant immune response. Cell 124:803–814

Chujo T, Takai R, Akimoto-Tomiyama C, Ando S, Minami E, Nagamura Y, Kaku H, Shibuya N, Yasuda M, Nakashita H (2007) Involvement of the elicitor-induced gene OsWRKY53 in the expression of defense-related genes in rice. Biochim Biophys Acta 1769:497–505

Ciolkowski I, Wanke D, Birkenbihl R, Somssich I (2008) Studies on DNA binding selectivity of WRKY transcription factors lend structural clues into WRKY-domain function. Plant Mol Biol 68:81–92

Cvetkovska M, Ramptitsch C, Bykova N, Xing T (2005) Genomic analysis of MAPK Cascade in *Arabidopsis* defense responses. Plant Mol Biol Rep 23:331–343

Dai Y, Wang H, Li B, Huang J, Liu X, Zhou Y, Mou Z, Li J (2006) Increased expression of MAP KINASE KINASE7 causes deficiency in polar auxin transport and leads to plant architectural abnormality in *Arabidopsis*. Plant Cell 18:308–320

Dang FF, Wang YN, Yu L, Eulgem T, Lai Y, Liu ZQ, Wang X, Qiu AL, Zhang TX, Lin J, Chen YS, Guan DY, Cai HY, Mou SL, He SL (2012) CaWRKY40, a WRKY protein of pepper, plays an important role in the regulation of tolerance to heat stress and resistance to *Ralstonia solanacearum* infection. Plant Cell Environ 36(4):757–774

Daxberger A, Nemak A, Mithofer A, Fliegmann J, Ligterink W, Hirt H, Ebel J (2006) Activation of members of a MAPK module in beta-glucan elicitor mediated non-host resistance of soybean. Planta 225 (6):1559–1571. doi:10.1007/s00425-006-0442-6

Del Pozo O, Pedley KF, Martin GB (2004) MAPKKKα is a positive regulator of cell death associated with both plant immunity and disease. EMBO J 15:3072–3082

Desikan R, Hancock TJ, Ichimura K, Shinozaki K, Neill ST (2001) Harpin induces activation of the *Arabidopsis* mitogen activated protein kinase AtMpk4 and AtMpk6. Plant Physiol 126:1579–1587

Deslandes L, Olivier J, Theulieres T, Hirsch J, Feng DX, Bittner-Eddy P, Beynon J, Marco Y (2002) Resistance to *Ralstonia solanacearum* in *Arabidopsis thaliana* is conferred by the recessive RRS1-R gene, a member of a novel family of resistance genes. Proc Natl Acad Sci USA 99:2404–2409

Doczi R, Brader G, Pettko-Szandtner A, Rajh I, Djamei A, Pitzschke A, Teige M, Hirt H (2007) The *Arabidopsis* mitogen-activated protein kinase kinase MKK3 is upstream of group C mitogen-activated protein kinases and participates in pathogen signaling. Plant Cell 19:3266–3279

Dong J, Chen C, Chen Z (2003) Expression profiles of the *Arabidopsis* WRKY gene superfamily during plant defense response. Plant Mol Biol 51:21–37

Eckardt NA (2002) Good things come in threes: a trio of triple kinases essential for cell division in *Arabidopsis*. Plant Cell 14:965–967

Ecker JR (2004) Reentry of the ethylene MPK6 module. Plant Cell 16:3169–3174

Eulgem T, Somssich IE (2007) Networks of WRKY transcription factors in defense signaling. Curr Opin Plant Biol 10:366–371

Eulgem T, Rushton PJ, Schmelzer E, Hahlbrock K, Somssich IE (1999) Early nuclear events in plant defense: rapid gene activation by WRKY transcription factors. EMBO J 18:4689–4699

Eulgem T, Rushton PJ, Robatzek S, Somssich IE (2000) The WRKY superfamily of plant transcription factors. Trends Plant Sci 5:199–206

Felix G, Duran JD, Volko S, Boller T (1999) Plants have a sensitive perception system for the most conserved domain of bacterial flagellin. Plant J 18:265–276

Fellbrich G, Romanski A, Varet A, Blume B, Brunner F, Engelhardt S, Felix G, Kemmerling B, Krzymowska M, Nürnberger T (2002) NPP1, a phytophthora-associated trigger of plant defense in parsley and *Arabidopsis*. Plant J 32:375–390

Fliegmann J, Mithofer A, Wanner G, Ebel J (2004) An ancient enzyme domain hidden in the putative beta-glucan elicitor receptor of soybean may play an active part in the perception of pathogen-associated molecular patterns during broad host resistance. J Biol Chem 279:1132–1140

Fliegmann J, Montel E, Djulic A, Cottaz S, Driguez H, Ebel J (2005) Catalytic properties of the bifunctional soybean beta-glucan-binding protein, a member of family 81 glycoside hydrolases. FEBS Lett 579:6647–6652

Fritz-Laylin LK, Krishnamurthy N, Tor M, Sjolander KV, Jones JD (2005) Phylogenomic analysis of the receptor-like proteins of rice and *Arabidopsis*. Plant Physiol 138:611–623

Frye CA, Tang D, Innes RW (2001) Negative regulation of defense responses in plants by a conserved MAPKK kinase. Proc Natl Acad Sci USA 98:373–378

Furman-Matarasso N, Cohen E, Du Q, Chejanovsky N, Hanania U, Avni A (1999) A point mutation in the ethylene-inducing xylanase elicitor inhibits the B-1-4-endoxylanase activity but not the elicitation activity. Plant Physiol 121:345–352

Gomez-Gomez L, Boller T (2000) An LRR receptor-like kinase involved in the perception of the bacterial elicitor flagellin in *Arabidopsis*. Mol Cell 5:1003–1011

Guo R, Yu F, Gao Z, An H, Cao X, Guo X (2011) GhWRKY3, a novel cotton (*Gossypium hirsutum* L.) WRKY gene, is involved in diverse stress responses. Mol Biol Rep 38:49–58

Hadiarto T, Nanmori T, Matsuoka D, Iwasaki T, Sato K (2006) Activation of *Arabidopsis* MAPK kinase kinase (AtMEKK1) and induction of AtMEKK1-AtMEK1 pathway by wounding. Planta 223:708–713

Hamel LP, Beaudoin N (2010) Chitooligosaccharide sensing and downstream signaling: contrasted outcomes in pathogenic and beneficial plant-microbe interactions. Planta 232:787–806

Han L, Li GJ, Yang KY, Mao G, Wang R, Liu Y, Zhang S (2010) Mitogen-activated protein kinase 3 and 6 regulate Botrytis cinerea-induced ethylene production in *Arabidopsis*. Plant J 64:114–127

Hann DR, Rathjen JP (2007) Early events in the pathogenicity of *Pseudomonas syringae* on *Nicotiana benthamiana*. Plant J 49:607–618

Hashimoto M, Komatsu K, Maejima K, Okano Y, Shiraishi T et al (2012) Identification of three MAPKKKs forming a linear signaling pathway leading to programmed cell death in. BMC Plant Biol 12(1):103

He ZH, Fujiki M, Kohorn BD (1996) A cell wall associated receptor-like kinase. J Biol Chem 271:19789–19793

He ZH, He D, Kohorn BD (1998) Requirement for the induced expression of a cell wall associated receptor kinase for survival during the pathogen response. Plant J 14:55–63

He P, Shan L, Lin NC, Martin GB, Kemmerling B, Nürnberger T, Sheen J (2006) Specific bacterial suppressors of PAMP signaling upstream of MAPKKK in *Arabidopsis* innate immunity. Cell 125:563–575

Higashi K, Ishiga Y, Inagaki Y, Toyoda K, Shiraishi T, Ichinose Y (2008) Modulation of defense signal transduction by flagellin-induced WRKY41 transcription factor in *Arabidopsis thaliana*. Mol Genet Genomics 279:303–312

Hirt H (2000) Connecting oxidative stress, auxin, and cell cycle regulation through a plant MAPK pathway. Proc Natl Acad Sci USA 97:2405–2407

Hirt H (2009) Connecting oxidative stress, auxin, and cell cycle regulation through a plant MAPK pathway. Proc Natl Acad Sci USA 97:2405–2407

Hord CLH, Sun Y-J, Pillitteri LJ, Torii KU, Wang H, Zhang S, Ma H (2008) Regulation of Arabidopsis early anther development by the mitogen-activated protein kinases, MPK3 and MPK6, and the ERECTA and related receptor-like kinases. Mol Plant 1:645–658

Huh SU, Choi LM, Lee GJ, Kim YJ, Paek KH (2012) *Capsicum annuum* WRKY transcription factor d (CaWRKYd) regulates hypersensitive response and defense response upon Tobacco mosaic virus infection. Plant Sci 197:50–58

Ichimura K, Casais C, Peck SC, Shinozaki K, Shirasu K (2006) MEKK1 is required for MPK4 activation and regulates tissue-specific and temperature dependent cell death in *Arabidopsis*. J Biol Chem 281 (48):36969–36976

Ichimura K, Mizoguchi T, Irie K, Morris P, Giraudat J, Matsumoto K, Shinozaki K (1998) Isolation of AtMEKK1 (A MAP kinase kinase kinase)-Interacting proteins and analysis of a MAPK kinase cascade in *Arabidopsis*. Biochem Biophys Res Commun 253:532–543

Ichimura K, Tena G, Henry Y, Zhang Z, Hirt H, Wilson C, Morris P, Mundy J, Innes R, Ecker J (2002) Mitogen-activated protein kinases in plants: a new nomenclature. Trends Plant Sci 7:301–308

Ishiguro S, Nakamura K (1994) Characterization of a cDNA encoding a novel DNA-binding protein, SPF1, that recognizes SP8 sequences in the 5' upstream regions of genes coding for sporamin and β -amylase from sweet potato. Mol Gen Genet 244:563–571

Ishihama N, Yoshioka H (2012) Post-translational regulation of WRKY transcription factors in plant immunity. Curr Opin Plant Biol 15:431–437

James TY, Litvintseva AP, Vilgalys R, Morgan JAT, Taylor JW et al (2009) Rapid global expansion of the fungal disease chytridiomycosis into declining and healthy amphibian populations. PLoS Pathog 5 (5):e1000458. doi:10.1371/journal.ppat.1000458

Jin H, Axtell MJ, Dahlbeck D, Ekwenna O, Zhang S, Staskawicz B, Baker B (2002) NPK1, an MEKK1-like mitogen-activated protein kinase kinase kinase, regulates innate immunity and development in plants. Dev Cell 2:291–297

Jin H, Liu Y, Yang KY, Kim CY, Baker B, Zhang S (2003) Function of a mitogen-activated protein kinase pathway in N gene-mediated resistance in tobacco. Plant J 33:719–731

Jonak C, Okresz L, Bogrw L, Hirt H (2002) Complexity, cross talk and integration of plant MAP kinase signalling. Curr Opin Plant Biol 5:415–424

Jones JD, Dangl JL (2006) The plant immune system. Nature 444:323–329

Joo S, Liu Y, Lueth A, Zhang S (2008) MAPK phosphorylation-induced stabilization of ACS6 protein is mediated by the non-catalytic C-terminal domain, which also contains the cis-determinant for rapid degradation by the 26S proteasome pathway. Plant J 54:129–140

Journot-Catalino N, Somssich IE, Roby D, Kroj T (2006) The transcription factors WRKY11 and WRKY17 act as negative regulators of basal resistance in *Arabidopsis thaliana*. Plant Cell 18:3289–3302

Kaboshi MK, Okada K, Kurimoto L, Murakami S, Umezawa T et al (2010) A rice fungal MAMP-responsive MAPK cascade regulates metabolic flow to antimicrobial metabolite synthesis. Plant J 63:599–612

Kaku H, Nishizawa Y, Ishii-Minami N, Akimoto-Tomiyama C, Dohmae N, Takio K, Minami E, Shibuya N (2006) Plant cells recognize chitin fragments for defense signaling through a plasma membrane receptor. Proc Natl Acad Sci USA 103:11086–11091

Kalde M, Barth M, Somssich IE, Lippok B (2003) Members of the *Arabidopsis* WRKY group III transcription factors are part of different plant defense signaling pathways. Mol Plant Microbe Interact 16:295–305

Kim YS, Lee JH, Yoon GM, Cho HS, Park S-W, Suh MC, Choi D, Ha HJ, Liu JR, Pai H-S (2000) CHRK1, a Chitinase-related receptor-like kinase in tobacco. Plant Physiol 123:905–915

Kim KC, Lai Z, Fan B, Chen Z (2008) *Arabidopsis* WRKY38 and WRKY62 transcription factors interact with histone deacetylase 19 in basal defense. Plant Cell 20:2357–2371

Kim CY, Zhang S (2004) Activation of a mitogen-activated protein kinase cascade induces WRKY family of transcription factors and defense genes in tobacco. Plant J 38:142–151

Knoth C, Ringler J, Dangl JL, Eulgem T (2007) *Arabidopsis* WRKY70 is required for full RPP4-mediated disease resistance and basal defense against *Hyaloperonospora parasitica*. Mol Plant-Microbe Interact 20:120–128

Kohorn B (2000) Plasma membrane-cell wall contacts. Plant Physiol 124:249–262

Komjanc M, Festi S, Rizzotti L, Cattivelli L, Cervone V, De Lorenzo G (1999) A leucine-rich repeat receptor-like protein kinase (LRPKm1) gene is induced in *Malus domestica* by *Venturia inaequalis* infection and salicylic acid treatment. Plant Mol Biol 40:945–995

Kovtun Y, Chium WL, Tena G, Sheen J (2000) Functional analysis of oxidative stress-activated mitogen-activated protein kinase cascade in plants. Proc Natl Acad Sci USA 97:2940–2945

Kroj T, Rudd JJ, Nurnberger T, Gabler Y, Lee J, Scheel D (2003) Mitogen activated protein kinases play an essential role in oxidative burst independent expression of pathogenesis-related genes in parsley. J Biol Chem 278:2256–2264

Krysan PJ, Jester PJ, Gottwald JR, Sussman MR (2002) An *Arabidopsis* mitogen-activated protein kinase kinase kinase gene family encodes essential positive regulators of cytokinesis. Plant Cell 14:1109–1120

Kurusu T, Yagala T, Miyao A, Hirochika H, Kuchitsu K (2005) Identification of a putative voltage-gated Ca^{2+} channel as a key regulator of elicitor-induced hypersensitive cell death and mitogen-activated protein kinase activation in rice. Plant J 42:798–809

Lai Z, Vinod KM, Zheng Z, Fan B, Chen Z (2008) Roles of *Arabidopsis* WRKY3 and WRKY4 transcription factors in plant responses to pathogens. BMC Plant Biol 8:68

Lalle M, Visconti S, Marra M, Camoni L, Velasco R, Aducci P (2005) ZmMPK6, a novel maize MAP kinase that interacts with 14–3-3 proteins. Plant Mol Biol 59:713–722

Lampard GR, MacAlister CA, Bergmann DC (2008) Arabidopsis stomatal initiation Is controlled by MAPK-mediated regulation of the bHLH SPEECHLESS. Science 322:1113–1116

Lee SW, Han SW, Bartley LE, Ronald PC (2006) Unique characteristics of *Xanthomonas oryzae* pv. oryzae AvrXa21 and implications for plant innate immunity. Proc Natl Acad Sci USA 103:18395–18400

Lee HB, Kim CJ, Yu SH (2001) *Alternaria tenuissima* causes strawberry fruit rot in Korea. Plant Dis 85:563

Li J, Brader G, Palva ET (2004) The WRKY70 transcription factor: a node of convergence for jasmonate-mediated and salicylate-mediated signals in plant defense. Plant Cell 16:319–331

Li J, Brader G, Kariola T, Palva ET (2006) WRKY70 modulates the selection of signaling pathways in plant defense. Plant J 46:477–491

Lieberherr D, Thao NP, Nakashima A, Umemura K, Kawasaki T, Shimamoto K (2005) A sphingolipid elicitor-inducible mitogen-activated protein kinase is regulated by the small GTPase OsRac1 and heterotrimeric G-protein in rice. Plant Physiol 138:1644–1652

Ligterink W, Kroj T, zur Nieden U, Hirt H, Scheel D (1997) Receptor-mediated activation of a MAP kinase in pathogen defense of plants. Science 276:2054–2057

Lippok B, Birkenbihl RP, Rivory G, Brummer J, Schmelzer E, Logemann E, Somssich IE (2007) Expression of AtWRKY33 encoding a pathogen- or PAMP-responsive WRKY transcription factor is regulated by a composite DNA motif containing W box elements. Mol Plant Microbe Interact 20:420–429

Liu J, Coaker G (2008) Nuclear trafficking during plant innate immunity. Mol Plant 1:411–422

Liu Y, Ren D, Pike S, Pallardy S, Gassmann W, Zhang S (2007) Chloroplast-generated reactive oxygen species are involved in hypersensitive response-like cell death mediated by a mitogen-activated protein kinase cascade. Plant J 51:941–954

Liu Y, Zhang S (2004) Phosphorylation of 1-aminocyclopropane-1-carboxylic acid synthase by MPK6, a stress-responsive mitogen-activated protein kinase, induces ethylene biosynthesis in *Arabidopsis*. Plant Cell 16:3386–3399

Madsen EB, Madsen LH, Radutoiu S, Olbryt M, Rakwalska M, Szczyglowski K, Sato S, Kaneko T, Tabata S, Sandal N, Stougaard J (2003) A receptor kinase gene of the LysM type is involved in legume perception of rhizobial signals. Nature 425:637–640

Maleck K, Levine A, Eulgem T, Morgan A, Schmid J, Lawton KA, Dangl JL, Dietrich RA (2000) The transcriptome of *Arabidopsis thaliana* during systemic acquired resistance. Nat Genet 26:403–410

Mao P, Duan M, Wei C, Li Y (2007) WRKY62 transcription factor acts downstream of cytosolic NPR1 and negatively regulates jasmonate responsive gene expression. Plant Cell Physiol 48:833–842

Mao G, Meng X, Liu Y, Zheng Z, Chen Z, Zhang S (2011) Phosphorylation of a WRKY transcription factor by two pathogen-responsive MAPKs drives phytoalexin biosynthesis in *Arabidopsis*. Plant Cell 23:1639–1653

Marchive C, Léon C, Kappel C, Coutos-Thévenot P, Corio-Costet MF, Delrot S, Lauvergeat V (2013) Over-expression of VvWRKY1 in grapevines induces expression of jasmonic acid pathway-related genes and confers higher tolerance to the Downy mildew. PLoS One 8:e54185

Matsuoka D, Nanmori T, Sato K, Fukami Y, Kikkawa U, Yasuda T (2002) Activation of AtMEK1, an *Arabidopsis* mitogen-activated protein kinase kinase, in vitro and in vivo: analysis of active mutants expressed in *E. coli* and generation of the active form in stress response in seedlings. Plant J 29:637–647

Melech-Bonfil S, Sessa G (2010) Tomato MAPKKKε is a positive regulator of cell-death signaling networks associated with plant immunity. Plant J 64:379–391

Menges M, Doczi R, Okresz L, Morandini P, Mizzi L, Soloviev M, Murray J, Bogre L (2008) Comprehensive gene expression atlas for the *Arabidopsis* MAP kinase signalling pathways. New Phytol 179:643–662

Menke FL, Van Pelt JA, Pieterse CM, Klessig DF (2004) Silencing of the mitogen-activated protein kinase MPK6 compromises disease resistance in *Arabidopsis*. Plant Cell 16:897–907

Mészáros T, Helfer A, Hatzimasoura E, Magyar Z, Serazetdinova L, Rios G, Bardóczy V, Teige M, Koncz C, Peck S et al (2006) The *Arabidopsis* MAP kinase kinase MKK1 participates in defence responses to the bacterial elicitor flagellin. Plant J 48:485–498

Mishra NS, Tuteja R, Tuteja N (2006) Signaling through MAP kinase networks in plants. Arch Biochem Biophys 452:55–68

Mizoguchi T, Hayashida N, Yamaguchi-Shinozaki K, Kamada H, Schinozaki K (1993) AtMPKs: a gene family of MAP Kinase in *Arabidopsis thaliana*. FEBS Lett 336:440–444

Mizoguchi T, Irie K, Hirayama T, Hayashida N, Yamaguchi-Shinozaki K, Matsumoto K, Shinozaki K (1996) A gene encoding a mitogen-activated protein kinase kinase kinase is induced simultaneously with genes for a mitogen-activated protein kinase and an S6 ribosomal protein kinase by touch, cold, and water stress in *Arabidopsis thaliana*. Proc Natl Acad Sci USA 93:765–769

Morris PC (2001) MAP kinase signal transduction pathways in plants. New Phytol 151:67–89

Mzid R, Marchive C, Blancard D, Deluc L, Barrieu F, Corio-Costet MF, Drira N, Hamdi S, Lauvergeat V (2007) Overexpression of VvWRKY2 in tobacco enhances broad resistance to necrotrophic fungal pathogens. Physiol Plant 131:434–447

Nadarajah K, Kasirn NM, Fui VV (2009) Function of mitogen protein kinase gene in biotic stress. Asian J Plant Sci 85:191–198

Nakagami H, Kiegeri S, Hirt H (2004) OMTK1, a novel MAPKKK, channels oxidative stress signalling through direct MAPK interaction. J Biol Chem 279:26959–26966

Nakagami H, Pitzschke A, Hirt H (2005) Emerging MAP kinase pathways in plant stress signaling. Trends Plant Sci 10(7):1360–1385

Nakagami H, Soukupová H, Schikora A, Zarsky V, Hirt H (2006) A mitogen-activated protein kinase kinase kinase mediates reactive oxygen species homeostasis in *Arabidopsis*. J Biol Chem 281:38697–38704

Navarro L, Zipfel C, Rowland O, Keller I, Robatzek S, Boller T, Jones JDG (2004) The transcriptional innate immune response to flg22. Interplay and overlap with Avr gene-dependent defense responses and bacterial pathogenesis. Plant Physiol 135:1113–1128

Oh CS, Pedley KF, Martin GB (2010) Tomato 14-3-3 protein 7 positively regulates immunity-associated programmed cell death by enhancing protein abundance and signaling ability of MAPKKKa. Plant Cell 22:260–272

Pandey SP, Somssich IE (2009) The role of WRKY transcription factors in plant immunity. Plant Physiol 150:1648–1655

Peck SC, Nuhse TS, Hess D, Iglesias A, Meins F, Boller T (2001) Directed proteomics identifies a plant specific protein rapidly phosphorylated in response to bacterial and fungal elicitors. Plant Cell 13:1467–1475

Peng X, Hu Y, Tang X, Zhou P, Deng X, Wang H, Guo Z (2012) Constitutive expression of rice WRKY30 gene increases the endogenous jasmonic acid accumulation, PR gene expression and resistance to fungal pathogens in rice. Planta 236:1485–1498

Petersen M, Brodersen P, Naested H, Andreasson E, Lindhart U, Johansen B, Nielsen HB, Lacy M, Austin MJ, Parker JE (2000) *Arabidopsis* map kinase 4 negatively regulates systemic acquired resistance. Cell 103:1111–1120

Popescu SC, Popescu GV, Snyder M, Dinesh SP (2009) Integrated analysis of co-expressed MAP kinase substrates in *Arabidopsis thaliana*. Plant Signal Behav 4(6):524–527

Postel S, Kemmerling B (2009) Plant systems for recognition of pathogen-associated molecular patterns. Semin Cell Dev Biol 20:1025–1031

Qiu D, Xiao J, Ding X, Xiong M, Cai M, Cao Y, Li X, Xu C, Wang S (2007) OsWRKY13 mediates rice disease resistance by regulating defense related genes in salicylate- and jasmonate-dependent signaling. Mol Plant Microbe Interact 20:492–499

Qiu D, Xiao J, Xie W, Liu H, Li X, Xiong L, Wang S (2008a) Rice gene network inferred from expression profiling of plants overexpressing OsWRKY13, a positive regulator of disease resistance. Mol Plant 1:538–551

Qiu JL et al (2008b) *Arabidopsis* MAP kinase 4 regulates gene expression through transcription factor release in the nucleus. EMBO J 27:2214–2221

Qiu JL, Zhou L, Yun BW, Nielsen HB, Fiil BK, Petersen K, MacKinlay J, Loake GJ, Mundy J, Morris PC (2008c) *Arabidopsis* mitogen-activated protein kinase kinases MKK1 and MKK2 have overlapping functions in defense signaling mediated by MEKK1, MPK4, and MKS1. Plant Physiol 148:212–222

Qutob D et al (2006) Phytotoxicity and innate immune responses induced by Nep1-like proteins. Plant Cell 18:3721–3744

Radutoiu S, Madsen LH, Madsen EB, Felle HH, Umehara Y, Gronlund M, Sato S, Nakamura Y, Tabata S, Sandal N, Stougaard J (2003) Plant recognition of symbiotic bacteria requires two LysM receptor-like kinases. Nature 425:585–592

Ren D, Liu Y, Yang KY, Hanm L, Mao G (2008) A fungal-responsive MAPK cascade regulates phytoalexin biosynthesis in *Arabidopsis*. Proc Natl Acad Sci USA 105:5638–5643

Ren D, Yang H, Zhang S (2002) Cell death mediated by MAPK is associated with hydrogen peroxide production in *Arabidopsis*. J Biol Chem 277(1):559–565

Reyna NS, Yang YN (2006) Molecular analysis of the rice MAP kinase gene family in relation to *Magnaporthe grisea* infection. Mol Plant Microbe Interact 19:530–540

Rodríguez CS, Estévez JM, Llorente F, Blanco CH, Jordá L, Pagán I, Berrocal M, Marco Y, Somerville S, Molina A (2009) The ERECTA receptor-like kinase regulates cell wall–mediated resistance to pathogens in *Arabidopsis thaliana*. Am Phytopathol Soc 22:953–963

Rodriguez MC, Adams-Phillips L, Liu Y, Wang H, Su SH, Jester PJ, Zhang S, Bent AF, Krysan PJ (2007) MEKK1 is required for flg22-induced MPK4 activation in *Arabidopsis* plants. Plant Physiol 143:661–669

Rodriguez MCS, Petersen M, Mundy J (2010) Mitogen-activated protein kinase signaling in plants. Annu Rev Plant Biol 61:621–649

Romeis T (2001) Protein kinases in the plant defence response. Curr Opin Plant Biol 4:407–414

Ron M, Avni A (2004) The receptor for the fungal elicitor ethylene inducing xylanase is a member of a resistance-like gene family in tomato. Plant Cell 16:1604–1615

Ross CA, Liu Y, Shen QJ (2007) The WRKY gene family in rice (*Oryza sativa*). J Integr Plant Biol 49:827–842

Rudd JJ, Keon J, Kosack KEH (2008) The wheat mitogen-activated protein kinases TaMPK3 and TaMPK6 are differentially regulated at multiple levels during compatible disease interactions with *Mycosphaerella graminicola*. Plant Physiol 147:802–815

Rushton PJ, Somssich IE (1998) Transcriptional control of plant genes responsive to pathogens. Curr Opin Plant Biol 1:311–315

Rushton PJ, Torres JT, Parniske M, Wernert P, Hahlbrock K, Somssich IE (1996) Interaction of elicitor-induced DNA binding proteins with elicitor response elements in the promoters of parsley PR1 genes. EMBO J 15:5690–5700

Rushton PJ, Somssich IE, Ringler P, Shen QJ (2010) WRKY transcription factors. Trends Plant Sci 15:247–258

Schenk PM, Kazan K, Wilson I, Anderson JP, Richmond T, Somerville SC, Manners JM (2000) Coordinated plant defense responses in *Arabidopsis* revealed by microarray analysis. Proc Natl Acad Sci USA 97:11655–11660

Shen QH, Saijo Y, Mauch S, Biskup C, Bieri S, Keller B, Seki H, Ulker B, Somssich IE, Schulze-Lefert P (2007) Nuclear activity of MLA immune receptors links isolate-specific and basal disease-resistance responses. Science 315:1098–1103

Shi J, An HL, Zhang L, Gao Z, Guo XQ (2010) GhMPK7, a novel multiple stress-responsive cotton group C MAPK gene, has a role in broad spectrum disease resistance and plant development. Plant Mol Biol 74 (1–2):1–17. doi:10.1007/s11103-010-9661-0

Shi J, Zhang L, An H, Wu C, Guo X (2011) GhMPK16, a novel stress-responsive group D MAPK gene from cotton, is involved in disease resistance and drought sensitivity. BMC Mol Biol 12:22

Shim JS, Jung C, Lee S, Min K, Lee YW, Choi Y, Lee JS, Song JT, Kim JK, Choi YD (2013) AtMYB44 regulates WRKY70 expression and modulates antagonistic interaction between salicylic acid and jasmonic acid signaling. Plant J 73:483–495

Shimono M (2012) Rice WRKY45 plays important roles in fungal and bacterial disease resistance. Mol Plant Pathol 13:83–94

Shimono M, Sugano S, Nakayama A, Jiang CJ, Ono K, Toki S, Takatsuji H (2007) Rice WRKY45 plays a crucial role in benzothiadiazole-inducible blast resistance. Plant Cell 19:2064–2076

Shiu SH, Bleecker AB (2001) Receptor-like kinases from *Arabidopsis* form a monophyletic gene family related to animal receptor kinases. Proc Natl Acad Sci USA 98:10763–10768

Shu-Ling Z, Xing-Fen W, Yan Z, Jian-Feng L, Li-Zhu W, Dong-Mei Z, Zhi-Ying M (2012) GbWRKY1, a novel cotton (*Gossypium barbadense*) WRKY gene isolated from a bacteriophage full-length cDNA library, is induced by infection with *Verticillium dahliae*. Indian J Biochem Biophys 46:405–413

Sun C, Palmqvist S, Olsson H, Borén M, Ahlandsberg S, Jansson C (2003) A novel WRKY transcription factor, SUSIBA2, participates in sugar signaling in barley by binding to the sugar-responsive elements of the iso1 promoter. Plant Cell 15:2076–2092

Sturgill TW, Ray LB (1986) Muscle proteins related to microtubule associated protein-2 are substrates for an insulin-stimulatable kinase. Biochem Biophys Res Commun 134(2):565–571

Taguchi F, Shimizu R, Inagaki Y, Toyoda K, Shiraishi T, Ichinose Y (2003) Post translational modification of flagellin determines the specificity of HR induction. Plant Cell Physiol 44:342–349

Takahashi F, Riichiro Y, Ichimura K, Mizoguchi T, Seo S, Yonezawa M, Maruyama K, Yamaguchi-Shinozaki K, Shinozaki K (2007) The mitogen-activated protein kinase cascade MKK3–MPK6 is an important part of the jasmonate signal transduction pathway in *Arabidopsis*. Plant Cell 19:805–818

Teige M (2004) The MKK2 pathway mediates cold and salt stress signaling in *Arabidopsis*. Mol Cell 15:141–152

Teige M, Scheikl E, Eulgem T, Doczi F, Ichimura K, Shinozaki K, Dangl JL, Hirt H (2004) The MKK2 pathway mediates cold and salt stress signaling in *Arabidopsis*. Mol Cell 15:141–152

Tena G, Asai T, Chiu WL, Sheen J (2001) Plant mitogen activated protein kinase signaling cascades. Curr Opin Plant Biol 4:392–400

Tim X, Jordan M (2000) Genetic engineering of plant signal transduction mechanism. Plant Mol Biol Reprt 18:1–10

Torii KU, Mitsukawa N, Oosumi T, Matsuura Y, Yokoyama R, Whittier RF, Komeda Y (1996) The *Arabidopsis* ERECTA gene encodes a putative receptor protein kinase with extracellular leucine rich repeats. Plant Cell 8:735–746

Tsuji J, Jackson EP, Gage DA, Hammerschmidt R, Somerville SC (1992) Phytoalexin accumulation in *Arabidopsis thaliana* during the hypersensitive reaction to *Pseudomonas syingae* pv syringae. Plant Physiol 98:1304–1309

van Loon LC, Geraats BPJ, Linthorst HJM (2006) Ethylene as a modulator of disease resistance in plants. Trends Plant Sci 11:184–191

Van verk MC, Pappaioannou D, Neeleman L, Bol JF, Linthorst HJM (2008) A novel WRKY transcription factor is required for induction of PR-1A gene expression by salicylic acid and bacterial elicitors. Plant Physiol 146:1983–1995

Wagner T, Kohorn B (2001) Wall-associated kinases are expressed throughout plant development and are required for cell expansion. Plant Cell 13:303–318

Walia A, Lee JS, Wasteneys GO, Ellis BE (2009) *Arabidopsis* mitogen-activated protein kinase MPK18 mediates cortical microtubule functions in plant cells. Plant J 59(4):565–575

Wan J, Zhang S, Stacey G (2004) Activation of a mitogen-activated protein kinase pathway in *Arabidopsis* by chitin. Mol Plant Pathol 5:125–135

Wan J, Zhang XC, Neece D, Ramonell KM, Clough S, Kim SY, Stacey MG, Stacey G (2008) A LysM receptor-like-kinase plays a critical role in chitin signalling and fungal resistance in *Arabidopsis.* Plant Cell 20:471–481

Wang D, Amornsiripanitch N, Dong X (2006) A genomic approach to identify regulatory nodes in the transcriptional network of systemic acquired resistance in plants. PLoS Pathog 2(11):123

Wang H, Liu Y, Bruffett K, Lee J, Hause G, Walker JC, Zhang S (2008) Haplo-insufficiency of MPK3 in MPK6 mutant background uncovers a novel function of these two MAPKs in Arabidopsis ovule development. Plant Cell 20:602–613

Wang H, Avci U, Nakashima J, Hahn MG, Chen F, Dixon RA (2010) Mutation of WRKY transcription factors initiates pith secondary wall formation and increases stem biomass in dicotyledonous plants. Proc Natl Acad Sci USA 107:22338–22343

Wang N, Huabg HJ, Ren S, Sun Y, Sun D, Zhang SQ (2012) The rice wall-associated receptor-like kinase gene OsDEES1 plays a role in female gametophyte development. Plant Physiol 160(2):696–707

Wang Y, Dang F, Liu Z, Wang X, Eulgem T, Lai Y, Yu L, She J, Shi Y, Lin J, Chen C, Guan D, Qiu A, He S (2013) *CaWRKY58*, encoding a group I WRKY transcription factor of *Capsicum annuum*, negatively regulates resistance to *Ralstonia solanacearum* infection. Mol Plant Pathol 14:131–144

Widmann C, Gibson S, Jarpe MB, Johnson GL (1999) Mitogen-activated protein kinase: conservation of a three kinase module from yeast to human. Physiol Rev 79:143–180

Wrzaczek M, Hirt H (2001) Plant MAP kinase pathways: how many and what for? Biol Cell 93:81–87

Xing T, Ouellet T, Miki BL (2002) Towards genomic and proteomic studies of protein phosphorylation in plant-pathogen interactions. Trends Plant Sci 7:224–231

Xiong L, Yang Y (2003) Disease resistance and abiotic stress tolerance in rice are inversely modulated by an abscisic acid-inducible mitogen activated protein kinase. Plant Cell 15:745–759

Xu X, Chen C, Fan B, Chen Z (2006a) Physical and functional interactions between pathogen-induced *Arabidopsis* WRKY18, WRKY40 and WRKY60 transcription factors. Plant Cell 18:1310–1326

Xu WH, Wang YS, Liu GZ, Chen X, Tinjuangjun P, Pi LY, Song WY (2006b) The autophosphorylated Ser686, Thr688, and Ser689 residues in the intracellular juxtamembrane domain of XA21 are implicated in stability control of rice receptor-like kinase. Plant 45:740–751

Xu J, Li Y, Wang Y, Liu H, Lei L, Yang H, Liu G, Ren D (2008) Activation of MAPK kinase 9 induces ethylene and camalexin biosynthesis and enhances sensitivity to salt stress in *Arabidopsis.* J Biol Chem 283:26996–27006

Yamaguchi Y, Pearce G, Ryan CA (2006) The cell surface leucine-rich repeat receptor for AtPep1, an endogenous peptide elicitor in *Arabidopsis*, is functional in transgenic tobacco cells. Proc Natl Acad Sci USA 103:10104–10109

Yamaguchi Y, Huffaker A, Bryan AC, Tax FE, Ryana CA (2010) PEPR2 is a second receptor for the Pep1 and Pep2 peptides and contributes to defense responses in Arabidopsis. Plant Cell 22:508–522

Yang KY, Liu Y, Zhang S (2001) Activation of a mitogen-activated protein kinase pathway is involved in disease resistance in tobacco. Proc Natl Acad Sci USA 98:741–746

You MK, Oh SL, Ok SH, Cho SK, Shin HY, Jeung JU, Shin JS (2007) Identification of putative MAPK kinases in *Oryza minuta* and *O. sativa* Responsive to biotic stresses. Mol Cells 23(1):108–114

Yuasa T, Ichimura K, Mizoguchi T, Shinozaki K (2001) Oxidative stress activates ATMPK6, an *Arabidopsis* homologue of MAP kinase. Plant Cell Physiol 42:1012–1016

Yu F, Huaxia Y, Lu W, Wu C, Cao X, Guo X (2012) GhWRKY15, a member of the WRKY transcription factor family identified from cotton (*Gossypium hirsutum* L.) is involved in disease resistance and plant development. BMC Plant Biol 12:144

Zhang S, Du H, Klessig DF (1998) Activation of the tobacco SIP kinase by both a cell wall-derived carbohydrate elicitor and purified proteinaceous elicitins from *Phytophthora* spp. Plant Cell 10:435–450

Zhang S, Klessig DF (2001) MAPK cascades in plant defense signaling. Trends Plant Sci 6:520–527

Zhang Y, Wang L (2005) The WRKY transcription factor superfamily: its origin in eukaryotes and expansion in plants. BMC Evol Biol 5:1–12

Zhang L, Li Y, Lu W, Meng F, Wu C, Guo X (2012) Cotton GhMKK5 affects disease resistance, induces HR-like cell death, and reduces the tolerance to salt and drought stress in transgenic *Nicotiana benthamiana*. J Exper Bot 63:3935–3952

Zhang S, Liu Y, Klessig DF (2000) Multiple levels of tobacco WIPK activation during the induction of cell death by fungal elicitins. Plant J 23:339–347

Zhang X, Dai Y, Xiong Y, DeFraia C, Li J, Dong X, Mou Z (2007) Over expression of *Arabidopsis* MAP kinase kinase 7 leads to activation of plant basal and systemic acquired resistance. Plant J 52:1066–1079

Zhao J, Davis LC, Verpoorte R (2005) Elicitor signal transduction leading to production of plant secondary metabolites. Biotechnol Adv 23:283–333

Zheng Z, Qamar SA, Chen Z, Mengiste T (2006) *Arabidopsis* WRKY33 transcription factor is required for resistance to necrotrophic fungal pathogens. Plant J 48:592–605

Zhou QY, Tian AG, Zou HF, Xie ZM, Lei G, Huang J, Wang CM, Wang HW, Zhang JS, Chen SY (2008) Soybean WRKY-type transcription factor genes, GmWRKY13, GmWRKY21, and GmWRKY54, confer differential tolerance to abiotic stress in transgenic *Arabidopsis* plants. Plant Biotechnol J 6:486–503

Zhou C, Cai Z, Guo Y, Gan S (2009) An *Arabidopsis* mitogen-activated protein kinase cascade, MKK9-MPK6, plays a role in leaf senescence. Plant Physiol 150:167–177

Zipfel C, Robatzek S (2010) Pathogen-associated molecular pattern triggered immunity. Plant Physiol 154:551–554

Zipfel C, Robatzek S, Navarro L, Oakeley EJ, Jones JD, Felix G, Boller T (2004) Bacterial disease resistance in *Arabidopsis* through flagellin perception. Nature 428:764–767

Zipfel C, Kunze G, Chinchilla D, Caniard A, Jones JD, Boller T, Felix G (2006) Perception of the bacterial MAMP EF-Tu by the receptor EFR restricts Agrobacterium-mediated transformation. Cell 125:749–760

An Overview of Antiviral RNA Silencing in Plant: Biogenesis, Host–Virus Interaction and Potential Applications

Saumik Basu, Veerendra Kumar Sharma, Dhriti Bhattacharyya, and Supriya Chakraborty

Abstract

Small RNA molecules play a crucial regulatory role in maintaining genome stability as well as developmental regulations through a set of complex and partially overlapping pathways in a wide range of eukaryotic organisms. Active in both cytoplasm and nucleus, RNA interference regulates eukaryotic gene expression through transcriptional repression by epigenetic modification and interaction with transcription machinery. Small interfering RNAs (siRNAs/miRNAs) of 21–24 nucleotides constitute the innate defence arm against a variety of pathogens, especially viruses. Plant viruses with either DNA or RNA genomes are subjected to small RNA-directed RNA degradation. Additionally, DNA viruses are subjected to another line of defence through 'RNA-directed DNA methylations' (RdDM). On the other hand, viral-encoded proteins, called silencing suppressors (VSRs), are known to counter the defence machinery, and therefore the virus can evade the host surveillance system. Some plant viruses additionally adopt certain strategies like acquiring silencing resistant structures (some RNA virus) to evade the RNA silencing machinery and thereby shaping the viral as well as the host genome. Recently, it has been reported that particular viral proteins and viral siRNAs contribute directly to pathogenicity by interacting with certain host proteins or RNAs. Transcriptional regulation of host gene by small RNA of viral origin plays important role in pathogenesis and symptom development. Small regulatory RNAs of cellular rather than pathogen origin have also been found to play a broad role in improving the basal defence in the case of plant–virus interaction. This chapter provides key insights into the complex intricate machinery of diverse RNA silencing mechanisms, describes various evolutionary diverse strategies of viral

S. Basu • V.K. Sharma • D. Bhattacharyya
S. Chakraborty (✉)
Molecular Virology Laboratory, School of Life Sciences,
Jawaharlal Nehru University, New Delhi 110067, India
91-11-26704153 91-11-26742558
e-mail: supriyachakrasls@yahoo.com

R.K. Gaur and P. Sharma (eds.), *Approaches to Plant Stress and their Management*,
DOI 10.1007/978-81-322-1620-9_18, © Springer India 2014

silencing suppressors at various steps, offers a broader view of host recovery following virus infection and finally suggests the possible applications of RNA silencing to generate virus resistant plants.

Keywords
siRNAs • miRNAs • RdDM • VSRs • Pathogenesis • Host recovery

Introduction

RNA silencing is an evolutionary conserved gene regulatory mechanism active in a majority of eukaryotic system (e.g. plants, animals, yeast and insects). The versatile mechanism involves inhibition both at translational level (through the degradation of the target mRNA and/or by inhibiting its translation in a sequence-specific manner, i.e. posttranscriptional gene silencing, PTGS) and during transcription of transposons and repetitive DNA elements (transcriptional gene silencing, TGS). The concept of RNA silencing was born in early 1990 when transgenic plants expressing an extra copy of chalcone synthase (CHS) unexpectedly resulted in the suppression of both transgene and endogenous CHS mRNA giving birth to a phenomenon called co-suppression in scientific landscape (Napoli et al. 1990; van der Krol et al. 1990). Subsequently, in 1996, in the fungus *Neurospora crassa*, a similar phenomenon was noticed (Cogoni et al. 1996). The direct study of RNA interference was performed in *C. elegans,* where delivery of exogenous dsRNA resulted in sequence-specific degradation of cognate cellular mRNAs (Fire et al. 1998). Similar effects were observed in the majority of other eukaryotes including mouse, *Drosophila* and human (Elbashir et al. 2001; Billy et al. 2001). From the inception of the concept, the intricacy and importance of this process channelized persistent efforts in detailed exploration of the mechanism.

Three classes of small RNA regulate gene expression in cytoplasm. These are microRNAs (miRNAs), small RNAs which have imperfect complementarity with target and are generated from long RNA with hairpin structure causing translational repression of target mRNA; small interfering RNAs (siRNAs), with perfect complementarity to targets and cause transcript degradation; and PIWI-interacting RNAs (piRNAs) targeting transcripts in animal germ lines. Plants produce miRNA and siRNA but no piRNA. The term RNAi is conventionally used for siRNA-mediated silencing but convergence of different small RNA pathway prompted us to use RNAi as an umbrella term in this chapter to describe small RNA-dependent silencing. In plants, the RNA silencing machinery includes at least three complex yet partially overlapping pathways: (1) siRNA-mediated cytoplasmic RNA silencing (PTGS), (2) silencing mediated by miRNAs and (3) DNA methylation-dependent silencing at the transcriptional level (Baulcombe 2004).

Viruses, one of the most important causative agents of infectious diseases in both plants and animals, encode few multitasking proteins to support their life cycle. For successful replication of the genome, viruses use self-encoded replicase along with various replication factors (Mori et al. 1992) or reverse transcriptase (Laco and Beachy 1994) in case specific manner. Interestingly, viruses are also known to efficiently use host-encoded RNA-dependent RNA polymerase (Dalmay et al. 2000; Mourrain et al. 2000). Most plant viruses have a narrow host range and capable of manipulating developmental pathways of the hosts leading to striking and elaborate array of symptoms formation (Hull 2002). Plant viral symptoms were also observed and commercially exploited long before the concept of virus came into existence. For example, the flame-like streaks of tulip flower caused by infection with tulip-breaking virus (Dekker et al. 1993) achieved high prices. The autumnal yellow appearance of eupatorium plants caused by *Eupatorium yellow vein virus* and its cognate betasatellite was praised by a Chinese princess (Saunders et al. 2003).

Recent studies have demonstrated 'RNA silencing' as a major contributor of plant defence response against viruses (Wang and Metzlaff 2005; Pallas and Garcia 2011; Waterhouse et al. 2001). Plant viruses induce PTGS in a homology-dependent manner (Baulcombe 2004; Meister and Tuschl 2004; Eamens et al. 2008). During the initial stage of RNA silencing, long double-stranded RNAs (dsRNAs) (produced by the transcripts of inverted repeat sequence as in transposons or by the transcription from convergent promoters or by host RDRs (either in primer-dependent or primer-independent mechanism) (Dalmay et al. 2000; Mourrain et al. 2000; Sijen et al. 2001)) are cleaved into small interfering RNAs (siRNAs) of 21–24 nucleotides (Bernstein et al. 2001; Hamilton et al. 2002) which play diverse and redundant functions (Xie et al. 2004; Gasciolli et al. 2005; Blevins et al. 2006). Duplex siRNAs then undergo unwinding in ATP-dependent manner (Nykanen et al. 2001) and one of the two strands called guide strand gets incorporated into complex machinery of proteins called RNA-induced silencing complex (RISC) to carry out sequence-specific degradation of complementary target mRNA (Khvorova et al. 2003). A similar strategy has been adopted by both plants and animals to generate endogenous microRNAs (miRNAs). siRNA-directed transcriptional and heterochromatic gene silencing (Lippman and Martienssen 2004) constitute the third branch where the siRNAs (24–26 nt) of slightly larger size can be generated from transcripts of inverted repeats or tandem-repeated sequence and by ectopically expressed RNAs corresponding to the promoter region (Jones et al. 1999; Mette et al. 2000) and function in association with AGO4 and RDR2 (Hamilton et al. 2002; Qi et al. 2005; Xie et al. 2005; Zilberman et al. 2003). Methylation of cytosine residue in DNA (RNA-dependent DNA methylation, RdDM) or at the 9th position of histone H3 (H3K9) also results in suppression of gene expression (Blander and Guarente 2004). This RdDM pathway has been reported to maintain the genome integrity at both centromeric and telomeric repeat regions and suppress the transcription of transposons and other invasive DNAs (Matzke et al. 2009; Haag and Pikaard 2011). On the other hand, viruses in turn develop several strategies to interfere with host defence machinery to establish successful pathogenesis. One of the most fascinating strategies is the evolution of viral suppressors of RNA silencing which interfere with various steps of RNA silencing pathway of host (Anandalakshmi et al. 1998; Brigneti et al. 1998; Kasschau and Carrington 1998; Llave et al. 2000). Additionally, viruses also adopt different strategies to bypass RNA silencing machinery. Bromoviruses protect their RNA genome from host ribonucleases by accumulating inside the membrane vesicle (Schwartz et al. 2002). Members of the family *Avsunviroidae* undergo chloroplastic replication and thereby protect the viroid genome from RNA silencing (Tabler and Tsagris 2004). Again, the extensive intramolecular fold structures of viroids make them inaccessible to RISC complex (Wang et al. 2004). Defective interfering RNAs, which are devoid of target sequences also help to escape RNA silencing as observed in case of tombusviruses (Dalmay et al. 1995). Transfected siRNAs specific for either influenza A or HIV virus failed to target viral genome because of the occurrence of quasi species by spontaneous mutations in the target region (Ge et al. 2003; Boden et al. 2003; Das et al. 2004). Viruses (e.g. respiratory syncytial virus) also interact with certain cytoplasmic proteins (Biltko and Barik 2001) for encapsidation, high rate of replication and spread that may finally aid the viruses to escape RNA silencing machinery.

Components of RNA Silencing Machinery

RNA silencing is an ancient defence mechanism. While coming down the ladder of evolution, the process has incorporated several unique and functionally diverse proteins as important contributors in the complexity and specificity of the pathway. Proteins like Dicer-like enzymes (DCLs), ARGONAUTEs (AGOs), HYPONASTIC LEAVES (HYL1) and other dsRNA-binding proteins, HUA ENHANCER1 (HEN1) and RNA-DEPENDENT RNA POLYMERASEs are providing specificity to the RNA silencing machinery of plant.

Dicer-Like Proteins (DCLs): In comparison to other eukaryotes (mammals, worms, flies and fungi) *A. thaliana* encodes different DCLs. DCLs/DICERs are multidomain RNAse III-like ribonucleases which include evolutionary conserved N-terminal RNA helicase domain, central Piwi/Argonaute/Zwille (PAZ) domain and C-terminal dual catalytic and dsRNA-binding domains (Bernstein et al. 2001; Schauer et al. 2002). DCL1, the first RNAse III-like ribonucleases discovered, processes endogenous dsRNAs to miRNAs (21–22 nt) which in turn control diverse set of mRNAs of various transcription factors (Park et al. 2002; Xie et al. 2005). DCL2 produces less abundant 22-nt siRNAs population, and DCL3 produces 24-nt hc-siRNAs to carry out RdRM and to modify cis- and trans-elements of the gene, DNA repeats and transposons loci (Xie et al. 2004; Bouche et al. 2006). DCL4 is involved in the biogenesis of endogenous tasiRNAs (Reinhart et al. 2002; Gasciolli et al. 2005). Recently, DICER-LIKE 4 (DCL4) has been shown to terminate transcription of *Arabidopsis* endogenous FCA gene (a nuclear RNA-binding protein which controls the flowering time in Arabidopsis) by promoting cleavage of the aberrant RNA produced from the locus (Liu et al. 2012). In rice, DCL4 and DCL3 homolog DCL3b are likely to be involved in the generation of phased siRNAs of 21 and 24 nucleotides, respectively (Song et al. 2012). It has been proposed that formation of differently sized siRNAs is probably mediated by a PAZ domain of the DCLs, which configures the single-cleavage centre with respect to long N-terminal RNase III domain (Schauer et al. 2002).

Argonautes (AGOs): The uniqueness of RISC complex is provided by the AGO proteins. They can bind to both siRNA and miRNA. Ten different AGO proteins are encoded by *Arabidopsis* genome (Baulcombe 2004) of which AGO1, AGO2, AGO5 and AGO7 reportedly contribute to the antiviral defence in plant (Wang et al. 2011). Typically RISC-containing miRNAs and 21-nt siRNAs (produced by DCL1 and DCL4, respectively) associate with either AGO1, AGO2, AGO7 or AGO10 to cause post-transcriptional gene silencing (PTGS) of target mRNA by translational repression (Brodersen et al. 2008) or slicing (Baumberger and Baulcombe 2005). In contrast, 24-nt siRNAs, produced by DCL3, associate with AGO4, AGO6 or AGO9 and initiate transcriptional gene silencing (TGS) (Brosnan and Voinnet 2011).

AGO contains four functionally distinct domains to interact extensively with small RNA molecules. PAZ domain recognizes the 3′ end (Lingel et al. 2003, 2004; Ma et al. 2004); PIWI domain adopts an RNase-H fold and confers targeted endonucleolytic activity to certain AGOs to cleave the target mRNA in the region complementary to the guide RNA (Song et al. 2003; Yuan et al. 2005; Kawamura et al. 2008); MID domains of AGOs interact with the 5′ end of small RNAs and can direct the sorting of different classes of small RNAs into the appropriate AGO family members (Frank et al. 2012, 2010; Ma et al. 2005; Parker et al. 2005). Recently, the N-terminal domain of AGO has been proposed to be involved in unwinding of duplex siRNAs/miRNAs (Kwak and Tomari 2012). These AGOs vary in terms of catalytic triad present either in them or in residues that are involved in 5′ phosphate binding (Liu et al. 2004; Rivas et al. 2005). AGOs have been found to coimmunoprecipitate with viral small RNAs, but AGO1, largely considered as the principal slicer, has been found to bind to miRNAs and certain class of siRNAs of endogenous origin but not with the viral siRNAs (Hunter et al. 2003; Fagard et al. 2000; Baumberger et al. 2007), while AGO4 is required for RdDM mediated by 24-nt siRNAs (Zilberman et al. 2004). AGO1 and AGO5 preferentially bind to small RNAs containing 5′ terminal U or C residues, respectively whereas AGO2 and AGO4 have a strong strand bias for small RNAs with 5′ terminal A. 5′ terminal nucleotide of small RNAs determines strand selection into AGO complexes. Nevertheless, this 5′ end-dependent incorporation is not exclusive (Mi et al. 2008; Montgomery et al. 2008; Takeda et al. 2008; Havecker et al. 2010). DDH residues of AGO1 have been shown to possess cleavage activity (Elbashir et al. 2001; Mallory et al. 2004).

RNA-Dependent RNA Polymerases (RDRs): Host-encoded RNA-dependent RNA polymerase (RDR) uses viral primary siRNA molecules as primers to convert (aberrant) RNA target sequences into new long dsRNAs which in turn are processed into secondary siRNAs. These RDR-dependent amplified pools of viral siRNAs are originated from the entire target RNA sequence causing transitive silencing Sijen et al. (2001)). In *Arabidopsis* 6 RDRs have been reported. Three of them, i.e. RDR1, RDR2 and RDR6, belong to RDRα group containing a catalytic DLDGD motif. Arabidopsis RDR1, RDR*2* and RDR*6*, and orthologs of these genes, are involved in the amplification, and plants from which these genes have been knocked out are highly susceptible to various plant viruses. RDR3, RDR4 and RDR5 contain DFDGD motif and are characterized as members of RDRγ group (Wassenegger and Krczal 2006). RDR2 and RDR6 are found to be the most important members that contribute significantly in endogenous small RNA pathway by converting the ssRNA templates to dsRNA in a primer-independent manner (Curaba and Chen 2008). The role of RDR 3, 4 and 5 are not well explored yet. They are found as tandemly repeated clusters on chromosome II. RDR1 has been found to be induced in response to either viral infection or salicylic acid (Yu et al. 2003). Interplay of different RDRs is important in regulating antiviral response of host. RDR1 from *N. tabacum* suppresses RDR6-mediated antiviral silencing and enhances viral infection in *N. benthamiana* where it is reported to be truncated due to insertion of inframe mutation (Ying et al. 2010). RDR2 also antagonizes the production of RDR6-dependent siRNAs in sense PTGS (Jauvion et al. 2012).

Nonfamily dsRNA-Binding Proteins: HYPONASTIC LEAVES1 (HYL1) and dsRNA-binding proteins (DRB 2–4) bind to DCLs and assist cleavage of double stranded RNAs. HYL1 interacts with and DCL1 and colocalizes in the nuclear bodies along with C2H2 Zn finger protein, Serrate and this complex is required for miRNA processing (Han et al. 2004; Vazquez et al. 2004; Fang and Spector 2007; Song et al. 2007). DRB4 interacts specifically with DCL4 (Hiraguri et al. 2005). R2D2 in *Drosophila* (Liu et al. 2003) and RDE4 in *C. elegans* (Tabara et al. 2002) are two important DRBs helping Dicers to deliver duplex small RNAs to downstream effector complexes.

HUA ENHANCER 1 (HEN1): HEN1, a small methylase, is unique to plants which methylates the 2′ OH of the terminal nucleotide at 3′ end of the small RNAs (Yang et al. 2006; Yu et al. 2005). The small RNA duplexes with 3′ 2-nt overhangs are preferred substrate for HEN1 and get methylated immediately after DCL-mediated cleavage to provide stability and protection to the small RNAs against oligouridylation (Yang et al. 2006).

Origin of Viral siRNAs

Genome of plant viruses interestingly can serve as both the target and trigger of RNA silencing. Earlier it was speculated by the scientific fraternity that double-stranded RNA intermediate generated during replication of positive-strand RNA virus could trigger the production of vsiRNA (Ahlquist 2002) which, if true, would generate equal amount of siRNA from both positive and negative RNA strand. But Molnar and associates found that the genomic strand of the viruses gave rise to greater amount of vsiRNA. Consequently, it was proposed that highly structured, single-stranded viral RNAs could be processed into vsiRNAs to trigger RNA silencing (Molnar et al. 2005). Moreover, certain regions on the viral genome were identified as 'hot' which had greater potential of producing vsiRNA over 'non-hot' regions. It was further suggested that single-stranded viral RNA with stable secondary structure is more likely the probable source of vsiRNA than dsRNA replication intermediates (Szittya et al. 2010; Donaire et al. 2009). In the case of plant DNA viruses which replicate through dsDNA intermediate also produce vsiRNAs from foldback structures of RNA transcription units (Moissiard and Vionnet 2006; Vanitharani et al. 2005).

Role of vsiRNAs in Attenuating Expression of Host Transcript

Detailed studies with vsiRNA have indicated that vsiRNAs can posttranscriptionally regulate the host transcripts expression. Bioinformatics study with Potato spindle tuber viroid (PSTVd-RG1) revealed presence of stretches of 19–20-nt sequences from various plant species that share sequence identity with the viroid. Interestingly, most of these sequences corresponded to virulence-regulating region of the pathogen. Analysing the plant sequence divulged presence of number of transcription factors and chromodomain helicase DNA-binding protein that shared sequence homology with the viral sequences (Wang et al. 2004). This result suggested putative role of vsiRNA in regulating expression of host regulatory genes. Study with *Cauliflower mosaic virus* revealed that the CaMV infection greatly reduced the expression of one mRNA from *Arabidopsis* sharing 18–25nt microhomology with 35S RNA leader sequence (Moissiard and Vionnet 2006). The functionality and efficiency of vsiRNA in regulating host genes depends on many cellular factors including activity of the silencing suppressors of viral origin and abundance of vsiRNAs. Recently it has been reported that siRNAs derived from viral satellite RNA can directly regulate the expression of a host gene and hence attenuate the disease symptoms. A 24-nt region of CMV-Y satellite RNA (Y-Sat), called the 'yellow domain', was shown to be responsible for yellow symptoms induced in Y-Sat-infected tobacco plants (Kuwata et al. 1991). Smith et al. (2011) observed that a 22-nt complementary region of this yellow domain was present in the sequence of subunit I of magnesium chelatase, an enzyme involved in chlorophyll biosynthesis. Extensive study revealed that Y-Sat-induced symptoms are elicited by the vsiRNAs-mediated silencing of *CHLI* (2011). This result was further confirmed by Shimura et al. (2011), who showed that *N. benthamiana* plants expressing inverted repeat sequence of Y-Sat also develop yellow symptom mimicking the virus-infected phenotype. Downregulation of CHLI expression in both transgenic and Y-Sat-infected plants further proved the role of Y-Sat-derived vsiRNA in affecting host expression. Finally, it was demonstrated that Y-Sat-derived vsiRNAs could specifically target the 22-nt sequence in *CHLI* mRNA and therefore downregulate *CHLI* mRNA, thus inducing the yellowing symptom by impairing the chlorophyll biosynthesis pathway.

Cell-to-Cell and Long-Distance Movement of Virus-Derived siRNAs in Plants

In plants, RNAi acts non-cell autonomously and spreads in transacting manner. Grafting experiments with transgene-induced rootstocks with non-silenced target shoots or scion showed that a sequence-specific silencing signal is transmitted from rootstocks into shoots (Palauqui et al. 1997). Mobile RNA silencing was found to have two distinct arms in plants: cell-to-cell (through plasmodesmata) (Himber et al. 2003; Dunoyer et al. 2010) and long-distance movement through phloem (Palauqui et al. 1997; Voinnet and Baulcombe 1997; Yoo et al. 2004; Kalantidis et al. 2008). The siRNAs (21 to 24 nt) generated from the processing of the long dsRNAs act as mobile silencing signal for both the cases. Few of the components of this signal transduction path way including number of small RNAs, proteins and protein channels have been identified.

Local Movement of Silencing Signal From Cell to Cell: Local movement of silencing signal occurs through specialized intercellular channels called plasmodesmata (Lucas and Lee 2004; Oparka 2004; Kim and Zambryski 2005; Maule 2008). In the absence of signal amplification triggered by cellular RDRs, the spread of silencing signal is limited to 10–15 cells beyond the site of signal initiation. Plasmodesmata can allow the transfer of up to 27 kDa protein (Kobayashi and Zambryski 2007). However, plasmodesmata, upon binding to various virus-encoded proteins, can undergo significant change in their size exclusion limit (Imlau et al. 1999) and thereby allowing larger molecules like viral ribonucleoproteins and transcription factors (e.g. KNOTTED 1) to pass through it (Lucas et al. 1995; Carrington et al. 1996). Spread of 'local' or 'limited' cell-to-cell silencing depends on the 21-nt siRNAs generated by DCL4 in AGO1-dependent cleavage of target endogenous genes (Himber et al. 2003; Parizotto et al. 2004). When SULPHUR gene fragment was expressed using phloem companion cell-specific promoter, mutation in RDR6 failed to interfere with the yellowing of the companion and its adjacent 10–15 cells (Himber et al. 2003) indicating little role of RDR6 in the local silencing process. Mutation of DCL4 leads to loss of non-cell-autonomous silencing indicating that 21-nt but not 24-nt siRNAs are sufficient for non-cell-autonomous RNAi (Hamilton et al. 2002; Dunoyer et al. 2005, 2010).

Extensive Long-Distance Movement of Silencing Signals: Spread of silencing signal beyond 10–15 cells is termed as extensive silencing. The larger siRNAs (24–26 nt) rather than 21 nt are essential for spread of long-distance silencing signals (Himber et al. 2003) which is dependent on signal amplification by RDR6, SGS3 and a putative RNA helicase (SDE3) (Mourrain et al. 2000; Vaistij et al. 2002; Himber et al. 2003) either in primer-dependent (5′–3′ transitivity) or primer-independent way (3′–5′ transitivity). This amplification is predominantly carried out by the secondary siRNAs generated from cleaved dsRNAs that function as repetitive wave of local cell-to-cell signalling of 10–15 cells at a time (Himber et al. 2003).Two important proteins NRPD1a (a component of RNA Pol IV) (Herr et al. 2005; Kanno et al. 2005; Onodera et al. 2005; Pontier et al. 2005) and RDR2 (an RNA-dependent RNA polymerase 2) (Xie et al. 2004; Herr et al. 2005; Pikaard 2006) function as essential components of non-cell-autonomous RNA silencing (Dunoyer et al. 2007; Smith et al. 2007). Phloem acts as a specific highway for transport of long-distance systemic silencing signals through specialized sieve tube cells from source to sink (Palauqui et al. 1997). Additionally, long-distance spread of silencing signal requires high amount of target transcripts (Garcia-Perez et al. 2004; Schwach et al. 2005). Spreading of miR166 expression in phloem tissues during leaf development indicates its involvement in long-distance signal movement to act at distance (Juarez et al. 2004). Abundance of PHO2 and miR399 in the phloem with regard to inorganic phosphate (Pi) alteration and coexpression suggests their involvement in systemic silencing (Lin et al. 2008). miR172 was found to be present in sRNA libraries prepared from phloem exudates and likely to play important role in long-distance signalling (Zeevaart 2008). miR319 gets transported from leaves to roots where it targets a subset of the TCP family of transcription factors that regulates *LOX2* expression (Yoo et al. 2004; Schommer et al. 2008; Buhtz et al. 2010). Phloem small-RNA-binding protein 1 (PSRP1) was subsequently shown to bind and facilitate movement of single-stranded sRNA molecules between cells (Ham et al. 2009). CmPP16 protein from *Cucurbita maxima* was shown to possess properties similar to those of viral movement proteins (Aoki et al. 2005).

Antiviral Defence Pathways in Plants: RNA Silencing

In 1992, Lindbo and Dougherty observed that transgenic plant expressing non-translatable coat protein of tobacco etch virus (TEV) was resistant to cognate virus. Taking clue from this observation, it was rather sagaciously proposed that the resistant phenotype was the consequence of a mechanism active in cytoplasm which target and destroy the mRNA in sequence-specific manner (Lindbo et al. 1993). The hypothesis was the first step in building the concept of RNA silencing as antiviral defence. Later on it was also observed that an infectious viral cDNA clone engineered to carry a part of a host gene, when mobilized inside the plant, caused silencing of both the specific host gene and the viral sequence. Antiviral RNA silencing in plant has turned out to be an integrated network of at least 3 different mechanisms, namely, cytoplasmic RNA silencing, endogenous mRNA silencing by microRNAs and DNA methylation-dependent silencing at transcriptional level (Baulcombe 2004). These mechanisms not only provide antiviral resistance but also are crucial for cellular functions such as regulation of gene expression, maintenance of genome integrity and stress response. The basic processes of RNA silencing have been well documented in several reviews (Meister and Tuschl 2004; Eamens et al. 2008; Ruiz-Ferrer and Vionnet 2009; Ding 2010; Llave 2010).

Cytoplasmic RNA silencing (Fig. 1a) starts through a process known as virus-induced gene silencing (VIGS). dsRNAs or hpRNAs are targeted by DCLs to produce small RNAs of varying length (21–24 nt). The resulting small RNAs are unwound with the help of an ATP-dependent RNA helicase and subsequently incorporated into RISC-containing AGO1 to perform degradation of viral mRNA and translational repression or methylation of the homologous target genes. Transcriptional gene silencing (Fig. 1c) initiates in the nucleus following infection with viruses or subviral elements which are gradually subjected to inactivation through DNA methylation (TGS). RNA-directed DNA methylation (RdDM) plays a very important role in terms of silencing transposons as well as repetitive DNA elements to maintain genome integrity as well as stability (Matzke et al. 2009; Haag and Pikaard 2011). In RdDM, dsRNAs are synthesized by a DNA-dependent RNA polymerase called RNA polymerase IV (Pol IV) and RDR2 specific to plant and then processed by DCL3 to produce 24-nt siRNAs. These 24-nt siRNAs form an AGO4-containing RISC and interact with the nascent transcript prepared by RNA Pol V (another plant-specific RNA polymerase). This interaction facilitates recruitment of various methylation factors like DRM2, and ultimately de novo cytosine methylation of the target DNA takes place. Therefore, in general RdDM has been known to contribute to plant defence by transcriptional repression of genes from DNA viruses.

The miRNA pathway (Fig. 1b) starts when the miRNA genes are transcribed by RNA polymerase II, and the resulting transcripts contain complementary regions that form short imperfect hairpins. These imperfect hairpins are processed by DCL1 in the nucleus into 21-nt miRNAs with the aid of several other proteins like zinc-finger protein SERRATE and the dsRNA-binding proteins DRB1 and HYL1. The miRNAs play a decisive role in plant development by either repressing or optimizing the expression of various transcription factors associated with developmental processes. miRNAs in plants function through homology-dependent RNA degradation as well as through translational repression (Brodersen et al. 2008) unlike animal miRNAs which usually bind to 3′ UTR. In cytoplasmic siRNA-dependent RNA silencing pathway, the exogenous or endogenous long dsRNAs or short hpRNAs are degraded by either DCL4 or DCL2 to generate 21- and 22-nt siRNAs, respectively. These siRNAs then recruited onto AGO1-containing RISC and RISC-containing guide siRNA cleave target mRNAs. RDR6, one among the six reported RDRs of Arabidopsis, then synthesizes long double-stranded RNA using ssRNAs as template to give rise to trans-acting siRNAs of 21 nt which also have been found to play a role in various stress responses as well as plant developmental processes. Another type of siRNAs called natural antisense siRNAs have been reported in many eukaryotes including plants which are produced from cis-natural antisense transcripts (cis-NATs) in response to various biotic and abiotic stresses (Zhang et al. 2012).

Counter-Defence Response of Virus: Plant Viral Suppressors (VSRS)

Plant viral synergism is defined by a situation, where a plant infected with two or more unrelated viruses shows symptoms much severe than that caused by either of the virus alone. Majority

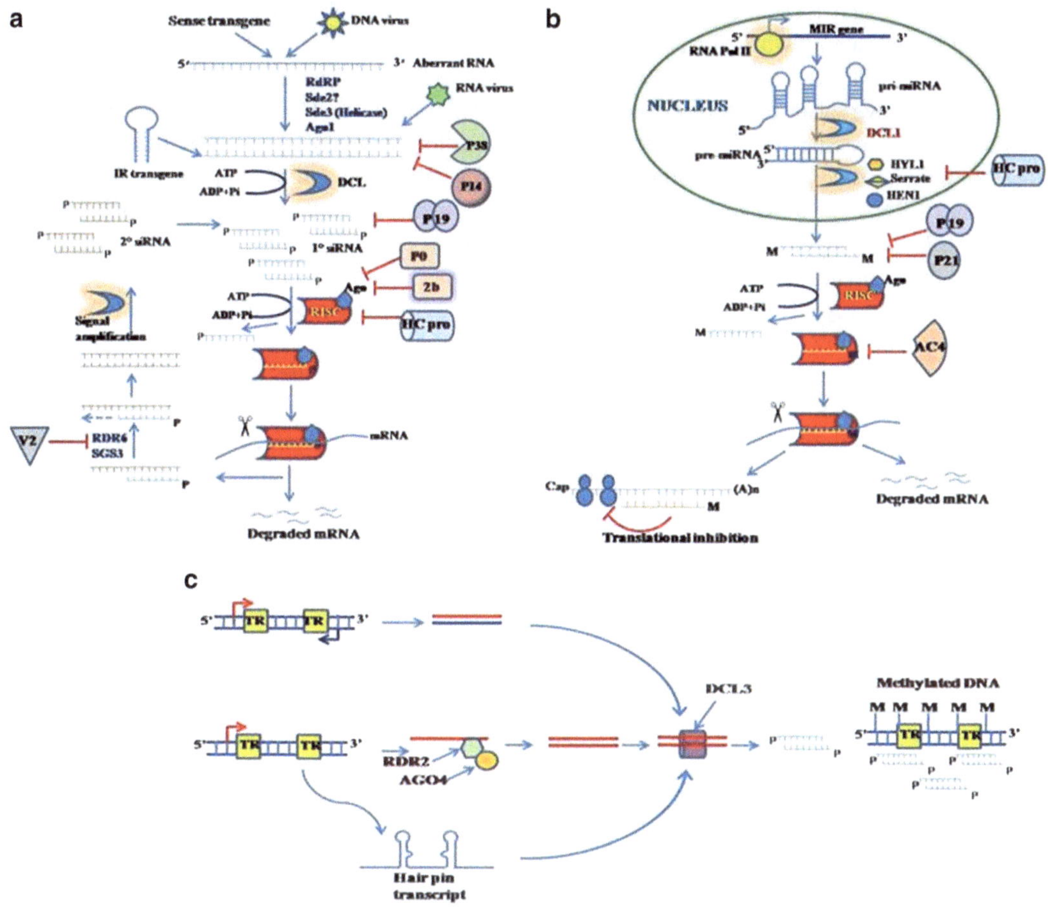

Fig. 1 Anti-viral RNA silencing pathways in plants and their suppression by plant viral encoded suppressors. Three major RNA silencing pathways include (**a**) cytoplasmic RNA silencing, (**b**) endogenous mRNA silencing by mRNAs and (**c**) transcriptional gene silencing through RdDM. Plant VSRs are represented by various shapes, while various steps of inhibition is represented by *block arrows*

of the synergistic interaction between two viruses usually involves potyvirus as one of the co-infecting virus. With the discovery of importance of helper component proteinase (Hc-Pro) of potyvirus in synergism, the concept of viral suppressor working as a counter-defence tool was born. Hc-Pro was designated as viral suppressor of gene silencing (Vance et al. 1995). Viral suppressors (VSRs) have emerged as one of the most potent tricks available with the viruses to invade hosts' defence for establishing successful pathogenesis. VSRs are found in almost all the plant viruses though few reports of suppressors from insects and mammalian viruses are also available at present. Plant VSRs have been found to evolve as diverse group of proteins and share very less sequence homology across the genera. VSRs are presumably evolved to counter the host silencing machinery-mediated defence response and therefore to suppress the host surveillance system. Different suppressors are reported to inhibit different steps of RNA silencing machinery by interacting with different effector components by any of the following ways: (i) through interaction with dsRNA to inhibit their processing by Dicers, (ii) binding to siRNAs and sequestering them to make them unavailable for the RISC, (iii) interacting either directly or indirectly with AGO to degrade or inactivate them and thus preventing functional

RISC assembly and (iv) inhibition of systemic silencing by interacting with either host RDR or DCL4 or DBR4. In addition, some other mechanisms have also been proposed. Some VSRs activate specific group of miRNAs and thereby inhibit some of the important effector molecule of RNA silencing machinery. In other cases, VSRs outcompetes HEN1 for binding to siRNA duplex having 2-nt overhang at 3′ end. There are VSRs, specially found in DNA viruses, which have been shown to inactivate TGS either by transactivating a set of host genes, which in turn act as suppressor or inactivate some of the host methyl transferases. Role of the individual group of suppressors and their mode of action is given in Table 1.

In addition to their primary role in suppression of antiviral RNA silencing, VSRs can act as potent mediator of plant viral diseases by affecting the intrinsic function of essential host factors through direct or indirect interactions. For example, P2 protein, encoded in *Rice dwarf virus* (RDV), interacts directly with rice ent-kaurene oxidases (Zhu et al. 2005) and interferes with gibberellin biosynthesis. Reduced hormone accumulation results in stunting of the infected rice plants. CMV-2b protein has been shown to interact with Arabidopsis catalase (CAT3) and interfere with its scavenging activity (Inaba et al. 2011). At least 10 host proteins were reported to contribute to pathogenicity of tombusviruses (Ishibashi et al. 2010).

The Plant Fights Back: Phenomenon of Host Recovery

The idea of 'host recovery' phenomenon came as a converging mechanism of both natural resistance and host PTGS where the plants infected by virus showed initial symptom on inoculated leaves, but the newly emerging leaves were completely asymptomatic. The systemic and newly emerged leaves provided complete sequence-specific resistance against the virus. The incident of virus-induced symptom recovery was observed for the first time in 1928 when the initial leaves of tobacco plants infected with tobacco ring spot virus showed necrosis and disease symptom. The upper systemic leaves were asymptomatic and consequently showed resistance to secondary infection by the homologous virus (Wingard 1928). Further study suggests that methylation-dependent gene silencing is also associated with host recovery. Such type of recovery has been well documented in geminivirus-infecting host plants. Wild-type *Arabidopsis* and *N. benthamiana* plants inoculated with Beet curly top virus (BCTV) L2⁻ mutant showed recovery (L2 interferes with methylation by interacting with host methyl transferase) owing to the recovery of host from the mutant virus infection (Hormuzdi and Bisaro 1995; Wang et al. 2003). Geminivirus-induced symptom recovery has also been reported in watermelon, cassava and pepper following infection with cucurbit leaf crumple virus, African cassava mosaic Cameroon virus [ACMV-CM] and pepper golden mosaic virus, respectively (Hagen et al. 2008; Chellappan et al. 2004; Rodriguez-Negrete et al. 2009). Transient expression of dsRNA corresponding to viral IR showed enhanced symptom recovery in Zucchini plants (Hagen et al. 2008). Recovery of plants from virus infection was linked with RNA silencing machinery, and particularly in geminiviruses recovery was correlated with reduced viral titre followed by increased viral siRNA accumulation (Chellappan et al. 2004). Natural recovery of host can also be observed following infection with nepovirus (Ratcliff et al. 1997) and caulimovirus (Covey et al. 1997). In contrary to previous studies as observed in DNA viruses, it was reported that the recovery of *N. benthamiana* carrying functional RDR1 orthologue of *Medicago truncatula* was associated with RNA silencing but not with reduced viral titre from a necrotic response induced by a nepovirus, Tomato ring spot virus (ToRSV). The disappearance of symptoms was not accompanied by reduction of viral mRNA (Jovel et al. 2007). Mutation in AV2 also leads to recovery (Basu et al. unpublished data) because of its inability to bind to SGS3, and therefore, allowing the

Table 1 Different plant viral suppressors and their mode of action

Mechanism of suppression	Virus genus	Name of the virus	VSRs	Other functions	Reference
Binding to dsRNA	Aureusvirus	Pothos latent aureusvirus	P14	Symptom determinant	Merai et al. (2005)
	Carmovirus	Turnip crinkle virus	P38	Coat protein	Thomas et al. (2003)
					Qu et al. (2003)
	Cucumovirus	Tomato aspermy virus	2b	Nuclear localization/	Brigneti et al. (1998)
		Cucumber mosaic virus	2b	Host-specific movement	
	Tenuivirus	Rice hoja blanca virus	NS3	Unknown	Bucher et al. (2003); Yang et al. (2011b)
	Nodavirus	Flock house virus	B2	Plaque formation	Li et al. (2002)
	Tospovirus	Groundnut bud necrosis virus (GBNV)	NSs	Interference in plant defence and development	Goswami et al. (2012)
Sequestering/binding siRNAs duplex	Tombusvirus	Cymbidium ring spot virus	P19	Movement	Silhavy et al. (2002)
		Tomato bushy stunt virus	P19		
	Closterovirus	Beet yellow virus	P21	Replication enhancer	Reed et al., (2003); Ye and Patel (2005)
	Potyvirus	Tobacco etch virus	HC-pro	Movement, polyprotein processing, aphid transmission, pathogenicity determinant	Anandalaxmi et al. (1998), Brigneti et al. (1998); Kasschau and Carrington (1998)
	Pecluvirus	Peanut clump virus	P15	Movement	Dunoyer et al. (2002)
	Hordei virus	Barley stripe mosaic virus	γB	Replication enhancer, movement, seed transmission and pathogenicity determinant	Yelina et al. (2002)
Binding to single-stranded miRNA/siRNA	Geminivirus	African cassava mosaic virus	AC4	Movement , virulence	Chellappan et al. (2005)
Interfering methylation of siRNA/miRNA	Tombusvirus	Carnation Italian ring spot virus	P19	Movement	Lozsa et al. (2008); Yu et al. (2005)
	Tombus	Tobacco mosaic virus	P122, P126 and P130	Replication protein	Kubota et al. (2003)
					Csorba et al. (2007)
					Vogler et al. (2007)
Degrading 21-, 22- and 24-nt siRNAs to 14 nts	Closterovirus	Sweet potato chloratic stunt virus	RNase3	dsRNA-specific endonuclease and helps in viral synergism	Cuellar et al. (2009)

(continued)

Table 1 (continued)

Mechanism of suppression	Virus genus	Name of the virus	VSRs	Other functions	Reference
Interaction with DRB4	Caulimovirus	Cauliflower mosaic virus	P6	Viral translational transactivator , motile inclusion formation and microtubule stabilization, inhibition of signalling responses to salicylic acid and regulation of innate immunity	Hass et al. (2008)
					Harries et al. (2009)
					Love et al. (2012)
Targeting AGO1 protein	Cucumovirus	Fny-CMV	2b	Nuclear localization and movement	Mayers et al. (2000)
					Zhang et al. (2006)
Degradation of AGO1 through SCF complex	Polerovirus	Beet western yellows virus and potato leaf roll virus	P0	Pathogenicity determinant	Pazhouhandeh et al. (2006), Bortolamiol et al. (2007); Baumberger et al. (2007)
Interfering RDR6-SGS3-mediated signal amplification	Begomovirus	Tomato yellow-leaf curl virus	V2	Pre-coat protein	Glick et al. (2008)
Inhibiting RDR6-dependent 2^0 siRNA production	Potexvirus	Potato virus X	P25	Movement, Nucleotide binding and RNA helicase	Vionnet et al. (2000)
					Kalinina et al. (2002)
		Turnip yellow mosaic virus	P69	Movement, pathogenicity determinant	Chen et al. (2004)
Blocking intercellular spread of silencing	Closterovirus	Citrus tristeza virus	P20, P23 and CP	Replication enhancer, nucleic acid binding and encapsidation, respectively	Lu et al. (2004); Chiba et al. (2006)
Suppressing local and systemic S-PTGS	Phytoreovirus	Rice dwarf virus (RDV)	Pns10	Actin binding, viroplasm assembly	Cao et al. (2005); Wei et al. (2006); Jia et al. (2012)
Interfering long-distance and systemic silencing	Hordeivirus	*Poa semilatent virus*	γb	Movement and virulence	Yelina et al. (2002)
Inactivation of adenosine kinase and sucrose non-fermenting 1 (SNF1)	Begomovirus	Tomato golden mosaic virus	AC2	Transcriptional activator	Bisaro (2006)
		Beet curly top virus			Wang et al. (2003)
					Hao et al. (2003)
Interaction and attenuation of S-adenosyl methionine (SAMDC1) and its degradation	Curtovirus	Beet severe curly top virus (BSCTV)	C2	Transcriptional activator	Zhang et al. (2011b)
Inhibiting S-adenosyl-L-homocysteine hydrolase (SAHH)	Begomovirus	Tomato yellow-leaf curl china virus [TYLCCV]	βC1	Movement, virulence	Yang et al. (2011a)

amplification of silencing signal in presence of RDR6–SGS3 interaction and the secondary siRNA so produced degrade the viral mRNA and ultimately led to recovery in the newly emerging systemic leaves.

The Role of Plant miRNA in Plant–Virus Interaction

Recently it has been reported that plant miRNAs are responsive to developmental cues and environmental stresses. Tomato plants after infection with Cucumovirus and Tobamovirus showed significant differential expression in 85 % of its total miRNA pool (Fang and Spector 2007). All the differentially expressed miRNA were classified into 25 families. Among all these families, miR159 and miR171 contained most number of miRNAs. Most of these miRNAs were targeted to control expression of transcription factors, plant flower and leaf and height development and reproductive growth. High-throughput sequencing revealed a set of conserved miRNAs. Earlier it was also shown that infection with *Tobamoviridae*, *Potyviridae*, and *Potexviridae* families caused altered accumulation of certain miRNA in *Nicotiana tabucum* in which miRNAs 156, 164, 165 and 167 accumulated to higher levels compared to noninfected tissues (Bazzini et al. 2007). Silencing suppressors of various plant viruses have been reported to change target mRNA level through directly altering the accumulation of endogenous miRNA levels inducing changes also in target mRNA accumulations (Kasschau et al. 2003; Dunoyer et al. 2004; Zhang et al. 2006). Other workers have established the correlation between enhanced expression of miR168 and *AGO1* mRNA in virus-infected plants (Zhang et al. 2006; Csorba et al. 2007; Havelda et al. 2008). These reports also make room to develop a novel strategy where manipulating the host miRNA level holds promise to combat with the viral stress.

Application of RNA Silencing Towards Plant Virus Resistance

Plant pathogens especially viruses are responsible for severe loss in crop production every year throughout the world. Earlier these pathogens were controlled using conventional measures including crop rotation, use of insecticides and breeding with resistant varieties. During 1986, Beachy and his associates demonstrated for the first time the use of pathogen-derived sequence (using TMV coat protein) to engineer resistance in the host (Powell et al. 1986). Since then various strategies based on either protein-mediated or RNA-mediated resistance have been developed. The actual mechanism of protein-mediated resistance is still not clear, and several pathways may be involved based on the type of gene used for engineering resistance. On the other hand, the mechanism of RNA silencing is well understood. During the last two decades, substantial effort has been channelized based on siRNA-mediated RNA silencing to engineer resistance in plants. These approaches differed in varied precursor sequence like pathogen-derived sequences in sense or antisense orientation, shRNA constructs, intron hairpin constructs and artificial miRNA sequences that were used to generate siRNA in plants. The use of intron hairpin RNAi constructs has been shown to be highly effective and caused nearly 100 % silencing of the target gene as compared to sense, antisense or hpRNAi constructs (Smith et al. 2000). It is also possible to target multiple viruses using single-RNAi constructs containing sequences from multiple viruses to generate broad-spectrum resistance (Jan et al. 2000; Bucher et al. 2006). One important hindrance to employ RNAi for engineering resistance is the selection of target and the minimum length of the target sequence for effective silencing. Hutvagner et al. (2000) showed that siRNAs generated by silencing of GUS gene mainly correspond to two-third region of 3′end of mRNA. Now a number of computational algorithms are freely available online for the rational design of siRNA and selection of target sequence to generate effective silencing of the target gene using several parameters.

Increased knowledge of microRNA (miRNA) biogenesis machinery and their role in regulation of transcript expression has helped to develop synthetic or artificial miRNAs (amiRNAs) to direct efficient silencing of any target transcript. amiRNA-mediated approach is one of the recently developed strategies with wide range of applications especially for conferring viral resistance in crop plants. Several studies have established potential of amiRNAs to target and degrade mRNAs of both viral and plant origin and thereby specifically degrading the target mRNA (Schwab et al. 2006, Niu et al. 2006, Qu et al. 2007, Zhang et al. 2012). Recently, amiRNAs targeting different ORFs of Tomato leaf curl virus AC1 along with AC2/AC4 (Yadava and Mukherjee 2010), the middle region of the AV1 (coat protein) transcript (amiR-AV1-3) and the overlapping region of the AV1 and AV2 (pre-coat protein) transcripts (amiR-AV1-1) (Vu et al. 2013) were designed and expressed in transgenic tomato plants to confer resistance and tolerance to ToLCV, respectively. Seemingly, amiRNA approach has several advantages over conventional siRNA-mediated strategy. In the hairpin RNAi approach, multiple siRNAs are formed from single precursor, and off-target genes are often silenced, while in amiRNA approach only one mature miRNA is produced targeting the specific gene. In amiRNAs mismatches can be introduced to avoid signal amplification and transitivity. siRNA-based gene silencing has been shown to be temperature dependent, while miRNA biogenesis has been shown to occur in various conditions and under extreme temperatures and therefore has wider scope of applications.

Conclusion

RNA silencing is an evolutionary conserved mechanism, which operates in several eukaryotic organisms across kingdoms and involves highly sequence-specific degradation of complementary RNA and transcriptional gene silencing. sRNAs of 21–24 nts in length are the key players of RNA silencing. The major components (players) of RNA silencing machinery in plants include AGO1, RDRs, DCLs, HEN1 and HYL1. These components are required for processing of dsRNA into siRNA and maintenance of RNA silencing. The mobile silencing signal moves from initiating cell to neighbouring cells through plasmodesmata and to long distance through phloem. Viruses are one of the most devastating pathogens of plants causing substantial crop loss every year. Viruses are both inducers and targets of RNA silencing. dsRNAs generated during replication of RNA viruses or transcription of overlapping sequences in DNA viruses induce RNA silencing which leads to sequence-specific degradation of target RNA into 21–24-nt siRNAs. Viruses in turn evolved suppressors of RNA silencing as powerful weapon to counter the host defence machinery. The suppressors encoded by different plant viruses act at different steps of RNA silencing thus inhibiting RNA silencing pathways in plants. Occasionally, infected plants show recovery from virus infection leading to remission of symptoms. Recovered plants remain immune to subsequent infection by a homologous virus through RNA silencing mechanism. Based upon the knowledge of RNA silencing mechanism, it is possible to engineer virus resistance in plants based on RNA silencing using viral-derived sequences as target.

Acknowledgement We acknowledge the support for the Department of Biotechnology (DBT), Govt. of India, for financial assistance.

References

Ahlquist P (2002) RNA-dependent RNA polymerases, viruses, and RNA silencing. Science 296:1270–1273

Anandalakshmi R, Pruss GJ, Ge X, Marathe R, Mallory AC, Smith TH, Vance VB (1998) A viral suppressor of gene silencing in plants. Proc Natl Acad Sci USA 95:13079–13084

Aoki K, Suzui N, Fujimaki S, Dohmae N, Yonekura-Sakakibara K, Fujiwara T, Hayashi H, Yamaya T,

Sakakibara H (2005) Destination-selective long-distance movement of phloem proteins. Plant Cell 17:1801–1814
Baulcombe D (2004) RNA silencing in plants. Nature 431:356–363
Baumberger N, Baulcombe DC (2005) Arabidopsis ARGONAUTE1 is an RNA Slicer that selectively recruits microRNAs and short interfering RNAs. Proc Natl Acad Sci USA 102:11928–11933
Baumberger N, Tsai CH, Lie M, Havecker E, Baulcombe DC (2007) The Polerovirus silencing suppressor P0 targets ARGONAUTE proteins for degradation. Curr Biol 17:1609–1614
Bazzini AA, Hopp HE, Beachy RN, Asurmendi S (2007) Infection and coaccumulation of tobacco mosaic virus proteins alter microRNA levels, correlating with symptom and plant development. Proc Natl Acad Sci USA 104:12157–12162
Bernstein E, Caudy AA, Hammond SM, Hannon GJ (2001) Role for a bidentate ribonuclease in the initiation step of RNA interference. Nature 409:363–366
Billy E, Brondani V, Zhang H, Muller U, Filipowicz W (2001) Specific interference with gene expression induced by long, double-stranded RNA in mouse embryonal teratocarcinoma cell lines. Proc Natl Acad Sci USA 98:14428–14433
Biltko V, Barik S (2001) Phenotypic silencing of cytoplasmic genes using sequence specific double stranded interfering RNA and its application in the reverse genetics of wild type negative strand RNA viruses. Biomed Center Microbiol 1:34–55
Bisaro DM (2006) Silencing suppression by geminivirus proteins. Virology 344:158–168
Blander G, Guarente L (2004) The Sir2 family of protein deacetylases. Ann Rev Biochem 73:417–435
Blevins T, Rajeswaran R, Shivaprasad PV, Beknazariants D, Si-Ammour A, Park HS, Vazquez F, Robertson D, Meins F, Hohn T, Pooggin MM (2006) Four plant Dicers mediate viral small RNA biogenesis and DNA virus induced silencing. Nucleic Acids Res 34:6233–6246
Boden D, Pusch O, Lee F, Tucker L, Shank PR, Ramratnam B (2003) Promoter choice affects the potency of HIV-1 specific RNA interference. Nucleic Acids Res 31:5033–5038
Bortolamiol D, Pazhouhandeh M, Marrocco K, Genschik P, Ziegler-Graff V (2007) The Polerovirus F box protein P0 targets ARGONAUTE1 to suppress RNA silencing. Curr Biol 17:1615–1621
Bouche N, Lauressergues D, Gascioli V, Vaucheret H (2006) An antagonistic function for *Arabidopsis* DCL2 in development and a new function for DCL4 in generating viral siRNAs. The EMBO J 25:3347–3356
Brigneti G, Voinnet O, Li WX, Ji LH, Ding SW, Baulcombe DC (1998) Viral pathogenicity determinants are suppressors of transgene silencing in Nicotiana benthamiana. EMBO J 17:6739–6746
Brodersen P, Sakvarelidze-Achard L, Bruun-Rasmussen M, Dunoyer P, Yamamoto YY, Sieburth L, Voinnet O (2008) Widespread translational inhibition by plant miRNAs and siRNAs. Science 320:1185–1190
Brosnan CA, Vionnet O (2011) Cell to cell and long-distance siRNA movement in plants: mechanisms and biological implications. Curr Opin Plant Biol 14:580–587
Bucher E, Sijen T, De Haan P, Goldbach R, Prins M (2003) Negative-strand tospoviruses and tenuiviruses carry a gene for a suppressor of gene silencing at analogous genomic positions. J Virol 77:1329–1336
Bucher E, Lohuis D, van Poppel PMJA, Geerts-Dimitriadou C, Goldbach R, Prins M (2006) Multiple virus resistance at a high frequency using a single transgene construct. J Gen Virol 87:3697–3701
Buhtz A, Pieritz J, Springer F, Kehr J (2010) Phloem small RNAs, nutrient stress responses, and systemic mobility. BMC Plant Biol 10:64–76
Cao XS, Zhou P, Zhang XM, Zhu SF, Zhong XH, Xiao Q, Ding B, Li Y (2005) Identification of an RNA silencing suppressor from a plant double-stranded RNA virus. J Virol 79:13018–13027
Carrington JC, Kasschau KD, Mahajan SK, Schaad MC (1996) Cell-to-cell and long-distance transport of viruses in plants. Plant Cell 8:1669–1681
Chellappan P, Vanitharani R, Pita J, Fauquet CM (2004) Short interfering RNA accumulation correlates with host recovery in DNA virus-infected hosts, and gene silencing targets specific viral sequences. J Virolol 78:7465–7477
Chellappan P, Vanitharani R, Fauquet CM (2005) MicroRNA-binding viral protein interferes with Arabidopsis development. Proc Natl Acad Sci USA 102:10381–10386
Chen J, Li WX, Xie DX, Peng JR, Ding SW (2004) Viral virulence protein suppresses RNA silencing-mediated defense but upregulates the role of MicroRNA in host gene expression. Plant Cell 16:1302–1313
Chiba M, Reed JC, Prokhnevsky AI, Chapman EJ, Mawassi M, Koonin EV, Carrington JC, Dolja VV (2006) Diverse suppressors of RNA silencing enhance agroinfection by a viral replicon. Virology 346:7–14
Cogoni C, Irelan JT, Schumacher M, Schmidhauser T, Selker EU, Macino G (1996) Transgene silencing of the al-1 gene in vegetative cells of Neurospora is mediated by a cytoplasmic effector and does not depend on DNA-DNA interactions or DNA methylation. EMBO J 15:3153–3163
Covey SN, AlKaff NS, Langara A, Turner DS (1997) Plants combat infection by gene silencing. Nature 385:781–782
Csorba T, Bovi A, Dalmay T, Burgyan J (2007) The p122 subunit of tobacco mosaic virus replicase is a potent silencing suppressor and compromises both small interfering RNA- and MicroRNA-mediated pathways. J Virol 81:11768–11780
Cuellar WJ, Kreuze JF, Rajamaki ML, Cruzado KR, Untiveros M, Valkonen JPT (2009) Elimination of

antiviral defense by viral RNase III. Proc Natl Acad Sci USA 106:10354–10358

Curaba JL, Chen XM (2008) Biochemical activities of Arabidopsis RNA-dependent RNA polymerase 6. J Biol Chem 283:3059–3066

Dalmay T, Szittya G, Burgyan J (1995) Generation of defective interfering RNA dimers of cymbidium ringspot tombusvirus. Virology 207:510–517

Dalmay T, Hamilton A, Rudd S, Angell S, Baulcombe DC (2000) An RNA-dependent RNA polymerase gene in *Arabidopsis* is required for posttranscriptional gene silencing mediated by a transgene but not by a virus. Cell 101:543–553

Das AT, Brummelkamp TR, Westerhout EM, Vink M, Madiredjo M, Bernards R, Berkhout B (2004) Human immunodeficiency virus type 1 escapes from RNA interference-mediated inhibition. J Virol 78:2601–2605

Dekker EL, Derks AF, Asjes CJ, Lemmers ME, Bol JF, Langeveld SA (1993) Characterization of potyviruses from tulip and lily which cause flower-breaking. J Gen Virol 74(Pt 5):881–887

Ding SW (2010) RNA-based antiviral immunity. Nat Rev Immunol 10:632–644

Donaire L, Wang Y, Gonzalez-Ibeas D, Mayer KF, Aranda MA, Llave C (2009) Deep-sequencing of plant viral small RNAs reveals effective and widespread targeting of viral genomes. Virology 392:203–214

Dunoyer P, Ritzenthaler C, Hemmer O, Michler P, Fritsch C (2002) Intracellular localization of the Peanut clump virus replication complex in tobacco BY-2 protoplasts containing green fluorescent protein-labelled endoplasmic reticulum or golgi apparatus. J Virol 76:865–874

Dunoyer P, Lecellier CH, Parizotto EA, Himber C, Voinnet O (2004) Probing the microRNA and small interfering RNA pathways with virus-encoded suppressors of RNA silencing. Plant Cell 16:1235–1250

Dunoyer P, Himber C, Voinnet O (2005) DICER-LIKE 4 is required for RNA interference and produces the 21-nucleotide small interfering RNA component of the plant cell-to-cell silencing signal. Nat Genet 37:1356–1360

Dunoyer P, Himber C, Ruiz-Ferrer V, Alioua A, Voinnet O (2007) Intra- and intercellular RNA interference in *Arabidopsis* thaliana requires components of the microRNA and heterochromatic silencing pathways. Nat Genet 39:848–856

Dunoyer P, Brosnan CA, Schott G, Wang Y, Jay F, Alioua A, Himber C, Voinnet O (2010) An endogenous systemic RNAi pathway in plants. EMBO J 29:1699–1712

Eamens A, Wang MB, Smith NA, Waterhouse PM (2008) RNA silencing in plants: yesterday, today, and tomorrow. Plant Physiol 147:456–468

Elbashir SM, Martinez J, Patkaniowska A, Lendeckel W, Tuschl T (2001) Functional anatomy of siRNAs for mediating efficient RNAi in *Drosophila melanogaster* embryo lysate. EMBO J 20:6877–6888

Fagard M, Boutet S, Morel JB, Bellini C, Vaucheret H (2000) AGO1, QDE-2, and RDE-1 are related proteins required for post-transcriptional gene silencing in plants, quelling in fungi, and RNA interference in animals. Proc Natl Acad Sci USA 97:11650–11654

Fang Y, Spector DL (2007) Identification of nuclear dicing bodies containing proteins for microRNA biogenesis in living *Arabidopsis* plants. Curr Biol 17:818–823

Fire A, Xu S, Montgomery MK, Kostas SA, Driver SE, Mello CC (1998) Potent and specific genetic interference by double-stranded RNA in *Caenorhabditis elegans*. Nature 391:806–811

Frank F, Sonenberg N, Nagar B (2010) Structural basis for 5′-nucleotide base-specific recognition of guide RNA by human AGO2. Nature 465:818–822

Frank F, Hauver J, Sonenberg N, Nagar B (2012) Arabidopsis Argonaute MID domains use their nucleotide specificity loop to sort small RNAs. EMBO J 31:3588–3595

Garcia-Perez RD, Houdt HV, Depicker A (2004) Spreading of post-transcriptional gene silencing along the target gene promotes systemic silencing. Plant J 38:594–602

Gasciolli V, Mallory AC, Bartel DP, Vaucheret H (2005) Partially redundant functions of *Arabidopsis* DICER-like enzymes and a role for DCL4 in producing trans-acting siRNAs. Curr Biol 15:1494–1500

Ge Q, McManus MT, Nguyen T, Shen CH, Sharp PA, Eisen HN, Chen J (2003) RNA interference of influenza virus production by directly targeting mRNA for degradation and indirectly inhibiting all viral RNA transcription. Proc Natl Acad Sci USA 100:2718–2723

Glick E, Zrachya A, Levy Y, Mett A, Gidoni D, Belausov E, Citovsky V, Gafni Y (2008) Interaction with host SGS3 is required for suppression of RNA silencing by tomato yellow leaf curl virus V2 protein. Proc Natl Acad Sci USA 105:157–161

Goswami S, Sahana N, Pandey V, Doblas P, Jain RK, Palukaitis P, Canto T, Praveen S (2012) Interference in plant defense and development by non-structural protein NSs of Groundnut bud necrosis virus. Virus Res 163(1):368–373

Haag JR, Pikaard CS (2011) Multisubunit RNA polymerases IV and V: purveyors of non-coding RNA for plant gene silencing. Nat Rev Mol Cell Biol 12:483–492

Hagen C, Rojas MR, Kon T, Gilbertson RL (2008) Recovery from *Cucurbit leaf crumple virus* (Family *Geminiviridae*, Genus *Begomovirus*) infection is an adaptive antiviral response associated with changes in viral small RNAs. Virology 98:1029–1037

Ham BK, Brandom JL, Xoconostle-Cazares B, Ringgold V, Lough TJ, Lucas WJ (2009) A polypyrimidine tract binding protein, pumpkin RBP50, forms the basis of a phloem-mobile ribonucleoprotein complex. Plant Cell 21:197–215

Hamilton A, Voinnet O, Chappell L, Baulcombe DC (2002) Two classes of short interfering RNA in RNA silencing. EMBO J 21:4671–4679

Han MH, Goud S, Song L, Fedoroff N (2004) The Arabidopsis double-stranded RNA-binding protein HYL1 plays a role in microRNA-mediated gene regulation. Proc Natl Acad Sci USA 101:1093–1098

Hao L, Wang H, Sunter G, Bisaro DM (2003) Geminivirus AL2 and L2 proteins interact with and inactivate SNF1 kinase. Plant Cell 15:1034–1048

Harries PA, Palanichelvam K, Yu W, Schoelz JE, Nelson RS (2009) The Cauliflower mosaic virus protein p6 forms motile inclusions that traffic along actin microfilaments and stabilize microtubules. Plant Physiol 149:1005–1016

Hass G, Azevedo J, Moissiard G, Geldreich A, Himber C, Bureau M, Fukuhara T, Keller M, Voinnet O (2008) Nuclear import of CaMV P6 is required for infection and suppression of the RNA silencing factor DRB4. EMBO J 27:2102–2112

Havecker ER, Wallbridge LM, Hardcastle TJ, Bush MS, Kelly KA, Dunn RM, Schwach F, Doonan JH, Baulcombe DC (2010) The Arabidopsis RNA-directed DNA methylation argonautes functionally diverge based on their expression and interaction with target loci. Plant Cell 22:321–334

Havelda Z, Varallyay E, Valoczi A, Burgyan J (2008) Plant virus infection-induced persistent host gene downregulation in systemically infected leaves. Plant J 55:278–288

Herr AJ, Jensen MB, Dalmay T, Baulcombe DC (2005) RNA polymerase IV directs silencing of endogenous DNA. Science 308:118–120

Himber C, Dunoyer P, Moissiard G, Ritzenthaler C, Voinnet O (2003) Transitivity-dependent and -independent cell-to-cell movement of RNA silencing. EMBO J 22:4523–4533

Hiraguri A, Itoh R, Kondo N, Nomura Y, Aizawa D, Murai Y, Koiwa H, Seki M, Shinozaki K, Fukuhara T (2005) Specific interactions between Dicer-like proteins and HYL1/DRB family dsRNA-binding proteins in Arabidopsis thaliana. Plant Mol Biol 57:173–188

Hormuzdi SG, Bisaro DM (1995) Genetic analysis of beet curly top virus: examination of the roles of L2 and L3 genes in viral pathogenesis. Virology 206:1044–1054

Hull R (2002) Mathews plant virology, 4th edn. Academic, San Diego, pp 47–74

Hunter C, Sun H, Poethig RS (2003) The Arabidopsis heterochronic gene ZIPPY is an ARGONAUTE family member. Curr Biol 13:1734–1739

Hutvagner G, Mlynarova L, Nap JP (2000) Detailed characterization of the posttranscriptional gene-silencing-related small RNA in a GUS gene-silenced tobacco. RNA 6:1445–1454

Imlau A, Truernit E, Sauer N (1999) Cell-to-cell and long-distance trafficking of the green fluorescent protein in the phloem and symplastic unloading of the protein into sink tissues. Plant Cell 11:309–322

Inaba J, Kim BM, Shimura H, Masuta C (2011) Virus-induced necrosis is a consequence of direct protein-protein interaction between a viral RNA-silencing suppressor and a host catalase. Plant Physiol 156:2026–2036

Ishibashi K, Nishikiori M, Ishikawa M (2010) Interactions between tobamovirus replication *proteins* and cellular factors: their impacts on virus multiplication. Mol Plant Microbe Interact 23:1413–1419

Jan FJ, Pang SZ, Tricoli DM, Gonsalves D (2000) Evidence that resistance in squash mosaic comovirus coat protein-transgenic plants is affected by plant developmental stage and enhanced by combination of transgenes from different lines. J Gen Virol 81:2299–2306

Jauvion V, Rivard M, Bouteiller N, Elmayan T, Vaucheret H (2012) RDR2 partially antagonizes the production of RDR6-dependent siRNA in sense transgene-mediated PTGS. PLoS One 7:e29785

Jia D, Guo N, Chen H, Akita F, Xie L, Omura T, Wei T (2012) Assembly of the viroplasm by viral non-structural protein Pns10 is essential for persistent infection of rice ragged stunt virus in its insect vector. J Gen Virol 93(Pt 10):2299–2309

Jones L, Hamilton AJ, Voinnet O, Thomas CL, Maule AJ, Baulcombe DC (1999) RNA–DNA interactions and DNA methylation in post-transcriptional gene silencing. Plant Cell 11:2291–2301

Jovel J, Walker M, Sanfacon H (2007) Recovery of *Nicotiana benthamiana* plants from a necrotic response induced by a nepovirus is associated with RNA silencing but not with reduced virus titer. J Virol 81:12285–12297

Juarez MT, Kui JS, Thomas J, Heller BA, Timmermans MCP (2004) microRNA-mediated repression of rolled leaf1 specifies maize leaf polarity. Nature 428:84–88

Kalantidis K, Schumacher HT, Alexiadis T, Helm JM (2008) RNA silencing movement in plants. Biol Cell 100:13–26

Kalinina NO, Rakitina DV, Solovyev AG, Schiemann J, Morozov SY (2002) RNA helicase activity of the plant virus movement proteins encoded by the first gene of the triple gene block. Virology 296:321–329

Kanno T, Huettel B, Mette MF, Aufsatz W, Jaligot E, Daxinger L, Kreil DP, Matzke M, Matzke AJ (2005) Atypical RNA polymerase subunits required for RNA-directed DNA methylation. Nat Genet 37:761–765

Kasschau KD, Carrington JC (1998) A counterdefensive strategy of plant viruses: suppression of posttranscriptional gene silencing. Cell 95:461–470

Kasschau KD, Xie Z, Allen E, Llave C, Chapman EJ, Krizan KA, Carrington JC (2003) P1/HC-Pro, a viral suppressor of RNA silencing, interferes with Arabidopsis development and miRNA function. Dev Cell 4:205–217

Kawamura Y, Saito K, Kin T, Ono Y, Asai K, Sunohara T, Okada TN, Siomi MC, Siomi H (2008) *Drosophila* endogenous small RNAs bind to Argonaute 2 in somatic cells. Nature 453:793–795

Khvorova A, Reynolds A, Jayasena SD (2003) Functional siRNAs and miRNAs exhibit strand bias. Cell 115:209–216

Kim I, Zambryski PC (2005) Cell-to-cell communication via plasmodesmata during *Arabidopsis* embryogenesis. Curr Opin Plant Biol 8:593–599

Kobayashi K, Zambryski P (2007) RNA silencing and its cell-to-cell spread during *Arabidopsis* embryogenesis. Plant J 50:597–604

Kubota K, Tsuda S, Tamai A, Meshi T (2003) Tomato mosaic virus replication protein suppresses virus-targeted posttranscriptional gene silencing. J Virol 77:11016–11026

Kuwata S, Masuta C, Takanami Y (1991) Reciprocal phenotype alterations between two satellite RNAs of cucumber mosaic virus. J Gen Virol 72(Pt 10):2385–2389

Kwak PB, Tomari Y (2012) The N domain of Argonaute drives duplex unwinding during RISC assembly. Nat Struct Mol Biol 19:145–151

Laco GS, Beachy RN (1994) Rice tungro bacilliform virus encodes reverse transcriptase, DNA polymerase, and ribonuclease H activities. Proc Natl Acad Sci USA 91:2654–2658

Li H, Li WX, Ding SW (2002) Induction and suppression of RNA silencing by an animal virus. Science 296:1319–1321

Lin SI, Chiang SF, Lin WY, Chen JW, Tseng CY, Wu PC, Chiou TJ (2008) Regulatory network of microRNA399 and PHO2 by systemic signaling. Plant Physiol 147:732–746

Lindbo JA, Silvarosales L, Proebsting WM, Dougherty WG (1993) Induction of a highly specific antiviral state in transgenic plants – implications for regulation of gene-expression and virus-resistance. Plant Cell 5:1749–1759

Lingel A, Simon B, Izaurralde E, Sattler M (2003) Structure and nucleic-acid binding of the *Drosophila* Argonaute 2 PAZ domain. Nature 426:465–469

Lingel A, Simon B, Izaurralde E, Sattler M (2004) Nucleic acid 3′-end recognition by the Argonaute2 PAZ domain. Nat Struct Mol Biol 11:576–577

Lippman Z, Martienssen R (2004) The role of RNA interference in heterochromatic silencing. Nature 431:364–370

Liu Q, Rand TA, Kalidas S, Du F, Kim HE, Smith DP, Wang X (2003) R2D2, a bridge between the initiation and effector steps of the *Drosophila* RNAi pathway. Science 301:1921–1925

Liu JD, Carmell MA, Rivas FV, Marsden CG, Thomson JM, Song JJ, Hammond SM, Joshua-Tor L, Hannon GJ (2004) Argonaute2 is the catalytic engine of mammalian RNAi. Science 305:1437–1441

Liu FQ, Bakht S, Dean C (2012) Cotranscriptional role for *Arabidopsis* DICER-LIKE 4 in transcription termination. Science 335:1621–1623

Llave C (2010) Virus-derived small interfering RNAs at the core of plant-virus interactions. Trends Plant Sci 15:701–707

Llave C, Kasschau KD, Carrington JC (2000) Virus-encoded suppressor of posttranscriptional gene silencing targets a maintenance step in the silencing pathway. Proc Natl Acad Sci USA 97:13401–13406

Love AJ, Geri C, Laird J, Carr C, Yun BW, Loake GJ, Tada Y, Sadanandom A, Milner JJ (2012) Cauliflower mosaic virus protein P6 inhibits signaling responses to salicylic acid and regulates innate immunity. PLoS One 7(10):e47535

Lozsa R, Csorba T, Lakatos L, Burgyan J (2008) Inhibition of 3′ modification of small RNAs in virus-infected plants require spatial and temporal co-expression of small RNAs and viral silencing-suppressor proteins. Nucleic Acids Res 36:4099–4107

Lu R, Folimonov A, Shintaku M, Li WX, Falk BW, Dawson WO, Ding SW (2004) Three distinct suppressors of RNA silencing encoded by a 20-kb viral RNA genome. Proc Natl Acad Sci USA 101:15742–15747

Lucas WJ, Lee JY (2004) Plant cell biology – Plasmodesmata as a supracellular control network in plants. Nat Rev Mol Cell Biol 5:712–726

Lucas WJ, Bouche-Pillon S, Jackson DP, Nguyen L, Baker L, Ding B, Hake S (1995) Selective trafficking of KNOTTED1 homeodomain protein and its mRNA through plasmodesmata. Science 270:1980–1983

Ma JB, Ye KQ, Patel DJ (2004) Structural basis for overhang-specific small interfering RNA recognition by the PAZ domain. Nature 429:318–322

Ma JB, Yuan YR, Meister G, Pei Y, Tuschl T, Patel DJ (2005) Structural basis for 5′-end-specific recognition of guide RNA by the *A. fulgidus* Piwi protein. Nature 434:666–670

Mallory AC, Reinhart BJ, Jones-Rhoades MW, Tang G, Zamore PD, Barton MK, Bartel DP (2004) MicroRNA control of PHABULOSA in leaf development: importance of pairing to the microRNA 59 region. EMBO J 23:3356–3364

Matzke M, Kanno T, Daxinger L, Huettel B, Matzke AJ (2009) RNA-mediated chromatin-based silencing in plants. Curr Opin Cell Biol 21:367–376

Maule AJ (2008) Plasmodesmata: structure, function and biogenesis. Curr Opin Cell Biol 11:680–686

Mayers CN, Palukaitis P, Carr JP (2000) Subcellular distribution analysis of the cucumber mosaic virus 2b protein. J Gen Virol 81:219–226

Meister G, Tuschl T (2004) Mechanisms of gene silencing by double-stranded RNA. Nature 431:343–349

Merai Z, Kerenyi Z, Molnar A, Barta E, Valoczi A, Bisztray G, Havelda Z, Burgyan J, Silhavy D (2005) Aureusvirus P14 is an efficient RNA silencing suppressor that binds double-stranded RNAs without size specificity. J Virol 79:7217–7226

Mette MF, Aufsatz W, van der Winden J, Matzke MA, Matzke AJM (2000) Transcriptional silencing and promoter methylation triggered by double-stranded RNA. EMBO J 19:5194–5201

Mi SJ, Cai T, Hu YG, Chen Y, Hodges E, Ni FR, Wu L, Li S, Zhou H, Long CZ et al (2008) Sorting of small RNAs into Arabidopsis argonaute complexes is directed by the 5′ terminal nucleotide. Cell 133:116–127

Moissiard G, Voinnet O (2006) RNA silencing of host transcripts by cauliflower mosaic virus requires coordinated action of the four *Arabidopsis* Dicer-like proteins. Proc Natl Acad Sci USA 103:19593–19598

Molnar A, Csorba T, Lakatos L, Varallyay E, Lacomme C, Burgyan J (2005) Plant virus-derived small interfering RNAs originate predominantly from highly structured single-stranded viral RNAs. J Virol 79:7812–7818

Montgomery TA, Yoo SJ, Fahlgren N, Gilbert SD, Howell MD, Sullivan CM, Alexander A, Nguyen G, Allen E, Ahn JH, Carrington JC (2008) AGO1-miR173 complex initiates phased siRNA formation in plants. Proc Natl Acad Sci USA 105:20055–20062

Mori M, Mise K, Okuno T, Furusawa I (1992) Expression of brome mosaic virus-encoded replicase genes in transgenic tobacco plants. J Gen Virol 73(Pt 1):169–172

Mourrain P, Beclin C, Elmayan T, Feuerbach F, Godon C, Morel JB, Jouette D, Lacombe AM, Nikic S, Picault N, Remoue K, Sanial M, Vo TA, Vaucheret H (2000) *Arabidopsis* SGS2 and SGS3 genes are required for posttranscriptional gene silencing and natural virus resistance. Cell 101:533–542

Napoli C, Lemieux C, Jorgensen R (1990) Introduction of a chimeric chalcone synthase gene into petunia results in reversible co-suppression of homologous genes in trans. Plant Cell 2:279–289

Niu QW, Lin SS, Reyes JL, Chen KC, Wu HW, Yeh SD, Chua NH (2006) Expression of artificial microRNAs in transgenic *Arabidopsis* thaliana confers virus resistance. Nat Biotechnol 24:1420–1428

Nykanen A, Haley B, Zamore PD (2001) ATP requirements and small interfering RNA structure in the RNA interference pathway. Cell 107:309–321

Onodera Y, Haag JR, Ream T, Costa Nunes P, Pontes O, Pikaard CS (2005) Plant nuclear RNA polymerase IV mediates siRNA and DNA methylation-dependent heterochromatin formation. Cell 120:613–622

Oparka KJ (2004) Getting the message across: how do plant cells exchange macromolecular complexes? Trends Plant Sci 9:33–41

Palauqui JC, Elmayan T, Pollien JM, Vaucheret H (1997) Systemic acquired silencing: transgene-specific post-transcriptional silencing is transmitted by grafting from silenced stocks to non-silenced scions. EMBO J 16:4738–4745

Pallas V, Garcia JA (2011) How do plant viruses induce disease? Interactions and interference with host components. J Gen Virol 92:2691–2705

Parizotto EA, Dunoyer P, Rahm N, Himber C, Voinnet O (2004) In vivo investigation of the transcription, processing, endonucleolytic activity, and functional relevance of the spatial distribution of a plant miRNA. Genes Dev 18:2237–2242

Park W, Li JJ, Song RT, Messing J, Chen XM (2002) CARPEL FACTORY, a Dicer homolog, and HEN1, a novel protein, act in microRNA metabolism in *Arabidopsis* thaliana. Curr Biol 12:1484–1495

Parker JS, Roe SM, Barford D (2005) Structural insights into mRNA recognition from a PIWI domain-siRNA guide complex. Nature 434:663–666

Pazhouhandeh M, Dieterle M, Marrocco K, Lechner E, Berry B, Brault V, Hemmer O, Kretsch T, Richards KE, Genschik P, Ziegler-Graff V (2006) F-box-like domain in the polerovirus protein P0 is required for silencing suppressor function. Proc Natl Acad Sci USA 103:1994–1999

Pikaard CS (2006) Cell biology of the Arabidopsis nuclear siRNA pathway for RNA-directed chromatin modification. Cold Spring Harbour Symp Quant Biol 71:473–480

Pontier D, Yahubyan G, Vega D, Bulski A, Saez-Vasquez J, Hakimi MA, Lerbs-Mache S, Colot V, Lagrange T (2005) Reinforcement of silencing at transposons and highly repeated sequences requires the concerted action of two distinct RNA polymerases IV in *Arabidopsis*. Genes Dev 19:2030–2040

Powell AP, Nelson RS, De B, Hoffmann N, Rogers SG, Fraley RT, Beachy RN (1986) Delay of disease development in transgenic plants that express the tobacco mosaic virus coat protein gene. Science 232:738–743

Qi Y, Denli AM, Hannon GJ (2005) Biochemical specialization within *Arabidopsis* RNA silencing pathways. Mol Cell 19:421–428

Qu F, Ren T, Morris TJ (2003) The coat protein of turnip crinkle virus suppresses posttranscriptional gene silencing at an early initiation step. J Virol 77:511–522

Qu J, Ye J, Fang R (2007) Artificial microRNA-mediated virus resistance in plants. J Virol 81:6690–6699

Ratcliff F, Harrison BD, Baulcombe DC (1997) A similarity between viral defense and gene silencing in plants. Science 276:1558–1560

Reed JC, Kasschau KD, Prokhnevsky AI, Gopinath K, Pogue GP, Carrington JC, Dolja VV (2003) Suppressor of RNA silencing encoded by beet yellows virus. Virology 306:203–209

Reinhart BJ, Weinstein EG, Rhoades MW, Bartel B, Bartel DP (2002) MicroRNAs in plants. Genes Dev 16:2313–2313

Rivas FV, Tolia NH, Song JJ, Aragon JP, Liu JD, Hannon GJ, Joshua-Tor L (2005) Purified Argonaute2 and an siRNA form recombinant human RISC. Nat Struc Mol Biol 12:340–349

Rodriguez-Negrete EA, Carrillo-Tripp J, Rivera-Bustamante RF (2009) RNA silencing against geminivirus: complementary action of posttranscriptional gene silencing and transcriptional gene silencing in host recovery. J Virol 83:1332–1340

Ruiz-Ferrer V, Voinnet O (2009) Roles of plant small RNAs in biotic stress responses. Annu Rev Plant Biol 60:485–510

Saunders K, Bedford ID, Yahara T, Stanley J (2003) Aetiology: the earliest recorded plant virus disease. Nature 422:831

Schauer SE, Jacobsen SE, Meinke DW, Ray A (2002) DICER-LIKE1: blind men and elephants in *Arabidopsis* development. Trends Plant Sci 7:487–491

Schommer C, Palatnik JF, Aggarwal P, Chetelat A, Cubas P, Farmer EE, Nath U, Weigel D (2008) Control of jasmonate biosynthesis and senescence by miR319 targets. PLoS Biol 6:1991–2001

Schwab R, Ossowski S, Riester M, Warthmann N, Weigel D (2006) Highly specific gene silencing by artificial microRNAs in *Arabidopsis*. Plant Cell 18 (5):1121–1133

Schwach F, Vaistij FE, Jones L, Baulcombe DC (2005) An RNA-dependent RNA polymerase prevents meristem invasion by potato virus X and is required for the activity but not the production of a systemic silencing signal. Plant Physiol 138:1842–1852

Schwartz M, Chen J, Janda M, Sullivan M, den Boon J, Ahlquist P (2002) A positive-strand RNA virus replication complex parallels form and function of retrovirus capsids. Mol Cell 9:505–514

Shimura H, Pantaleo V, Ishihara T, Myojo N, Inaba J, Sueda K, Burgyan J, Masuta C (2011) A viral satellite RNA induces yellow symptoms on tobacco by targeting a gene involved in chlorophyll biosynthesis using the RNA silencing machinery. PLoS Pathog 7: e1002021

Sijen T, Fleenor J, Simmer F, Thijssen KL, Parrish S, Timmons L, Plasterk RHA, Fire A (2001) On the role of RNA amplification in dsRNA-triggered gene silencing. Cell 107:465–476

Silhavy D, Molnar A, Lucioli A, Szittya G, Hornyik C, Tavazza M, Burgyan J (2002) A viral protein suppresses RNA silencing and binds silencing-generated, 21- to 25-nucleotide double-stranded RNAs. EMBO J 21:3070–3080

Smith NA, Singh SP, Wang MB, Stoutjesdijk PA, Green AG, Waterhouse PM (2000) Gene expression – total silencing by intron-spliced hairpin RNAs. Nature 407:319–320

Smith LM, Pontes O, Searle I, Yelina N, Yousafzai FK, Herr AJ, Pikaard CS, Baulcombe DC (2007) An SNF2 protein associated with nuclear RNA silencing and the spread of a silencing signal between cells in *Arabidopsis*. Plant Cell 19:1507–1521

Smith NA, Eamens AL, Wang MB (2011) Viral small interfering RNAs target host genes to mediate disease symptoms in plants. PLoS Pathog 7:e1002022

Song JJ, Liu JD, Tolia NH, Schneiderman J, Smith SK, Martienssen RA, Hannon GJ, Joshua-Tor L (2003) The crystal structure of the Argonaute2 PAZ domain reveals an RNA binding motif in RNAi effector complexes. Nat Struct Biol 10:1026–1032

Song L, Han MH, Lesicka J, Fedoroff N (2007) *Arabidopsis* primary microRNA processing proteins HYL1 and DCL1 define a nuclear body distinct from the Cajal body. Proc Natl Acad Sci USA 104:5437–5442

Song XW, Li PC, Zhai JX, Zhou M, Ma LJ, Liu B, Jeong DH, Nakano M, Cao SY, Liu CY, Chu CC, Wang XJ, Green PJ, Meyers BC, Cao XF (2012) Roles of DCL4 and DCL3b in rice phased small RNA biogenesis. Plant J 69:462–474

Szittya G, Moxon S, Pantaleo V, Toth G, Rusholme Pilcher RL, Moulton V, Burgyan J, Dalmay T (2010) Structural and functional analysis of viral siRNAs. PLoS Pathog 6:e1000838

Tabara H, Yigit E, Siomi H, Mello CC (2002) The dsRNA binding protein RDE-4 interacts with RDE-1, DCR- 1, and a DexH-box helicase to direct RNAi in *C. elegans*. Cell 109:861–871

Tabler M, Tsagris M (2004) Viroids: petite RNA pathogens with distinguished talents. Trends Plant Sci 9:339–348

Takeda A, Iwasaki S, Watanabe T, Utsumi M, Watanabe Y (2008) The mechanism selecting the guide strand from small RNA duplexes is different among Argonaute proteins. Plant Cell Physiol 49:493–500

Thomas CL, Leh V, Lederer C, Maule AJ (2003) Turnip crinkle virus coat protein mediates suppression of RNA silencing in *Nicotiana benthamiana*. Virology 306:33–41

Vaistij FE, Jones L, Baulcombe DC (2002) Spreading of RNA targeting and DNA methylation in RNA silencing requires transcription of the target gene and a putative RNA-dependent RNA polymerase. Plant Cell 14:857–867

van der Krol AR, Mur LA, Beld M, Mol JN, Stuitje AR (1990) Flavonoid genes in petunia: addition of a limited number of gene copies may lead to a suppression of gene expression. Plant Cell 2:291–299

Vance VB, Berger PH, Carrington JC, Hunt AG, Shi XM (1995) 5′ proximal potyviral sequences mediate potato virus X/potyviral synergistic disease in transgenic tobacco. Virology 206:583–590

Vanitharani R, Chellappan P, Fauquet CM (2005) Geminiviruses and RNA silencing. Trends Plant Sci 10:144–151

Vazquez F, Gasciolli V, Crete P, Vaucheret H (2004) The nuclear dsRNA binding protein HYL1 is required for microRNA accumulation and plant development, but not posttranscriptional transgene silencing. Curr Biol 14:346–351

Vogler H, Akbergenov R, Shivaprasad PV, Dang V, Fasler M, Kwon MO, Zhanybekova S, Hohn T, Heinlein M (2007) Modification of small RNAs associated with suppression of RNA silencing by tobamovirus replicase protein. J Virol 81:10379–10388

Voinnet O, Baulcombe DC (1997) Systemic signalling in gene silencing. Nature 389:553

Vionnet O, Lederer C, Baulcombe DC (2000) A viral movement protein prevents spread of the gene silencing signal in *Nicotiana benthamiana*. Cell 103:157–167

Vu TV, Choudhury NR, Mukherjee SK (2013) Transgenic tomato plants expressing artificial microRNAs for silencing the pre-coat and coat proteins of a begomovirus, tomato leaf curl New Delhi virus, show tolerance to virus infection. Virus Res 172 (1–2):35–45

Wang MB, Metzlaff M (2005) RNA silencing and antiviral defense in plants. Curr Opin Plant Biol 8:216–222

Wang H, Hao LH, Shung CY, Sunter G, Bisaro DM (2003) Adenosine kinase is inactivated by geminivirus AL2 and L2 proteins. Plant Cell 15:3020–3032

Wang MB, Bian XY, Wu LM, Liu LX, Smith NA, Isenegger D, Wu RM, Masuta C, Vance VB, Watson JM, Rezaian A, Dennis ES, Waterhouse PM (2004) On the role of RNA silencing in the pathogenicity and evolution of viroids and viral satellites. Proc Natl Acad Sci USA 101:3275–3280

Wang XB, Jovel J, Udomporn P, Wang Y, Wu Q, Li WX, Gasciolli V, Vaucheret H, Ding SW (2011) The 21-nucleotide, but not 22-nucleotide, viral secondary small interfering RNAs direct potent antiviral defense by two cooperative argonautes in *Arabidopsis* thaliana. Plant Cell 23:1625–1638

Wassenegger M, Krczal G (2006) Nomenclature and functions of RNA-directed RNA polymerases. Trends Plant Sci 11:142–151

Waterhouse PM, Wang MB, Lough T (2001) Gene silencing as an adaptive defence against viruses. Nature 411:834–842

Wei T, Kikuchi A, Moriyasu Y, Suzuki N, Shimizu T, Hagiwara K, Chen H, Takahashi M, Ichiki-Uehara T, Omura T (2006) The spread of Rice dwarf virus among cells of its insect vector exploits virus-induced tubu*lar struc*tures. J Virol 80:8593–8602

Wingard SA (1928) Hosts and symptoms of ring spot, a virus disease of plants. J Agric Res 37:127–153

Xie Z, Johansen LK, Gustafson AM, Kasschau KD, Lellis AD, Zilberman D, Jacobsen SE, Carrington JC (2004) Genetic and functional diversification of small RNA pathways in plants. PLoS Biol 2:E104

Xie ZX, Allen E, Wilken A, Carrington JC (2005) DICER-LIKE 4 functions in trans-acting small interfering RNA biogenesis and vegetative phase change in *Arabidopsis* thaliana. Proc Natl Acad Sci USA 102:12984–12989

Yadava P, Mukherjee SK (2010) Engineering geminivirus resistance in tomatoes using artificial microRNAs. Keystone symposium on RNA silencing mechanisms in plants, Santa Fe, 21–26 Feb 2010

Yang Z, Ebright YW, Yu B, Chen X (2006) HEN1 recognizes 21–24 nt small RNA duplexes and deposits a methyl group onto the 29 OH of the 39 terminal nucleotide. Nucleic Acids Res 34:667–675

Yang XL, Xie Y, Raja P, Li SZ, Wolf JN, Shen QT, Bisaro DM, Zhou XP (2011a) Suppression of methylation-mediated transcriptional gene silencing by beta C1-SAHH protein interaction during geminivirus-betasatellite infection. PLoS Pathog 7(10):e1002329

Yang X, Tan SH, The YJ, Yuan YA (2011b) Structural implications into dsRNA binding and RNA silencing suppression by NS3 protein of Rice Hoja Blanca Tenuivirus. RNA 17(5):903–911

Ye K, Patel DJ (2005) RNA silencing suppressor p21 of Beet yellows virus forms an RNA binding octameric ring structure. Structure 13:1375–1384

Yelina NE, Savenkov EI, Solovyev AG, Morozov SY, Valkonen JP (2002) Long-distance movement, virulence, and RNA silencing suppression controlled by a single protein in hordei- and potyviruses: complementary functions between virus families. J Virol 76:12981–12991

Ying XB, Dong L, Zhu H, Duan CG, Du QS, Lv DQ, Fang YY, Garcia JA, Fang RX, Guo HS (2010) RNA-dependent RNA polymerase 1 from *Nicotiana tabacum* suppresses RNA silencing and enhances viral infection in *Nicotiana benthamiana*. Plant Cell 22:1358–1372

Yoo BC, Kragler F, Varkonyi-Gasic E, Haywood V, Archer-Evans S, Lee YM, Lough TJ, Lucas WJ (2004) A systemic small RNA signaling system in plants. Plant Cell 16:1979–2000

Yu DQ, Fan BF, MacFarlane SA, Chen ZX (2003) Analysis of the involvement of an inducible *Arabidopsis* RNA-dependent RNA polymerase in antiviral defense. Mol Plant Microbe Interact 16:206–216

Yu B, Yang Z, Li J, Minakhina S, Yang M, Padgett RW, Steward R, Chen X (2005) Methylation as a crucial step in plant microRNA biogenesis. Science 307:932–935

Yuan YR, Pei Y, Ma JB, Kuryavyi V, Zhadina M, Meister G, Chen HY, Dauter Z, Tuschl T, Patel DJ (2005) Crystal structure of A-aeolicus Argonaute, a site-specific DNA-guided endoribonuclease, provides insights into RISC-mediated mRNA cleavage. Mol Cell 19:405–419

Zeevaart JA (2008) Leaf-produced floral signals. Curr Opin Plant Biol 11:541–547

Zhang X, Yuan YR, Pei Y, Lin SS, Tuschl T, Patel DJ, Chua NH (2006) Cucumber mosaic virus-encoded 2b suppressor inhibits Arabidopsis Argonaute1 cleavage activity to counter plant defense. Genes De 20:3255–3268

Zhang X, Li H, Zhang J, Zhang C, Gong P, Ziaf K, Xiao F, Ye Z (2011a) Expression of artificial microRNAs in tomato confers efficient and stable virus resistance in a cell-autonomous manner. Transgenic Res 20(3):569–581

Zhang Z, Chen H, Huang X, Xia R, Zhao Q, Lai J, Teng K, Li Y, Liang L, Du Q, Zhou X, Guo H, Xie Q (2011b) BSCTV C2 attenuates the degradation of SAMDC1 to suppress DNA methylation-mediated gene silencing in *Arabidopsis*. Plant Cell 23:273–288

Zhang X, Xia J, Lii YE, Barrera-Figueroa BE, Zhou X, Gao S, Lu L, Niu D, Chen Z, Leung C, Wong T, Zhang H, Guo J, Li Y, Liu R, Liang W, Zhu JK, Zhang W, Jin H (2012) Genome-wide analysis of plant nat-siRNAs reveals insights into their distribution, biogenesis and function. Genome Biol 13:R20

Zhu S, Gao F, Cao X, Chen M, Ye G, Wei C, Li Y (2005) The rice dwarf virus P2 protein interacts with entkaurene oxidases in vivo, leading to reduced biosynthesis of gibberellins and rice dwarf symptoms. Plant Physiol 139:1935–1945

Zilberman D, Cao X, Jacobsen SE (2003) ARGONAUTE4 control of locus-specific siRNA accumulation and DNA and histone methylation. Science 299:716–719

Zilberman D, Cao XF, Johansen LK, Xie ZX, Carrington JC, Jacobsen SE (2004) Role of *Arabidopsis* ARGONAUTE4 in RNA-directed DNA methylation triggered by inverted repeats. Curr Biol 14:1214–1220

Fungal Disease Management in Plants

Deepali Singh and Sachin Teotia

Abstract

Fungal diseases damage crop plants and affect agricultural production. The defence strategy of plants against stress factors involves a multitude of tools, including various types of stress proteins with putative protective functions. Recent molecular advancements in understanding plant–pathogen studies have led to the identification of various host genes involved in the plant's defence against pathogen attack. This knowledge has paved path for a number of options and strategies that can be and have been developed to make plants resistant to pathogens. These genes may involve resistance gene–avirulence gene interaction, antimicrobial peptides, enzymes for phytoalexin production, proteins involved in defence-signalling cascades and hydrolytic enzymes or pathogenesis-related proteins that are directly or indirectly responsible for the plant's defence responses following a pathogen attack. Recently small RNAs have been identified as key players of many pathways they are important transcriptional and post-transcriptional regulators of gene expression. RNA interference (RNAi) is an emerging strategy for control of fungal pathogens, through silencing of pathogen-associated genes. miRNAs also play an important role in plant defence responses to pathogen attack. Certain microRNAs (miRNAs) are up- or downregulated during pathogen attack, indicating that these miRNAs could play important roles in biotic stress tolerance. All this information has been/or is being used to produce fungus-resistant transgenic plants in different crop species.

Keywords

Pathogenesis-related proteins • Hypersensitive response • RNA interference • microRNAs • Fungal pathogens • Disease resistance

D. Singh (✉) • S. Teotia
School of Biotechnology, Gautam Buddha University,
Greater Noida, Uttar Pradesh, India
e-mail: deepali.singh@gmail.com

R.K. Gaur and P. Sharma (eds.), *Approaches to Plant Stress and their Management*,
DOI 10.1007/978-81-322-1620-9_19, © Springer India 2014

Introduction

Fungal diseases cause major yield losses and major economic and social problems in the affected countries. Phytopathogenic fungi and oomycetes are the major causative agents of infectious crop plant diseases (Kamoun 2001).

During any attempted fungal infection, the penetration and growth on host plants is arrested by a complex, multifaceted plant defence response comprising of rigidifications of the plant cell wall at the infection site, the production of antimicrobial plant protein enzymes and phytoalexins as well as programmed plant cell death, termed the hypersensitive response. Activation of disease-resistance responses in plants can be at non-cultivar-specific level or cultivar specific (gene-for-gene hypothesis) (Cohn et al. 2001; Dangl and Jones 2001; Heath 2000; Kamoun 2001). Cultivar specific is genetically determined by complementary pairs of pathogen-encoded avirulence (Avr) genes and plant resistance (R) genes (gene-for-gene hypothesis) (Cohn et al. 2001; Dangl and Jones 2001) and lack or non-functional products of either gene would result in disease.

Scientists have used agronomic practices to control plant diseases including use of agrochemicals, crop rotation and breeding disease-resistant varieties. However, extensive use of fungicides is environmentally not safe and breeding for disease resistance is a time-consuming process. Though breeders have successfully incorporated disease-resistance genes into plants, and almost every agricultural crop grown today has some form of genetic resistance, generally against a number of diseases. However, breeding has to be a continuous process as pathogens evolve very fast and crops become susceptible. A need for alternative strategies is the need of the time, and genetic engineering allows expression of new or modified traits. With the development of transformation and regeneration protocols for most of the crop plants as well as sequence information being continuously available through genome sequencing projects, identifying and cloning new disease-resistance genes has become very convenient.

We will review here three strategies for the production of fungus-resistant transgenics: (i) pathogenesis-related proteins, (ii) hypersensitive response and (iii) RNA interference.

PR Proteins

The concept of pathogenesis-related protein (PR) was first identified by Van Loon and Van Kammen (1970) to designate any protein expressed by the host plant but induced only in pathological or related situations (Antoniw et al. 1980), including viral, fungal or bacterial infections, parasitic attack by nematodes, phytophagous insects and other higher forms of animals such as herbivores. Recently it has been shown that colonization by non-pathogenic/beneficial fungi and bacteria also leads to induction of the PR proteins (Blilou et al. 2000; Coventry and Dubery 2001). The PR proteins consist of different groups of structurally and functionally unrelated proteins that are classified according to coding sequence similarities, serological relationships and enzymatic activities (Fritig et al. 1998; Somssich and Hahlbrock 1998). Initially, only four proteins were detected as part of hypersensitive response in tobacco plants infected with Tobacco Mosaic Virus (TMV). They were designated I, II, III and IV based on their increasing order of electrophoretic mobility (van Loon and van Kammen 1970). Later, these proteins were classified into five groups, PR-1 to PR-5. Each of these five classical groups of PR proteins comprised two subclasses: an acidic subclass, extracellular in occurrence and whose members are induced by salicylic acid (Chen et al. 1993); and a basic subclass, found in the plant cell vacuole, whose members are induced by ethylene or jasmonic acid (Boller et al. 1983; Selitrennikoff 2001; Thomma et al. 1998). At present there are 17 classes of PR protein, numbered in the order in which they were discovered, from PR-1 to PR-17 (Table 1). Following infection, they can account to more than 10 % of the total soluble protein. The term PR-like protein was coined to describe proteins that are present in healthy plants and are induced in a developmentally controlled and

Table 1 Families of pathogenesis-related proteins

Family	Representative protein	Biochemical properties	Molecular size range
PR-1	Tobacco PR-1a	Unknown	15–17 kDa
PR-2	Tobacco PR-2	Glucanase	30–41 kDa
PR-3	Tobacco P, Q	Chitinase type I, II, IV, V, VI, VII	35–46 kDa
PR-4	Tobacco R	Chitinase type I, II	13–14 kDa
PR-5	Tobacco S	Thaumatin-like	16–26 kDa
PR-6	Tomato inhibitor I	Proteinase inhibitor	8–22 kDa
PR-7	Tomato P69	Endoproteinase	69 kDa
PR-8	Cucumber chitinase	Chitinase type III	30–35 kDa
PR-9	Tobacco "lignin-forming peroxidase"	Peroxidase (POX)	50–70 kDa
PR-10	Parsley "PR1"	Ribonuclease-like	18–19 kDa
PR-11	Tobacco class V chitinase	Chitinase, type I	40 kDa
PR-12	Radish Rs-AFP3	Ion transport (defensins)	5 kDa
PR-13	*Arabidopsis* TH12.1	Thionins	5–7 kDa
PR-14	Barley LTP4	Lipid transfer proteins	9 kDa
PR-15	Barley OxOa (germin)	Oxalate oxidases	22–25 kDa
PR-16	Barley OxOLP	Oxalate oxidase-like protein	100 kDa (hexamer)
PR-17	Tobacco PRp27	Unknown	

Source: modified from van Loon et al. (2006)

tissue-specific manner. These proteins are basic in nature and localized intracellularly in the vacuole (van Loon et al. 1994). Recently, van Loon et al. (2006) introduced the general term 'inducible defence-related proteins' to describe proteins that are mostly non-detectable in healthy tissues and but are induced after pathogen infection.

Some PR proteins are induced under pathological conditions suggesting a role for these proteins in plant defence. Therefore, PR proteins are generally considered as defence proteins, acting as a barrier in preventing or limiting pathogen invasion and spread. PR proteins are produced by plants during normal development and also as part of an induced defence from fungal pathogens. Thus, their biosynthesis and accumulation is considered a major defence mechanism of plants against fungal pathogens (Odjakova and Hadjiivanova 2001; Somssich and Hahlbrock 1998). Some members of PR-protein family have been shown to exhibit antifungal properties *in vitro*; they are also induced *in vivo* in response to fungal attack. A large group of PR proteins has been shown to be rapidly and massively induced both locally around infection sites and systemically. PR proteins have also been shown to be induced in response to various environmental stress factors, such as drought, salinity, wounding, heavy metals, endogenous and exogenous elicitor treatment and plant growth regulators (Derckel et al. 1996; Xie et al. 1999; Yu et al. 2001). Developmentally induced PRs accumulate in an organ- and tissue-specific manner (Van Loon and Van Strien 1999). PRs in plants are encoded by a small multigene family. It has been proven that PR synthesis is regulated at transcriptional level (Hooft van Huijsduijnen et al. 1985; Cornelissen et al. 1986).

The function of many PR proteins is unknown. However, members of several of these families have shown damaging actions on the pathogens, thus exhibiting antifungal activity, in *in vitro* bioassays, and supporting a possible role for these proteins in plant defence (Kombrink and Somssich 1997; Odjakova and Hadjiivanova 2001). Chitinase activity was detected in PR-4, PR-8 and PR-11. Proteinase, peroxidase, ribonuclease and lysozyme activities were established in PR-7, PR-9, PR-10 and PR-8, respectively. The peroxidase activity of the PR-9 family may act in cell wall reinforcement by catalysing lignification, leading to enhanced resistance against multiple pathogens (Passardi et al. 2004). PR-6 has shown proteinase-inhibitory properties. Membrane-permeabilizing functions are characteristic of defensins, thiols and lipid-transfer proteins (LTPs), referred to as PR-12, PR-13 and PR-14, respectively. PR-15 (oxalate oxidases) and PR-16 (oxalate oxidase-like proteins) proteins generate hydrogen peroxide that may be toxic to attackers or stimulate plant defence responses (Bernier and Berna 2001). PR-1 and PR-5 (osmotins and thaumatin-like proteins) create transmembrane pores and have therefore been termed permatins (Vigers et al. 1992; Abad et al. 1996). Members of the PR-1 family have been associated with activity against oomycetes (van Loon et al. 2006). Plasma membrane-permeabilizing ability proper to PR-5, PR-12, PR-13 and PR-14 contributes to plasmolysis and damage of fungal and bacterial pathogens, inhibiting their growth and development (El-kereamy et al. 2011). PR-2 (β-1,3-glucanases) and PR-3, 4, 8 and 11 (chitinases) attack components of the cell walls in most higher fungi β-1,3-glucans and chitin, respectively, (Honée 1999). These oligosaccharides act as elicitors and induce a chain of defence reactions in the host plant. PR-17 proteins, as yet uncharacterized, have been detected in infected tobacco, wheat and barley (Christensen et al. 2002).

PR proteins have shown synergism. Since chitin and β-1,3-glucan are synthesized simultaneously in the apex of growing hyphae of filamentous fungi, the effectiveness of a hydrolytic enzyme may depend on the simultaneous action of another one to hydrolyse mixed chitin–glucan fibres (Stintzi et al. 1993). Thus, for example, class II β-1,3-glucanases only show antifungal activity *in vitro* when they are applied in combination with chitinases or class I β-1,3-glucanases (Theis and Stahl 2004). Thus, constitutive, high-level expression of combinations of PR proteins with different modes of action against target organisms may provide broad-spectrum, durable resistance to a variety of diseases and pathogens. Several reports have demonstrated that transgenic plants constitutively overexpressing some PR genes show enhanced resistance to fungal pathogens (Table 2) (Wally and Punja 2010). Examples include overexpression of chitinases, glucanases and ribosome-inactivating proteins (RIPs) (Alexander et al. 1993; Hong and Hwang 2006) and osmotin (Liu et al. 1994). Constitutive co-expression of PR proteins provides the host plant the required level of proteins for effective resistance before the pathogen attack (Table 2). For example, transgenic tomato plants expressing only a chitinase transgene or a β-1,3-glucanase transgene were susceptible to *Fusarium oxysporum*, but plants expressing both genes had significantly higher resistance than the plants expressing only one of these two enzymes (Jongedijk et al. 1995; Wally and Punja 2010).

Hypersensitive Response

Plants have sophisticated and efficient defence mechanisms to prevent the colonization of their tissues by microbial pathogens and parasites. Preformed physical and chemical barriers constitute the first line of defence. Second line of defence are inducible defence responses that are initiated after successful recognition of the invading pathogen. The induced defence responses can be assigned to three major categories, according to their distinct temporal and spatial expression patterns (Kombrink and Somssich 1995).

Hypersensitive response is manifested as rapid collapse and death of host cells. Recent genetic, biochemical and morphological evidence suggests that HR cell death in plants is controlled by endogenous genetic mechanisms and hence is a kind of programmed cell death (PCD), similar to apoptosis of mammalian cells (Danon et al. 2000). Disease resistance in plants is often mediated by specific interactions between plant resistance (*R*) genes and corresponding avirulence (*Avr*) genes of the pathogen (Dangl and Jones 2001). When corresponding *R* and *Avr* genes are present in both host and pathogen, respectively, the result is disease resistance; if either gene is inactive or absent, the result is disease. This model is explained by gene-for-gene interaction, this interaction activates a signal-transduction cascade leading to hypersensitive response. Many R-genes have been isolated from model and crop plants (Dangl and Jones 2001), and they encode five classes of proteins depending on the presence of typical structural motifs, such as a nucleotide-binding domains (NBs), leucine-rich repeats

Table 2 PR proteins (in chronological order) used for making fungus-resistant transgenic plants

Name of the gene	Source	Host plant	Reference
Chitinase (chi1)	*Rhizopus oligosporus*	Tobacco	Terakawa et al. (1997)
Chitinase (RCC2)	Rice	Chrysanthemum	Takatsu et al. (1999)
TLP	Rice	Wheat	Chen et al. (1999)
TLP	Rice	Rice	Datta et al. (1999)
Glucanase (SGN1)	Soybean	Tobacco	Cheong et al. (2000)
Chitinase (RCC2)	Rice	Grapevine	Yamamoto et al. (2000)
Chitinase (OsChia)	Rice	Rice	Takakura et al. (2000)
Chitinase (Chi)	Tobacco	Peanut	Rohini and Rao (2001)
Chitinase (RC7)	Rice	Rice	Datta et al. (2001)
TLP	Tomato	Orange	Fagoaga et al. (2001)
Chitinase (RCC2)	Rice	Cucumber	Kishimoto et al. (2002)
Chitinase like (Chs2)	American elm	Creeping bentgrass	Chai et al. (2002)
β-1,3-glucanase and chitinase genes	Pea	Potato	Chang et al. (2002)
Chitinase	*Saccharomyces cerevisiae*	Tobacco	Carstens et al. (2003)
Ribosome-inactivating protein (MOD1); Chitinase (RCH10)	RCH10 from rice; MOD1 from maize	Rice	Kim et al. (2003)
β-glucanase (Gns1)	Rice	Rice	Nishizawa et al. (2003)
Chitinase (RCH10); Glucanase (ALG)	RCH10 from rice; ALG from alfalfa	Creeping bentgrass	Wang et al. (2003)
Chitinase (chi11)	Rice	Rice	Kumar et al. (2003)
Cationic peptide (msrA3)	Synthetic preparation	Potato	Osusky et al. (2004)
Glucanase (Bglu)	Potato	Flax	Wro'bel-Kwiatkowska et al. (2004)
Chitinase (ech42); Chitinase (nag70); Glucanase (gluc78)	*Trichoderma atroviride*	Rice	Mei et al. (2004)
Chitinase (Chi); Ribosome-inactivating protein (Rip)	Chi from bean; Rip from barley	Soya bean	Li et al. (2004)
Chitinase (CHIT); Glucanase (GLUC)	GLUC from tobacco; CHIT from cucumber	Potato	Moravčíková et al. (2004)
Glucanase (OsGLN2)	Rice	Rice	Akiyama et al. (2004)
Antifungal protein (Afp)	*Aspergillus giganteus*	Rice	Coca et al. (2004)
Chitinase (BjCHI1); Glucanase (HbGLU)	HbGLU from rubber tree; BjCHI1 from mustard	Potato	Chye et al. (2005)
Antifungal protein (AFP-PIN)	Prawn (synthetic preparations)	Finger millet	Latha et al. (2005)
Thaumatin	*Thaumatococcus daniellii*	Strawberry	Schestibratov and Dolgov (2005)

(continued)

Table 2 (continued)

Name of the gene	Source	Host plant	Reference
Chitinase (Chi)	Bean	Cotton	Tohidfar et al. (2005)
Thaumatin-like protein	Rice	Carrot	Punja (2005)
Chitinase (RCC2)	Rice	Trifoliate orange	Mitani et al. (2006)
Chitinase (ch5B); Glucanase (gln2); Antifungal protein (ap24)	gln2 and ap24 from tobacco; ch5B from beans	Strawberry	Vellicce et al. (2006)
Chitinase; Glucanase	Barley	Oilseed rape	Melander et al. (2006)
Glucanase (GLU); Antifungal protein Glucanase (GLU-AFP)	alfAFP from Alfalfa (alfAFP); GLU from tobacco	Tomato	Chen et al. (2006)
AFP	*Aspergillus giganteus*	Pearl Millet	Girgi et al. (2006)
α-1-purothionin; β-1,3-glucanase gene; tlp-1 gene;	tlp-1 and β-1,3-glucanase from barley; a-1-purothionin from wheat	Wheat	Mackintosh et al. (2007)
Chitinase (CHIT); Glucanase (GLUC)	GLUC from tobacco; CHIT from cucumber	Potato	Moravcikova et al. (2007)
Chitinases (RCH10 and RAC22); Glucanase (β-Glu); Ribosome-inactivating protein (B-RIP)	β-Glu from alfalfa; RCH10 and RAC22 from rice; B-RIP from barley	Rice	Zhu et al. (2007)
Chitinase (chi11); Thaumatin-like protein (tlp)	Rice	Barley	Tobias et al. (2007)
β-1,3-glucanase	Soybean	Banana	Maziah et al. (2007)
Thaumatin II	*Thaumatoccous daniellii*	Tobacco	Rajam et al. (2007)
Glucanase	Tomato	Indian mustard	Mondal et al. (2007)
Chitinase (chi11); Thaumatin-like protein (tlp); *Xa21*	Rice	Rice	Maruthasalam et al. (2007)
Chitinase	Rice	Taro	He et al. (2008)
Chitinase	*Streptomyces griseus*	Potato	Raham et al. (2008)
Mustard defensin	Mustard	Tobacco, Peanut	Anuradha et al. (2008)
Chitinase (chi11); Glucanase (gluc)	Chi11 from rice; gluc from Tobacco	Rice	Sridevi et al. (2008)
Chitinase383; Glucanase638; Cationic peroxidise (POX1)	Chitinase and glucanase from wheat; POX1 from Rice	Carrot	Wally et al. (2009)
Chitinase (Chit30)	*Streptomyces olivaceoviridis*	Pea	Hassan et al. (2009)
β-1,3-glucanase	Tobacco	Groundnut	Sundaresha et al. (2010)
Endochitinase	Trichoderma virens	Tobacco and Tomato	Shah et al. (2010)
β-1,3-glucanase gene (gluc); chitinase gene (Chit30)	Chit30 from *S. olivaceoviridis*; Gluc from barley	Pea	Amian et al. (2011)
Endochitinase	Wheat	Tomato	Girhepuje and Shinde (2011)
Chitinase (Chi11)	Rice	Finger millet	Ignacimuthu and Ceasar (2012)
Thaumatin-like protein	Rice	Banana	Mahdavi et al. (2012)

(continued)

Table 2 (continued)

Name of the gene	Source	Host plant	Reference
Thaumtin-like protein	Rice	Rice	Naseri et al. (2012)
Thaumatin-like protein	Secale cereal	Canola	Zamani et al. (2012)
Chitinase (Rchit)	Rice	Peanut	Prasad et al. (2013)
tlp-D34 and chi11	chi11 from Rice; tlp-D34 from	Rice	Shah et al. (2013)
Thaumatin II	*Thaumatococcus daniellii*	Carrot	Sidorova et al. (2013)
Thaumatin-like protein (CsTLP)	*Camelia sinensis*	Potato	Acharya et al. (2013)
Chitinase RC24	Rice	Wheat	Huang et al. (2013)
Chitinase	Rice	Banana	Kovacs et al. (2013)

(LRRs), transmembrane domains (TMs) and serine/threonine protein kinase domains (PKs). There are several examples of the expression of R-genes in transgenic plants. Transgenic tobacco and potato expressing Cf-9 gene of tomato (confers resistance in tomato to races of *Cladosporium fulvum*) showed hypersensitive response when challenged with Avr9 peptide (Hammond-Kosack et al. 1998). Similarly, expression of the Cf-g gene in oilseed rape enhanced resistance to *Leptosphaeria maculans* (Hennin et al. 2001). The R-gene Rxol from maize was successfully introduced into rice and conferred resistance against bacterial streak disease caused by *Xanthomonas oryzae* pv. *oryzicola* (Zhao et al. 2005). The R-gene RCT1 from *Medicago truncatula* was expressed in alfalfa and conferred resistance to *Colletotrichum trifolii* (Yang et al. 2008), and RPI-BLB2 from wild potato, *Solanum bulbocastanum* conferred resistance to *Phytophthora infestans* in cultivated potato (van der Vossen et al. 2005). The *Xa21* gene isolated from indica rice strain IRBB21 (conferring resistance to *Xanthomonas oryzae*) when introduced into a susceptible variety IR72 resulted in excellent field resistance against the pathogenic bacteria (Tu et al. 2000). Transgenic flax plants expressing three alleles of flax rust resistance genes, namely, L2, L6 and L10, were shown to be resistant to strains of flax rust which had corresponding *Avr* genes (Ellis et al. 2000).

The overexpression of the HRT gene, which controls the hypersensitive response in Arabidopsis to turnip crinkle virus, did not confer enhanced resistance to *Peronospora tabacina* (Cooley et al. 2000). This could happen possibly due to two reasons – one, there could be multiple factors involved in determining the resistance response and two, the resistance might be HR independent. Results to date suggest that the expression of cloned R-genes in heterologous transgenic plants is unlikely by itself to enhance tolerance to fungal pathogens, due to the complexity of the interacting signalling pathways. A combination of several interacting genes (gene pyramiding), similar to that for the antifungal proteins, will likely be required. There are several other potential problems with genetic engineering of resistance through the use of R-genes. These include the potential for spontaneous activation leading to cell death via HR-like response, the development of pathogens with an alternative *Avr* gene and reduced overall fitness.

RNA Interference

RNA silencing, or RNA interference (RNAi), is a nucleotide sequence-specific conserved regulatory mechanism of gene expression that has been widely characterized in eukaryotic organisms. Silencing can occur either through transcriptional gene silencing (TGS) or posttranscriptional gene silencing (PTGS). Both the cases, long ds-RNA are cleaved into small RNAs (siRNAs or miRNAs) of 21–24 nt in length by Dicer or a Dicer-like (DCL) protein, in an ATP-dependent step. The siRNAs become integrated into a multi-subunit protein complex, commonly known as the RNAi-induced silencing complex (RISC), which guides the siRNAs to the target RNA sequence; the antisense siRNA guides RISC complex to the complimentary mRNA and targets it for degradation (Meister and Tuschl 2004). Alternatively generated siRNAs or miRNAs can be recruited into a microRNA ribonucleoprotein complex (miRNP) which mediate translational repression of mRNA targets (Bartel 2004). miRNAs are produced from imperfectly base-paired hairpin loop structures, whereas siRNAs are produced from perfectly paired double-stranded RNA (dsRNA) precursors (Katiyar-Agarwal and Jin 2010).

RNAi regulates endogenous genes involved in plant development (Vazquez et al. 2004) and stress adaptation (Ruiz-Ferrer and Voinnet 2009) besides functioning as a natural antiviral defence mechanism, a process named virus-induced gene silencing (VIGS). Recently RNAi has been used in crop pest management. This is generally achieved by ectopic expression of long dsRNA hairpin molecules encoding the transcribed regions of gene (or genes) of interest. This results in targeting and sequence-specific degradation of homologous mRNAs. Thus, in principle every gene is amenable to RNAi-based silencing where the efficiency and specificity can be controlled by varying the length and sequence of the dsRNA (Jan et al. 2000).

Using the (RNAi) mechanism, transgenic plants have been generated that target the pathogen genomes. Recently there have been few reports about successful generation of transgenics against viral pathogens using varied approaches to generate siRNAs including long/small hairpin RNA, sense/antisense RNA and artificial miRNA precursors (Prins et al. 2008; de Alba AE Martinez et al. 2002; Simon-Mateo and Garcia 2011). Plant fungal pathogens interact via haustoria, which is an interface used for exchange of material between fungal pathogen and plant host, and this can also be used to facilitate the uptake of dsRNA or siRNA from the host plant cells into the fungal pathogens to create RNA silencing-mediated resistance. This concept has been proved for the barley powdery mildew *Blumeria graminis*. Barley transgenics were generated expressing dsRNA directed against *B. graminis* effector gene *Avra*10. There was significant reduction of disease symptoms of a *B. graminis* infection, whereas transgenic control that had lost the hairpin RNAi cassette was as susceptible as wild-type control plants (Nowara et al. 2010), indicating trafficking of dsRNA or siRNA from host plants into *B. graminis.* This may lead to an RNA silencing-based crop protection strategy against fungal pathogens (Duan et al. 2012). Khatri and Rajam (2007) had shown silencing of fungal *ornithine decarboxylase* (ODC) *in vitro* using antisense RNA. Recently, Tinoco et al. (2010) have reported that GUS-specific siRNAs expressed in the transgenic tobacco could silence GUS gene in GUS expressing *Fusarium verticillioides*' transformants. This could be a result of movement of silencing signal through the germinating spores into the fungal cells. Transgenic lettuce plants expressing a GUS dsRNA could induce specific gene silencing in the parasitizing plant *Triphysaria versicolor* expressing GUS gene (Tomilov et al. 2008; Rajam 2012).

miRNAs are evolutionally conserved in the kingdom-specific manner and are differentially expressed during cell differentiation and development as well as stress responses. Novel miRNAs have been identified in the last few years that are intimately associated with various important cellular functions, including stress responses and disease development in plants.

The expression profile of miRNA genes has been studied in plants in response to biotic stresses. These investigations have found that there are miRNAs that are over- or underexpressed to withstand the stress conditions (Jones-Rhoades and Bartel 2004; Guo et al. 2011; Khraiwesh et al. 2012). There are also studies on the expression profile of host plant miRNAs following fungal infections. Guo et al. (2011) identified miRNAs involved in the response of soybean to *Phytophthora sojae*. They studied expressional patterns of miRNAs upon infection by *P. sojae* by microarray analysis in a susceptible, a quantitative-resistant and a qualitative-resistant soybean cultivars. Expression of a number of miRNAs was significantly altered upon infection and (or) in the different genotypes. Lu et al. (2007) have found that miRNAs were differentially expressed in response to infection by the rust fungus, *Cronartium quercuum* f. sp. *fusiforme*, in pines. Some of the conserved miRNAs were found to be differentially expressed in wheat infected with the powdery mildew fungus (Xin et al. 2010), for example miR156, miR159, miR164, miR171 and miR396 were repressed and miR393, miR444 and miR827 were upregulated. Jin et al. (2012) have identified two microRNAs miR160 and miR171a, that were downregulated and one microRNA miR169 that was upregulated in tomato infected with *Botrytis cinerea*. Similarly, Zhao et al. (2012) identified 12 miRNAs, including miR156, miR159, miR160, miR164 and miR168 to be upregulated in the stem bark of *Populus trichocarpa* following stem canker disease, caused by *Botryosphaeria dothidea*. Host miRNAs, regulated by fungal elicitors, target endogenous genes for various defense responses. miRNA from rice, osa-miR7695, is expressed in response to the infection by *Magnaporthe oryzae* and negatively regulates an alternatively spliced transcript of OsNramp6 (Natural resistance-associated macrophage protein 6). The overexpression of osa-miR7695 in rice confers fungal resistance (Campo et al. 2013). Similar strategies can be deployed for other plant-fungal pathosystems, where other miRNAs can be used to induce disease resistance.

Thus host plant miRNA genes that respond to pathogen or pest attack can be over-expressed or the miRNA gene of the pathogen can be suppressed by plant expression of an antisense RNA specific to pathogen miRNA sequence.

Conclusion

Due to increasing human population and public concern over the use of chemical fungicides, development of resistance in pathogens over a period of time demands alternative ways for disease control. A detailed understanding of the molecular events that take place during a plant–pathogen interaction is a prerequisite for identifying the genes/proteins/factors associated with inherent defences of plants and then to transfer these defensive traits into the genome of economically important crops. PR proteins are natural defences of plants, and although combinations of PR proteins may not result in complete resistance to a given disease (as resistance genes may), they provide and offer a broad range of protection against a large assortment of pathogens and pests. Targeting of the most effective antifungal PR proteins to appropriate cellular locations may also prove effective in control of pathogens. Genetic engineering and molecular breeding will help in identifying and cloning R-genes which can then be introduced more rapidly into elite germplasm. The engineering of both R-genes and downstream defence responses will also enhance disease resistance. Future research that focuses on R-genes, pathogen-inducible promoters and expression of traits associated with broad-spectrum resistance will further enhance their impact in crop improvement programmes.

References

Abad LR, D'Urzo MP, Liu D, Narasimhan ML, Reuveni M, Zhu JK, Niu X, Singh NK, Hasegawa PM, Bressan RA (1996) Antifungal activity of tobacco osmotin has specificity and involves plasma membrane permeabilization. Plant Sci 118:11–23

Acharya K, Pal AK, Gulati A, Kumar S, Singh AK, Ahuja PS (2013) Overexpression of *Camellia sinensis* thaumatin-like protein, CsTLP in potato confers enhanced resistance to *Macrophomina phaseolina* and *Phytophthora infestans* infection. Mol Biotechnol 54:609–622

Akiyama T, Pillai MA, Sentoku N (2004) Cloning, characterization and expression of OsGLN2, a rice endo-1,3-betaglucanase gene regulated developmentally in flowers and hormonally in germinating seeds. Planta 220:129–139

Alexander D, Goodman RM, Gut-Rella M, Glascock C, Weymann K, Friedrich L, Maddox D, Ahl-Goy P, Luntz T, Ward E, Ryals J (1993) Increased tolerance to two oomycete pathogens in transgenic tobacco expressing pathogenesis-related protein 1a. Proc Natl Acad Sci U S A 90:7327–7331

Amian AA, Papenbrock J, Jacobsen HJ, Hassan F (2011) Enhancing transgenic pea (*Pisum sativum* L.) resistance against fungal diseases through stacking of two antifungal genes (Chitinase and Glucanase). GM Crops 2:104–109

Antoniw JF, Ritter CE, Pierpoint WS, Van Loon LC (1980) Comparison of the pathogenesis-related proteins from plants of two cultivars of tobacco infected with TMV. J Gen Virol 47:79–87

Anuradha TS, Divya K, Jami SK, Kirti PB (2008) Transgenic tobacco and peanut plants expressing a mustard defensin show resistance to fungal pathogens. Plant Cell Rep 27:1777–1786

Bartel DP (2004) MicroRNAs: genomics, biogenesis, mechanism and function. Cell 116:281–297

Bernier F, Berna A (2001) Germins and germin-like proteins: plant do-all proteins. But what do they do exactly? Plant Physiol Biochem 39:545–554

Blilou I, Juan A, Ocampo JA, Carcia-Garrido JM (2000) Induction of LTP (lipid transfer protein) and PAL (phenylalanine ammonia-lyase) gene expression in rice roots colonized by the arbuscular mycorrhizal fungus *Glomus mosseae*. J Exp Bot 51:1969–1977

Boller T, Gehri A, Mauch F, Vögeli U (1983) Chitinase in bean leaves: induction by ethylene, purification, properties, and possible function. Planta 157:22–31

Campo S, Peris-Peris C, Siré C, Moreno AB, Donaire L, Zytnicki M, Notredame C, Llave C, San Segundo B (2013) Identification of a novel microRNA (miRNA) from rice that targets an alternatively spliced transcript of the Nramp6 (Natural resistance-associated macrophage protein 6) gene involved in pathogen resistance. New Phytol 199(1):212–227

Carstens M, Vivier MA, Pretorius IS (2003) The *Saccharomyces cerevisiae* chitinase, encoded by the CTS1–2 gene, confers antifungal activity against *Botrytis cinerea* to transgenic tobacco. Transgenic Res 12:497–508

Chai B, Maqbool SB, Hajela RK, Green D, Vargas JM, Warkentin D, Sabzikar R, Sticklen MB (2002) Cloning of a chitinase-like cDNA (hs2), its transfer to creeping bentgrass (*Agrostis palustris* Huds.) and development of brown patch (*Rhizoctonia solani*) disease resistant transgenic lines. Plant Sci 163:183–193

Chang M, Culley D, Choi JJ, Hadwiger LA (2002) Agrobacterium-mediated co-transformation of a pea b-1,3-glucanase and chitinase genes in potato (*Solanum tuberosum* L. c.v. Russet Burbank) using a single selectable marker. Plant Sci 163:83–89

Chen Z, Ricigliano JW, Klessig DF (1993) Purification and characterization of a soluble salicylic acid-binding protein from tobacco. Proc Natl Acad Sci U S A 90 (20):9533–9537

Chen SC, Liu AR, Zou ZR (2006) Overexpression of glucanase gene and defensin gene in transgenic tomato enhances resistance to *Ralstonia solanacearum*. Plant Sci 5:2134–2140

Chen WP, Chen PD, Liu DJ, Kynast R, Friebe B, Velazhahan R, Muthukrishnan S, Gill BS (1999) Development of wheat scab symptoms is delayed in transgenic wheat plants that constitutively express a rice thaumatin-like protein gene. Theor Appl Genet 99:755–760

Cheong YH, Kim CY, Chun HJ, Moon BC, Park HC, Kim JK, Lee SY, Cho MJ (2000) Molecular cloning of a soybean class III b-1,3-glucanase gene that is regulated both developmentally and in response to pathogen infection. Plant Sci 154:71–81

Christensen A, Ho Cho B, Naesby M, Gregersen P, Brandt J, Madriz-Ordeñana K, Collinge D, Thordal-Christensen H (2002) The molecular characterization of two barley proteins establishes the novel PR-17 family of pathogenesis-related proteins. Mol Plant Pathol 3:135–144

Chye M, Zhao K, He Z, Ramalingam S, Fung K (2005) An agglutinating chitinase with two chitin-binding domains confers fungal protection in transgenic potato. Planta 220:717–730

Coca M, Bortolotti C, Rufat M, Peñas G, Eritja R, Tharreau D, Martinez del Pozo A, Messeguer J, Segundo SB (2004) Transgenic rice plants expressing the antifungal AFP protein from *Aspergillus giganteus* show enhanced resistance to the rice blast fungus *Magnaporthe grisea*. Plant Mol Biol 54:245–259

Cohn J, Sessa G, Martin GB (2001) Innate immunity in plants. Curr Opin Immunol 13:55–62

Cooley MB, Pathirana S, Wu H-J, Kachroo P, Klessig DF (2000) Members of the *Arabidopsis* HRT/RPP8 family of resistance genes confer resistance to both viral and oomycete pathogens. Plant Cell 12:663–676

Cornelissen BJ, Hooft van Huijsduijnen RA, Van Loon LC, Bol JF (1986) Molecular characterization of messenger RNAs for 'pathogenesis related' proteins la, lb and lc, induced by TMV infection of tobacco. EMBO J 5(1):37–40

Coventry HS, Dubery IA (2001) Lipopolysaccharides from *Burkholderia cepacia* contribute to an enhanced defensive capacity and the induction of pathogenesis-related proteins in *Nicotiana tabacum*. Physiol Mol Plant Pathol 58:149–158

Dangl JL, Jones JDG (2001) Plant pathogens and integrated defence responses to infection. Nature 411:826–833

Danon A, Delorme V, Mailhac N, Gallois P (2000) Plant programmed cell death: a common way to die. Plant Physiol Biochem 38:647–655

Datta K, Velazhahan R, Oliva N, Ona I, Mew T, Khush GS, Muthukrishnan S, Datta SK (1999) Over-expression of the cloned rice thaumatin-like protein (PR-5) gene in transgenic rice plants enhances environmental friendly resistance to *Rhizoctonia solani* causing sheath blight disease. Theor Appl Genet 98:1138–1145

Datta K, Tu J, Oliva N, Ona I, Velazhahan R, Mew TW, Muthukrishnan S, Datta SK (2001) Enhanced resistance to sheath blight by constitutive expression of infection- related rice chitinase in transgenic elite indica rice cultivars. Plant Sci 160:405–414 de Alba Martinez AE, Flores R, Hernandez C (2002) Two chloroplastic viroids induce the accumulation of small RNAs associated with posttranscriptional gene silencing. J Virol 76:13094–13096

Derckel JP, Legendre L, Audran J-C, Haye B, Lambert B (1996) Chitinases of the grapevine (*Vitis vinifera* L.): five isoforms induced in leaves by salicylic acid are constitutively expressed in other tissues. Plant Sci 119:31–37

Duan CG, Wang CH, Guo HS (2012) Application of RNA silencing to plant disease resistance. Silence 3:5

El-kereamy A, El-sharkawy I, Ramamoorthy R, Taheri A, Errampalli D, Kumar P, Jayasankar S (2011) *Prunus domestica* pathogenesis-related protein-5 activates the defense response pathway and enhances the resistance to fungal infection. PLoS One 6:e17973

Ellis J, Dodds P, Pryor T (2000) The generation of plant disease resistance gene specificities. Trends Plant Sci 5:373–379

Fagoaga C, Rodrigo I, Conejero V, Hinarejos C, Tuset JJ, Arnau J, Pina JA, Navarro L, Pena L (2001) Increased tolerance to *Phytophthora citrophthora* in transgenic orange plants constitutively expressing a tomato pathogenesis related protein PR-5. Mol Breed 7:175–185

Fritig B, Heitz T, Legrand M (1998) Antimicrobial proteins in induced plant defense. Curr Opin Immunol 10:16–22

Girgi M, Breese WA, Lorz H, Oldach KH (2006) Rust and downy mildew resistance in pearl millet (*Pennisetum glaucum*) mediated by heterologous expression of the afp gene from *Aspergillus giganteus*. Transgenic Res 15:313–324

Girhepuje PV, Shinde GB (2011) Transgenic tomato plants expressing a wheat endochitinase gene demonstrate enhanced resistance to *Fusarium oxysporum* f. sp. *Lycopersici*. Plant Cell Tiss Organ Cult 105:243–251

Guo N, Ye WW, Wu XL, Shen DY, Wang YC, Xing H, Dou DL (2011) Microarray profiling reveals microRNAs involving soybean resistance to *Phytophthora sojae*. Genome 54(11):954–958

Hammond-Kosack KE, Tang S, Harrison K, Jones JDG (1998) The tomato Cf-9 disease resistance gene functions in tobacco and potato to confer responsiveness to the fungal avirulence gene product Avr 9. Plant Cell 10:1251–1266

Hassan F, Meens J, Jacobsen H, Kiesecker H (2009) A family 19 chitinase (Chit30) from Streptomyces olivaceoviridis ATCC 11238 expressed in transgenic pea affects the development of *T. harzianum in vitro*. J Biotechnol 143:302–330

He X, Miyasaka SC, Fitch MMM, Moore PH, Zhu YJ (2008) *Agrobacterium tumefaciens* mediated transformation of taro [*Colocasia esculenta* (L.) Schott] with a rice chitinase gene for improved tolerance to a fungal pathogen *Sclerotium rolfsii*. Plant Cell Rep 27:903–909

Heath MC (2000) Nonhost resistance and nonspecific plant defenses. Curr Opin Plant Biol 3:315–319

Hennin C, Hofte M, Diederichsen E (2001) Functional expression of Cf9 and Avr9 genes in *Brassica napus* induces enhanced resistance to *Leptosphaeria maculans*. Mol Plant Microbe Interact 14:1075–1085

Honée G (1999) Engineered resistance against fungal plant pathogens. Eur J Plant Pathol 105:319–326

Hong JK, Hwang BK (2006) Promoter activation of pepper class II basic chitinase gene, CAChi2, and enhanced bacterial disease resistance and osmotic stress tolerance in the CAChi2-overexpressing *Arabidopsis*. Planta 223:433–448

Hooft van Huijsduijnen RA, Cornelissen BJ, van Loon LC, van Boom JH, Tromp M, Bol JF (1985) Virus-induced synthesis of messenger RNAs for precursors of pathogenesis-related proteins in tobacco. EMBO J 4 (9):2167–2171

Huang X, Wang J, Du Z, Zhang C, Li L, Xu Z (2013) Enhanced resistance to stripe rust disease in transgenic wheat expressing the rice chitinase gene RC24. Transgenic Res 22(5):939–947

Ignacimuthu S, Ceasar SA (2012) Development of transgenic finger millet (*Eleusine coracana* (L.) Gaertn.) resistant to leaf blast disease. J Biosci 37:135–147. doi:10.1007/s12038-011-9178-y

Jan FJ, Fagoaga C, Pang SZ, Gonsalves D (2000) A minimum length of N gene sequence in transgenic plants is required for RNA-mediated tospovirus resistance. J Gen Virol 81:235–242

Jin W, Fangli WU, Liang X, Guangwang L, Yuxian Z (2012) Microarray based analysis of tomato miRNA regulated by *Botrytis cinerea*. J Plant Growth Regul 31:38–46

Jones-Rhoades MW, Bartel DP (2004) Computational identification of plant microRNAs and their targets, including a stress-induced miRNA. Mol Cell 14:787–799

Jongedijk E, Tigelaar H, Roekel JSC, Bres-Vloemans SA, Dekker I, Elzen PJM, Cornelissen BJC, Melchers LS (1995) Synergistic activity of chitinases and beta-1,3-glucanases enhances fungal resistance in transgenic tomato plants. Euphytica 85:173–180

Kamoun S (2001) Nonhost resistance to *Phytophthora*: novel prospects for a classical problem. Curr Opin Plant Biol 4:295–300

Katiyar-Agarwal S, Jin H (2010) Role of small RNAs in host-microbe interactions. Annu Rev Phytopathol 48:225–246

Khatri M, Rajam MV (2007) Targeting polyamines of *Aspergillus nidulans* by siRNA specific to fungal ornithine decarboxylase gene. Med Mycol 45(3):211–220

Khraiwesh B, Zhu JK, Zhu J (2012) Role of miRNAs and siRNAs in biotic and abiotic stress responses of plants. Biochem Biophys Acta 1819:137–148

Kim JK, Jang IC, Wu R, Zuo WN, Boston RS, Lee YH, Ahn IP, Nahm BH (2003) Co-expression of a modified maize ribosome-inactivating protein and a rice basic chitinase gene in transgenic rice plants confers enhanced resistance to sheath blight. Transgenic Res 12:475–484

Kishimoto K, Nishizawa Y, Tabei Y, Hibi T, Nakajima M, Akutsu K (2002) Detailed analysis of rice chitinase gene expression in transgenic cucumber plants showing different levels of disease resistance to gray mold (*Botrytis cinerea*). Plant Sci 162:655–662

Kombrink E, Somssich IE (1995) Defense responses of plants to pathogens. Adv Bot Res (Formerly Adv Plant Pathol) 21:1–34

Kombrink E, Somssich IE (1997) Pathogenesis-related proteins and plant defense. In: Carroll G, Tudzynski P (eds) The mycota, vol 5, Part A, Plant relationships. Springer, Berlin/Heidelberg, pp 107–128

Kovacs G, Sagi L, Jacon G, Arinaitwe G, Busogoro JP, Thiry E, Strosse H, Swennen R, Remy S (2013) Expression of a rice chitinase gene in transgenic banana ('Gros Michel', AAA genome group) confers resistance to black leaf streak disease. Transgenic Res 22:117–130

Kumar KK, Poovannan K, Nandakumar R, Thamilarasi K, Geetha C, Jayashree N, Kokiladevi E, Raja JAJ, Samiyappan R, Sudhakar D, Balasubramanian P (2003) A high throughput functional expression assay system for a defense gene conferring transgenic resistance on rice against the sheath blight pathogen, *Rhizoctonia solani*. Plant Sci 165:969–976

Latha MA, Rao KV, Reddy VD (2005) Production of transgenic plants resistant to leaf blast disease in finger millet (*Eleusine coracana* (L.) Gaertn.). Plant Sci 169:657–667

Li HY, Zhu YM, Chen Q, Conner RL, Ding XD, Zhang BB (2004) Production of transgenic soybean plants with two anti-fungal protein genes via *Agrobacterium* and particle bombardment. Biol Plant 48:367–374

Liu D, Raghothama KG, Hasegawa PM, Bressan RA (1994) Osmotin overexpression in potato delays development of disease symptoms. Proc Natl Acad Sci U S A 91:1888–1892

Lu S, Sun YH, Amerson H, Chiang VL (2007) MicroRNAs in loblolly pine (*Pinus taeda* L.) and their association with fusiform rust gall development. Plant J 51:1077–1098

Mackintosh C, Lewis J, Radmer L, Shin S, Heinen S, Smith L, Wyckoff M, Dill-Macky R, Evans C, Kravchenko S, Baldridge G, Zeyen R, Muehlbauer G (2007) Overexpression of defense response genes in transgenic wheat enhances resistance to *Fusarium* head blight. Plant Cell Rep 26:479–488

Mahdavi F, Sariah M, Maziah M (2012) Expression of rice thaumatin-like protein gene in transgenic banana plants enhances resistance to *Fusarium* wilt. Appl Biochem Biotechnol 166:1008–1019

Maruthasalam S, Kalpana K, Kumar KK, Loganathan M, Poovannan K, Raja JA, Kokiladevi E, Samiyappan R, Sudhakar D, Balasubramanian P (2007) Pyramiding transgenic resistance in elite indica rice cultivars against the sheath blight and bacterial blight. Plant Cell Rep 26:791–804

Maziah M, Saraih M, Sreeramanan S (2007) Transgenic banana Rastali (AAB) with b-1,3-glucanase gene for tolerance to *Fusarium* wilt race 1 disease via *Agrobacterium*-mediated transformation system. Plant Pathol J 6:271–282

Mei L, Zong-xiu S, Jei Z, Tong X, Gary EH, Matteo L (2004) Enhancing rice resistance to fungal pathogens by transformation with cell degrading enzyme genes from *Trichoderma atroviride*. J Zhejiang Uni Sci 5:133–136

Meister G, Tuschl T (2004) Mechanisms of gene silencing by double-stranded RNA. Nature 431:343–350

Melander M, Kamnert I, Happstadius I, Liljeroth E, Bryngelsson T (2006) Stability of transgene integration and expression in subsequent generations of doubled haploid oilseed rape transformed with chitinase and b-1,3-glucanase genes in a double-gene construct. Plant Cell Rep 25:942–952

Mitani N, Kobayashi S, Nishizawa Y, Kuniga T, Matsumoto R (2006) Transformation of trifoliate orange with rice chitinase gene and the use of the transformed plant as a rootstock. Sci Hortic 108:439–441

Mondal K, Bhattacharya R, Koundal K, Chatterjee S (2007) Transgenic Indian mustard (*Brassica juncea*) expressing tomato glucanase leads to arrested growth of *Alternaria brassicae*. Plant Cell Rep 26:247–252

Moravčíková J, Matušíková I, Libantová J, Bauer M, Mlynárová L (2004) Expression of cucumber class III chitinase and *Nicotiana plumbaginifolia* class I glucanase genes in transgenic potato plants. Plant Cell Tissue Organ Cult 79:161–168

Moravczkova J, Libantova J, Heldak J, Salaj JM, Matuszkova I, Galova Z, Mlynarov L (2007) Stress-induced expression of cucumber chitinase and *Nicotiana plumbaginifolia* b-1,3- glucanase genes in transgenic potato plants. Acta Physiol Plant 29:133–141

Naseri G, Sohani MM, Pourmassalehgou A, Allahi S (2012) In planta transformation of rice (*Oryza sativa*) using thaumatin-like protein gene for enhancing resistance to sheath blight. Afr J Biotechnol 11:7885–7893

Nishizawa Y, Saruta M, Nakazono K, Nishio Z, Soma M, Yoshida T, Nakajima E, Hibi T (2003) Characterization of transgenic rice plants over-expressing the stress-inducible b-glucanase gene Gns1. Plant Mol Biol 51:143–152

Nowara D, Gay A, Lacomme C, Shaw J, Ridout C, Douchkov D, Hensel G, Kumlehn J, Schweizer P (2010) HIGS: host-induced gene silencing in the obligate biotrophic fungal pathogen *Blumeria graminis*. Plant Cell 22:3130–3141

Odjakova M, Hadjiivanova C (2001) The complexity of pathogen defence in plants. Bulg J Plant Physiol 27:101–109

Osusky M, Osuska L, Hancock RE, Kay W, Misra S (2004) Transgenic potatoes expressing a novel cationic peptide are resistant to late blight and pink rot. Trans Res 13:181–190

Passardi F, Penel C, Dunand C (2004) Performing the paradoxical: how plant peroxidases modify the cell wall. Trends Plant Sci 9:534–540

Prasad K, Bhatnagar-Mathur P, Waliyar F, Sharma KK (2013) Overexpression of a chitinase gene in transgenic peanut confers enhanced resistance to major soil borne and foliar fungal pathogens. J Plant Biochem Biotechnol 22:222–233

Prins M, Laimer M, Noris E, Schubert J, Wassenegger M, Tepfer M (2008) Strategies for antiviral resistance in transgenic plants. Mol Plant Pathol 9:73–83

Punja ZK (2005) Transgenic carrots expressing a thaumatin-like protein display enhanced resistance to several fungal pathogens. Can J Plant Pathol 27:291–296

Raham SK, Rinaldi S, Ikuo N, Masahiro M (2008) Production of transgenic potato exhibiting enhanced resistance to fungal infections and herbicide applications. Plant Biotechnol Rep 2:13–20

Rajam MV (2012) Host induced silencing of fungal pathogen genes: an emerging strategy for disease control in crop plants. Cell Dev Biol 1(6):e118, http://dx.doi.org/10.4172/2168-9296.1000e118

Rajam MV, Chandola N, SaiPrasad Goud P, Singh D, Kashyap V, Choudhary ML, Sihachakr D (2007) Thaumatin sweet protein gene confers tolerance to fungal pathogens as well as abiotic stresses in transgenic tobacco plants. Biol Plant 51:135–141

Rohini VK, Rao KS (2001) Transformation of peanut (*Arachis hypogaea* L.) with tobacco chitinase gene: variable response of transformants to leaf spot disease. Plant Sci 160:889–898

Ruiz-Ferrer V, Voinnet O (2009) Roles of plant small RNAs in biotic stress responses. Annu Rev Plant Biol 60:485–510

Schestibratov KA, Dolgov SV (2005) Transgenic strawberry plants expressing a thaumatin II gene demonstrate enhanced resistance to *Botrytis cinerea.* Sci Hortic 106:177–189

Selitrennikoff CP (2001) Antifungal proteins. Appl Environ Microbiol 67:2883–2894

Shah MR, Mukherjee PK, Eapen S (2010) Expression of a fungal endochitinase gene in transgenic tomato and tobacco results in enhanced tolerance to fungal pathogens. Physiol Mol Biol Plants 16(1):39–51

Shah JM, Singh R, Veluthambi K (2013) Transgenic rice lines constitutively co-expressing tlp-D34 and chi11 display enhancement of sheath blight resistance. Biol Plant 57:351–358

Sidorova T, Miroshnichenko D, Dolgov S, Tjukavin G (2013) Transgenic carrot expressing Thaumatin II gene has enhanced resistance against *Fusarium avenaceum.* Acta Hort (ISHS) 974:123–130

Simon-Mateo C, Garcia JA (2011) Antiviral strategies in plants based on RNA silencing. Biochim Biophys Acta 1809:722–731

Somssich IE, Hahlbrock K (1998) Pathogen defence in plants – a paradigm of biological complexity. Trends Plant Sci 3:86–90

Sridevi G, Parameswari C, Sabapathi N, Raghupathy V, Veluthambi K (2008) Combined expression of chitinase and b-1,3-glucanase genes in indica rice (*Oryza sativa* L.) enhances resistance against *Rhizoctonia solani.* Plant Sci 175:283–290

Stintzi A, Heitz T, Prasad V, Wiedemann-Merdinoglu S, Kauffmann S, Geoffroy P, Legrand M, Fritig B (1993) Plant 'pathogenesis-related' proteins and their role in defense against pathogens. Biochimie 75:687–706

Sundaresha S, Manoj Kumar A, Rohini S, Math S, Keshamma E, Chandrashekar S, Udayakumar M (2010) Enhanced protection against two major fungal pathogens of groundnut, *Cercospora arachidicola* and *Aspergillus flavus* in transgenic groundnut over-expressing a tobacco b-1,3-glucanase. Eur J Plant Pathol 126:497–508

Takakura Y, Ito T, Saito H, Inoue T, Komari T, Kuwata S (2000) Flower-predominant expression of a gene encoding a novel class I chitinase in rice (*Oryza sativa* L.). Plant Mol Biol 42:883–897

Takatsu Y, Nishizawa Y, Hibi T, Akutsu K (1999) Transgenic chrysanthemum (*Dendranthema grandiflorum* (Ramat.) Kitamura) expressing a rice chitinase gene shows enhanced resistance to gray mold (*Botrytis cinerea*). Sci Hortic 82:13–123

Terakawa T, Takaya N, Horiuchi H, Koike M, Takagi M (1997) A fungal chitinase gene from *Rhizopus oligosporus* confers antifungal activity to transgenic tobacco. Plant Cell Rep 16:439–443

Theis T, Stahl U (2004) Antifungal proteins: targets, mechanisms and prospective applications. Cell Mol Life Sci 61:437–455

Thomma B, Eggermont K, Penninckx I, Mauch-Mani B, Vogelsang R, Cammue BPA, Broekaert WF (1998) Separate jasmonate-dependent and salicylate-dependent defense-response pathways in *Arabidopsis* are essential for resistance to distinct microbial pathogens. Proc Natl Acad Sci U S A 95:15107–15111

Tinoco ML, Dias BB, Dall'Astta RC, Pamphile JA, Aragão FJ (2010) *in vivo* trans-specific gene silencing in fungal cells by in planta expression of a double-stranded RNA. BMC Biol 8:27

Tobias DJ, Manoharan M, Pritsch C, Dahleen LS (2007) Co-bombardment, integration and expression of rice chitinase and thaumatin-like protein genes in barley (*Hordeum vulgare* cv.Conlon). Plant Cell Rep 26:631–639

Tohidfar MM, Mohammadi T, Ghareyazie B (2005) Agrobacterium- mediated transformation of cotton (*Gossypium hirsutum*) using a heterologous bean chitinase gene. Plant Cell Tissue Organ Cult 83:83–96

Tomilov AA, Tomilova NB, Wroblewski T, Michelmore R, Yoder JI (2008) Trans-specific gene silencing between host and parasitic plants. Plant J 56:389–397

Tu J, Datta K, Khush GS, Zhang Q, Datta SK (2000) Field performance of *Xa21* transgenic indica rice (*Oryza sativa* L.), IR72. Theor Appl Genet 101:15–20

van der Vossen EAG, Gros J, Sikkema A, Muskens M, Wouters D, Wolters P, Pereira A, Allefs S (2005) The Rpi-blb2 gene from *Solanum bulbocastanum* is an

Mi-1 gene homolog conferring broad-spectrum late blight resistance in potato. Plant J 44:208–222

Van Loon LC, Van Kammen A (1970) Polyacrylamide disc electrophoresis of the soluble leaf proteins from *Nicotiana tabacum* var. 'Samsun' and 'Samsun NN'. II. Changes in protein constitution after infection with tobacco mosaic virus. Virology 40:199–211

van Loon L, van Strien E (1999) The families of pathogenesis-related proteins, their activities, and comparative analysis of PR-1 type proteins. Physiol Mol Plant Pathol 55:85–97

van Loon LC, Pierpoint WS, Boller T, Conejero V (1994) Recommendations for naming plant pathogenesis-related proteins. Plant Mol Biol Rep 12:245–264

van Loon LC, Rep M, Pieterse CM (2006) Significance of inducible defense-related proteins in infected plants. Annu Rev Phytopathol 44:135–162

Vazquez F, Vazquez F, Vaucheret H, Rajagopalan R, Lepers C, Gasciolli V et al (2004) Endogenous trans-acting siRNAs regulate the accumulation of *Arabidopsis* mRNAs. Mol Cell 16:69–79

Vellicce GR, Ricci JCD, Herna'ndez L, Castagnaro AP (2006) Enhanced resistance to *Botrytis cinerea* mediated by the transgenic expression of the chitinase gene ch5B in strawberry. Transgenic Res 15:57–68

Vigers A, Wiedemann S, Roberts WK, Legrand M, Selitrennikoff CP, Fritig B (1992) Thaumatin-like pathogenesis- related proteins are antifungal. Plant Sci 83:155–161

Wally O, Jayaraj J, Punja Z (2009) Comparative resistance to foliar fungal pathogens in transgenic carrot plants expressing genes encoding for chitinase, b-1,3-glucanase and peroxidise. Eur J Plant Pathol 123:331–342

Wally O, Punja ZK (2010) Genetic engineering for increasing fungal and bacterial disease resistance in crop plants. GM Crops 1(4):199–206

Wang Y, Kausch AP, Chandlee JM, Luo H, Ruemmele BA, Browning M, Jackson N, Goldsmith MR (2003) Co-transfer and expression of chitinase, glucanase, and bar genes in creeping bentgrass for conferring fungal disease resistance. Plant Sci 106:497–506

Wrobel-Kwiatkowska M, Lorenc-Kukula K, Starzycki M, Oszmianski J, Kepczynska E, Szopa J (2004) Expression of b-1,3-glucanase in flax causes increased resistance to fungi. Physiol Mol Plant Pathol 65:245–256

Xie Z, Staehelin C, Wiemken A, Broughton W, Muller J, Boller T (1999) Symbiosis-stimulated chitinase isoenzymes of soybean (Glycine max (L.) Merr.). J Exp Bot 50:327–333

Xin M, Wang Y, Yao Y, Xie C, Peng H et al (2010) Diverse set of microRNAs are responsive to powdery mildew infection and heat stress in wheat (*Triticum aestivum* L.). BMC Plant Biol 10:123

Yamamoto T, Iketani H, Ieki H, Nishizawa Y, Notsuka K, Hibi T, Hayashi T, Matsuta N (2000) Transgenic grapevine plants expressing a rice chitinase with enhanced resistance to fungal pathogens. Plant Cell Rep 19:639–646

Yang SM, Gao MQ, Xu CW, Gao JC, Deshpande S, Lin S, Roe BA, Zhu H (2008) Alfalfa benefits from *Medicago truncatula*: the RCT1 gene from *M. truncatula* confers broad spectrum resistance to anthracnose in alfalfa. Proc Natl Acad Sci U S A 105:12164–12169

Yu XM, Griffith M, Wiseman SB (2001) Ethylene induces antifreeze activity in winter rye leaves. Plant Physiol 126:1232–1240

Zamani A, Motallebi M, Jonoubi P, Ghafarian-Nia NS, Zamani MR (2012) Heterologous expression of the *Secale* cereal thaumatin-like protein in transgenic canola plants enhances resistance to stem rot disease. Iran J Biotechnol 10:87–95

Zhao BY, Lin XH, Poland J, Trick H, Leach J, Hulbert S (2005) A maize resistance gene functions against bacterial streak disease in rice. Proc Natl Acad Sci U S A 102:15383–15388

Zhao JP, Jiang XL, Zhang BY, Su XH (2012) Involvement of microRNAmediated gene expression regulation in the pathological development of stem canker disease in *Populus trichocarpa*. Plos One 7:1–13

Zhu H, Xu X, Xiao G, Yuan L, Li B (2007) Enhancing disease resistances of super hybrid rice with four antifungal genes. Sci China C Life Sci 50:31–39

Systemic Infection of Potyvirus: A Compatible Interaction Between Host and Viral Proteins

Rakesh Kumar Verma, Ritesh Mishra, Pooja Sharma, D.K. Choudhary, and Rajarshi Kumar Gaur

Abstract

Viruses profoundly depend on endogenous host transport system and interact with preexisting host cellular factors during movement. Potyviral movement is directed by several movement proteins that are HC-Pro, CP, VPg, and CI and newly discovered P3N-PIPO. CP and HC-Pro facilitate movement of virus by increasing size exclusion limit (SEL) of plasmodesmata (PD). These movement proteins serve many functions: binding the viral genome, transporting the viral genome to plasmodesmata, gating plasmodesmata, trafficking through plasmodesmata, and then transporting through phloem. TuMV P3N-PIPO is a PD-localized protein and mediates the targeting of CI to PD. The P3 protein was not previously associated with potyvirus movement, but it was known to interact with the P1 protein; it is co-localized with 6K2 vesicles (site of potyviral replication). This points out a link between virus replication complexes and intracellular movement. CP has the ability to increase SEL of PD and interact with host RTM factors and suppress RTM resistance of plants. HC-Pro is crucial for long-distance movement of potyvirus by suppressing gene silencing mechanism of host plant. Interaction with host factors and chaperones is also required for efficient spread of potyvirus; presumably interaction of the viral CP with a plant Dna J-like protein NtCPIP (capsid protein interacting proteins) provides a strong in vivo confirmation for the essential role of plant chaperones in potyvirus movement. In this chapter, we are concerned on potyvirus intracellular, intercellular, and long-distance movement, focusing on the host cellular factors' interaction with movement proteins involved.

R.K. Verma • R. Mishra • P. Sharma • D.K. Choudhary
R.K. Gaur (✉)
Department of Science, Faculty of Arts, Science & Commerce, Mody Institute of Technology & Science, Laxmangarh, Sikar 332311, Rajasthan, India
e-mail: gaurrajarshi@hotmail.com

R.K. Gaur and P. Sharma (eds.), *Approaches to Plant Stress and their Management*,
DOI 10.1007/978-81-322-1620-9_20, © Springer India 2014

Keywords
Potyvirus • Cell-to-cell movement • Plasmodesmata • Chaperones • Movement proteins

Introduction

Viruses are obligate parasites that can multiply genomes in their host organisms. A successful infection requires counteracting host defenses as well as intimate interactions between the viral genomes/genome-encoded products and host cellular factors. The infection of plants by viruses is a result of series of compatible interactions between viral and host factors to complete their life cycle. For infection, first, the infected cells must support viral replication for the supply of infectious material. Second, the virus or viral genome moves from cell to cell through the plasmodesmata, within the initially infected leaf. Third, the virus must move through several vasculature-associated cell types and enter the sieve elements, where movement occurs passively over long distances within the same leaf and between organs. Finally, the virus must exit the sieve elements and reestablish replication and cell-to-cell movement in tissues distant from the initial infection site (Whitham and Wang 2004). Potyvirus is the largest genus of plant viruses, with 180 definite or possible members causing significant losses in a wide range of crop plants. The viruses are aphid transmitted in a nonpersistent manner, and some of them are also seed transmitted (Shukla et al. 1994). The organization of the potyvirus single-stranded RNA genome is shown in Fig. 1 (Cuevas et al. 2012). The genome has terminal untranslated (UTR) regions flanking a single large open reading frame. The viral polyprotein is cleaved by three virus-encoded proteases (P1, HC-Pro, and NIa-Pro) into ten products (P1, HC-Pro, P3, 6K1, CI, 6K2, VPg, NIa-Pro, NIb, and CP). An additional peptide, P3N-PIPO, is translated from an overlapping ORF after +2 frame shifting of the P3 cistron (Chung et al. 2008).

Systemic infection is a result of compatible interaction between viral and host factors. This interaction controls viral translation, replication, viral assembly, cell-to-cell movement, and long-distance movement (Fig. 2).

Genome Amplification

Once the virion has entered into the host cell, it must be unencapsidated to expose the viral RNA.

Genome amplification of virus requires two fundamental processes, viral RNA translation for the synthesis of virus-specific proteins, including the viral replicase, and RNA replication itself. Potyviral replicases include core proteins, CI, NIb, and NIa, and accessory factors are P1, HC-Pro, and P3 (Kasschau et al. 1997). The potyvirus RNA lacks a 5′cap structure, but it has been observed in TuMV using Yeast *in-situ* Hybridization (YTH) analysis that the VPg at the 5′end of potyviral RNA binds with the host translation factors, i.e., eIF4E and eIF (iso)4E (Leonard et al. 2000). The RNA-dependent RNA polymerase (RdRp) is the core polypeptide that catalyzes the synthesis of RNA chains of potyviruses. The recruitment of the RdRp to

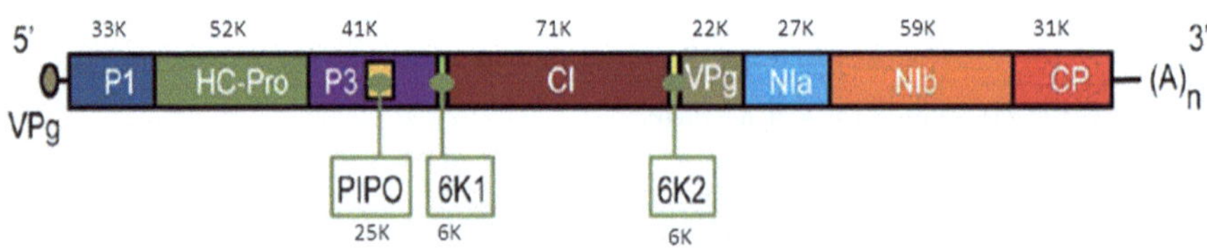

Fig. 1 Schematic representation of potyvirus genome including UTR regions and gene distribution

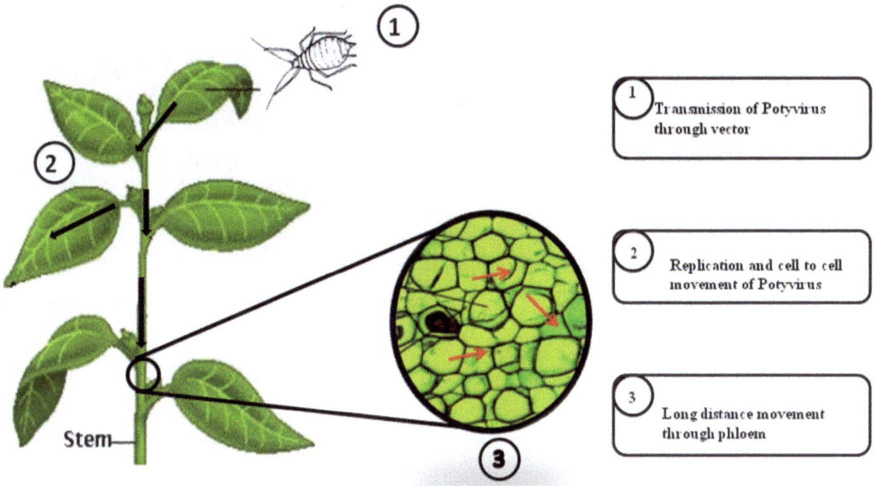

Fig. 2 Systemic infection of virus in plant. Plant viruses accomplish four main steps to complete their infection cycle: (i) transmission through the vector, (ii) replication inside the cell, (iii) cell-to-cell movement through plasmodesmata, and (iv) long-distance movement through the vascular tissue. The *black arrows* show systemic movement through the phloem, and *red arrows* show cell-to-cell movement through plasmodesmata

membranes and its interaction with viral and host factors are critical for the efficiency, specificity, and regulation of viral replication (Buck 1996).

RdRp protein localizes to ER membranes where RNA synthesis takes place (Schaad et al. 1997). RdRp interact with 6K-VPg-Pro polypeptide for association with ER. The 6K domain of the 6K-VPg-Pro polyprotein has been shown to be necessary for ER membrane targeting of VPg-Pro. The 6-kDa protein may serve as anchor for RNA replication complexes to membranous sites of synthesis because it is always found tightly associated with membranes (Li et al. 1997).

Translation initiation factors eIF(iso) 4E (eukaryotic initiation factor (iso)4E) and PABP (poly(A)-binding protein) interact with potyviral VPg for the recruitment of translation initiation factors for viral RNA translation (Cotton et al. 2006). This interaction suggests that VPg may play a role in the assembly of viral translation initiation complex (Léonard et al. 2004). As well as, it has been suggested that the role of the VPg in the initiation of translation could be comparable to that of 5′cap structure and subsequent recruitment of 40S ribosomal subunits (Tavert et al. 2007). HC-Pro functions in replication as a suppressor of posttranscriptional gene silencing (PTGS) as well as plays a role in long-distance movement of potyvirus by interaction with other proteins, and these functions of HC-Pro are independent to its proteolytic activity (Dufresne et al. 2008).

Cell-to-Cell Movement Within the Cell

Depending on the virus species, cell-to-cell movement may occur either in virion or nonvirion form. In plant system, plasmodesmata (PD) provide a whole body macromolecular transport network and work as gateway of plant viruses. Cell-to-cell movement involves the use of the host endoplasmic reticulum (ER)/actin network as an intracellular transport pathway, recognition of adhesion sites at the cell periphery, and modification of PD by alteration of the cell wall structure.

In mesophyll cells, PD have size exclusion limit of approximately 1000 Da for passive transport of molecules; without changing the SEL of PD, it is very difficult to transport the viral particle (Rojas et al. 1997). In case of potyvirus, both CP and HC-Pro were shown to increase plasmodesmal SEL and mediate viral RNA transport from cell to cell. From the model for potyvirus cell-to-cell movement proposed by

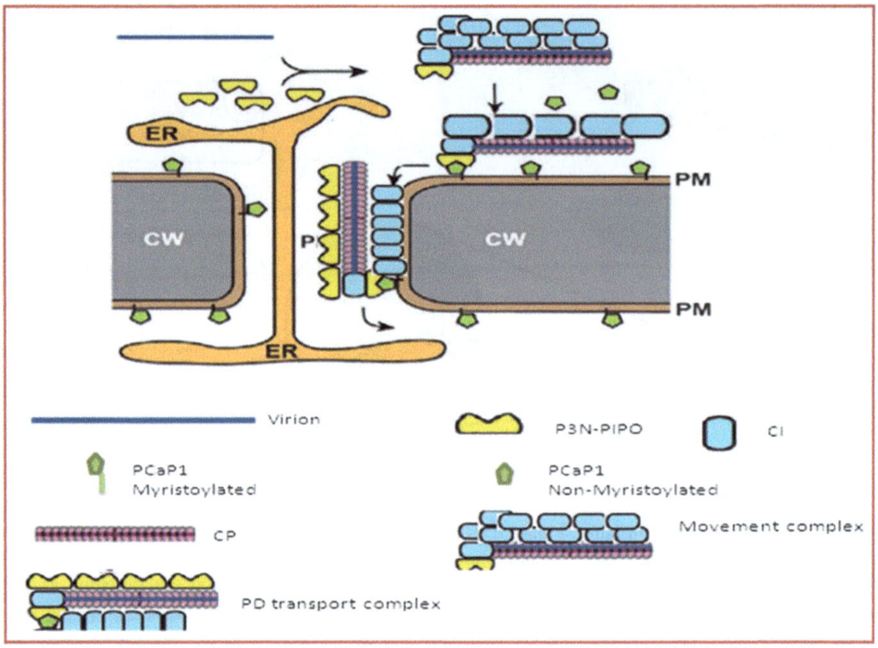

Fig. 3 Model for potyvirus intercellular transport through PD. The virion–CI movement complex is intracellularly transported to the modified PD where CI forms conical structures anchored by the PD-located P3N-PIPO. The virion is then fed through the CI structures and PD to enter the adjacent cell (Wei et al. 2010a; Vijayapalani et al. 2012)

Carrington et al. (1998), CI protein may direct intracellular translocation of a viral transport complex that includes the CP. Then, CP may interact with the plasmodesmata to increase the SEL, and CI protein may function to position the viral complex for translocation through the CI structures, into the plasmodesmata, and finally into the adjacent cells (Fig. 3). The strong correlation between competence for virion assembly and for cell-to-cell movement in the case of TEV (Dolja et al. 1994, 1995), and the fibrillar material (similar to PSbMV particles) observed within plasmodesmata in pea (Roberts et al. 1998), may be taken as an indication that potyviruses move from cell to cell as virions.

Long-Distance Movement from Cell to Phloem

Long-distance movement (also known as phloem-dependent movement) of a virus within a plant generally refers to translocation from the inoculated leaf through the stem to uninoculated leaves and occurs through the plant vasculature. It involves the movement of the infectious agent from the mesophyll via the bundle sheath cells, phloem parenchyma, and companion cells into phloem sieve elements, passive translocation in the phloem, and unloading at a remote site to establish further infection (Carrington et al. 1996). Viruses are passively transported from source to sink tissues along with the photoassimilates at a rate similar to the rate of nutrient movement (Santa Cruz 1999). In a source-to-sink pattern of translocation, mature leaves that are photosynthetically active produce and supply photoassimilates, i.e., serve as sources of photoassimilates, while young leaves that are not photosynthetically active serve as sinks for photoassimilates. Vascular movement of most viruses is thus dependent on the source–sink relationship and seems to be restricted only by the as yet unknown mechanisms by which viruses are loaded and unloaded from the phloem transport system (Silva et al. 2002). Potyviral proteins HC-Pro and CP play a key role in long-distance movement, as well as the central part of the VPg is a domain with universal importance to virus–host interactions required for systemic invasion of plants with potyviruses (Rajamaki and Valkonen 2002). Spetz and Jari (2004) showed that the PVA 6K2 protein affects viral long-distance movement and symptom induction independently and in a host-specific manner.

Potyviral Movement Proteins

For most plant virus groups, the movement process involves one or more specialized virus-encoded proteins, termed movement proteins (MPs). These proteins are usually characterized by mutagenesis when cell-to-cell movement from the primary infected cell is altered without affecting virus replication. Some viruses, such as tobacco mosaic virus and red clover necrotic mosaic virus, encode single dedicated MPs that modify plasmodesmata and facilitate the transport of themselves and nucleic acids through the modified channel. Several families of viruses contain a set of movement proteins called the triple gene block, which encodes three proteins that are proposed to function coordinately for transport (Carrington et al. 1998).

Potyviruses do not encode a dedicated MP, but movement functions have been allocated to several proteins, including the coat protein (CP), HC-Pro, the cylindrical inclusion (CI) protein, and the genome-linked protein (VPg). Movement proteins are multifunctional; they are required for the movement of viral particle as well as for localizing the plasmodesmata and interact with host factors.

HC-Pro: The HC-Pro (50–53 kDa) was initially identified as an accessory helper factor required during the plant-to-plant transmission process of potyviruses by aphid vectors (Govier and Kassanis 1974), a function that gave the original name to this protein (HC, helper component). HC-Pro is a multifunctional protein that is involved in many functions during the virus life cycle.

The N-terminal part of HC-Pro affects virulence (symptom severity), genome amplification, and virus accumulation (Kasschau and Carrington 1995). KITC and PTK amino acid motifs of HC-Pro are required for transmissibility by aphid vectors (Huet et al. 1994). The central region of HC-Pro affects long-distance movement of virus (Cronin et al. 1995), whereas the C-terminal part is needed for cell-to-cell movement (Rojas et al. 1997). HC-Pro has plasmodesmatal-gating (Rojas et al. 1997) and nucleic acid-binding properties (Maia and Bernardi 1996). Recently HC-Pro has been identified as a suppressor of posttranscriptional gene silencing (Kasschau et al. 2003).

The role of HC-Pro in long-distance movement is reported by Kasschau et al. by mutant-defective phenotype; they developed defective TEV-GUS/CCCE mutant virus (substitution of the Cys293, Cys294, Cys295, and Glu299), highly conserved within the HC-Pro central region (Cronin et al. 1995) infection in a series of grafted plants composed of various combinations of HC-Pro transgenic and nontransgenic scions and rootstocks. Systemic infection was only when both the stock and the scion could provide a complementing function from a wild-type HC-Pro transgene. This indicates that HC-Pro is required in both inoculated and noninoculated tissues for efficient long-distance movement and, hence, presumably for both entry into, and exit from, the host plant vascular system (Kasschau et al. 1997).

CP Protein: The CP is a three-domain protein with variable N- and C-terminal domains exposed on the virion surface and a core region that binds RNA. The N-terminal domain is required for transmission of potyviruses by aphids (Atreya et al. 1991). Mutations in the core region of TEV CP revealed an essential role for virus assembly and cell-to-cell movement, suggesting that intercellular transport involves virions (Dolja et al. 1994). A distinct MP-like function for potyvirus CPs in cell-to-cell transport has been proposed from microinjection studies with recombinant CPs from *bean common mosaic necrosis virus* and *lettuce mosaic virus* demonstrating that CPs are able to modify plasmodesmal SEL and to mediate their own trafficking, as well as the transport of viral

RNA from cell to cell (Rojas et al. 1997). For CP, TEV-GUS mutants with deletions in the CP N- or C-terminal domains produced virions in vivo, but the virus exhibited defects in long-distance movement in plants (Dolja et al. 1994, 1995). Also, mutational analysis demonstrated that changes to Ser47 of the PSbMV CP and Asp5 in the DAG motif of the TVMV CP N-terminal domain (López-Moya and Pirone 1998) can modulate the ability of the virus to move systemically in *Chenopodium quinoa* and tobacco plants, respectively, suggesting strong evidence of the role of CP in long-distance movement.

CI Protein: CI protein, one of 11 known potyviral proteins, is associated with cone-shaped structures at plasmodesmata (PD) and is involved in viral cell-to-cell movement. The CI protein, an RNA helicase, is required for genome replication (Klein et al. 1994) and for potyvirus cell-to-cell movement. By electron microscopy, CI protein is seen to form aggregates (called pinwheel or cylindrical inclusions [CIs]) in the cytoplasm of infected cells. These inclusions are frequently seen positioned over the plasmodesmal aperture (Langenberg 1986).

P3N-PIPO: P3N-PIPO is a PD-located protein and directs the CI protein to PD, facilitating the deposition of the cone-shaped structures of CI at PD by interacting with CI protein (Wei et al. 2010a, b). PIPO is a translational fusion of the N-terminus of P3 and PIPO, and hence the name P3N-PIPO. Mutations in the PIPO coding region impeded the virus cell-to-cell movement, allowed virus accumulation in cells localized at the inoculation site, and rendered the virus non-infectious or nearly so in whole plants, suggesting that P3N-PIPO-GFP facilitates its own cell-to-cell movement (Wen and Hajimorad 2010; Vijayapalani et al. 2012).

VPg Protein: VPg is also multifunctional and needed for virus replication (Shahabuddin et al. 1988). It is also needed for virus cell-to-cell and long-distance movement (Rajamaki and Valkonen 2002). It contains a nuclear localization signal that is important for virus replication (Schaad et al. 1996) and a sequence-nonspecific RNA-binding domain (Merits et al. 1998).

Compatible Interaction Between Viral and Host Factors Is Required for Potyvirus Movement

The identification and investigation of protein–protein interactions comprise an important step in understanding the virus infection cycle and the interplay between virus and host. Several methods have been developed to identify and examine protein–protein interactions. In addition to different in vitro methods, the yeast two-hybrid (YTH) system is the most popular in vivo method for the detection of protein interactions (Fields and Song 1989); Fig. 4 describes the protein interaction map of potyviral proteins with itself and with host factors.

- The RTM resistance genes restrict the long-distance movement of several potyviruses in Arabidopsis thaliana (Cosson et al. 2010); these are firstly identified as specific to tobacco etch virus (TEV; Whitham et al. 2000), but recently RTM resistance has been shown to be active against two other unrelated potyviruses, lettuce mosaic virus (LMV) and plum pox virus (PPV; Decroocq et al. 2006). RTM1 (jacalin lectin protein family) and RTM2 (small HSP domain) are expressed in phloem-associated tissues, that the corresponding proteins localize to phloem sieve elements (Chisholm et al. 2001); RTM3 belongs to new plant gene family encoding a meprin and TRAF homology domain-containing protein. The N-terminal end of the potyviral capsid protein (CP) is involved in breaking of the RTM resistance (Decroocq et al. 2009), suggesting a direct or indirect interaction between the potyvirus CP and the RTM factors.
- DnaJ proteins are assumed to function as co-chaperones and regulators of heat shock protein 70 (HSP70) proteins by stimulating their ATPase activity via interaction of the J domain (Kelley 1998), and recruitment of CP-mediated host chaperones is required for efficient spread of viruses. Hofius et al. (2007) identified a novel subset of DnaJ-like proteins from *N. tabacum*, designated capsid protein

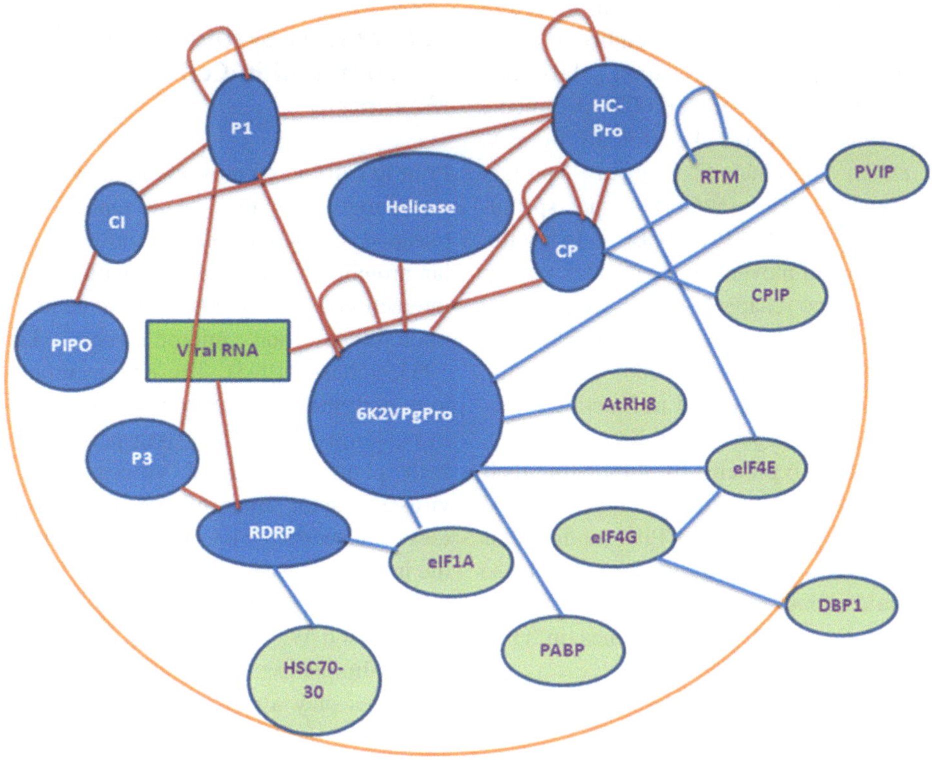

Fig. 4 Interaction map of potyvirus proteins; *blue boxes* show the viral proteins and *green boxes* show plant proteins. Interaction connecting nodes are represented in color code. *Red line* shows interaction between potyviral proteins with itself, and *blue line* shows interaction between host and viral factors

interacting proteins (NtCPIPs), that specifically bind to PVY CP in yeast and in vitro. NtCPIPs are found as novel potyviral susceptibility factors and also provide a strong in vivo confirmation for the essential role of plant chaperones in virus movement (Hofius et al. 2007).

- The eukaryotic translation initiation factor eIF4E has been identified as susceptibility factor supporting potyvirus movement through interaction with the virus genome-linked protein VPg (Gao et al. 2004). The eIF(iso)4E–VPg interaction could anchor viral genomic RNA to translation initiation complexes . This interaction provides viral genome stability by protecting viral RNA from DCP1p-like "decapping" enzymes, thereby blocking degradation of the viral genome by XRN-like host-encoded exonucleases. eIF(iso)4E interacts with eIF(iso)4G, a protein with microtubule-binding activity, and a kinesin-like domain interaction between VPg and eIF (iso)4E may engage the potyviral genome with the intracellular trafficking machinery (Lellis et al. 2002).
- The yeast two- and three-hybrid systems are used to detect direct protein interactions between host and viral factors (Bilgin et al. 2003), and for identification of associated host factors, purification of viral-protein complexes from infected cells has been used (Serva and Nagy 2006). HSP70s represent a conserved family of cellular chaperones. These proteins function as core components in the cellular chaperone network and participate in a wide variety of processes, including the folding of newly synthesized protein refolding of misfolded or aggregated proteins,

translocation of organellar and secretory proteins, protein complex assembly or disassembly, and protein degradation (Mayer and Bukau 2005). Hofius et al. (2007) identified a novel subset of DnaJ-like proteins from *N. tabacum*, designated capsid protein interacting proteins (NtCPIPs), that specifically bind to PVY CP in yeast and in vitro. For cell-to-cell movement, heat shock protein (Hsp) 70-class chaperones act as potential translocation factors and regulation of movement (Boevenik and Oparka 2005). Generally, DnaJ proteins are assumed to function as co-chaperones and regulators of heat shock protein 70 (HSP70) proteins by stimulating their ATPase activity via interaction of the J domain (Kelley 1998). NtCPIPs are found as novel potyviral susceptibility factors and also provide a strong in vivo confirmation for the essential role of plant chaperones in virus movement (Hofius et al. 2007).

- PCaP1 is a cation-binding protein attached to the plasma membrane via myristoylation of a glycine residue and interacts to the P3N-PIPO protein of turnip mosaic virus (TuMV). It links potyviral movement complex to the plasma membrane by binding P3N-PIPO and enables localization of viral movement complex to the plasmodesmata and cell-to-cell movement. Interaction of P3N-PIPO with PCaP1 may directly affect the calcium levels at the plasmodesmata owing to the $Ca2^{+}$-CAM-binding activity of PCaP1. This, in turn, may increase the size exclusion limit (SEL) of the plasmodesmata by reducing callose accumulation (Vijayapalani et al. 2012).
- CK2 (casein kinase II) is a plant protein kinase responsible for PVA CP phosphorylation both in vivo and in vitro (Ivanov et al. 2001). Dynamic balance between CP phosphorylation and dephosphorylation is crucial for PVA infectivity because this phosphorylation regulates the binding of PVA CP to RNA. The confirmation of this mechanism was obtained by the movement-deficient phenotype of the GFP-tagged mutant of CK2 (Ivanov et al. 2003).

Role of Host System in Potyvirus Infection and in Cell-to-Cell Movement

Prior to cell-to-cell movement, the virus must be transported from the site of replication to the plasmodesmata. Intracellular trafficking of cellular proteins and mRNAs is facilitated by their interactions with cytoskeletal elements such as microtubules and microfilaments (Langford 1995; Johnston1995). The endomembrane system and the cytoskeleton of host cooperate in numerous intracellular transport processes in both plant and animal viruses. For animal viruses, f-actin (microfilaments) and microtubules have been shown to be important throughout the infection process, from virus entry and intracellular transport to virus egress and budding (Greber and Way 2006; Radtke and O'Riordan 2006). In plants, there is evidence that these structures play a pivotal role in viral infection (Laporte et al. 2003). The movement of tobacco mosaic virus (TMV) has been particularly well studied and represents a unique situation where both the microtubule and actin cytoskeleton have been implicated in supporting its movement. MP association with the ER also has been reported for viruses whose MPs form tubules. For example, the *alfalfa mosaic virus* MP behaves as an integral membrane protein and localizes to the ER (Huang and Zhang 1999). In the case of CPMV and *cauliflower mosaic virus*, tubule assembly was independent of microtubules or microfilaments; it requires a functional secretory pathway (COP-II) (Huang et al. 2000; Pouwels et al. 2002). At present it remains unclear whether viral RNA transport complex along the ER depends on the ER-associated actin system or whether the ER alone could provide sufficient membrane-associated motility for virus movement (Eduardo et al. 2012). Both 6K2 and P3 proteins encoded by either tobacco etch virus or turnip mosaic virus (TuMV) have been shown to produce mobile granules when ectopically expressed in cells (Wei et al. 2010b; Cui et al. 2010). These researchers also determined that this mobility was associated with the actomyosin network. The 6K2 granules were determined to associate with the ER exit sites (ERES) through their co-localization with Sec23 and Sec24, known markers for this location (Wei and Wang 2008).

Conclusion

There is so much loss occurring because of biotic and abiotic stress in agriculture field. Potyvirus group is main part of biotic stress, which causes heavy loss in important crops. Potyviruses are mainly transmitted through vectors in plants. Despite their importance, the cell-to-cell spread of potyviruses remains poorly understood (Wei et al. 2010a). Potyviruses move from cell to cell through modification of size exclusion limit of plasmodesmata and infect whole plant through phloem in source to sink pattern by interacting with host proteins and several chaperones. Potyviral movement proteins interact with host proteins like JDP CPIP, eukaryotic translation factor eIF4e, PCaP1, protein kinase CK2, and chaperone HSP70 and co-chaperones for successful infection. Analysis of these interactions is very important for understanding the basic mechanism behind the involvement of host proteins during potyvirus infection and also suggests clues about sites and structures within cell at which they function. Many scientists concern that the limitation of host range is the result of disruption in cell-to-cell and long-distance movement of virus rather than the inability to replicate within cell. Information about these interactions may help to improve our understanding of diseases and also can provide some therapeutic approaches like knockout of PCaP1 for inhibition of TuMV cell-to-cell movement and lectin-mediated resistance (Yamaji et al. 2012) by RTM gene. Disruption of compatible viral-protein interactions may provide a new way for effective resistance of plants against viruses.

Acknowledgement The authors are thankful to Department of Biotechnology, New Delhi, India for providing the financial support (ref no. BT/PR14902/BRB/10/889/2010) during this study.

References

Atreya PL, Atreya CD, Pirone TP (1991) Amino acid substitutions in the coat protein result in loss of insect transmissibility of a plant virus. Proc Natl Acad Sci U S A 88:7887–7897

Bilgin DD, Liu Y, Schiff M, Dinesh-Kumar SP (2003) P58 (IPK), a plant ortholog of double-stranded RNA-dependent protein kinase PKR inhibitor, functions in viral pathogenesis. Dev Cell 4:651–661

Boevenik P, Oparka K (2005) Virus-host interactions during movement process. Plant Physiol 138:1815–1821

Buck KW (1996) Comparison of the replication of positive-stranded RNA viruses of plants and animals. Adv Virus Res 47:159–251

Carrington JC, Kasschau KD, Mahajan SK, Schaad MC (1996) Cell-to-cell and long-distance transport of viruses in plants. Plant Cell 8:1669–1681

Carrington JC, Jensen PE, Schaad MC (1998) Genetic evidence for an essential role for potyvirus CI protein in cell-to-cell movement. Plant J 14:393–400

Chisholm ST, Parra MA, Anderberg RJ, Carrington JC (2001) Arabidopsis RTM1 and RTM2 genes function in phloem to restrict long-distance movement of tobacco etch virus. Plant Physiol 127:1667–1675

Chung BY, Miller WA, Atkins JF, Firth AE (2008) An overlapping essential gene in the Potyviridae. Proc Natl Acad Sci U S A 105:5897–5902

Cosson P, Sofer L, Quang Hien Le, Leger V, Schurdi-Levraud V, Whitham SA, Yamamoto ML, Gopalan S, Le Gall O, Candresse T, Carrington JC, Revers F (2010) RTM3, which controls long-distance movement of potyviruses, is a member of a new plant gene family encoding a Meprin and TRAF homology domain-containing protein. Plant Physiol 154:222–232

Cotton S, Dufresne PJ, Thivierge K, Ide C, Fortin MG (2006) The VPgPro protein of Turnip mosaic virus: in vitro inhibition of translation from a ribonuclease activity. Virology 351:92–100

Cronin S, Verchot J, Haldeman-Cahill R, Schaad MC, Carrington JC (1995) Long-distance movement factor: a transport function of the potyvirus helper component-proteinase. Plant Cell 7:549–559

Cuevas JM, Delaunay A, Visser JC, Bellstedt DU, Jacquot E, Elena SF (2012) Phylogeography and molecular evolution of potato virus Y. PLoS One 7(5):e37853

Cui X, Wei T, Chowda-Reddy RV, Sun G, Wang A (2010) The tobacco etch virus P3 protein forms mobile inclusions via the early secretory pathway and traffics along actin microfilaments. Virology 397:56–63

Decroocq V, Sicard O, Alamillo JM, Lansac M, Eyquard JP, Garcia JA, Candresse T, Le Gall O, Revers F (2006) Multiple resistance traits control plum pox virus infection in Arabidopsis thaliana. Mol Plant Microbe Interact 19:541–549

Decroocq V, Salvador B, Sicard O, Glasa M, Cosson P, Svanella-Dumas L, Revers F, García JA, Candresse T (2009) The determinant of potyvirus ability to overcome the RTM resistance of Arabidopsis thaliana maps to the N-terminal region of the coat protein. Mol Plant Microbe Interact 22:1302–1311

Dolja VV, Haldeman R, Robertson NL, Dougherty WG, Carrington JC (1994) Distinct functions of capsid protein in assembly and movement of tobacco etch potyvirus in plants. EMBO J 13:1482–1491

Dolja VV, Haldeman-Cahill R, Montgomery AE, VandenBosch KA, Carrington JC (1995) Capsid protein determinants involved in cell-to-cell and long distance movement of tobacco etch potyvirus. Virology 206:1007–1016

Dufresne PJ, Thivierge K, Cotton S, Beauchemin C, Ide C, Ubalijoro E, Laliberte JF, Fortin MG (2008) Heat shock 70 protein interaction with Turnip mosaic virus RNA-dependent RNA polymerase within virus-induced membrane vesicles. Virology 374:217–227

Eduardo I, Chietera G, Pirona R, Pacheco I, Troggio M, Banchi E, Bassi D, Rossini L, Vecchietti A, Pozzi C (2012) Genetic dissection of aroma volatile compounds from the essential oil of peach fruit: QTL analysis and identification of candidate genes using dense SNP maps. Tree Genetics Genomes 9 (1):189–204. http://link.springer.com/journal/11295

Fields S, Song O (1989) A novel genetic system to detect protein-protein interactions. Nature 340:245–246

Gao Z, Johansen E, Eyers S, Thomas CL, Noel Ellis TH, Maule AJ (2004) The potyvirus recessive resistance gene, sbm1, identifies a novel role for translation initiation factor eIF4E in cell-to-cell trafficking. Plant J 40:376–385

Govier DA, Kassanis B (1974) A virus-induced component of plant sap needed when aphids acquire potato virus Y from purified preparations. Virology 61:420–426

Greber UF, Way M (2006) A superhighway to virus infection. Cell 124:741–754

Hofius D, Maier AT, Dietrich C, Jungkunz I, Bo¨rnke F, Maiss E, Sonnewald U (2007) Capsid protein-mediated recruitment of host DnaJ-Like proteins is required for *Potato Virus Y* infection in tobacco plants. J Virol 81:11870–11880

Huang M, Zhang L (1999) Association of the movement protein of alfalfa mosaic virus with the endoplasmic reticulum and its trafficking in epidermal cells of onion bulb scales. Mol Plant Microbe Interact 12:680–690

Huang Z, Han Y, Howell SH (2000) Formation of surface tubules and fluorescent foci in *Arabidopsis thaliana* protoplasts expressing a fusion between the green fluorescent protein and the *Cauliflower mosaic virus* movement protein. Virology 271:58–64

Huet H, Gal-on A, Meir E, Lecoq H, Raccah B (1994) Mutations in the helper component protease gene of zucchini yellow mosaic virus affect its ability to mediate aphid transmissibility. J Gen Virol 75:1407–1414

Ivanov KI, Puustinen P, Merits A, Saarma M, Mäkinen K (2001) Phosphorylation down-regulates the RNA binding function of the coat protein of potato virus A. J Biol Chem 276:13530–13540

Ivanov KI, Puustinen P, Gabrenaite R, Vihinen H, Rönnstrand L, Valmu L, Kalkkinen N, Mäkinen K (2003) Phosphorylation of the potyvirus capsid protein by protein kinase CK2 and its relevance for virus infection. Plant Cell 15:2124–2139

Kasschau KD, Carrington JC (1995) Requirement for HC-Pro processing during genome amplification of tobacco etch potyvirus. Virology 209:268–273

Kasschau KD, Cronin S, Carrington JC (1997) Genome amplification and long-distance movement functions associated with the central domain of tobacco etch potyvirus helper component-proteinase. Virology 228:251–262

Kasschau KD, Xie ZX, Llave AE, Chapman EJ, Krizan KA, Carrington JC (2003) P1/HC-Pro, a viral suppressor of RNA silencing, interferes with Arabidopsis development and miRNA function. Dev Cell 4:205–217

Kelley WL (1998) The J-domain family and the recruitment of chaperone power. Trends Biochem Sci 23:222–227

Klein PG, Klein RR, RodrõÂguez-Cerezo E, Hunt AG, Shaw JG (1994) Mutational analysis of the tobacco vein mottling virus genome. Virology 204:759–769

Langenberg WG (1986) Virus protein association with cylindrical inclusion of two viruses that infect wheat. J Gen Virol 67:1161–1168

Langford GM (1995) Actin- and microtubule-dependant organelle motors: interrelationships between the two motility systems. Curr Opin Cell Biol 7:82–88

Laporte C, Vetter G, Loudes AM, Robinson DG, Hillmer S et al (2003) Involvement of the secretory pathway and the cytoskeleton in intracellular targeting and tubule assembly of grapevine fanleaf virus movement protein in tobacco BY-2 cells. Plant Cell 15:2058–2075

Lellis AD, Kasschau KD, Whitham SA, Carrington JC (2002) Loss-of-susceptibility mutants of *Arabidopsis thaliana* reveal an essential role for eIF(iso)4E during potyvirus infection. Curr Biol 12:1046–1051

Leonard S, Plante D, Wittmann S, Daigneault N, Fortin MG, Laliberte JF (2000) Complex formation between potyvirus VPg and translation eukaryotic initiation factor 4E correlates with virus infectivity. J Virol 74:7730–7737

Léonard S, Viel C, Beauchemin C, Daigneault N, Fortin MG, Laliberté JF (2004) Interaction of VPg-Pro of turnip mosaic virus with the translation initiation factor 4E and the poly (A)-binding protein in planta. J Gen Virol 85:1055–1063

Li XH, Valdez P, Olvera RE, Carrington JC (1997) Functions of the tobacco etch virus RNA polymerase (NIb): subcellular transport and protein-protein interaction with VPg/proteinase (NIa). J Virol 71:1598–1607

Lopez-Moya JJ, Pirone TP (1998) Charge changes near the N terminus of the coat protein of two potyviruses affect virus movement. J Gen Virol 79:161–165

Maia IG, Bernardi F (1996) Nucleic acid-binding properties of a bacterially expressed potato virus Y helper component-proteinase. J Gen Virol 77:869–877
Mayer MP, Bukau B (2005) Hsp70 chaperones: cellular functions and molecular mechanism. Cell Mol Life Sci 62:670–684
Merits A, Guo D, Järvekülg L, Saarma M (1998) Biochemical and genetic evidence for interactions between potato a potyvirus-encoded proteins P1 and P3 and proteins of the putative replication complex. Virology 263:15–22
Peña EJ, Heinlein M (2012) RNA transport during TMV cell-to-cell movement. Front Plant Sci 3:193
Pouwels J, Van Der Krogt GNM, Van Lent J, Bisseling T, Wellink J (2002) The cytoskeleton and the secretory pathway are not involved in targeting the cowpea mosaic virus movement protein to the cell periphery. Virology 297:48–56
Radtke AL, O'Riordan MXD (2006) Intracellular innate resistance to bacterial pathogens. Cell Microbiol 8:1720–1729. doi:10.1111/j.1462-5822.2006.00795.x
Rajamaki M-L, Valkonen JPT (2002) Viral genome-linked protein (VPg) controls accumulation and phloem-loading of a potyvirus in inoculated potato leaves. Mol Plant-Microbe Interact 15:138–149
Roberts IM, Wang D, Findaly K, Maule AJ (1998) Ultrastructural and temporal observations of the potyvirus cylindrical inclusions (CIs) show that the CI protein acts transiently in aiding virus movement. Virology 244:173–181
Rojas MR, Zerbini FM, Allison RF, Gilbertson RL, Lucas WJ (1997) Capsid protein and helper component-proteinase function as potyvirus cell-to-cell movement proteins. Virology 237:283–295
Roudet-Tavert G, Michon T, Walter J, Delaunay T, Redondo E, Le Gall O (2007) Central domain of a potyvirus VPg is involved in the interaction with the host translation initiation factor eIF4E and the viral protein HcPro. J Gen Virol 88:1029–1033
Santa Cruz S (1999) Perspective: phloem transport of viruses and macromolecules-what goes in must come out. Trends Microbiol 7:237–241
Schaad MC, Haldeman-Cahill R, Cronin S, Carrington JC (1996) Analysis of the VPg-proteinase (NIa) encoded by tobacco etch potyvirus: effects of mutations on subcellular transport, proteolytic processing, and genome amplification. J Virol 70:7039–7048
Schaad MC, Jensen PE, Carrington JC (1997) Formation of plant RNA virus replication complexes on membranes: role of an endoplasmic reticulum-targeted viral protein. EMBO J 16:4049–4059
Serva S, Nagy PD (2006) Proteomics analysis of the tombusvirus replicase: Hsp70 molecular chaperone is associated with the replicase and enhances viral RNA replication. J Virol 80:2162–2169
Shahabuddin M, Shaw JG, Rhoads RE (1988) Mapping of the tobacco vein mottling virus VPg cistron. Virology 163:635–637
Shukla DD, Ward CW, Brunt AA (1994) The Potyviridae. CAB International, Oxford
Silva MS, Wellink J, Goldbach RW, van Lent JWM (2002) Phloem loading and unloading of *Cowpea mosaic virus* in *Vigna unguiculata.* J Gen Virol 83:1493–1504
Spetz C, Jari PT (2004) Valkonen potyviral 6K2 protein long-distance movement and symptom-induction functions are independent and host-specific. Mol Plant-Microbe Interact 17:502–510
Vijayapalani P, Maeshima M, Nagasaki-Takekuchi N, Miller WA (2012) Interaction of the trans-frame potyvirus protein P3N-PIPO with host protein PCaP1 facilitates potyvirus movement. PLoS Pathog 8(4): e1002639. doi:10.1371/journal.ppat.1002639
Wei T, Wang A (2008) Biogenesis of cytoplasmic membranous vesicles for plant potyvirus replication occurs at the endoplasmic reticulum exit sites in a COPI- and COPII-dependent manner. J Virol 82:12252–12264
Wei T, Zhang C, Hong J, Xiong R, Kasschau KD, Zhou XP, Carrington JC, Wang AM (2010a) Formation of complexes at plasmodesmata for potyvirus intercellular movement is mediated by the viral protein P3N-PIPO. PLoS Pathog 6:e1000962
Wei TY, Huang TS, McNeil J, Laliberte JF, Jong J, Nelson RS, Wang AM (2010b) Sequential recruitment of the endoplasmic reticulum and chloroplasts for plant potyvirus replication. J Virol 84:799–809
Wen RH, Hajimorad MR (2010) Mutational analysis of the putative PIPO of soybean mosaic virus suggests disruption of PIPO protein impedes movement. Virology 400:1–7
Whitham S, Wang Y (2004) Roles for host factors in plant viral pathogenicity. Curr Opin Plant Biol 7:1–7
Whitham SA, Anderberg RJ, Chisholm ST, Carrington JC (2000) Arabidopsis RTM2 gene is necessary for specific restriction of tobacco etch virus and encodes an unusual small heat shock-like protein. Plant Cell 12:569–582
Yasuyuki Yamaji, Kensaku Maejima, Ken Komatsu, Takuya Shiraishi, Yukari Okano, Misako Himeno, Kyoko Sugawara, Yutaro Neriya, Nami Minato, Chihiro Miura, Masayoshi Hashimoto, Shigetou Namba (2012) Lectin-mediated resistance impairs plant virus infection at the cellular level. Plant Cell 24:778–793

Bioinformatics Resources for the Management of Biological Information on Plant Responses Towards Stresses

Ratna Prabha, Dhananjaya P. Singh, Vivek Keshri, Sevyaa, and Arun K. Sharma

Abstract

In natural and agricultural conditions, plants are exposed to multiple stresses that impose severe impact on growth and development. Stress in plants can be considered as a change in growth phage which leads to disturbance of metabolic homeostasis. Plant pathogens result in biotic stresses, whereas environmental stimuli generate multiple abiotic stresses like temperature extremes, drought, chemical toxicity, salinity, heavy metals, oxidative stress and radiation. Plants respond to these stresses and develop better adaptation by activating their intrinsic machinery. Currently, biological research is witnessing increasing development of methods, technologies and implementations for a better mechanistic representation of biological systems. In the era of multi-omics approaches addressing multiple aspects of biological mechanisms shown by plants in response to biotic or abiotic stresses, huge data is emerging out from research work. This data needs proper management, analysis and interpretation in order to decipher plant strategies to combat stresses. In the past decades, a large number of databases, software, tools and web resources were developed to make ease of access for researchers working to decipher plant responses towards stresses. In this chapter we are describing bioinformatics resources which need to describe information on plant responses against stresses. Bioinformatics resources like database of annotated tentative orthologs from crop abiotic stress transcripts, MIPS PlantsDB, GreenPhylDB, Gramene, GCP Comparative Stress Gene Catalog, Plant Stress Gene Database, PASmiR, QlicRice, Rice Stress Gene Catalog, Arabidopsis Stress Responsive Gene Database, STIFDB and STIFDB2 are available to facilitate multi-omics research in this field.

R. Prabha • D.P. Singh (✉) • V. Keshri • Sevyaa
A.K. Sharma
National Bureau of Agriculturally Important Microorganisms, Indian Council of Agricultural Research, Kushmaur, Maunath Bhanjan 275101, India
091-547-2530125 091-547-2530358
e-mail: dpsfarm@rediffmail.com

R.K. Gaur and P. Sharma (eds.), *Approaches to Plant Stress and their Management*,
DOI 10.1007/978-81-322-1620-9_21,

Some of these resources are specific to particular plants, whereas others are generalized and contain information about multiple species.

Keywords
Bioinformatics • Plant stress • Databases • Genomics • Proteomics • Metabolomics

Introduction

In the entire life cycle, plants in natural habitat are exposed to multiple biotic and abiotic factors which lead to stresses in plants. Thus, plants require stringent biological mechanisms for fighting against such stresses and gaining better adaptation. Plants recognize any unfavourable changes in their surrounding environment and respond to it. Plants are exposed to these stresses in natural and agricultural conditions, both, and this imposes severe impact on development and growth of plants. Multiple environmental conditions like light, temperature, soil and water availability and salt have influence on growth and development of plants, and thus they represent the most limiting factors that strongly affect agricultural productivity and significant economic losses across the globe (Hirt and Shinozaki 2004). However, a clear-cut definition for plant stress is not available (Kranner et al. 2010).

Plant pathogens result in biotic stress, whereas there are multiple abiotic stresses like temperature extremes, drought, chemical toxicity, salinity, heavy metals, oxidative stress and radiation causing loss of more than 50 % productivity of important crops worldwide. Abiotic stress is a fundamental component of such climate change and thus poses serious risk to agriculture. They are major limiting factors on plant growth and productivity and possess injurious effects (Bray et al. 2000; Mantri et al. 2012).

Biotic Stress

Plants are in regular contact with various biotic agents, like bacterial, viral and fungal pathogens and even parasitic plants and insect herbivores which regularly attack and impose negative impact (Bilgin et al. 2010). These factors are reducing huge amount of crop yield since long time. It is a well-accepted fact that the increasing number of pests and pathogens are progressively reducing crop yield. Current science and agriculture is focusing on the study of damaging effect of biotic stress agents on yield (Peterson and Higley 2000). Plants have evolved inducible defences against these microbial pathogens, herbivores and parasitic plants. These defences involve activation of specific gene regulatory networks which initiates gene expression for the synthesis of specific proteins and defensive secondary metabolites. In systemic defence signalling, various secondary metabolites including plant hormones play essential role and protect plants from damage (Dorantes-Acosta et al. 2012).

Abiotic Stress

Environment is a key factor in growth and development of plants. Unfavourable environmental conditions pose negative impacts on plant physiology and phenotype (Osakabe et al. 2013). Plants are exposed to numerous abiotic stresses which are result of multifactor environmental conditions, e.g. temperature variation (high and low both), UV, light, drought, salinity, heavy metals and hypoxia. Better understanding of plant responses towards abiotic stress is currently hot topic in plant research as plants are sessile organisms and require tolerance towards these stresses. In fact, these stresses are expected to increase in near future due to global climate changes (Intergovernmental Panel of Climate Change, http://www.ipcc.ch) (Hirayama and

Shinozaki 2010). Plants have developed numerous responses towards abiotic stresses through which they can tolerate and survive in adverse conditions (Knight and Knight 2001). There are cascades of molecular networks which are triggered in plants in response to environmental stresses leading to stress sensitivity, signal transduction, expression of genes and metabolites involved in specific stress conditions (Vinocur and Altman 2005).

Impact of Stresses on Plant Growth and Productivity

Stress in plants can lead to disturbance of metabolic homeostasis and thus requires proper adjustment of it through acclimation (Shulaev et al. 2008). In response to stress, plants involve biochemical, physiological, morphological and developmental changes (Caliskan 2011; Kranner et al. 2010). Abiotic stress leads to amendment in soil-plant-atmosphere continuum. Thus, the study of metabolism, productivity and sustainability in plants which are subjected to various abiotic stresses is a significant subject of research. Long-term introduction to these abiotic stresses results in distorted biomolecules and transformed metabolism that affects vegetative as well as reproductive growth and development (Hirt and Shinozaki 2004; Wang et al. 2003; Djilianov et al. 2005).

Plant Responses Towards Environmental and Biotic stresses

For plants, natural environment is a complex set of stress conditions, and the similar is also true for plant responses towards these stresses. Although, fundamentally, plants require energy (light), water, carbon and mineral nutrients for their optimal growth and development, these conditions sometimes limit plant growth due to being below the optimal levels or excessiveness. Water stress reduced the leaf size, stunted plant growth, suppressed root growth, delayed flowering and fruits and reduced seed number, size and viability. Low or excess temperature affects physiological processes and adversely influences growth and development of plants. Freezing injury is a major cause of crop losses. Likewise light intensity and atmospheric gases also badly affect plant development. Edaphic conditions like salinity, alkalinity, acidity pollutant contaminations and many other anthropogenic causes are severely affecting plant development and adversely managing crop production (Haferkamp 1988).

During evolution, plants have developed a wide range of mechanisms to deal with stresses. Cellular and molecular responses of plants to environmental stresses are complex and have been studied intensively (Thomashow 1999; Hasegawa et al. 2000; Xiong et al. 2002). Evidences suggest that plants resist to abiotic stresses in a complex but integrated manner and adapt to the existing constraints in due course of time (Lata et al. 2011). Plant molecular responses to various biotic and abiotic stresses routinely involve interactive crosstalk with diverse biosynthetic networks and pathways (Takahashi et al. 2004). Tolerance or susceptibility to these stresses is a dynamic event involving multiple stages of plant development (Chinnusamy et al. 2008). Defence, repair, acclimation and adaptation are the major components of resistance responses towards stresses (Kranner et al. 2010).

Plant adaptations to stresses involves series of reaction which implies at each stage of cellular, biochemical and molecular organization. This multipart response requires an extensive molecular regulation of gene expression. In the course of evolution, plants have developed mechanisms of survival and adaptation to these various stresses and other crisis caused due to repeatedly adverse environment (Fraire-Velázquez et al. 2011). Studying the mechanisms of stress adaptation helps to address environmental, toxicological and physiological troubles (Borkotoky et al. 2013). In recent years, extensive research has been carried out to understand the mechanisms through which plants recognize environmental signals and pass on the signals to cellular machinery to trigger adaptive responses (Prabha et al. 2011; Mantri et al. 2012). Interconnections have also been searched out to find commons between biotic and abiotic stress signalling pathways, but it is basically in initial phase as much of research in the area of stress molecular mechanisms has been done in fragmented and independent manner (Fujita et al. 2006). As plants receive any stimulus from the environment, manifold pathways of cellular signalling having complex interactions or crosstalk are activated (Fraire-Velázquez et al. 2011). This leads to the activation and regulation of particular stress-related genes involved in the stress response mechanisms which include transcriptional control, signalling, membrane and protein protection and scavenging of free radical and toxic compound (Wang et al. 2003). Current research has shown that transcription factors, kinases and other molecules played important role in crosstalk between stress signalling pathways. Similarly, pathways like hormone as well as ROS signalling pathways take part in the crosstalk between biotic and abiotic stress signalling (Fujita et al. 2006). One of the widely studied stress impacts on plants result in the accumulation of reactive oxygen species (ROS) and reactive nitrogen species (RNS) which regulate enzyme activity and gene regulation (Wilkinson and Davies 2009; Molassiotis and Fotopoulos 2011; Mittler et al. 2011). It is widely accepted that ROS and RNS constitute a coordinated network which regulates plant responses to the environment. Plant defences in the form of the accumulation of such secondary metabolites like phenolics, flavonoids, abscisic acid (ABA) etc., have also come out to minimize the effects of the ROS or RNS damage inside the cells (Wilkinson and Davies 2009). Hormones like ABA and ethylene are important regulators of plant responses to abiotic stress (Goda et al. 2008). ABA is a regulator of plant responses to osmotic stress stimuli (Chinnusamy et al 2008; Hubbard et al. 2010; Kim et al 2010). ABA-mediated signalling is fast and controls stomatal aperture through the biochemical regulation of ion and water transport processes (Kim et al. 2010). Under water-deficient conditions, cellular dehydration induces increased accumulation of ABA to trigger downstream genes encoding signalling and transcription factors and metabolic enzymes (Yamaguchi-Shinozaki and Shinozaki 2006). It is reported that the activation of ABA signalling cascades results in enhanced plant tolerance to drought stress. Transcriptional regulation of dehydration and salinity has shown both ABA-dependent and ABA-independent pathways (Yamaguchi-Shinozaki and Shinozaki 2006). Ethylene is also involved in stress responsiveness including drought, ozone, flooding (hypoxia and anoxia), heat, freezing, wounding and UV-B light (Morgan and Drew 1997). Ethylene signalling is well studied and interactions exist between ethylene and ABA during drought (Stepanova and Alonso 2009; Yoo et al. 2009). All of these interactions make the plant response to stresses very complex (Pinheiro and Chaves 2011). Understanding stress signal transduction is moreover imperative for constant progress of rational breeding and transgenic strategies which aims at developing stress tolerance in crops (Xiong et al. 2002). In recent times, exploration of molecular mechanisms of stress responses along with genetic modification of stress tolerance has shown promising results which may be further applied to agriculturally and ecologically important plants (Wang et al. 2003).

Abiotic stresses and losses in crop production have become a hot topic in plant research in recent years. Plant stress has been recognized as major constituent of the outcome of global warming on worldwide food production. Studies to plant responses to stress involve genomics, proteomics, transcriptomics and metabolomics approaches. Research in this aspect is a complex issue and requires integration of multiple disciplines such as soil science, plant physiology, biochemistry, plant breeding and molecular biology. It also entails close interdisciplinary group efforts and integration which can comprehensively exchange understandings and information pertaining to plants under stress and needs bioinformatics resources to analyse, interpret and integrate data and manage biological data coming out of interaction studies and system biology.

Genomic Approaches

Research interest in the area of plant stress management and its implications gained strength due to their impact on plant productivity and sustainability that was harshly affected by stress factors (Aarts and Fiers 2003; Sreenivasulu et al. 2007). In the past, focus was on breeding and genetic engineering for improving tolerance of plants towards abiotic stress factors which had very restricted success due to the genetic complexity of stress responses (Cushmana and Bohnertb 2000). Presently focus has been given on large-scale genomics approaches which are facilitating identification of target gene or genes involved in this complex process of adaptation (Aarts and Fiers 2003). Transcriptional and translational levels also involve multiple genes and their products. Better understanding of these stress-inducible genes and their functions will unravel the probable system of stress tolerance (Sreenivasulu et al. 2007). Comparative genomic studies of an evolutionarily diverse set of model organisms and high-throughput techniques such as analysis of expressed sequence tags, microarray, targeted or random mutagenesis are helping in getting better understanding. The background of effective engineering strategies for greater stress tolerance is lying in novel gene(s) discovery and analysis and better understanding of their expression patterns in stress adaptation (Cushmana and Bohnertb 2000). As there is increase in number of genomic resources, there are mainly two methods used for the identification of probable gene pool conferring abiotic stress tolerance: firstly, functional genomics approach involving gene identification and transgenesis, and, secondly, the identification of QTLs/genes conferring tolerance to stress in germplasm collections and marker-assisted breeding programmes. Functional genomics approaches have opened new avenues from a single-gene discovery up to thousands of genes that can be identified via this approach (Sreenivasulu et al. 2007). Genomics-based approaches related to quantitative trait loci (QTLs) and marker-assisted selections are also permitting researchers to promote research in area of stress tolerance (Tuberosa and Salvi 2006). Genome-wide profiling of stress-induced expression and their post-transcriptional events also enable identification of candidate genes. Trait gene discovery via functional genomics necessitates use of multiple genetic resources, phenotyping and genomics tools along with bioinformatics (Ishitania et al. 2004). In plants a little number of whole genome sequences are available in comparison to microorganisms and mammalian species (Neilson et al. 2010). Next-generation approach in plant genomics towards abiotic stress management research will involve perceptive of stress response gene networks (Rowley and Mockler 2011).

Transcriptomics Approaches

In the recent years, genome- and transcriptome-wide investigations have provided insight of transcript abundance profiles under abiotic stresses. Transcriptome analysis also provides glimpse of alternative splicing patterns and upregulation of key transcription factors which are involved in stress-induced signalling cascades. DNA microarrays are extremely used for the analysis of plant transcriptomes and provide information about their response and tolerance towards stresses. These studies also assist in the identification of stress-related genes which improve our understanding towards biotic and abiotic stress tolerance in plants (Öktem et al. 2008; Rowley and Mockler 2011).

Proteomic Approaches

Much emphasis is given on transcriptomics and proteomics for the identification of stress-responsive genes and proteins (Neilson et al. 2010). First line of plant cell defence against stress involves changes in gene expression which consequently leads to change in their proteome. Changes in gene expression depend on particular stress and its rigorousness along with the developmental stage of the plant and proteome analysis and provide potent means for connecting gene expression to cell metabolism (Nouri et al. 2011). Significant and intense changes are observed in proteome composition of plants while acclimation towards stresses. As proteins contribute directly in plant stress response, it might be expected that proteomics studies can enhance existing knowledge about the association between protein abundance and plant stress acclimation. Most studies in this aspect are carried out on model plant *Arabidopsis thaliana* and rice as large protein sequence databases are available for them. However, progress is also made for performing proteomic analyses on other plant species (Kosová et al. 2011). Through proteomic approaches general stress-related genes and proteins involved in several pathways like those involved in abscisic acid and jasmonic acid biosynthesis, redox homeostasis, energy metabolism, etc., have been recognized (Neilson et al. 2010). A number of techniques are being developed nowadays for extraction, identification and protein-interaction study, allowing a deeper insight into the mechanism that is involved in the cell responses towards abiotic stresses. All these approaches finally aim at attaining an improved understanding and justification of the morphological behaviours of plants in stress conditions (Nouri et al. 2011).

Proteomics has witnessed technological development, particularly with the entry of mass spectroscopy- and nuclear magnetic resonance (NMR)-based high-end techniques. This has enhanced our knowledge in crop plant abiotic stress tolerance (Barkla et al. 2013). However, despite of the particular techniques available, complete genome sequence information is very important in proteomics as it is the major source of protein identification (Neilson et al. 2010).

Metabolomics Approaches

Metabolomics have emerged as post-genomic era technologies (Ishitania et al. 2004). Metabolome of any organism represents the complete set of metabolites found in the organism. Metabolome is dynamic and, therefore, makes available a snapshot of the processes going on in the organism at specific time. In combination with other "omics" data, it facilitates a complete chemical understanding of the organism (Grennan 2009). Plant metabolomics has developed into a very important tool for the analysis of primary and secondary metabolism in plants (Sumner 2010). Although this area is facing certain technical problems as presently no single technique is available to detect all of the metabolites and therefore, variety of analytical techniques are being used. Presently metabolomic databases have developed progressively to facilitate the identification of the compounds. Metabolomics is an emerging field with many challenges to be met and possesses the potential for comparing an ecotype or species for study of effect of stresses upon a plant. Metabolic data needs combination with other genomic data to become most useful and informative, and this can be facilitated by the creation of high-quality, publicly available databases (Grennan 2009).

Database for Information on Plant Stress Management

In the last few years, there have been tremendous increases in the quantity of existing biological databases (Stein 2003) that provide users the capacity to process, integrate and interpret massive amounts of biological data. For several decades, databases are used as a model system of managing and processing large quantities of information (Rhee and Crosby 2005), biological databases have been very useful for management of data and providing accessibility in an informative way. Databases execute different functions that depend on the data they hold (Stein 2003).

Basically, there are three main types of biological databases which were developed or still in developmental phase:

(i) Large-scale public repositories developed and maintained by government authorities or international groups, e.g. GenBank, EMBL, DDBJ, UniProt and ArrayExpress
(ii) Community-specific database resources involving model organism databases,

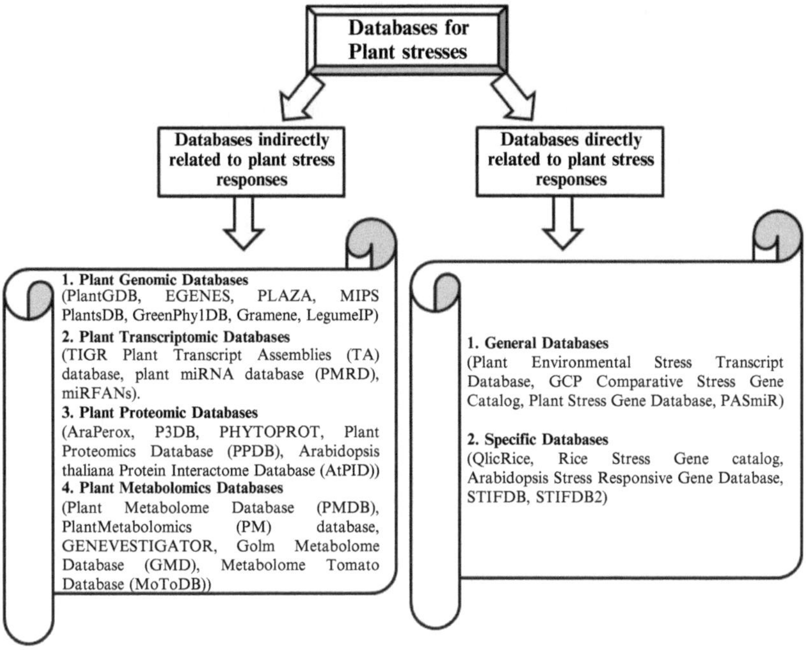

Fig. 1 Bioinformatics resources for the information on plant responses to stresses

databases of taxonomically related species or databases focused towards specific kind of data, genome annotation, metabolism, orthologous relationships or like plant stresses

(iii) Project-specific databases including smaller-scale and habitually short-lived databases which manage data of projects during their running period (Rhee and Crosby 2005)

Currently biological research is in transition phase and witnessing increasing development of methods, technologies and implementations for a better mechanistic perceptive of biological systems (Rhee and Crosby 2005). In the past few decades, a large number of databases, software, tools and web resources were developed to make ease of access to bioinformatics-related tasks and provide broad relevance for researchers (Singh et al. 2012). The sudden increase of biological data along with auxiliary abundance of dispersed databases sometimes creates difficulty for biologists and bioinformaticians to identify and select the best and concise data resources for their requirements and the most proficient way to access them (Ali et al. 2011). Plant science has witnessed this development, and a number of databases were developed to facilitate researchers. We are providing information about resources from where data or information related to plant stresses, either directly or indirectly, can be accessed (Fig. 1).

Databases Directly Related to Plant Stress

We are providing information about those databases which specifically holds data and information related to plant stresses. We identified nine databases (database of annotated tentative orthologs from crop abiotic stress transcripts, GCP Comparative Stress Gene Catalog, Plant Stress Gene Database, PASmiR, QlicRice, Rice Stress Gene Catalog, Arabidopsis Stress Responsive Gene Database, STIFDB, STIFDB2) which possess information regarding plant stresses (Table 1). Some of these resources are plant specific, whereas rest are generalized and contain information about multiple species.

Table 1 Database resources for biological information on plant responses to stresses

S. no	Name of the database	Description	Weblink	References
1.	Plant Environmental Stress Transcript Database	Contains information about annotated tentative orthologous sequences of 16 crop species across 4 abiotic stress situations	http://intranet.icrisat.org/gt1/tog/homepage.htm	Balaji et al. (2006)
2.	GCP Comparative Stress Gene Catalog	Provides comparative information for stress-responsive genes across plant species	http://dayhoff.generationcp.org	Wanchana et al. (2008)
3.	Plant Stress Gene Database	Includes information about 259 stress-related genes from 11 different plant species	http://ccbb.jnu.ac.in/stressgenes/frontpage.html	Prabha et al. (2011)
4.	PASmiR	Database for miRNA molecular regulation in abiotic stress in plants	http://hi.ustc.edu.cn:8080/PASmiR or http://pcsb.ahau.edu.cn:8080/PASmiR	Zhang et al. (2013)
5.	QlicRice	Contains information of abiotic stress-responsive quantitative trait loci (QTL) and loci interaction channels in rice	http://nabg.iasri.res.in:8080/qlic-rice/	Smita et al. (2011)
6.	Rice Stress Gene Catalog	Integrated data source of pre-existing rice stress gene families	Not available	Ali et al. (2011)
7.	Arabidopsis Stress Responsive Gene Database	Contains 637 potential stress-responsive genes along with 44 types of different stress factors	http://srgdb.bicpu.edu.in/	Borkotoky et al. (2013)
8.	STIFDB	Contains information about putative abiotic stress-responsive transcription factor binding sites (TFBS) for 2,269 abiotic stress-responsive genes	http://caps.ncbs.res.in/stifdb/	Shameer et al. (2009)
9.	STIFDB2	STIFDB2 is a recent version of STIFDB and is information of stress-responsive genes and stress-inducible transcription factor for both *Arabidopsis thaliana* and *Oryza sativa*	http://caps.ncbs.res.in/stifdb2	Naika et al. (2013)

Most of these databases are focused towards abiotic stresses rather than biotic stresses. A major reason of this biasness might be the fact that abiotic stresses are responsible for a loss of more than 50 % productivity of important crops worldwide (Bray et al. 2000; Mantri et al. 2012).

Plant Environmental Stress Transcript Database

Plant Environmental Stress Transcript Database is a database of annotated tentative orthologs from crop abiotic stress transcripts. This database contains information about annotated tentative orthologous sequences of 16 crop species like *Arabidopsis thaliana*, *Glycine max*, barley, Medicago, rice, rye, sorghum, chickpea, potato, wheat, maize, tomato, pennisetum, phaseolus, cowpea and groundnut across four abiotic stress situations and includes ESTs from stress cDNA libraries. This also carried out bioinformatics analysis like clustering, assemblage of tentative orthologous sets and annotated information under stress conditions and putative functions. This database allows the user to assess multiple information like annotated transcripts expressed in stress conditions, microsatellites in these transcripts for purpose of conserved functional markers, conserved hypothetical genes having orthologs in multiple species but function unpredicted, ortholog sets with sequence alignment related to stress conditions in these species. The information is useful for users working in the area of molecular evolution of genes and genomics of stress responses in plants. For particular organism, it includes information for total number of sequences, number of contigs, number of singletons, number of cDNA libraries, number of sequences in orthologous groups and number of SSRs. It is available online at http://intranet.icrisat.org/gt1/tog/homepage.htm (Balaji et al. 2006).

GCP Comparative Stress Gene Catalog

Comparative Stress Gene Catalog is developed by Generation Challenge Programme (GCP) which provides comparative information for stress-responsive genes across plant species and act as a summarized resource of protein families, phylogenetic trees, multiple sequence alignments (MSA) and related facts for genes which are involved in various environmental stresses, specifically abiotic. GCP Comparative crop Stress Gene Catalog is designed with the expectations of assisting comparative genomics and bioinformatics analysis and elucidation of research results. Comparative biology across multiple crop species is the main feature in the GCP for identifying stress-responsive gene loci and their corresponding alleles which can be further used in plant breeding programmes for stress tolerance (Wanchana et al. 2008). GCP Comparative Stress Gene Catalog can be accessed worldwide at http://dayhoff.generationcp.org

Plant Stress Gene Database

Plant Stress Gene Database includes information about 259 stress-related genes from 11 different plant species, viz. *Arabidopsis thaliana*, *Arachis hypogaea*, *Glycine max*, *Triticum aestivum*, *Hordeum vulgare*, *Saccharum officinarum*, *Solanum lycopersicum*, *Oryza sativa*, *Pennisetum*, *Phaseolus vulgaris* and *Zea mays*. Plant Stress Gene Database focuses basically on genes and includes information about genes which are either up- or down-regulated in stress conditions. It attempts to provide maximum information available for each gene along with their references. Along with the detailed information about gene, its protein product, ortholog and paralog and bibliography about each gene are also available in this database. Database can be searched through many different keywords. Links to other useful databases are also provided on a separate web page. This database is valuable for researchers because of wealth of information it contains on plant genes involved in stress conditions and thus influencing crop productivity and defence system of plants (Prabha et al. 2011). Database is publicly accessible at http://ccbb.jnu.ac.in/stressgenes/frontpage.html.

PASmiR

PASmiR is database for microRNAs (miRNA) molecular regulation of abiotic stress in plants. miRNAs are reported to have a role in regulation of plant responses towards abiotic stresses. PASmiR is a literature-curated database which provides descriptions of miRNA molecular regulation in detailed and searchable way. PASmiR database includes information of 715 miRNAs and 1,085 miRNA-abiotic stress regulatory entries of 33 plant species across 35 different conditions of abiotic stresses collected from ~200 published studies. Users are able to search PASmiR for plant species, abiotic stress and miRNA identifier. Results include specific miRNA, its regulation information, identifier, expression pattern, species name, stress type, reference literature and other related information. Users can also submit entries through a web-based submission page. It is a valuable resource for research of miRNA regulatory mechanisms involved in plant reactions to abiotic stresses (Zhang et al. 2013). The PASmiR database is online accessible from http://hi.ustc.edu.cn:8080/PASmiR or http://pcsb.ahau.edu.cn:8080/PASmiR.

QlicRice

QlicRice is a database aimed towards abiotic stress-responsive quantitative trait loci (QTL) and loci interaction channels in rice. It contains information about 974 abiotic stress-related QTLs from 53 different traits. These QTLs were further annotated for their physical position on rice chromosomes by using Rice Genome Sequencing and Annotation Project. QlicRice provides information for novel candidate genes underlying specific QTLs, corresponding biochemical pathways responding to particular abiotic stresses, location on genome, GenBank accessions, associated expressed sequence tags (ESTs), protein structure, tandem repeats and link to other related resources like Gene Ontology (GO), Kyoto Encyclopedia of Genes and Genomes (KEGG) and other related information. Results are provided in graphical manner including physical and genetic map with statistics for illustrating the location of QTLs on different chromosomes. This database is useful for researchers working in the area of plant breeding especially on functional genomics of rice plants in respect to abiotic stress and will help to get information about agronomically important QTLs in rice responding to abiotic stresses (Smita et al. 2011). This database is online accessible via weblink http://nabg.iasri.res.in:8080/qlic-rice/ .

Rice Stress Gene Catalog

Rice Stress Gene Catalog (RSGC) facilitates users as an integrated data source of pre-existing rice stress gene families, its protein members and sequences in FASTA format. RSGC tool provides global classification of rice genes and includes functionally defined gene families for comparative analyses of rice genome. It also provides opportunities for identification of new gene families, initial annotation errors and best gene models along with the structures of existing gene families as well as tools for alignments, multiple sequence comparison and database search. RSGC is designed to facilitate users with genome-scale analysis of multiple gene families, annotation on the basis of homology and identification of genes with common domain architectures (Ali et al. 2011).

Arabidopsis Stress Responsive Gene Database

Arabidopsis Stress Responsive Gene Database (ASRGD) contains 637 potential stress-responsive genes for *Arabidopsis thaliana* along with 44 types of different stress factors. For each of the genes, associated information like gene ID, sequences and cross-response is also available. It also includes BLAST search interface for sequences. The information available in database is gathered solely from literature available for stress tolerance genes associated with plants. The database is online available via http://srgdb.bicpu.edu.in/. This database grants a potent approach for management, assessment, exploration and retrieval of information of various stress-responsive genes in *A. thaliana* (Borkotoky et al. 2013).

STIFDB: Arabidopsis Stress Responsive Transcription Factor Database

STIFDB is also dedicated to *Arabidopsis thaliana* plant. It contains information about putative abiotic stress-responsive transcription factor binding sites (TFBS) for 2,269 abiotic stress-responsive genes (identified through numerous stress-responsive microarray experiments) which makes it different from Arabidopsis Stress Responsive Gene Database which contains information about genes only (identified through publications). Transcription factor binding sites are identified through STIF (Sundar et al. 2008) and implement Hidden Markov Models for the prediction of the TFBSs, whereas further validation is carried out through Jackknifing method (Sundar et al. 2008). STIFDB makes available comprehensive information about genes, transcription factors and transcription factor binding sites which are induced by stress in *A. thaliana*. STIFDB points out that a number of stress-responsive genes are almost similar on all chromosomes. It offers analysis of the promoters of these stress-responsive genes and information about regulation by upstream transcription factors. STIFDB is useful source for scientists working on abiotic stress responses in plants due to information regarding stress regulome of abiotic stress-responsive genes in plants and stress signalling affecting the transcription of these genes (Shameer et al. 2009). It is online available at http://caps.ncbs.res.in/stifdb/.

STIFDB2: Stress Responsive Transcription Factor Database

STIFDB2 is a recent version of STIFDB (Arabidopsis Stress Responsive Transcription Factor Database) and, along with *A. thaliana*, it also includes information of stress-responsive genes and stress-inducible transcription factor for *Oryza sativa*. It also has options to identify probable transcription factor binding sites in the promoters of abiotic stress-responsive genes. In case of *A. thaliana*, 10 specific transcription factor families are known to be stress responsive, whereas for *O. sativa* only five specific families of transcription factors are known. The database is useful for researchers working in the area of transcriptional regulation of genes in response to abiotic stresses (Naika et al. 2013). STIFDB2 can be accessed online via http://caps.ncbs.res.in/stifdb2.

Databases Indirectly Related to Plant Stress

Multiple databases are also available for different areas of plant science. Here, we are providing information about databases which can be used in analysis related to plant stresses. These databases are categorized in genomic, transcriptomic, proteomic and metabolomic areas (Table 2) and are briefly described herewith.

Plant Genomic Databases

PlantGDB

PlantGDB contains molecular sequence data for all plant species and organizes EST sequences into contigs which are annotated and linked to particular genomic DNA. PlantGDB aims at identifying genes which are common to all plants and those which are specific to particular species so that it can help in gene prediction and cross variety comparisons. PlantGDB provides genome browsing capability for species for which large-scale genome sequencing efforts are going on (Dong et al. 2004). It is online accessible via http://www.plantgdb.org/.

EGENES

EGENES is an integrated resource consisting of genomic, chemical and network information for multiple species so that cellular functions can be represented. EGENES is a knowledge-based database focusing on the analysis of plant expressed sequence tags (ESTs) present in the KEGG suite of databases. It connects plant genomic information with functional information. For each genome, it connects a set of genes/transcripts with a network of interacting molecules in the cell. EGENES is publicly available via http://www.genome.jp/kegg-bin/create_kegg_menu?category5plants_egenes through the KEGG's navigation system (Masoudi-Nejad et al. 2007).

PLAZA

PLAZA is an online database for plant comparative genomics and helps in the analysis of genome organization, gene function and regulatory pathways. PLAZA extracts structural and functional annotation for plant genomes which are published and then incorporates the same with a large set of interactive tools which make the study of gene, gene function and genome evolution easy (Proost et al. 2009). It is online available at http://bioinformatics.psb.ugent.be/plaza/.

MIPS PlantsDB

MIPS PlantsDB (Spannagl et al. 2007) provides data and information resources for tomato, *Medicago, Arabidopsis, Brachypodium*, sorghum, maize, rice, barley and wheat as well as support for integrative and comparative analysis. It incorporates CrowsNest for synteny analysis between monocots, MIPS Repeat Element Database (mips-REdat) and Catalog (mips-REcat) along with links to other databases (Nussbaumer et al. 2013). It is online available at http://mips.helmholtz-muenchen.de/plant/genomes.jsp.

GreenPhylDB

GreenPhylDB is focused on *Oryza sativa* and *Arabidopsis thaliana* genomes and assists comparative functional genomics study of these two species. It encompasses the largest number of manually curated plant gene families. It incorporates information from numerous other databases including KEGG, InterPro, UniProt, TAIR and TIGR (Conte et al. 2008). It can be accessed via link http://greenphyl.cirad.fr.

Table 2 Resources useful for analysis relevant to plant stresses

S. no	Name	Description	Weblink	References
Plant genomic databases				
1.	PlantGDB	Contains molecular sequence data for all plant species	http://www.plantgdb.org/	Dong et al. (2004)
2.	EGENES	Provides analysis of plant expressed sequence tags (ESTs) present in the KEGG	http://www.genome.jp/kegg-bin/create_kegg_menu?category5plants_egenes	Masoudi-Nejad et al. (2007)
3.	PLAZA	Online database for plant comparative genomics	http://bioinformatics.psb.ugent.be/plaza/	Proost et al. (2009)
4.	MIPS PlantsDB	Provide data and information resources for tomato, *Medicago*, *Arabidopsis*, *Brachypodium*, sorghum, maize, rice, barley and wheat	http://mips.helmholtz-muenchen.de/plant/genomes.jsp	Nussbaumer et al. (2013)
5.	GreenPhylDB	Assists in comparative functional genomics study of *Oryza sativa* and *Arabidopsis thaliana*	http://greenphyl.cirad.fr	Conte et al. (2008)
6.	Gramene	Curated resource intended for genetic, genomic and comparative genomics statistics predominantly for multiple plant species	www.gramene.org	Liang et al. (2008)
7.	LegumeIP	Integrative database for comparative genomics and transcriptomics of model legumes	http://plantgrn.noble.org/LegumeIP/	Li et al. (2012)
Plant transcriptomic databases				
8.	TIGR Plant Transcript Assemblies (TA) database	Provides transcript assemblies for plants	http://plantta.tigr.org	Childs et al. (2007)
9.	Plant miRNA database (PMRD)	Contains information about 8,433 miRNAs collected from 121 plant species	http://bioinformatics.cau.edu.cn/PMRD	Zhang et al. (2010)
10.	miRFANs	Provides miRNA function annotations for *Arabidopsis thaliana*	http://www.cassava-genome.cn/mirfans	Liu et al. (2012)
Plant proteomic databases				
11.	AraPerox	Database of novel proteins of plant peroxisomes carrying putative peroxisome targeting signals (PTSs) identified from Arabidopsis genome	http://www.araperox.uni-goettingen.de/	Reumann et al. (2004)
12.	P^3DB	Database of protein phosphorylation data from many plants	http://www.p3db.org/	Gao et al. (2009)
13.	PHYTOPROT	Database of protein cluster of *Arabidopsis thaliana* with all the existing sequences from other plants	http://genoplante-info.infobiogen.fr/phytoprot	Mohseni-Zadeh et al. (2004)
14.	Plant Proteomics Database (PPDB)	Source of proteins experimentally identified via mass spectrometry (MS) in *Arabidopsis* and maize (*Zea mays*)	http://ppdb.tc.cornell.edu	Sun et al. (2009)
15.	*Arabidopsis thaliana* Protein Interactome Database (AtPID)	AtPID offers system-level understanding of gene function and biological processes in *Arabidopsis thaliana*	http://atpid.biosino.org/	Cui et al. (2008)

(continued)

Table 2 (continued)

S. no	Name	Description	Weblink	References
Plant metabolomics databases				
16.	Plant Metabolome Database (PMDB)	Database of a structurally and functionally annotated metabolites found in plants	http://www.sastra.edu/scbt/pmdb	Udayakumar et al. (2012)
17.	Plant Metabolomics (PM) database	Possesses broad targeted and untargeted mass spectrum metabolomics data for *Arabidopsis* mutants	http://www.plantmetabolomics.org	Bais et al. (2012)
18.	GENEVESTIGATOR	Assign contextual information to gene expression data	https://www.genevestigator.ethz.ch	Zimmermann et al. (2004)
19.	Golm Metabolome Database (GMD)	Makes custom mass spectral libraries, metabolite profiling experiments publicly available	http://csbdb.mpimp-golm.mpg.de/gmd.html	Kopka et al. (2005)
20.	The Metabolome Tomato Database (MoTo DB)	Liquid chromatography–mass spectrometry (LC–MS)-based metabolome database for tomato fruit	http://appliedbioinformatics.wur.nl	Moco et al. (2006)

Gramene

Gramene (www.gramene.org) is a curated resource intended for genetic, genomic and comparative genomics statistics predominantly for crop species (rice, maize, wheat, etc.) and several other plant (mainly grass) species. It provides information regarding assembly and annotations of particular genomes, genes, nucleotide sequences, proteins, genetic and physical maps, quantitative trait loci (QTLs), markers, comparative mappings, ontologies, pathways and genetic diversity data from rice, creation of orthologous gene sets and phylogenetic trees and literature (Liang et al. 2008).

LegumeIP

LegumeIP (http://plantgrn.noble.org/LegumeIP/) is an integrative database for comparative genomics and transcriptomics of model legumes. It contains complete genomic sequences of *Medicago truncatula*, *Glycine max*, *Lotus japonicus*, *A. thaliana* and *Populus trichocarpa* where last two are for reference purposes. LegumeIP assembles information for gene function, gene and gene family information, tissue-specific transcriptomic profiles, synteny and phylogeny. For annotation, it relies on other databases, i.e. UniProt, InterProScan, Gene Ontology, KEGG and Genomes databases. LegumeIP also contains large-scale microarray and RNA-Seq-based gene expression data (Li et al. 2012).

Plant Transcriptomic Databases

TIGR Plant Transcript Assemblies (TA) database

The TIGR Plant Transcript Assemblies (TA) database (http://plantta.tigr.org) collects expressed sequences from the GenBank and constructs transcript assemblies from it. Sequences collected are expressed sequence tags (ESTs), full-length and partial cDNAs (exclusive of in silico predicted gene). The TA database includes a total of 215 plant species, each of which contains more than 1,000 EST or cDNA sequences (Childs et al. 2007).

MicroRNAs databases

MicroRNAs (miRNAs) are 21 nucleotide-long non-coding small RNAs having role in plant growth and development and response to environmental stresses. Presently, two databases are available for miRNA in plants. The plant miRNA database (PMRD) contains information about 8,433 miRNAs collected from 121 plant species such as Arabidopsis, rice, wheat, soybean, maize, sorghum and barley. It collects data from public databases and recent literature, and data is generated by in-house projects. The database incorporates information about sequence, secondary structure, target genes and expression profiles with genome browser. The PMRD is freely obtainable at http://bioinformatics.cau.edu.cn/PMRD (Zhang et al. 2010). miRFANs is an online database of miRNA function annotations for *Arabidopsis thaliana* only. For each miRNA, it provides integrated information including miRNA–target interactions, expression profiles, genomic annotations and pathways, gene ontology and target site information (Liu et al. 2012). miRFANs is online available at http://www.cassava-genome.cn/mirfans.

Plant Proteomic Databases

AraPerox

Plant peroxisomes are involved in numerous processes including primary and secondary metabolism, development and responses to abiotic and biotic stresses (Hu et al. 2012). AraPerox (http://www.araperox.uni-goettingen.de/) is a database of novel proteins of plant peroxisomes carrying putative peroxisome targeting signals (PTSs) identified from Arabidopsis genome. Information regarding prediction of subcellular localization is also provided to sort out peroxisomal and nonperoxisomal proteins. AraPerox is likely to smoothen the progress of the identification of new utility and complex biochemical pathways of plant peroxisomes (Reumann et al. 2004).

P^{3}DB

Protein phosphorylation is the major casual post-translational modification regulating dynamic behaviours and decision processes in cells of numerous organisms. P^{3}DB (http://www.p3db.org/) is a database of protein phosphorylation data from many plants. P^{3}DB was initially constructed with dataset from oilseed rape, but presently it includes information for *A. thaliana* and soybean. It also incorporates phosphorylation data from current literature (Gao et al. 2009).

PHYTOPROT

This is a database of protein cluster. Complete proteome of *A. thaliana* was compared with all the existing sequences from other plants and was then grouped into clusters. Resulting clusters were stored in PHYTOPROT and can be queried at http://genoplante-info.infobiogen.fr/phytoprot. Users can also perform BLAST search for any query sequence against all the clusters (Mohseni-Zadeh et al. 2004).

Plant Proteomics Database

The Plant Proteomics Database (PPDB; http://ppdb.tc.cornell.edu) is a source of proteins experimentally identified via mass spectrometry (MS) in Arabidopsis and maize (*Zea mays*). Other features of PPDB include information of posttranslational modifications, experimental proteotypic peptides and spectral count which can be searched also (Sun et al. 2009).

AtPID

Arabidopsis thaliana Protein Interactome Database (AtPID) contains data appropriate to protein–protein interaction, domain attributes, gene regulation, protein subcellular location and ortholog maps which is obtained from GO annotation, ortholog interactome, microarray profiles and genome contexts. For *A. thaliana*, AtPID offers system-level understanding of gene function and biological processes (Cui et al. 2008). It is online accessible at http://atpid.biosino.org/.

Plant Metabolomics Databases

Plant Metabolome Database

Plant Metabolome Database (PMDB) is database of a structurally and functionally annotated metabolites found in plants. Presently, it contains more than 1,000 metabolites. PMDB includes secondary metabolites possessing three-dimensional structures in the biological data banks and databases. It integrates with JME Editor and Jmol for output in textual and graphical format both. It also provides links to other databases such as KEGG, PUBCHEM and CAS NUMBER for providing additional information about the metabolites (Udayakumar et al. 2012). PMDB is freely available to all users at http://www.sastra.edu/scbt/pmdb.

PlantMetabolomics (PM) Database

PlantMetabolomics.org is aimed towards developing metabolomics as a functional genomics tool so that functions of *A. thaliana* genes can be exposed devoid of visible phenotype. The PlantMetabolomics (PM) database possesses broad targeted and untargeted mass spectrum metabolomics data for Arabidopsis mutants.

This permits researcher to create models of the metabolic network of *A. thaliana* (Bais et al. 2012). PM is publicly available at http://www.plantmetabolomics.org.

GENEVESTIGATOR

GENEVESTIGATOR is database cum tool which allows users to analyse the data in respect to plant development, organ and environmental conditions. It consists of a gene expression database and multiple facilities for query and analysis which aimed at gene functional discovery. Main purpose of GENEVESTIGATOR is to assign contextual information to gene expression data (Zimmermann et al. 2004). The database and analysis toolbox is online available at https://www.genevestigator.ethz.ch.

Golm Metabolome Database

The Golm Metabolome Database (GMD) makes possible the search for and distribution of mass spectra in biologically active metabolites which are quantified using gas chromatography (GC) united with mass spectrometry (MS). It makes custom mass spectral libraries, metabolite profiling experiments publicly available. GMD aims to represent an exchange platform for experimental research activities and bioinformatics (Kopka et al. 2005). It is publicly available via http://csbdb.mpimp-golm.mpg.de/gmd.html.

Metabolome Tomato Database

The Metabolome Tomato Database (MoTo DB) is an open-access liquid chromatography–mass spectrometry (LC–MS) based metabolome database for tomato fruit. MoTo DB contains semipolar metabolites along with metabolite information from the literature biologically (Grennan 2009; Moco et al. 2006).

Conclusion

Plant responses to stress conditions for management of their growth and development are complex and chemically diverse phenomenon. Since plant losses caused due to biotic and abiotic stresses are almost irreversible in environmental or biotic condition above or below optimal levels, the study of plant responses towards the management of physiological, biochemical and molecular status within the cell remains an ever-challenging research area. In the era of multi-omics approaches towards all kind of biological problems, data generation is a huge task and so are its analysis, interpretation, storage and retrieval. Bioinformatics provides such resources to access, analyse and interpret what message biological data means to convey.

The above-described databases are useful in terms of knowledge they hold in the area of plant stress management. Most of these databases contain information of plant genes involved in stress mechanisms. However, some also contain information about miRNA molecular regulation, abiotic stress-responsive quantitative trait loci (QTL) and loci interaction channels in rice and putative abiotic stress-responsive transcription factor binding sites (TFBS) in plants under stress condition. We are sure that this information will benefit researchers working in the area of plant biology of biotic and abiotic stress management.

Acknowledgement Authors are thankful to National Agricultural Innovation Project (NAIP), India sponsored *Establishment of National Agricultural Bioinformatics Grid (NABG)* project from Indian Council of Agricultural Research (ICAR), INDIA for financial support.

References

Aarts MG, Fiers MW (2003) What drives plant stress genes? Trends Plant Sci 8:99–102

Ali M, Khalid RR, Nawaz M, Qamar N (2011) Development of a tool for the analysis of plant stress proteins. Int J Bioautomation 15:261–266

Bais P, Moon-Quanbeck SM, Nikolau BJ, Dickerson JA (2012) Plantmetabolomics.org: mass spectrometry-based Arabidopsis metabolomics–database and tools update. Nucleic Acids Res 40:D1216–D1220. doi:10.1093/nar/gkr969

Balaji J, Crouch JH, Petite PV, Hoisington DA (2006) A database of annotated tentative orthologs from crop abiotic stress transcripts. Bioinformation 1:225–227

Barkla BJ, Vera-Estrella R, Pantoja O (2013) Progress and challenges for abiotic stress proteomics of crop plants. Proteomics. doi:10.1002/pmic.201200401

Bilgin DD, Zavala JA, Zhu J, Clough SJ, Ort DR, DeLucia EH (2010) Biotic stress globally

downregulates photosynthesis genes. Plant Cell Environ 33:1597–1613. doi:10.1111/j.1365- 3040.2010.02167.x
Borkotoky S, Saravanan V, Jaiswal A, Das B, Selvaraj S, Murali A, Lakshmi PTV (2013) The Arabidopsis Stress Responsive Gene Database. Int J Plant Genomics. http://dx.doi.org/10.1155/2013/949564
Bray EA, Bailey-Serres J, Weretilnyk E (2000) Responses to abiotic stresses. In: Gruissem W, Buchannan B, Jones R (eds) Biochemistry and molecular biology of plants. American Society of Plant Physiologists, Rockville, pp 1158–1249
Caliskan M (2011) The roles of germin gene products in plants under salt stress. In: Shanker A, Venkateswarlu B (eds) Abiotic stress response in plants – physiological, biochemical and genetic perspectives.. ISBN 978-953-307-672-0
Childs KL, Hamilton JP, Zhu W, Ly E, Cheung F, Wu H, Rabinowicz PD, Town CD, Buell CR, Chan AP (2007) The TIGR plant transcript assemblies database. Nucleic Acids Res 35:D846–D851
Chinnusamy V, Gong Z, Zhu JK (2008) Abscisic acid-mediated epigenetic processes in plant development and stress responses. J Integr Plant Biol 50 (10):1187–1195
Conte MG, Gaillard S, Lanau N, Rouard M, Périn C (2008) GreenPhylDB: a database for plant comparative genomics. Nucleic Acids Res 36:D991–D998
Cui J, Li P, Li G, Xu F, Zhao C, Li Y, Yang Z, Wang G, Yu Q, Li Y, Shi T (2008) AtPID: *Arabidopsis thaliana* protein interactome database–an integrative platform for plant systems biology. Nucleic Acids Res 36: D999–D1008
Cushmana JC, Bohnertb HJ (2000) Genomic approaches to plant stress tolerance. Cur Opin Plant Biol 3:117–124
Djilianov D, Georgieva T, Moyankova D, Atanassov A, Shinozaki K, Smeeken SCM, Verma DPS, Murata N (2005) Improved abiotic stress tolerance in plants by accumulation of osmoprotectants – gene transfer approach. Biotechnol Biotechnoll Equip 19(3):63–71
Dong Q, Schlueter SD, Brendel V (2004) PlantGDB, plant genome database and analysis tools. Nucleic Acids Res 32:D354–D359
Dorantes-Acosta AE, Sánchez-Hernández CV, Arteaga-Vázquez MA (2012) Biotic stress in plants: life lessons from your parents and grandparents. Front Genet 3:256. doi:10.3389/fgene.2012.00256
Fraire-Velázquez S, Rodríguez-Guerra R, Sánchez-Calderón L (2011) Abiotic and biotic stress response crosstalk in plants. In: Shanker A (ed) Abiotic stress. InTech, Croatia, pp 3–26. ISBN 979-953-307-195-3
Fujita M, Fujita Y, Noutoshi Y, Takahashi F, Narusaka Y, Yamaguchi-Shinozaki K, Shinozaki K (2006) Crosstalk between abiotic and biotic stress responses: a current view from the points of convergence in the stress signalling networks. Cur Opin Plant Biol 9:436–442
Gao J, Agrawal GK, Thelen JJ, Xu D (2009) P3DB: a plant protein phosphorylation database. Nucleic Acids Res 37:D960–D962. doi:10.1093/nar/gkn733
Goda H, Sasaki E, Akiyama K, Maruyama-Nakashita A, Nakabayashi K, Li W, Ogawa M, Yamauchi Y, Preston J, Aoki K, Kiba T, Takatsuto S, Fujioka S, Asami T, Nakano T, Kato H, Mizuno T, Sakakibara H, Yamaguchi S, Nambara E, Kamiya Y, Takahashi H, Hirai MY, Sakurai T, Shinozaki K, Saito K, Yoshida S, Shimada Y (2008) The AtGenExpress hormone and chemical treatment data set: experimental design, data evaluation, model data analysis and data access. Plant J 55(3):526–542
Grennan AK (2009) MoTo DB: a metabolic database for tomato. Plant Physiol 151:1701–1702
Haferkamp MR (1988) Environmental factors affecting plant productivity. In: White RS, Short RE (eds) Achieving efficient use of rangeland resources. Fort Keogh Research symposium, Miles City, pp 27–36
Hasegawa PM, Bressan RA, Zhu J-K, Bohnert HJ (2000) Plant cellular and molecular responses to high salinity. Annu Rev Plant Physiol Plant Mol Biol 51:463–499
Hirayama T, Shinozaki K (2010) Research on plant abiotic stress responses in the post-genome era: past, present and future. Plant J 61:1041–1052
Hirt H, Shinozaki K (eds) (2004) Plant responses to abiotic stress, vol 4, Topics in current genetics. Springer, Berlin/New York. ISBN 978-3-540-20037-6
Hu J, Baker A, Bartel B, Linka N, Mullen RT, Reumann S, Zolman BK (2012) Plant peroxisomes: biogenesis and function. Plant Cell 24:2279–2303. doi:10.1105/tpc.112.096586
Hubbard KE, Nishimura N, Hitomi K, Getzoff ED, Schroeder JI (2010) Early abscisic acid signal transduction mechanisms: newly discovered components and newly emerging questions. Genes Dev 24 (16):1695–1708
Ishitania M, Raoa I, Wenzlb P, Beebea S, Tohme J (2004) Integration of genomics approach with traditional breeding towards improving abiotic stress adaptation: drought and aluminum toxicity as case studies. Field Crops Res 90:35–45
Kim TH, Bohmer M, Hu H, Nishimura N, Schroeder JI (2010) Guard cell signal transduction network: advances in understanding abscisic acid, CO_2, and Ca^{2+} signalling. Annu Rev Plant Biol 61:561–591
Knight H, Knight MR (2001) Abiotic stress signalling pathways: specificity and cross-talk. Trends Plant Sci 6:262–267
Kopka J, Schauer N, Krueger S, Birkemeyer C, Usadel B, Bergmüller E, Dörmann P, Weckwerth W, Gibon Y, Stitt M, Willmitzer L, Fernie AR, Steinhauser D (2005) GMD@CSB.DB: the Golm Metabolome Database. Bioinformatics 21:1635–1638
Kosová K, Vítámvás P, Prášil IT, Renaut J (2011) Plant proteome changes under abiotic stress–contribution of proteomics studies to understanding plant stress response. J Proteomics 74:1301–1322. doi:10.1016/j.jprot.2011.02.006
Kranner I, Minibayeva FV, Beckett RP, Seal CE (2010) What is stress? Concepts, definitions and applications in seed science. New Phytol 188:655–673. doi:10.1111/j.1469-8137.2010.03461.x

Lata C, Yadav A, Prasad M (2011) Role of plant transcription factors in abiotic stress tolerance. In: Shanker A, Venkateswarlu B (eds) Abiotic stress response in plants – physiological, biochemical and genetic perspectives. InTech, Croatia. ISBN 978-953-307-672-0

Li J, Dai X, Liu T, Zhao PX (2012) LegumeIP: an integrative database for comparative genomics and transcriptomics of model legumes. Nucleic Acids Res 40:D1221–D1229

Liang C, Jaiswal P, Hebbard C, Avraham S, Buckler ES, Casstevens T, Hurwitz B, McCouch S, Ni J, Pujar A, Ravenscroft D, Ren L, Spooner W, Tecle I, Thomason J, Tung CW, Wei X, Yap I, Youens-Clark K, Ware D, Stein L (2008) Gramene: a growing plant comparative genomics resource. Nucleic Acids Res 36:D947–D953

Liu H, Jin T, Liao R, Wan L, Xu B, Zhou S, Guan J (2012) miRFANs: an integrated database for *Arabidopsis thaliana* microRNA function annotations. BMC Plant Biol 12:68. doi:10.1186/1471-2229-12-68

Mantri N, Patade V, Penna S, Ford R, Pang E (2012) Abiotic stress responses in plants: present and future. In: Ahmad P, Prasad MNV (eds) Abiotic stress responses in plants: metabolism to productivity. Springer, New York, pp 1–19

Masoudi-Nejad A, Goto S, Jauregui R, Ito M, Kawashima S, Moriya Y, Endo TR, Kanehisa M (2007) EGENES: transcriptome-based plant database of genes with metabolic pathway information and expressed sequence tag indices in KEGG. Plant Physiol 144:857–866. doi:10.1104/pp. 106.095059

Mittler R, Vanderauwera S, Suzuki N, Miller G, Tognetti VB, Vandepoele K, Gollery M, Shulaev V, Van BF (2011) ROS signalling: the new wave? Trends Plant Sci 16(6):300–309

Moco S, Bino RJ, Vorst O, Verhoeven HA, de Groot J, van Beek TA, Vervoort J, de Vos CH (2006) A liquid chromatography-mass spectrometry-based metabolome database for tomato. Plant Physiol 141:1205–1218

Mohseni-Zadeh S, Louis A, Brézellec P, Risler JL (2004) PHYTOPROT: a database of clusters of plant proteins. Nucleic Acids Res 32:D351–D353. doi:10.1093/nar/gkh040

Molassiotis A, Fotopoulos V (2011) Oxidative and nitrosative signalling in plants: two branches in the same tree? Plant Signal Behav 6(2):210–214

Morgan PW, Drew MC (1997) Ethylene and plant responses to stress. Physiol Plant 100(3):620–630

Naika M, Shameer K, Mathew OK, Gowda R, Sowdhamini R (2013) STIFDB2: an updated version of plant stress-responsive transcription factor database with additional stress signals, stress-responsive transcription factor binding sites and stress-responsive genes in *Arabidopsis* and rice. Plant Cell Physiol 54 (2):e8. doi:10.1093/pcp/pcs185

Neilson KA, Gammulla CG, Mirzaei M, Imin N, Haynes PA (2010) Proteomic analysis of temperature stress in plants. Proteomics 10:828–845

Nouri M, Toorchi M, Komatsu S (2011) Proteomics approach for identifying abiotic stress responsive proteins in soybean. In: Sudaric A (ed) Soybean-molecular aspects of breeding.. ISBN 978-953-307-240-1

Nussbaumer T, Martis MM, Roessner SK, Pfeifer M, Bader KC, Sharma S, Gundlach H, Spannagl M (2013) MIPS PlantsDB: a database framework for comparative plant genome research. Nucleic Acids Res 41:D1144–D1151

Öktem HA, Eyidoğan F, Selçuk F, Tufan ÖM, da Silva JAT, Yücel M (2008) Revealing response of plants to biotic and abiotic stresses with microarray technology. Genes Genomes Genomics 14:48

Osakabe Y, Yamaguchi-Shinozaki K, Shinozaki K, Tran LS (2013) Sensing the environment: key roles of membrane-localized kinases in plant perception and response to abiotic stress. J Exp Bot 64:445–458. doi:10.1093/jxb/ers354

Peterson RKD, Higley LG (2000) Illuminating the black box: the relationship between injury and yield. CRC Press, Boca Raton, FL. ISBN-978-0-8493-1145-1

Pinheiro C, Chaves MM (2011) Photosynthesis and drought: can we make metabolic connections from available data? J Exp Bot 62(3):869–882

Prabha R, Ghosh I, Singh DP (2011) Plant Stress Gene Database: a collection of plant genes responding to stress condition. ARPN J Sci Technol 1:28–31

Proost S, Van Bel M, Sterck L, Billiau K, Van Parys T, Van de Peer Y, Vandepoele K (2009) PLAZA: a comparative genomics resource to study gene and genome evolution in plants. Plant Cell 21:3718–3731. doi:10.1105/tpc.109.071506

Reumann S, Ma C, Lemke S, Babujee L (2004) AraPerox. a database of putative Arabidopsis proteins from plant peroxisomes. Plant Physiol 136:2587–2608

Rhee SY, Crosby B (2005) Biological databases for plant research. Plant Physiol 138(1):1–3. doi:10.1104/pp. 104.900158

Rowley ER, Mockler TC (2011) Plant abiotic stress: insights from the genomics era. In: Shanker A, Venkateswarlu B (eds) Abiotic stress response in plants – physiological, biochemical and genetic perspectives.. ISBN 978-953-307-672-0

Shameer K, Ambika S, Varghese SM, Karaba N, Udayakumar M, Sowdhamini R (2009) STIFDB – Arabidopsis stress responsive transcription factor dataBase. Int J Plant Genomics. doi: 10.1155/2009/583429

Shulaev V, Cortes D, Miller G, Mittler R (2008) Metabolomics for plant stress response. Physiol Plant 132:199–208. doi:10.1111/j.1399-3054.2007.01025.x

Singh DP, Prabha R, Rai A, Arora DK (2012) Bioinformatics-assisted microbiological research: tasks, developments and upcoming challenges. Am J Bioinform 1:10–19. doi:10.3844/ajbsp.2012.10.19

Smita S, Lenka SK, Katiyar A, Jaiswal P, Preece J, Bansal KC (2011) QlicRice: a web interface for abiotic stress responsive QTL and loci interaction channels in rice. Database 2011. doi:10.1093/database/bar037

Spannagl M, Noubibou O, Haase D, Yang L, Gundlach H, Hindemitt T, Klee K, Haberer G, Schoof H, Mayer KF (2007) MIPSPlantsDB–plant database resource for

integrative and comparative plant genome research. Nucleic Acids Res 35:D834–D840
Sreenivasulu N, Sopory SK, Kavi Kishor PB (2007) Deciphering the regulatory mechanisms of abiotic stress tolerance in plants by genomic approaches. Gene 388:1–13
Stein LD (2003) Integrating biological databases. Nat Rev Genet 4:337–345. doi:10.1038/nrg1065
Stepanova AN, Alonso JM (2009) Ethylene signalling and response: where different regulatory modules meet. Curr Opin Plant Biol 12(5):548–555
Sumner LW (2010) Recent advances in plant metabolomics and greener pastures. F1000 Biol Rep 2:7. doi:10.3410/B2-7
Sun Q, Zybailov B, Majeran W, Friso G, Olinares PD, van Wijk KJ (2009) PPDB, the Plant Proteomics Database at Cornell. Nucleic Acids Res 37:D969–D974. doi:10.1093/nar/gkn654
Sundar AS, Varghese SM, Shameer K, Karaba N, Udayakumar M, Sowdhamini R (2008) STIF: identification of stress-upregulated transcription factor binding sites in *Arabidopsis thaliana*. Bioinformation 2:431–437
Takahashi S, Seki M, Ishida J, Satou M, Sakurai T, Narusaka M, Kamiya A, Nakajima M, Enju A, Akiyama K, Yamaguchi-Shinozaki K, Shinozaki K (2004) Monitoring the expression profiles of genes induced by hyperosmotic, high salinity, and oxidative stress and abscisic acid treatment in Arabidopsis cell culture using a full-length cDNA microarray. Plant Mol Biol 56:29–55
Thomashow MF (1999) Plant cold acclimation: freezing tolerance genes and regulatory mechanisms. Annu Rev Plant Physiol Plant Mol Biol 50:571–599
Tuberosa R, Salvi S (2006) Genomics-based approaches to improve drought tolerance of crops. Trends Plant Sci 11(8):405–412
Udayakumar M, Prem Chandar D, Arun N, Mathangi J, Hemavathi K, Seenivasagam R (2012) PMDB: plant metabolome database – a metabolomic approach. Med Chem Res 21:47–52
Vinocur B, Altman A (2005) Recent advances in engineering plant tolerance to abiotic stress: achievements and limitations. Curr Opin Biotechnol 16:123–132
Wanchana S, Thongjuea S, Ulat VJ, Anacleto M, Mauleon R, Conte M, Rouard M, Ruiz M, Krishnamurthy N, Sjolander K, van Hintum T, Bruskiewich RM (2008) The generation challenge programme comparative plant stress responsive gene catalogue. Nucleic Acids Res 36:D943–D946
Wang W, Vinocur B, Altman A (2003) Plant responses to drought, salinity and extreme temperatures: towards genetic engineering for stress tolerance. Planta 218:1–14. doi:10.1007/s00425-003-1105-5
Wilkinson S, Davies WJ (2009) Drought, ozone, ABA and ethylene: new insights from cell to plant to community. Plant Cell Environ 33:510–525. doi:10.1111/j.1365-3040.2009.02052.x
Xiong L, Schumaker KS, Zhu JK (2002) Cell signalling during cold, drought, and salt stress. Plant Cell 14: S165–S183
Yamaguchi-Shinozaki K, Shinozaki K (2006) Transcriptional regulatory networks in cellular responses and tolerance to dehydration and cold stresses. Annu Rev Plant Biol 57:781–803
Yoo SD, Cho Y, Sheen J (2009) Emerging connections in the ethylene signalling network. Trends Plant Sci 14 (5):270–279
Zhang Z, Yu J, Li D, Zhang Z, Liu F, Zhou X, Wang T, Ling Y, Su Z (2010) PMRD: plant microRNA database. Nucleic Acids Res 38:D806–D813. doi:10.1093/nar/gkp818
Zhang S, Yue Y, Sheng L, Wu Y, Fan G, Li A, Hu X, ShangGuan M, Wei C (2013) PASmiR: a literature-curated database for miRNA molecular regulation in plant response to abiotic stress. BMC Plant Biol 13:33. doi:10.1186/1471-2229-13-33
Zimmermann P, Hirsch-Hoffmann M, Hennig L, Gruissem W (2004) GENEVESTIGATOR. Arabidopsis microarray database and analysis toolbox. Plant Physiol 136:2621–2632

Genomics-Based Analyses of Environmental Stresses in Crop Plants

Meenakumari Muthuramalingam, Yong-Fang Li, and Ramamurthy Mahalingam

Abstract

The obligate sessile nature of plants imposes a considerable challenge with regard to their ability to adapt and thrive in the wake of rapid climate change disasters. This includes extremes of moisture stress – drought and flooding, temperature extremes – heat and cold, and atmospheric pollutants such as ozone. These climate change factors can significantly retard the phenology, physiology, and molecular programs of crop plants that in turn adversely impact the crop yields and hence represent a significant threat for global food security. On the brighter side, plants being resilient have evolved a gamut of adaptive mechanisms to thwart such catastrophes. In this chapter, we focus our attention on the molecular aspects of adaptive mechanisms in plants. In particular, we attempt to highlight the major findings from omics-based studies in response to climate change factors. We offer some perspectives on the need for integrated omics approaches and realistic field-level studies of stresses.

Keywords

Climate change • Cold • Drought • Genomics • Heat • miRNA • Ozone • Proteome • Stress • Transcriptome

Introduction

Food security is a major issue in the global policy agenda (Rosegrant and Cline 2003). In a world where population growth exceeds food supply (Malthus 1817), a second green revolution is necessary (Dhalmini et al. 2005). Opportunities for plant biotechnologies to contribute to second green revolution have been widely recognized (Fedoroff and Brown 2004). Three major yield-increasing strategies have been suggested: (1). increasing harvest index (ratio of grain to total crop biomass), (2). increasing plant biomass, and (3). improving stress tolerance (Evans 1998; Cassman 1999). Since greatest losses in crop yields are due to abiotic stresses, we consider improving stress tolerance in plants as the most important yield-increasing strategy.

M. Muthuramalingam • Y.-F. Li • R. Mahalingam (✉)
Department of Biochemistry and Molecular Biology, Oklahoma State University, Stillwater, OK 74078, USA
e-mail: ramamurthy.mahalingam@okstate.edu

R.K. Gaur and P. Sharma (eds.), *Approaches to Plant Stress and their Management*,
DOI 10.1007/978-81-322-1620-9_22, © Springer India 2014

Among abiotic stresses, water deprivation or drought accounts for more than 50 % of the estimated losses (Araus et al. 2002; Boyer 1982). Plant water status is a central component of several abiotic stresses including drought, freezing, and salinity (Verslues et al. 2006). Moisture status prevailing in soil environment in turn depends on ambient environmental conditions such as temperature, light, and humidity. Increase in earth's surface temperature is a major indicator of global warming (Van Vuuren et al. 2008). This raise in mean global temperature is attributed to increases in greenhouse gases such as CO_2 and air pollutants such as ozone. Thus plants not only have to deal with dwindling water resources but also have to deal with increasing temperatures, CO_2, and ozone (http://www.ipcc.ch).

The advent of genomics technologies has ushered in a new era for unraveling the plant responses to stresses. The availability of whole genome microarrays not only in model plants but also in crops led to a plethora of studies identifying stress-responsive gene networks. Recent advances in sequencing, especially RNA-seq technologies are becoming popular tools for gene expression studies especially in response to environmental perturbations in plants. The role of post-transcriptional gene regulation via small RNAs gained importance during the last decade. MicroRNAs play important roles in regulating plants growth, development, differentiation, and metabolism. Increasing evidence demonstrate that miRNAs are involved in biotic and abiotic stress responses (Sunkar 2010; Sunkar et al. 2012; Lu et al. 2008; Sunkar and Zhu 2004; Ruiz-Ferrer and Voinnet 2009; Khraiwesh et al. 2012; de Lima et al. 2012). Concomitant developments in the field of mass spectrometry and two-dimensional gel electrophoresis have led to whole proteome analysis of plants subjugated to various environmental stresses (Barkla et al. 2013). In this chapter we provide an overview of the omics-based studies on abiotic stresses in crop plants. In particular, drought, heat, cold, and ozone stress will be discussed.

Transcriptome Profiling

A comprehensive review of the transcriptomic changes in response to climate change factors, especially drought and heat, has been reported (Ahuja et al. 2010). Here we will summarize studies on cold and ozone stress in plants of economic importance.

Cold Stress

In rice seedlings, comparative transcriptomics of a chilling tolerant (LTH) and a sensitive line (IR29) revealed that upregulation of genes was a marked feature of the former. Furthermore, the tolerant line was able to rapidly return to basal level of gene expression when the cold stress was removed, suggesting plasticity of the transcriptome as an important cold-adaptive trait (Zhang et al. 2012). Using a subtractive hybridization approach, it was shown in oranges that genes involved in secondary metabolism especially anthocyanin biosynthesis, osmoregulation, lipid desaturation, and defense mechanisms against oxidative stress were upregulated in response to cold (Crifo et al. 2011). Using a similar subtractive hybridization technique, a cold-induced dehydrin was identified as a marker for cold stress in tomato fruits (Weiss and Egea-Cortines 2009). In an Affymetrix GeneChip-based analysis of cold stress in grapefruits, it was reported that genes associated with lipid, sterol, and carbohydrate metabolism and hormone biosynthesis were upregulated while photosynthesis, respiration, nucleic acid, and secondary metabolism were downregulated (Maul et al. 2008). Transcriptome analyses of barley chloroplast mutants, *albina* and *xantha*, demonstrated that chloroplasts play a pivotal role in the cold stress adaptation (Svensson et al. 2006). Using two wheat cultivars with contrasting phenotypes to cold stress, 65 genes with opposite expression patterns were identified. This included transcription factors, calcium-binding proteins, protein kinases, pyrophosphatase and cell wall-associated hydrolase (Gulick et al. 2005).

Ozone Stress

A large number of studies related to ozone-induced oxidative stress have been reported in the model plant Arabidopsis. An overview of the key biological processes affected by ozone is summarized below (Fig. 1, modified from Ludwikow and Sadowski 2008). Comparative analysis of ozone-tolerant JE154 and sensitive Jemalong cultivar of *Medicago truncatula* showed rapid changes within 1 h after treatment initiation in the transcriptome of the former (Puckette et al. 2008). This suggested that initial signals activated in response to this oxidant were rapidly perceived in the tolerant line while this may not be occurring in the sensitive line. Nearly 120 genes were differentially expressed in a transcriptome analysis of ozone-tolerant pepper cultivar Buchon and ozone-sensitive cultivar Dabotop (Lee and Yun 2006). Interestingly, many of the ozone-responsive genes were specifically upregulated in the sensitive pepper cultivar. Ozone induced unique transcriptome signatures in panicles and grains of rice plants. Majority of the ozone-responsive genes were associated with hormonal signaling, proteolysis, transcription, cell wall and defense signaling (Cho et al. 2013).

Posttranscriptional Gene Regulation via Small RNAs

MicroRNAs are approximately 21 nucleotides-long noncoding RNAs, which are generated from processing of imperfectly folded hairpin-like single-strand RNAs by the Dicer-like 1 complex (Jones-Rhoades et al. 2006; Ramachandran and Chen 2008). In plants miRNAs regulate their mRNA targets (mainly transcription factors) based on imperfect sequence complementarity and suppress expression of the target gene by Argonaute (AGO)-mediated mRNA cleavage and/or inhibit translation (Jones-Rhoades and Bartel 2004; Voinnet 2009; Brodersen et al. 2008; Carrington and Ambros 2003; Lanet et al. 2009; Chen 2005). miRNAs responsive to drought, heat, cold, and ozone are discussed below.

Drought-Responsive miRNAs

miRNAs responsive to drought stress have been thoroughly studied in various plant species and several of these have been summarized in Table 1. A few miRNAs display the same pattern of regulation in response to drought in different plant species. For example, miR159 is induced

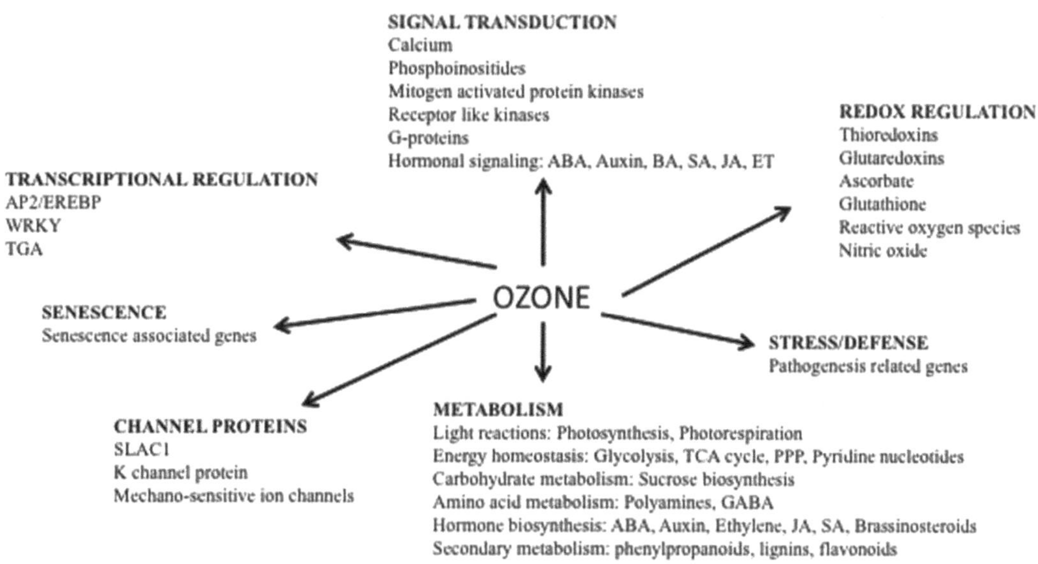

Fig. 1 Various biological processes affected in response to ozone exposure in plants

Table 1 Abiotic stress-responsive miRNAs in various plant species

miRNAs	Drought	Cold	Heat
miR156/157	Ath+, Ttu+, Hvu+, Osa−, Peu+	Ptc−, Mes#	Tae+, Bra+
miR159	Ath+, Peu+, Ptc +	Mes#	Tae+
miR160	Mes#, Peu+, Ptc−	Mes#	Tae+
miR162	Peu+	Mes#	
miR164	Ptc−		
miR165/166	Hvu + leaf, Ttu−, Hvu-root, Mes#, Gma#, Peu + &−	Ath+, Mes#	Tae+
miR167	Ath+, Mes#, Peu+	Mes#	Pto+
miR168	Ath+, Osa−, Peu−	Ptc+, Ath+	Tae+
miR169	Ath−, Osa+, Mtr−, Peu+	Ath+, Bdi+	Tae+, Pto−
miR170/171	Ath+, Hvu + leaf, Ttu−, Osa + &−, Peu + &−	Mes#, Ptc−, Ath+	Ptc−, Pto−
miR172	Osa−, Peu + &−	Ath+, Bdi+	Tae−
miR319	Ath+, Osa + &−, Peu + &−	Ath+	
miR393	Ath+, Osa+, Mtr+, Pvu+, Peu + &−	Ath+	Tae+
miR395	Peu + &−, Osa+	Mes#	
miR396	Ath+, Osa−, Ttu−, Peu + &−	Ath+, Mes#	
miR397	Ath+, Osa−, Gma#, Peu+	Ath+, Bdi+, Mes#	
miR398	Mtr+, Ttu+, Peu+	Ath−	Ath+, Bra−
miR408	Ath+, Mtr+, Hvu + leaf, Osa−, Ptc−	Ath+	
miR390	Peu−		

Abbreviations and related literature: + upregulated, − downregulated, +&− some members were upregulated, some were downregulated, # two varieties displayed opposite expression pattern, *Ttu Triticum turgidum* (wild emmer wheat) (Kantar et al. 2011), *Ptc Populus trichocarpa* (Lu et al. 2005; Shuai et al. 2013), *Peu Populus euphratica* (Li et al. 2011), *Pto Populus tomentosa* (Chen et al. 2012), *Gma Glycine max* (Kulcheski et al. 2011), *Ath Arabidopsis thaliana* (Liu et al. 2008; Reyes and Chua 2007; Li et al. 2008; Sunkar and Zhu 2004; Zhou et al. 2008; Guan et al. 2013), *Osa Oryza sativa* (Zhou et al. 2010; Zhao et al. 2007), *Mtr Medicago truncatula* (Trindade et al. 2010), *Bdi Brachypodium distachyon* (Zhang et al. 2009), *Pvu Phaseolus vulgaris* (Kantar et al. 2010), *Hvu Hordeum vulgare* (Kantar et al. 2010), *Bra Brassica rapa* (Yu et al. 2012), *Tae Triticum aestivum* (Xin et al. 2010), *Mes Manihot esculenta* (Zeng et al. 2010)

by drought in Arabidopsis and Populus (Reyes and Chua 2007; Li et al. 2011; Shuai et al. 2013), and miR398 is induced in Medicago, wheat, and Populus (Kantar et al. 2011; Li et al. 2011; Trindade et al. 2010). While most miRNAs respond to drought stress differently in different plant species, a few exhibited opposite expression pattern (Table 1). For example, miR156 was induced by drought in Arabidopsis, wild emmer wheat, barley, and Populus, but was repressed in rice (Kantar et al. 2011; Li et al. 2011; Liu et al. 2008, Zhou et al. 2010, Kantar et al. 2010). Interestingly, different miRNA expression profile has been observed even among different cultivars of the same species under drought stress. In soybean, majority of the tested miRNAs were upregulated during water-deficit stress in drought-sensitive genotype, while most of the miRNAs were downregulated in the drought-tolerant genotype (Kulcheski et al. 2011). This opposite patterns of miRNA expression has also been reported in two cassava varieties (Zeng et al. 2010). These studies further exemplify the importance of miRNAs as a resource for developing drought-resistant cultivars.

Differences in spatial and temporal expression patterns of miRNAs have been observed in different tissues and at different time points during drought regime. Studies in barley showed that Hvu-miR156a, miR166, miR171, and miR408 were induced in leaf under dehydration stress. Interestingly, miR166 was downregulated in roots, while expression of miR156a, miR171, and miR408 was not altered in roots (Kantar et al. 2010). Dynamic regulation pattern of miRNAs were reported in drought-resistant wild emmer wheat (Kantar et al. 2011). Rice drought-responsive miR169g and miR393 exhibited different spatial and temporal expression patterns (Zhao et al. 2007). A thorough understanding of the spatial and temporal expression patterns of drought-responsive miRNAs will provide a valuable resource for engineering crop plants with improved drought tolerance.

Heat Stress-Responsive miRNAs

Several reports have examined the heat stress-responsive miRNAs in several different plant species (Table 1). miR156/157 is induced by heat stress in wheat and Brassica (Xin et al. 2010; Yu et al. 2012). In wheat miR159, miR160, miR165/166, miR168, miR169, and miR393 were heat induced, while miR172 was downregulated (Xin et al. 2010). Interestingly, miR398 was upregulated by heat stress in Arabidopsis (Guan et al. 2013) but was downregulated in Brassica (Yu et al. 2012). Downregulation of miR170/171 and 14 other miRNA families including miR168 and miR169 was reported in two Poplar species in response to heat stress (Lu et al. 2008; Chen et al. 2012). miR167 and miR1450 were upregulated only in *P. tomentosa* (Chen et al. 2012).

Cold Stress-Responsive miRNAs

Similar to the drought studies, numerous reports have explored the role of miRNAs in response to cold stress in a number of different plant species (Table 1). In Arabidopsis several miRNAs (miR165/166, miR169, miR172, miR393, miR396, miR397, miR402, miR408) are significantly upregulated by cold stress, while others (miR156/157, miR159/319, miR164, miR394, miR398) show either transient or small difference in expression in Arabidopsis (Sunkar and Zhu 2004; Zhou et al. 2008; Liu et al. 2008). Using microarray platform, 18 miRNAs were found rapidly responding to cold in rice, most of which were downregulated. Intriguingly, members of miR171 family showed diverse expression patterns (Lv et al. 2010). In *Brachypodium*, a total of 25 miRNAs were found to be cold stress responsive and majority was downregulated, while only three conserved miRNAs (miR169, miR172, and miR397) were upregulated (Zhang et al. 2009). Repression of six miRNAs and a tasiRNA3 in spike tissues of wheat thermosensitive genic male-sterile line during the early stages of cold stress was reported (Tang et al. 2012).

As observed in drought and heat stress, members of the same miRNA family exhibited different response patterns to cold stress. Members of wheat miR166 family showed different expression patterns; tae-miR167 and tae-miR167e showed completely reversed expression profiles in spikes of genic male-sterile line (Tang et al. 2012). Differential miRNA responsive pattern was even observed between two cassava cultivars (SC124 and C4) under cold stress. Majority of miRNAs were downregulated in SC124. On the contrary in cultivar C4, only four miRNAs were downregulated and 31 miRNAs were upregulated (Zeng et al. 2010). These results indicate miRNAs not only at the species level but also at the level of variety or cultivars that are crucial for mediating stress responses. These observations strongly suggest that miRNA family members can be carefully manipulated in extant germplasm to overcome temperature extremes.

Oxidative Stress-Responsive miRNAs

A direct link between miRNA and oxidative stress was uncovered by miR398, which targets two different gene families: Cu-Zn SODs (*CSD1* and *CSD2*) and a copper chaperone for superoxide dismutase (*CCS1*) (Beauclair et al. 2010; Li et al. 2010; Sunkar et al. 2006). miR398 levels gradually receded in rosette leaves during acute ozone stress and even continued to decrease following 6 h of recovery but was gradually restored to normal level after 18 h of recovery from this oxidant (Jagadeeswaran et al. 2009). Using plant miRNA atlas, 22 miRNA families were differentially expressed within 1 h of ozone fumigation, and most of these miRNAs were upregulated by ozone stress (Iyer et al. 2012). These studies demonstrate that miRNAs play a crucial role in regulating the oxidative signaling pathway, a central coordinator of stress/defense signaling in plants.

Proteomics Studies of Environmental Stresses in Crop Plants

A recent comprehensive review has described the advances in proteomics technologies and their applications in plant stress proteome analysis (Barkla et al. 2013).

Drought Stress

Proteome analysis of rice leaves during drought stress and after recovery resulted in identification of 15 reversibly responding proteins. Among them, RNA-binding protein and EF-Tu proteins involved in chloroplastic protein synthesis were upregulated. Proteins involved in photosynthesis and carbon metabolism, such as RuBisCo activases, fructose 1,6-bisphosphate aldolase, and cytosolic triosephosphate isomerase were upregulated, while chloroplastic Rieske FeS proteins were downregulated. In response to ROS burst caused by drought stress, the Cu-Zn SOD and glutathione dehydroascorbate reductase were upregulated (Salekdeh et al. 2002). Later Choudhary et al. 2009 compared the nuclear proteome of rice seedlings and identified dehydration responsive proteins involved in transcriptional regulation and chromatin remodeling. Four days of drought stress on soybean roots altered abundances of proteins participating in signal transduction, structural organization, carbohydrate, and nitrogen metabolism (Alam et al. 2010). Some of the cell wall proteins were modified to reduce water loss indicating the plant tolerance to stress. Similarly studies in maize and sugar beet reported increased levels of oxidation-protective enzymes upon dehydration (Benesova et al. 2012; Hajheidari et al. 2005). Analysis of the proteome of drought-tolerant wheat genotype showed upregulation of cytosolic Trx h (Hajheidari et al. 2007). Nuclear proteome of chickpea crop revealed accumulation of proteins involved in cell signaling, antioxidant defense-related enzymes, Ran GTPase, HSPs, and chaperones. Out of 205 dehydration-responsive proteins identified in this study, 46 proteins were downregulated, which included mainly the subunits of ATPase, transcription factors (AP2/EREBP), and transketolase (Pandey et al. 2008).

Heat Stress

Heat stress in plants leads to elevated expression of a class of heat shock proteins (HSPs) (Baniwal et al. 2004). Besides HSPs and chaperonins, proteins involved in redox regulation such as glutathione S-transferase, thioredoxin h, and dehydroascorbate reductase are upregulated in rice (Lee et al. 2007). Other heat stress-responsive proteins such as UDP-glucose pyrophosphorylase, pyruvate dehydrogenase, and transketolase were reported in rice seedlings. Similar studies in wheat reported accumulation of enzymes involved in starch synthesis and degradation. Additionally some translation initiation factors and elongation factors were enhanced upon heat stress in wheat (Majoul et al. 2004). In soybean seedlings heat stress led to differential expression of 54, 35, and 61 proteins in leaves, stems, and roots, respectively (Ahsan et al. 2010). This study revealed that soybeans operate tissue-specific defenses and adaptive mechanisms in response to heat stress, while common defense mechanism may be responsible for induction of HSPs in all the tissues (Ahsan et al. 2010).

Cold Stress

Differential abundance of a hypothetical Osl_33098, mitochondrial malate dehydrogenase, and legumin-like protein were reported in a comparison between two wheat cultivars and four reciprocal substitution lines during long-term cold acclimation (Vitamvas et al. 2012). Comparison of wheat leaf proteomes exposed to cold (4 °C) or normal temperatures (20 °C) showed upregulation of ascorbate peroxidase, dehydroascorbate reductase, cysteine proteinase, proteosome subunit, and glutamate semialdehyde aminomutase. Downregulation of Krebs cycle enzymes and photosynthesis-related proteins were also observed (Rinalducci et al. 2011). In rice seedlings exposed to cold stress, 19 proteins were upregulated and 20 proteins were downregulated. In the leaf blades the upregulated proteins were associated with energy metabolism, while downregulated proteins were related to defense (Hashimoto and Komatsu

2007). A large number of proteins were identified in cold-tolerant sunflower lines when compared with cold-sensitive genotypes. As observed in cold-responsive proteomes of other plant species, the identified proteins in sunflower were associated with metabolism, protein synthesis, energy, and defense-related processes (Balbuena et al. 2011).

Ozone Stress

Studies in soybean seedlings exposed to moderate O_3 dose exhibited only microscopic leaf damage after 3 days. However, the impact on the protein level was highly visible in total as well in chloroplastic proteome. In 2D gel analysis of the total protein extract, 23 out of more than 500 protein spots showed more than 1.5-fold difference between the control and treated samples. In contrast to that on 2-D gels loaded with the chloroplastic protein extract, a total of 35 proteins showed more than 1.5-fold change in abundance. Downregulation of 15 proteins in total protein extract and 24 proteins from chloroplast extract was reported. The O_3-responsive proteins can be categorized in eight functional groups involved in photosynthesis (45 %), energy metabolism (16 %), protein synthesis, and assembly (7 %). Proteins with unknown functions were represented by 11 % and 13 % of the O_3 – responsive proteins are assigned to miscellaneous processes. Proteins involved in antioxidant defense, such as peroxidases and Cu-Zn SOD, as well as proteins associated in carbon metabolism, secondary metabolism, and nitrogen metabolism were mostly upregulated, whereas proteins involved in photosynthesis were downregulated under O_3 stress (Agrawal et al. 2002; Sharma and Davis 1994). Additionally there are indications for an alteration in starch and sucrose concentration caused by O_3 stress (Ahsan et al. 2010). Galant et al. (2012) showed changes in protein abundance and oxidation state of several redox regulated proteins upon exposure of soybean to ozone stress. Glycosyl hydrolase revealed high susceptibility to oxidation under high ozone. As distinct protein markers responding to O_3 treatment, four proteins were identified in cultivated bean and maize utilizing the gel-based proteomics, namely, the small and big RuBisCo subunit, superoxide dismutase (SOD), heat shock protein (HSP30), and naringenin 7-O-methyltransferase (Torres et al. 2007). Overall elevated levels of atmospheric ozone concentration cause reduction in plant growth, due to impairment of photosynthesis, lowered CO_2 fixation, and cell death (Table 2).

Conclusion

The omics-based studies in four major climate change factors have provided a wealth of information regarding the various biological processes that are altered in response to these environmental perturbations. Several reviews have been published on integrating data from multiple omic strategies to gain a holistic view of an organism's responses to changing environment (Liberman et al. 2012; Fukushima et al. 2009). The reason behind this being information from each "omic" approach is valuable; however, it is incomplete. Further, their predictive quality and usefulness are compromised by false-negatives and false-positives depending on the technology adapted. Each data source offers a different, partial view of – transcripts, proteins, or miRNAs or metabolites. Hence their integration is necessary for reconstructing biological networks that will aid in understanding the "phenotype" and for identifying reliable functional predictors. Such integrative studies are necessary for selecting rational targets for transgenics- and/or cisgenics-based improvement of stress tolerance in crops.

Secondly, many of the studies reported above are based on plants grown in greenhouse or growth chambers. It is important to undertake studies on stress responses using plants grown under field conditions. Furthermore, most of these studies were considering a single stressor at a time. In reality, these stresses can occur in nature together, e.g., drought and chronic ozone, and at different times during a growing season,

Table 2 Protein classes showing significant regulation in abundance in crop plants in response to abiotic stresses

Climatic stress	Plants	Functional classification of stress-responsive proteins	References
		Upregulated	
Drought	Rice	Cellular metabolism	Biehler and Fock (1996)
	Maize	Protein processing	Gorantla et al. (2007)
	Sugar beet	Cell communication	Salekdeh et al. (2002)
	Wheat	Energy transfer proteins	Benesova et al. (2012)
	Chickpea	Transcription factors	Hajheidari et al. (2005)
		Defense proteins	Hajheidari et al. (2007)
		Proteins in carbon fixation	Pandey et al. (2008)
		Heat shock proteins	
		Downregulated	
	Rice	Rieske FeS proteins	
	Chickpea	NADPH oxidase	
		AP2 class of transcription factors	
Heat	Wheat	Heat shock proteins	Baniwal et al. (2004)
	Rice	Chaperonins	Lee et al. (2007)
	Seedling	Antioxidant defense proteins	Majoul et al. (2004)
		Calvin cycle enzymes	
		Proteins of translational machinery	
Cold	Rice	Chaperones	Cui et al. (2005)
	Seedlings	Protein biosynthesis	Balbuena et al. (2011)
	Sunflower	Antioxidative/detoxifying enzymes	
		Energy pathway	
		Cell wall components	
		Proteases	
	Wheat	Ascorbate recycling	
		Protein recycling	Rinalducci et al. (2011)
		Tetrapyrrole biosynthesis	
		Krebs cycle enzymes	
		Photosynthesis	
Ozone	Rice	*Upregulated*	Agrawal et al. (2002)
	Maize	Antioxidant defense proteins	Ahsan et al. (2010)
	Soybean	Carbon and nitrogen metabolism	Galant et al. (2012)
		Downregulated	
		Photosynthetic proteins	

e.g., heat and drought (Tester and Bacic 2005). There is an imminent need for a thorough understanding of crop responses to combined stresses that usually co-occur in nature (Mittler 2006; Mittler and Blumwald 2010). Analyzing combined stress scenarios in plants growing under field conditions in conjunction with multiple omics platforms will pave the way for engineering novel germplasm that can tolerate and thrive well under adverse environmental conditions.

References

Agrawal GK, Rakwal R, Yonekura M, Kubo A, Saji H (2002) Proteome analysis of differentially displayed proteins as a tool for investigating ozone stress in rice (*Oryza sativa* L.) seedlings. Proteomics 2:947–959

Ahsan N, Nanjo Y, Sawada H, Kohno Y, Komatsu S (2010) Ozone stress-induced proteomic changes in leaf total soluble and chloroplast proteins of soybean reveal that carbon allocation is involved in adaptation in the early developmental stage. Proteomics 10:2605–2619

Ahuja I, De Vos RC, Bones AM, Hall RD (2010) Plant molecular stress responses face climate change. Trends Plant Sci 15:664–674

Alam I, Lee DG, Kim KH et al (2010) Proteome analysis of soybean roots under waterlogging stress at an early vegetative stage. J Biosci 35:49–62

Araus JL, Slafer GA, Reynolds MP, Royo C (2002) Plant breeding and drought in C-3 cereals: what should we breed for? Ann Bot 89:925–940

Balbuena TS, Salas JJ, Martinez-Force E, Garces R, Thelen JJ (2011) Proteome analysis of cold acclimation in sunflower. J Proteome Res 10:2330–2346

Baniwal SK, Bharti K, Chan KY et al (2004) Heat stress response in plants: a complex game with chaperones and more than twenty heat stress transcription factors. J Biosci 29:471–487

Barkla BJ, Vera-Estrella R, Pantoja O (2013) Progress and challenges for abiotic stress proteomics of crop plants. Proteomics. doi: 10.1002/pmic.201200401

Beauclair L, Yu A, Bouche N (2010) microRNA-directed cleavage and translational repression of the copper chaperone for superoxide dismutase mRNA in Arabidopsis. Plant J 62:454–462

Benesova M, Hola D, Fischer L et al (2012) The physiology and proteomics of drought tolerance in maize: early stomatal closure as a cause of lower tolerance to short-term dehydration? PLoS One 7:e38017

Biehler K, Fock H (1996) Evidence for the contribution of the Mehler-peroxidase reaction in dissipating excess electrons in drought-stressed wheat. Plant Physiol 112:265–272

Boyer JS (1982) Plant productivity and environment. Science 218:443–448

Brodersen P, Sakvarelidze-Achard L, Bruun-Rasmussen M et al (2008) Widespread translational inhibition by plant miRNAs and siRNAs. Science 320:1185–1190

Carrington JC, Ambros V (2003) Role of microRNAs in plant and animal development. Science 301:336–338

Cassman KG (1999) Ecological intensification of cereal production systems: yield potential, soil quality, and precision agriculture. Proc Natl Acad Sci U S A 96:5952–5959

Chen XM (2005) microRNA biogenesis and function in plants. FEBS Lett 579:5923–5931

Chen L, Ren YY, Zhang YY et al (2012) Genome-wide identification and expression analysis of heat-responsive and novel microRNAs in *Populus tomentosa*. Gene 504:160–165

Cho K, Shibato J, Kubo A et al (2013) Genome-wide mapping of the ozone-responsive transcriptomes in rice panicle and seed tissues reveals novel insight into their regulatory events. Biotechnol Lett 35:647–656

Choudhary MK, Basu D, Datta A, Chakraborty N, Chakraborty S (2009) Dehydration-responsive nuclear proteome of rice (*Oryza sativa* L.) illustrates protein network, novel regulators of cellular adaptation, and evolutionary perspective. Mol Cell Proteomics 8:1579–1598

Crifo T, Puglisi I, Petrone G, Recupero GR, Lo Piero AR (2011) Expression analysis in response to low temperature stress in blood oranges: implication of the flavonoid biosynthetic pathway. Gene 476:1–9

Cui S, Huang F, Wang J, Ma X, Cheng Y, Liu J (2005) A proteomic analysis of cold stress responses in rice seedlings. Proteomics 5:3162–3172

De Lima JC, Loss-Morais G, Margis R (2012) MicroRNAs play critical roles during plant development and in response to abiotic stresses. Genet Mol Biol 35:1069–1077

Dhalmini Z, Spillane C, Moss JP, Ruane J, Urquia N, Sonnino A (2005) Status of research and application of crop biotechnologies in developing countries. Food and Agricultural Organization, Rome, pp 1–62. Available at http://www.fao.org/docrep/008/y5800e/Y5800E09.htm

Evans LT (1998) Feeding the ten billion: plants and population growth. Cambridge University Press, Cambridge

Fedoroff N, Brown NM (2004) Mendel in the kitchen. Joseph Henry Press, Washington, DC

Fukushima A, Kusano M, Redestig H, Arita M, Saito K (2009) Integrated OMICS approaches in plant systems biology. Curr Opin Chem Biol 13:532–538

Galant A, Koester RP, Ainsworth EA, Hicks LM, Jez JM (2012) From climate change to molecular response: redox proteomics of ozone-induced responses in soybean. New Phytol 194:220–229

Gorantla M, Babu PR, Lachagari VB et al (2007) Identification of stress-responsive genes in an indica rice (*Oryza sativa* L.) using ESTs generated from drought-stressed seedlings. J Exp Bot 58:253–265

Guan Q, Lu X, Zeng H, Zhang Y, Zhu J (2013) Heat stress induction of miR398 triggers a regulatory loop that is critical for thermotolerance in *Arabidopsis*. Plant J 74:840–851

Gulick PJ, Drouin S, Yu Z et al (2005) Transcriptome comparison of winter and spring wheat responding to low temperature. Genome 48:913–923

Hajheidari M, Abdollahian-Noghabi M, Askari H et al (2005) Proteome analysis of sugar beet leaves under drought stress. Proteomics 5:950–960

Hajheidari M, Eivazi A, Buchanan BB, Wong JH, Majidi I, Salekdeh GH (2007) Proteomics uncovers a role for redox in drought tolerance in wheat. J Proteome Res 6:1451–1460

Hashimoto M, Komatsu S (2007) Proteomic analysis of rice seedlings during cold stress. Proteomics 7:1293–1302

Iyer NJ, Jia X, Sunkar R, Tang G, Mahalingam R (2012) MicroRNAs responsive to ozone-induced oxidative stress in *Arabidopsis thaliana*. Plant Signal Behav 7:484–491

Jagadeeswaran G, Saini A, Sunkar R (2009) Biotic and abiotic stress down-regulate miR398 expression in *Arabidopsis*. Planta 229:1009–1014

Jones-Rhoades MW, Bartel DP (2004) Computational identification of plant microRNAs and their targets, including a stress-induced miRNA. Mol Cell 14:787–799

Jones-Rhoades MW, Bartel DP, Bartel B (2006) MicroRNAs and their regulatory roles in plants. Annu Rev Plant Physiol Plant Mol Biol 57:19–53

Kantar M, Unver T, Budak H (2010) Regulation of barley miRNAs upon dehydration stress correlated with target gene expression. Funct Integr Genomics 10:493–507

Kantar M, Lucas SJ, Budak H (2011) miRNA expression patterns of *Triticum dicoccoides* in response to shock drought stress. Planta 233:471–484

Khraiwesh B, Zhu JK, Zhu J (2012) Role of miRNAs and siRNAs in biotic and abiotic stress responses of plants. Biochim Biophys Acta 1819:137–148

Kulcheski FR, De Oliveira LF, Molina LG et al (2011) Identification of novel soybean microRNAs involved in abiotic and biotic stresses. BMC Genomics 12:307

Lanet E, Delannoy E, Sormani R et al (2009) Biochemical evidence for translational repression by *Arabidopsis* microRNAs. Plant Cell 21:1762–1768

Lee S, Yun SC (2006) The ozone stress transcriptome of pepper (*Capsicum annuum* L.). Mol Cells 21:197–205

Lee DG, Ahsan N, Lee SH et al (2007) A proteomic approach in analyzing heat-responsive proteins in rice leaves. Proteomics 7:3369–3383

Li WX, Oono Y, Zhu JH et al (2008) The *Arabidopsis* NFYA5 transcription factor is regulated transcriptionally and posttranscriptionally to promote drought resistance. Plant Cell 20:2238–2251

Li YF, Zheng Y, Addo-Quaye C et al (2010) Transcriptome-wide identification of microRNA targets in rice. Plant J 62:742–759

Li B, Qin Y, Duan H, Yin W, Xia X (2011) Genome-wide characterization of new and drought stress responsive microRNAs in *Populus euphratica*. J Exp Bot 62:3765–3779

Liberman LM, Sozzani R, Benfey PN (2012) Integrative systems biology: an attempt to describe a simple weed. Curr Opin Plant Biol 15:162–167

Liu HH, Tian X, Li YJ, Wu CA, Zheng CC (2008) Microarray-based analysis of stress-regulated microRNAs in *Arabidopsis thaliana*. RNA 14:836–843

Lu SF, Sun YH, Shi R, Clark C, Li LG, Chiang VL (2005) Novel and mechanical stress-responsive microRNAs in *Populus trichocarpa* that are absent from *Arabidopsis*. Plant Cell 17:2186–2203

Lu S, Sun YH, Chiang VL (2008) Stress-responsive microRNAs in *Populus*. Plant J 55:131–151

Ludwikow A, Sadowski J (2008) Gene networks in plant ozone stress response and tolerance. J Integr Plant Biol 50:1256–1267

Lv DK, Bai X, Li Y et al (2010) Profiling of cold-stress-responsive miRNAs in rice by microarrays. Gene 459:39–47

Majoul T, Bancel E, Triboi E, Ben Hamida J, Branlard G (2004) Proteomic analysis of the effect of heat stress on hexaploid wheat grain: characterization of heat-responsive proteins from non-prolamins fraction. Proteomics 4:505–513

Malthus TR (1817) An essay on the principle of population. Murray, London

Maul P, Mccollum GT, Popp M, Guy CL, Porat R (2008) Transcriptome profiling of grapefruit flavedo following exposure to low temperature and conditioning treatments uncovers principal molecular components involved in chilling tolerance and susceptibility. Plant Cell Environ 31:752–768

Mittler R (2006) Abiotic stress, the field environment and stress combination. Trends Plant Sci 11:15–19

Mittler R, Blumwald E (2010) Genetic engineering for modern agriculture: challenges and perspectives. Annu Rev Plant Physiol Plant Mol Biol 61:443–462

Pandey A, Chakraborty S, Datta A, Chakraborty N (2008) Proteomics approach to identify dehydration responsive nuclear proteins from chickpea (*Cicer arietinum* L.). Mol Cell Proteomics 7:88–107

Puckette MC, Tang Y, Mahalingam R (2008) Transcriptomic changes induced by acute ozone in resistant and sensitive *Medicago truncatula* accessions. BMC Plant Biol 8:46

Ramachandran V, Chen X (2008) Degradation of microRNAs by a family of exoribonucleases in *Arabidopsis*. Science 321:1490–1492

Reyes JL, Chua NH (2007) ABA induction of miR159 controls transcript levels of two MYB factors during *Arabidopsis* seed germination. Plant J 49:592–606

Rinalducci S, Egidi MG, Karimzadeh G, Jazii FR, Zolla L (2011) Proteomic analysis of a spring wheat cultivar in response to prolonged cold stress. Electrophoresis 32:1807–1818

Rosegrant MW, Cline SA (2003) Global food security: challenges and policies. Science 302:1917–1919

Ruiz-Ferrer V, Voinnet O (2009) Roles of plant small RNAs in biotic stress responses. Annu Rev Plant Biol 60:485–510

Salekdeh GH, Siopongco J, Wade LJ, Ghareyazie B, Bennett J (2002) Proteomic analysis of rice leaves during drought stress and recovery. Proteomics 2:1131–1145

Sharma YK, Davis KR (1994) Ozone-induced expression of stress-related genes in *Arabidopsis thaliana*. Plant Physiol 105:1089–1096

Shuai P, Liang D, Zhang Z, Yin W, Xia X (2013) Identification of drought-responsive and novel *Populus trichocarpa* microRNAs by high-throughput sequencing and their targets using degradome analysis. BMC Genomics 14:233

Sunkar R (2010) MicroRNAs with macro-effects on plant stress responses. Semin Cell Dev Biol 21:805–811

Sunkar R, Zhu JK (2004) Novel and stress-regulated microRNAs and other small RNAs from *Arabidopsis*. Plant Cell 16:2001–2019

Sunkar R, Kapoor A, Zhu JK (2006) Posttranscriptional induction of two Cu/Zn superoxide dismutase genes in *Arabidopsis* is mediated by downregulation of miR398 and important for oxidative stress tolerance. Plant Cell 18:2051–2065

Sunkar R, Li YF, Jagadeeswaran G (2012) Functions of microRNAs in plant stress responses. Trends Plant Sci 17:196–203

Svensson JT, Crosatti C, Campoli C et al (2006) Transcriptome analysis of cold acclimation in barley albina and xantha mutants. Plant Physiol 141:257–270

Tang Z, Zhang L, Xu C et al (2012) Uncovering small RNA-mediated responses to cold stress in a wheat thermosensitive genic male-sterile line by deep sequencing. Plant Physiol 159:721–738

Tester M, Bacic A (2005) Abiotic stress tolerance in grasses. From model plants to crop plants. Plant Physiol 137:791–793

Torres NL, Cho K, Shibato J et al (2007) Gel-based proteomics reveals potential novel protein markers of ozone stress in leaves of cultivated bean and maize species of Panama. Electrophoresis 28:4369–4381

Trindade I, Capitao C, Dalmay T, Fevereiro MP, Santos DM (2010) miR398 and miR408 are up-regulated in response to water deficit in *Medicago truncatula*. Planta 231:705–716

Van Vuuren DP, Meinshausen M, Plattner GK et al (2008) Temperature increase of 21st century mitigation scenarios. Proc Natl Acad Sci U S A 105:15258–15262

Verslues PE, Agarwal M, Katiyar-Agarwal S, Zhu JH, Zhu JK (2006) Methods and concepts in quantifying resistance to drought, salt and freezing, abiotic stresses that affect plant water status. Plant J 45:523–539

Vitamvas P, Prasil IT, Kosova K, Planchon S, Renaut J (2012) Analysis of proteome and frost tolerance in chromosome 5A and 5B reciprocal substitution lines between two winter wheats during long-term cold acclimation. Proteomics 12:68–85

Voinnet O (2009) Origin, biogenesis, and activity of plant microRNAs. Cell 136:669–687

Weiss J, Egea-Cortines M (2009) Transcriptomic analysis of cold response in tomato fruits identifies dehydrin as a marker of cold stress. J Appl Genet 50:311–319

Xin M, Wang Y, Yao Y et al (2010) Diverse set of microRNAs are responsive to powdery mildew infection and heat stress in wheat (*Triticum aestivum* L.). BMC Plant Biol 10:123

Yu X, Wang H, Lu YZ et al (2012) Identification of conserved and novel microRNAs that are responsive to heat stress in *Brassica rapa*. J Exp Bot 63:1025–1038

Zeng CY, Wang WQ, Zheng Y et al (2010) Conservation and divergence of microRNAs and their functions in Euphorbiaceous plants. Nucleic Acids Res 38:981–995

Zhang JY, Xu YY, Huan Q, Chong K (2009) Deep sequencing of *Brachypodium* small RNAs at the global genome level identifies microRNAs involved in cold stress response. BMC Genom 10:449. doi: 10.1186/1471-2164-10-449

Zhang T, Zhao X, Wang W et al (2012) Comparative transcriptome profiling of chilling stress responsiveness in two contrasting rice genotypes. PLoS One 7:e43274

Zhao B, Liang R, Ge L et al (2007) Identification of drought-induced microRNAs in rice. Biochem Biophys Res Commun 354:585–590

Zhou X, Wang G, Sutoh K, Zhu JK, Zhang W (2008) Identification of cold-inducible microRNAs in plants by transcriptome analysis. Biochim Biophys Acta 1779:780–788

Zhou L, Liu Y, Liu Z, Kong D, Duan M, Luo L (2010) Genome-wide identification and analysis of drought-responsive microRNAs in Oryza sativa. J Exp Bot 61:4157–4168

About the Editors

Dr. Rajarshi Kumar Gaur is presently working as Head and Associate Professor, Department of Science, Mody Institute of Technology and Science (Deemed University), Laxmangarh, Sikar, Rajasthan, India. He did his Ph.D. on molecular characterization of sugarcane viruses of India. He has partially characterized three sugarcane viruses, namely, sugarcane mosaic virus, sugarcane streak mosaic virus, and sugarcane yellow luteovirus. He received the MASHAV Fellowship of the Government of Israel in 2004 for his postdoctoral studies and joined the Volcani Centre, Israel, and then shifted to Ben-Gurion University, Negev, Israel. In 2007, he received the Visiting Scientist Fellowship from the Swedish Institute, Sweden, for one year to work at Umeå University, Umeå, Sweden. He also received the ICGEB (Italy) Postdoctoral Fellowship in 2008. He worked on the development of marker-free transgenic plants against cucumber viruses. He has made significant contributions on sugarcane viruses and published 80 national and international papers and presented nearly 45 papers in national and international conferences. He was elected Fellow of the Linnean Society, London, UK. He has also visited Thailand, New Zealand, London, Canada, the USA, and Italy for attending conferences/workshops. Currently, he is handling different projects on plant viruses and disease management funded by the Government of India.

Dr. Pradeep Sharma is working as Senior Scientist (Biotechnology) at the Directorate of Wheat Research, Karnal, India. He did his Ph.D. in Plant Pathology (Molecular Virology) from Haryana Agricultural University, Hisar, on cotton leaf curl begomovirus. Dr. Sharma has made significant contributions on geminiviruses from South East Asia; published more than 80 national and international research papers, invited chapters, and reviews; and edited three books on biotechnology and geminivirology. He has studied the role of suppressors encoded by monopartite Tomato leaf curl Java and Ageratum yellow vein begomoviruses. Dr. Sharma also has the distinction of receiving numerous honors, international fellowships, and national awards in recognition to his excellent academic and research contributions. Some of these include the Young Scientist's Award (biannual 2005–2006) of the National Academy of Agricultural Sciences and the Pran Vohra Award (2008–2009) of the Indian Science Congress Association. As a recipient of

R.K. Gaur and P. Sharma (eds.), *Approaches to Plant Stress and their Management*,
DOI 10.1007/978-81-322-1620-9, © Springer India 2014

the JSPS Fellowship, he worked at the Tohoku University, Japan, during 2006–2008. He received a postdoctoral fellowship from the Ministry of Agriculture and Foreign Affairs, Israel, and joined the Volcani Centre, ARO, Israel, in 2005–2006. He was elected Fellow of the Indian Virological Society, Delhi, India. He has worked at and visited many pioneering laboratories of the USA, the UK, Japan, France, China, the Netherlands, Indonesia, Turkey, and Israel. He is member of and reviewer at several national and international scientific societies. Currently, he is working on functional genomics, small RNAs, and bioinformatics of wheat.

MIX
Papier aus verantwortungsvollen Quellen
Paper from responsible sources
FSC® C105338

If you have any concerns about our products,
you can contact us on
ProductSafety@springernature.com

In case Publisher is established outside the EU,
the EU authorized representative is:
Springer Nature Customer Service Center GmbH
Europaplatz 3, 69115 Heidelberg, Germany

Printed by Libri Plureos GmbH
in Hamburg, Germany